Handbook of
Adult Development

The Plenum Series in Adult Development and Aging

SERIES EDITOR:
Jack Demick, *University of Massachusetts Medical School, Worcester, Massachusetts*

ADULT DEVELOPMENT, THERAPY, AND CULTURE
A Postmodern Synthesis
Gerald D. Young

AGING AND HUMAN MOTIVATION
Ernest Furchtgott

THE AMERICAN FATHER
Biocultural and Developmental Aspects
Wade C. Mackey

THE CHANGING NATURE OF PAIN COMPLAINTS OVER THE LIFESPAN
Michael R. Thomas and Ranjan Roy

THE DEVELOPMENT OF LOGIC IN ADULTHOOD
Postformal Thought and Its Applications
Jan D. Sinnott

HANDBOOK OF ADULT DEVELOPMENT
Edited by Jack Demick and Carrie Andreoletti

HANDBOOK OF AGING AND MENTAL HEALTH
An Integrative Approach
Edited by Jacob Lomranz

HANDBOOK OF CLINICAL GEROPSYCHOLOGY
Edited by Michel Hersen and Vincent B. Van Hasselt

HANDBOOK OF PAIN AND AGING
Edited by David I. Mostofsky and Jacob Lomranz

HUMAN DEVELOPMENT AND THE SPIRITUAL LIFE
How Consciousness Grows Toward Transformation
Ronald R. Irwin

HUMAN DEVELOPMENT IN ADULTHOOD
Lewis R. Aiken

A Continuation Order Plan is available for this series. A continuation order will bring delivery of each new volume immediately upon publication. Volumes are billed only upon actual shipment. For further information please contact the publisher.

Handbook of
Adult Development

Edited by

Jack Demick

University of Massachusetts Medical School
Worcester, Massachusetts

and

Carrie Andreoletti

Brandeis University
Waltham, Massachusetts

Kluwer Academic /Plenum Publishers
New York • Boston • Dordrecht • London • Moscow

Library of Congress Cataloging-in-Publication Data

Handbook of adult development/edited by Jack Demick, Carrie Andreoletti.
    p. cm. — (The Plenum series in adult development and aging)
    Includes bibliographical references and index.
    ISBN 0-306-46758-5
      1. Adulthood—Psychological aspects. 2. Adulthood—Social aspects. 3.
Aging—Psychological aspects. 4. Developmental psychology. I. Demick, Jack. II.
Andreoletti, Carrie, 1967– III. Series.

BF724.5 .H36 2002
155.6—dc21

2002025686

ISBN: 0-306-46758-5

©2003 Kluwer Academic / Plenum Publishers, New York
233 Spring Street, New York, New York 10013

10 9 8 7 6 5 4 3 2 1

A C.I.P. record for this book is available from the Library of Congress

Printed in Great Britain by IBT Global, London

Contributors

CARRIE ANDREOLETTI, Department of Psychology, Brandeis University, Waltham, Massachusetts, 02454-9110

CHERYL ARMON, Education Programs, Antioch University Southern California, Los Angeles, Marina del Rey, California, 90292-7008

ROBERT ATKINS, Department of Nursing, Temple University, Philadelphia, Pennsylvania, 19122-6072

SANDRA T. AZAR, Department of Psychology, The Pennsylvania State University, University Park, Pennsylvania, 16802-3104

CATHERINE E. BARTON, Lynch School of Education, Program in Counseling Psychology, Boston College, Chestnut Hill, Massachusetts, 02467-3813

MICHAEL BASSECHES, Bureau of Study Counsel, Harvard University, Cambridge, Massachusetts, 02138-5091

CYNTHIA A. BERG, Department of Psychology, The University of Utah, Salt Lake City, Utah, 84112-1102

MARY M. BRABECK, Lynch School of Education, Boston College, Chestnut Hill, Massachusetts, 02467-3813

LAURA HESS BROWN, Department of Psychology, State University of New York, at Oswego, Oswego, New York, 13126-3599

JANE ALLIN BYBEE, Department of Psychology, Suffolk University, Boston, Massachusetts, 02114-2770

JOHN C. CAVANAUGH, Office of the President, University of West Florida, Pensacola, Florida, 32514-5750

ALISON H. CLIMO, Department of Social Work, Warren Wilson College, Ashville, North Carolina, 28815-9000

MICHAEL LAMPORT COMMONS, Department of Psychiatry, Harvard Medical School, Boston, Massachusetts, 02115-6196

STEVEN W. CORNELIUS, Department of Human Development, Cornell University, Ithaca, New York, 14853-4401

THEO LINDA DAWSON, Graduate School of Education, University of California at Berkeley, Berkeley, California, 94720-1670

JAMES M. DAY, Faculty of Psychology and Educational Sciences, Universite Catholique de Louvain, Louvain-la-Neuve, Belgium

JACK DEMICK, Center for Adoption Research, University of Massachusetts Medical School, Worcester, Massachusetts, 01605-2397

KURT W. FISCHER, Department of Human Development and Psychology, Harvard Graduate School of Education, Cambridge, Massachusetts, 02138-3752

CARL GOLDBERG, Albert Einstein College of Medicine, Bronx, New York, 10461-1975

HOWARD E. GRUBER, Teacher's College, Columbia University, New York, 10027-6696

DANIEL HART, Camden College of Arts and Sciences, Rutgers University, Camden, New Jersey, 08102-1405

WILLIAM J. HOYER, Department of Psychology, Syracuse University, Syracuse, New York, 13244-2340

RUTHELLEN JOSSELSON, School of Psychology, The Fielding Graduate Institute, Santa Barbara, California, 93105-3538 and Department of Psychology, The Hebrew University of Jerusalem, Jerusalem, Israel

EDITH F. KAPLAN, Department of Psychology, Suffolk University, Boston, Massachusetts, 02114-2770

MAUREEN E. KENNY, Lynch School of Education, Program in Counseling Psychology, Boston College, Chestnut Hill, Massachusetts, 02467-3813

DEIRDRE KRAMER, Department of Psychology, Rutgers, The State Universtiy of New Jersey, Piscataway, New Jersey, 08854-8040

OTTO E. LASKE, Personnel Development Consultation, Inc., West Medford, Massachusetts, 02155-3643

ALICE LOCICERO, Department of Psychology, Suffolk University, Boston, Massachusetts, 02114-2770

GREG MICHAUD, The Fielding Institute, Santa Barbara, California, 93105-3538

DAVID MOSHMAN, Department of Educational Psychology, University of Nebraska, Lincoln, Nebraska, 68588-0345

MARGARET G. O'CONNOR, Division of Behavioral Neurology, Beth Israel Deaconess Medical Center, Boston, MA, 02215-5501

ELLEN PRUYNE, Human Development and Psychology, Harvard Graduate School of Education, Cambridge, Massachusetts, 02138-3752

FRANCIS A. RICHARDS, Office of School Improvement, Rhode Island Department of Education, Providence, Rhode Island, 02903-3400

PAUL A. ROODIN, Office of Experience-Based Education, State University of New York at Oswego, Oswego, New York, 13126-3599

DAWN E. SCHRADER, Department of Education, Cornell University, Ithaca, New York, 14853-4203

DOROTHY J. SHEDLOCK, Department of Psychology, State University of New York at Oswego, Oswego, New York, 13126-3599

ERIKA L. SHORE, Lynch School of Education, Program in Counseling Psychology, Boston College, Chestnut Hill, Massachusetts, 02467-3813

JAN D. SINNOTT, Department of Psychology, Towson University, Towson, Maryland, 21252-0002

NANCY SOUTHERLAND, Graduate School of Education, University of Pennsylvania, Philadelphia, Pennsylvania, 19104-6216

ROBERT J. STERNBERG, Department of Psychology, Yale University, New Haven, Connecticut, 06520-6614

ABIGAIL J. STEWART, Institute for Research on Women and Gender, University of Michigan, Ann Arbor, Michigan, 48109-1318

LAURA TAHIR, Director of Psychology, Garden State Youth Correctional Facility, Highbridge Road, Yardville, New Jersey, 08620-9632

SEYMOUR WAPNER, Heinz Werner Institute for Developmental Analysis, Clark University, Worcester, Massachusetts, 01610-1477

YVONNE V. WELLS, Department of Psychology, Suffolk University, Boston, Massachusetts, 02114-2770

SUSAN KRAUSS WHITBOURNE, Department of Psychology, University of Massachusetts, Amherst, Massachusetts, 01003-9271

DEBORAH J. YOUNGMAN, Department of Developmental Studies and Counseling, Boston University, Boston, Massachusetts, 02215-1300

Preface

This volume is an outgrowth of contemporary research on development over the adult lifespan, which by now has burgeoned and developed both nationally and internationally. However, for us, the impetus to be involved in this area was spawned and nurtured by our initial association with the Society for Research in Adult Development (SRAD) with its origins some 15 years ago by Michael Commons and his associates in Cambridge, Massachusetts. Through the good will and support of this society, we also became, and are still, heavily involved with the *Journal of Adult Development* and the *Kluwer-Plenum Monograph Series on Adult Development and Aging*, of which this volume is a companion.

Many of the contributions in the volume are from SRAD members, who consistently adhere to a focus on positive adult development. Their chapters have been complemented by pieces from other researchers, who have adopted more mainstream approaches to adult development and/or aging. Regardless of the particular approach and/or focus of the chapter, all the work reported herein supports the relatively recent idea that development is not restricted to children and adolescents but continues throughout the adult lifespan in ways that we never envisioned some 20 years ago. Thus, the volume represents state-of-the-art theory, research, and practice on adult development, which has the potential to occupy us all for some time to come.

When one writes or edits a book on adult development, it is inevitable that one thinks about those individuals and/or forces that have shaped our current lives as adults. In this way, we are no different. We would therefore like to express our sincere gratitude to those family members and friends who have contributed to this volume in more ways than they might ever know. Our collective list includes: Joan Kellerman, Katie Kellerman-Demick, Dan Kellerman-Demick, Seymour Wapner, Joel and Jayne Demick, Judy Miller, Teri-Leigh Boutot, Gail Horowitz, Susan Brand, Jack Horowitz, David Ash, Helena Andreoletti, Gary Andreoletti, Sandy Forrest, Mary and Edward Berlin, Leslie Zebrowitz, and Margie Lachman. We also wish to thank the people and processes associated with our intellectual homes of Clark University and Brandeis University, respectively. Without these individuals and institutions, our development as adults would surely be less advanced.

JACK DEMICK
University of Massachusetts Medical School

CARRIE ANDREOLETTI
Brandeis University

Contents

Introduction

A brief synopsis of the chapters contained in this volume follows to help orient the reader. Organizationally, we have divided the volume into three interrelated categories. Specifically, Part I (Chapters 1–5) treats introductory issues related to *theory and method* in the study of adult development, Part II (Chapters 6–18) examines aspects of *biocognitive development* in adulthood, Part III (Chapters 19–30) considers aspects of *social development* in adulthood, and the Epilogue attempts to integrate the preceding chapters against the backdrop of a holistic, developmental, systems-oriented approach to human functioning (e.g., Wapner & Demick, 1998, 1999, 2000, 2002, Chapter 4).

Part I, which covers *introductory issues relevant to the theories and methods of adult development*, treats the material through both general and specific approaches. It begins with Judith Stevens-Long and Greg Michaud's (Chapter 1) stage-setting presentation on *the state of theory building in the field of adult development*. Referencing and making overt some of the underlying assumptions of many of the contributors to this volume, they conclude that future advances in theory may lie in *dynamic systems models* of human development. On the most general level, such models posit systems in which elements are interrelated so that changes in one sector of the system systematically affect the rest of the system; several more specific variants have appeared in the literature, for example, Fischer and Bidell's (1998) dynamic systems model (Chapter 10); Thelen and Smith's (1998) ecological model, and Wapner and Demick's (1998) holistic, developmental, systems-oriented approach (Chapter 3). As the authors note, such models "... appear to hold promise for many kinds of integrations... [they] encourage us to analyze human activities in all their complexity and variability and yet maintain our interest in order and patterning within that variation... this paradigm calls for complex patterns of variables in the laboratory. It suggests statistical methods that permit pattern analysis and the identification of clusters of both skills and conditions... It reminds us that no one developmental growth curve may be sufficient to describe a complex skill... they may predict movement toward greater complexity, but they don't tell us whether that complexity is good or mature or wise" (p. 19).

William Hoyer and Dayna Touron (Chapter 2) discuss *the concept of learning in adulthood*. They begin by placing learning in a developmental context, which leads them to distinguish *learning* and *development* along two dimensions. As they write: "First, in terms of the inclusiveness or scope of the behavior and of the antecedents of change, learning refers to the effects of practice or experience on behavior whereas development refers to a wider variety of influences that are associated with time-related change ... It is generally accepted that developmental change is multidetermined ... and multidirectional ... Second, development and learning differ in regard to the specificity and durability of change. That is, the learning of a set of relatively specific cognitive skills might endure or remain asymptotic for minutes or maybe weeks, whereas the time-related processes associated with the development of a set of general cognitive skills might endure for years or decades ... Learning refers to practice-based improvements in the speed, accuracy, or efficiency of performing specific tasks and skills whereas development refers to changes in the repertoire of capabilities and strategies for assimilation and accommodation in understanding and interacting with structured environments ... " (pp. 23–24). They then demonstrate the ways in which the study of learning and development in adulthood have recently changed in fundamental ways "... because learning is now seen not only as the acquisition of domain knowledge, but also as the storage and use of knowledge ... The processes of learning and memory serve to produce dynamic and diverse contents and abilities, from perceptual phenomena to the representation of abstract domains of knowledge" (p. 24).

In a related vein, David Moshman (Chapter 3) examines the issue of *developmental change in adulthood*. After addressing a series of conceptual and empirical questions about developmental change across the lifespan (using epistemic cognition, morality, and identity as examples), he concludes that development continues past childhood but in fundamentally different ways (e.g., adult developmental change as qualitative, progressive, and internally directed rather than universal and subject to genetic explanation). Integrating these ideas, he then presents a *pluralistic rational constructive perspective*, which may serve as a paradigm for the study of advanced development as obtains in adults.

Seymour Wapner and Jack Demick (Chapter 4) present their *holistic, developmental, systems-oriented approach to person-in-environment functioning*, which is an elaboration and extension of Heinz Werner's (1940/1957) comparative–developmental theory. They then demonstrate its usefulness for conceptualizing and integrating the field of adult development. For example, they have reframed traditional developmental tasks of adulthood in terms of tasks or goals that serve as the basis for critical person-in-environment transitions that many adults face and illustrate the ways in which their approach empirically shapes the study of these transitions. The transitions they discuss include: parental development (including adoptive parenting), family functioning in general, divorce, empty nest, grandparenting, and retirement.

Finally, Part I concludes with John Cavanaugh and Susan Whitbourne's (Chapter 5) discussion of *research methods in adult development*. Following the presentation of general issues in research design (e.g., validity, selection and measurement equivalence, basic variables such as age, cohort, and time of measurement), they describe both quantitative and qualitative methodologies and designs for the study of adult development and delineate the implications of each (e.g., ethics in research with adults).

Part II discusses multiple aspects of *biocognitive development in adulthood*. Specifically, it treats such topics at the interface of biological and psychological functioning by focusing primarily on adult cognitive processes (e.g., memory, wisdom, intelligence, postformal thought) and issues related to age and gender. Part II begins with the work of Cynthia Berg and Robert Sternberg (Chapter 6), which presents multiple perspectives on the *development of adult intelligence* most generally. Specifically, they compare and contrast psychometric, cognitive, neo-Piagetian, and contextual theories of adult intelligence and make recommendations for future research in this area.

Margaret O'Connor and Edith Kaplan (Chapter 7) discuss specific *adult age-related changes in memory functioning*. Employing a neuropsychological approach, they describe the components of memory that are particularly vulnerable to aging (e.g., working memory, supraspan learning, memory for context) as well as those that remain impervious to age (e.g., implicit memory, procedural learning, well rehearsed information). While they note that our understanding of age-related memory loss has increased substantially over the last several decades, they nonetheless caution that work in this area continues to be confounded by cohort effects—a theme that is echoed throughout several other chapters in the volume.

The next two chapters address the issue of *wisdom and its development*. In an overview, Deirdre Kramer (Chapter 8) outlines the basic principles underlying *psychological models of wisdom*, discusses the empirical research on the construct, and speculates as to the processes underlying its development. She notes that at least two variations of wisdom, each with its own developmental path, emerge. The first (relatively rare) type is characterized by postformal thinking while the second (more common) type is characterized by the cognitive deduction of truths from experience. In a related manner, Dorothy Shedlock and Steven Cornelius (Chapter 9) consider the strengths and weakness of existing *psychological approaches to wisdom* and then posit their own integrative theoretical model of wisdom, which employs measures from the other major approaches, namely, subjective social judgment, personality, and objective wisdom performance. They report that their empirical work has found theoretically consistent relationships between their model of wisdom and its hypothetical correlates of experiential factors, personality, and cognitive abilities.

The next three chapters in Part II deal with *metatheories of cognitive development in adulthood*. Kurt Fischer and Ellen Pruyne (Chapter 10), Michael Commons and Francis Richards (Chapter 11), and Jan Sinnott (Chapter 12) all adhere to postformal models of cognitive development in adulthood, but indeed there are major differences as well. For example, Fischer and Pruyne (Chapter 10) treat the interface between biological and psychological levels of functioning in their work on the emergence, development, and variation of more specific *reflective thinking in adulthood*. They argue that reflective thinking, which they define as the consideration of beliefs in terms of the evidence and arguments for them is a complex form of cognition that requires an advanced level of skill involving the construction of abstract systems and principles. As they note, the capacity for advanced abstract thinking, which serves as the foundation for reflective thinking, emerges in late adolescence or early adulthood with a growth spurt in the brain that reorganizes neural networks and strengthens the connection between the frontal cortex and other parts of the brain. However, whether the capacity for reflective thinking becomes realized depends upon the actual experience of the individual.

In contrast, Commons and Richards (Chapter 11) focus on the psychological level of functioning (cognitive development) through their *model of hierarchical complexity*, which generates one sequence that addresses all cognitive tasks in all domains and is based on a contentless, axiomatic theory. Sinnott (Chapter 12) addresses the interface between functioning at the psychological and sociocultural levels. Specifically, her aim is to understand *adult psychosocial development from the perspective of cognitive development, which is linked to emotion, spirituality, community, and existential meaning.* As previously stated: "Spurred by everyday social encounters, fresh from the everyday problem-solving tasks of creating a marriage, a long-term friendship, a parent–child relationship, an organization, a social role, a self…the midlife adult seems to use assimilation and accommodation to become skilled in new ways of filtering life with a new postformal logic that combines subjectivity and objectivity" (Sinnott, 1998, p. 55).

Related to the convergent processes of cognition and intelligence, Laura Tahir and Howard Gruber (Chapter 13) focus on those *divergent processes associated with creativity.* Specifically, through the lens of the evolving systems approach to creativity that focuses on the development of the creator, of his or her work, and of the contextual frame in which this work occurs, the authors present case studies of developmental trajectories and creative work in later life. They employ illuminative examples from both the arts and the (physical, social, political) sciences.

The next three chapters deal with the interface of the psychological part-processes of cognition and valuation. Jane Bybee and Yvonne Wells (Chapter 14) examine the issue of *the development of possible selves during adulthood.* From a cognitive perspective, they address developmental changes in the discrepancy between the real (or current) self and the ideal self (the self as one would like to be) as well as developmental changes in the nightmare self (the self as one does not want to be), the moral (or ought) self, and the fantasy self (the self as one would like to be if anything were possible). The adaptability of these different selves at different adult life stages is also considered.

Cheryl Armon (Chapter 15) then presents findings from a longitudinal study of *adult value reasoning.* Employing a structural–developmental approach (e.g., Kohlberg, 1984; Piaget, 1968), her findings lend support to the notion of an invariant sequence of stages to reasoning about the good life (e.g., what is the good life? good work? good friendship? good person?). Implications for the study of adult development are also delineated. In a related vein, Dawn Schrader (Chapter 16) provides three case studies of adolescent and adult metacognitive awareness of thought processes when resolving ethical dilemmas. Using a metacognitive interview (in which participants construct a process for resolving moral dilemmas and then resolve hypothetical and real life moral dilemmas and reflect on their thinking processes), she: (1) distinguishes different forms of metacognition (e.g., elementary reflection, self-reflective monitoring, identification of processes, explanation of processes, evaluation of processes); (2) differentiates metacognition, moral judgment stage, and postformal operational thinking; and (3) draws out the implications of her work for the field of education.

The last two chapters in Part II deal with issues related to sex and gender. First, Alice LoCicero (Chapter 17) presents her exploratory work on *women's health activists using the internet.* Beginning with the assumption that disease cannot be understood outside of its social and cultural context, she describes a technologically driven model of care for women's health, which requires ongoing constructivist activities that integrate the intuitive experiential knowledge processes of the subjectivist perspective with the logical, rational processes of the procedural perspective.

Perhaps in opposition to such an approach, Mary Brabeck and Erika Shore (Chapter 18) conclude Part II by providing evidence that refutes the claim of *gender differences in intellectual* (e.g., Belenky, Clinchy, Goldberger, & Tarule, 1986) *and moral* (e.g., Gilligan, 1982) *development*. They suggest that advances might be made if researchers were to explore more complex questions such as those concerning the relationship between gender and different epistemological stances as well as the relationships among gender, type of dilemma, and ego development.

Part III treats aspects of *adult social development*. It covers the adult life stages and concludes with selections concerning practice and/or intervention. Part III begins with Maureen Kenny and Catherine Barton's (Chapter 19) discussion of the implications of *attachment theory and research for understanding late adolescent and young adult development*. They present findings from their original research, which demonstrates the importance and function of attachment in late adolescent/young adult samples (e.g., secure parental attachment is positively related to academic, emotional, and social adjustment to college in both black and white students). They conclude with directions for further clarity and integration of findings for these age groups.

The next two chapters deal with parenting. Sandra Azar (Chapter 20) presents *a social–cognitive perspective on adult development and parenthood*. Drawing on her earlier work that documented a positive association between illogical thought processes and maladaptive parenting in abusing mothers, here she delineates her approach for parenthood more generally and asks whether parenting is a "necessary" and unique stage of development or just one of many stressors that can lead to growth. Yvonne Wells (Chapter 21) then discusses the phenomenon of *superwomen raising superdaughters*. Specifically, she applies adult developmental theory to black mothers and daughters and raises significant questions about the appropriateness of such to them both individually and dyadically.

Ruthellen Josselson (Chapter 22) presents findings from an intensive study of midlife women (born in 1950), who were interviewed when they were 20 years old (college graduation), 32 years old, and 43 years old. Presenting case vignettes of some of these women, Josselson highlights *the psychological realities of midlife* by demonstrating shifts in these women's expression of competence and connection. Continuing with the theme of middle age, Alison Climo and Abigail Stewart (Chapter 23) discuss *elder care and personality development in middle age*. Using memoirs and autobiographical accounts, they "...show how the stresses of eldercare might indeed pose some challenges for middle-aged caregivers (both male and female), while there might also be understudied potential for eldercare to enhance and promote their personality development" (p. 443).

Laura Hess Brown and Paul Roodin (Chapter 24) outline the importance of *grandparent–grandchild relationships*, review previous research on this problem, and discuss grandparent–grandchild relationships from a life course perspective.

Toward a more integrated and comprehensive study of grandparent–grandchild relationships, they highlight: temporal contexts ("... the relationships, like individuals, are not static, but dynamic; they change and are re-conceptualized over time as grandchildren mature and grandparents age ... "; p. 464); the (microstemic, mesosystemic, and exosystemic) social ecologies of families; and diversity issues including gender, lineage, race, and ethnicity.

Using a strategy similar to Brown and Hess, Jack Demick (Chapter 25) discusses the ways in which the study of *adoption and foster care may illuminate adult development and the life course more generally*. Arguing that adoption may well serve as a microcosm for life issues (e.g., attachment, control, identity, guilt and shames, values, concept of families) with which we must all deal, he delineates a holistic, developmental, systems-oriented program of research with implications for both child and adult development.

Goldberg (Chapter 26) discusses the issue of whether adults who commit destructive acts are or are not psychologically and/or legally responsible for these acts. On the basis of clinical evidence, he examines both positions, discusses the tasks and responsibilities of virtue, and draws out the implications for *a theory of recovery ethics* for the field of adult development.

The remaining four chapters have implications for practice. James Day and Deborah Youngman (Chapter 27) begin by outlining the major features of *a cognitive–developmental approach to religious development*. They then demonstrate how research on epistemological and religious styles as well as more recent narrative approaches to the psychology of religion raise questions about the cognitive–developmental paradigm and suggest alternative approaches for theory, research, and ultimately practice. Considering religious languages as relational performatives, they argue that moral action as embodied in religion "... is something that cannot be understood apart from the narrative forms in which it is imagined, explicated, played out, and that understanding this involves an appreciation of the speaker's location in relationship to those to whom she speaks, or imagines herself speaking, to whom she expects she will (have to) be accountable. These contexts inform the sense of possibilities she has for action, and indeed become part of the action, insofar as it becomes impossible to act in ways she cannot imagine describing to those who constitute the community in which she has her being as a speaker" (p. 524).

Michael Bassesches (Chapter 28) begins by asking the question: "In what ways does using life span developmental psychology as one's primary conceptual frame of reference lead to differences in approaches to *psychotherapy practice with adult clients and/or to the training and supervision of adult psychotherapy practitioners?*" (p. 533). He then goes on to delineate the ways in which a lifespan developmental psychological approach has implications for conceptualizing a variety of phenomena, including the philosophical justification of psychotherapy practice, psychotherapy availability/accessibility (for whom psychotherapy is or is not), the nature of therapeutic processes, case formulation, therapist functioning, training and supervision, and why psychotherapy sometimes fails. In a related manner, Otto Laske (Chapter 29) considers *executive development in the workplace* as one form of adult development. Drawing on theories of executive development per se, career theory, and constructivist theories of adult development, he—similar to Bassesches on psychotherapy—provides a conceptualization of (here, executive) development that emphasizes human agency, dialectical processes, and self-construction over the lifespan.

Part III concludes with Daniel Hart, Nancy Southerland, and Robert Atkins (Chapter 30), who explore the connections between *community service and adult development*. After reviewing recent theoretical accounts for community service involvement, the authors present some findings from research that utilizes data from the National Survey of Midlife Development. Specifically, they demonstrate that community service can produce identity change, but caution that the processes through which this change occurs remain unclear. They conclude with a much-needed call for research on the individual and cumulative effects of community service on the life course.

Finally, in the Epilogue, Jack Demick and Carrie Andreoletti *review and synthesize the contributions* to this volume. Employing categories from the holistic, developmental, system-oriented approach to person-in-environment functioning across the lifespan (see Chapter 4), they develop the implications of the volume for the interrelated areas of problem, theory, method, and practice in the field of adult development.

REFERENCES

Belenky, M. F., Clinchy, M., Goldberger, N. R., & Tarule, J. M. (1986). *Women's ways of knowing: The development of self, voice, and mind.* New York: Basic Books.

Fischer, K. W., & Bidell, T. R. (1998). Dynamic development of psychological structures in action and thought. In R. M. Lerner (Ed.), & W. Damon (Series Ed.), *Handbook of child psychology, Vol. 1: Theoretical models of human development,* 5th ed. (pp. 467–561). New York: John Wiley & Sons.

Gilligan, C. (1982). *In a different voice: Psychological theory and women's development.* Cambridge, MA: Harvard University Press.

Kohlberg, L. (1984). *The psychology of moral development.* New York: Harper & Row.

Piaget, J. (1968). *Six psychological studies.* New York: Random House.

Sinnott, J. D. (1998). *The development of logic in adulthood: Postformal thought and its applications.* New York: Plenum Press.

Thelen, E., & Smith, L. B. (1998). Dynamic systems theory. In W. Damon & R. R. Lerner (Eds.), *Handbook of Child Psychology, Vol. 1* (pp. 807–863). New York: John Wiley & Sons.

Wapner, S., & Demick, J. (1998). Developmental analysis: A holistic, developmental, systems-oriented perspective. In W. Damon (Series Ed.) & R. M. Lerner (Vol. Ed.), *Handbook of child psychology, Vol. 1: Theoretical models of human development,* 5th ed. (pp. 761–805). New York: John Wiley & Sons.

Wapner, S., & Demick, J. (1999). Developmental theory and clinical practice: A holistic, developmental, systems-oriented approach. In W. K. Silverman & T. H. Ollendick (Eds.), *Developmental issues in the clinical treatment of children* (pp. 3–30). Boston, MA: Allyn & Bacon.

Wapner, S., & Demick, J. (2000). Person-in-environment psychology: A holistic, developmental, systems-oriented perspective. In W. B. Walsh, K. H. Craik, & R. H. Price (Eds.), *Person-environment psychology: New directions and perspectives,* 2nd ed. (pp. 25–60). Hillsdale, NJ: Lawrence Erlbaum.

Wapner, S., & Demick, J. (2002). The increasing *context* of *context*: in the study of environment-behavior relations. In R. B. Bechtel & A. Churchman (Eds.), *Handbook of environmental psychology* (pp. 3–14). New York: John Wiley & Sons.

Wapner, S., & Demick, J. (in press). Critical person-in-environment transitions across the life span: a holistic, development, systems-oriented program of research. In J. Valsiner (Ed.), *Differentiation and integration of a developmentalist: Heinz Werner's ideas in Europe and America.* New York: Kluwer Academic/Plenum Publishers.

Werner, H. (1940/1957). Comparitive psychology of mental development. New York: International Universities Press.

PART I

Introductory Theory and Method

CHAPTER 1

Theory in Adult Development

The New Paradigm and the Problem of Direction

JUDITH STEVENS-LONG AND GREG MICHAUD

Over the past 20 years, Stevens-Long (1979, 1990; Stevens-Long & Commons, 1992) has been reviewing the state of theory building in the field of adult development. These efforts have produced "a fascinating account of the complex contradictions that have coexisted in adult developmental theory over the last twenty years. Her conclusion that the contradictions were not all bad, since dialectics cast theoretical conflict in the role of change agent, is a meaningful one ..." (Nemiroff & Colarusso, 1990, p. 165). Although these fascinating contradictions have certainly not been resolved in the past 10 years, we conclude in this chapter that some progress has been made toward framing a model of human development that offers a promise, if not of synthesis, of meaningful coexistence.

Other recent reviews have been less sanguine. In *Handbook of Theories of Aging*, for instance, Bengston, Rice, and Johnson (1999) suggest that many researchers in gerontology "seem to have abandoned any attempt at building theory" (p. 3) and assert that the lack of theory has retarded the "process of connecting findings to explanations, thereby undermining the enrichment of knowledge about phenomenon of aging" (p. 9). In the same volume, James Birren asserts that "The contemporary picture of theories of aging is obviously a very fragmented one. If one adds to this theories of life span development which attempt to embrace both developments and senescence then a further broadening of the scope is often made without any apparent means of integration" (p. 466).

Birren does note, however, that "It seems desirable to adopt an ecological point of view of aging, and that theory should embrace many forces not commonly grouped together as a result of disciplinary specialization" (p. 466). This is precisely where we find reason to be optimistic. Over the last 5 years, many writers have adopted an overarching ecological paradigm that might, indeed, provide a

JUDITH STEVENS-LONG AND GREG MICHAUD • The Fielding Institute, Santa Barbara, California 93109.

Handbook of Adult Development, edited by J. Demick and C. Andreoletti. Plenum Press, New York, 2002.

basis for integration across the disciplinary fragmentation that Birren as well as Hendricks and Achenbaum (1998) decry.

This interdisciplinary model has been variously dubbed a dynamic systems model (Fischer & Bidell, 1998); a holistic, developmental system-oriented approach (Wapner & Demick, 1998); and an ecological model (Thelen & Smith, 1998). In this chapter, we propose to look at the common assumptions and metaphors that emerge over various descriptions and to see how several domains of developmental theory fit within such a paradigm.

THE ECOLOGICAL PARADIGM

In 1990, Stevens-Long wrote that a growing awareness of the historical and societal forces, along with an appreciation for the politics of thought, had led many thinkers in adult development to describe a dialectical model as the clearest reflection of the complex interactions they studied. Dialectical interactions imply movement from one state to another through a process characterized by opposition and/or contradiction and governed by internal logic. Dialectics was said to assume an intense transaction between the organism and the environment. Dialectical theorists acknowledge that humans, although shaped by environmental conditions, are, at the same time, central to shaping those conditions (Reese, 1983; Riegel, 1975, 1977). As Paul Baltes and his colleagues have recently written, "Dialectical conceptions of development were at the core of early work in life-span development theory" (Baltes, Lindenberger, & Staudinger, 1998, p. 1041).

However, even the complexity of the dialectical interpretation has proved unable to capture current thinking about development. Somehow, dialectics fails to suggest the many levels at which these mutual interactions take place. As Thelen and Smith argue, "Development can only be understood as the multiple, mutual, and continuous interaction of all levels of the developing system, from the molecular to the cultural." Furthermore, "Development can only be understood as nested processes that unfold over many time scales, from milliseconds to years" (p. 563).

Working within a dialectical framework, Stevens-Long (1990) proposed that adult development can best be understood as a system of systems and offered the matrix shown in Fig. 1.1 as a beginning place for thinking about how research might be integrated and where work needs to be done. At the time, she suggested that the matrix might be conceived in three dimensions at the individual level, as a kind of topography of development.

The height of the landscape along any row or column might represent the degree of individual development that has emerged in one arena or domain. There is no reason to think that everyone will present the same landscape. Some people may develop only toward the end of the lifespan; some will develop early. "In this way, everyone's development can be described by a unique topography determined, to some extent, by logical sequences of growth, but also shaped by the unique events that influence a particular life course" (p. 160). The labels for each square were based on a summary of theory and data over the previous decade.

The conceptualization of development as a "topography" and a "system of systems" seemed fairly sophisticated at the time, but now appears static when compared to the lively, complex metaphors in more recent theoretical work.

	Young Adulthood	Middle Adulthood	Later Adulthood
Motivations	*Self-Actualized Intimacy* The need to resolve the conflict between individuation and fusion in the context of close relationships; to be intimate and self-sufficient	*Self-Actualized Generativity* The need to develop and maintain the social system and continue to individuate in the context of pressure; to be stable and responsible	*Self-Actualized Integrity* The need to accept one's past, one's life history as meaningful, and to continue to develop or individuate
Emotion	*Mature Love* The ability to identify completely with another and maintain a strong sense of self	*Responsibility* The ability to maintain a sense of self and exercise judgment in spite of personal and social disequilibrium; to exhibit both compassion and control	*Patience* The ability to tolerate conflict; to identify with opposition
Cognition	*Insight* The ability to analyze relationships within a system and to find logical solutions	*Perspective* The ability to compare relationships across systems, and to find adequate solutions	*Autonomy* The ability to see one's own role in the experience of reality; to mediate between emotions and cognition
Behavior	*Ethical/Committed* Behavior becomes ethical, driven by personal principles rather than conformity; interests deepen	*Effective/Enabling* One is able to meet one's own needs and to assist others without wasted effort; behavior becomes productive	*Reciprocity* One is able to meet one's own needs without using another person instrumentally

Figure 1.1. Development across the adult lifespan. (Note: Dark outline indicates cells where the most research is available.) (From Judith Stevens-Long, *Adult life*. Palo Alto, CA: Mayfield, 1988, p. 66.)

Consider, for example, the illustration from Thelen and Smith (1998) that human development is like a mountain stream that "shows shape and form and dynamic changes over time" (p. 587). In this conceptualization, human development is an ecological system consisting of many different subsystems. Some will oscillate, others reach stable asymptotes, and others appear to jump about randomly, but are deterministically chaotic (Van Geert, 1991, 1993, 1994).

Such complex formulations are based, in part, on work taken from general systems theory (Bertalanffy, 1968), open systems theory, and chaos and complexity theory (Prigogine & Stengers, 1984). Smith and Thelen refer to multiple, heterogeneous components exhibiting various degrees of stability and change. According to a dynamic systems view, then, development can be envisioned as a series of patterns evolving and dissolving over time and, at any point in time, possessing particular degrees of stability. They offer the adaptation of Waddington's epigenetic landscape in Fig. 1.2 as a way to depict these patterns. In this figure, development progresses from past (background) to present (foreground) and each of the lines represents a particular moment in time. The configuration of a line is the history of that subsystem. Dips and valleys represent the stability of the system at that time.

Still not content with the level of complexity illustrated in Fig. 1.2, the authors go on to note that there are many such landscapes interacting with one another over the course of human development. Each component that develops has a history and the components interact with one another in multidirectional ways. As an example, they describe the growth of the stepping response in infants. Stepping demands the coordination of locomotion and spatial cognition, the evaluation of feedback, motor planning, and rapid adjustments to unexpected events. They offer a further depiction of a developmental landscape as a multilayered system in which the components can influence each other in changing ways.

Many other authors have presented similar ideas. Kurt Fischer (1999) notes that many components contribute to the development of any activity. Development takes many shapes. It is more like a web with different strands that have different developmental ranges depending on contextual conditions. As Fischer and Bidell (1998) write, "Dazzled by all this complexity, scientists often retreat to oversimplification and stereotyping. Most approaches to explaining human activity start with

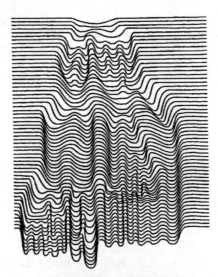

Figure 1.2. An adaptation of Waddington's epigenetic landscape. This version depicts behavioral development as a series of evolving and dissolving attractors of different stability. [From Thelin, E. and Smith, L. B. (1998). Dynamic systems theory. In W. Damon & R. Lerner (Eds.), *Handbook of child psychology, Vol. 1*. New York: John Wiley & Sons, p. 592.]

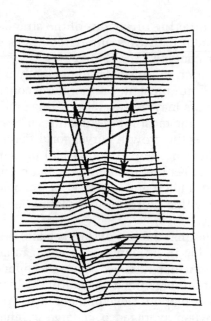

Figure 1.3. The epigenetic landscape as a multilayered system where the components mutually influence each other in changing ways. [From Thelin, E. and Smith, L. B. (1998). Dynamic systems theory. In W. Damon & R. Lerner (Eds.), *Handbook of child psychology, Vol. 1.* New York: John Wiley & Sons, p. 595].

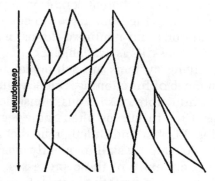

Figure 1.4. Idealized constructive web. The lines or strands represent potential skill domains. The connections between strands represent possible relations among skill domains, and the differing directions of the strands indicate possible variations in developmental pathways and outcomes as skills are constructed for use in different contexts. [Fischer, K. and Bidell, T. (1998). Dynamic development of psychological structures in action and thought. In W. Damon & R. Lerner (Eds.), *Handbook of child psychology, Vol. 1.* New York: John Wiley & Sons, p. 474.]

one limited aspect, such as stage of development … reinforcement contingency, emotional distortion, competitive advantage, or cultural role. They then attempt to analyze all human complexity in terms of that one aspect" (p. 468).

Fischer and Bidell choose to depict this complexity as an "idealized constructive web," in which the "lines or strands represent potential skill domains. In Fig. 1.4, the connections between strands represent possible relationships among skill domains, and the differing directions of the strands indicate possible variations in the developmental pathways and outcomes as skills are constructed for participation in diverse contexts" (p. 474).

Now, imagine that each of those strands is really a plane that looks like the landscape Thelen and Smith present. Furthermore, consider that for each component of development, there may be several possible landscapes that develop over different contexts. One of the most interesting lines of research in the past 10 years has been work that shows the growth curves for different skills may change over contexts. For instance, Fischer and Bidell (1998) present evidence that different levels of social support affect the apparent growth curves for various social skills including the ability of adolescents to describe their self-concepts in complex ways and the probability that children will include describe social situations in complex ways.

Fischer (1999) argues that the use of static metaphors have led us to focus our attention on what is stable in human development rather than on what changes and the conditions under which changes occur. He believes that the skills reflected in human development are organized in multilevel hierarchies and constructed through a process of coordination. Development of any particular skill may appear to be a smooth curve when averaged over large number of instances. However, both analysis of individual growth curves and larger studies that focus on the coordination of multiple actions into integrated complexes often show nonlinear growth, that is, stagelike jumps in development. Furthermore, the development of skills may produce a smooth growth curve under one set of conditions, and a less linear, more stagelike curve given different conditions. An example is the work of Fischer and Biddell (1998), showing smooth growth curves for certain skills under the usual laboratory conditions, but nonlinear jumps in skill construction where social support for higher level skills is available.

Fischer notes that living systems are self-organizing, that is, they change and adapt; they show a range of functioning. Under optimal conditions, one may see stagelike advances while more continuous, nonstagelike growth is typical in ordinary circumstances. Growth curves often show regression or even reversal while new skills develop. Fischer and Bidell argue that researchers need to sample the performance of a developing skill frequently over reasonably long periods of time to capture the complexity of these processes. Moreover, they need to measure the growth of skills in a variety of contexts to reflect the complexity of dynamic systems.

Fischer and Bidell's work is reminiscent of an earlier distinction made by Dannefer and Perlmutter (1990) between physical ontogeny, habituation, and cognitive generativity. *Physical ontogeny* represents development that is largely dependent on biological regulation, such as learning to walk or sensory acuity, whereas habituation represents types of development that are largely dependent on environmental experience. *Habituation* includes most of what is called learning in laboratory research with animals. It probably includes behaviors that are acquired through classical and operant conditioning, and perhaps modeling or observation.

Cognitive generativity refers to the kind of development that occurs when the individual is actively recombining experience with logic, memory, and imagination. "It allows the individual to interpret the past and present, as well as to envision alternative lives and alternative futures" (p. 110). Figure 1.5 suggests how

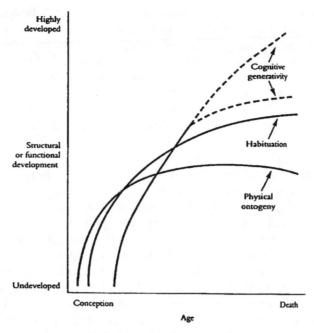

Figure 1.5. Age trajectories of development for three developmental processes. [From Dannefer, D. and Perlmutter, M. (1990). Development as a multidimensional process: Individual and social constituents. *Human Development, 33*, 111.]

these three capacities may develop over time and how effective each might be at different points of the lifespan. *Functional effectiveness* refers to the relative effectiveness of each process, and as you might note from the figure, these authors propose that habituation becomes less effective in later life.

Figure 1.5 depicts different curves for cognitive generativity and shows how it might develop and how effective it might be under favorable and unfavorable conditions. Dannefer and Perlmutter suggest a further complexity as well. Cognitive generativity may, in fact, give humans some control over habituation such that the curve for cognitive generativity may influence the height of the curve for habituation.

Ford and Lerner (1992) argue that dynamic systems theory allows for great diversity of development among humans. The model allows one to study growth in terms of both emergence and reduction. That is, while events at one level (such as the growth of vocabulary) may appear to proceed through linear processes, in some cases constraints at one level (for instance, biology prevents humans from flying) can be overcome at another (humans can build planes). Emergent properties (such as the use of language) can change all the interrelationships at other levels. Someday, as Ford and Lerner remind us, we may even change our own genetic makeup.

Ford and Lerner go on to argue that mutual transactions and mutual causality create great diversity and ensure the relative plasticity of human development. Their version of dynamic systems theory assumes no unity of direction in the course of development, but does allow for the appearance of qualitative changes and discontinuities. Development is defined as incremental and transformational, structural or functional relatively enduring change that "elaborates or increases the diversity of the person's structural or functional characteristics and the patterns of his or her environmental interactions while maintaining coherent organization and functional unity as a person" (p. 92). The task of the scientist, from this viewpoint, is to identify the organismic variables under which various kinds of change occur and to describe change over context.

Magnusson and Stattin (1998) present a thorough review of the issue of context. They point out that context includes not only the conditions that researchers can control, but also much larger variables from the social environment of the family, to the historical and cultural setting. Certain interactions are much clearer in some contexts than others. For instance, among Swedish females the early menarche of females is related to early marriage, lower educational attainment, more children, and lower job status (Stattin and Magnusson, 1990), but these relationships are not nearly as striking among young women in the United States. We must begin to look at environmental patterns rather than differences in single variables. As Magnusson and Stattin put it, environmental variables occur in contexts too and can be taken out of context when re-created in laboratory conditions.

Magnusson and Stattin (1998) believe that "The acceptance of a common theoretical framework for research in natural sciences, can be regarded as a prerequisite for continuous progress in these fields … the holistic integrated view … may serve that purpose" (p. 732). They define the holistic view as a paradigm that emphasizes the continuous, ongoing multidirectional interaction of person and environment as well as the ongoing reciprocal interactions between and among mental, biological, and behavioral factors. They maintain that research in cognitive science, neuropsychology, developmental biology, chaos theory, and other developmental sciences has converged on the dynamic systems approach.

As Wapner and Demick (1998) iterate, the holistic approach is nonreductionist. It assumes that developing systems are mutually interactive and that affecting one part of a developing system affects another. It assumes that all parts and processes are interrelated and are affected by context. At the level of the whole organism, however, Wapner and Demick (see also Chapter 4, this volume) believe that development can be characterized by the orthogenetic principle (Werner, 1957). That is, they see development as proceeding from lack of differentiation to differentiation and hierarchical integration. In this view, development is ordered so that goals become distinct and subordinated, perception and behavior become articulate and flexible, and the structure of thought and behavior becomes stable over time. However, Wapner and Demick also emphasize the need to analyze these relationships in context. Their work describes how development is affected by major shifts in context, such as the transition to school, the entrance into marriage or parenthood, and the like.

In fact, the issue of context has become so salient in the last 10 years that some authors have suggested that there are multiple, diverse psychologies rather than a single psychology (Shweder, Goodnow, Hatano, LeVine, Markus, & Miller, 1998). Here, it is argued, the formal universals of the mind interact with any culture of way of life to produce many mentalities. Culture and psyche are said to "make up each other" (p. 871). In stripping the psyche down to those formal universals, Shweder and his colleagues suggest the following: all human minds appear to have the capacities of representation and intentionality, knowing, wanting, feeling, valuing and deciding, a general interest in active construction, personal worth, and communication. In particular, these authors emphasize the cross-cultural viability of knowing, thinking, feeling, wanting, and valuing. In the study of adult development, the development of knowing and thinking has clearly been of great interest to researchers, feeling and wanting less so, and valuing of very little interest. In the next few pages, we shall take a brief look at the state of theory in each of these areas. Although we shall see some convergence on the general features of this emerging systems paradigm, we shall also find a lack of consensus as to whether a general direction of development can be detected in the data. Unfortunately, we will also see that this new paradigm itself does not resolve the issue of direction.

KNOWING AND THINKING

Most of the work in adult development involving knowing and thinking can be categorized as work on intelligence and cognition. This is certainly one area in which great differences exist between researchers as to whether there is a direction in development. Baltes, Lindenberger, and Staudinger (1998) capture one pole of the argument when they venture that an open systems view of the "incomplete biological and cultural architecture of life-span development and the multiple ecologies of life also made it obvious that the postulation of a single endstate to development was inappropriate" (p. 1046).

Like Ford and Lerner (1992), who argue for the importance of diversity and flexibility instead of a single endstate, Baltes and his colleagues believe that the capacity to move between levels of knowledge and skills is crucial for effective individual development. They (Smith & Baltes, 1999) posit that beyond this, all human development can best be understood in terms of gains and losses. No gain occurs without a loss and no loss without a gain. Over the lifespan, there is a changing ratio of resources allocated to growth maintenance vs. loss management. Successful lifespan adaptation requires the interplay of selection, compensation, and optimization, and the demand for coordination of these components increases in late adulthood as losses begin to mount. Figure 1.6 outlines the lifespan model proposed by Baltes and his associates.

Most of the research that inspired the Baltes model is found in the domain of intelligence (Baltes, Dittmann-Kohli, & Dixon, 1984) and, specifically, in work based on the psychometric theory of fluid and crystallized intelligence (Cattell, 1963; Horn, 1982). Data from these studies clearly show that the mechanics of intelligence, including basic skills such as perceiving relationships, classification, and logical reasoning, peaks in young adulthood. On the other hand, the pragmatics of intelligence, the application of social wisdom or good judgment, may increase with age throughout the normal adult lifespan. Furthermore, the rise in crystallized intelligence often matches the decline in fluid intelligence.

Baltes has also argued (Baltes, Lindenberger, & Staudinger, 1998) that this two-factor model of intelligence accurately predicts differential ontogenetic sources of explanation. "In adulthood and old age, for instance, the mechanics are primarily regulated by biological factors, whereas the cognitive pragmatics evince … a substantial link to culture-based experiential factors" (p. 1081).

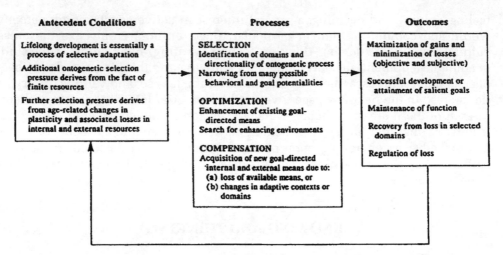

Figure 1.6. The lifespan model of selective optimization with compensation. The essentials of the model are proposed to be universal, but specific phenotypic manifestations will vary by domain, individual, sociocultural context, and theoretical perspective. [From Baltes, P. and Staudinger, U. (1998). Life-span theory in developmental psychology. In W. Damon & R. Lerner (Eds.), *Handbook of child psychology, Vol. 1.* New York: John Wiley & Sons, p. 1055.]

Baltes' work on thinking provides strong contrast to the work of neo-Piagetian or post-Piagetian thinkers like Michael Commons (Commons, 1999; Commons & Richards, 1982), Gisela Labouvie-Vief (1982a, 1982b), Michael Basseches (1984), and others (Pascual-Leone, 1983; Riegel, 1973). These authors argue that there are clear stagelike changes in cognition over the lifespan moving toward a particular endstate and that these can be found cross-culturally (Commons, Armon, Richards, & Schrader, 1989; Galaz-Fontes & Commons, 1989).

According to Commons (1999), the three most common stages of adult reasoning are abstract operations, formal operations, and *systematic operations*. Formal operators can think logically about the possible as well as the real and can use concepts such as probability and chance to explain events. Adults can systematically generate a number of alternative solutions to problems and then test each of those solutions in turn, using any negative information to eliminate more alternatives.

To go beyond formal operations to systematic operations, Commons argues (Stevens-Long & Commons, 1992), people must be willing to deal with contradiction and paradox. At the systematic stage, adults are able to compare whole systems of relationships and to see how sets of relationships might affect other whole sets of relationships. We are discussing here the notion that human development is a system of mutually interactive systems and might not be wholly predictable, but can be described and understood. This statement must be considered in relation to its differences and similarities to more static models of development. Systematic thinking is probably required to appreciate the problems and advantages of an ecological model or a dynamic systems model of human development.

Basseches (1984) describes a stage he calls *dialectical thinking*. Like systems operational thinking, dialectical thinking allows one to analyze competing systems of relationships. Basseches maintains that a basic attribute of dialectal thinking is the dynamic of relativity. Formal thinkers, Basseches writes, focus on universals. They assume a single best answer to a problem exists. Postformal thinkers assume that because change is a basic characteristic of reality, there may be no finite truths. The dialectical thinker is aware that interactions between ideas and fact, as well as interactions between people and information, create what we regard as truth within a particular system of thought or a particular historical period. Such thinkers understand that every effort to organize or systematize knowledge omits something—something that will eventually threaten the system with contradictions and create change.

Despite the strong emphasis on change, however, Basseches also argues that the dialectical thinker is able to identify universal forms within a changing context. For example, although paradox or contradiction between systems may create a constantly changing view of the universe, the significance of contradiction is constant. Similarly, the existence of paradox at the edges of our ability to reason is a constant.

Gisela Labouvie-Vief (1982a, 1982b, 1997) has argued that formal operations are appropriate only when a single, correct answer can be deduced from the premises of a problem. Life, however, is rarely this logical. The best solutions to social problems are seldom clear to everyone, and often the most rational solutions are impractical. Reality is fraught with uncertainty and paradox. As a function of experience in an illogical universe, some adults progress not only beyond formal operations, but also beyond dialectics (she calls dialectical thinking *intersystemic thought*) to the stage of *autonomous thought*.

Labouvie-Vief contends that in the final or autonomous stage of development, there is an "increasingly complex understanding of real-life system(s)..." which finally subordinates the "buoyant and naïve if brilliant thought derived from 'pure logic'" (1982b, p. 76). Autonomous thinkers are able to formulate decisions that integrate logic and the irrational aspects of experience. Autonomous thinkers see not only how truth can be a product of a particular system, but also how the thinker participates in creating truth. They understand that social and personal motivations influence one's formulation of truth and are able to discriminate between personal and universal reality. Performance on tests of formal operations may decline with age, in part, because older people no longer see problems of pure logic as relevant or interesting.

Similarly, John Rybash, William Hoyer, and Paul Roodin (1986) contend that the "formal thinker is infatuated with ideas, abstractions, and absolutes" (p. 31). Through young adulthood and middle age, they maintain, people learn to reconnect reason with the socioemotional realities of life. Whereas formal operations allow the thinker to analyze a closed system (one that is characterized by a finite knowable number of variables), problems in real life are influenced by an unbounded number of interacting variables. Rybash, Hoyer, and Roodin believe that over the span of adult life, thought becomes increasingly relative. Postformal thinkers understand that knowledge and reality are only temporarily true, not universally fixed, and that contradiction is an aspect of reality. Postformal thinkers possess the ability to synthesize contradictory thoughts, emotions, and experiences.

Labouvie-Vief's conceptualization of an autonomous stage, along with similar work by Chinen (1984), Kohlberg (1973), and Van den Dale (1975), suggest a further stage shift back from the highly relativistic, contextual thinking of early postformal thought to a renewed appreciation of the universal, perhaps in the form of aesthetic or affective properties. Whether we are looking at real, qualitatively different stages that appear in nonlinear, irreversible leaps or not, the overarching theme in later stages appears to be a synthesis of motivation, emotion, thought, and values in the struggle to understand reality. This may be one reason why Baltes and his associates (Baltes et al., 1998) have classified the work of Labouvie-Vief, along with that of Kohlberg, Loevinger, and Erikson, as theories of personality development or ego development.

In fact, Baltes, Lindenberger, and Staudinger admit that "It may seem surprising that cognitive models such as that of Labouvie-Vief are included [as theories of self and personality development]. However, if one assumes that self-reflexivity and aspects of intersubjectivity and social cognition are important components of self and personality, cognitive development has its natural place in any developmental model of self and personality" (p. 1093).

If it is true that there is an emergent dynamic systems model of development that might be depicted as interacting planes (see Fig. 1.4) or a web (Fig. 1.5), theories of intelligence or cognition or knowing will have to refer to the mutual interactions of other subsystems, such as emotion, perception, and motivation. This seems a predictable consequence of adopting a model that assumes mutual causality and multidirectional influence across many levels of development.

FEELING AND WANTING

Work that clearly posits a direction to development has been done in the areas of motivation and emotion. Beginning with Maslow (1968, 1970) and Erikson (1963) and continuing in works by Jane Loevinger (1976) and George Vaillant (1977, 1993), researchers in the area of ego development have emphasized motivation and emotion, positing a mature personality.

Erikson (1963), for example, suggests that age-related crises in development produce autonomy, love, care, integrity and, ultimately, wisdom. Moreover, Erikson argued there were strong interactions, not only between the person and the external environment, but among developmental systems as well: the biological, the psychological, and the social systems. He also assumed that the organism both transforms the environment and is transformed by it. Finally, cross-cultural evidence had convinced him that every society affects the particular manifestation of pivotal developments such as intimacy or generativity; in other words, he emphasized the role of conflict.

Extending Erikson's ideas, Vaillant (1993) argues that maturity proceeds through the maturing of the ego defenses. He sees the theories of Maslow, Kolberg, and Loevinger as "characterized by an increasing differentiation of self from others and progressive freeing of self from contextual and social restraint" (p. 161). By contrast, he argues that Erikson (1963) defines maturity as movement toward more, rather than less, attachment. While issue has been taken with view this of Loevinger (Levenson & Crumpler, 1996), all of these theories clearly suggest a direction of development moves toward concern or care for that which lies beyond the self.

Jane Loevinger (1976) and Robert Kegan (1982, 1994) have also offered two quite ambitious versions of ego development. In both of these constructivist theories, Werner's orthogentic principle is assumed: as individuals develop they can simultaneously process and integrate elements from more diverse sources at higher levels of synthesis. People who demonstrate high levels of ego development are said to consider multiple sources of information from others (perceptions of feelings, behaviors, thought), and from the self (motivation, primary process and secondary ideation, emotions, biases, dreams and intuitions, bodily states, and altered states of consciousness). "At the most advanced levels, whole systems of human behavior and thought—for instance moral reasoning, physics as a discipline, or language as an automatic habit—are taken as objects of reflections. Individuals can become simultaneously aware of the most concrete and the most ephemeral aspects of their experience" (Cook-Greuter, 1999).

Both Loevinger and Kegan subscribe to the view that there are general stages or levels of development that tend to emerge over many domains at approximately the same time. Kegan's (1984) scheme emphasizes the development of consciousness which includes emotional, cognitive, interpersonal, and intrapersonal experiencing. He believes all of these domains are strongly constellated. According to Kegan (1994), over time, people organize experience according to higher order principles such that they can take as the subject of reflection what used to exist as context. We come to be able to reflect on our own experience, interactions, and development. Growth of consciousness means development toward more inclusive and complex principles for organizing experience.

Loevinger (1976) emphasizes the attainment of an objective, integrated identity as the goal of development and equates great maturity with the set of characteristics described as "self-actualized" by Maslow (Cook-Greuter, 1999). For this reason, Erikson and Loevinger have been criticized as Euro-centric or Western, along with most versions of ego development that posit a mature endstate (Stevens-Long, 2000). There is little cross-cultural work in either theory as empirical evidence is derived from complex instruments that require elaborate schemes for scoring.

Simpler measures that reflect the coordination of the emotions and cognitive systems over the course of development form a base for work on differential emotion theory (DET) (Izard & Ackerman, 1997). DET defines the experience of an emotion as "a feeling or motivational state that is often accompanied by a readiness for cognitive or motor activity" (p. 9). This kind of research on emotions relates the dependability and regularity of emotional experience to self-identity, self-concept, and personality. The stability of our emotional experiences over time gives us a growing knowledge of how we will think and act in a situation that elicits an emotion and contributes to feelings of self-worth and control.

DET suggests that what develops in emotional development is mainly connections among systems, particularly interconnections between the emotion and cognitive systems in patterns that anchor self-identity, self-knowledge, and self-regulation. These developments are not simply the product of maturity, but of "interactions and transactions among biosocial and environmental systems" (p. 11). There is evidence that complexity of the self-concept is accompanied by fewer, less intense mood swings (Linville, 1982) and that the expression of emotion over the adult years becomes more differentiated and includes the experience of sustained mixed feelings along with greater feelings of control (Magai & Passman, 1997).

In fact, Shulz and Heckhausen (1997) have argued that "Attaining a measure of control over emotional states, once instigated, developing self-regulated emotions such as pride, shame and guilt, using emotions as a mechanism for developing and regulating goal achievement are important milestones in development" (p. 197). They claim cross-sectional findings indicate self-regulation increases with age and that people are able to modulate their own affective responsiveness with age.

Labouvie-Vief (1997) has argued that cognitive complexity, coping, and emotional complexity are strongly related. Cognition, she maintains, does not just control emotion, but also informs it and is informed by it. Cognition develops enough to "bear affect," so that mature adults can better mediate between their outer and inner worlds. Her research demonstrates that younger, less mature, people use more conventional language when speaking about emotions and refer to external rules and standards more often to define what should or ought to be felt. Younger people also forget, ignore, and redirect their thoughts more than middle-aged adults who appear more able to fully tolerate conflict and articulate complex emotional responses.

According to Labouvie-Vief and Diehl (1999), increased cognitive development and complexity give the older adult greater capacity for greater emotional experiencing. Older adults are generally more effective at emotional regulation. They are consistently less likely to regulate their behavior through acting out, projection, and turning against others. It may be that older adults learn more regulated behavior to gate out negative experiences or restrict their focus to a narrower, more self-protective set of strategies. However, Labouvie-Vief believes regulation may also reflect a widening concern for well-being between individuals and institutions. These competing views raise the basic issues of direction. Do people simply become more diverse over time or is there increasing integration, a widening horizon of thought and feeling? And how does this widening or restriction manifest itself in behavior (if at all)?

VALUING AND DECIDING (AND ACTING?)

Shweder and his associates (Shweder et al., 1998) maintain that the mental concepts of knowing, thinking, feeling wanting, and valuing occur in the lexicons of all the languages of the world and are used to explain what people do. Valuing and wanting are particularly important, they believe, in choosing or deciding what to do. It is interesting, in this light, that there is so little research and theory on the issue of values, and so little of what we know about any of these dimensions is based on the observation of behavior in real-life settings. The majority of developmental research is about what people say they think or feel, how they solve abstract problems, respond to laboratory tests or, at best, how they relate to others in controlled settings.

One of the few theorists to address behavioral development directly is Jochen Brandtstaedter (1984a, 1984b, 1998). The *action perspective* is particularly focused on how it may be possible to understand and explain *intentional behavior*. Action theorists are interested in how beliefs, plans, and expectations influence subsequent behaviors. In particular, they are interested in what intentional behavior means to the actor; how feeling affects actions; and how goals, evaluations, memories, and other cognitive events influence action (Eckensberger & Meacham, 1984).

Brandtstaedter (1998) defines actions as behavior that can be predicted and explained by reference to goals, values, and beliefs that are at least partly under personal control and are constrained by convention. As suggested by the ecological paradigm, situations constrain and potentiate actions and actions transform situations in accord with desired future states. Brandtstaedter maintains that actions are overdetermined (they often serve more than one intention) and they often have unexpected or unintended effects that lead to changes in goals and beliefs.

Expectations about age, he writes, are powerful predictors of action and, therefore, of development. Furthermore, all of our ideas about human development are important motivators of action. People begin, sometime during adolescence and young adulthood, to monitor their own development, to make choices and plans, and execute processes over long periods of time that are designed to improve the self.

Emotions are linked to or mediate between cognition and action, according to this view. Emotions signal mismatches between hopes and states of development and lead to actions. Thus, worry leads to planning and education, guilt leads to self-punishment or attempts at reparation, and so on. Actions are also strongly related to feelings of self-efficacy. If people believe that they are in control of their own development, they are more likely to create the conditions that enable change. Moveover, their behavior will remain flexible and adaptive over the lifespan. Brandtstaedter's view raises many concerns for the most empirically minded.

As Baltes and Brim (1984) caution, "For many behavior scientists the use of action and intention may initially spell a return to a 'philosophical' rather than a 'natural-science' approach to human behavior" (p. 137). On the other hand, it has been difficult for even the most dedicated behavioral scientists to talk about the interaction between person and environment without considering how people's plans, expectations, and beliefs influence subsequent behavior (Bandura, 1981). In fact, there has been a fascinating growth of the literature in transpersonal psychology that is devoted to the study of beliefs, values, and practices under the rubric of spirituality.

Transpersonal psychologists such as Ken Wilber (1995) make the strongest claims about the direction of development. In *Sex, Ecology and Spirituality*, Wilber has argued that development moves from egocentricity to ethnocentricity to world culture, from narcissism, toward autonomy, decentering, reflection, and a wider, deeper perspective. He presents many cross-cultural and historical exemplars of human maturity and argues that particular actions (such as meditation and participation in a spiritual community) must be undertaken to achieve the further reaches of development. He describes a stage of cognition, *vision-logic*, in which the individual can hold a variety of perspectives without privileging one over the other; can maintain paradoxes in mind, and is able to see the unity in opposites through a dialectical process not unlike what others have described as systems thinking.

In more mainstream developmental theory, work on human wisdom is the clearest about behavior and the issue of direction. Diedre Kramer (in press and this volume) presents a vision of wisdom not far removed from the transpersonal perspective. She writes of the ability to see and accept multiple perspectives, an awareness of the relativistic and uncertain nature of reality. She believes there are three to six dimensions in most of the work on wisdom including reflective-ness, emotional understanding, social unobtrusiveness, judicious understanding and action, skillful communication, and compassionate concern. Wisdom, she argues, stands at the intersection of cognition, emotion, and behavior.

Achenback and Orwoll (1991) discuss wisdom as the interaction of personal-ity, cognition, and conation (commitment to the philosophical and spiritual). They claim that wisdom allows one to transcend the self and gain access to ordi-narily unconscious information about the self and others. These insights, they argue, manifest themselves in behaviors that communicate understanding and caring. Labouvie-Vief (1997) and Pascual-Leone (1990) have proposed that wis-dom allows one to individuate from the conventional norms that govern adult behavior in order to integrate the inner self and internal, affective concerns with outer, conventional reality.

Perhaps because wisdom is generally thought to occupy a position at the intersection of multiple psychological dimensions, the topic of behavior is fre-quently addressed. Wisdom, it appears, manifests in wise actions. Kramer (in press) argues that wise people interact with others in ways that allow others to remain open rather than to respond defensively. Orwell and Perlmutter (1990) talk about compassionate action and Pascual-Leone (1990) argues for wisdom as the active exertion of will in order to counter and change automatic processes and achieve greater self-actualization. Even Baltes and Staudinger (1993), who treat wisdom primarily as a cognitive domain based in crystallized knowledge, suggest that wise judgment will produce wise action.

The work of Brandtstaedter on intentional self-development therefore seems quite relevant to the development of wisdom. Through intentional actions aimed at self-development, one might come to counter and change automatic maladap-tive processes. Because many aspects of wisdom, such as greater reflectivity and growth of empathy (Pascual-Leone, 1990), appear closely related to the breadth of life experience, it is not surprising that research on wise performance identifies older adults among the top scorers (Staudinger & Baltes, 1997; Staudinger, Lopez, & Baltes, 1997). Furthermore, people who are nominated as wise persons tend to hold positions of leadership, be employed in human services, or have exceptional experiences, such as being a Nazi resistor during the Third Reich (Staudinger, 1996). They are consistently seen as open to experience (Bacelar, 1998).

Kramer (in press) concludes that wisdom appears to "stem from a capacity to reflect on and grapple with difficult existential life issues" (p. 24). Wise people, she writes, are willing to explore the shadow side of existence and to express a wide array of human emotions. Wisdom may require a highly developed form of dialectical and relativistic thinking, and appears profoundly related to emotion and action. However, Kramer notes, wisdom is at least partly a reflection of the biases of the beholder. Much work on wisdom is done using samples that have been nominated as wise by others. There is little cross-cultural work on wisdom outside Euro-American populations and, of course, that is one of the primary con-cerns of those who grapple with the problem of direction in human development. To what extent can we, as Westerners, or even we as a group of developmental sci-entists across cultures, step outside our own thoughts to observe whether there are human universals?

CONCLUSIONS

Dynamic system models of human development appear to hold promise for many kinds of integrations, across domains of study within developmental psychology and the wider discipline of psychology, as well as across a variety of disciplines from neuropsychology, biology, and physiology to sociology and anthropology. Dynamic systems models encourage us to analyze human activities in all their complexity and variability and yet maintain our interest in order and patterning within that variation. This way of conceptualizing development calls for complex procedures and methods of analysis. Dynamic system models dictate, for instance, greater care about the frequency of sampling in longitudinal data, concern for cohort and historical effects, as well as study over a wider variety of conditions and contexts. They suggest statistical methods that permit pattern analysis and the identification of clusters of both skills and conditions. They encourage us to observe people in their interactions with others and with their environments, rather than in controlled laboratory conditions.

Dynamic systems models, however, leave open the question of direction. They may predict movement toward greater complexity, but they don't tell us whether that complexity is good or mature or wise. There are certainly ways in which it might be said that the question of direction is an empirical one, that with research over generations and various populations, we shall see whether development proceeds toward an endstate or simply toward greater diversity. However, we are also becoming sophisticated enough to know that the observer changes the observed in complex and unconscious ways we may never fully appreciate. Moreover, the development of science changes what we see and how we interpret those findings in radical and most probably unpredictable ways. Finally, perhaps, even the truth changes over time as our own history and context change. Alas, it seems unlikely that progress in developmental science will prevent us from having to grapple with the profound existential questions of human existence by mapping out the answers.

REFERENCES

Achenback, W. A., & Orwoll, L. (1991). Becoming wise: A psycho-gerontological interpretation of the book of Job. *International Journal of Aging and Human Development, 32*, 21–39.

Baltes, P. B., & Brim, O. G., Jr. (1984). Discussion: Some constructive caveats on action psychology and the study of intention. *Human Development, 27*, 135–139.

Baltes, P. B., Dittmann-Kohli, R., & Dixon, R. A. (1984). New perspectives on the development of intelligence in adulthood: Toward a dual-process conception and a model of selective optimization with compensation. In P. B. Baltes & O. G. Brim, Jr. (Eds.), *Life-span development and behavior, Vol. 6* (pp. 4–21). New York: Academic Press.

Baltes, P. B., Lindenberger, U., & Staudinger, U. M. (1998). Life-span theory in developmental psychology. In W. Damon & R. Lerner (Eds.), *Handbook of child psychology, Vol. 1* (pp. 1029–1143). New York: John Wiley & Sons.

Baltes, P. B., & Staudinger, U. M. (1993). The search of a psychology of wisdom. *Current Directions in Psychological Science, 2*, 1–6.

Bandura, A. (1981). Self-referent thought: A developmental analysis of self-efficacy. In P. B. Baltes & O. G. Brim, Jr. (Eds.), *Social cognitive development: Frontiers and possible futures.* New York: Academic Press.

Basseches, J. (1984). *Dialectical thinking and adult development*. Norwood, NJ: Ablex.

Bacelar, W. T. (1998). Age differences in adult cognitive complexity: The role of life experiences and personality. Doctoral Dissertation, Rutgers University, New Brunswick, NJ.

Bengston, V. L., Rice, C. I. J., & Johnson, M. L. (1999). Are theories of aging important? Models and explanations in gerontology at the turn of the century. In V. L. Bengston & K. W. Schaie (Eds.), *Handbook of theories of aging* (pp. 3–20). New York: Springer.

Bertalanffy, L. von (1968). *General system theory*. New York: Braziller.

Birren, J. (1999). Theories of aging: A personal perspective. In V. L. Bengston & K. W. Schaie (Eds.), *Handbook of theories of aging* (pp. 459–471). New York: Springer.

Brandtstaedter, J. (1984a). Personal and social control over development: Some Implications of an action perspective in life-span developmental psycholgy. In P. B. Baltes & O. G. Brim, Jr. (Eds.), *Life-span development and behavior, Vol. 6*. New York: Academic Press.

Brandtstaedter, J. (1984b). Action development and development through action. *Human Development, 27*, 11–29.

Brandtstaedter, J. (1998). Action perspectives on human development. In W. Damon & R. Lerner (Eds.), *Handbook of child psychology, Vol. 1* (pp. 467–561). New York: John Wiley & Sons.

Cattell, R. B. (1963). Theory of fluid and crystallized intelligence: A critical experiment. *Journal of Educational Psychology, 54*, 1–22.

Chinen, A. B. (1984). Modal logic: A new paradigm of development and late life potential. *Human Development, 27*, 52–56.

Cook-Greuter, S. R. (1999). *Postautonomous ego development: A study of it's nature and measurement*. Unpublished dissertation, Harvard University, Cambridge, MA.

Commons, M. L. (1999). Producing and measuring transition to higher stages in individuals, organizations and developing cultures. Paper presented at the meeting of the Society for Research in Adult Development, Salem, MA, June 18–20.

Commons, M. L., Armon, C., Richards, F. A., & Schrader, D. E. (1989). A multidomain study of adult development. In M. L. Commons, J. Sinnott, F. A. Richards, & C. Armon (Eds.), *Adult development, Vol. 1: Comparisons and applications of adolescent and adult development models*. New York: Praeger.

Commons, M. L., & Richards, F. A. (1982). A general model of stage theory. In M. L. Commons, F. S. Richards, & C. Armon (Eds.), *Beyond formal operations: Late adolescent and adult cognitive development*. New York: Praeger.

Dannefer, D., & Perlmutter, M. (1990). Development as a multidimensional process: Individual and constituents. *Human Development, 33*, 108–137.

Eckensberger, L. H., & Meacham, J. A. (1984). The essentials of action theory: A framework for discussion. *Human Development, 27*, 166–173.

Erikson, E. H. (1963). *Childhood and society*, 2nd edit. New York: W. W. Norton.

Fischer, K. W. (1999). Dynamics of adult cognitive-emotional development. Paper presented at the annual symposium of the Society for Research in Adult Development, Salem, MA, June 19–21.

Fischer, K. W., & Bidell, T. R. (1998). Dynamic development of psychological structures in action and thought. In W. Damon & R. M. Lerner (Eds.), *Handbook of child psychology, Vol. 1* (pp. 807–863). New York: John Wiley & Sons.

Ford, D. H., & Lerner, R. M. (1992). *Developmental systems theory: An integrative approach*. Newbury Park, CA: Sage.

Galaz-Fontes, J. F., & Commons, M. L. (1989). Desarrollow moral y educasion. *Revista Travesia, 18*, 5–8.

Hendricks, J., & Achenbaum, A. (1999). Historical development of theories of aging. In V. L. Bengston & K. W. Schaie (Eds.), *Handbook of theories of aging* (pp. 21–39). New York: Springer.

Horn, J. L. (1982). The theory of fluid and crystallized intelligence in relation to concept of cognitive psychology and aging in adulthood. In F. I. M. Craik & S. E. Truhub, *The 1980 Erindate Symposium*. Beverly Hills, CA: Sage.

Izard, C., & Ackerman, B. (1997). Emotions and self-concept across the life-span. In K. Schaie & M. Lawton (Eds.), *Annual review of gerontology and geriatrics, Vol. 17* (pp. 1–26). New York: Springer.

Kegan, R. (1982). *The evolving self-problem and process in human development*. Cambridge, MA: Harvard University Press.

Kegan, R. (1994). *In over our heads: The demands of modern life*. Cambridge, MA: Harvard University Press.

Kohlberg, L. (1973). The claim to moral adequacy of a higher stage of moral development. *Journal of Philosophy, 70*, 630–646.

Kramer (in press). Wisdom as a classical source of human strength: Conceptualization and empirical inquiry. *Journal of Social and Clinical Psychology*.

Labouvie-Vief, G. (1982a). Dynamic development and mature autonomy. *Human Development, 25,* 161–191.

Labouvie-Vief, G. (1982b). Growth and aging in life-span perspective. *Human Development, 25,* 65–78.

Labouvie-Vief, G. (1997). Cognitive-emotional integration in adulthood. In K. Schaie & M. Lawton (Eds.), *Annual review of gerontology and geriatrics, Vol. 17* (pp. 206–237). New York: Springer.

Labouvie-Vief, G., & Diehl, M. (1999). Self and personality development. In J. Cavanaugh & S. Whitbourne (Eds.), *Gerontology.* New York: Oxford University Press.

Levenson, M. R., & Crumpler, C. A. (1996). Three models of adult development. *Human Development, 39,* 135–149.

Linville, P. W. (1982). Affective consequences of complexity regarding the self and others. In M. S. Clark & S. T. Fiske (Eds.), *Affect and cognition.* Hillsdale, NJ: Lawrence Erlbaum.

Loevinger, J. (1976). *Ego development: Conceptions and theories.* San Francisco: Jossey-Bass.

Magai, C., & Passman, V. (1997). The interpersonal basis of emotional behavior and emotional regulation in adulthood. In K. Shcaie & M. Lawton (Eds.), *Annual review of gerontology and geriatrics, Vol. 17* (pp. 104–137). New York: Springer.

Magnusson, D., & Stattin, H. (1998). Person-context interaction theories. In W. Damon & R. Lerner (Eds.), *Handbook of child psychology, Vol. 1* (pp. 685–759). New York: John Wiley & Sons.

Maslow, A. H. (1968). *Toward a psychology of being.* New York: Van Nostrand.

Maslow, A. H. (1970). *Motivation and personality,* 2nd edit. New York: Harper & Row.

Nemiroff, R. A., & Colarusso, C. A. (Eds.) (1990). Discussion. In *New dimensions in adult development (pp. 165–169). New York: Basic Books.*

Orwell, L., & Perlmutter, M. (1990). The study of wise persons: Integrating a personality perspective. In R. J. Sternberg (Eds.), *Wisdom: Its nature, origins, and development* (pp. 160–177). Cambridge: Cambridge University Press.

Pascual-Leone, J. (1983). Growing into human maturity: Toward a metasubjective theory of adult stages. In P. B. Baltes & O. Brim (Eds.), *Life-span development and behavior, Vol. V* (pp. 117–156). New York: Academic Press.

Pascual-Leone, J. (1990). An essay on wisdom: Toward organismic processes that make it possible. In R. J. Sternberg (Ed.), *Wisdom: Its nature, origins, and development* (pp. 244–278). Cambridge, MA: Cambridge University Press.

Prigogine, I., & Stengers, I. (1984). *Order out of chaos: Man's new dialogue with nature.* New York: Bantam Books.

Reese, H. W. (1983). Some notes on the meaning of the dialectic. *Human Development, 26,* 315–320.

Riegel, K. F. (1973). Dialectic operations: The final period of cognitive development. *Human Development, 16,* 346–370.

Riegel, K. F. (1977). History of psychological gerontology. In J. E. Birren & K. W. Shaie (Eds.), *Handbook of the psychology of aging,* 2nd edit. New York: Van Nostrand.

Rybash, J. M., Hoyer, W., & Roodin, P. (1986). *Adult cognition and aging.* New York: Pergamon Press.

Schulz, R., & Heckhausen, J. (1997). Emotion and control: A life-span perspective. In K. Schaie & M. Lawton (Eds.), *Annual review of gerontology and geriatrics, Vol. 17* (pp. 185–205). New York: Springer.

Shweder, R. A., Goodnow, J., Hatano, G., LeVine, R. A., Markus, H., & Miller, P. (1998). The cultural psychology of development: One mind, many mentalities. In W. Damon & R. Lerner (Eds.), *Handbook of child psychology, Vol. 1* (pp. 865–937). New York: John Wiley & Sons.

Smith, J., & Baltes, P. B. (1999). Differential psychological aging: Profiles of the old and the very old. *Ageing and Society, 13,* 551–587.

Stattin, H., & Magnusson, D. (1990). Pubertal maturation in female development. In D. Magnusson (Ed.), *Paths through life, Vol. 2.* Hillsdale, NJ: Lawrence Erlbaum.

Staudinger, U. (1996). Wisdom and the social-interactive foundation of the mind. In P. B. Baltes & U. M. Staudinger (Eds.), *Interactive minds: Life-span perspectives on the social foundation of cognition* (pp. 276–315). New York: Cambridge University Press.

Staudinger, U., & Baltes, P. B. (1997). Interactive minds: A facilitative setting for wisdom-related performance? *Journal of Personality and Social Psychology, 71,* 746–762.

Staudinger, U., Lopez, D. F., & Baltes, P. B. (1997). The psychometric location of wisdom-related performance: Intelligence, personality and more? *Personality and Social Psychology Bulletin, 23,* 1200–1214.

Stevens-Long, J. (1979). *Adult life*. Palo Alto, CA: Mayfield.

Stevens-Long, J. (1990). Adult development: Theories past and future. In R. A. Nemiroff & C. A. Colarusso (Eds.), *New dimension in adult development*. New York: Basic Books.

Stevens-Long, J., & Commons, M. L. (1992). *Adult life*, 4th edit. Palo Alto, CA: Mayfield.

Stevens-Long, J. (2000). The prism self: Personality and transcendence. In M. Miller & P. Young-Eisendrath (Eds.), *The psychology of mature spirituality: Integrity, wisdom and transcendence*. London: Psychology Press LTD (Taylor and Francis).

Thelen, E., & Smith, L. B. (1998). Dynamic systems theory. In In W. Damon & R R. Lerner (Eds.), *Handbook of child psychology, Vol. 1* (pp. 807–863). New York: John Wiley & Sons.

Vaillant, G. E. (1977). *Adaptation to life*. Boston: Little, Brown.

Vaillant, G. E. (1993). *The wisdom of the ego*. Cambridge, MA: Harvard University Press.

Van den Dale, L. (1975). Ego development and preferential judgment in life-span perspective. In N. Datan and L. Ginsberg (Eds.), *Life-span developmental psychology: Normative life crisis*. New York: Academic Press.

van Geert, P. (1991). A dynamic system model of cognitive and language growth. *Psychological Review, 98*, 3–53.

van Geert, P. (1993). A dynamic systems model of cognitive growth: Competition and support under limited resource conditions. In L. B. Smith & E. Thelen (Eds.), *A dynamic systems approach to development: Applications*. Cambridge, MA: The MIT Press.

van Geert, P. (1994). *Dynamic systems of development*. London: Harvester Wheatsheaf.

Wapner, S., & Demick, J. (1998). Development analysis: A holistic, developmental systems-oriented approach. In W. Damon & R. Lerner (Eds.), *Handbook of child psychology, Vol. 1* (pp. 761–806). New York: John Wiley & Sons.

Werner, H. (1957). The concept of development from a comparative and organismic point of view. In D. B. Harris (Ed.), *The concept of development: An issue in the study of human behavior*. Minneapolis, MN: University of Minnesota Press.

Wilber, K. (1995). *Sex, ecology and spirituality: The spirit of evolution*. Boston: Shambala Press.

CHAPTER 2

Learning in Adulthood

WILLIAM J. HOYER AND DAYNA R. TOURON

In recent years, a number of theories and frameworks have emerged that try to address both the potentials and limitations of effective cognitive and social functioning during the adult years (e.g., Baltes & Staudinger, 2000; Baltes, Staudinger, & Lindenberger, 1999; Hoyer & Rybash, 1994; Lemme, 1999; Rowe & Kahn, 1997; Seligman & Csikszentmihalyi, 2000; Staudinger & Pasupathi, 2000; Wapner & Demick, this volume). Such frameworks have aided the articulation of the characteristics of adult development by integrating observations that would otherwise have been disconnected pieces of a puzzle and less meaningful. In the study of adult cognitive development, for example, much of the available data and theory address the description and explanation of age-related cognitive decline. However, everyday observations and the results from some studies suggest that there are improvements or stability as well as declines in cognitive function during the adult years. The aim of the present chapter is to describe what is known about the potentials as well as limits of learning during the adult years.

The processes and outcomes of learning influence the nature and course of adult development, and reciprocally, developmental variables influence the processes and products of learning. Although it is important to consider learning as an antecedent of developmental change, the focus of this chapter is on the description and explanation of age-related differences in the products and processes of learning. Because the term *learning* is used in reference to a wide variety of levels of behavior change, and because the concepts of learning and development overlap in that both refer to behavior change over time, we begin by examining some of the differences between concepts of learning and development.

LEARNING IN A DEVELOPMENTAL CONTEXT

The concepts of learning and development can be distinguished along two dimensions. First, in terms of the inclusiveness or scope of the behavior and of

WILLIAM J. HOYER • Department of Psychology, Syracuse University, Syracuse, New York, 13244-2340.
DAYNA R. TOURON • School of Psychology, Georgia Institute of Technology, Atlanta, Georgia 30332.

Handbook of Adult Development, edited by J. Demick and C. Andreoletti. Plenum Press, New York, 2002.

the antecedents of change, learning refers to the effects of practice or experience on behavior whereas development refers to a wider variety of influences that are associated with time-related change (Baltes et al., 1999; Bronfenbrenner & Ceci, 1994). It is generally accepted that developmental change is multidetermined (e.g., by genetic programs and neurobiological clocks as well as by external influences) and multidirectional (i.e., there are gains as well as losses). A quote from Talland (1965) illustrates the multidimensionality of adult development, as follows:

> I am still puzzled by the contrast of the athlete, who at thirty, is too old for the championship and the maestro, who at eighty, can treat us to a memorable performance on the concert stage ... Are our aged masters freaks of nature, paragons of self-discipline, or do they demonstrate the inadequacy of our present notions about the effects of age on human capacities? (p. 558)

Learning is also multidetermined and multidirectional, but the outcomes of learning are measured in terms of gains or improvements in knowledge or skills that occur at different rates and to different degrees at different ages (e.g., Ruch, 1934).

Second, the concepts of development and learning differ in regard to the specificity and durability of change. That is, the learning of a set of relatively specific cognitive skills might endure or remain asymptotic for minutes or maybe weeks, whereas the time-related processes associated with the development of a set of general cognitive skills might endure for years or decades (but see Feldman, 1995). Learning refers to practice-based improvements in the speed, accuracy, or efficiency of performing specific tasks and skills, whereas cognitive development refers to changes in task-general abilities (e.g., see Baltes, 1987; Bronfenbrenner & Ceci, 1994; Hoyer & Rybash, 1994, 1996; Karmiloff-Smith, 1992; Lerner, 1991).

PRODUCTS AND PROCESSES OF LEARNING

Study of the relationships between learning and development have changed fundamentally in recent years because learning is now seen not only as the acquisition of domain knowledge, but also as the storage and use of that knowledge (e.g., see Halford, 1995; Pascual-Leone, 1995). The processes of learning and memory serve to produce dynamic and diverse contents and abilities, from perceptual phenomena to the representations of abstract domains of knowledge. Recognition of the fact that there are different developmental trends for different cognitive abilities or products can be traced to the early work of Hollingworth (1927), Jones and Conrad (1933), and Willoughby (1927). Jones and Conrad's data showing age trends for measures of general information, vocabulary, and analogies from the Army Alpha test battery are illustrated in Fig. 2.1. As pointed out by Salthouse (2002), these data illustrate that there are different age trends for measures of the products (general information, vocabulary) and the processes of cognition (analogies), and that cognitive functioning is more likely to rely on accumulated information with advancing age (e.g., see also Hoyer, 1980, 1985, 1986). Research findings bearing on the distinction between the processes and products of learning and age differences in the use of acquired skills are discussed later in the chapter.

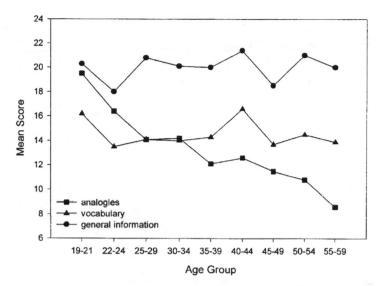

Figure 2.1. Mean scores for measures of General Information, Vocabulary, and Analogies from the Army Alpha Test for different age groups between 20 and 55 years. The data are from Jones and Conrad (1933). Note that age differences are negligible for the measure of general information, and that there is an age-related decline in performance on analogies.

AGE-RELATED DIFFERENCES IN THE PROCESSES OF LEARNING

Except in situations where there is opportunity to rely on previously acquired skills or products of learning, research findings are consistent in demonstrating that there are age-related declines on measures of basic learning (e.g., for excellent reviews, see Arenberg & Robertson-Tchabo, 1977; Hultsch & Dixon, 1990; Kausler, 1994, MacKay & Burke, 1990). The consistency of the data is illustrated in Fig. 2.2. Data from these early studies examining the relationship between age and performance on associative learning tasks clearly indicate a reliable pattern of decline (Gladis & Braun, 1958; Hulicka, 1966; Monge, 1971; Smith, 1975; Thorndike, Bregman, Tilton, & Woodyard, 1928).

There are some results to suggest that adult age differences are larger for some measures of learning than for others (e.g., for reviews see Howard & Wiggs, 1993; Kausler, 1994). In an early study by Ruch (1934), for example, the amount of age differences in paired associate learning for equations that were false or that were meaningless was larger than the amount of age differences for paired associate learning of word pairs. These data are illustrated in Fig. 2.3.

One important factor that differentiates tasks that show different amounts of performance differences for different age groups is the complexity of the task requirements. Generally, tasks and the conditions of tasks that are more complex produce larger age differences. The effect is ubiquitous. Across many different kinds of task conditions, the performance of older adults is almost completely described as a ratio of the performance of younger adults (e.g., Cerella, 1990, 1991). Such a general finding might seem trivial or uninteresting at first glance,

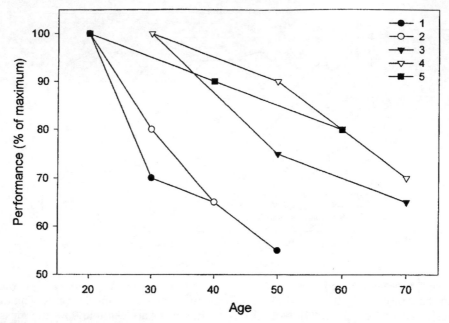

Figure 2.2. The results of five studies showing a consistent relationship between adult age and performance on measures of paired-associate learning. The circles refer to studies by (1) Monge, 1971; (2) Thorndike et al., 1928; (3) Gladis & Braun, 1958; (4) Smith, 1975; and (5) Hulicka, 1966. For each study, performance is shown as a percentage of the maximum score for the different age groups.

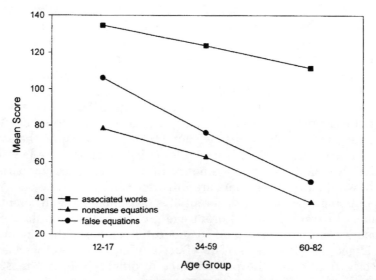

Figure 2.3. Age differences in mean correct responses for paired-associate learning of three different kinds of materials. (Adapted from Ruch, 1934.)

but its ubiquity gives it substantial descriptive power, as well as import for the proper interpretation of findings that appear to indicate age by task interactions in learning. Observed age-related deficits in measures of learning processes are largely the result of the effects of age-related slowing on the performance of each component or each step of a complex learning task, and not the result of an age deficit associated with particular types of learning. Even the findings of studies suggesting that age differences are larger for visual–spatial tasks than for lexical tasks can be accounted for by age-complexity effects (e.g., Jenkins, Myerson, Joerding, & Hale, 2000; Myerson & Hale, 1993; Myerson, Hale, Rhee, & Jenkins, 1999; Zheng, Myerson, & Hale, 2000).

The most prominent age deficits in learning have to do with a slowing in the speed of performance across task conditions. The performance of older adults can be precisely described in terms of a ratio of about 1.4 to 1.7 of the performance of young adults. To quote Cerella (1991):

> The effects in some 288 experimental conditions are primarily determined by a single aspect of the information processing requirement, namely, task duration. The evidence is near-to-overwhelming that age is experienced, at least to a first approximation, as some sort of generalized slowing... success over such a diversity of data suggests that aging effects stem from some elementary aspect of the biology of the nervous system. (pp. 220–221)

The consequences of age-related slowing can be readily observed in tasks that entail the active and simultaneous processing and storing of information (e.g., Salthouse, 1996). The term *working memory* is used to refer to the kind of processing that occurs when some information needs to be held in memory while computations are carried out using other information. It is well known that age differences in performance are particularly large for tasks that seem to require working memory (e.g., Myerson, Hale, Rhee, & Jenkins, 1999; Verhaeghen, Kliegl, & Mayr, 1997; Zachs, Hasher, & Li, 2000). Research bearing on the neural substrates underlying age-related deficits in working memory suggests that the coordination of computational and storage processes requires cooperation among scattered areas of the brain (Raz, 2000). Which specific areas are activated depends on whether the task requires spatial or semantic memory. Working memory has been associated with the prefrontal cortex. Structural imaging studies indicate shrinkage of the prefrontal cortex with aging. However, functional imaging studies indicate increased activation in the prefrontal areas for older adults than for younger adults (e.g., see Raz, 2000, for a review). These findings can be interpreted as suggesting that older adults use more cognitive resources to carry out demanding cognitive tasks.

It is interesting to note that individuals who are middle-aged show a relatively variable or differentiated pattern of performance across measures of the products of learning compared with inter-task differences in performance for younger adults and older adults. Differences within-individuals in variability across tasks have been attributed to differential patterns of age-related deficits associated with particular learning processes or mechanisms (e.g., Rogers, Hertzog, & Fisk, 2000), and to individual differences in deliberate skill development (e.g., Charness, Krampe, & Mayr, 1996; Clancy & Hoyer, 1994; Hoyer & Ingolfsdottir, in press; Krampe & Ericsson, 1996). It is possible that differentiated patterns of cognitive skills and reliance on the products of learning are most apparent during the middle years because the age-ordered constraints associated with the neurobiological mechanisms of learning are likely to be least influential at this period (e.g., Flavell, 1970; Hoyer & Rybash, 1994).

Individual differences in speed of processing and working memory are general factors that have been used to explain age-related differences in many aspects of cognition including learning (Salthouse, 1994; Salthouse & Coon, 1994). Developmental changes in learning and cognitive skill acquisition have also been attributed to differences in task strategies (e.g., Kramer, Larish, & Strayer, 1995; Rogers & Gilbert, 1997; Rogers et al., 2000; Strayer & Kramer, 1994). That is, it has been shown that older adults often use different strategies in cognitive tasks, and that some of the age-related difference observed in learning are due to the use of non-optimal strategies. In terms of the development of automaticity, for example, the effects of practice on performance can be attributed to improved computational efficiency, to the use of item memory in lieu of computation, or to other factors such as increased familiarity with the task characteristics (e.g., Charness & Campbell, 1988; Fisk & Rogers, 1990; Fisk, Cooper, Hertzog, Anderson-Garlach, & Lee, 1995; Touron, Hoyer, & Cerella, 2001; Jenkins & Hoyer, 2000).

AGING AND THE DEVELOPMENT OF AUTOMATICITY

Generally performance improves with practice, and performance sometimes becomes automatic as a result of consistent practice. Consistent practice refers to task conditions that involve using the same response to repeated presentations of the same stimulus. Learning goes from being a deliberate computational process to an automatic process with consistent practice. Early in practice, performance is controlled—typically slow and effortful. As practice continues in a consistent format, performance eventually becomes fast, effortless, and automatic. However, older subjects do not reach young levels of automatic performance (Fisk & Rogers, 1991; Jenkins & Hoyer, 2000; Plude et al., 1983; Fisk & Rogers, 1991). Theories of skill learning and automaticity emphasize two processes, practice-related changes in the speed-of-processing and repetitions-based accumulation of information about particular stimuli (e.g., Logan, 1988; Rickard, 1997; Strayer & Kramer, 1994).

For some kinds of learning tasks, age differences in the initial level of performance are large, there are age differences in the rate of learning across training, and the age difference in performance remains after extensive training. For example, in an early study by Thorndike et al. (1928), right-handed young adults between the ages of 20 and 25 years and right-handed older adults between the ages of 35 and 57 years were given 15 hours of practice writing left-handed. Large age differences were found in the rate at which writing speed improved with practice, and age differences did not diminish with practice. Similar patterns of results have been reported in recent studies (Czaja & Sharit, 1993; Kramer et al., 1995; Salthouse, Hambrick, Lukas, & Dell, 1996).

For other kinds of tasks, age differences in the initial level of performance are large, but there are equivalent rates of learning across training, and performance reaches an equivalent level for younger and older adults by the end of training. In the execution of well-learned, real-world skilled tasks, for example, it has been reported that older adults perform without noticeable deficits (e.g., Salthouse, 1984).

Precise descriptions of age differences in the rate of learning or in the rate of development of automaticity can be derived from fitting power functions to individual data, and then analyzing the parameters of the power function fits for different age groups. A power function depicts learning as a negatively accelerated improvement in performance as a function of practice (e.g., see Fig. 2.4). Performance on many different types of learning tasks is well described by fitting a power function to the data (e.g., Touron, Hoyer, & Cerella, 2001; Lincourt, Cerella, & Hoyer, 2000; Logan, 1988; Newell & Rosenbloom, 1981). For the power function, $RT = a + bN^{-c}$, where RT is response time and N is the number of practice trials, the a parameter represents asymptotic performance, the b parameter represents improvement span, and the c parameter represents rate of change or learning rate. In a recent study, Touron, Hoyer, and Cerella (2001) examined the rates of learning, the amount of practice required to reach asymptote, and asymptotic performance for younger and older adults by analyzing the parameters of a power function representing these aspects of acquisition. The parameter values were derived from individual data and were averaged for the purposes of age group comparisons. For illustration, Fig. 2.5 shows two hypothetical power function curves that differ in terms of the values of the a, b, and c parameters. Curve 1 resembles the performance of a group of older adults and curve 2 resembles the performance of a group of younger adults. A difference in the a parameter indicates that the groups reach different levels of performance as a result of training. Differences in the b parameter (and/or the a parameter) are produced by differences in starting values and indicate the amount of improvement that can be brought about by practice. The starting value is calculated by adding the b parameter and the a parameter. Obtained differences in the exponents of the power function, c, indicate differences in the learning rate (e.g., $c_{old} < c_{yng}$). In the Green, Hoyer, and Cerella study, analyses of the parameters derived from the power function fits revealed age differences favoring young adults in improvement span, rate of learning, and asymptotic performance. Differences in the development of automaticity seemed to be due to the efficiency of shifting from computation to instance-based retrieval.

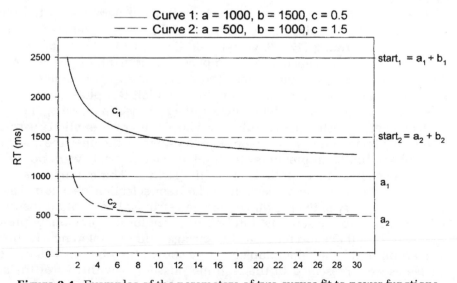

Figure 2.4. Examples of the parameters of two curves fit to power functions.

Fisk, Hertzog, Rogers, and their colleagues have also reported that older adults and younger adults do not show equivalent gains as a result of consistent practice (e.g., Fisk, Cooper, Hertzog, Anderson-Garlach, & Lee, 1995; Fisk & Rogers, 1991; Hertzog, Cooper, & Fisk, 1996). These investigators suggested that there is an age-related deficit in the development of an *automatic attention response* to consistently mapped items in visual search, and that there is no age-related deficit in *memory set unitization* in memory search.

IMPLICIT LEARNING VS. EXPLICIT LEARNING

Learning in older adults is less likely to be impaired on measures of non-intentional acquisition than on measures of explicit learning. Evolutionary considerations provide some justification for a distinction between nonintentional learning and explicit learning (Reber, 1993). Reber argued that unconscious, implicit, covert cognitive functions predated the emergence of the conscious, explicit functions of mind. Because implicit learning is a phylogenetically more primitive system than explicit learning, implicit learning might be *less vulnerable* than explicit learning to the neurological insults of aging. In contrast to measures of the individual's deliberate efforts to learn, implicit learning measures presumably tap the nonintentional aspects of the acquisition of knowledge about the structure of a complex stimulus environment (Frensch, 1998).

One of the key issues in evaluating the usefulness of the distinction between explicit and implicit learning vis-à-vis aging has to do with what information is learned. Two kinds of tasks have been used to investigate adult age differences in implicit learning: grammar learning and serial reaction time. There are pros and cons associated with the use of each of these kinds of tasks, but the serial reaction time generally provides a cleaner measure of nonintentional rule acquisition and the learning of simple covariations or associations inherent in the tasks (e.g., see Perruchet, 1994).

D'Eredita and Hoyer (1999) examined adult age differences in the implicit learning and explicit learning of abstract rules and of item-to-item covariations in figural sequences. Young (17–23 years), middle-aged (35–45 years), and older adults (55–65 years) participated in a two-phase experiment. In phase 1, research participants were given sufficient practice so as to learn strings of figural sequences to the same criterion. In phase 2, the implicit/explicit test phase, subjects made forced-choice judgments about the sequences they learned in phase 1, under either explicit or implicit instructions. Subjects who were given explicit instructions were told that some of the strings would be the same as the ones they saw in the learning phase. Unlike in previous studies, however, subjects were not told that there was a grammar that applied across the strings. The explicit instructions were intended to direct the subjects to make item selection decisions based on their explicit memory of the previously learned strings. The implicit instructions were intended to tap the subjects' implicit memory for the previously presented strings by asking them to complete the strings without reference to previous string learning or rule learning. Subjects were shown the same strings in the testing phase as were presented in the learning phase, but with two of the shapes removed. Subjects were asked to complete each string by selecting one of two choices. Thus, instead of making judgments about the rule-based nature of each string, subjects were completing each string, either to make it the same as a previously seen string in the explicit instructions condition or to make a "better fit" in the implicit instructions condition. Analyses revealed evidence for the acquisition of item-to-item covariations. There was no evidence of implicit learning of the abstract rules of the artificial grammar. As expected, an age-related deficit was found for explicit relearning of grammar-following sequences.

D. V. Howard and J. H. Howard (1992) used a serial reaction time task to examine adult age differences in response times that occurred over a series of blocks containing a repeated pattern for the locations of asterisks on a computer screen. Response times improved with repetitions of the sequence for both younger and older adults. This increase in the speed of responding was due primarily to subjects' learning the pattern rather than to a general practice-based improvement in visual–motor speed, because it was found that response times nearly returned to their original level on blocks of trials in which the locations of the asterisks occurred in a random pattern. The increase in response times when the pattern was removed reflects the extent to which response times were influenced by the presence of the repeating pattern apparently without the subjects being aware of learning any pattern. The lack of age differences on implicit measures of serial reaction time contrasts with the data obtained from a comparable situation in which subjects attempted to generate or predict in advance the locations of asterisks; these explicit measures revealed reliable age-related deficits.

An earlier study by Howard and Howard (1989) also used the serial reaction time task to assess implicit and explicit learning in younger and older adults. Participants responded to asterisks presented in four different locations on a computer screen whose positions followed a repeating pattern. Subjects were instructed to press a button that corresponded to one of four possible item locations. After a series of repetitions of the pattern, a random series of items was presented. An increase in response times associated with the switch from the repeating sequence to the random sequence was used as a measure of implicit learning. Overall, older adults produced longer response times, but the switch from the patterned blocks to the random blocks was equally disruptive for young adults and older adults. As an explicit measure of sequence learning, subjects were asked to predict the location of the next item to be presented based on previous learning. The explicit learning measure revealed an age-related deficit.

Howard and Howard (1997) also examined pattern learning in younger and older adults under conditions designed to be more difficult for serial pattern learning. In this study, the sequences of the asterisks alternated between a repeating pattern and a random pattern. Subjects representing three age groups, including a group of six old–old adults ranging in age from 76 to 80 years, were able to learn this relatively difficult sequence of covariations of locations. In contrast to earlier findings by Howard and Howard (1989, 1992) in which younger and older adults showed equal amounts of pattern learning on an implicit measure, there were age differences in the magnitude of pattern learning and in the sensitivity to patterns. The data suggested that only the younger adults showed sensitivity to the higher order dependencies in the sequence.

Cherry and Stadler (1995) examined the implicit serial learning of relatively nonobvious sequences in a sample of young adults and samples of low-ability and high-ability older adults. The two samples of older adults differed in educational attainment, occupational status, and verbal ability, and these groups performed differently on implicit sequence learning. That is, the amount of disruption produced by changing from a repeated sequence to a random sequence using the Nissen and Bullemer task was not different for the samples of younger adults and high-ability older adults, but the sample of low-ability older adults showed less evidence of implicit learning. The samples of younger adults and the high-ability older adults were also more accurate on an explicit learning measure than the sample of low-ability older adults.

In addition to age differences in the ability of discern a structure in the sequence, it has been reported that the complexity of the task affects the degree of implicit learning obtained by older adults. In one study, Frensch and Minor (1994) reported age differences in implicit learning under dual-task conditions, but not under single-task conditions. In the French and Minor study (experiment 1), implicit learning was found at short response–stimulus intervals (500 ms), but not at long response–stimulus intervals (1500 ms); explicit learning was found at both response–stimulus intervals, although it was less at the longer response–stimulus intervals. This finding suggests that implicit sequence learning depends on the coactivation of stimuli for associations to develop in short-term memory. Harrington and Haaland (1992) reported adult age differences in implicit learning for highly educated participants under single-task conditions when the task required complex hand movements to make the responses instead of simple button presses. Thus, age deficits do emerge in implicit learning tasks when the task is relatively difficult or when the older subjects are of relatively low ability.

Related to these findings, it is consistently reported that there are age-related differences in discerning higher order dependencies in studies of the effects of adult age on explicit detection of relatively complicated event covariations (e.g., Kay, 1951; Mutter & Pliske, 1996). Kay (1951) showed that age differences in the accurate performance of an explicit sequence learning task depended on unlearning, then rebuilding a correct schema for performing the sequence. Kay presented subjects with a row of five buttons and five lights. There was a correct sequence for pressing the buttons (2, 4, 3, 1, 5), and the task was to explicitly learn the sequence by trial and error. Feedback was signaled by a change in one of the five lights whenever a correct button was pressed. Older adults performed poorly on this task compared with younger adults, and the effects of age on reaching the criterion of two correct sequences increased from the 20s through the 30s, 40s, 50s, and 60s. In terms of current descriptive frameworks, Kay's findings are compatible with an account that emphasizes age-related declines in working memory because the correct response at each position had to be found before proceeding to the next, and the products of learning had to be held in memory. The results suggested that schema correction was required, and that learning depended not just on a process of gradually eliminating errors.

Relative sparing of implicit learning, compared with deficits in explicit learning of a series, have been reported in studies comparing healthy and impaired older adults (e.g., Ferraro, Balota, & Connor, 1993; Knopman & Nissen, 1987; Mutter, D. V. Howard, J. H. Howard, & Wiggs, 1990; Mutter, D. V. Howard, & J. H. Howard, 1994). Dissociations between implicit and explicit measures of learning have been found in studies with clinically impaired samples of

Korsakoff's patients (Nissen & Bullemer, 1987), Huntington's disease patients (Knopman & Nissen, 1991), and Alzheimer's disease patients (Ferraro et al., 1993; Knopman & Nissen, 1987). Also, Nissen, Knopman, and Schacter (1987) observed a similar dissociation in college students when they were administered the drug scopolamine so as to examine sequence learning under conditions of suppressed awareness and to simulate the effects of aging on learning systems. A connection between the distinctive characteristics of learning systems and the effects of aging and insult on particular brain functions is an active area of research in cognitive neuroscience (e.g., Knowlton, Ramus, & Squire, 1992; Pascual-Leone, Grafman, & Hallatt, 1994; Rybash, 1996). There is some evidence to suggest that the neural substrates that support implicit learning and implicit memory may be relatively unaffected by aging and by the kinds of disorders that compromise explicit learning (e.g., see Rybash, 1996; Squire, Knowlton, & Musen, 1993). The available neural evidence suggests that implicit learning and explicit learning depend on nonidentical neural mechanisms (e.g., Gabrieli, 1994; Squire et al., 1993). Implicit learning seems to depend on the integrity of the striatum, cerebellum, amygdala, and neocortex, whereas explicit learning is associated with the integrity of limbic and diencephalic brain structures.

AGE AND UTILIZATION OF THE
PRODUCTS OF LEARNING

William James (1890) was probably one of the first to call attention to the distinction between acquisition processes and the products of learning. He stated:

> When we are learning to walk, to ride, to swim, skate, fence, write, play, or sing, we interrupt ourselves at every step by unnecessary movements and false notes. When we are proficient, on the contrary, the results not only follow with the very minimum of muscular action requisite to bring them forth, they also follow from a single instantaneous "cue." The marksman sees the bird, and before he knows it, he has aimed and shot. A gleam in his adversary's eye, a momentary pressure from his rapier, and the fencer finds he has instantly made the right parry and return. A glance at the musical hieroglyphics, and the pianist's fingers have rippled through a cataract of notes. (p. 86)

That is, the processes of acquisition may not be evident in measures of the utilization of the products of acquisition. The distinction between acquisition and the utilization of cognitive skills can be especially poignant when observing the performance of middle-aged and older adults in skilled tasks. Frequently, middle-aged adults perform skilled tasks without any noticeable deficit even though the processes that were involved in skill learning have been affected by aging.

Ericsson and Kintsch (1995) suggested that maintenance of chunks, units, associations, and instance-based representations may underlie the maintenance of expert performance in skilled tasks, and that access to expert knowledge may be triggered by a cue. Thus, the availability of formed chunks, units, and instances may enable an older individual to bypass or circumvent speed of processing limitations. In other words, learning as well as the products of learning (i.e., the development of skilled performance) depends on memory retrieval supported by an associative memory system, such that quick access to learned knowledge is triggered by cues, bypassing limitations age-sensitive computational processes.

Such an account does not apply to understanding expert performance in domains that appear to require effortful computation and/or integration of novel information; the manifestation of extraordinary skills of this type probably depends heavily on the status of the individual's repertoire of fluid mental abilities. Certainly, a comprehensive account of effective performance in real-world learning tasks depends on the person's knowledge as it applies to the task at hand, and on the functioning of the person's fluid abilities as they apply to computational or processing demands of the task (e.g., Hoyer, 1985, 1986; Krampe & Ericsson, 1996; Salthouse, 1984).

Some researchers have examined the skilled performance of adults of different ages for the purpose of trying to describe how it is that middle-aged and older adults seem to perform as well as younger adults in skilled cognitive tasks, despite substantial evidence indicating age-related deficits on novel or nonskilled cognitive tasks. Findings from a number of recent investigations of the interactive effects of age and experience suggest that particular work environments or life experiences contribute to the maintenance of a relatively high level of cognitive performance in the later years (e.g., Charness, 1981; Charness & Bosman, 1990; Clancy & Hoyer, 1994; Hoyer & Ingolfsdottir, in press; Krampe & Ericsson, 1996; Morrow, Leirer, Altieri, & Fitzsimmons, 1994; Salthouse, 1984; Shimamura, Berry, Mangels, Rusting, & Jurica, 1995). These studies as well as everyday observations of maintained cognitive expertise in selected domains for some older adults support the view that access to the products of learning can be maintained in at least some domains during the middle-adult years.

However, in contrast to these reports, most studies fail to demonstrate a clear advantage of experience in attenuating age-related deficits in cognitive performance (e.g., Czaja & Sharit, 1998; Halpern, Bartlett, & Dowling, 1995; Hambrick, Salthouse, & Meinz, 1999; Hardy & Parasuraman, 1997; Meinz, 2000; Meinz & Salthouse, 1998; Salthouse & Mitchell, 1990). There are several possible reasons for the paucity of evidence demonstrating the developmental benefits of domain-specific experience. First, for many kinds of laboratory and real-world tasks, there is relatively little spare variance to be accounted for by experience effects because of the pervasive influences of age-related slowing on measures of cognitive performance.

A second issue related to the selection and sensitivity of dependent measures has to do with realism or naturalness and the extent to which selected measures tap the mechanism or skill that is the locus of maintained expertise. It is difficult to bring the real world into the research lab without losing what very well might be the source of the performance advantages inherent in real-world tasks.

LEARNING AND BRAIN PLASTICITY

Considering the inevitability of age-related cognitive slowing and its negative consequences for the efficiency of learning during the adult years, how and to what extent it is possible to improve, sustain, or restore effective behavioral and cognitive function? The term *brain plasticity* refers to an individual's potential for change and especially the potential for growth or maintenance or restoration of function in response to loss or disease. Recent conceptions of brain plasticity are based on a number of findings showing that some aspects of neural circuitry and synaptic connectivity are capable of growth and repair throughout life. Brain plasticity and effective cognitive function may be interdependent in that continued active involvement in cognitive activities might serve to facilitate brain plasticity or at least lead to the use of alternative strategies for effective cognitive and behavioral functioning. That is, it is likely that brain plasticity itself is partially an outcome of deliberate cognitive efforts to restore, protect, or promote function, and partially an innate reserve.

There is evidence demonstrating that associative learning, especially deliberate or explicit learning, produces changes in cortical sensory and motor function. Among the many important questions still to be answered is how intraneuron and interneuron function is altered in response to stimulation. Recent work examining long-term potentiation, for example, shows how changes in the sensitivity of neural transmission are affected by repeated stimulation and how such changes in sensitivity are maintained across time. Repeated stimulation produces alterations in the neurochemical functions of presynaptic and postsynaptic neurons.

Although relatively little is known about the potential for learning across the adult lifespan, the extent to which age-related deficits can be remedied might provide an index to cognitive potential and cognitive plasticity in the adult years (e.g., see Baltes, 1997; Baltes & Kliegl, 1992; Kliegl, Smith, & Baltes, 1990; Lindenberger & Baltes, 1995). Figure 2.3 shows the results of the Kliegl and Baltes' study in which younger adults and older adults were given training and extensive practice in using the Method of Loci to improve recall of word lists. Individuals were trained to form visual associations between to-be-remembered words and a familiar series of locations. Using this technique, recall is cued by mentally travelling through the familiar sequence, and retrieving the associated word at each locus. Figure 2.5 shows that the recall performance of both age groups improved with training and practice across sessions, but that the performance of the younger group improved more with practice. These data tell us about the extent to which memory can be improved with training. That training serves to enlarge age differences suggests that there are age differences in the plasticity of cognitive function (see also Verhaeghen & Marcoen, 1996).

Across many studies, the age difference between younger adults and older adults in recall and recognition from episodic memory is about one standard

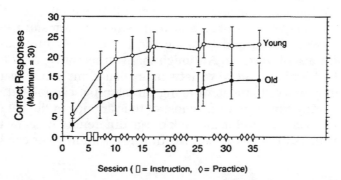

Figure 2.5. Data from Baltes and Kliegl (1992) showing that age differences in word recall persist even after extensive training. These data suggest that there are age limitations in cognitive plasticity as measured by level of improvement with practice. (Adapted from Baltes & Kliegl, 1992.)

deviation (see Verhaeghen, Marcoen, & Goossens, 1993). The negative effects of aging on memory have been attributed to limitations in the recruitment and activation of the brain mechanisms responsible for memory. Alternatively, the age differences might be due to differences in the strategies that younger and older individuals use for remembering and retrieving information. Training studies demonstrate that some portion of the memory decline of older adults has to do with the use of nonoptimal learning and memory strategies, and that older adults can learn to improve memory by using more effective strategies. In a meta-analysis of the results of 32 studies, based on the data from 1539 persons, Verhaeghen, Marcoen, and Goossens (1992) reported that memory training boosted the performance of older adults by .73 standard deviations. The effects of training on performance were larger than the effects of retesting (.38 standard deviations) or placebo treatments (.37 standard deviations).

Examination of the studies in the Verhaeghen et al. meta-analysis revealed that no one type of training procedure was any more effective than other procedures. Similarly, in a study directly comparing the effectiveness of a microcomputer-based memory training program, a commercially available audiotape memory improvement program, and a group-based memory course, Rasmusson, Rebok, Bylsma, and Brandt (1999) reported that there was no evidence to suggest that one type of training was superior to others.

Examination of the results of memory training studies reveals that improvements associated with memory training effects are specific to the type of training provided. That is, there is little or no evidence to suggest that general-purpose memory function can be improved by training. In other words, memory training probably does not improve learning or memory per se, but instead yields its beneficial effects by instilling specific strategies for the effective retrieval of specific kinds of information.

SUMMARY AND CONCLUSIONS

Our purpose in writing this chapter was to review what is known about learning during the adult years and to try to expose the main sources of age effects on learning. We presented evidence to suggest that age-related differences in the processes of learning are probably best understood in terms of age-related differences in processing speed and task-specific factors that correspond to different forms or conditions of learning. Although many different types of tasks have been used to study learning, and a variety of explanations have been advanced to account for the observed effects of aging on learning, age-related slowing in the speed of processing can account for most of the variance in the data on age-related differences across tasks, and for the lack of age differences in using the products of learning. The effects of aging on learning during the adult years can be summarized as follows:

1. The research findings from many kinds of speeded cognitive tasks suggest that there are noticeable age-related declines in the efficiency of the processes of learning throughout the adult years. Age-related changes in the speed of processing and in the efficiency of working memory are probably

immutable despite practice and prior knowledge. Training and prior knowledge aid performance by virtue of providing alternative strategies carrying out task demands.

2. If the processing of task materials can be automated in an instance-based fashion, the performance of older and younger adults will benefit from consistent practice. Two mechanisms seem to be responsible for the development of automaticity, computational speed-up and a shift from computation to the retrieval of instances. Performance improvements for older adults seem to involve a qualitative shift in the strategies for performing the task rather than computational speedup per se (Hoyer, Cerella, & Lincourt, 2000; Hoyer & Lincourt, 1998).

3. The cognitive performance of different aged adults is unaffected by aging to the extent that domain-specific strategies or heuristics can be deployed to support effective performance. That is, if the individual knows how to approach a particular kind of problem or has already learned or knows how to perform a particular task, age limitations in the speed of processing or in learning new associations are likely to have negligible influences on performance.

4. Age-related changes in learning processes during adult years and in later life depend on the interplay of brain aging and external factors (i.e., regularly occurring adaptive organism–environment transactions). Age-related changes in brain function affect the processes of learning. Concepts of brain plasticity give emphasis to the potential of individuals to continue to improve, maintain, or optimize development during the adult years.

ACKNOWLEDGMENT. Preparation of this chapter was supported by research grant AG11451 to W. J. H.

REFERENCES

Arenberg, D., & Robertson-Tchabo, E. A. (1977). Learning and aging. In J. E. Birren & K. W. Schaie (Eds.), *Handbook of the psychology of aging* (pp. 421–449). New York: Van Nostrand Reinhold.

Baltes, P. B. (1987). Theoretical propositions of life-span developmental psychology: On the dynamics between growth and decline. *Developmental Psychology, 23*, 611–626.

Baltes, P. B. (1997). On the incomplete architecture of human ontogeny. *American Psychologist, 52*, 366–380.

Baltes, P. B., & Kliegl, R. (1992). Further testing of limits of cognitive plasticity: Negative age differences in a mnemonic skill are robust. *Developmental Psychology, 28*, 121–125.

Baltes, P. B., & Staudinger, U. M. (2000). Wisdom: a metaheuristic (pragmatic) to orchestrate mind and virtue toward excellence. *American Psychologist, 55*, 122–135.

Baltes, P. B, Staudinger, U. M., & Lindenberger, U. (1999). Lifespan psychology. Theory and application to intellectual functioning. *Annual Review of Psychology, 50*, 471–507.

Bronfenbrenner, U., & Ceci, S. J. (1994). Nature-nurture reconceptualized in developmental perspective: A bioecological model. *Psychological Review, 101*, 568–586.

Cerella, J. (1990). Aging and information-processing rate. In J. E. Birren & K. W. Schaie (Eds.), *Handbook of the psychology of aging*, 3rd edit. (pp. 201–221). New York: Academic Press.

Cerella, J. (1991). Age deficits may be global, not local: Comment on Fisk and Rogers (1991). *Journal of Experimental Psychology: General, 120*, 215–223.

Charness, N. (1981). Aging and skilled problem solving. *Journal of Experimental Psychology: General, 110*, 21–38.

Charness, N., & Bosman, E. A. (1990). Expertise and aging: Life in the lab. In T. M. Hess (Ed.), *Aging and cognition: Knowledge organization and utilization* (pp. 343–385). Amsterdam: Elsevier.

Charness, N., & Campbell, J. I. D. (1988). Acquiring skill at mental calculation in adulthood: A task decomposition. *Journal of Experimental Psychology: General, 117*, 115–129.

Charness, N., Krampe, R., & Mayr, U. M. (1996). The role of practice and coaching in entrepreneurial skill domains: An international comparison of life-span chess skill acquisition. In K. A. Ericsson (Ed.), *The road to excellence* (pp. 51–80). Mahwah, NJ: Lawrence Erlbaum.

Cherry, K. E., & Stadler, M. A. (1995). Implicit learning of a nonverbal sequence in younger and older adults. *Psychology and Aging, 10*, 379–394.

Clancy, S. M., & Hoyer, W. J. (1994). Age and skill in visual search. *Developmental Psychology, 30*, 545–552.

Czaja, S. J., & Sharit, J. (1993). Age differences in performance of computer-based work. *Psychology and Aging, 8*, 59–67.

Czaja, S. J., & Sharit, J. (1998). Ability-performance relationships as a function of age and task experience for a data entry task. *Journal of Experimental Psychology: Applied, 4*, 332–351.

D'Eredita, M. A., & Hoyer, W. J. (1999). An examination of the effects of adult age on explicit and implicit learning of figural sequences. *Memory & Cognition, 27*, 890–895.

Ericsson, K. A., & Kintsch, W. (1995). Long term working memory. *Psychological Review, 102*, 211–245.

Feldman, D. H. (1995). Learning and development in nonuniversal theory. *Human Development, 38*, 315–321.

Ferraro, F. R., Balota, D. A., & Connor, L. T. (1993). Implicit memory and the formation of new associations in non-demented Parkinson's disease individuals and individuals with dementia of the Alzheimer's type: A serial reaction time (SRT) investigation. *Brain and Cognition, 21*, 163–180.

Fisk, A. D., Cooper, B. P., Hertzog, C., & Anderson-Garlach, M. (1995). Age-related retention of skilled memory search: Examination of associative learning, interference, and task-specific skills. *Journal of Gerontology: Psychological Sciences, 50B*, P150–P161.

Fisk, A. D., Cooper, B. P., Hertzog, C ., Anderson-Garlach, M., & Lee, M. D. (1995). Understanding performance and learning in consistent memory search: An age-related perspective. *Psychology and Aging, 10*, 255–268.

Fisk, A. D., & Rogers, W. A. (1991). Toward an understanding of age-related memory and visual search effects. *Journal of Experimental Psychology: General, 120*, 131–149.

Flavell, J. H. (1970). Cognitive changes in adulthood. In L. R. Goulet & P. B. Baltes (Eds.), *Life-span developmental psychology* (pp. 247–253). New York: Academic Press.

Frensch, P. A. (1998). One concept, multiple meanings: On how to define the concept of implicit learning. In M. A. Stadler & P. A. Frensch (Eds.), *Implicit learning: Representation and process* (pp. 47–104). Thousand Oaks, CA: Sage.

Frensch, P. A., & Minor, C. S. (1994). Effects of presentation rate and individual differences in short-term memory capacity on an indirect measure of serial learning. *Memory & Cognition, 22*, 95–110.

Gabrieli, J. (1994). Contributions of the basal ganglia to skill learning and working memory in humans. In J. Houk, J. L. Davis, & D. G. Beiser (Eds.), *Information processing in the basal ganglia* (pp. 277–294). Cambridge, MA: The MIT Press.

Gladis, M., & Braun, H. (1958). Age differences in transfer and retention as a function of intertask response similarity. *Journal of Experimental Psychology, 55*, 25–30.

Halford, G. (1995). Learning processes in cognitive development: A reassessment with some unexpected implications. *Human Development, 38*, 295–301.

Halpern, A. R., Bartlett, J. C., & Dowling, W. J. (1995). Aging and experience in the recognition of musical transpositions. *Psychology and Aging, 10*, 325–342.

Hambrick, D. Z., Salthouse, T. A., & Meinz, E. J. (1999). Predictors of crossword puzzle proficiency and moderators of age-cognition relations. *Journal of Experimental Psychology: General, 128*, 131–164.

Hardy, D. J., & Parasuraman, R. (1997). Cognition and flight performance in older pilots. *Journal of Experimental Psychology: Applied, 3*, 313–348.

Harrington, D. L., & Haaland, K. Y. (1992). Skill learning in the elderly. Diminished implicit and explicit memory for a motor sequence. *Psychology and Aging, 7*, 425–434.

Hertzog, C., Cooper, B. P., & Fisk, A. D. (1996). Age and individual differences in the development of skilled memory search. *Psychology and Aging, 11*, 497–520.

Hollingworth, H. L. (1927). *Mental growth and decline.* New York: Appleton.

Howard, D. V., & Howard, J. H., Jr. (1989). Age differences in learning serial patterns: Direct versus indirect measures. *Psychology and Aging, 4,* 357–364.

Howard, D. V., & Howard, J. H., Jr. (1992). Adult age differences in the rate of learning serial patterns: Evidence from direct and indirect tests. *Psychology and Aging, 7,* 232–241.

Howard, D. V., & Wiggs, C. L. (1993). Aging and learning: Insights from implicit and explicit tests. In J. Cerella, J. M. Rybash, W. J. Hoyer, & M. L. Commons (Eds.), *Adult information processing: Limits on loss* (pp. 511–527). San Diego: Academic Press.

Howard, J. H., Jr., & Howard, D. V. (1997). Age differences in implicit learning of higher-order dependencies in serial patterns. *Psychology and Aging, 12,* 634–656.

Hoyer, W. J. (1980). Information processing, knowledge acquisition, and learning. *Human Development, 23,* 389–399.

Hoyer, W. J. (1985). Aging and the development of expert cognition. In T. M. Schlecter & M. P. Toglia (Eds.), *New directions in cognitive science* (pp. 69–87). Norwood, NJ: Ablex.

Hoyer, W. J. (1986). On the growth of knowledge and the decentralization of *g* in adult intellectual development. In C. Schooler & K. W. Schaie (Eds.), *Cognitive functioning and social structures over the life course* (pp. 120–141). Norwood, NJ: Ablex.

Hoyer, W. J., Cerella, J., & Lincourt, A. E. (2000). A re-examination of the skill acquisition curve: Adult age differences. Unpublished manuscript, Syracuse University.

Hoyer, W. J., & Ingolfsdottir, D. (in press). Age, skill, and contextual cueing in target detection. *Psychology and Aging.*

Hoyer, W. J., & Lincourt, A. E. (1998). Aging and the development of learning. In M. A. Stadler & P. A. Frensch (Eds.), *Handbook of implicit learning* (pp. 445–470). Thousand Oaks, CA: Sage.

Hoyer, W. J., & Rybash, J. M. (1994). Characterizing adult cognitive development. *Journal of Adult Development, 1,* 7–12.

Hoyer, W. J., & Rybash, J. M. (1996). Life-span theory. In J. E. Birren (Ed.), *Encyclopedia of gerontology* (pp. 65–71). San Diego: Academic press.

Hulicka, I. M. (1966). Age differences in Wechsler Memory Scale scores. *Journal of Genetic Psychology, 109,* 134–145.

Hultsch, D. F., & Dixon, R. A. (1990). Learning and memory in aging. In J. E. Birren & K. W. Schaie (Eds.), *Handbook of the psychology of aging,* 3rd edit. (pp. 258–274). San Diego: Academic Press.

James, W. (1890). *Principles of psychology.* New York: Dover.

Jenkins, L., & Hoyer, W. J. (2000). Memory-based automaticity and aging: Acquisition, re-acquisition, and retention. *Psychology and Aging, 15,* 551–565.

Jenkins, L., Myerson, J. Joerding, J. A., & Hale, S. (2000). Converging evidence that visual-spatial cognition is more age-sensitive than verbal cognition. *Psychology and Aging, 15,* 157–175.

Jones, H. E., & Conrad, H. (1933). The growth and decline of intelligence: A study of a homogenous group between the ages of ten and sixty. *Genetic Psychology Monographs, 13,* 223–298.

Karmiloff-Smith, A. (1992). *Beyond modularity: A developmental perspective on cognitive science.* Cambridge, MA: The MIT Press.

Kausler, D. H. (1994). *Learning and memory in normal aging.* San Diego: Academic Press.

Kay, H. (1951). Learning of a serial task by different age groups. *Quarterly Journal of Experimental Psychology, 3,* 166–183.

Kliegl, R., Smith, J., & Baltes, P. B. (1990). On the locus and process of magnification of age differences during mnemonic training. *Developmental Psychology, 26,* 894–904.

Knopman, D., & Nissen, M. J. (1987). Implicit learning in patients with probable Alzheimer's disease. *Neurology, 37,* 784–788.

Knopman, D., & Nissen, M. J. (1991). Procedural learning is impaired in Huntington's disease. Evidence from the serial reaction time task. *Neuropsychologia, 29,* 245–254.

Knowlton, B. J., Ramus, S. J., & Squire, L. R. (1992). Intact artificial grammar learning in amnesia. *Psychological Science, 3,* 172–179.

Kramer, A. F., Larish, J. F., & Strayer, D. L. (1995). Training for attentional control in dual task settings: A comparison of young and old adults. *Journal of Experimental Psychology: Applied, 1,* 50–76.

Krampe, R., & Ericsson, K. A. (1996). Maintaining excellence: Deliberate practice and elite performance in young and older pianists. *Journal of Experimental Psychology: General, 125,* 331–359.

Lemme, B. H. (1999). *Development in adulthood,* 2nd edit. Boston: Allyn & Bacon.

Lerner, R. M. (1991). Changing organism-context relations as the basic process of development: A developmental contextual perspective. *Developmental Psychology, 27,* 27–32.

Lincourt, A., Cerella, J., & Hoyer, W. J. (2000). Attention and learning in the development of instance-based learning: Adult age differences. Unpublished manuscript, Syracuse University.

Lindenberger, U., & Baltes, P. B. (1995). Testing the limits and experimental simulation: Two methods to explicate the role of learning in development. *Human Development, 38*, 349–360.

Logan, G. D. (1988). Toward an instance theory of automatization. *Psychological Review, 95*, 492–528.

MacKay, D., & Burke, D. (1990). Cognition and aging: A theory of new learning and the use of old connections. In T. Hess (Ed.), *Aging and cognition: Knowledge organization and utilization* (pp. 213–263). Amsterdam: North Holland.

Meinz, E. J. (2000). Experience based attenuation of age-related differences in music cognition tasks. *Psychology and Aging, 15*, 297–312.

Meinz, E. J., & Salthouse, T. A. (1998). The effects of age and experience on memory for visually presented music. *Journal of Gerontology: Psychological Sciences, 53B*, P60–P69.

Monge, R. H. (1971). Studies of verbal learning from the college years through middle age. *Journal of Gerontology, 26*, 324–329.

Morrow, D. G., Leirer, Von O., Altieri, P., & Fitzsimmons, C. (1994). When expertise reduces age differences in performance. *Psychology and Aging, 9*, 134–148.

Mutter, S. A., Howard, J. H., & Howard, D. V. (1994). Serial pattern learning after head injury. *Journal of Clinical and Experimental Neuropsychology, 16*, 271–288.

Mutter, S. A., Howard, D. V., Howard, J. H., & Wiggs, C. L. (1990). Performance on direct and indirect tests of memory after mild closed-head injury. *Cognitive Neuropsychology, 7*, 329–346.

Mutter, S. A., & Pliske, R. M. (1996). Judging event covariation: Effects of age and memory demand. *Journal of Gerontology: Psychological Sciences, 51B*, P70–P80.

Myerson, J., & Hale, S. (1993). General slowing and age invariance in cognitive processing: The other side of the coin. In J. Cerella, J. M. Rybash, W. J. Hoyer, & M. L. Commons (Eds.), *Adult information processing: Limits on loss* (pp. 115–141). San Diego, CA: Academic Press.

Myerson, J., Hale, S., Rhee, S. H., & Jenkins, L. (1999). Selective interference with verbal and spatial working memory in young and older adults. *Journal of Gerontology: Psychological Sciences, 54B*, P161–P164.

Newell, A., & Rosenbloom, P. S. (1981). Mechanisms of skill acquisition and the law of practice. In J. R. Anderson (Ed.), *Cognitive skills and their acquisition* (pp. 1–56). Hillsdale, NJ: Lawrence Erlbaum.

Nissen, M. J., & Bullemer, P. (1987). Attentional requirements of learning: Evidence from performance measures. *Cognitive Psychology, 19*, 1–32.

Nissen, M. J., Knopman, D. S., & Schacter, D. L. (1987). Neurochemical dissociation of memory systems. *Neurology, 37*, 789.

Pascual-Leone, J. (1995). Learning and development as dialectical factors in cognitive growth. *Human Development, 38*, 338–348.

Pascual-Leone, J., Grafman, J., & Hallatt, M. (1994). Modulation of cortical motor output maps during development of implicit and explicit knowledge. *Science, 263*, 1287–1289.

Perruchet, P. (1994). Learning from complex rule-governed environments: On the proper functions of nonconscious and conscious processes. In C. Umilta & M. Moscovitch (Eds.), *Attention and performance XV* (pp. 811–836). Cambridge, MA: The MIT Press.

Plude, D. J., Kaye, D. B., Hoyer, W. J., Post, T. A., Saynisch, M. J., & Hahn, M. V. (1983). Aging and visual search under consistent and varied mapping. *Developmental Psychology, 19*, 508–512.

Rasmusson, D., Rebok, G. W., Bylsma, F. W., & Brandt, J. (1999). Effects of three types of memory training in normal elderly. *Aging, Neuropsychology, and Cognition, 6*, 56–66.

Raz, N. (2000). Aging of the brain and its impact on cognitive performance: Integration of structural and functional findings. In F. I. M. Craik & T. A. Salthouse (Eds.), *The handbook of aging and cognition*, 2nd edit. (pp. 1–90). Mahwah, NJ: Lawrence Erlbaum.

Reber, A. S. (1993). *Implicit learning and tacit knowledge.* New York: Oxford University Press.

Rickard, T. C. (1997). Bending the power law: A CMPL theory of strategy shifts and the automatization of cognitive skills. *Journal of Experimental Psychology: General, 126*, 288–310.

Rogers, W. A., & Gilbert, D. K. (1997). Do performance strategies mediate age-related differences in associative learning? *Psychology and Aging, 12*, 620-633.

Rogers, W. A., Hertzog, C., & Fisk, A. D. (2000). An individual differences analysis of ability and strategy influences: Age-related differences in associative learning. *Journal of Experimental Psychology: Learning, Memory, and Cognition, 26*, 359–394.

Rowe, J. W., & Kahn, R. L. (1997). Successful aging. *The Gerontologist, 37*, 433–440.

Ruch, F. L. (1934). The differentiative effects of age upon human learning. *Journal of General Psychology, 11*, 261–286.

Rybash, J. M. (1996). Implicit memory and aging: A cognitive neuropsychological perspective. *Developmental Neuropsychology, 12*, 127–179.

Salthouse, T. A. (1984). Effects of age and skill in typing. *Journal of Experimental Psychology: General, 113*, 345–371.

Salthouse, T. A. (1994). Aging associations: Influence of speed on adult age differences in associative learning. *Journal of Experimental Psychology: Learning, Memory, and Cognition, 20*, 1486–1503.

Salthouse, T. A. (1996). The processing speed theory of adult age differences in cognition. *Psychological Review, 103*, 403–428.

Salthouse, T. A. (2002). Interrelations of aging, knowledge, and cognitive performance. In U. M. Staudinger & U. Lindenberger (Eds.), *Understanding human development: Life-span psychology in exchange with other disciplines.* New York: Kluwer Academic.

Salthouse, T. A., & Coon, V. E. (1994). Interpretation of differential deficits: The case of aging and mental arithmetic. *Journal of Experimental Psychology: Learning, Memory, and Cognition, 20*, 1172–1182.

Salthouse, T. A., Hambrick, D. Z., Lukas, K. E., & Dell, T. C. (1996). Determinants of adult age differences on synthetic work performance. *Journal of Experimental Psychology: Applied, 2*, 305–329.

Salthouse, T. A., & Mitchell, D. R. D. (1990). Effect of age and naturally occurring experience on spatial visualization performance. *Developmental Psychology, 26*, 845–854.

Seligman, M. E. P., & Csikszentmihalyi, M. (2000). Positive psychology: An introduction. *American Psychologist, 55*, 5–14.

Shimamura, A. P., Berry, J. M., Mangels, J. A., Rusting, C. L., & Jurica, P. J. (1995). Memory and cognitive abilities in university professors: Evidence for successful aging. *Psychological Science, 6*, 271–277.

Smith, A. D. (1975). Partial learning and recognition memory in the aged. *International Journal of Aging and Human Development, 6*, 359–365.

Squire, L. R., Knowlton, B., & Musen, G. (1993). The structure and organization of memory. *Annual Review of Psychology, 44*, 453–495.

Staudinger, U. M., & Pasupathi, M. (2000). Life-span perspectives on self, personality, and social cognition. In F. I. M. Craik & T. A. Salthouse (Eds.), *The handbook of aging and cognition*, 2nd edit. (pp. 633–688). Mahwah, NJ: Lawrence Erlbaum.

Strayer, D. L., & Kramer, A. F. (1994). Aging and skill acquisition: Learning-performance distinctions. *Psychology and Aging, 9*, 589–605.

Talland, G. A. (1965). Initiation of response, and reaction time in aging, and with brain damage. In A. T. Welford & J. E. Birren (Eds.), *Behavior, aging, and the nervous system* (pp. 526–561). Springfield, IL: Charles C Thomas.

Thorndike, E. L., Bregman, E. O., Tilton, J. W., & Woodyard, E. (1928). *Adult learning.* New York: Macmillan.

Touron, D. R., Hoyer, W. J., & Cerella, J. (2001). Cognitive skill acquisition and transfer in younger and older adults. *Psychology and Aging, 16*, 555–563.

Verhaeghen, P., Kliegl, R., & Mayr, U. (1997). Sequential and coordinative complexity in time-accuracy functions for mental arithmetic. *Psychology and Aging, 12*, 555–564.

Verhaeghen, P., & Marcoen, A. (1996). On the mechanisms of plasticity in young and older adults after instruction in the method of loci: Evidence for an amplification model. *Psychology and Aging, 11*, 164–178.

Verhaeghen, P., Marcoen, A., & Goossens, L. (1992). Improving memory performance in the aged through mnemonic training: A meta-analytic study. *Psychology and Aging, 7*, 242–251.

Verhaeghen, P., Marcoen, A., & Goossens, L. (1993). Fact and fiction about memory aging: A quantitative integration of research findings. *Journal of Gerontology: Psychological Sciences, 48*, P157–P171.

Willoughby, R. R. (1927). Family similarities in mental test abilities (with a note on the growth and decline of these abilities). *Genetic Psychological Monographs, 2*, 235–277.

Zacks, R. T., Hasher, L., & Li, K. Z. H. (2000). Human memory. In F. I. M. Craik & T. A. Salthouse (Eds.), *The handbook of aging and cognition* (pp. 293–358). Mahwah, NJ: Lawrence Erlbaum.

Zheng, Y., Myerson, J., & Hale, S. (2000). Age and individual differences in visuospatial processing speed: Testing the magnification hypothesis. *Psychonomic Bulletin and Review, 7*, 113–120.

CHAPTER 3

Developmental Change in Adulthood

DAVID MOSHMAN

The traditional distinction between children and adults is that children are developing, whereas adults have attained a relatively stable state of maturity. The existence of a *Handbook of Adult Development*, however, casts doubt on this distinction. Presumably, I needn't convince the readers of this *Handbook* that development continues into adulthood.

To say that adults develop, however, is not to say that they change in precisely the same ways that children change. In the present chapter, I address a series of conceptual and empirical questions about developmental change across the lifespan. In the first section, I propose a strict and narrow conception of development and argue that development, thus defined, is largely a phenomenon of childhood, as suggested in the traditional distinction between children and adults. In the second section, I propose a broader and more flexible conception of development and review research indicating that development, thus defined, continues well into adulthood. In the third section, drawing these two perspectives together, I demonstrate that development does indeed continue long beyond childhood but that advanced development differs in fundamental ways from early development. Then, in the final section, I present *pluralist rational constructivism* as a metatheory that highlights the sorts of developmental changes associated with adolescence and adulthood.

THE CONCEPT OF DEVELOPMENT: A TRADITIONAL VIEW

Prototypical examples of developmental change in early childhood include the emergence of (1) language, (2) elementary number concepts, (3) tacit knowledge of mechanics, (4) intuitive understandings of the nature of life, and (5) a basic theory of

DAVID MOSHMAN • Department of Educational Psychology, University of Nebraska, Lincoln, Nebraska 68588-0345.

Handbook of Adult Development, edited by J. Demick and C. Andreoletti. Plenum Press, New York, 2002.

mind. Developmental theorists have long marveled at the extraordinary linguistic competence achieved by children in the first 5 years of life (Maratsos, 1998). Over the last quarter of the 20th century, developmentalists became increasingly impressed with the levels achieved by young children in their understandings of mathematics, physics, biology, and psychology (Case, 1998; Flavell & Miller, 1998; Gelman & Williams, 1998; Karmiloff-Smith, 1992; Rosser, 1994; Wellman & Gelman, 1998). Why do we see the emergence of such competencies as a process of development?

Development as Structural Transformation

One reason we see young children as developing, not just learning or changing, is their achievement of competencies that do not appear to be mere collections of various facts and skills. The grammatical competence of a 5-year-old, for example, constitutes a structure so complex that it defies scientific efforts to characterize it (Maratsos, 1998). Similarly, young children routinely show highly structured forms of mathematical knowledge (Case, 1998).

Since the 1970s, developmentalists have been increasingly dubious of orthodox Piagetian claims about general stages of development such as concrete operations. Development is now seen as more specific to foundational domains such as language, mathematics, physics, biology, and psychology (Gelman & Williams, 1998; Karmiloff-Smith, 1992; Maratsos, 1998; Rosser, 1994; Wellman & Gelman, 1998). Such domains, however, are still quite broad, involving networks of concepts, actions, and inferences sufficiently structured to warm a Piagetian's heart (Case, 1998). To the extent that change in some domain involves a succession of such structures, we are likely to see such change as developmental.

Development as Progress Toward Maturity

Child development in foundational domains involves not only transitions to new structures but also the emergence of *better* structures. In the domain of mathematics, for example, mature thinkers understand that counting a row of objects from left to right will generate the same result as counting it from right to left. Similarly, they understand that $5 + 3$ necessarily yields the same result as $3 + 5$. Moreover, they understand the relationship of these two additions to the subtractions $8 - 5 = 3$ and $8 - 3 = 5$, and the relationship of these subtractions to each other. Structural transformations in early numerical and arithmetic knowledge and reasoning can be evaluated with respect to whether they constitute progress toward the sort of comprehension that constitutes mathematical maturity (Case, 1998).

Mathematics is only one of several major domains in which theorists readily agree on fundamental aspects of psychological maturity. To the extent that we observe progress toward maturity, we are likely to deem a change developmental.

Development as Universal Maturation

Implicit in the standard conception of maturity is an assumption of universality. Maturity is defined in a manner that transcends particular contexts and cultures, with the implication that child development in foundational domains is to a large degree universal. With regard to mathematics, for example, it is difficult to question the logical necessities that undergird mature conceptions of number, addition, and subtraction. Despite substantial variation in mathematical terms and strategies, there is no evidence that the developmental sequence for basic conceptions of number and arithmetic varies across cultures or across educational environments. The same can be said for the emergence of basic conceptions about matters of physics and biology (Karmiloff-Smith, 1992; Rosser, 1994; Wellman & Gelman, 1998). Even with regard to people's intuitive conceptions about psychological phenomena, where profound cultural differences have been documented (Lillard, 1998), there are likely to be universal sequences in the early emergence of a basic understanding of persons as beings with perceptions, thoughts, feelings, and intentions (Flavell & Miller, 1998; Wellman, 1998). In fact, children's intuitive theories of mind arguably provide the basis for any human culture.

Progress toward universal aspects of psychological maturity, moreover, is commonly regarded as a process of maturation rather than a process of learning, with the implication that this process is strongly guided by the genes. Not all child developmentalists hold genetic determinist views, but many see the human genome, common to all normal children, as an important basis for universal pathways and endpoints (Gelman & Williams, 1998; Karmiloff-Smith, 1992; Rosser, 1994; Spelke & Newport, 1998).

Do Adults Develop?

A focus on age-related changes in young children, then, may lead us to define development as a universal process of structural transformation that constitutes progress toward maturity. Given this strict definition of development, the sorts of changes typical of adulthood are not, for the most part, developmental changes. As seen throughout this volume, adult changes are largely a function of specific experiences; it is difficult to identify universal endpoints against which later changes can be evaluated as progressive or not (Lillard, 1998). Thus it may be misleading to speak of adult development. Nevertheless, many adults believe that they are, or should be, developing, and many theorists share this assumption. Perhaps we need to reconsider our conception of development.

THE CONCEPT OF DEVELOPMENT: A BROADER VIEW

Rather than allow the phenomena of early childhood to determine what sorts of changes we construe as developmental, perhaps we need a broader conception of development. Such a conception may enable us to identify some changes that take place during adulthood as similar in some important respects to the sorts of changes we typically refer to as child development. An excessively broad and flexible definition of development, however, may render the concept of development useless. If all changes are deemed to be developmental, then it is redundant to speak of developmental change. Unless there is a distinction to be made between developmental and nondevelopmental change, the term *development* serves no purpose.

The challenge, then, is to formulate a conception of development strict enough to meaningfully distinguish some types of change from others but broad enough to encompass what can reasonably be construed as the developmental aspects of change in adulthood. In this section, I propose a flexible conception in which, to be construed as developmental, a change must be (1) qualitative, but not necessarily stagelike; (2) progressive, but not necessarily headed toward some specifiable endpoint; and (3) internally directed, but not necessarily universal or genetic.

Development as Qualitative Change

In classical Piagetian theory (Inhelder & Piaget, 1958, 1964), the transition to concrete operations at about age 7 is a transition to a new stage of development and the transition to formal operations at about age 11 or 12 marks the beginning of another developmental stage. The concrete and formal stages are not collections of facts or skills. Rather, they are defined by cognitive structures so general that these structures radically transform our understanding of the world and our most fundamental modes of reasoning. The transition to a new stage does not occur instantaneously, nor are the structures associated with a given stage applied with complete consistency. Piaget (1972) was well aware that development is gradual and that we all make mistakes. Nevertheless, orthodox Piagetian stages involve forms of reasoning and levels of understanding that are potentially applicable to all domains of knowledge.

As noted earlier, however, research has shown the importance of cognitive structures specific to foundational domains such as language, mathematics, physics, biology, and psychology (Karmiloff-Smith, 1992; Rosser, 1994; Wellman & Gelman, 1998). As a result, developmentalists over the past several decades have increasingly questioned the importance, and even the existence, of the sorts of domain-general cognitive structures highlighted by Piaget. This has led to doubts about the reality of developmental stages, at least with regard to stages defined on the basis of domain-general Piagetian structures. Stagelike transformations are, moreover, even more difficult to document in adulthood than in childhood (Moshman, 1998).

Even the domain-specific structures favored by most current child developmentalists, however, show qualitative changes over time. The 4-year-old's theory of mind, for example, constitutes an understanding of human cognition, behavior, and social interaction fundamentally different from the understanding of a typical 2-year-old (Flavell & Miller, 1998). Adult changes may typically be less revolutionary than those of early childhood but often involve the emergence of genuine novelty. Even changes that seem to be something less than stagelike transformations may be qualitative rather than merely quantitative.

Whether we construe development as continuing into adulthood, then, depends on what we mean by development. If we limit our concept of development to transformations that are structural in the strictest sense—stagelike changes in general and fundamental forms of understanding and reasoning—development may be largely a phenomenon of childhood, if it occurs at all. If we relax our definition to include less encompassing sorts of qualitative change, however, we can make a stronger case that development continues across the lifespan.

Development as Progressive Change

In classical Piagetian theory, formal operations, the final stage of development, constitutes an endpoint against which cognitive changes can be evaluated. Concrete operational reasoning is superior to preoperational reasoning because it involves logical structures than can, through further intercoordination and reflection, be transformed into the mature logical structure of formal operations.

In the context of adult development, however, maturity is not so readily identified. Although various theorists have posited postformal stages, there is no consensus on what constitutes the highest stage of cognitive development (Commons, Richards, & Armon, 1984; see other chapters in this volume). Proposals for developmental endpoints within specific domains are likewise open to question. Kohlberg (1984) proposed a highest stage of moral development (stage 6) but others have questioned whether this adequately represents the most advanced form of morality (Campbell & Christopher, 1996; Eisenberg & Fabes, 1998; Gilligan, 1982; Moshman, 1995, 1999). King and Kitchener (1994) proposed a highest stage of reflective judgment (stage 7) but it is unclear what makes this the highest possible stage. Empirically, moreover, it is well established that relatively few individuals attain any of these developmental endpoints (King & Kitchener, 1994; Kohlberg, 1984; Moshman, 1998, 1999). We may be able to specify states of maturity relative to the acquisitions of early childhood, such as the structure of basic arithmetic, but endpoints for adult development are not so readily identified.

Even if maturity is a problematical concept, however, we may be able to distinguish progressive changes from those that are neutral or regressive. Change may, for example, be regarded as progressive if it involves a dialectical synthesis of what earlier appeared to be contradictory ideas, even if that synthesis is not a final achievement or a step toward some identifiable endpoint (Basseches, 1984; Moshman, 1998, 1999). Similarly, without positing a state of maturity, we may identify progress in a process of differentiation and hierarchic integration (Werner, 1957), a transition from disequilibrium to equilibrium (Piaget, 1985; Rawls, 1971), or increasingly reflective awareness of cognitive processes and perspectives (Campbell & Bickhard, 1986; Moshman, 1994, 1999). Whether any given change represents progress may be controversial, but a plausible argument for progress need not depend on postulating a specific endpoint for the developmental sequence in question.

With regard to the criterion that developmental change must involve progress toward maturity, then, a less stringent standard is that it must involve progress in some sense but that one need not posit a state of maturity to propose that change is progressive. By relaxing the definition in this way, we may be able to count as developmental a variety of advanced transitions that are arguably progressive without having to specify in each case where such development is headed or where it will end.

Development as Internally Directed Change

In addition to being qualitative and progressive, developmental change is typically seen as coming from the inside rather than from the outside, as a process of maturation rather than a process of learning. With regard to young children, this is commonly taken to mean that development, as distinct from learning, is guided by the genes rather than by the environment. Contemporary developmentalists generally acknowledge that genes exert their influence within environments and that development is thus influenced by cultural and other environmental factors. Nevertheless, we are most likely to speak of development with respect to universal changes that appear to be guided by the human genome, and more likely to speak of learning when we seek to explain why children in different environments come to have differing skills and beliefs.

Strong claims of genetic determination are controversial, however, and those who emphasize the role of genes tend to focus on early childhood (Gelman & Williams, 1998; Karmiloff-Smith, 1992; Rosser, 1994, Spelke & Newport, 1998). With older individuals, it becomes increasingly difficult to make the case for universal changes guided by a human genetic program. Does this mean that whatever changes take place over the course of adulthood are externally directed? If so, it may be highly misleading to construe adults as developing. Perhaps it would be better to recognize adulthood as a period of learning in which most changes are driven by our various environments.

There is another possibility, however. Constructivist metatheory suggests a sense in which change may be internally directed without being genetic. In its account of change, constructivism recognizes the emergent organism as a factor increasingly distinct from genes and environment. Especially at advanced levels, the organism may be construed as an autonomous agent, constructing new ideas, perspectives, and modes of reasoning through its own interpretations, actions, coordinations, and reflections. This is not to say that genes and environments become irrelevant. The point is that genes are not the only force that can be said to be internal to the organism. To the extent that organisms transform themselves through their own actions, such changes can be said to be internally directed, and are thus potentially developmental.

Changes in adulthood, then, may be less universal and less attributable to genes than are changes in early childhood, but may nevertheless be internally directed in the sense that they are constructed by the individual rather than caused by the environment. If we restrict our conception of development to universal outcomes of genetic programs, psychological development, if it occurs at all, may be largely a phenomenon of early childhood. A broader conception of development recognizes the constructive role of the organism and thus the ways that internally directed change continues well into adulthood.

Adult Development

As noted earlier, a focus on young children suggests a conception of development as a universal process of structural transformation that constitutes progress toward maturity. Given this strict definition, development may be largely a phenomenon of childhood. The present section of the chapter, however, has suggested a broader conception: Developmental changes are those that are qualitative, progressive, and internally directed. Examples of such changes abound throughout this volume.

Looking across the first two sections of this chapter, then, it appears that adults do indeed develop, in a broad sense of the term, but that adult development differs in some important respects from child development. In the next section, I consider the nature of advanced development with regard to epistemic cognition, morality, and identity.

ADVANCED DEVELOPMENT IN THREE ILLUSTRATIVE DOMAINS

Three domains in which development clearly continues long beyond childhood, at least in some individuals and social contexts, are epistemic cognition, morality, and identity (Moshman, 1999). An overview of development in each of these domains will illustrate how advanced development resembles and differs from early development.

The Development of Epistemic Cognition

Consider the following claims:

1. Whales are bigger than germs.
2. $5 + 3 = 8$.
3. Chocolate is better than vanilla.
4. Einstein's theory is better than Newton's.
5. Mozart's music is better than Madonna's.

In each case, one can ask whether the claim is true. An epistemologist would ask how we justify such judgments. Research on epistemic cognition has shown that people have a variety of epistemological beliefs—beliefs about the nature and justification of knowledge (Hofer & Pintrich, 1997). On the basis of such research, theorists have identified three general epistemic perspectives (Boyes & Chandler, 1992; Chandler, Boyes, & Ball, 1990; King & Kitchener, 1994; Kuhn, 1991); terminologies vary, but I have labeled these *objectivist*, *subjectivist*, and *rationalist* (Moshman, 1998, 1999).

An individual with an objectivist epistemology sees truth as unproblematic. Claims 1 and 2 above would be seen as prototypical examples of knowledge. One can easily establish that each of these claims is true and that alternative claims, such as *germs are bigger than whales* or *5 + 3 = 6*, are false. Claim 4 may be a more difficult matter because it involves technical knowledge but, if appropriate authorities determine that Einstein's theory is consistent with relevant evidence and Newton's is not, we may also accept this as a true claim. Claim 3 might be dismissed as a matter of opinion, not a matter of knowledge. Claim 5 might also be simply a matter of opinion, though perhaps an expert in music could establish its truth.

A subjectivist, in contrast, sees truth as relative to one's point of view. Claim 3 would be seen as a prototypical example: No flavor is better than any other—this is literally a matter of taste. And for that matter, who is to say that one form of music is better than another (claim 5)? Even an expert musician evaluates music from his or her own musical perspective, which is no better than anyone else's perspective. With regard to claim 4 it may be true that most contemporary physicists prefer Einstein's theory to Newton's, but there was a time when Newton's theory was widely accepted and there may come a time when Einstein's falls into disfavor. Even in science, a subjectivist would argue, our "facts" are a function of our theoretical perspectives, and such perspectives are ultimately subjective, neither true nor false. Claims 1 and 2 may seem indisputable, but knowledge is rarely this simple, and even in these cases the claims are true only within a shared network of concepts about whales, germs, size, number, addition, and equality.

Finally, a rationalist epistemology might take claim 4 as a prototypical example of knowledge. Einstein's theory may not be true in the same simple sense that whales are bigger than germs or $5 + 3 = 8$, but preferring it to Newton's theory is not just a matter of taste, like preferring one flavor to another. In complex domains of knowledge we may use justifiable criteria to evaluate various judgments and justifications and, as a result, have rational grounds for preferring some beliefs to others even if we cannot prove them true. Whether and how musical preferences, such as claim 5, can be justified may be unclear, but this doesn't mean all knowledge is subjective any more than the existence of some relatively clear-cut truths, such as claims 1 and 2, means all knowledge is objective.

Does epistemic cognition develop? This is in part an empirical question, but the answer also depends on how we define development. Research on adolescents and adults has shown a trend from objectivist to subjectivist to rationalist epistemologies, but the relation of epistemic cognition to age is modest and the trend is far from universal. Is this a developmental change?

The answer depends, I think, on our definition of development. If by development we mean a universal process of structural transformation that constitutes progress toward maturity, with the implication that all members of the species are genetically programmed to move through some universal sequence of structures toward a genetically determined endpoint, transitions in epistemic cognition beyond childhood are not developmental changes because they are neither universal nor genetically driven. If developmental changes are simply those that are qualitative, progressive, and internally directed, however, research shows that many adolescents and adults change in ways that constitute epistemic development.

Epistemic development is generally seen as a process in which a rational agent constructs a series of increasingly sophisticated epistemic perspectives (Moshman, 1999). The direction of change is from objectivist to subjectivist to rationalist epistemologies, though the rate and extent of change is highly variable across individuals (Boyes & Chandler, 1992; Chandler et al., 1990; Hofer & Pintrich, 1997; King & Kitchener, 1994; Kuhn, 1991). The sequence is developmental, in the broad sense of the term, in that it is (1) qualitative, (2) progressive, and (3) internally directed.

First, the sequence involves qualitative changes from one epistemic perspective to another. This does not mean that people are entirely consistent in their epistemic reasoning or that the shift to a new epistemic perspective takes place suddenly. What it means is that a subjectivist perspective is not simply a larger, faster, or better version of an objectivist perspective, and a rationalist perspective is in turn a different way of thinking about epistemic matters than either of the others. Whether or not this sequence is presented as a series of structures (Kitchener & Fischer, 1990), it is clearly a series of qualitatively distinct viewpoints.

Second, the sequence is arguably progressive in that, at least from a rational constructivist point of view, each step in the sequence represents an advance over the previous one. An objectivist epistemology is transformed into a subjectivist epistemology through processes of coordination and reflection that enable increasing awareness of cognitive diversity and our inevitable subjectivity. Further reflections and coordinations may enable a dialectical synthesis of objectivity and subjectivity to create a rationalist epistemology that overcomes the limits of each and thus constitutes a higher level of equilibrium. Without positing the existence of a mature epistemology or proposing that some version of a rationalist epistemology constitutes the endpoint of epistemic development, we can nevertheless make a case for progress and thus for developmental change. The theoretical argument that people construct what they themselves recognize as increasingly justifiable epistemologies is supported by empirical evidence that change, when it occurs, is from objectivism to subjectivism to rationalism.

Finally, the sequence is internally directed in that it involves a rational agent who constructs new epistemologies out of earlier ones through internal processes of reflection and coordination. These processes take place in the course of interacting with others, but they are not simply a matter of learning what one sees or is told. Without suggesting that the constructive process is universal, much less genetic, we can construe it as internally directed to a sufficient degree to refer to it as developmental.

How does the late emergence of the sorts of explicit epistemologies observed in adolescents and adults compare to the early emergence of tacit theories of mind during the preschool years? The early developmental trends are not only qualitative, progressive, and internally directed but also appear to be universal (Flavell & Miller, 1998; Wellman, 1998), and a case can even be made for their genetic basis (Gelman & Williams, 1998; Spelke & Newport, 1998). Thus the late development of epistemic cognition may differ in some important ways from the earlier development of intuitive theories of mind but is similar in ways that justify considering this a continuation of a developmental process.

Moral Development

Morality is another domain in which development may continue long beyond childhood but later development differs in important ways from earlier development. The best case for advanced moral development is the research generated by Kohlberg's (1984) theory. Kohlberg proposed six moral frameworks within which moral dilemmas can be addressed. He labeled these stages 1 through 6, reflecting his philosophical assessment of their relative adequacy, and research in diverse cultures has supported his view that people move through these stages in the order postulated (Boyes & Walker, 1988; Kohlberg, 1984; Snarey, 1985; Walker, 1989). This research has indicated, however, that many adults reason mostly at the stage 2 level, that most individuals never move beyond stage 4, and that stage 6 reasoning is extremely rare.

With regard to the criterion of qualitative change, Kohlberg's stages clearly represent qualitatively distinct perspectives on morality. The stages are not defined on quantitative grounds, such as how quickly a moral conclusion is generated or what proportion of a person's judgments are morally correct. Rather, the stages represent distinct sets of assumptions about what sorts of considerations are relevant to moral judgment and how these considerations should be coordinated. It has been argued that Kohlberg's stages focus on abstract issues of justice and do not adequately address moral considerations of personal virtue (Campbell & Christopher, 1996), concern for others (Eisenberg & Fabes, 1998), and responsibility in close relationships (Gilligan, 1982). If so, Kohlberg's stages may be something less than general stages of moral development (Moshman, 1995, 1999). Even if one questions the breadth of the moral structures proposed by Kohlberg, however, there is no doubt that a transition from one of these structures to another represents a qualitative change.

A strong case can be made, moreover, that these qualitative changes represent progress in moral reasoning and understanding. Kohlberg's stage 2 involves an awareness of diverse perspectives impossible at stage 1. Stage 3 coordinates such perspectives within a new framework of human relationships, and stage 4 reconstrues human relationships from the abstract perspective of the social system that gives them meaning. Stage 5, in turn, enables one to evaluate social systems from the more abstract perspective of a justifiable social contract, and stage 6 provides a still more abstract conception of the rationale for a stage 5 social contract morality. There is room for argument as to whether Kohlberg's conception of stage 6 constitutes a philosophically defensible endpoint for moral development (Campbell & Christopher, 1996; Eisenberg & Fabes, 1998; Gilligan, 1982; Moshman, 1995, 1999). Even if it doesn't, however, Kohlbergian theory and research demonstrate that many adolescents and adults make moral progress.

Finally, there is the question of whether transitions to advanced forms of morality are internally directed. It is clear that such transitions are far from universal, and there is no reason to think they are genetically directed. There is substantial agreement, however, that individuals construct advanced moralities out of earlier moralities (Moshman, 1995, 1999). Social interaction among individuals with diverse moral perspectives surely plays an important role in such construction, and may be a necessary condition for it, but advanced moralities are not simply learned from our environments. Rather, they emerge from internal processes of reflection and coordination.

Does morality develop? There may be norms of cooperation and mutual respect that virtually all children come to understand and that can be considered endpoints of early moral development (Piaget, 1932/1965). Such endpoints are more difficult to establish at advanced levels, however. Even if we accept Kohlberg's stage 6 as a moral ideal, it is an endpoint virtually no one achieves. If by moral development we mean progress toward a universally achieved state of moral maturity, it is difficult to make a case for advanced moral development. If by moral development we simply mean qualitative, internally directed progress in moral understanding and reasoning, however, we can conclude that morality continues to develop, in many individuals, well into adulthood.

Identity Formation

Identity formation has traditionally been viewed as the major developmental task of adolescence (Erikson, 1968). It has long been clear, however, that identity formation often continues well into adulthood and is far from a universal achievement (Marcia, Waterman, Matteson, Archer, & Orlofsky, 1993; Moshman, 1999). Is the formation of identity a developmental process? How does it relate to the earlier attainment of foundational self-conceptions—conceptions of oneself as a person with a name, as a person distinct from others, with experiences and attributes of one's own?

An identity may be defined as an explicit theory of oneself as a person (Moshman, 1999). Thus, my identity is not just any self-concept. It is a highly organized and reflective structure of self-conceptions that enables me to construe myself as a person—as a rational agent responsible for my beliefs and actions.

With regard to the criterion of qualitative change, identities are fundamentally qualitative phenomena. An identity is not the sum of one's ratings on, say, some set of personality dimensions. My identity gets to issues of who I perceive myself to be. Am I a Christian, a Muslim, a humanist, a pagan? Am I a liberal? A radical? A conservative? Do I define myself in terms of my role as a parent, a spouse, a worker, a soldier, a student, a teacher, a citizen, an activist? Constructing an identity involves creating a unique structure of self-understanding that enables me to explain my most fundamental commitments and actions. Thus identity formation is clearly a process of qualitative change.

Does such change represent progress? In general, we should be hesitant to suggest that one identity is better than another. In forming an initial identity, however, one generates an explicit self-understanding that incorporates and transcends the self-conceptions of childhood, thus moving to a higher level of reflection, responsibility, and commitment. Regardless of the specific identity constructed, this may be seen as psychological progress. Later changes in identity would not necessarily be progressive but in some cases we might identify reorganizations that incorporate and transcend earlier self-understandings and thus represent progress. Although a particular identity may turn out to be a stable and final state for a given person, there is no highest stage in the process of identity formation, no mature identity toward which all changes of identity are striving. Nevertheless the construction of identity often appears to make progress.

Finally, the process of identity formation is generally seen as an active process guided to a substantial extent by the individual (Marcia et al., 1993; Moshman, 1999). Identities are formed in social contexts, of course, and are highly constrained by the social and vocational roles available in those contexts, but they are not simply internalized from one's environment. Rather, in the course of our social interactions, we construct our identities. Thus, despite individual and cultural variability in the construction of identity, there is an important sense in which identities are internally generated.

Early self-conceptions appear to be fundamental human attainments, emerging through a universal, and perhaps genetically guided, sequence of increasingly sophisticated structures (Harter, 1998). The later construction of identity may be much less universal and may owe much less to any sort of genetic program, but identity formation nevertheless appears sufficiently qualitative, progressive, and internally directed to be seen as a continuation of the developmental process.

Conclusion

Looking across three domains of potential development, I have considered changes in epistemic cognition, morality, and identity over the course of adolescence and adulthood. These changes are much less universal, and less subject to genetic explanation, than are corresponding aspects of child development. Nevertheless, later changes can be seen as developmental in that they are qualitative, progressive, and internally directed. It thus appears that development continues long beyond childhood, but that advanced development differs, in important ways, from child development.

PLURALIST RATIONAL CONSTRUCTIVISM: A METATHEORY OF ADVANCED DEVELOPMENT

Development is a concept with roots in biology, in the way embryos transform themselves into mature organisms. In the traditional view, development is a universal process directed by the genes that leads toward a state of maturity. Today, this would be called a modernist view.

Early psychological development is in many respects an extension of anatomical and physiological development. Although its genetic basis is less clear, it still seems to fit our core notion of a developmental change in that it posits endpoints toward which development is directed. As we proceed through later childhood, however, and especially in adolescence and adulthood, change is less closely tied to age. People change in different ways and to different degrees in diverse personal and cultural circumstances. Is change ultimately idiosyncratic and unpredictable? If so, perhaps adolescents and adults are best understood from a postmodern perspective.

A postmodern perspective, as opposed to a modernist conception of maturation, might note that adolescents and adults in varied social, cultural, and linguistic contexts construct a wide variety of epistemic conceptions, forms of reasoning, moral perspectives, and identities. To say some of these are better than others is to privilege some beliefs, forms of reasoning, moral perspectives, or identities over others. Such judgments inevitably reflect our biases and thus, a postmodernist would argue, cannot be justified.

More generally, postmodernism views knowledge as radically collective and subjective and thus rejects the modernist image of rational agents constructing higher levels of understanding and reasoning. There is much more to postmodernism than this, and much to be said about its relationship to issues of psychology, development, and education (Chandler, 1997; Kahn & Lourenco, 1999; Prawat, 1999; Ryan, 1999; Siegel, 1997). For present purposes, the central concern is that the radical collectivism and subjectivism of postmodern perspectives appear inconsistent with the possibility of progress. Postmodernism thus rules out development.

Rational Constructivism

Major theories of advanced development, such as those addressing the development of epistemic cognition, morality, and identity, view development as a rational process, directed by rational agents, rather than a process directed by genes and/or environments. We can call this modernist view *rational constructivism* (Moshman, 1998, 1999). In its Piagetian forms, rational constructivism sees cognitive development as a sequence of structures constructed by all normal individuals, each more equilibrated than the one it replaces, culminating in a state of cognitive maturity such as formal operations.

Postmodernism, in contrast, suggests a more radical constructivism. Perhaps construction is a free act of creation, unconstrained by considerations of objectivity or justifiability. In that case, individuals and/or cultures construct whatever they construct, and there is no defensible basis for evaluating such constructions or suggesting that any particular restructuring represents progress.

Constructivist views of psychological development highlight the role of persons as agents in their understanding and reasoning. *Rational constructivism* highlights the role of persons as rational agents making progress in their understanding and reasoning, and thus developing. *Radical constructivism*, in contrast, rejects the possibility of progress, and thus of development.

Is progress impossible? Can change never be justified? All forms of postmodernism face the general philosophical problem of justifying a doctrine that denies justifiability. If no idea is true, why should I believe postmodernist ideas? If no perspective is better than any other, why should I prefer postmodernism to modernism? If no cognitive transition represents progress, then a transition from modernism to postmodernism is no more justifiable than the reverse.

Postmodernism thus appears to be self-contradictory at a most fundamental level. It is committed to denying the possibility that it can possibly be justified. Its radical subjectivism rules out the possibility that postmodern beliefs might represent or enable progress.

Lemke (1999), for example, defending postmodernism against a critique by Prawat (1999), argues that postmodern theorists have made substantial progress in addressing a variety of psychological phenomena. Turning to the specific concerns of developmentalists, Lemke adds, "I believe there is also substantial progress afoot on better ways of modeling change" (p. 89). But if postmodernists are only making progress from some idiosyncratic point of view that is no more justifiable than any other point of view, why should I be impressed? If, on the other hand, postmodernists are in some general way making progress toward higher levels of understanding, doesn't that run counter to the postmodernist assertion that such progress is impossible?

With regard to developmental theory, radical constructivism cannot explain development because it denies the possibility of progressive change. Genetic explanations of development are also inadequate, however, especially at advanced levels. If adolescents and adults develop, we need a rational constructivist account to explain this.

Rational constructivism highlights the roles of reflection and coordination in developmental change (Case, 1998), especially at advanced levels (Moshman, 1998, 1999). Reflection is a developmental process in which we become conscious of our previously tacit assumptions and perspectives, thus achieving a higher level of understanding (Campbell & Bickhard, 1986; Moshman, 1994, 1998, 1999). Reflective understanding may enable a dialectical synthesis of two apparently contradictory views, thus coordinating and transcending them (Basseches, 1984; Moshman, 1998, 1999). Complex coordinations of multiple assumptions and perspectives may generate higher levels of equilibrium (Piaget, 1985; Rawls, 1971). Peer interaction may be especially likely to encourage reflection, coordination, dialectical synthesis, and equilibration, and thus be a critical context for advanced development (Moshman, 1995, 1998, 1999; Moshman & Geil, 1998).

Rational constructivism thus proposes constructive processes that can generate progress, and thus development. Rational constructivist theories have traditionally assumed a single sequence of structures, such that progress beyond any given structure necessarily consists of a transition to the next. The only way to progress beyond concrete operations, in classical Piagetian theory, is to construct formal operations. Similarly, the only way to transcend a Kohlbergian stage is to move to the immediately following stage, which is defined in terms of a specific structure.

Pluralist Rational Constructivism

There may indeed be progressive changes, especially in early childhood, that are universal across the human species. Rational constructivism does not, however, require universality. There could be two distinct structures that transcend concrete operations, for example; formal operations might only be one of these. Similarly, it may be possible to achieve a higher level of moral reflection without moving toward Kohlberg's stage 6 or any other final moral structure (Moshman, 1995). A commitment to progress need not assume the existence of a highest stage.

It is thus useful to distinguish pluralist from universalist versions of rational constructivism (Moshman, 1995, 1998, 1999). Rational constructivism proposes that the construction of knowledge is a rational process leading to increasingly reflective and/or equilibrated forms of understanding and reasoning. *Universalist rational constructivism* assumes that this process involves a single sequence of structures culminating in a mature and thus final state. *Pluralist rational constructivism* does not rule out the possibility of universal sequences and outcomes but does not assume them either. It leaves open the possibility that change may proceed in more than one justifiable direction.

Pluralist rational constructivism can thus accommodate claims that adolescents and adults justifiably construct multiple forms of reasoning and understanding, varied epistemic and moral perspectives, and diverse identities. The nature and extent of such diversity is an empirical question.

The locus of diversity is also an empirical question. There could, for example, be important differences across cultures or between demographic groups such as men and women. There could also be important differences among individuals that have little or no relation to categories such as those of culture or gender.

Current research indicates that individual differences with regard to reasoning, morality, and identity are substantial (Marcia et al., 1993; Stanovich, 1999). There are also findings of statistically significant group differences, but these are generally small compared to the variability among individuals within groups. Claims of categorical differences between women and men, and among distinct cultures, are largely undermined by evidence that psychological diversity *within* such demographic groups is generally greater than diversity *across* those groups (Moshman, 1999).

Perhaps most importantly, current research demonstrates substantial diversity *within individuals* (Kuhn, Garcia-Mila, Zohar, & Andersen, 1995; Siegler,

1996; Turiel, 1998). A given person may manifest multiple perspectives or forms of reasoning. Whatever tensions exist between care and justice as moral orientations, for example, may not be differences between women and men, or between distinct cultures, or between some individuals and others. Rather, they may be tensions most of us, regardless of gender or culture, experience between two forms or aspects of human morality. Advanced reasoning and development may be largely a matter of consciously coordinating multiple frameworks and perspectives, many of which may be widely shared across individuals and groups (Moshman, 1998, 1999).

If this complex picture of psychological and cultural diversity leads us to assume that all perspectives and constructions are equally good, we have lapsed into a radical constructivism that denies the possibility of developmental change. If, however, we maintain the possibility of progress, we may find ourselves with a pluralist rational constructivism that questions universal sequences but is nevertheless developmental.

Modernity Reinvigorated

Development, with its assumption of potential progress, is a modernist concept. Postmodernism teaches that progress is an illusion, because our assessment of it is always and utterly subjective. Claims of progressive change, and thus of development, are always relative to some unjustifiable perspective, and thus unjustifiable. With the rise of postmodernism, then, we may have reached "the end of the age of development" (Kessen, 1984).

As the study of development finds itself unable to take the postmodern turn, Chandler (1995, p. 1) pleads, "Please wait a minute, Mr. Post-Man." Rising "against mounting odds" to the defense of developmental theory and research, he provides a manifesto

> in opposition to a contemporary current of so-called *post-modern thought* that, if left unchecked, would not only work to collapse the usual common language distinction that ordinarily divides the notions of *change* and *development*...but would also serve, in the bargain, to carry away the essential foundations upon which the scientific study of human development necessarily rests. (1997, p. 1, italics in original)

As Chandler has been sounding the alarm among developmentalists, Siegel (1997) has been warning educational philosophers that postmodernism also undermines any conception of education as a justifiable effort to promote rationality. If no idea or form of reasoning can possibly be more or less justifiable than any other, what is the purpose of education? "Gimme that Old-Time Enlightenment Metanarrative," proclaims Siegel (p. 129) in the title of one of his chapters. As the subtitle explains, "Radical Pedagogies (and Politics) Require Old-Fashioned Epistemology (and Moral Theory)."

Prawat (1999) agrees that postmodernism is deeply problematic but insists that postmodern critiques of modernism, as Chandler and Siegel both acknowledge, are in some important respects on target. What we need, Prawat argues, is a middle ground between modernism and postmodernism. Efforts to find such a middle ground with regard to issues of psychological development, he suggests, date back to the work of Dewey and Vygotsky in the first third of the 20th century:

> Dewey, and this is true of Vygotsky as well, sought to discover what a more modest brand
> of epistemology might achieve, one that was relieved of its traditional responsibility to
> lay bare the grounds of true belief. Dewey rejected the notion that truth is timeless and
> universal. He also rejected the alternative position adopted by postmodernists like Rorty
> [1979], which equates truth claims to moves in a language game. Truth, according to this
> perspective, is nothing more 'than what our peers will let us get away with saying'
> [p. 176]. Dewey and Vygotsky both believed that there is more to truthful belief than
> social consensus. (p. 62)

Thus Dewey and Vygotsky independently found their ways to "a middle ground position that takes into account the individual and social construction of knowledge, and the individual and social verification of that knowledge" (p. 67).

Kahn and Lourenco (1999) agree that modernism is subject to legitimate critique, but firmly reject the postmodern assumption

> that fundamentally there is little of importance that people share psychologically, and
> that epistemologically there is little that transcends culture and context by which we can
> judge the intellectual or moral merits of such difference. This view seems to us not only
> empirically wrong and philosophically inadequate, but politically unworthy in that it
> increasingly fragments people from one another, and promotes a view that power itself is
> the only legitimate regulator. In other words, when postmodern theory—and particularly
> deconstruction—is taken seriously, it leads to contradictions in epistemology, to frag-
> mentation in knowledge, to opportunism in interpersonal relationships, and to nihilism
> in moral action and commitment. (p. 105)

What we need, they conclude, is not a vaguely defined middle ground between modernism and postmodernism but a reinvigorated version of modernism that responds to postmodern concerns without lapsing into the radical subjectivism and relativism of a postmodern perspective.

In the study of adolescent and adult development, consistent with this suggestion, there appears to be a trend toward pluralist versions of rational constructivism. Pluralist rational constructivism does not insist that universal sequences are necessarily the most fundamental but neither does it assume that all sequences are equally justifiable. It recognizes the diversity of constructive processes and outcomes within a framework that preserves the modernist commitment to rationality and progress. Thus pluralist rational constructivism may be a metatheory especially suited to the study of development beyond childhood.

CONCLUSION

Development has traditionally been construed as a universal process of structural transformation that constitutes progress toward maturity. Research on advanced forms of reasoning and understanding indicates that diversity among adolescents and adults is substantial and that universal sequences and outcomes are difficult to specify. If adolescents and adults do not show universal progress toward universal states of maturity, it may be misleading to speak of adult development. It might be suggested that development is a modernist concept best suited for understanding the universal, genetically directed changes of early childhood. Perhaps the study of adolescence and adulthood, in contrast, requires a postmodern perspective.

There is substantial evidence, however, that qualitative, internally directed progress in reasoning and understanding continues long into adulthood in domains such as epistemic cognition, morality, and identity. It thus appears that adults do develop, in a broad sense of that term, though later development differs from earlier development in the increasing diversity of pathways and outcomes. With this in mind, I have suggested pluralist rational constructivism as an emerging paradigm of advanced development. Without assuming universal sequences and endpoints, pluralist rational constructivism preserves a modernist conception of developmental change as qualitative, progressive, and internally directed. Thus it enables us to see how development continues into adulthood and to understand how developmental change in adulthood both resembles and differs from developmental change in early childhood.

ACKNOWLEDGMENT.   I am grateful to Rick Lombardo for feedback on an earlier draft.

REFERENCES

Basseches, M. (1984). *Dialectical thinking and adult development*. Norwood, NJ: Ablex.

Boyes, M. C., & Chandler, M. (1992). Cognitive development, epistemic doubt, and identity formation in adolescence. *Journal of Youth and Adolescence, 21*, 737–763.

Boyes, M. C., & Walker, M. (1988). Implications of cultural diversity for the universality claims of Kohlberg's theory of moral reasoning. *Human Development, 31*, 44–59.

Campbell, R. L., & Bickhard, M. H. (1986). *Knowing levels and developmental stages*. Basel: Karger.

Campbell, R. L., & Christopher, J. C. (1996). Moral development theory: A critique of its Kantian presuppositions. *Developmental Review, 16*, 1–47.

Case, R. (1998). The development of conceptual structures. In W. Damon (Series Ed.), D. Kuhn & R. Siegler (Vol. Eds.), *Handbook of child psychology, Vol. 2: Cognition, perception, and language*, 5th edit. (pp. 745–800). New York: John Wiley & Sons.

Chandler, M. (1995). Is this the end of "The Age of Development," or what? Or: Please wait a minute, Mr. Post-Man. *Genetic Epistemologist, 23*(1), 1, 3–11.

Chandler, M. (1997). Stumping for progress in a post-modern world. In E. Amsel & K. A. Renninger (Eds.), *Change and development: Issues of theory, method, and application* (pp. 1–26). Mahwah, NJ: Lawrence Erlbaum.

Chandler, M., Boyes, M., & Ball, L. (1990). Relativism and stations of epistemic doubt. *Journal of Experimental Child Psychology, 50*, 370–395.

Commons, M. L., Richards, F. A., & Armon, C. (1984). Beyond formal operations: Late adolescent and adult cognitive development. New York: Praeger.

Eisenberg, N., & Fabes, R. A. (1998). Prosocial development. In W. Damon (Series Ed.) & N. Eisenberg (Vol. Ed.), *Handbook of child psychology, Vol. 3: Social, emotional, and personality development*, 5th edit. (pp. 701–778). New York: John Wiley & Sons.

Erikson, E. H. (1968). *Identity: Youth and crisis*. New York: W. W. Norton.

Flavell, J. H., & Miller, P. H. (1998). Social cognition. In W. Damon (Series Ed.), D. Kuhn & R. Siegler (Vol. Eds.), *Handbook of child psychology, Vol. 2: Cognition, perception, and language*, 5th edit. (pp. 851–898). New York: John Wiley & Sons.

Gelman, R., & Williams, E. M. (1998). Enabling constraints for cognitive development and learning: Domain specificity and epigenesis. In W. Damon (Series Ed.), D. Kuhn & R. Siegler (Vol. Eds.), *Handbook of child psychology, Vol. 2: Cognition, perception, and language*, 5th edit. (pp. 575–630). New York: John Wiley & Sons.

Gilligan, C. (1982). *In a different voice: Psychological theory and women's development*. Cambridge, MA: Harvard University Press.

Harter, S. (1998). The development of self-representations. In W. Damon (Series Ed.) & N. Eisenberg (Vol. Ed.), *Handbook of child psychology, Vol. 3: Social, emotional, and personality development* 5th edit. (pp. 553–617). New York: John Wiley & Sons.

Hofer, B. K., & Pintrich, P. R. (1997). The development of epistemological theories: Beliefs about knowledge and knowing and their relation to learning. *Review of Educational Research, 67,* 88–140.

Inhelder, B., & Piaget, J. (1958). *The growth of logical thinking from childhood to adolescence.* New York: Basic Books.

Inhelder, B., & Piaget, J. (1964). *The early growth of logic in the child.* London: Routledge.

Kahn, P. H., Jr., & Lourenco, O. (1999). Reinstating modernity in social science research—or the status of Bullwinkle in a post-postmodern era. *Human Development, 42,* 92–108.

Karmiloff-Smith, A. (1992). *Beyond modularity: A developmental perspective on cognitive science.* Cambridge, MA: The MIT Press.

Kessen, W. (1984). Introduction: The end of the Age of Development. In R. J. Sternberg (Ed.), *Mechanisms of cognitive development* (pp. 1–17). New York: Freeman.

King, P. M., & Kitchener, K. S. (1994). *Developing reflective judgment.* San Francisco: Jossey-Bass.

Kitchener, K. S., & Fischer, K. W. (1990). A skill approach to the development of reflective thinking. In D. Kuhn (Ed.), *Developmental perspectives on teaching and learning thinking skills* (pp. 48–62). Basel: Karger.

Kohlberg, L. (1984). *The psychology of moral development.* San Francisco: Harper & Row.

Kuhn, D. (1991). *The skills of argument.* Cambridge, UK: Cambridge University Press.

Kuhn, D., Garcia-Mila, M., Zohar, A., & Andersen, C. (1995). Strategies of knowledge acquisition. *Monographs of the Society for Research in Child Development, 60,* Serial No. 245.

Lemke, J. L. (1999). Meaning-making in the conversation: Head spinning, heart winning, and everything in between. *Human Development, 42,* 87–91.

Lillard, A. (1998). Ethnopsychologies: Cultural variations in theories of mind. *Psychological Bulletin, 123,* 3–32.

Maratsos, M. (1998). The acquisition of grammar. In W. Damon (Series Ed.), D. Kuhn & R. Siegler (Vol. Eds.), *Handbook of child psychology, Vol. 2: Cognition, perception, and language,* 5th edit. (pp. 421–466). New York: John Wiley & Sons.

Marcia, J. E., Waterman, A. S., Matteson, D. R., Archer, S. L., & Orlofsky, J. L. (Eds.) (1993). *Ego identity: A handbook for psychosocial research.* New York: Springer-Verlag.

Moshman, D. (1994). Reason, reasons, and reasoning: A constructivist account of human rationality. *Theory & Psychology, 4,* 245–260.

Moshman, D. (1995). The construction of moral rationality. *Human Development, 38,* 265–281.

Moshman, D. (1998). Cognitive development beyond childhood. In W. Damon (Series Ed.), D. Kuhn & R. Siegler (Vol. Eds.), *Handbook of child psychology, Vol. 2: Cognition, perception, and language,* 5th edit. (pp. 947–978). New York: John Wiley & Sons.

Moshman, D. (1999). *Adolescent psychological development: Rationality, morality, and identity.* Mahwah, NJ: Lawrence Erlbaum.

Moshman, D., & Geil, M. (1998). Collaborative reasoning: Evidence for collective rationality. *Thinking & Reasoning, 4,* 231–248.

Piaget, J. (1965). *The moral judgment of the child.* New York: Free Press. (Orig. pub. 1932.)

Piaget, J. (1972). Intellectual evolution from adolescence to adulthood. *Human Development, 15,* 1–12.

Piaget, J. (1985). *The equilibration of cognitive structures.* Chicago: University of Chicago Press.

Prawat, R. S. (1999). Cognitive theory at the crossroads: Head fitting, head splitting, or somewhere in between? *Human Development, 42,* 59–77.

Rawls, J. (1971). *A theory of justice.* Cambridge, MA: Harvard University Press.

Rorty, R. (1979). *Philosophy and the mirror of nature.* Princeton, NJ: Princeton University Press.

Rosser, R. (1994). *Cognitive development: Psychological and biological perspectives.* Boston: Allyn and Bacon.

Ryan, B. A. (1999). Does postmodernism mean the end of science in the behavioral sciences, and does it matter anyway? *Theory & Psychology, 9,* 483–502.

Siegel, H. (1997). *Rationality redeemed? Further dialogues on an educational ideal.* London: Routledge.

Siegler, R. S. (1996). *Emerging minds: The process of change in children's thinking.* Oxford, UK: Oxford University Press.

Snarey, J. (1985). Cross-cultural universality of social-moral development: A critical review of Kohlbergian research. *Psychological Bulletin, 97,* 202–232.

Spelke, E. S., & Newport, E. L. (1998). Nativism, empiricism, and the development of knowledge. In W. Damon (Series Ed.) & R. M. Lerner (Vol. Ed.), *Handbook of child psychology: Vol. 1. Theoretical models of human development,* 5th edit. (pp. 275–340). New York: John Wiley & Sons.

Stanovich, K. E. (1999). *Who is rational? Studies of individual differences in reasoning.* Mahwah, NJ: Lawrence Erlbaum.

Turiel, E. (1998). The development of morality. In W. Damon (Series Ed.) & N. Eisenberg (Vol. Ed.), *Handbook of child psychology, Vol. 3: Social, emotional, and personality development,* 5th edit. (pp. 863–932). New York: John Wiley & Sons.

Walker, L. J. (1989). A longitudinal study of moral reasoning. *Child Development, 60,* 157–166.

Wellman, H. M. (1998). Culture, variation, and levels of analysis in folk psychologies: Comment on Lillard (1998). *Psychological Bulletin, 123,* 33–36.

Wellman, H. M., & Gelman, S. A. (1998). Knowledge acquisition in foundational domains. In W. Damon (Series Ed.), D. Kuhn & R. Siegler (Vol. Eds.), *Handbook of child psychology, Vol. 2: Cognition, perception, and language,* 5th edit. (pp. 523–573). New York: John Wiley & Sons.

Werner, H. (1957). The concept of development from a comparative and organismic point of view. In D. B. Harris (Ed.), *The concept of development* (pp. 125–147). Minneapolis, MN: University of Minnesota Press.

CHAPTER 4

Adult Development

The Holistic, Developmental, and Systems-Oriented Perspective

SEYMOUR WAPNER AND JACK DEMICK

OVERVIEW OF OUR APPROACH

The study of adult development has had an extensive and variegated coverage. For example, consider the work of Baltes (1979), Birren and Schaie (1990), Erikson (1950), Havighurst (1953), Jung (1933), Neugarten (1968), and Schaie (1983). The breadth and extensity of the work in this area is readily seen in Lemme's (1999) comprehensive survey of research and theory on adult development. The enormity of these contributions points to the need for their organization by a comprehensive theoretical perspective. Toward this end, the present chapter describes the holistic, developmental, systems-oriented perspective (Wapner, 1981, 1987; Wapner & Demick, 1990, 1992, 1998, 1999, 2000, 2002, in press) that is an extension of some facets of Werner's (1940/1957, 1957) classic work on comparative psychology of mental development.[1] For the most part, the research on adult development has involved theoretical approaches restricted to ontogenesis and aging. The approach presented here includes *ontogenesis and aging*, but also has a broader conceptualization of development that encompasses *phylogenesis, microgenesis, psycho- and neuro-pathology, and genesis* and *development of operations*

---

[1] For a review of Werner's work, with associates and students, on which the holistic, developmental, systems-oriented approach is based, see: Barten and Franklin (1978); Franklin (1990); Kaplan (1966, 1967, 1983); Wapner, Cirillo, and Baker (1969, 1971); Wapner and Werner (1957, 1965); Werner and Kaplan (1956, 1963); Werner and Wapner (1949, 1952); and Witkin (1965).

SEYMOUR WAPNER • Heinz Werner Institute for Developmental Analysis, Clark University, Worcester, Massachusetts, 01610.   JACK DEMICK • Center for Adoption Research, University of Massachusetts Medical School, Worcester, Massachusetts, 01605.

Handbook of Adult Development, edited by J. Demick and C. Andreoletti. Plenum Press, New York, 2002.

varying in degree of optimal functioning (e.g., stress, drugs). Thus, it includes progressive as well as regressive development. These diverse areas are analyzed by use of the *orthogenetic principle* that formally describes development as change from dedifferentiated to differentiated and hierarchically integrated person-in-environment system states.

Along with the developmental aspect of the perspective, one must consider its *holistic* aspect, which is directly concerned with the living organism as an integrated system where parts are considered in relation to the functioning whole. The parts of the systems are designated in the *systems-oriented* feature that refers to the *person-in-environment system state* as the unit of analysis. This unit includes three aspects of the *person* (physical/biological, e.g., health; intra-psychological, e.g., stress; and sociocultural, e.g., role), and three aspects of the *environment* (physical, e.g., natural or built environment; interpersonal; e.g., friends, relatives; and sociocultural, e.g., rules, laws of society).

In addition to these main categories of the perspective, a number of other features should be considered. These include: (1) *transactions* of the organism with the environment including *experience* (cognitive, affective, valuative) and *action*; and (2) modes of analysis including *structural analysis* (part–whole relationships) and *dynamic analysis* (means–ends relationships). Further, assumptions underpinning the analysis of transactions (experience and action) include the ideas of: *constructivism* (the organism actively constructs or construes its experience of the environment); *multiple intentionality* (the organism adopts different intentions with respect to self–world relationships, i.e., toward self or world out-there); *multiple worlds* (the organism lives in different spheres of existence, e.g., home and work); *directedness and planning* (the person-in-environment system is directed toward goals related to the capacity to plan); *complementarity of qualitative and quantitative analyses* (concern with both description of the nature of experience and action and factors underlying change in the nature of that experience and action); and, finally, *preference for process rather than achievement analysis*.

OTHER APPROACHES TO ADULT DEVELOPMENT

The variety of investigators concerned with adult development mentioned earlier and others (e.g., Harevan, 1994; Havighurst, 1953; Kitrell, 1998; Labouvie-Vief, Orwell, & Manion, 1995; Lemme, 1999; Levenson, 1986; O'Connor & Chamberlain, 1996; Roberts & Newton, 1987; Winter & Samuels, 1998) have largely conceptualized the period during ontogenesis and aging presumed to cover adult development to begin around the end of adolescence (18–20 years) and to end with death. Further, many of them have focused their analysis on delineating the developmental tasks of adulthood, that is, the abilities, skills, and/or responsibilities that, if accomplished at the appropriate stage of adult life, contribute to future development and happiness (cf. Havighurst's, 1953, tasks of childhood and adolescence). These tasks—that, from our perspective, may be reframed as goals and categorized according to those that involve the physical/biological, intra-personal/psychological, and sociocultural aspects of the person and physical, interpersonal, and sociocultural aspects of the environment—are presented in Table 4.1.

Table 4.1. Some Tasks or Goals to Be Fulfilled at Different Periods During the Course of Human Development with Respect to Various Aspects of the Person-in-Environment System

Category	Tasks or Goals
PERSON	
Physical/biological	18–35 years: Optimal physical development; desire for physical fitness; fulfillment of sexual desires
	35–60 years: Coping with and adjusting to decreasing physical strength and physical changes of middle age
	60 years and on: Continued coping with physical changes
Psychological	18–35 years: Develop and maximize cognitive (sensorimotor, perceptual, and conceptual) skills; affective functions; appropriate values; develop independence; self-concept; self-esteem; self-control; self-efficacy; self-identity; cope with midlife crises; objectivity; individually oriented dignity; sense of security; develop and work toward occupational goals; find life work
	35–60 years: Develop leisure time activities
	60 years and on: Deal with changes in cognitive, affective, and valuative functioning and in action
Sociocultural	18–35 years: Marriage; parenthood; serving as a wife and mother; managing a home and family; household provider; getting started in an occupation and achieving success; optimizing relation between work and home worlds
	35–60 years: Grandparenting; coping with possible divorce; adjusting to remarriage; stepparenting; caring for aging parents
	60 years and on: Disengagement from work; retirement; care for aging parents; coping with possible loss of partner
ENVIRONMENT	
Physical	18–35 years: Leave home and establish satisfactory physical living arrangements; adapt to physical location and nature of physical work place
	35–60 years: Deal with common changes in residences and/or physical work places
	60 years and on: Develop a deeper appreciation of nature and the physical surrounds; possibly adapt to supportive living communities and/or nursing homes
Interpersonal	18–35 years: Seek affiliation; social support; social network; attachment; friendship; love; emotional bonds; select a mate; intimacy with others; develop an optimal spousal relationship; start family; childbirth; raising children and possibly stepchildren; possibility of coping with divorce; develop social milieu
	35–60 years: Coping with empty nest
	60 years and on: Find a comfortable social network; adjust to loss of one's significant other; establishing an explicit affiliate relationship with one's age group; coping with social interaction that declines in older age groups
Sociocultural	18–35 years: Develop educationally; develop social networks that conform to cultural setting; adopt civic and societal responsibility; adhere to societal rules and regulations; getting started in an occupation
	35–60 years: Achieve adult and societal responsiblity
	60 years and on: Deal with new rules and regulations of older adulthood; develop new leisure activities (e.g., Edlerhostel)

Inspection of Table 4.1 reveals a variety of tasks and goals of the developing adult. It is important to appreciate that the items presented here are not rigidly fixed with respect to time of occurrence or relevance to all humans and to all cultures. Such a characterization is, of course, not precise because the time of occurrence is largely an individual matter depending on the rate of development of the individual as well as the environmental (physical and social) situation in which he or she finds himself or herself. Thus, although there are marked differences that exist with respect to these factors, the areas delineated clearly indicate the diversity and breadth of scope of problems, issues, and concerns that the developing adult must face.

RELATIONSHIP OF GOALS TO CRITICAL PERSON-IN-ENVIRONMENT TRANSITIONS

The tasks or goals that require action during adult development are inextricably linked to our research program on critical person-in-environment transitions across the lifespan conducted from the holistic, developmental, systems-oriented perspective for more than two decades. Though it is recognized that every moment of person-in-environment functioning involves change, our concern has been with those transitions that we regard as critical, that is, where a perturbation to any aspect of the person-in-environment system is experienced as so potent that the ongoing modes of transacting with the physical, interpersonal, and sociocultural features of the environment no longer suffice. Powerful changes in the person, in the environment, and in the relations between them may make for *developmental regression* in functioning, which may in turn—depending on conditions promoting more optimal functioning—make for *developmental progression* as characterized by the orthogenetic principle. Such changes are expected to be manifest in various aspects of the transactions (experience and action) of the person with the environment.

Now, not all of the tasks or goals listed in Table 4.1 would, from our perspective, operate as critical person-in-environment transitions for all people. Whether they operate in that manner depends both on the particular capacities and interests (physically, psychologically, and socioculturally) of the focal person and the characteristics of the physical, interpersonal, and sociocultural environments in which he or she is embedded. Consider some examples.

Tasks or goals that do not serve as the basis for critical transitions. Of course, if the focal person under consideration does not consider a particular task or goal as personally important (e.g., goal of physical fitness), lack of satisfaction of that goal would have little or no impact on the person-in-environment system state, and thereby make neither for developmental regression nor for developmental progression. This would, of course, hold in parallel fashion for other nonpotent goals.

Tasks or goals that serve as the basis for critical transitions. Many of the tasks or goals listed in Table 4.1 can operate as critical person-in-environment transitions. Indeed, some of items listed have been or are currently being studied as part of the research program on critical person-in-environment transitions from the holistic, developmental, systems-oriented perspective. Here, to illustrate our methodological approach, we shall restrict ourselves to problems relevant to adult development: birth of the first child; adoption; family functioning in general; divorce; "empty nest"; transitions to new environments; returning to school; transition to grandparenthood; retirement; coping with loss; aging and cognitive development; and adaptation to the nursing home. Before describing these studies, it would be of value to consider, in slightly greater detail, some general features of the mode of analysis of the perspective.

MODES OF ANALYSIS

Structural Analysis

While the tasks or goals have been categorized in terms of six aspects of the person-in-environment system in keeping with the holistic (part–whole) assumption, these tasks or goals are not independent of each other. The interrelationships among parts are clearly evident both within and among categories.

In relation to Table 4.1, for example, within the person/physical level, desire for physical fitness is related to adjusting to physiological changes of middle age; within the person/intra-psychological level, cognitive, affective, and valuative functions are clearly interrelated within the normal functioning adult; within the person/ sociocultural level, marriage is related to success as a household provider; within the environmental/physical level, establishing satisfactory living arrangements is related to adaptation to the workplace; within the environmental/interpersonal level, social support is related to coping with the empty nest; and within the environment/sociocultural level, development of social networks that conform to the cultural setting is linked to adherence to societal rules and regulations. The interrelationships of parts among categories is readily exemplified. Consider the linkage among the goals of fulfillment of physical development (person/physical/biological), achievement of self-esteem (physical, academic, social, and occupational) and self-identity (person/intra-psychological), and managing a home and family (person/sociocultural). Another example involves interrelationships among the goals of establishing satisfactory living arrangements (environment/physical), finding a comfortable social network (environment/interpersonal), and developing educationally (environment/sociocultural).

A critical question is related to the part–whole relationship. What is the impact of a given goal on the total functioning of a given human being? One item may have a powerful impact on the characterization of the whole. For example, given an individual's desire to become a professional athlete, the goal in his or her younger years of achieving optimal physical development is critical.

Given the variety of components and subcomponents of the person-in-environment system, the structure is described and analyzed—using the orthogenetic principle—in formal, comparative terms where the parts of the system are characterized as more or less differentiated parts or subsystems that are more or less integrated with one another in specifiable ways.

Dynamic (Means–Ends) Analysis

Focusing on the dynamics of the person-in-environment system entails a determination of the means whereby a characteristic structure is achieved or

maintained. It is here that *planning* and the relationship between *experience/
intentionality* and *action* enter into our approach.

Planning is used in the sense of plotting a course of future action that moves
the person-in-environment system from some initial state of functioning to some
end state. Such movement involves transactions (cognitive, affective, and evaluative experience and action—organized activities in relation to some object—with
the environment on the part of the person). The formulation of a plan involves a
set of acts that are preparatory for carrying out a more complex set of concrete
actions directed toward the achievement of some goal (Wapner & Cirillo, 1973).
Further, in keeping with Werner's (1937) methodological approach, a process
analysis is preferred over an achievement analysis.

Developmental Analysis

The developmental mode of analysis is based on the orthogenetic principle,
which characterizes development in formal terms with respect to the degree of
organization of the person-in-environment system. This applies to the components (person, environment) as well as to the relationships among components
(e.g., means–ends) and among part-processes (e.g., cognition, affection, valuation,
and action). The more differentiated and hierarchically integrated a system is, in
terms of its parts and its means and ends, the more advanced in development it is
said to be. Optimal development entails a differentiated and hierarchically integrated person-in-environment system with flexibility, freedom, self-mastery, and
the capacity to shift from one mode of person-in-environment organization to
another depending on the goals, the demands of the situation, and the instrumentalities available (B. Kaplan, 1966, 1967; Wapner, 1987; Wapner, Ciottone, Hornstein,
McNeil, & Pacheco, 1983; Wapner & Demick, 1990, 1998, 1999, 2000, 2002, in press;
Wapner, Kaplan, & Ciottone, 1981; Werner, 1940/1957; Werner & Kaplan, 1956).

Polarities. The orthogenetic principle has also been specified with respect
to a number of polarities, which at one extreme represent developmentally less
advanced and at the other more advanced functioning (Werner & Kaplan, 1956).
Consider these with respect to examples from adult development:

1. *Interfused to subordinated.* In the former, ends or goals are not sharply differentiated; in the latter, functions are differentiated and hierarchized
   with drives and momentary states subordinated to more long-term goals.
   For example, for the adult with less developmentally advanced experience and action, fulfillment of sexual desires is not differentiated from the
   need to be concerned with the long-term feelings of one's spouse (each is
   viewed as a short-term goal); in contrast, the more developmentally
   advanced adult differentiates and subordinates the short-term goal (fulfillment of sexual desires) to the long-term goal of maintaining an optimal
   relationship with his or her spouse.
2. *Syncretic to discrete.* Syncretic refers to the merging of several mental
   phenomena, whereas discrete refers to experience, functions, meanings,
   and action that represent something specific and unambiguous. Syncretic
   thinking is represented, for example, by the less developmentally advanced

adult (e.g., mother, father, teacher) being unable to differentiate between his or her feelings and those of others out-there (e.g., child). In contrast, discrete is exemplified by the more advanced adult who is able to differentiate accurately between one's feeling and those of others out-there.

3. *Diffuse to articulate.* Diffuse represents a relatively uniform, homogeneous structure with little differentiation of parts, whereas articulate refers to a structure in which differentiated parts make up the whole. The law of *pars pro toto* (a part stands for the whole) provides a good example. The less developmentally advanced adult is exhibiting diffuse behavior when a minor disagreement between self and other impacts the nature of the whole self–other relationship; in contrast, articulate behavior is indicated by his or her capacity to cope with that part and yet separate it from the whole.

4. *Rigid to flexible.* Rigid refers to experience and action that is fixed and not readily changeable whereas flexible refers to experience and action that is readily changeable and plastic. For example, the less developmentally advanced adult shows fixed ways of coping with the child leaving home and the change in household duties and behaviors, whereas the developmentally advanced adult changes depending on that change in context.

5. *Labile to stable.* Labile refers to the fluidity and inconsistency that goes along with changeability; stable refers to the consistency or unambiguity that occurs with fixed properties. For example, lability is evident in the less developmentally advanced adult rapidly changing from feelings of love to hate and vice versa, by shifts of attention from other to self, and so forth; in contrast, stability is represented by the avoidance of stimulus-bounded shifts of attention from other to self.

Modes of Coping.   Utilizing the orthogenetic principle, four modes of coping have been described to characterize individual differences developmentally as follows (e.g., Wapner & Demick, 1998):

1. *Dedifferentiated person-in-environment system state.* For example, a wife (husband) immediately, unequivocally, and unquestioningly goes along with her husband's (his wife's) wishes.

2. *Differentiated and isolated person-in-environment system state.* Withdrawal and removal of self from the desires of one's spouse is a classic example.

3. *Differentiated and in conflict.* For instance, there is ongoing conflict between the desires of parent and child.

4. *Differentiated and hierarchically integrated person-in-environment relationship.* An example might be when the parent distinguishes between short- and long-term goals for the child and the capacity to subordinate the former to the latter occurs (e.g., when the parent insists on the child's studying for tomorrow's examination rather than permitting him or her to play with friends this afternoon).

This developmental mode of analysis can be applied to the variety of tasks or goals for the three aspects of the person and the three aspects of the environment at various periods during adulthood, as listed in Table 4.1.

Relationships Between Experience and Action.   Some initial inroads have been made in our analysis on the problem of the relationship between experience and action by utilizing Turvey's (1973) musical instrument metaphor for understanding neurological mechanisms. On the analogy between tightening strings (*tuning*) as affecting music produced and string plucking *activation* that causes the sound (see Gallistel, 1980), we have introduced the notion of *general factors* preparing the individual for some action (e.g., advertisements about the importance for health of a fit physical body) and *specific precursors, precipitating events*, or *triggers* (e.g., a close friend who asks one to join him or her working out in a gym; a person does not have the strength to pick up a small package). General factors and precipitating events represent *means* toward achieving the goal of moving toward optimal fitness. Given these considerations concerning methodology, we may now turn to some studies relevant to critical person-in-environment transitions.

SOME RESEARCH ON CRITICAL PERSON-IN-ENVIRONMENT TRANSITIONS

Birth of the First Child

This problem of the differential changes in experience and action of first-time parents is of critical significance in adult development. It was examined in a pilot study by Coltrera (1978) and has been examined more systematically from the holistic, developmental, systems-oriented perspective by Clark (2001) in a Ph.D. dissertation (see Demick &Wapner, 1991).

Coltrera's study was directed at assessing changes in the transactions (experience and action) of the parents-to-be with respect to their experience of self and of the world out-there. Here, we restrict ourselves to some findings on the body percept as assessed by the Draw-a-Person technique (Machover, 1949; Witkin, Lewis, Hertzman, Machover, Meisner, & Wapner, 1954). The Draw-a-Person technique was administered on four occasions (3–4 weeks and 3–4 days before birth; 3–4 days and 3–4 weeks after birth) with the following instructions: (1) draw a person; (2) draw a person of the opposite sex; (3) draw your spouse and yourself before birth; and (4) after birth, draw yourself, your spouse, and your child. Consider some findings for one typical wife and husband.

For the wife, the first spontaneous drawing of a person 3 to 4 weeks before birth is an adult male with a moustache, and so forth; in contrast, just before birth, the moustache is no longer present and the drawing has childlike, boyish qualities. After birth, to the request to draw a person, there is a dramatic shift to drawing a baby rather than an adult. The baby, conforming to actuality, is female. The second drawings, which are self sex in this case, also undergo dramatic change from a woman with a rounded middle to a staid figure sitting and "waiting for birth." Finally after the birth, the drawings show a woman with pronounced breasts and a small stomach. The husband's drawings also show the impact of the transition. Here, the major shift is from representation of an adult to representation of a child's body, especially in the third and fourth drawings following birth. The third drawings of husband, wife, and family show the stressful impact of the transition. One point relevant to stress is that the first drawings by the wife all have mouths with a smile represented, whereas in the fourth drawing there is a mouth half downward expressing sadness and also half upward with a smile.

Clark's (2001) more extensive study again collected data on two occasions prior to birth (6–9 and 1–2 weeks) and two following birth (2–3 and 8–10 weeks) with the following instruments: a demographic questionnaire; a neutral, open-ended interview (e.g., "What has life been like for you during the past months?"); a Psychological Distance Map (PDM) for "people" to indicate the people most important in the lives of the husband and wife; a PDM for "places" (Wapner, 1977); a Pie Chart of Important Roles (PIE) (Cowan & Cowan, 1988, 1990) to assess the roles important to the focal person during pregnancy (person, co-worker, friend, partner, lover, etc.); a family assessment device that assesses individual and family perspectives on problem solving, communication, rule following behavior, and so forth; a Multiple Worlds Questionnaire, which provides a description of life in the home world vs. the work world, and so forth; and the Draw-a-Person Test (Machover, 1949).

In Clark's (2001) study, there are highly suggestive findings. Some, for example, are in keeping with those described earlier in the pilot study by Coltrera (1978). That is, consistent with Coltrera's results, Clark's data from the Draw-a-Person Test suggest, as expected, that the transition to parenthood has an impact on the drawings of both husband and wife. For some participants, the stressful nature of the transition to parenthood is evident in the contrast between the drawings from approximately 2 months before the birth of the child and the subsequent drawings after birth. For others, the drawings provide evidence of the development of a closer connection between the couple over the transition to parenthood.

Further, findings from multiple facets of Clark's (2001) study show promise of moving our description of development in the transition to parenthood beyond Coltrera's focus on the experience of stress or well-being to a holistic description of the person's multifaceted experience within a person-in-environment system. That is, results from the PDM for "people" and "places" suggest a systematic progression in the differentiation and integration of the focal person's experience over the transition to parenthood. Psychological distance appears to be related to the person's experience of people or places in the environment as sources of more or less support as he or she transitions from childlessness to first-time parenthood. The greater the support in terms of type and intensity, the more likely it is that the person or place will be less distant from the focal person. It is expected that the analysis of data from a person-in-environment questionnaire (open-ended and family assessments) and other instruments will further enhance our understanding of the person's experience and action as well as his or her development under this transition.

Adoption

In line with our work on parental development (see: Demick, 2002; Demick & Wapner, 1991; Wapner, 1993), Demick and his associates have developed a research program (see Demick, chapter 25) on the adaptation of individuals who were adopted and their families (adoptive parents, birthparents). As clearly multiple parties are involved (e.g., several sets of parents, agency workers, legal personnel), our transactionally oriented, systems approach has been particularly helpful. Further, our characteristic focus on both structural (e.g., how does the individual who was adopted organize the various parts of identity into a coherent whole?) and dynamic (e.g., how do preadoptive couples plan for a pending adoption?) analyses has likewise suggested open empirical problems.

For example, based on the assumption that some of the same processes relevant to the general problem of the transition to parenthood may become more focal when one examines the contrasts between the transition to biological vs. adoptive parenthood, Demick (1993) has assessed the differential effects of open vs. closed adoption (communication vs. no communication between biological and adoptive parents) on children and their families. Speculating that open adoption represents a differentiated and integrated self–world relationship and traditional closed adoption either a dedifferentiated, a differentiated and isolated, or a differentiated and in conflict self–world relationship, Demick has found no differences in life satisfaction, stress, and control in parents practicing open vs. closed adoption. However, relative to those practicing closed adoption, those practicing open adoption worried less about attachment to their child and felt confident about the "story" that they would ultimately tell their child. Such developmentally based individual differences have relevance not only for adoptive families, but also for all families irrespective of the way(s) in which the children have been acquired.

As a second example, Aronson, Ronayne, Hayaki, and Demick (1994) have found that, relative to nonadopted young adults, young adults who were adopted: exhibited a more internal locus of control; made more positive attributions about story characters; had more confidence in their judgments; and did not differ with respect to perceptions of their parents or to standard defense mechanisms. Taken with more general findings from developmental psychology on resilient children (e.g., E. Werner, 1993), such research has the potential to lead to a "new look" in adoption and foster care (i.e., adoptive status leading to psychological strength rather than disability).

Family Functioning in General

Related to the larger issue of parental development, our approach has also framed several treatments of family functioning, more generally. First, Kaden (1958) has suggested that structural analyses—for example, dominance of one partner to the other, independence of partners, mutuality—are useful in couple and family conceptualization. In line with this, McAvay (1977) has posited developmentally based modes of conflict resolution that include fusion (one partner yields to the other), differentiation-incongruity (each goes his or her own separate way), and integration (respective viewpoints are integrated). More recently, Melito (1985, 1988) has presented an integrative framework for combining some central aspects of individual psychodynamics with structural family therapy. The framework derives from our approach and relies on the concept of integrative levels of organization and on the goals of therapy as viewed from a developmental perspective, namely, the differentiation and hierarchic integration of the family system characterized by flexibility, stability, and adaptability.

Divorce

As Wapner (1993) has noted, "Divorce operates as a powerful perturbation to the state of parenthood. With the change in relationship between the two parents, there are profound effects on the relations between each parent and the child" (p. 26). Drawing on this notion, several pilot studies conducted from our approach (e.g., Grossman, 1999; Minian, 1996; Plante, 1992) have reported that parental divorce has an effect on every aspect of the person-in-environment relationship, namely, biological, psychological, and sociocultural levels of the person and physical, interpersonal, and sociocultural aspects of the environment. When considering the impact of parental divorce on the individual, there has been a tendency to focus mainly on relationships with parents and the loss of contact with one parent. In contrast, these studies have suggested the importance of considering all of the other areas of a person's life that are also influenced by the disruption. Participants also indicated that not all of the effects of divorce are negative: many suggested that with the dissolution of parents' marriages come opportunities for growth.

The "Empty Nest"

A pilot study was conducted on the impact of the departure of adult children on the changes that take place within the home setting and on the personal experience of middle- to upper-class mothers with two to four children (Capobianco, 1992). A questionnaire covering the six aspects of the person-in-environment system and an interview including open-ended questions were administered to 12 mothers ranging in age from 45 to 70 years. While there have been studies concerned with the effect of the empty nest on the self experience of mothers, there has been little or no research on the relationship between the impact of the empty nest on changes in the physical aspect of the environment and interpersonal aspects of the environment.

Accordingly, to obtain a more holistic picture, information was obtained not only on the experience of the mothers, but also on the actions taken on the children's rooms (were they changed or preserved?), treatment of the children's possessions, and spatial arrangements for staying over when the children visited. There is evidence, albeit preliminary, that: (1) whereas a third of the rooms, and so forth, were left unchanged, others were changed most frequently into guestrooms, dens, sitting rooms, and so forth; (2) the majority of mothers stored some or all of their children's possessions that were left behind; (3) a high proportion of those children who do not live close by stayed in what had been their own bedrooms; (4) there was a significant relationship between the length of time since the child left and whether the room was still designated by the child's name, that is, a greater proportion of children's names were retained for children who left 3 or fewer years ago, while a greater proportion was not used for children who left

4 or more years ago; (5) the majority of mothers responded that there was a positive change in the interpersonal relationships with their children including a feeling of closeness, more communication, relating to children as adults, and experiencing the children as now independent and responsible, while a few of the mothers experienced no change or a change in herself; (6) the main changes included a quieter and cleaner/neater physical environment; fewer commitments and responsibilities; and feeling closer to children, husbands, and others in the interpersonal environment; (7) the role of the mother was reported as changed from caretaker, disciplinarian, and mediator to that of being less responsible and less of a housekeeper and defender; and (8) with respect to mothers' physical and mental well-being, a majority noticed a change including feelings of loss and of not being needed as well as feeling more relaxed and less worried.

More generally, it was found that the transition to the empty nest is neither as unhappy nor as painful as some studies have concluded (e.g., Pillay, 1988). While many mothers reported a positive change in the relationship with their children and the physical environment, none of the mothers reported a change in their physical well-being. For those who noticed a change in their interpersonal relationships, it was generally positive and occurred most frequently with the husband. These findings are consonant with those of Harkins (1978) and White and Edwards (1990). Specifically, Harkins found that empty nest mothers experience no physical changes in themselves, while White and Edwards (1990) concluded that there are improvements in the marital happiness of empty nesters.

Although the person-in-environment category system (three aspects of person and three of environment) has been helpful in suggesting a variety of interesting experiential findings that occur with the empty nest, the pilot study has also suggested other directions that might be profitably explored in other studies with larger samples. Such studies should include the father as well as the mother and should consider relationships among the physical, intra-personal/psychological, and interpersonal factors. Moreover, it is essential from our perspective to include a developmental analysis utilizing the orthogenetic principle to assess the modes of coping of both mothers and fathers as well as the relationship to each other.

Transitions to New Environments

Life continually involves transitions to new environments and adulthood is no exception (e.g., vacations, new jobs). Relevant here is a series of studies that has dealt with similarities and differences in perception of necessities, amenities, and luxuries among young, middle-aged, and older adults. For example, Cool (1991) has reported that there is much commonality among the three groups: for example, members of all groups included food, shelter, and clothing as physical necessities; cars and television sets as physical amenities; and close emotional friends as interpersonal luxuries. There were also, however, significant age differences. For instance, relative to young adults, middle-aged and older adults more often included heat and air and less often mentioned food as physical necessities. Relative to the other two groups, young adults more often considered parents as interpersonal necessities. Relative to young adults, middle-aged adults and to a lesser extent older adults considered education and civil rights as sociocultural necessities. These age differences appear linked to values that are representative of the particular age groups (see Wapner, Quilici-Matteucci, & Cool, 1991; Wapner, Quilici-Matteucci, Yamamoto, & Ando, 1990).

Returning to School

Many adults often return to school after completing other life tasks such as raising one's family. For the most part, the open empirical question remains: Are there similarities and/or differences in the behavior and experience of traditional aged and older returning college students? Based on our approach, Abe (1990) has demonstrated that: (1) returning older students use more developmentally advanced modes of coping with conflict (e.g., less financial aid than expected) than traditional aged students, who characteristically respond with non-constructive ventilation and cognitive accommodation; and (2) relative to traditional aged students, returning older students are more likely to be concerned about the study environment, to desire more individual attention, and to perceive courses as not challenging enough. Such findings have both theoretical and practical import with potential to ameliorate the lives of adults during critical person-in-environment transitions.

Transition to Grandparenthood

From our perspective, Wapner (1993) has conducted a holistic, developmental, systems-oriented analysis of grandparenting. Specifically, he has suggested that researchers describe the nature and course of grandparent development, a problem that has received almost no attention in the literature. Further, he has suggested that research be directed at understanding changes over time in the meaning of grandparenthood—centrality, valued elder, immortality through clan, reinvolvement with personal past, and indulgence (Kivnik, 1988)—as well as differences in these meanings between biological and adoptive grandparents. Finally, he has recommended that:

> ... it would be of great value to conduct a longitudinal study of this transition parallel to Coltrera's (1978) previously mentioned pilot study on the transition to parenthood. Assessment of the grandparent's cognitive, affective, and valuative experience as well as the actions to be with his or her child prior and following birth of the grandchild would throw light on the underlying process of becoming a grandparent as well as the changes that take place in the course of the grandchild's growth ... For each of Cherlin and Furstenberg's (1986) stages, it would be of value to characterize the individuals with respect to the developmentally-ordered person-in-environment system states described earlier for parenthood, namely, dedifferentiated, differentiated and isolated, differentiated and in conflict, and differentiated and hierarchically integrated. (p. 26)

Retirement

A study utilizing phenomenological methods was conducted to assess the experience and action of people about to undergo retirement and the role of the social network in the individual's adaptation to his or her new person-in-environment status. Four types of retirement experience were evident: (1) transition to old age—retirement was experienced as a transition to the last phase of life and the importance of putting things in order; (2) new beginning—a new time in life when one can do what one wants; (3) continuation—retirement is not an event of major importance but rather valued activities of work, and so forth can be continued; and (4) imposed disruption—devastated by the loss of a highly valued sphere of activity (Hornstein & Wapner, 1985). The social networks of the retirees showed either: (1) low involvement in planning and support, high congruence, and minimal support for the network member; (2) low to moderate involvement in planning and support, moderate incongruence, and some—often negative—support for the network member; or (3) high involvement in planning and support, high degree of congruence, and moderate change in the network member's life (Hornstein & Wapner, 1984).

The findings of the Hornstein and Wapner studies (1984, 1985) were duplicated in a study by Hanson and Wapner (1994) and, in addition, revealed differences between men and women. Women, for example, had more positive attitudes toward work and old age, retired at younger ages than men, reported greater financial worries, and reported greater degrees of preretirement planning. A follow-up study (Wapner, Demick, & Damrad, 1988) 8 years later with 17 of the original retirees revealed that retirees tend to move to a person-in-environment state in which their lives are very busy, filled with working activities including earlier employment, highly active volunteer work, and active involvement in hobbies and other pursuits. In other words, they are acting in a parallel fashion to the continuation group noted earlier.

Loss of a Loved One

With ontogenesis, one must negotiate the inevitable losses associated with the deaths of close friends and/or loved ones. While no one is exempt from potential losses at any stage of development, middle-aged adults are often required to care for their aging parents, who often become more sick and die. In older adulthood, the loss of peers and older family members is often a common experience. How do adults face and deal with such losses? While we do not have data on this problem per se, a pilot study generated from our approach (Caputo, 1997) has suggested that, for college students, the death of a parent—similar to parental divorce—has the potential to impact the individual holistically at all levels of the person-in-environment system. Most notable was the fact that the death affects participants particularly at the interpersonal level of the environment (e.g., fear of abandonment by others). Whether, and if so how, such generalizations obtain for adults at various stages in the life cycle appear worthy of future empirical investigation and have the potential to lead to practical programs aimed at helping individuals cope with such situations.

Aging and Cognitive Development

It is almost a cliché that with increasing age come changes in cognitive functioning. While much of the empirical evidence has been conceptualized as indicating a state of cognitive deterioration in aging, research generated from within our approach has alternatively argued for the position of cognitive reorganization in aging. For example, early studies in the Clark University laboratories (e.g., Comalli, Wapner, & Werner, 1962) have indicated that, relative to younger adults, the total time scores of older adults on the Stroop Color-Word Test are impaired only on Card C (color names printed in different colored inks where the task is to identify the color of the ink). In line with this, process-oriented analyses of performance on this task (e.g., Demick & Wapner, 1985) have also revealed modifications of cognitive strategies at later developmental stages (adults 70 years and older). For instance, older adults typically emphasize boundaries between items, use more sharp and articulate pronunciation, and employ nonlinguistic insertions rather than the inserted linguistic phrases characteristic of younger participants. In addition, participants' verbalized strategies on completion of the task (e.g., the common response of "I'm no good at concentrating") have suggested exploration into illogical thought processes—rather than actual cognitive deficits—as contributing factors to older individuals' less than optimal cognitive functioning.

Additional evidence for the notion of reorganization (rather than of deterioration) of cognitive strategies in older adults has come from the more recent work of Demick and Harkins (1999), which examined relationships between field dependence–independence cognitive style and driving behaviors across the adult lifespan. Specifically, we have found that factors other than age (e.g., cognitive style, personality) are generally the best predictors of driving behavior. While our participants were admittedly in relatively good physical and psychological health and hence not representative of all older adults, it has become clear that overly broad generalizations about older adulthood in general and about cognitive changes in aging more specifically are unwarranted and fail to capture the diversity and the potential of older adults under optimal conditions as they negotiate the inevitable transitions of adult life.

Adaptation to a Nursing Home

One hundred older people (75 females and 25 males) responded to a questionnaire on cherished possessions and on adaptation to the nursing home (Wapner, Demick, & Redondo, 1990). It was found that the presence of cherished possessions is positively related to adaptive status in a nursing home, a new environment. There remains, of course, the question of the directionality of this relationship. Does the presence of cherished possessions serve as an aid in adaptation to the nursing home? In contrast to the findings of Csikszentmihalyi and Rochberg-Halton (1981) that, with age, there is a shift from an emphasis on objects of action to objects of contemplation, we found the suggestion of a reversion to objects of action (e.g., musical instruments, silverware, tools, sports equipment, cameras) when one enters a nursing home in later adulthood. Whether such reversion implies developmental regression or progression is worthy of inquiry.

Our findings have also suggested that cherished possessions largely function to provide historical continuity, comfort, and a sense of belongingness. There remains the question of whether cherished possessions also function to provide control, which is in keeping with Furby's (1978a, 1978b) notion that possessions offer the individual a sense of control over his or her environment. It is also of interest that there are gender differences, namely, that, relative to men, women have significantly more cherished possessions and attribute less utilitarian meaning to them, greater comforter functions, and associate the possessions with self–other relationships. These findings, if replicated, have practical implications such as suggesting to men entering a nursing home that they bring their cherished possessions with them.

CONDITIONS FACILITATING DEVELOPMENTAL ADVANCE

Conditions concerning facilitation of developmental progression have recently been described by Wapner and Demick (1998). They are briefly described below with implications for adult development.

Self–World Distancing

A " ... decrease in self–world distancing between the individual and the consequences of his or her negative actions may lead to safer, more optimal person-in-environment functioning" (p. 795). For instance, parents who allow themselves the opportunity to distance themselves from the problematic behavior of their children may ultimately be able to reconnect and work through such behavior. Further, whether optimal parenting and self–world distancing are positively correlated is a problem worthy of future empirical investigation.

Reculer Pour Mieux Sauter (Draw Back to Leap)

"A negative experience (e.g., loss of self-esteem) may serve the function of fostering greater self-insight and providing the formal condition of dissolution of a prior organization of the self, thereby permitting a creative organization of the self" (p. 795). In line with this, dealing with stressful life events that occur routinely during adulthood (e.g., transition to parenthood, coping with illness) may actually lead to developmental advances in self experience and the like.

Triggers to Action

"There is evidence to suggest that making people aware of the precipitating events or triggers to their own or others' actions leads to a heightened awareness, which in turn leads to appropriate action and hence more optimal person-in-environment functioning" (p. 795). Thus, as adults become more health conscious as they enter middle age, it might be important for them to consider the triggers to more developmentally advanced action (e.g., initiating a diet regime, using automobile safety belts) as outlined in our previous research.

Individual Differences

"Various studies from within our approach have suggested that consideration of individual differences may be a route for fostering developmental progression"

(p. 795). As noted previously, if individual differences in the retiree's needs are taken into account in making a match with the post-retirement environment, developmental advances may result.

Planning

"There is also evidence that a simple request for the person to verbalize plans about actions to be taken to advance to a new, more ideal person-in-environment system state may bring the state into effect (see Neuhaus, 1988; Shapiro, 1973; Shapiro, Rierdan, & Wapner, 1972; Wapner, 1987)" (p. 796). In line with this, planning is a useful cognitive process that should be nurtured at every juncture of adult life (e.g., planning for parenthood, planning for vacations, planning for retirement).

Anchor Point

"Both physical (Schouela, Steinberg, Leviton, & Wapner, 1980) and social anchor points (Kaplan, Pemstein, Cohen, & Wapner, 1974; Minami, 1985a, 1985b; see also Winnicott, 1958, 1971) have been shown to play a role in the development of a spatial organization and social network in a adapting to a new environment" (p. 795). Thus, the developing adult—regardless of whether he or she is visiting a new country, starting a new job, beginning his or her post-retirement life, or entering a nursing home—would do well to consider this strategy of adopting various anchor points for orientation to the environment as he or she negotiates the myriad of transitions inherent in adult life.

CONCLUSIONS

The problem areas discussed here indicate the relevance of our holistic, developmental, systems-oriented approach to the field of adult development. While most authors have focused primarily on the developmental tasks of adulthood, we have alternatively examined critical person-in-environment transitions across the adult lifespan, which have included the transitions to parenthood (with an emphasis on adoptive parenthood); retirement; aging more generally; and the nursing home. Over the course of our journey, we have also identified numerous open research problems—including, though not limited to, the relationships among levels of integration in adult development; the roles of individual differences and processes such as planning in adult development; and developmental analyses of the status of the individual who was adopted, grandparent, retiree, and so forth—and have attempted to demonstrate that the original Wernerian approach coupled with our recent theoretical elaboration has the potential to integrate the prevalent diversity within contemporary psychology.

REFERENCES

Abe, J. A. (1990, April). *The experiences and modes of handling conflicts of older returning students and traditional age students.* Unpublished manuscript, Clark University.

Aronson, E., Ronayne, M., Hayaki, J., & Demick, J. (1994, April). *Adopted does not mean seeing the world through rose-colored glasses.* Paper presented at the annual meeting of the Eastern Psychological Association, Providence, RI.

Baker, A. H., Cirillo, L., & Wapner, S. (1969). Perceived body position under lateral body tilt. *American Psychological Association Proceedings, 4, Part 1*, 41–42.

Baltes, P. B. (1979). Life span developmental psychology: Some converging observations on history and theory. In P. B. Baltes & O. G. Brim, Jr. (Eds.), *Life-span development and behavior, Vol. 2* (pp. 255–279). New York: Academic Press.

Barten, S. S., & Franklin, M. B. (1978). *Developmental processes: Heinz Werner's selected writings.* New York: International Universities Press.

Birren, J. E., & Schaie, K. W. (Eds.) (1990). *Handbook of the psychology of aging,* 3rd ed. San Diego: Academic Press.

Capobianco, N. (1992). *Changes in the home setting following the departure of adult children: A preliminary study.* Unpublished manuscript, Clark University.

Caputo, B. (1997). *The effects of death of a parent on college students: A pilot study.* Unpublished manuscript, Clark University.

Cherlin, A. J., & Furstenberg, F. F. (1986). *The new American grandparent: A place in the family, a life apart.* New York: Basic Books.

Clark, W. (2001). Transition to parenthood: The experience and action of first-time parents. Unpublished doctoral dissertation, Clark University, Worcester, MA.

Coltrera, D. (1978). *Experiential aspects of the transition to parenthood.* Master's thesis proposal, Clark University, Worcester, MA.

Comalli, P. E., Jr., Wapner, S., & Werner, H. (1962). Interference effects of Stroop Color-Word Test in childhood, adulthood, and aging. *Journal of Genetic Psychology, 100*, 47–53.

Cool, K.L. (1991). *Age differences (young, middle-age, and older adults) in necessities, amenities, and luxuries in various aspects of the environment.* Unpublished manuscript. Clark University, Worcester, MA.

Cowan, P. A., & Cowan, C. P. (1988). Changes in marriage during the transition to parenthood: Must we blame the baby? In G. Y. Michaels & W. A. Goldberg (Eds.), *The transition to parenthood: Current theory and research.* Cambridge, UK: Cambridge University Press.

Cowan, P. A., & Cowan, C. P. (1990). Becoming a family: Research and intervention. In I. Sigel & E. Brody (Eds.), *Family research, Vol. 1.* Hillsdale, NJ: Lawrence Erlbaum.

Csikszentmihalyi, M., & Rochberg-Halton, E. (1981). *The meaning of things: Domestic symbols of the self.* New York: Cambridge University Press.

Demick, J. (1993). Adaptation of marital couples to open versus closed adoption: A preliminary investigation. In J. Demick, K. Bursik, & R. DiBiase (Eds.), *Parental development* (pp. 175–201). Hillsdale, NJ: Lawrence Erlbaum.

Demick, J. (2002). Stages of parental development. In M.H. Borstein (Ed.), *Handbook of Parenting* (2nd ed.): *Vol. B. Being and becoming a parent* (pp. 389–413). Mahwah, NJ: Lawrence Erlbaum Associates.

Demick, J., & Harkins, D. (1999). Cognitive style and driving skills in adulthood: Implications for licensing of older adults. *International Association of Traffic Science and Safety (IATSS) Research, 23*(1), 1–16.

Demick, J., & Wapner, S. (1991). Field dependence-independence in adult development and aging. In S. Wapner & J. Demick (Eds.), *Field dependence-independence: Cognitive style across the life span* (pp. 245–268). Hillsdale, NJ: Lawrence Erlbaum.

Demick, J., & Wapner, S. (1985). *Age differences in processes underlying sequential activity (Stroop Color-Word Test).* American Psychology Association annual meetings, Los Angeles, CA.

Demick, J., & Wapner, S. (1992). Transition to parenthood: Developmental changes in experience and action. In T. Yamamoto & S. Wapner (Eds.), *Developmental psychology of life transitions* (pp. 243–265). (In Japanese. English translation available.) Kyoto, Japan: Kitaoji.

Erikson, E. (1950). *Childhood and society.* New York: W. W. Norton.

Franklin, M. B. (1990). Reshaping psychology at Clark: The Werner era. *Journal of the Histroy of the Behavioral Sciences, 26*, 176–189.

Furby, L. (1978a). Possessions: Toward a theory of their meaning and function throughout the life cycle. In P. Baltes (Ed.), *Life span development and behavior, Vol. 1* (pp. 297–336). New York: Academic Press.

Furby L. (1978b). Possessions in humans: An exploratory study of its meaning and motivation. *Journal of Social Behavior and Personality, 6*, 49–52.

Gallistel, C. R. 1980). *The organization of action: A new synthesis.* Hillsdale, NJ: Lawrence Erlbaum.

Grossman, J. L. (1999, April). *The effects of parental divorce on college students.* Unpublished manuscript, Clark University.

Hanson, K., & Wapner, S. (1994). Transition to retirement: Gender differences. *The International Journal of Aging and Human Development, 39*(3), 189–208.

Harevan, T. K. (1994). Aging and generational relations: A historical and life course perspective. *Annual Review of Sociology, 20*, 437–461.

Harkins, E. B. (1978). Effects of empty nest transition of self-report of psychological and physical well-being. *Journal of Marriage and the Family, 40*, 549–556.

Havighurst, R. (1953). *Human development and education.* New York: Longmans Green.

Hornstein, G. A., & Wapner, S. (1984). The experience of the retiree's social network during the transition to retirement. In C. M. Aanstoos (Ed.), *Exploring the lived world readings in phenomenological psychology* (pp. 119–136). Georgia College.

Hornstein, G. A., & Wapner, S. (1985). Modes of experiencing and adapting to retirement. *International Journal on Aging & Human Development, 21*(4), 291–315.

Jung, C. (1933). *Modern man in search of a soul* (W. S. Dell & C. F. Baynes, Trans). New York: Harcourt.

Kaden, S. E. (1958). A formal-comparative analysis of the relationship between the structuring of marital interaction and Rorschach blot stimuli. Unpublished doctoral dissertation, Clark University, Worcester, MA.

Kaplan, B. (1966). The study of language in psychiatry: The comparative developmental approach and its application to symbolization and language in psychopathology. In S. Arieti (Ed.), *American handbook of psychiatry, Vol. 3* (pp. 659–668). New York: Basic Books.

Kaplan, B. (1967). Meditations on genesis. *Human Development, 10*, 65–87.

Kaplan, B. (1983). Genetic-dramatism: Old wine in new bottles. In S. Wapner & B. Kaplan (Eds.), *Toward a holistic developmental psychology* (pp. 53–74). Hillsdale, NJ: Lawrence Erlbaum.

Kaplan, B., Pemstein, D., Cohen, S. B., & Wapner, S. (1974). *A new methodology for studying the processes underlying the experience of a new environment: Self-observation and group debriefing.* Working notes, Clark University, Worcester, MA.

Kitrell, D. (1998). A comparison of the evolution of men's and women's dreams in Daniel Levinson's theory of adult development. *Journal of Adult Development, 5*(2), 105–115.

Kivnik, H. Q. (1988). Grandparenthood, life review, and psychosocial development. *Journal of Gerontological Social Work, 12*, 63–81.

Labouive-Vief, G., Orwell, L., & Manion, M. (1995). Narratives of mind, gender and the life. *Human Development, 38*, 239–257.

Lemme, B. H. (1999) *Development in adulthood*, 2nd edit. Boston: Allyn and Bacon.

Levenson, D. J. (1986). A conception of adult development. *American Psychologist, 41*, 3–13.

Machover, K. (1949). *Personality projection in the drawing of the human figure.* Springfield, IL: Charles C Thomas.

McAvay, D. G. (1977). *An organismic-developmental analysis of marital interaction.* Unpublished doctoral dissertation, Clark University, Worcester, MA.

Melito, R. (1985). Adaptation in family systems: A developmental perspective. *Family Process, 24*, 89–100.

Melito, R. (1988). Combining individual psychodynamics with structural family therapy. *Journal of Marital and Family Therapy, 14*, 29–43.

Minami, H. (1985a). *Establishment and transformation of personal networks during the first year of college: A developmental analysis.* Unpublished doctoral dissertation, Clark University, Worcester, MA.

Minami, H. (1985b). *Development of personal networks during the first year of college.* Paper presented at the annual meeting of the Eastern Psychological Association, Boston, MA.

Minian, N. (1996, April). *Cross-cultural study on effects of parental divorce.* Unpublished manuscript, Clark University, Worcester, MA.

Neugarten, B. (Ed.). (1968). *Middle age and aging.* Chicago: University of Chicago Press.

Neuhaus, E. C. (1988). *A developmental approach to children's planning.* Unpublished doctoral dissertation, Clark University, Worcester, MA.

O'Connor, K., & Chamberlain, K. (1996). Dimensions of life meaning: Qualitative investigation at mid-life. *British Journal of Psychology, 89*, 461–477.

Pillay, A. L. (1988). Midlife depression and the "empty nest" syndrome in Indian women. *Psychological Reports, 63*(2), 591–594.

Plante, W. A. (1992, May). *Effects of parental divorce on college students: Study two.* Paper presented at the annual Academic Spree Day, Clark University, Worcester, MA.

Roberts, P., & Newton, P. (1987). Levinsonian studies of women's adult development. *Psychology and Aging, 2,* 154–163.

Schaie, K. W. (Ed.) (1983). *Longitudinal studies of adult psychological development.* New York: Guilford Press.

Schouela, D. A., Steinberg, L. J., Leviton, L. B., & Wapner, S. (1980). Development of the cognitive organization of an environment. *Canadian Journal of Behavioral Science, 12,* 1–16.

Shapiro, E. (1973, April). *The effect of verbalization of plans on sequence learning by children and adults.* Paper presented at the Psychology Conference, University of New Hampshire, Durham, NH.

Shapiro, E., Rierdan, J., & Wapner, S. (1972). *Effect of articulating plans of how to solve a sequence learning task on the learning of this task by children and adults.* Unpublished manuscript, Clark University, Worcester, MA.

Turvey, M. T. (1973). Preliminaries to a theory of action with reference to vision. In R. Shaw & J. Bransford (Eds.), *Perceiving, acting and knowing.* Hillsdale, NJ: Lawrence Erlbaum.

Wapner, S. (1977). Environmental transition: A research paradigm deriving from the organismic-developmental systems approach. In L. van Ryzin (Ed.), *Wisconsin Conference on Research Methods in Behavior Environment Studies Proceedings* (pp. 1–9). Madison, WI: University of Wisconsin Press.

Wapner, S. (1981). Transactions of persons-in-environments: Some critical transitions. *Journal of Environmental Psychology, 1,* 223–239.

Wapner, S. (1987). A holistic, developmental, systems-oriented environmental psychology: Some beginnings. In D. Stokols & I. Altman (Eds.), *Handbook of environmental psychology* (pp. 1433–1465). New York: John Wiley & Sons.

Wapner, S. (1993). Parental development: A holistic, developmental, systems-oriented perspective. In J. Demick, K. Bursik, & R. DiBiase (Eds.), *Parental development* (pp. 3–37). Hillsdale, NJ: Lawrence Erlbaum.

Wapner, S., & Cirillo, L. (1973). *Development of planning* (Public Health Service Grant Application). Worcester, MA: Clark University.

Wapner, S., Cirillo, L., & Baker, A.H. (1969). Sensory-tonic theory: Toward a reformulation. *Archivo di Psicologia e Psichiatria, XXX,* 493–512.

Wapner, S., Cirillo, L., & Baker, A. H. (1971). Some aspects of the development of space perception. In J. P. Hill (Ed.), *Minnesota Symposia on Child Psychology, Vol. 5* (pp. 162–204). Minneapolis, MN: University of Minnesota Press.

Wapner, S., Ciottone, R., Hornstein, G., McNeil, O., & Pacheco, A. M. (1983). An examination of studies of critical transitions through the life cycle. In S. Wapner & B. Kaplan (Eds.), *Toward a holistic developmental psychology* (pp. 111–132). Hillsdale, NJ: Lawrence Erlbaum.

Wapner, S., & Demick, J. (1990). Development of experience and action: Levels of integration in human functioning. In G. Greenberg & E. Tobach (Eds.), *Theories of the evolution of knowing: The T. C. Schneirla conference series* (pp. 47–68). Hillsdale, NJ: Lawrence Erlbaum.

Wapner, S., & Demick, J. (1992). The organismic-developmental, systems approach to the study of critical person-in-environment transitions through the life span. In T. Yamamoto & S. Wapner (Eds.), *Developmental psychology of life transitions* (pp. 25–49). Tokyo, Japan: Kyodo Shuppan.

Wapner, S., & Demick, J. (1998). Developmental analysis: A holistic, developmental, systems-oriented perspective. In W. Damon (Series Ed.) & R. M. Lerner (Vol. Ed.), *Handbook of child psychology, Vol. 1: Theoretical models of human development,* 5th ed. (pp. 761–805). New York: John Wiley & Sons.

Wapner, S., & Demick, J. (1999). Developmental theory and clinical practice: A holistic, developmental, systems-oriented approach. In W. K. Silverman & T. H. Ollendick (Eds.), *Developmental issues in the clinical treatment of children* (pp. 3–30). Boston: Allyn & Bacon.

Wapner, S., & Demick, J. (2000). Person-in-environment psychology: A holistic, developmental, systems-oriented approach. In W. B. Walsh, K. H. Craik, & R. H. Price (Eds.), *New directions in person-environment psychology,* 5th ed. (pp. 25–60) Hillsdale, NJ: Lawrence Erlbaum.

Wapner, S., & Demick, J. (2002). The increasing *contexts* of *context* in the study of environment-behavior relations. In R.B. Bechtel & A. Churchman (Eds.) *Handbook of environmental psychology* (pp. 3–14), New York: John Wiley & Sons.

Wapner, S., & Demick, J. (in press). Critical person-in-environment transitions across the life span: A holistic, developmental, systems-oriented program of research. In J. Valsiner (Ed.), *Differentiation and integration of a developmentalist: Heinz Werner's ideas in Europe and America*. New York: Kluwer Academic/Plenum Publishers.

Wapner, S., Demick, J., & Damrad, R. (1988, April). Transition to retirement: Eight years after. Presented at the annual meeting of the Eastern Psychological Association, Buffalo, NY.

Wapner, S., Demick, J., & Redondo, J. P. (1990). Cherished possessions and adaptation of older people to nursing homes. *The International Journal of Aging and Human Development, 31*(3), 299–315.

Wapner, S., Kaplan, B., & Ciottone, R. (1981). Self–world relationships in critical environmental transitions: Childhood and beyond. In L. Liben, A. Patterson, & N. Newcombe (Eds.), *Spatial representation and behavior across the life span* (pp. 251–282). New York: Academic Press.

Wapner, S., Quilici-Matteucci, F., & Cool, K. (1991). Cross-cultural comparison of necessities, amenities, and luxuries in the physical, interpersonal, and sociocultural aspects of the environment. In T. Niit, M. Raudsepp, & K. Liik (Eds.), *Environment and social development. Proceedings of the East–West Colloquium in Environmental Psyhcology, Tallinn, Estonia, May 16–19, 1991* (pp. 47–56). Tallinn, Estonia: Tallinn Pedagogical Institute.

Wapner, S., Quilici-Matteucci, F., Yamamoto, T., & Ando, T. (1990). Cross-cultural comparison of the concept of necessity, amenity, and luxury. In Y. Yoshitake, R. B. Bechtel, T. T. Akahashi, & M. Asai (Eds.), *Current issues in environment-behavior research. Proceedings of the Third Japan–United States Seminar, Kyoto, Japan, July 19–20, 1990* (pp. 21–32). Tokyo: University of Tokyo.

Wapner, S., & Werner, H. (1957). *Perceptual development*. Worcester, MA: Clark University Press.

Wapner, S., & Werner, H. (1965). An experimental approach to body perception from the organismic-developmental point of view. In S. Wapner & H. Werner (Eds.), *The body percept* (pp. 9–25). New York: Random House.

Werner, E. E. (1993). Risk and resilience in individuals with learning disabilities: Lessons learned from the Kauai longitudinal study. *Learning Disabilities Research and Practice, 8*, 28–34.

Werner, H. (1937). Process and achievement: A basic problem of education and developmental psychology. *Harvard Educational Review, 7*, 353–368.

Werner, H. (1940/1957). *Comparative psychology of mental development*. New York: Harper (2nd ed., Chicago: Follett; 1948 3rd ed., International Universities Press).

Werner, H., & Kaplan, B. (1956). The developmental approach to cognition: Its relevance to the psychological interpretation of anthroplogical and ethnolinguistic data. *American Anthropologist, 58*, 866–880.

Werner, H., & Kaplan, B. (1963). *Symbol formation*. New York: John Wiley & Sons.

Werner, H., & Wapner, S. (1949). Sensory-tonic field theory of perception. *Journal of Personality, 18*, 88–107.

Werner, H., & Wapner, S. (1952). Toward a general theory of perception. *Psychological Review, 59*, 324–338.

White, L., & Edwards, J. N. (1990). Emptying the nest and parent well being: An analysis of national panel data. *American Sociological Review, 55*, 235–242.

Winnicott, D. W. (1958). *Through pediatrics to psychoanalysis*. New York: Basic Books.

Winnicott, D. W. (1971). *Playing and reality*. New York: Basic Books.

Winter, L. E., & Samuels, C. A. (1998). The impact of "The Dream" on women's experience of the midlife transition. *Journal of Adult Development, 5*(1), 31–43.

Witkin, H. A. (1965). Heinz Werner: 1890–1964. *Child Development, 30*, 307–328.

Witkin, H. A., Lewis, H. B., Hertzman, M., Machover, K., Meisner, P. B., & Wapner, S. (1954). *Personality through perception*. New York: Harper.

Research Methods in Adult Development

JOHN C. CAVANAUGH AND
SUSAN KRAUSS WHITBOURNE

The study of adult development is grounded in the principles of scientific inquiry. Information concerning aging is gathered in the same ways as in other sciences, such as biology, psychology, sociology, anthropology, and the medical and allied health fields. Adult developmentalists have the same problems as other scientists: finding appropriate control or comparison groups, limiting generalizations to the types of groups included in the research, and finding adequate means of measurement (Kausler, 1982).

GENERAL ISSUES IN RESEARCH DESIGN

Before we consider specific types of research methods, it is necessary to examine several general issues regarding research design. The most important of these are issues involving validity in research design and selection and measurement equivalence.

Validity and Research Design

The chain of inferences in scientific research rests on the assumption that there are no rival explanations other than the substantive hypothesis for the set of empirical observations one has made. However, there are several threats to the validity of inferences in this chain that raise the possibility that one or more of the rival explanations may be plausible. With this in mind, Cook and Campbell

JOHN C. CAVANAUGH • President, University of West Florida, Pensacola, Florida, 32514-5750. SUSAN KRAUSS WHITBOURNE • Department of Psychology, University of Massachusetts at Amherst, Amherst, Massachusetts, 01003.

Handbook of Adult Development, edited by J. Demick and C. Andreoletti. Plenum Press, New York, 2002.

(1979) provided a taxonomy of threats to validity: construct validity, internal validity, external validity, and statistical conclusion validity. For a researcher to make valid substantive inferences, the study must be designed to guard against these sources of rival explanations to the study's hypotheses. Because the perfect study is impossible to design, the art of research is to identify the major sources of reasonable validity threats and then attempt to eliminate them by using appropriate research design techniques.

Construct validity is the degree to which the measurement method actually assesses the hypothetical constructs of interest. Construct validity is essential for linking substantive and empirical hypotheses. As Hertzog and Dixon (1996) note, though, behavioral and social science research is often deficient in providing an adequate test of construct validity. The literature is replete with examples of inadequately defined concepts and measurement assumptions. The inability of researchers to replicate their results is typically due to these failures (Hertzog, Hultsch, & Dixon, 1989).

Internal validity refers to the validity of substantive hypotheses within the context of a particular research design regarding the relationship among the variables of interest (Hertzog & Dixon, 1996). When threats to internal validity exist, it is not possible to determine whether the relationship among variables truly exists or whether there would be equally viable alternative explanations. Internal validity is particularly an issue in the case of designs other than those that fit the criteria for true experiments, as discussed later. In an experiment, the researcher attempts to rule out alternative explanations through the precise control of variables and conditions.

External validity represents the degree to which the findings from a particular study can be appropriately generalized beyond the sample and observational context of the investigation to other people, settings, and points in time (Hertzog & Dixon, 1996). Threats to external validity increase as the differences increase between the context of the investigation and the context to which the researcher attempts to generalize. For example, results from research in artificial laboratory environments suffer from low external validity if the desire is to generalize to the phenomena as they exist in the "real world."

Finally, statistical conclusion validity involves the appropriate use of techniques of statistical inference. Appropriate techniques are needed to ensure the correct interpretation of a failure to reject the null hypothesis. It is particularly important to use statistical tests that have adequate statistical power so that significant effects will be detected (Hertzog & Dixon, 1996).

Selection and Measurement Equivalence

Not all threats to validity can be eliminated, but it is possible to reduce them substantially through careful planning of the study's method. Most researchers focus their attention in the planning of methods on selection and measurement. Selection refers to the process by which the units of observation are chosen to be in the sample. Selection also refers to the process through which these units of observation are assigned within an investigation to conditions, such as being placed in different groups. Because no two units of observation are identical, concern over selection reflects the possibility of systematic differences between these units that will influence the outcome of the observation process. Selection factors go beyond the choice of units of observation and assignment to conditions. In addition, selection refers to the specific exemplars of a phenomenon chosen for examination, such as which variables to measure. The times at which observations are taken are another selection factor.

Measurement equivalence refers to the need to ensure that the measurement properties of an index used to assess a construct are the same across individuals and time (Labouvie, 1980). Without measurement equivalence, the researchers cannot determine what the responses meant, and any differences across groups would be impossible to interpret.

BASIC VARIABLES: AGE, COHORT, AND TIME OF MEASUREMENT

The research designs used in studies on adult development involve manipulation of the variables of age, cohort, and time of measurement. These three factors are thought to influence jointly the individual's performance on any given psychological measure at any point in life. As we shall see, these variables are highly related to each other, making the task of conducting research in gerontology a challenging enterprise.

Age

The study of development is inherently made difficult by the fact that age cannot be experimentally manipulated. As shown later, it is only when a variable can be manipulated that cause-and-effect relationships can be inferred. The status of age as a variable is comparable to the status of gender, ethnicity, place of birth, occupation, and other characteristics of the respondent that cannot be altered by the experimenter. At another level, age presents a complication as a variable because its meaning is not altogether clear. Age is a time-based measure, not a measure of features intrinsic to the individual. The older a person is, the more calendar years that person has experienced. However, there is no direct connection between the movement of the calendar and changes going on within the person. Developmentalists use age as a convenient shorthand but understand that age is an imperfect index of the phenomena being investigated.

Cohort

The term "cohort" signifies the general era in which a person was born. Conceptually, the term cohort may represent the more familiar term "generation," in that it is intended to refer to people who were born (and hence lived through) some of the same social influences. For example, members of the 1950 cohort were in college during the Vietnam War era, and shared certain experiences specific to this period of history. By contrast, the college experiences of people born in the 1960 cohort were far more quiescent. There is considerable evidence in the field of adult development and aging that the era in which a person reached maturity has a lasting influence on many aspects of functioning throughout the remaining years of that person's life.

Time of Measurement

Simply put, "time of measurement" is the year (or period) when testing has taken place. As such, it is a convenient way of representing the social and historical influences on the individual at the point when data are collected. Time of measurement is linked to cohort, in that people of a certain age being tested at a particular time were obviously born within the same cohort. As shown later, the inherent connection between time of measurement and cohort creates logical difficulties when investigators attempt to disentangle these indices of social and historical context.

DESCRIPTIVE RESEARCH DESIGNS

Three basic designs are used most often in adult developmental research: cross-sectional, longitudinal, and sequential. Each is considered in turn.

Cross-Sectional Designs

The most direct and quickest way to examine the possible effects of aging on a particular variable of interest is to compare people who are currently of different ages. This research design is called "cross-sectional" because the researcher is taking a "cross-section" of the population by age. The fact that the cross-sectional design does not take long to complete is its main (and perhaps, only) advantage. An investigator can test a hypothesis regarding possible age differences in as long a period it takes to gather the data. There is no need to wait until people actually "age."

Although adult developmentalists often rely heavily on the cross-sectional design, they realize that to do so involves a high degree of risk. It is never possible to know whether observed differences between people within different age groups reflect the aging process or the fact that the people being compared have lived through different historical periods. The best that the researcher can hope for is that as many alternative explanations as possible have been eliminated through careful selection of respondents. Furthermore, through repeated demonstration of the phenomenon of interest on different samples, researchers can establish the validity of cross-sectional findings with greater confidence.

Longitudinal Designs

The longitudinal design is one in which the investigator selects a sample and follows them over a designated period. At the end of that time, or at points along the way, the researcher can describe changes over the years on the particular set of variables being investigated. This design has many obvious advantages as a study of the aging process. Its main advantage is that it is chronicling the effects of aging as it is actually occurring. Rather than the artificiality of the cross-sectional study, in which conclusions about aging must be inferred, the evidence being accumulated from a longitudinal study captures the phenomenon in real time. However, as desirable as the longitudinal design seems to be, it is fraught with many practical and some theoretical difficulties.

Here are just some of the practical problems involved in longitudinal research. To embark on a longitudinal study is a major commitment of time and resources on the part of the researcher. There is inevitable "attrition" or loss of participants from the original sample, and so the numbers become smaller and smaller as the years go on. Although researchers can take measures to minimize dropouts, these will not necessarily resolve all difficulties. There are theoretical problems caused by attrition that are even more of a challenge to handle. The main theoretical problem is the fact that the "survivors" of a longitudinal investigation are not fully representative of the entire group of people who began the study. The survivors tend to be healthier, cognitively more capable, and perhaps different in personality than those who drop out of the study due to death, disinterest, or illness.

Schaie's General Developmental Model: Sequential Designs

It should be clear by now that the perfect study on aging is virtually impossible to conduct. Age is problematic as a variable of investigation. Furthermore, age is inherently linked with time and so personal aging can never be separated from social aging. However, researchers have turned to what are called the "sequential" designs in gerontology as a way of teasing apart at least some of the effects of personal aging from the effects of social time.

In a landmark paper, Schaie (1965) outlined the problems involved in traditional cross-sectional and longitudinal designs, as discussed in the preceding. His strategy for overcoming these problems was to design "sequential" studies that were intended to separate the effects of age, cohort, and time of measurement. Each design may be thought of as a 2×2 analysis of variance, crossing two of the three factors (age, cohort, time of measurement) at a time. A complete analysis requires arranging data collections that fit the criteria for all three designs. A layout for such an analysis is shown in Fig. 5.1.

Cohort-Sequential Design. In the cohort-sequential design, two cohorts are compared at two different ages (e.g., cells F + C vs. J + G in Fig. 5.1). In either case, if the cohort difference is significant, the researcher may conclude that time of birth has an effect on later adult performance. If both cohorts show similar scores, or similar patterns of change or stability across the ages they are studied, the researcher can conclude that the results reflect the influence of intrinsic aging processes. A third possibility is that members of the two cohorts show different patterns of performance over the two ages tested. Such findings would lead the researcher to look for differential early life experiences between the two cohorts that would become manifest in later life.

Year of birth (cohort)

Years of testing (Time of measurement)	1940	1930	1920	1910
1980	Cell A 40 years old	Cell B 50 years old	Cell C 60 years old	Cell D 70 years old
1990	Cell E 50 years old	Cell F 60 years old	Cell G 70 years old	Cell H 80 years old
2000	Cell I 60 years old	Cell J 70 years old	Cell K 80 years old	Cell L 90 years old

Figure 5.1. Layout of developmental research designs.

Time-Sequential Design. The time-sequential research design involves comparing the effects of age with the effects of time of measurement (e.g., cells F + C vs. G + D). Two or more age groups are compared at two or more times of measurement. The effects of time of measurement would be indicated if people in the sample, regardless of age, differed in their scores at one testing date compared to another testing date. Age differences across the two times of measurement would lead to the conclusion that age is a significant factor. Finally, the combined effect of age and time of measurement may result in different patterns of age differences for two times of testing.

Cross-Sequential Design. The third sequential design, called "cross-sequential," involves crossing the two factors of cohort and time of measurement (cells B + F vs. C + G). Two or more birth cohorts are compared at two or more times of measurement in this design, and as a result, age is not even a factor in the analysis. The purpose of this design is to compare these social time effects directly.

The "Sequences" Designs of Baltes

Baltes (1968) proposed an alternative model to Schaie's that directly pits personal aging against social time by repeating either a cross-sectional or a longitudinal study at different points of measurement.

Longitudinal Sequences Design. In the longitudinal sequences design, more than one longitudinal study is conducted at two or more periods of time (cells F–J vs. C–G). If the results comparing one age to the other differ, this would suggest that social time interacts with personal aging. Without this check, there would be no way of knowing whether the longitudinal results generalize over time or not.

Cross-Sectional Sequences Design.   In the cross-sectional sequences design, the researcher conducts more than one cross-sectional study at two or more different time points (cells B–C–D vs. E–F–G). The researcher can determine whether cross-sectional findings observed at one time of measurement are comparable to those obtained at other times. If the results are not consistent for the two cross-sectional studies, then the researcher must suspect that social time is influencing the results.

Implications

Schaie proposed that the General Developmental Model could separate personal from social aging. How successful is this model? The answer is, its success is limited. No matter what data collection strategy is used, the fact remains that age, cohort, and time of measurement are inherently related. Once two of these values have been set, the third is automatically determined. There is no way to manipulate all three in a totally independent fashion, which would be the case if, for instance, age and cohort could vary independently of time of measurement. Unfortunately, this solution might take place only in science fiction. Given the constraints of the real world, and the interdependence of these factors, interpretations derived from the sequential studies are always open to alternative explanations. The designs described by Baltes avert this issue as they do not attempt to settle the question of whether time of birth (cohort) is more important than time of measurement. This reduces the number of factors being manipulated to two, and they are not dependent on each other.

Although the concepts of age, cohort, and time of measurement can readily be translated into numerical values, it is easy to see how quickly they become interwoven. Researchers may lose track of what exactly is being measured and whether the findings pertain to age, cohort, time of measurement, or some combination of the three. Just as researchers must attempt to keep these distinctions straight, it is also important when reading published research to look for possible threats to the validity of a study's findings. A finding based on research conducted in the 1970s may not hold up to testing in the 2000s. The necessity for researchers to test their findings on various populations is another factor to consider. Many of the results about the "aging process" turn out, on closer inspection, to be descriptive of processes taking place within a culturally narrow group. As the field of gerontology matures, it will be increasingly important for researchers to think "sequentially" with respect not only to social time, but also to socially and culturally diverse groups.

EXPERIMENTAL, QUASIEXPERIMENTAL, AND CORRELATIONAL DESIGN

When people think of science and the scientific process, they tend to think of experiments. Indeed, many would argue that experimentation is the foundation of scientific inquiry. However, conducting experiments in the behavioral and social sciences is a process open to criticism. Some of the charges against experimental research involve the ethics involving manipulation and deception of participants. The artificiality of the experimental setting is another area of criticism regarding whether results can be generalized to the world outside the laboratory. In this section we consider the necessary elements for experimental design, and consider two common alternatives (quasiexperiments and correlational designs) that are used when at least one essential criterion for an experiment is lacking.

Experimental Design

The general definition of an experiment is that it is an investigation "in which at least one variable is manipulated and units are randomly assigned to the different levels or categories of the manipulated variable(s)" (Pedhazur & Schmelkin, 1991). Thus, the two necessary components of an experiment are manipulation (discussed in more detail later) and random assignment. Random assignment means that the units of analysis in the experiment must each have the same odds of being selected for assignment to any of the groups being examined in the study. These units may be, for example, individuals, families, communities, nursing homes, or social service agencies. The strategy of random assignment is used to control for extraneous factors influencing the results of the manipulation. These extraneous factors include aspects of the units being studied that are not being observed or manipulated. In short, randomization is used to create the situation in which "all other things being equal" any observed differences across groups would be most likely the result of the manipulation.

A lack of either necessary component (manipulation or randomization) results in a nonexperimental design. If no variable is manipulated, creating a situation in which the question of interest is how two or more variables covary, then the research is said to be correlational. If random assignment is lacking but one or more variables are manipulated, then the research takes the form of quasiexperimentation.

In an experimental design, there is at least one independent variable and one dependent variable. The independent variables are the variables manipulated by the experimenter. The dependent variables are the behaviors or outcomes that are measured. As mentioned previously, one problem with age is that it can never qualify as an independent variable because we can neither manipulate it nor randomly assign it to people. Consequently, it is impossible to conduct true experiments to examine the effects of age on a particular person's behavior. At best, we can only find age-related effects of an independent variable on dependent variables.

Quasiexperimental Design

The major difference between experimental and quasiexperimental designs is randomization. In quasiexperiments, units are not randomly assigned to groups (Campbell & Stanley, 1963). However, one or more variables may be manipulated. Without randomization, researchers face the difficult task of identifying and isolating the effects of the independent variable and all other variables that could potentially affect the dependent variable. Approximations of this isolation of effects can be made through various statistical adjustments and techniques. Such adjustments permit some insight into the causal effects of the independent variable, but none of these measures is as effective as randomization.

Despite the limitations of quasiexperiments, they play an invaluable role in adult developmental research. Given that age will never achieve the status of an independent variable, they are, in fact, the only viable alternative.

Correlational Design

In correlational studies, the goal is to uncover a relationship between two or more observed variables. The strength and direction of a correlation is expressed by a correlation coefficient, abbreviated as r, which can range from -1.0 to 1.0. When $r = 0$, the two variables are unrelated. When r is >0, the variables are positively related; when the value of one variable increases (or decreases), the other variable also increases (or decreases). When r is <0, the variables are inversely related; when one variable increases (or decreases), the other variable decreases (or increases).

Correlational studies do not give definitive information concerning cause-and-effect relationships. However, correlational studies do provide important information about the strength of relationship between variables, which is reflected in the absolute value of the correlation coefficient. Furthermore, the inclusion of multiple variables in more complex correlational designs makes it possible to observe the joint effects of multiple factors. Although correlational designs do not allow the same precise control as experimental designs, they make it possible for developmental researchers to examine in depth variables that are very difficult, if not impossible, to manipulate.

QUALITATIVE METHODS

There are often instances in which researchers wish to explore a phenomenon of interest in an open-ended fashion. The investigation of contextual factors in adult development may demand the researcher use a method that makes it possible to identify potentially relevant factors within a broad spectrum of possible influences. Qualitative methods allow for the exploration of such complex relationships without the restrictions and assumptions of the scientific model. In other cases, researchers may be working in an area in which conventional methods are neither practical nor appropriate for the problem under investigation. Qualitative methods are also used in the analysis of life history information, which is likely to be highly varied from person to person and not easily translated into numbers. The main point in using qualitative methods is that they provide researchers with alternative ways to test their ideas and that the method can be adapted in a flexible manner to the nature of the problem at hand.

Interviews

Face-to-face interviews may be regarded by the investigator as the best way to gather data that are of a highly personal nature, in which self-reports may be distorted by the individual's desire to appear "good" (a problem known as social desirability). Follow-up questions (called probes) can be inserted into the interview as needed to reach greater clarification of a point that the respondent has not addressed. Interviews are also of great use in clinical settings when the investigator is interested in aspects of personality functioning that are outside the interviewee's conscious awareness.

Some interview methods are highly structured, in which the respondent chooses an answer from a series of preset categories. Others are highly unstructured, and take the form of an extensive conversation about the topic of interest between the interviewer and respondent. The data from an unstructured interview can be difficult to analyze, however, because it can take many forms. Therefore, many investigators use the unstructured interview as a basis for formulating a semistructured interview.

In a semistructured interview, the respondent is asked a question, the answer to which is followed with a probe designed to elicit more specific information. To administer this interview properly, the interviewer should be familiar with how the responses will ultimately be scored. This ensures that the responses can be coded into the desired categories.

Focus Groups

As a preliminary step in the identification of variables of interest within a certain area, a researcher may arrange for the meeting of a group of respondents to discuss the topic of interest. The researcher attempts to identify important themes in the group's discussion and keep the conversation oriented to this theme. For example, the researcher may be interested in the caregiving experiences of elderly minority persons; focus groups would provide a way to get firsthand information. At the end of the focus group meeting, the researcher will have identified some concrete questions to pursue in subsequent studies.

Observational Methods

As a qualitative method, the observational study involves the careful examination of the behavior of individuals as noted by individuals trained to watch and record their actions. An observational study may be conducted with the assistance of videos or through the live presence of the observer. Videos have the advantage of remaining a permanent repository of data, but at the sacrifice of validity if the respondents are reacting in atypical ways to the presence of the camera.

In the participant-observation method, employed particularly in sociology and anthropology, the researcher actually takes part in the activities of the individuals being studied. The report of such an investigation includes the subjective experience of the researcher as well as descriptions of the actions of those being observed.

Developmentalists working from a behavioral perspective have also developed elaborate procedures for conducting research using observational methods, but these rely on quantitative methods for analysis. A behaviorist using such methods would define precisely the behavior to be observed (such as the number of times a person speaks) and then set up a viewing station to observe the behavior, preferably unobtrusively. The observer would examine behavior during predetermined period such as 5 minutes every 2 hours for a week. The data would be analyzed by connecting responses to "antecedents" or presumable causes in the environment.

Data Analytic Techniques

There are two fundamental approaches to the analysis of qualitative data. In the "pure" use of qualitative methods, the investigator attempts to describe underlying relationships in the data by using the circular method. In this approach, the investigator reads the data, which may consist of taped or written accounts of interviews or observations, and attempts to discern the presence of themes, issues, or consistencies across respondents. The investigator deliberately attempts to enter the subjective world of the respondents in this procedure rather than adopt the scientific hypothesis-testing mode. Throughout this process, the researcher becomes virtually immersed in the data until a clear picture emerges.

The second analytic method, usually derived from semistructured interviews, involves constructing coding categories of open-ended data and then counting the number of responses that fit these categories. The results may be reported as percentages of people who give a particular response, or perhaps as expressed in nonparametric statistics, such as chi-square.

Researchers who adopt the second, quantitative version of the qualitative procedure may move to the next step of constructing a closed-ended questionnaire with preset categories based on the responses given by people in the semistructured interview study. Some areas in adult development may lend themselves to such a transition. However, there may be good reasons for the investigator to stick with qualitative methods throughout the life of the study owing to the flexibility of the semistructured interview and its appropriateness for emotionally or personally sensitive areas of investigation.

STRUCTURAL EQUATION MODELING

Structural equation modeling (SEM) involves a comprehensive statistical approach to testing hypotheses about relationships among observed and latent variables (Hoyle, 1995). The emergence of SEM as a method of choice for testing complex hypotheses in the behavioral and social sciences is a direct result of the development of reasonably user-friendly software in the late 1980s (e.g., Bentler, 1992; Jöreskog & Sörbom, 1993).

Basic Tenets of SEM

SEM begins with the specification of a model, which is a statistical statement about the relationships among variables (Hoyle, 1995). These relationships involve a set of parameters called "fixed" and "free." Fixed variables are not estimated from the data and their value is fixed by the researcher. Free variables are ones in which the parameters are estimated from the data. Statistical tests of the adequacy of the model involve examining the goodness-of-fit. These tests indicate the degree to which the pattern of the fixed and free parameters specified in the model is consistent with the pattern of variances and covariances from the observed data (Hoyle, 1995).

Structural equation models are most easily communicated in a path diagram (see Fig. 5.2 for an example). Path diagrams have three principal components: rectangles, ellipses, and arrows. Rectangles represent observed or measured variables. Ellipses represent latent or unobserved variables, as well as errors of prediction and of measurement. Arrows indicate associations between variables: straight arrows point in one direction and represent the direction of prediction (from predictor to outcome); curved arrows point in two directions and represent nondirectional associations (i.e., correlations). Path diagrams are typically oriented so that the overall flow of the model is from left to right. Although errors of measurement and prediction are always present, they are often omitted from the depiction for clarity.

The structural equation model itself has two components defined by the pattern of fixed and free parameters: the measurement model and the structural model (Hoyle, 1995). The measurement model specifies relationships between the observed indicators (such as scores on tests) and unobserved latent variables (such as intelligence or personality characteristics). The structural model describes the relationships among the latent variables and any observed variables that are not indicators of latent variables. When combined, the measurement and structural models provide a comprehensive statistical model that can be used to evaluate relationships among variables that are free of measurement error (Hoyle, 1995).

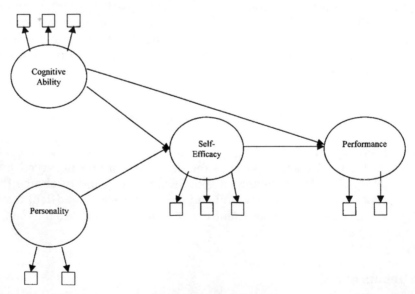

Figure 5.2. Hypothetical structural equation model relating cognitive ability, personality, self-efficacy, and performance. Circles represent latent variables and squares represent measures of the latent variables (see text).

The relationships among variables in a structural equation model can be of three types: association, direct effect, and indirect effect. An association represents a nondirectional relationship analogous to the correlation between variables. The building block of structural equation models is the direct effect, which is a directional relationship between two variables (an independent variable and a dependent variable) usually evaluated by analysis of variance or multiple regression. The dependent variable in one direct effect may be the independent variable in another direct effect, and one independent variable may be related to multiple dependent variables (or vice versa). An indirect effect is the effect of an independent variable on a dependent variable through one or more mediating variables. The combination of all direct and indirect effects of an independent variable on a dependent variable is called the total effect of the independent variable.

Once the model has been specified, the next step is to compute estimates of the free parameters from a set of observed data. The preferred approach for achieving these estimates is an iterative method such as maximum likelihood or generalized least squares (Hoyle, 1995). Each iteration creates an implied (estimated) covariance matrix that is compared to the actual covariance matrix observed in the data set. Because the estimation process only rarely produces an exact duplicate of the observed covariance matrix, the goal of the analysis is to minimize the difference between the estimated and observed matrices. This difference is referred to as the residual matrix. Iterations (i.e., repetitions of the analysis) continue until the residual matrix cannot be reduced any further. At this point, the estimation procedure is said to have converged on a solution, which becomes the final model.

How good a given estimation is defines the fit of the model to the observed data. This determination is a statistical one that takes into account features of the data, the model, and aspects of the estimation method (Hoyle, 1995). The latter point is important; for example, sampling error becomes increasingly problematic as sample size decreases, and the likelihood of reasonable fit increases with the number of free parameters estimated because these are derived from the data. Common statistical fit indexes include the χ^2 goodness-of-fit test, as well as several adjunct fit tests (e.g., incremental fit indexes, absolute fit indexes) (Hu & Bentler, 1995). Each of these statistical indexes has limitations, resulting in the relatively common practice of reporting multiple indicators of fit. An extension of tests of fit involves comparing two or more models of the same data. Such comparisons can be key in testing competing theories, for example, and are similar conceptually to comparing different models in hierarchical regression. The computation of estimated parameters and tests of fit is achieved most often through the use of specialized computer programs such as LISREL and EQS. Although both programs provide parameter estimates and tests of fit, they do so based on different mathematical approaches, especially when the data are not normally distributed (Byrne, 1995). Unfortunately, these different approaches can result in dramatically different solutions. Thus, researchers should be familiar with these differences to evaluate properly the results of SEM.

In sum, SEM is both similar and different from the related approaches of correlation, multiple regression, and analysis of variance (ANOVA) (Hoyle, 1995). It is similar to these methods in that all of are based on linear models. Furthermore, the statistical tests are true only if certain assumptions about the observed data are met. However, SEM differs from these other approaches in three key ways.

First, SEM requires a formal specification of the model to be estimated and tested. Thus, the researcher must state explicitly all the hypothesized relationships among the variables of interest prior to embarking on the study. Second, SEM provides the capacity to test relationships among latent variables isolated from the effects of unreliability and uniqueness. Third, the statistical indicators obtained in SEM do not have clear outcomes or interpretations, compared to those obtained in other approaches.

Overall, SEM is a more comprehensive and flexible approach to research design and data analysis than any other single approach in common use (Hoyle, 1995). Indeed, ANOVA, multiple regression, and factor analysis are all special instances of SEM. Clearly, SEM provides a way to test more complex and specific hypotheses, thereby providing an extremely powerful research tool.

ETHICS IN RESEARCH

The individuals who participate in research are asked to give of their time and, on some occasions, to provide information about themselves that is very private or potentially embarrassing. In some types of experimental studies, the researchers may need to use deception to study the variables of interest. It is important to protect the rights of respondents because there is information that they do not have about the study prior to agreeing to participate in it.

Institutional Review Boards

Starting in the mid-1970s, all institutions receiving federal funding for research are required to establish Institutional Review Boards (IRBs) that review all proposed studies to be carried out at that institution or by anyone employed by that institution. The purpose of such reviews is to ensure that the rights of the subjects are adequately protected and that the benefits of being in the study outweigh the risks to the individual. The American Psychological Association also has a set of ethical guidelines to ensure that studies specifically in psychology meet predetermined criteria for protection of human (and animal) subjects.

Informed Consent

The most important ethical guideline to be followed in conducting research is to ensure that, prior to being in the study, participants have as much information as possible about what they are being asked to do. Although this information might be limited by the nature of the study, the respondent should at least be told what to expect will happen during the course of the research.

Informed consent is usually accomplished by giving the respondent a consent form that describes the study in as much detail as possible. Clear information must be given about risks and benefits of participation. Furthermore, respondents should be given the right to refuse to participate, so that they do not feel that they are being coerced into participation. The consent form also specifies that the names of participants in the study will either remain anonymous or confidential (to be known only by the researchers). In the case of anonymity, the respondent is told that the responses are going to summed in the form of group data and there will be no reason to examine the respondent's data specifically. In some cases, confidentiality but not anonymity is guaranteed, as in a longitudinal study in which it is necessary to know people's names. In these cases, the investigator makes it clear that the data will be coded or in other ways made inaccessible to other parties. In cases where the person is cognitively impaired, consent for participation in research must be obtained from a family member or other individual who has the power of attorney.

Following the conclusion of the study, ethical guidelines dictate that respondents should be fully informed about the study's purpose. This debriefing should include the variables of interest and the expected relationships. If deception was involved, the debriefing form should state what that deception was and why it was needed. Respondents should also be given the opportunity to find out where to read the study's results when it is published. Generally, both debriefing and information about the results are accomplished by giving the respondent a written sheet of paper that fully describes the study's purpose and names of the investigator. In the case of a longitudinal investigation, the debriefing process becomes more complicated because the investigator may not want to reveal completely what the study was about or provide information on possible results.

Finally, researchers may choose to or be advised by the IRB that they should have some type of backup or referral agency to send respondents to should the study lead to emotional or medical problems. In other cases, the researchers may be encouraged to provide respondents automatically with the names of educational or other referral services at the time of debriefing.

CONCLUDING COMMENTS

The issues raised in this chapter form the very foundation for all of the data on substantive issues presented in the remainder of this book. Asking good, clear, and insightful research questions is only the beginning. Unless the investigator uses sound methodologies, the data that are gathered will not further understanding of the issues being studied. As you continue through this book, reflect on the techniques used to provide the data being discussed, and keep a healthy skepticism about the results.

REFERENCES

Baltes, P. B. (1968). Longitudinal and cross-sectional sequences in the study of age and generation effects. *Human Development, 11*, 145–171.

Bentler, P. M. (1992). *EQS structural equation program manual.* Los Angeles: BMDP Statistical Software.

Byrne, B. M. (1995). One application of structural equation modeling from two perspectives: Exploring the EQS and LISREL strategies. In R. H. Hoyle (Ed.), *Structural equation modeling: Concepts, issues, and applications* (pp. 138–157). Thousand Oaks, CA: Sage.

Campbell, D. T., & Stanley, J. C. (1963). Experimental and quasi-experimental designs for research on teaching. In N. L. Gage (Ed.), *Handbook of research on teaching* (pp. 171–246). Chicago: Rand McNally.

Cook, T. D., & Campbell, D. T. (1979). *Quasi-experimentation: Design and analysis issues for field settings.* Chicago: Rand McNally.

Hertzog, C., & Dixon, R. A. (1996). Methodological issues in research on cognition and aging. In F. Blanchard-Fields & T. M. Hess (Eds.), *Perspectives on cognitive change in adulthood and aging* (pp. 66–121). New York: McGraw-Hill.

Hertzog, C., Hultsch, D. F., & Dixon, R. A. (1989). Evidence for the convergent validity of two self-report metamemory questionnaires. *Development Psychology, 25,* 687–700.

Hoyle, R. H. (1995). The structural equation modeling approach: Basic concepts and fundamental issues. In R. H. Hoyle (Ed.), *Structural equation modeling: Concepts, issues, and applications* (pp. 1–15). Thousand Oaks, CA: Sage.

Hu, L., & Bentler, P. M. (1995). Evaluating model fit. In R. H. Hoyle (Ed.), *Structural equation modeling: Concepts, issues, and applications* (pp. 76–99). Thousand Oaks, CA: Sage.

Jöreskog, K. G., & Sörbom, D. (1993). *LISREL 8: User's reference guide.* Chicago: Scientific Software.

Kausler, D. H. (1982). *Experimental psychology and human aging.* New York: John Wiley & Sons.

Labouvie, E. W. (1980). Identity versus equivalence of psychological measures and constructs. In L. W. Poon (Ed.), *Aging in the 1980's: Psychological issues* (pp. 493–502). Washington, DC: American Psychological Association.

Pedhazur, E. J., & Schmelkin, L. P. (1991). *Measurement, design, and analysis: An integrated approach.* Hillsdale, NJ: Lawrence Erlbaum.

Schaie, K. W. (1965). A general model for the study of developmental change. *Psychological Bulletin, 64,* 92–107.

PART II

Biocognitive Development in Adulthood

CHAPTER 6

Multiple Perspectives on the Development of Adult Intelligence

Cynthia A. Berg and Robert J. Sternberg

In this chapter we explore the development of adult intelligence from four different perspectives: the psychometric, cognitive, neo-Piagetian, and contextual. These four perspectives were chosen to review the literature on adult intelligence because they offer a fairly diverse representation of guiding theories to adult intelligence that are dominant in the field at the present time (see Sternberg & Berg, 1992, for a review). Other perspectives on intelligence (e.g., comparative, biological, artificial intelligence) have not had the same presence in the field of adult intelligence as have these four perspectives.

These four perspectives on adult intelligence offer different answers to two questions that have guided the field of adult intelligence: (1) What is intelligence throughout adult development? and (2) How does intelligence develop across the adult lifespan? The psychometric and cognitive perspectives define intelligence to be largely the same throughout the lifespan, comprising broad mental abilities that are believed to characterize intelligence during childhood and adolescence. The neo-Piagetian and contextual perspectives hold that intelligence may change in its composition across the adult lifespan as individuals integrate the emotional and nonrational into thinking systems and traverse different contexts in middle and late life that afford different opportunities and constraints. In part because of these different starting points for the definition of intelligence, these four different perspectives chart different developmental trajectories for adult intelligence (i.e., decline, maintenance, and improvement). However, all struggle with the potential for both gains and losses at any point during adult development (see Baltes, Lindenberger, & Staudinger, 1998).

Cynthia A. Berg • Department of Psychology, The University of Utah, Salt Lake City, Utah, 84112. Robert J. Sternberg • Department of Psychology, Yale University, New Haven Connecticut 06520.

Handbook of Adult Development, edited by J. Demick and C. Andreoletti. Plenum Press, New York, 2002.

A general theme throughout all of these perspectives is the variability that exists across adult development in the expression of intelligence (Schaie, 1996b). Variability exists in the trajectory of adult intelligence (with some abilities showing increase, some decline, and some maintenance of function; Baltes et al., 1998; Horn & Hofer, 1992; Schaie, 1996a). Variability exists across individuals, with some individuals showing increase, decrease, or maintenance across time in their abilities (Schaie & Willis, 1996). Variability also exists within individuals, such that tremendous plasticity of functioning occurs and is evidenced in training studies (see Willis, 1987 for a review). Variability exists as well across contexts, with some contexts supporting superior intelligent functioning.

Our exploration of these four perspectives follows their chronological order of appearance in the field of adult intellectual development. We explore the major tenets of each perspective, present illustrative research from each perspective, and set forth the primary conclusions of each perspective regarding adult intellectual development from the relevant research. We conclude with a call for broad theories that can encompass the diverse findings concerning adult intellectual development that emerge from these four perspectives.

THE PSYCHOMETRIC PERSPECTIVE

The psychometric perspective has dominated the field of adult intelligence, such that many reviews of the development of adult intelligence include only this perspective (e.g., Schaie, 1996a; Schaie & Willis, 1996). The psychometric perspective focuses on how to characterize intellectual differences between individuals at various developmental periods. The view of intelligence from this perspective relies heavily on the conventional tests used to measure intelligence (e.g., vocabulary, inductive reasoning, simple arithmetic, spatial reasoning). In fact, this perspective comes the closest to defining intelligence as "how well one scores on an intelligence test." Research within this perspective begins with the administration of a large number of intelligence tests to individuals of a variety of ages. Statistical procedures such as factor analysis then summarize the data by extracting dimensions that illuminate the underlying structure or organization of individual differences in intelligence.

One question that dominated this field early on was whether the same number of dimensions or factors were needed to capture intelligence across the adult lifespan. Early research identified that a smaller set of factors was needed to capture intelligence in late life than in young adulthood, a process termed dedifferentiation (see Reinert, 1970, for a review). Current theorists hold that a small number (2–12) of different factors characterize intelligence and are largely unchanged in their organization across the adult lifespan (see Horn & Hofer, 1992, for a review).

A Model of Intelligence Based on Fluid and Crystallized Abilities

The psychometric perspective has uncovered two basic types of intellectual abilities that characterize adult intelligence and that demonstrate different developmental trajectories: fluid and crystallized intelligence (Carroll, 1993; Horn, 1994;

Horn & Cattell, 1967). Fluid intelligence is defined as abilities requiring adaptation to new situations and for which prior education or learning provide relatively little advantage. Measures of fluid intelligence include tests of abstract reasoning, spatial orientation, and perceptual speed. Crystallized intelligence is defined as abilities for which education and acculturation provide one with an advantage. Measures of vocabulary and arithmetic abilities would be examples of crystallized intelligence.

A similar distinction to the fluid-crystallized one has been advanced by Baltes and his colleagues (see Baltes et al., 1998), between the mechanics and pragmatics of intelligence. Baltes makes the analogy to a computer in defining the mechanics as the hardware of the intellectual system and the pragmatics as the software. Mechanics involve the basic processes of the intellectual system such as speed of processing and coordination of elementary operations as might be assessed on tests of simple discrimination and selective attention. The pragmatics of intelligence involve experience based declarative and procedural knowledge that one acquires during the course of socialization.

Of course, not all psychometric theorists accept the fluid-crystallized model. Jensen (1998) is one of many who believe that general ability accounts for most of what needs to be accounted for in terms of intellectual performance. This model, which dates back to Spearman (1927), remains a popular one, although perhaps not the ideal one for studying intelligence in adulthood.

Variability Across Components of Intelligence, Individuals, and Within Individuals

These two distinctions (fluid-crystallized and mechanics-pragmatics) have been useful as they characterize abilities that show different developmental trajectories across age. As seen in Fig. 6.1, fluid intelligence (i.e., reasoning, spatial orientation, perceptual speed) shows more rapid decline across age in cross-sectional (comparing people of different ages at one point in time) designs, whereas crystallized intelligence (i.e., verbal ability, number) is largely stable across age until the 60s. Similar findings have been reported for abilities constituting the mechanics vs. pragmatics of intelligence. Longitudinal research (in which the same individuals are followed across time) suggests a similar distinction between these two abilities, although the decline in both fluid and crystallized intelligence is not evident until much later (e.g., for fluid intelligence decline is not seen until the 60s; see Schaie, 1996b). The difference in results between cross-sectional and longitudinal designs reveals an important methodological issue that has been uncovered by Schaie (1984), namely, the influence of cohort effects. Cross-sectional designs, in which individuals are compared at one point in time, confound age and cohort (generational membership). Through comparing age differences uncovered through a variety of different methodological designs (cross-sectional, longitudinal, and cross-sequential), Schaie (1984) estimated that a substantial proportion of the age-related differences uncovered in cross-sectional designs could be accounted for by cohort effects, effects that vary by the ability under examination.

Related to the cohort effects uncovered by Schaie and colleagues, Flynn (1984, 1987) observed that throughout much of the 20th century, IQs rose in a monotonically increasing pattern. The level of increase is large—about 9 points

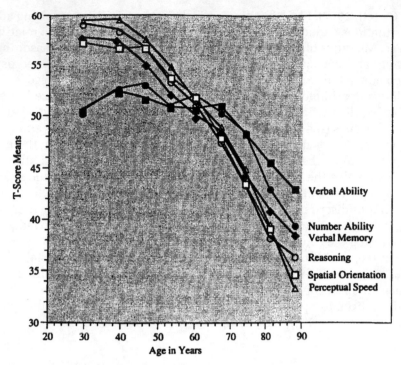

Figures 6.1. Cross-sectional trends in six primary mental abilities. [From Baltes, P. B., Lindenberger, U., & Staudinger, U. M. (1998). In *Handbook of Child Psychology, Vol. 1.* New York: John Wiley & Sons. Reprinted with permission.]

of IQ per generation—and is larger for fluid abilities than for crystallized abilities. One could argue that today's technologically driven world requires levels of intelligence that are higher than in the past. Indeed, this kind of consideration led Hunt (1995) to query whether the workforce of the future would have the cognitive competence to meet the demands of the rapidly changing work environment. Apparently, changes in the environment to which people are exposed over time affect their ability to cope with this environment. Although no one knows the reason for these increases, they are clearly environmental, given that the time frame is not sufficient for genetic mutations to have taken hold. The consensus view is that the increasing complexity of the environment, coupled perhaps with better nutrition, is creating smarter children and adults than was the case in generations past (Neisser, 1998).

Research from the psychometric perspective has also indicated great variability across individuals within a specific age group. Schaie's (1988) analysis of individuals in the Seattle Longitudinal Study indicated that a majority of individuals, even in their late 80s, perform within the distribution of young adults. Furthermore, Schaie and Willis (1986) found that the majority of individuals remain stable in measures of fluid intelligence (i.e., spatial and reasoning tests) over a 14-year period. In the Seattle Longitudinal Study, Schaie (1996b) found that this variability is due to a wide range of factors including genetic endowment, incidence of chronic disease, educational background, occupational pursuits, the stimulating vs. passive nature of daily life activities, and personality styles such as rigidity and flexibility.

The psychometric perspective has also revealed tremendous plasticity in intellectual functioning within individuals. Intervention studies have largely targeted aspects of fluid intelligence, as these abilities show the greatest cross-sectional decline. Results from the Seattle Longitudinal Study indicate that rather limited training (five 1-hour sessions) on tests such as spatial ability and reasoning returns performance back to a level demonstrated some 14 years previously. Training effects are maintained for as long as 7 years (Neely & Backman, 1993; Willis & Nesselroade, 1990), although training is limited to the trained tests and does not generally extend to other abilities (see Baltes & Willis, 1982).

Intelligence in Very Old Age

Recently researchers have turned their attention to issues of intelligence in what has been termed the "old–old," typically individuals over the age of 75. Two issues that are relevant to the basic issues addressed by the psychometric perspective include: Is intelligence still comprised of two separate factors of intelligence and do these two types of abilities continue to demonstrate different developmental trajectories? Much of this work has been conducted within the Berlin Aging Study (Baltes, Mayer, Helmchen, & Steinhagen-Thiessen, 1993), which consists of a representative stratified random sample of 516 individuals aged 70 to 103. Using the mechanics-pragmatics distinction, Lindenberger and Baltes (1997) found evidence that the intercorrelations among abilities reflective of mechanics and pragmatics are considerably higher in the old–old than in other age groups. Their data indicate that not only does this covariation increase among intellectual abilities but also among sensory abilities. Measures of sensory and sensorimotor functioning (e.g., basic hearing and vision, gait, and balance) also are related more to intelligence in the old–old than in other age groups. In addition, there is evidence in the old–old that the developmental trajectories for abilities reflective of pragmatics vs. mechanics are more similar than at other age groups (Lindenberger & Baltes, 1997). However, there still are continued differential relationships between the mechanics of intelligence, on the one hand, and the pragmatics of intelligence, on the other hand, and measures of biological (sensor:motor) and sociocultural variables (e.g., education, income). These differential relationships are supportive of the distinction between pragmatics and mechanics. Thus, the two-factor model of intelligence is somewhat less characteristic of intelligence in the old–old.

In sum, the psychometric perspective has indicated that a two-factor model of intelligence captures important individual differences in intelligence across the adult lifespan. Fluid and crystallized intelligence evidence different developmental trajectories across the lifespan. However, variability occurs in intellectual functioning across individuals within a specific age group and within individuals depending on supportive training conditions. The psychometric perspective has been useful in mapping out important individual differences in broad mental abilities. However, it has been less helpful in understanding why age differences are most pronounced on measures of fluid intelligence and less significant on measures of crystallized intelligence. The next perspective, the cognitive perspective, has been helpful in addressing this later question.

THE COGNITIVE PERSPECTIVE

The cognitive perspective takes as its starting point the intellectual products examined within the psychometric perspective and addresses at a microanalytic level the processes, representations, and strategies that individuals use to perform intellectual tasks (see Anderson, 1983, 1991, 1996; Sternberg, 1999). Cognitive theorists have been extremely influenced by the three-store model of Atkinson and Shiffrin (1968), which likened the intellectual system to a serial processing computer. The three-store model outlined the process that individuals go through in solving intellectual problems, such as encoding information, retrieving information from memory, using strategies to remember information, and metacognitive processes used in monitoring and allocating attention. Cognitive aging researchers enthusiastically adopted the information-processing model in an effort to localize the age differences found in measures of psychometric intelligence to a few of these component processes. This approach has been termed by Salthouse (1992b) as identifying "which box is broken?" (p. 267), given this approach's reliance on flow diagrams to capture the processes involved in intellectual functioning. However, as researchers began investigating task analyses of the component processes involved in tasks such as spatial ability (see Berg, Hertzog, & Hunt, 1982; Hertzog, Vernon, & Rypma, 1993) and inductive reasoning (Salthouse, 1991), it became clear that age differences could not be localized to a few component processes but were present in many of these.

This observation led investigators to search for a limited number of processing resources that were thought to be responsible for age-related changes in many processing components. Three such processing resources have been identified: speed of processing, working memory capacity, and loss of inhibition. Speed of processing has long been viewed as a resource that might be responsible for age-related differences in intellectual performance (Birren, 1974). The central slowing hypothesis advanced by Birren (1974) was used to explain why older adults' intellectual performance slows at a relatively constant rate, as a function of younger adults' performance (Cerella, 1990; Hale, Myerson, Smith, & Poon, 1988). The results from several large-scale studies have indicated that speed of processing is a resource that may be responsible for some of the age-related differences in measures of fluid intelligence (Hertzog, 1989; Salthouse, Kausler, & Saults, 1988).

Working memory capacity is another processing resource that has been shown to be related to age differences in fluid intelligence. Working memory capacity refers to the amount of information that one can hold in memory while performing some type of operation on the information (see Baddeley, 1992, 1995; Engle & Oransky, 1999). For example, working memory capacity has been measured by Salthouse (1992a) via the computation span task in which individuals are asked to complete arithmetic problems and simultaneously remember the final digits to be recalled at the end of a session. Age differences in working memory capacity have been shown to be involved in age differences in nearly every type of mental ability including language and comprehension (Cohen, 1988; McDowd, Oseas-Kreger, & Filion, 1995) and spatial relations and abstract reasoning (Salthouse, Mitchell, Skovronek, & Babcock, 1989).

An alternative view to working memory capacity is the inhibitory deficit theory advanced by Hasher and Zacks (1988). From this perspective, age differences in many components of language processing, memory, and cognition are due to a failure in attentional control, specifically, the failure to inhibit from working memory nonrelevant information. Such inhibition deficits may result in reductions in working memory capacity for older adults, as their working memory space may be allocated to both irrelevant and relevant information. An example of older adults' inhibition deficits is that their cognitive performance is differentially impaired (relative to young adults) in a reading task that involves distracting words and phrases that are related to the content of the read passage (Connelly, Hasher, & Zacks, 1991). Older adults have also been found to be impaired (relative to younger adults) on a range of tasks thought to reflect deficits in inhibitory control such as the production of off-task talk (Arbuckle & Gold, 1993), tasks of directed forgetting (Zacks, Radvansky, & Hasher, 1996), and interference effects (Stolzfus, Hasher, & Zacks, 1996). However, debate exists as to how general the inhibitory deficit is in accounting for many cognitive differences between young and older adults (Burke, 1997; McDowd, 1997) and how much of the age differences in intelligence inhibition may account for.

In sum, the cognitive perspective began as an effort to localize the age differences found on measures of fluid intelligence to a small number of component processes. However, this localization effort revealed that age differences exist in numerous component processes. Recent work indicates that age differences in measures of fluid intelligence may be due to a more limited number of processing resources (namely speed of processing, working memory capacity, and inhibitory control) that are responsible for age differences in component processes.

THE NEO-PIAGETIAN PERSPECTIVE

The neo-Piagetian (or postformal) approach in adult development arose in response to various criticisms of Piagetian accounts of intellectual development and critiques oriented around Piaget's exclusive focus on logical and rational systems of thought characterizing the end state of intellectual development. The neo-Piagetian perspective in adult development has involved identifying structures of reasoning that go beyond formal operational structures and that capture the development of reasoning beyond adolescence (see Labouvie-Vief, 1992, for a review). Thus, the approach has focused on showing that postformal thought is qualitatively different and superior to formal operational thought. The impetus for these qualitative changes in thought is related to the life tasks and social relationships that adults navigate (e.g., marriage, occupational decision-making, Havighurst, 1972) and the pressure of multiple social roles (Riegel, 1976; Sinnott, 1996). From this perspective intelligence is viewed as progressing during adult development toward a higher and superior state, rather than toward the decline often described by the psychometric and cognitive perspectives. Given the assumption that intelligence develops toward a higher state during adulthood, this perspective has focused much effort on intellectual developments that occur during young as well as middle adulthood, in contrast to the psychometric and cognitive perspectives, which largely have viewed young adulthood as the time of optimal intellectual functioning.

The Nature of Postformal Thought

Postformal thought has often been described as involving an integration and synthesis of the rational with the emotive, interpersonal, and nonrational (see Labouvie-Vief, 1992), although different theorists emphasize different components and offer varying stages beyond formal operations (Sinnott, 1996). Much of the work has traced developments from an objective separation of the knower and the known in adolescence to a more subjective or relativistic view of truth. Sinnott (1996) uses the example of choosing a particular model or perspective of intellectual development, much as the ones that are discussed in this chapter. To make judgments about the value of particular data, you must choose a model that is internally consistent and has some merit. There is, however, no "correct" model to use; rather these models are imposed because of logical, emotional (you subjectively like one perspective over another in terms of its aesthetics), and perhaps interpersonal concerns (i.e., researchers who you value also adopt this perspective).

Kitchener and King (1981; also King, Kitchener, Wood, & Davison, 1989; Kitchener, King, Wood, & Davison, 1989) have traced the developmental progression empirically from objective views of reality to subjective views. Participants were presented with ill-structured problems (e.g., the safety of adding chemicals to foods) and asked a variety of questions about the problems that assessed their reasoning about the state of knowledge. Participants' responses were then coded into one of several stages. Lower stages reflect a concrete, black-and-white way of knowing (e.g., chemical additives are wrong because they cause cancer), middle stages reflect that knowledge is uncertain and must be understood within specific contexts (e.g., there is evidence on both sides of the issue and people look at the evidence in a different way depending on their own perspective), to a final state of understanding in which knowledge develops in a probabalistic fashion that becomes generalizable across domains (e.g., the weight of the evidence coming in from experts in the field is that food additives are more likely to be bad than good). Their work consistently finds evidence for a developmental progression from adolescence into young adulthood in people's conceptions about the nature of knowledge and reality (see Kitchener, Lynch, Fischer, & Wood, 1993; see also Perry, 1968). These developments appear to be marked by discontinuities in cognitive growth, supportive of a stagelike notion of development.

These developments extend to the ways that individuals understand the self and their emotions. For instance, Labouvie-Vief, Chiodo, Goguen, Diehl, and Orwoll (1995) have traced the development of representations of the self from a relatively undifferentiated stage (where self and significant others are fused), to a more differentiated stage where the self and others are differentiated, yet tied to institutional roles (e.g., I am an effective mother), to higher stages that involve a more contextual, relativistic, and dynamic sense of the self (e.g., "At this point in time in my life and the lives of my parents, ... I am struggling to remain my adult self but getting pulled back to my younger self," p. 407). Self-representations were most advanced in middle age and least advanced among preadolescents and older adults. Higher levels of self-understanding were related to higher scores on measures of crystallized and fluid intelligence, and Loevinger's measures of ego-involvement, supporting the view that these levels of self-representation indicate higher, more advanced, levels of thinking about the self. Labouvie-Vief, DeVoe, and Bulka (1989) found similar results with respect to emotional understanding in adult development. Middle-aged individuals displayed more complex and integrated ways of thinking about emotion, synthesizing conventional, psychological, mind, and body experiences of various emotions. In Labouvie-Vief's work age has not been linearly related to cognitive growth.

Failure to recognize these advances in reasoning across adult development can, according to postformal theorists, result in misinterpretations of adults' intellectual functioning. For instance, Adams, Labouvie-Vief, Hobart, and Dorosz (1990) reported that older adults' story recall was more elaborate and integrative with somewhat less text recall, but with more additional filler material than that shown by young adults. Although the information-processing perspective might interpret such results as reflecting older adults' reduced ability to inhibit irrelevant thoughts, the neo-Piagetian approach would interpret such results as reflective of older adults' ability to integrate emotive, interpersonal, and logical material. In fact, Pascuale-Leone (1984) indicates that the lack of inhibiting emotive and interpersonal thoughts from logical thoughts leads to advances in postformal thought, such as transcendental thinking and wisdom. Similarly, on logical combination problems, reflective of those used by Piaget, older adults often perform at a level that indicates they are reasoning at the concrete-operational stage, rather than at the formal operational stage (Sinnott, 1989). However, Labouvie-Vief (1984) and Sinnott (1989) indicate that such results may reflect that older adults are more likely to interpret logical problems outside of the confines of logic and to integrate more pragmatic and nonrational ways of understanding.

In sum, evidence is mounting that supports the view that adult intellectual development may be characterized by qualitative shifts in the structure that underlies intelligent thought. Although there is great variability among theorists in the specific structures that are postulated to exist (see Sinnott, 1996, for a review), there is a growing consensus of the developmental shift from more concrete and undifferentiated ways of thinking to more contextualized and dynamic systems of thought that integrate objective and subjective ways of understanding. This approach has the potential for providing a broad-based theoretical account of developmental changes that occur in intellectual development. However, the present work is less clear on the mechanism for change and for clear-cut ways of identifying that a given structure is more complex or a higher stage of thought than a previous one.

THE CONTEXTUAL PERSPECTIVE

Intelligence as Adapting to Changing Life Contexts Across Adult Development

The three perspectives reviewed in the preceding have focused on intelligence as residing largely within the mind of the developing individual. The contextual perspective focuses on intelligence as it is expressed through a transaction of the individual with his or her context. This transaction is a dynamic one in which the context may offer opportunities or place constraints on an individual's functioning and simultaneously the individual may in his or her efforts to adapt and shape his or her environment change the context (see Berg, 2000). A central notion within the contextual perspective is that the intellectual processes and products that are required to adapt to one's environment may change throughout the lifespan as the contexts that are traversed vary in their opportunities and constraints. Shifts in the contexts of development from formal schooling and structured work environments in young adulthood, multiple roles in midlife, to retirement, and leisure activities in late life are thought to have important consequences for intellectual performance (Berg & Calderone, 1994; Laboratory of Comparative Human Cognition, 1982; Rogoff, 1982). For instance, individuals who occupy work contexts that vary in their level of complexity have been found to differ with respect to intellectual functioning such that individuals in more complex and stimulating work environments show more incremental gains in intellectual performance over a 20-year period than those in less stimulating work environments (Schooler, Mulatu, & Oates, 1999).

Berg and Sternberg (1992) examined laypeople's beliefs of what constitutes intelligence during adult development as a way to uncover the abilities that might be required to adapt to the changing context of adult development. This work drew from Neisser's (1979) suggestion that intelligence is the extent to which one resembles the culture's prototype of an intelligent individual. Berg and Sternberg found that adults across the lifespan perceived three characteristics to underlie an intelligent adult: interest in and ability to deal with novelty, verbal ability, and everyday competence. The importance of these three abilities to characterizing an intelligent young, middle-aged, and older adult, however, varied. The ability to deal with novelty was perceived as most important for characterizing intelligent young adults and decreased in importance for characterizing intelligent middle-aged and older adults. Characteristics associated with verbal and everyday competence, however, were perceived to be more important in characterizing intelligent middle-aged and older adults than young adults. The relationship between the dimension of interest in and ability to deal with novelty and measures of fluid intelligence suggest that the characteristic for which the most decline is seen in performance across adult development is the characteristic that individuals perceive to be least important in characterizing intelligent functioning in late life.

Given the importance of everyday competence to laypeople's views of intelligence during adult development, we will limit our review to how individuals use their mental abilities to adapt to everyday life environments. Research on everyday problem solving has typically focused on how individuals solve ill-structured problems, ones that do not have a single solution or a single means of solving the problem. Such work is often identified with the contextual perspective, although important exceptions exist (Willis and colleagues' research is conducted on everyday problem solving within the psychometric perspective; see Marsiske & Willis, 1995). Such problems might include making a decision (e.g., regarding a home purchase, a financial investment, or medical treatment), solving an interpersonal problem (a conflict between family members), or making long-terms plans with a complicated set of contingencies (e.g., planning a vacation). The focus of much of this research has been to compare the cognitive performance of individuals across contexts that vary in ways that may influence cognitive performance.

Variation in Cognitive Performance Across Contexts

Research indicates that individuals' everyday problem-solving strategies vary as a function of the domain or context (e.g., work, family, friends, health) in which everyday problems take place (Berg, Strough, Calderone, Sansone, & Weir, 1998; Blanchard-Fields, Chen, & Norris, 1997; Cornelius & Caspi, 1987). Berg et al. and Blanchard-Fields et al. found that the strategies that adults use to deal with everyday problem-solving situations and the attributions they make for the problem characters' behavior varied depending on the context in which everyday problems were couched. For instance, individuals used strategies that involved the use of others for problems that were in interpersonally relevant domains such as family and friends and used strategies that involved one's own behavior for domains such as work and school. Berg et al. found that different contexts may make salient certain goals (e.g., contexts such as family and friends may make salient interpersonally relevant goals) that guide the selection of particular problem-solving strategies.

Several investigators have examined how contexts in which tasks are quite familiar may support intellectual performance as compared to less familiar contexts. Smith and Baltes (1990; see also Baltes & Staudinger, 2000) operationalized wisdom as expert knowledge about fundamental life problems and explored wisdom through how adults approach life-planning problems. Problems were constructed to be normative for one's own age vs. normative for a different age. Participants responded to life-planning problems such as planning whether to accept a new promotion or begin a family (normative for a young adult) or to take early retirement or move to a new office in response to a company closing their headquarters (normative for an older adult). Individuals were asked to formulate a plan that covered what the target individual should consider in the next 3 to 5 years. Wise responses were characterized as those that reflected rich factual and procedural knowledge, knowledge about the contexts of life and relativism concerning values, goals, and the unpredictability of life. Smith and Baltes (1990) found that adults performed best when the problems were more normative for one's age group.

Sternberg (1998) has taken a different approach to wisdom. He has argued in his balance theory that wisdom involves the application of tacit knowledge for a common good. This application involves two kinds of balances. The first is among three interests: self (intrapersonal)-interests; other (interpersonal) interests; and institutional (extrapersonal) interests. The second is among three actions on the environment: adaptation, or changing oneself to suit the environment; shaping, or changing the environment to suit oneself; and selection, or changing environments entirely.

Research within the expertise literature has also found that intellectual operations performed within a familiar or overly practiced context are superior to the same operations that are taken out of their routine context. For instance, Salthouse's (1984) work comparing the speed of operations among typists varying in age indicates that the speed of typing does not vary by age. However, the typical age differences in speed of response are found among these typists when the task is a typical reaction time task (where individuals press a button as quickly as possible in response to a symbol). Older typists adopt a strategy (looking further down the visual field while typing) that seems to compensate for their reduction in speed of response. Such compensation seems to occur on tasks with which older adults have extensive experience (Charness, 1981; see Backman & Dixon, 1992, for a review).

Research regarding the tacit knowledge that individuals across the lifespan gain in their jobs also reveals dissociations between more standard measures of intelligence (decontextualized measures of intelligence) and measures of tacit knowledge. Wagner and Sternberg (1985) defined tacit knowledge as the knowledge that is not typically learned through formal instruction and is often not explicitly expressed. Tacit knowledge is typically measured by the extent to which individuals' ratings of strategies appropriate for solving job-relevant problems resemble those of acknowledged experts in one's field. In general, research examining the relationship between tacit knowledge in one's occupation (e.g., academic psychologists, salespeople, military officers, managers, and teachers) and conventionally measured psychometric intelligence reveals relatively small correlations (range across studies of $-.12$ to $-.3$; see Sternberg et al., 2000; Sternberg, Wagner, Williams, & Horvath, 1995, for a reviews). However, tacit knowledge typically predicts job performance as well as or better than does performance on conventional tests of intellectual abilities (Sternberg et al., 2000). Colonia-Willner (1998), in a study of bank managers ranging in age from 24 to 59 years, found typical negative relationships between age and psychometric measures of fluid intelligence. However, the relationship of age to tacit knowledge was much less strong.

In sum, research coming from the contextual perspective indicates that the context in which intelligence is expressed affects intelligent functioning. Cognitive performance is enhanced under conditions in which the cognitive operations are familiar, normative for one's age group, and grounded in daily life experience. Future research is needed to understand the ways that such contexts optimize cognitive performance and the specific adaptations and compensations that older adults perform to gain a better fit between themselves and their environment.

CONCLUSIONS AND FUTURE DIRECTIONS

This chapter has used four different perspectives that currently guide the field of adult intellectual development: the psychometric, the cognitive, the neo-Piagetian, and the contextual. Each of these perspectives adds a new dimension to our understanding of adult intelligence. The psychometric perspective advances a view of intelligence oriented around two factors (crystallized vs. fluid intelligence) that yield different trajectories of lifespan change (one oriented more toward stability and the other toward decline across adult development). The cognitive perspective further examines the mental processes that underlie the decline in fluid intellectual abilities through a focus on broad processing resources such as speed of response, working memory capacity, and attentional deficits in inhibiting irrelevant information. The neo-Piagetian perspective expands our view of intelligence beyond traditional realms of logical, rational thought to an understanding of how adults may integrate the rational with emotional and interpersonal ways of understanding. Such an expansion of the realm of intelligence led to an examination of how individuals use their intelligence to understand their own self and emotional development. From such a perspective, intelligence develops in an upward fashion, at least through middle age and perhaps into late life. Finally, the contextual perspective explores how the different contexts that individuals occupy across the adult lifespan may relate to the expression of different intellectual processes and products. The contextual perspective allows for intelligence to denote different characteristics across development and helps to appreciate how intellectual performance differs across contexts.

The challenge to current researchers and theorists of adult intellectual development is to encompass the findings revealed through these different theoretical perspectives. Currently, no single theoretical view has been advanced that takes into account all of the empirical findings uncovered through the different theoretical perspectives. Two broad theories of intellectual development, however, encompass three of the four. The triarchic theory of intelligence proposed by Sternberg (1985, 1997) and extended to adult development and aging by Berg and Sternberg (1985) views intelligence as comprising three parts: a contextual part, indicating that intelligence is the mental activity involved in providing a more optimal fit between the person and his or her context; a componential part, which specifies the elementary information-processing components that may be involved in providing a more optimal fit between the person and the context; and an experiential part, which specifies particular contexts that may be most optimally suited to the assessment of intelligence. This theory argues that what constitutes intelligence may change across adult development as the contexts that individuals occupy may draw for different abilities and mental processes.

A second theoretical view advanced by Baltes and colleagues is that of selective optimization with compensation. Individuals are posited to select across the lifespan from a variety of domains and goal possibilities those that can be enhanced and compensate for abilities and processes that may be showing elements of decline. For example, individuals may select out of some contexts and into others that can maximize their strengths and minimize their weaknesses in abilities and processes (Baltes et al., 1998, for a review). This theory contains an explicit self-regulatory notion of intellectual development such that individuals actively select contexts in which a better fit between mental abilities and processes occur. Recent work indicates that the use of selective optimization with compensation relates to better job performance differentially for older adults than young adults (Abraham & Hansson, 1995) and that greater use of such self-regulatory strategies may relate to measures of successful aging (e.g., life satisfaction, positive emotions, Freund & Baltes, 1998).

Such broad theoretical perspectives are useful in that they take into account a much broader array of the findings concerning adult intellectual development than any one single theoretical perspective reviewed in the preceding. The challenge for future research in adult intellectual development will be to address how intelligence as a construct may change across the adult lifespan as individuals adapt to different intellectual demands. Such contextual work on intelligence must include an examination of the abilities and processes that are necessary for such adaptation. Such work will lead us to a more complete understanding of the ways that individuals use their intelligence across the lifespan as they adapt to diverse contexts. Such work may also help to understand the tremendous variability that exists in the expression of intelligence across individuals and within individuals across contexts.

ACKNOWLEDGMENT. Robert Sternberg's contribution to this chapter was supported in part by Grant 206R50001 from the Office of Educational Research and Improvement, U.S. Department of Education. This support does not imply endorsement of any of the positions taken in the chapter.

REFERENCES

Abraham, J. D., & Hansson, R. O. (1995). Successful aging at work: An applied study of selection, optimization, and compensation through impression management. *Journal of Gerontology, 50B*, P94–P103.

Adams, C., Labouvie-Vief, G., Hobart, C. J., & Dorosz, M. (1990). Adult age group differences in story recall style. *Journal of Gerontology: Psychological Sciences, 45*, P17–P25.

Anderson, J. R. (1983). *The architecture of cognition.* Cambridge, MA: Harvard University Press.

Anderson, J. R. (1991). The adaptive nature of human categorization. *Psychological Review, 98*, 409–429.

Anderson, J. R. (1996). ACT: A simple theory of complex cognition. *American Psychologist, 51*, 355–365.

Arbuckle, T. Y., & Gold, D. P. (1993). Aging, inhibition, and verbosity. *Journal of Gerontology: Psychological Sciences, 48*, 225–232.

Atkinson, R. C., & Shiffrin, R. M. (1968). Human memory: A proposed system and its control processes. In K. W. Spence & J. T. Spence (Eds.), *The psychology of learning and motivation: Advances in research and theory, Vol. 2.* New York: Academic Press.

Backman, L., & Dixon, R. A. (1992). Psychological compensation: A theoretical framework. *Psychological Bulletin, 112*, 1–125.

Baddeley, A. D. (1992). Working memory. *Science, 255*, 556–559.

Baddeley, A. D. (1995). *Working memory.* In M. S. Gazzaniga (Ed.), *The cognitive neurosciences* (pp. 755–764). Cambridge, MA: The MIT Press.

Baltes, P. B., & Staudinger, U. M. (2000). Wisdom: A metaheuristic (pragmatic) to orchestrate mind and virtue toward excellence. *American Psychologist, 55*, 122–136. *Wisdom: The orchestration of mind and virtue.* Boston: Blackwell.

Baltes, P. B., Lindenberger, U., & Staudinger, U. M. (1998). Life-span theory in developmental psychology. In W. Damon (Series Ed., R. M. Lerner, Vol. Ed.), *Handbook of child psychology, Vol. 1* (pp. 1029–1143). New York: John Wiley & Sons.

Baltes, P. B., Mayer, K. U., Helmchen, H., & Steinhagen-Thiessen, E. (1993). The Berlin Aging Study (BASE): Overview and design. *Aging and Society, 13*, 483–515.

Baltes, P. B., & Willis, S. L. (1982). Plasticity and enhancement of intellectual functioning in old age: Penn State's adult development and enrichment project (ADEPT). In F. I. M. Craik & S. E. Trehub (Eds.), *Aging and cognitive processes* (pp. 353–389). New York: Plenum Press.

Berg, C. A. (2000). The development of adult intelligence. In R. J. Sternberg (Ed.), *Handbook of human intelligence.* (pp. 117–137). New York: Cambridge University Press.

Berg, C. A., & Calderone, K. S. (1994). The role of problem interpretations in understanding the development of everyday problem solving. In R. J. Sternberg & R. K. Wagner (Eds.), *Mind in context: Interactionist perspectives on human intelligence* (pp. 105–132). New York: Cambridge University Press.

Berg, C. A., Hertzog, C., & Hunt, E. (1982). Age differences in the speed of mental rotation. *Developmental Psychology, 18*, 95–107.

Berg, C. A., & Sternberg, R. J. (1985). A triarchic theory of intellectual development during adulthood. *Developmental Review, 5*, 334–370.

Berg, C. A., & Sternberg, R. J. (1992). Adults' conceptions of intelligence across the life span. *Psychology and Aging, 7*, 221–231.

Berg, C. A., Strough, J., Calderone, K. S., Sansone, C., & Weir, C. (1998). The role of problem definitions in understanding age and context effects on strategies for solving everyday problems. *Psychology and Aging, 13,* 29–44.

Birren, J. E. (1974). Translations in gerontology-From lab to life: Psychophysiology and speed of response. *American Psychologist, 29,* 808–815.

Blanchard-Fields, F., Chen, Y., & Norris, L. (1997). Everyday problem solving across the life span: The influence of domain-specificity and cognitive appraisal. *Psychology and Aging, 12,* 684–693.

Burke, D. M. (1997). Language, aging, and inhibitory deficits: Evaluation of a theory. *Journal of Gerontology: Psychological Sciences,* P254–264.

Carroll, J. B. (1993). *Human cognitive abilities: A survey of factor-analytic studies.* New York: Cambridge University Press.

Cerella, J. (1990). Aging and information processing rate. In J. E. Birren & K. W. Schaie (Eds.), *Handbook of the psychology of aging* (pp. 201–221). San Diego: Academic Press.

Charness, N. (1981). Age and skilled problem solving. *Journal of Experimental Psychology: General, 110,* 21.

Cohen, G. (1988). Age differences in memory for texts: Production deficiency or processing limitations? In L. L. Light & D. M. Burke (Eds.), *Language, memory and aging* (pp. 171–190). New York: Cambridge University Press.

Colonia-Willner, R. (1998). Practical intelligence at work: Relationships between aging and cognitive efficiency among managers in a bank environment. *Psychology and Aging, 13,* 45–57.

Connelly, S. L., Hasher, L., & Zacks, R. T. (1991). Age and reading: The impact of distraction. *Psychology and Aging, 6,* 533–541.

Cornelius, S. W., & Caspi, A. (1987). Everyday problem solving in adulthood and old age. *Psychology and Aging, 2,* 144–153.

Engle, R. W., & Oransky, N. (1999). Multi-store versus dynamic models of temporary storage in memory. In R. J. Sternberg (Ed.), *The nature of cognition* (pp. 515–556). Cambridge, MA: The MIT Press.

Flynn, J. R. (1984). The mean IQ of Americans: Massive gains 132 to 1978. *Psychological Bulletin, 95,* 29–51.

Flynn, J. R. (1987). Massive IQ gains in 14 nations: What IQ tests really measure. *Psychological Bulletin, 101,* 171–191.

Freund, A. M., & Baltes, P. B. (1998). Selection, optimization, and compensation as strategies of life management: Correlations with subjective indicators of successful aging. *Psychology and Aging, 13,* 531–543.

Hale, S., Myerson, J., Smith, G. A., & Poon, L. W. (1988). Age, variability, and speed: Between-subjects diversity. *Psychology and Aging, 3,* 407–410.

Hasher, L., & Zacks, R. T. (1988). Working memory, comprehension, and aging: A review and a new view. *The Psychology of Learning and Motivation, 22,* 193–225.

Havighurst, R. (1972). *Developmental tasks and education.* New York: Van Nostrand.

Hertzog, C. (1989). Influences of cognitive slowing on age differences in intelligence. *Developmental Psychology, 25,* 636–651.

Hertzog, C., Vernon, M. C., & Rypma, B. (1993). Age differences in mental rotation task performance: The influence of speed/accuracy tradeoffs. *Journal of Gerontology: Psychological Sciences, 48,* P150–P156.

Horn, J. L. (1994). Theory of fluid and crystallized intelligence. In R. J. Sternberg (Ed.), *Encyclopedia of human intelligence, Vol. 1* (pp. 443–451). New York: Macmillan.

Horn, J. L., & Cattell, R. B. (1967). Age differences in fluid and crystallized intelligence. *Acta Psychologica, 26,* 107–129.

Horn, J. L., & Hofer, S. M. (1992). Major abilities and development in the adult period. In R. J. Sternberg & C. A. Berg (Eds.), *Intellectual development* (pp. 44–99). New York: Cambridge University Press.

Hunt, E. B. (1995). *Will we be smart enough? A cognitive analysis of the coming workforce.* New York: Russell-Sage Foundation.

Jensen, A. R. (1998). *The g factor.* Westport, CT: Praeger.

King, P. M., Kitchener, K. S., Wood, P. K., & Davison, M. L. (1989). Relationships across developmental domains: A longitudinal study of intellectual, moral, and ego development. In M. L. Commons, J. D. Sinnott, F. A. Richards, & C. Armon (Eds.), *Adult development, Vol. 1: Comparisons and applications of developmental models* (pp. 57–72). New York: Praeger.

Kitchener, K. S., & King, P. M. (1981). Reflective judgement: Concepts of justification and their relationship to age and education. *Journal of Applied Developmental Psychology, 2,* 89–116.

Kitchener, K. S., King, P. M., Wood, P. K., & Davison, M. L. (1989). Consistency and sequentiality in the development of reflective judgement: a six year longitudinal study. *Journal of Applied Developmental Psychology, 10,* 73–95.

Kitchener, K. S., Lynch, C. L., Fischer, K. W., & Wood, P. K. (1993). Developmental range of reflective judgement: The effect of contextual support and practice on developmental stage. *Developmental Psychology, 29,* 893–906.

Laboratory of Comparative Human Cognition (1982). Culture and intelligence. In R. J. Sternberg (Ed.), *Handbook of human intelligence* (pp. 642–719). Cambridge: Cambridge University Press.

Labouvie-Vief, G. (1984). Logic and self-regulation from youth to maturity: A model. In M. L. Commons, F. A. Richards, & C. Armon (Eds.), *Beyond formal operations.* New York: Praeger.

Labouvie-Vief, G. (1992). A Neo-Piagetian perspective on adult cognitive development. In R. J. Sternberg & C. A. Berg (Eds.), *Intellectual developmental.* New York: Cambridge University Press.

Labouvie-Vief, G., Chiodo, L. M., Goguen, L. A., Diehl, M., & Orwoll, L. (1995). Representations of self across the life span. *Psychology and Aging, 10,* 404–415.

Labouvie-Vief, G., DeVoe, M., & Bulka, D. (1989). Speaking about feelings: Conceptions about emotion across the life span. *Psychology and Aging, 4,* 425–437.

Lindenberger, U., & Baltes, P. B. (1997). Intellectual functioning in old age and very old age: Cross-sectional results from the Berlin Aging Study. *Psychology and Aging, 12,* 410–432.

Marsiske, M., & Willis, S. L. (1995). Dimensionality of everyday problem solving in older adults. *Psychology and Aging, 10,* 269–283.

McDowd, J. (1997). Inhibition in attention and aging. *Journal of Gerontology: Psychological Sciences,* P265–273.

McDowd, J. M., Oseas-Kreger, D. M., & Filion, D. L. (1995). Inhibitory processes in cognition and aging. In F. N. Dempster & C. J. Brainerd (Eds.), *Interference and inhibition in cognition* (pp. 363–400). San Diego: Academic Press.

Neely, A. S., & Backman, L. (1993). Long-term maintenance of gains from memory training in older adults: Two $3\frac{1}{2}$-year follow-up studies. *Journal of Gerontology, 48,* 233–237.

Neisser, U. (1979). The concept of intelligence. In R. J. Sternberg & D. K. Detterman (Eds.), *Human Intelligence: Perspectives on its theory and measurement* (pp. 179–189). Norwood, NJ: Ablex.

Neisser, U. (Ed.) (1998). *The rising curve.* Washington, DC: American Psychological Association.

Pascual-Leone, J. (1984). Attentional, dialectic, and mental effort: Toward an Organismic theory of life stages. In M. L. Commons, F. A. Richards, & C. Armon (Eds.), *Beyond formal operations* (pp. 182–215). New York: Praeger.

Perry, W. G. (1968). *Forms of intellectual and ethical development in the college years.* New York: Holt, Rinehart & Winston.

Reinert, G. (1970). Comparative factor analytic studies of intelligence throughout the human life-span. In L. R. Goulet & P. B. Baltes (Eds.), *Life-span developmental psychology: Research and theory* (pp. 467–484). New York: Academic Press.

Riegel, K. F. (1976). The dialectics of human development. *American Psychologist, 31,* 689–700.

Rogoff, B. (1982). Integrating context and development. In M. E. Lamb & A. L. Brown (Eds.), *Advances in developmental psychology, Vol. 2* (pp. 125–170). Hillsdale, NJ: Lawrence Erlbaum.

Salthouse, T. A. (1984). Effects of age and skill in typing. *Journal of Experimental Psychology: General, 113,* 345–371.

Salthouse, T. A. (1991). *Theoretical perspectives on cognitive aging.* Hillsdale, NJ: Lawrence Erlbaum.

Salthouse, T. A. (1992a). *Mechanisms of age-cognition relations in adulthood.* Hillsdale, NJ: Lawrence Erlbaum.

Salthouse, T. A. (1992b). The information-processing perspective on cognitive aging. In R. J. Sternberg & C. A. Berg (Eds.), *Intellectual development* (pp. 261–277). New York: Cambridge University Press.

Salthouse, T. A. (1996). The processing-speed theory of adult age differences in cognition. *Psychological Review, 103,* 403–428.

Salthouse, T. A., Kausler, D. H., & Saults, J. S. (1988). Utilization of path analytic procedures to investigate the role of processing resources in cognitive aging. *Psychology and Aging, 3,* 158–166.

Salthouse, T. A., Mitchell D. R., Skovronek, E., & Babcock, R. L. (1989). Effects of adult age and working memory on reasoning and spatial abilities. *Journal of Experimental Psychology: Learning, Memory, and Cognition, 15,* 507–516.

Schaie, K. W. (1984). Historical time and cohort effects. In K. A. McCloskey & H. W. Reese (Eds.), *Life-span developmental psychology: Historical and generational effects* (pp. 1–15). New York: Academic Press.

Schaie, K. W. (1988). Variability in cognitive functioning in the elderly: Implications for societal participation. In A. D. Woodhead, M. A. Bender, & R. C. Leonard (Eds.), *Phenotypic variation in populations: Relevance to risk assessment* (pp. 191–212). New York: Plenum Press.

Schaie, K. W. (1996a). Intellectual development in adulthood. In J. E. Birren & K. W. Schaie (Eds.), *Handbook of the psychology of aging* (pp. 266–286). New York: Academic Press.

Schaie, K. W. (1996b). *Intellectual development in adulthood: The Seattle Longitudinal Study.* New York: Cambridge University Press.

Schaie, K. W., & Willis, S. L. (1986). Can decline in adult intellectual functioning be reversed? *Developmental Psychology, 22,* 223–232.

Schaie, K. W., & Willis, S. L. (1996). Psychometric intelligence and aging. In F. Blanchard-Fields & T. Hess (Eds.), *Perspectives on cognitive change in adulthood and aging.* New York: McGraw-Hill.

Schooler, C., Mulatu, M. S., & Oates, G. (1999). The continuing effects of substantively complex work on the intellectual functioning of older workers. *Psychology and Aging, 14,* 483–506.

Sinnott, J. D. (Ed.) (1989). *Everyday problem solving: Theory and applications.* New York: Praeger.

Sinnott, J. D. (1996). Post-formal intelligence. In F. Blanchard-Fields & T. M. Hess (Eds.), *Perspectives on cognition in adulthood and aging.* New York: McGraw-Hill.

Smith, J., & Baltes, P. B. (1990). Wisdom-related knowledge: Age/cohort differences in response to life-planning problems. *Developmental Psychology, 26,* 494–505.

Spearman, C. (1927). *The abilities of man: Their nature and measurement.* New York: Macmillan.

Sternberg, R. J. (1985). *Beyond IQ: A triarchic theory of human intelligence.* New York: Cambridge University Press.

Sternberg, R. J. (1997). *Successful intelligence.* New York: Plume.

Sternberg, R. J. (1998). A balance theory of wisdom. *Review of General Psychology, 2,* 347–365.

Sternberg, R. J. (Ed.) (1999). *The nature of cognition.* Cambridge, MA: The MIT Press.

Sternberg, R. J., & Berg, C. A. (1992). *Intellectual development.* Cambridge, UK: Cambridge University Press.

Sternberg, R. J., Forsythe, G. B., Hedlund, J., Horvath, J., Snook, S., Williams, W. M., Wagner, R. K., & Grigorenko, E. L. (2000). *Practical intelligence in everyday life.* New York: Cambridge University Press.

Sternberg, R. J., Wagner, R. K., Williams, W. M., & Horvath, J. A. (1995). Testing common sense. *American Psychologist, 50,* 912–927.

Stolzfus, E., Hasher, L., & Zacks, R. T. (1996). Working memory and aging: Current status of the inhibitory view. In J. T. E. Richardson, R. W. Engle, L. Haher, R. H. Logie, E. R. Stolzfus, & R. T. Zacks (Eds.), *Working memory and human cognition.* New York: Oxford University Press.

Wagner, R. K., & Sternberg, R. J. (1985). Practical intelligence in real-world pursuits: The role of tacit knowledge. *Journal of Personality and Social Psychology, 48,* 436–458.

Wagner, R. K., & Sternberg, R. J. (1986). Tacit knowledge and intelligence in the everyday world. In R. J. Sternberg & R. K. Wagner (Eds.), *Practical intelligence.* New York: Cambridge University Press.

Willis, S. L. (1987). Cognitive training and everyday competence. In K. W. Schaie (Ed.), *Annual Review of Gerontology and Geriatrics, Vol. 7* (pp. 159–188). New York: Springer.

Willis, S. L., & Nesselroade, C. S. (1990). Long-term effects of fluid ability training in old-old age. *Developmental Psychology, 26,* 905–910.

Zacks, R. T., Radvansky, G. A., & Hasher, L. (1994). Studies of directed forgetting in older adults. *Journal of Experimental Psychology: Learning Memory, and Cognition, 22,* 143–156.

CHAPTER 7

Age-Related Changes in Memory

Margaret G. O'Connor and Edith F. Kaplan

We begin this chapter with a review of theoretical and practical issues relevant to studies of age-related memory decline. Next we discuss the neural substrates and psychological components of memory in relation to memory changes that occur as part of the normal aging process. In the third section biological changes that occur with aging and that provide the backdrop against which memory decline occurs are discussed. In the final section neuropsychological studies of the relationship between aging and dementia are reviewed.

THE DEFINITION OF "NORMAL" IS ELUSIVE

Questions have arisen as to who should be included in studies of normal aging. Are individuals with mild memory loss abnormal? Or are individuals without any evidence of memory decline the abnormal ones? Many studies have shown that memory loss is common among elderly individuals. Nonetheless, some researchers interested in normal age-related memory loss routinely exclude individuals with performance that is one standard deviation below the mean on formal tests of memory while including individuals who perform one standard deviation above the mean. As a result of this practice, "normative" data from these studies may overestimate the average abilities and underestimate the variability in the memory capabilities of normal elders. The opposite problem occurs in studies of memory in which patients with preclinical Alzheimer's disease (AD) are included. These investigations tend to underestimate the mean and overestimate the variance in the memory abilities of this population (Sliwinski, Lipton, Buschke, & Stewart, 1996). Another common practice in studies of age-related

Margaret G. O'Connor • Division of Behavioral Neurology, Beth Isreal-Deaconess Medical Center, Boston, MA, 02215.   Edith F. Kaplan • Department of Psychology, Suffolk University, Boston, Massachusetts, 02114.

Handbook of Adult Development, edited by J. Demick and C. Andreoletti. Plenum Press, New York, 2002.

memory loss involves exclusion of elders as participants in memory studies if they have medical problems (e.g., hypertension, alcoholism, diabetes, etc.) that could negatively impact on memory. Again, this bias toward higher functioning elders limits the generalizability of the data obtained in these studies.

Investigations of age-related decline use either cross-sectional or longitudinal approaches. In cross-sectional studies older and younger adults are compared with respect to task performance. Such comparisons are confounded by cohort effects. Differences in nutrition, socioeconomic levels, education, and health status may confound cross-generational comparisons (Schaie & Willis, 1991). Selection differences in the individuals who volunteer to participate in memory investigations should also be considered. It is often the case that older individuals who participate in memory studies do so because they have observed memory problems in everyday life (Rentz, personal communication, 2000). Hence, this may result in a bias toward inclusion of elders with mild memory deficits.

Longitudinal studies are the preferred approach although there are also problems with this method that influence findings. The length of interval between test sessions may have a significant effect on whether decline is detected. Studies using short intertest intervals may not detect cognitive decline in normal elders (Flicker, Ferris, & Reisberg, 1993) whereas age-related decline may emerge when the intervals between test sessions are longer (Zelinski & Burnight, 1997). Longitudinal studies are also affected by possible inclusion of patients who develop dementia over the course of the study. One way of avoiding this problem involves the retrospective analysis of data. Only those individuals who are identified as "dementia free" at follow-up are included in the normative sample. This approach results in the exclusion of preclinical patients; however, patients with lifelong low abilities may also be inadvertently excluded (Sliwinski et al., 1996). Even more worrisome is the issue of selective attrition: healthier subjects with higher test scores are more likely to be followed up, particularly as the interval between test sessions increases (Brayne et al., 1999).

Diverse methodological approaches and selection biases preclude our ability to derive formulae for the prediction of age-related decline for the individual person. However, there is consensus of increased age-based heterogeneity with respect to cognitive abilities (Schaie, 1990). The effects of age on variability in cognitive performance were well illustrated in a study of 1000 physicians that revealed that the top 10% of elder participants were as cognitively adept as the top 10% of the younger group. At the same time the elder adults were far more variable than the younger adults in their overall performance (Weintraub, Powell, & Whitla, 1994). It appears to be the case that some elder adults are predisposed to "successful aging" during their latter years whereas others experience progressive memory loss associated with dementia. A number of factors have been studied in relation to cognitive fitness over the lifespan. In a study of 1192 individuals, 22 demographic variables were considered with respect to cognitive change over a 2-year period of time (Albert et al., 1995). Educational background was the best predictor of cognitive change in this study. It may be that educational reserve enables the individual to successfully circumvent age-related change and thereby mitigate the consequences of mild memory loss. Other factors that have also been shown as influential in the relationship between age and cognitive decline include extent of physical activity and feelings of self-efficacy (Albert et al., 1995).

NEUROPSYCHOLOGICAL STUDIES OF
MEMORY AND AGING

Aging is accompanied by changes in a broad array of cognitive abilities including attention (Somberg & Salthouse, 1982), speed of information processing (Salthouse & Litchy, 1985), capacity for abstraction (Albert, Wolfe, & Lafleche, 1990), naming (Albert, Heller, & Milbrerg, 1988), and overall intellectual abilities (Schaie, 1990), although some abilities are disproportionately sensitive to the aging process. Numerous investigations have revealed age-related memory decline that impacts on some components of memory more than others. Research over the past few decades has shown that memory is not a unitary system but is based on a collection of functional systems, each of which contributes to the analysis, storage, and subsequent retrieval of new information. Working memory refers to processes that are dedicated to the maintenance and manipulation of new information on line (Baddeley, 1992). Working memory is composed of "slave" systems responsible for temporary maintenance of new (verbal and nonverbal) information and executive control processes, which are involved in the manipulation of the new information (Baddeley, 1992). Long-term memory encompasses a variety of processes as diverse as memory for procedures and skills, semantic memory (i.e., knowledge of generic facts and concepts), and contextually embedded episodic material. Memory is viewed along a temporal continuum whereby the stages of encoding, consolidation, and retrieval are distinguished from one another (Cermak, 1986).

The neural substrates of various memory processes and stages are different, but overlapping. Damage to frontal brain regions or to pathways connecting frontal and anterior temporal areas results in diminished working memory capacities as well as in problems with encoding and retrieval (Buckner & Koutstaal, 1998; Shimamura, 1995). Damage to medial temporal circuitry, as seen in association with Alzheimer's disease and in the amnesic syndrome, results in consolidation deficits (Gabrieli, Brewer, Desmond, & Glover, 1997; McClelland, McNaughton, & O'Reilly, 1995; Squire, 1992). Semantic memory deficits are seen in the context of damage to association cortices (DeRenzi, Liotti, & Nichelli, 1987). Procedural learning problems arise as a result of basal ganglia or cerebellar lesions (Friston, Frith, Passingham, Liddle, & Frackowiak, 1992; Jenkins, Brooks, Nixon, Frackowiak, & Passingham, 1994).

Because aging is accompanied by selective changes in frontal and temporal brain regions (Raz, Gunning-Dixon, Head, Dupuis, & Acker, 1998; West, 1996), it is not surprising that memory processes subserved by these areas decline with age whereas other aspects of memory, such as procedural learning, fare well across the lifespan. Quite early in the aging process, adults become aware of changes in their working memory abilities. That is, they find it increasingly difficult to process new information particularly when the quantity (Light & Capps, 1986) or complexity of the stimuli increases (Light, Zelinski, & Moore, 1982; Salthouse, 1990; Salthouse, Mitchell, & Palmon, 1989) or in the context of interference (Light & Albertson, 1988). Indeed there is substantial evidence that age is accompanied by significant decrements in working memory that may undermine the integration of new material (Light et al., 1982) and thereby contribute to more generalized learning difficulties. Some researchers have speculated that working memory deficits account for deficits in long-term memory (Raz et al., 1998; Salthouse, 1992); however, not all studies have shown a significant relationship between working memory and long-term memory (Light, 1991).

In terms of long-term memory, encoding and retrieval, capacities that are mediated by frontal brain regions, are especially vulnerable to age-related decline (Poon, 1985; Small, Stern, Tang, & Mayeux, 1999). Elder adults have problems learning new information that exceeds attention span limitations and they may not form distinctive associations to new material at time of initial exposure (Rabinowitz, Ackerman, Craik, & Hinchey, 1982). Furthermore, their retrieval of previously stored knowledge declines even though their recognition is intact (Verhaeghen, Marcsen, & Goosens, 1993). The disparity between poor performance on tasks of free recall vs. intact recognition is consistent with retrieval deficits. Elder adults have more difficulty than do younger adults with the initiation of a strategic search of memory to access desired information. Problems with encoding and retrieval are invariant features of normal age-related memory decline. However, the psychological processes contributing to these deficits are not well understood. It may be that these deficits are a consequence of inefficient use of strategies (e.g., semantic clustering), failures of "metamemory," or reduced attentional resources (see Light, 1991 for a discussion of these issues).

Another aspect of memory that is adversely affected by age is memory for contextual information. There is abundant evidence showing that older adults have difficulties recollecting the time sequences or spatial locations of prior events (Kausler & Puckett, 1981; McIntyre & Craik, 1987). These problems with temporal and spatial processing may be due to the fact that contextual processing is mediated by frontal brain regions (Sagar, Sullivan, Gabrieli, Corkin, & Growdon, 1988; Shimamura, Janowski, & Squire, 1990).

In contrast with the above, some components of memory appear resistant to the effects of age. Aging does not impact on procedural learning abilities so that older adults are often able to play instruments or perform other skills quite proficiently. Furthermore, performance on tasks of implicit memory is robust across the lifespan (LaVoie & Light, 1994; Mitchell, 1993). Retention of new material, particularly if that material is well rehearsed and conceptually integrated (O'Connor, Sieggreen, Bachna, Kaplan, Cermak, & Ransil, 2000), is not affected by age, perhaps due to the relative preservation of critical hippocampal subfields in normal aging (Gomez-Isla et al., 1996, 1997). In fact, intact retention is one of the cardinal features differentiating normal aging from dementia (Morrison & Hof, 1997).

STRUCTURAL AND FUNCTIONAL CHANGES WITH AGING: THE BIOLOGICAL BACKDROP

Aging affects the structural and functional integrity of the brain. Several pivotal studies (Brody, 1955) gave rise to the idea that neuronal death was inevitable in aging. However, with the advent of new stereological techniques (which provide more accurate estimates of neuronal number) this view changed over the past few years (Morrison & Hof, 1997). Using stereological procedures, recent studies have not shown age-related decline in terms of absolute numbers of neurons (Morrison & Hof, 1997). With regard to memory, the structural health of the hippocampal subfields and the entorhinal cortex (a major pathway connecting the neocortex and the hippocampus) are of particular concern. Unlike the dramatic neuronal loss in these regions occurring in association with AD, neuropathological studies have shown that these regions are not vulnerable to decreased number of neurons in normal aging (Gomez-Isla et al., 1996, 1997).

Regardless as to whether there is a reduction in neuronal number with age, many studies have demonstrated an association between increased age and reduced cortical volume (Coffey, Saxton, Ratcliff, Bryan, & Lucke, 1999). Those brain regions mediating memory (e.g., temporal and frontal brain areas) appear susceptible to the deleterious effects of age. A significant decrease in medial temporal volume was noted in a study of normal elderly adults (ages 51–89), although this cross-sectional investigation included participants with a variety of medical conditions (Jack et al., 1997). Hence, it is not entirely clear whether volume loss would occur in those individuals without medical illnesses. Although some investigators have demonstrated significant associations between measures of medial temporal volume loss and performance on memory tasks (Petersen et al., 2000), others indicate that there is not a significant relationship between decreased volume in neocortical, hippocampal, and entorhinal cortices and memory performance in nondiseased elders (Raz et al., 1998).

Of interest, several structural imaging studies have revealed that gender affects patterns of brain atrophy. Men tend to have greater volume loss than women and the male pattern involves asymmetric loss (Cowell, Turetsky, Gur, Grossman, Shtasel, & Gur, 1991; Gur et al., 1991). Other studies have underscored gender-based regional specificity in terms of how the brain ages. Men are prone to greater reduction in frontal and temporal regions whereas women are vulnerable to increased atrophy in hippocampal and parietal regions (Coffey et al., 1999; Murphy et al., 1996). A positive correlation between education and cortical atrophy has also been noted (Coffey, 1999). The latter finding is viewed by some as evidence for the "cognitive reserve" hypothesis whereby individuals with more education tolerate greater atrophy while their less educated peers are not able to compensate in this manner (Coffey et al., 1999).

Functional neuroimaging studies provide resting and dynamic measures of brain activity. Resting studies have shown that aging is accompanied by reduced physiological activity in frontal and temporal brain regions, areas thought to mediate memory (Meyer, Kawamura, & Terayama, 1994). To date there have been only a few functional imaging studies examining the relationship between age and memory performance, and these have provided mixed results. Several studies have demonstrated that young and old individuals engage similar brain areas during retrieval of information from episodic memory (Backman et al., 1997) while others have shown that increased age results in different patterns of neural activity during tasks of encoding and retrieval. Whereas young adults activate left frontal brain regions when encoding new material and right frontal areas during retrieval (Cabeza, McIntosh, Tuluing, Nyberg, & Grady, 1997), older adults show enhanced frontal activation during encoding and bilateral activation during retrieval (Cabeza et al., 1996; Grady et al., 1995; Hazlett et al., 1998; Madden et al., 1999).

There is no consensus as to the behavioral significance of these different activation patterns although there is speculation that increased activation in the older adults may reflect increased attention or effort during task performance (Madden et al., 1999). Also noteworthy is that functional imaging studies have shown that older adults differ from patients with AD who have markedly reduced activity in prefrontal and hippocampal cortices during tasks of episodic memory (Sperling, 2000; Sperling et al., 2000).

NEUROPATHOLOGICAL DISTINCTIONS
BETWEEN NORMAL AGING AND DEMENTIA

Some of the brain alterations that occur with aging resemble those associated with AD, although there are differences in the quantity and distribution of these changes. Aging is accompanied by an accumulation of senile plaques (SPs), histopathological lesions that arise as a result of β-amyloid deposition. SPs occur frequently in the brains of nondemented elders and there is disagreement as to whether they are pathological or benign (Schmitt, Davis, Wekstein, Smith, Ashford, & Markesberg, 2000a). In one study autopsies were performed on 21 subjects who had been followed longitudinally for 13 years and who had been considered nondemented over that time period. A retrospective analysis of neuropsychological data indicated that the cognitive abilities of individuals with abundant SPs were relatively deficient in comparison to those who had few or no SPs. These data were viewed as support for the idea that diffuse SPs may not be normal but that they may represent incipient AD (Morris et al., 1996). Other studies have downplayed the clinical significance of SPs while emphasizing the importance of neuritic plaques as markers of early AD (Haroutunian, Perl, & Purohit, 1998).

Another neuropathological change that occurs with aging and in the context of AD involves increased accumulation of neurofibrillary tangles (NFTs) that emerge in memory circuitry such as the entorhinal and perirhinal cortices, hippocampal subfields, and temporal association cortex (Price & Morris, 1999; Price, 2000). While there is not an obligatory relationship between NFT and SP presence and cognitive impairment (Schmitt, Davis, Wekstein, Smith, Ashford, & Markesbury, 2000b), a number of recent studies have underscored a significant correlation between an increased accumulation of these changes and cognitive decline (Hartounian, 1999).

BEHAVIORAL DISTINCTIONS BETWEEN
NORMAL AGING AND DEMENTIA

There are a variety of names for the memory problems that exist in our later years. For many years the term benign senescent forgetfulness (BSF) was used to describe the mild and relatively stable memory loss occurring with age (Kral, 1962). By definition, the individual with BSF was aware of his or her memory problems and he or she forgot only minor details of daily events. In other words, he or she did not demonstrate memory problems that compromised capacity for independent living. Other terminology was proposed to denote the memory problems facing nondemented elderly individuals. Age-associated memory impairment (AAMI) referred to individuals with subjective and objective evidence of memory loss in the context of otherwise preserved mental abilities. Psychometric criteria for AAMI included performance one standard deviation below the mean on tests of memory (Crook, Bartus, Ferris, Whitehouse, Cohen, & Gershon, 1986).

More recently the term mild cognitive impairment (MCI) has been used to refer to age-related abnormal memory difficulties that do not compromise other cognitive abilities or one's capacity for independent living (Petersen, Smith, Waring, Ivnik, Tangalos, & Kokmen, 1999).

It is believed that elders with memory problems but otherwise preserved mental abilities are at increased risk for developing dementia (Petersen et al., 1999). To date, a number of standards have been used to identify incipient dementia. Close friends' and family members' perceptions of memory functions have been useful (Tierney, Szalai, Snow, & Fisher, 1996) as has baseline performance on neuropsychological tests (Bondi, Monsch, Galasko, Butters, Salmon, & Delis, 1994; Linn et al., 1995; Masur, Sliwinski, Lipton, Blau, & Crystal, 1994; Tierney et al., 1996). Performance on tasks of immediate recall, a reflection of initial encoding, has distinguished early AD patients from normal elders in several studies (Chapman, White, & Storandt, 1997; Jacobs, Sano, Dooneief, Marder, Bell, & Stern, 1995). In other studies delayed recall performance, reflecting capacity for consolidation, has been more important in distinguishing between elders who go on to develop dementia vs. those who do not (Masur et al., 1994; Small et al., 1999; Welsh, Butters, Hughes, Mohs, & Heyman, 1992). Identification of patients who are at risk for developing dementia takes on added importance in the context of new developments in the treatment of AD (Hanson & Galvez-Jimenez, 2000; Prasad, Nahreini, & Edwards-Prasad, 2000; Schenk, Seubert, Lieberg, & Wallace, 2000).

SUMMARY

Normal aging is associated with decrements in many aspects of cognition including memory. Although not all aspects of memory are affected by age, some components of memory appear vulnerable to the aging process. Relatively subtle memory decline is noted by some elder individuals whereas others go on to develop severe memory loss in conjunction with dementia. Of note, the memory loss that occurs with aging takes place in the context of brain changes that affect frontal and temporal brain regions. Qualitative features of "normal" memory decline are different from, but overlap with, those associated with dementia. In like manner, there are differences in neuropathological features of aging vs. dementia. To date, numerous investigators have attempted to identify risk factors for dementia. Such information may be useful in attempts to develop effective preventative and treatment strategies.

ACKNOWLEDGMENT.   Preparation of this chapter was supported by NINDS program project grant NS26985.

REFERENCES

Albert, M., Heller, H., & Milbrerg, W. (1988). Changes in naming ability with age. *Psychology and Aging, 3*, 173–178.

Albert, M., Savage, C., Blazer, D., Jones, K., Berkman, L., Seeman, T., & Rowe, J. (1995). Predictors of cognitive change in older persons: MacArthur studies of successful aging. *Psychology and Aging, 10*(4), 578–589.

Albert, M., Wolfe, J., & Lafleche, G. (1990). Differences in abstraction with age. *Psychology and Aging* (5)1, 94–100.

Backman, L., Almkvist, O., Anderson, J., Nordberg, A., Windblad, B., Rineck, R., & Lagstrom, B. (1997). Brain activation in young and older adults during implicit and explicit memory retrieval. *Journal of Cognitive Neuroscience, 9*(3), 378–391.

Baddeley, A. (1992). Working memory. *Science, 255,* 556–559.

Bondi, M., Monsch, A., Galasko, D., Butters, N., Salmon, D., & Delis, D. (1994). Preclinical markers of dementia of the Alzheimer's type. *Neuropsychology, 8*(3), 374–384.

Brayne, C., Spiegelhalter, D., Dufouil, C., Chi, L., Dening, T., Paykel, E., O'Connor, D., Ahmed, A., McGee, M., & Huppert, F. (1999). Estimating the true extent of cognitive decline in the old old. *Journal of the American Geriatric Society, 47,* 183–188.

Brody, H. (1955). Organization of the cerebral cortex: III A study of aging in the human cerebral cortex. *Journal of Comparative Neurology, 102,* 511.

Buckner, R. L., & Koutstaal, W. (1998). Functional neuroimaging studies of encoding, priming and explicit memory retrieval. *Proceedings of the National Academy of Sciences USA, 95,* 891–898.

Cabeza, R., McIntosh, A., Tulving, E., Nyberg, L., & Grady, C. (1997). Age-related differences in effective neural connectivity during encoding and retrieval. *NeuroReport, 8,* 3479–3483.

Cabeza, R., Tulving, E., Kapur, S., McIntosh, A. R., Jennings, J. M., Nyberg, L., Houle, S., & Craik, F. I. M. (1996). Functional compensation in the aging brain: A PET study of memory encoding and retrieval. *Rotman Research Institute Conference.*

Cermak, L. S. (1986). Amnesia as a processing deficit. In G. Goldstein & R. E. Tarter (Eds.), *Advances in Clinical Neuropsychology, Vol. 3* (pp. 265–290). New York: Plenum Press.

Chapman, L., White, D., & Storandt, M. (1997). Prose recall in dementia: A comparison of delay intervals. *Archives of Neurology, 54,* 1501–1504.

Coffey, C., Saxton, J., Ratcliff, G., Bryan, R., & Lucke, J. (1999). Relation of education to brain size in normal aging: Implications for the reserve hypothesis. *Neurology, 53,* 189–196.

Cowell, P., Turetsky, B., Gur, R., Grossman, R., Shtasel, D., & Gur, R. (1991). Sex differences in aging of the human frontal and temporal lobes. *Journal of Neuroscience, 14,* 4748–4756.

Crook, T., Bartus, R., Ferris, S., Whitehouse, P., Cohen, G., & Gershon, S. (1986). Age-associated memory impairment: Proposed diagnostic criteria and measures of clinical change-report of a National Institute of Mental Health Work Group. *Developmental Neuropsychology, 2,* 261–276.

DeRenzi, E., Liotti, M., & Nichelli, P. (1987). Semantic amnesia with preservation of autobiographical memory: A case report. *Cortex, 23,* 575–597.

Flicker, C., Ferris, S. H., & Reisberg, B. (1993). A Two year longitudinal study of cognitive function in normal aging and Alzheimer's disease. *Journal of Geriatric Psychiatry and Neurology, 6,* 84–96.

Friston, K. J., Frith, C. D., Passingham, R. E., Liddle, P. F., & Frackowiak, R. S. J. (1992). Motor practice and neurophysiological adaptation in the cerebellum: A positron tomography study. *Proceedings of the Royal Society of London B, 248,* 223–228.

Gabrieli, J. D. E., Brewer, J. B., Desmond, J. E., & Glover, G. H. (1997). Separate neural bases of two fundamental memory processes in the human medial temporal lobe. *Science, 276,* 264–266.

Gomez-Isla, T., Hollister, R., West, H., Mui, S., Growdon, J., Petersen, R., Parisi, J., & Hyman, B. (1997). Neuronal loss correlates with but exceeds neurofibrillary tangles in Alzheimer's disease. *Annals of Neurology, 41,* 17–24.

Gomez-Isla, T., Price, J., McKeel, D., Morris, J., Growdon, J., & Hyman, B. (1996). Profound loss of layer II entorhinal cortex neurons occurs in very mild Alzheimer's disease. *The Journal of Neuroscience, 16,* 4491.

Grady, C. L., McIntosh, A. R., Horwitz, B., Maisog, J. M., Ungerleider, L. G., Mentis, M. J., Pietrini, P., Schapiro, M. B., & Haxby, J. V. (1995). Age-related reductions in human recognition memory involve altered cortical activation during encoding. *Science, 269,* 218–220.

Gur, R., Mozley, P., Resnick, S., Gottlieb, G., Kohn, M., Herman, G., Atlas, S., Grossman, R., Berretta, D., Erwin, R., & Gur, R. (1991). Gender differences in age effect on brain atrophy measured by magnetic resonance imaging. *Proceedings for the National Academy of Sciences USA, 88,* 2845–2849.

Hanson, M., & Galvez-Jimenez, N. (2000). Effective treatment of Alzheimer's disease and its complications. *Cleveland Clinic Journal of Medicine, 67*(6), 441–448.

Haroutunian, V., Purohit, D. P., Perl, D. P., Marin, D., Lantz, M., Davis, K. L., Mohs, R. C. (1999) Neurofibrillary tangles in nondemented elderly subjects and mild Alzheimer's Disease. *Arch Neurol, 56*(6), 713–718.

Haroutunian, V., Perl, D., Purohit, D., Marin, D., Khan, K., Lantz, M. Davis, K. L., & Mohs, R. C. (1998). Regional distribution of neuritic plaques in the nondemented elderly and subjects. *Archives of Neurology, 55,* 1185–1191.

Hazlett, E., Buchsbaum, M., Mohs, R., Spiegel-Cohen, J., Wei, T., Azueta, R., Haznedar, M., Singer, M., Shihabuddin, L., Luu-Hsia, C., & Harvey, P. (1998). Age-related shift in brain region activity during successful memory performance. *Neurobiology of Aging, 19*(5), 437–445.

Jack, C., Petersen, R., Xu, Y. C., Waring, S., O'Brien, P., Tangalos, E., Smith, G., Ivnik, R., & Kokmen, E. (1997). Medial temporal atrophy on MRI in normal aging and very mild Alzheimer's disease. *Neurology, 49*, 786–794.

Jacobs, D., Sano, M., Dooneief, G., Marder, K., Bell, K., & Stern, Y. (1995). Neuropsychological detection and characterization of preclinical Alzheimer's disease. *Neurology, 45*, 957–962.

Jenkins, I. H., Brooks, D. J., Nixon, P. D., Frackowiak, R. S. J., & Passingham, R. E. (1994). Motor sequence learning: A study with positron emission tomography. *Journal of Neuroscience, 14*, 3775–3790.

Kausler, D., & Puckett, J. (1981). Adult age differences in memory for modality attributes. *Experimental Aging Research, 7*, 117–125.

Kral, V. (1962). Senescent forgetfulness: Benign or malignant. *Journal of the Canadian Medical Association, 86*, 257–260.

LaVoie, D., & Light, L.L. (1994). Adult age differences in repetition priming: A meta-analysis. *Psych and Aging, 9*, 539–553.

Light, L. L. (1991). Memory and aging: Four hypotheses in search of data. *Annual Review of Psychology, 42*, 333–376.

Light, L., & Albertson, S. (1988). Comprehension of pragmatic implications in young and older adults. In L. Light & D. Burke (Eds.), *Language, Memory and Aging* (pp. 133–153). New York: Cambridge University Press.

Light, L., & Capps, J. (1986). Comprehension of pronouns in young and older adults. *Developmental Psychology, 22*, 580–585.

Light, L., Zelinski, E., & Moore, M. (1982). Adult age differences in reasoning from new information. *Journal of Experimental Psychology: Learning, Memory and Cognition, 8*, 435–447.

Linn, R., Wolf, P., Bachman, D., Knoefel, J., Cobb, J., Belanger, A., Kaplan, E., & D'Agnostino, R. (1995). The preclinical phase of probable Alzheimer's disease: A 13-year prospective study of the Framingham cohort. *Archives of Neurology, 52*, 485–490.

Madden, D., Turkington, T., Provensale, J., Denny, L., Hawk, T., Gottlob, L., & Coleman, R. (1999). Adult age differences in the functional neuroanatomy of verbal recognition memory. *Human Brain Mapping, 7*, 115–135.

Masur, D., Sliwinski, M., Lipton, R., Blau, A., & Crystal, A. (1994). Neuropsychological prediction of dementia and the absence of dementia in healthy elderly persons. *Neurology, 44*, 1427–1432.

McClelland, J. L., McNaughton, B. L., & O'Reilly, R. C. (1995). Why there are complementary learning systems in the hippocampus and neocortex: Insights from the successes and failures of connectionist models of learning and memory. *Psychological Review, 102*, 419–457.

McIntyre, J., & Craik, F. (1987). Age differences in memory for item and source information. *Canadian Journal of Psychology, 41*, 175–192.

Meyer, J., Kawamura, J., & Terayama, Y. (1994). Cerebral blood flow and metabolism with normal and abnormal aging. In M. Albert & J. Knoefel (Eds.), *Clinical neurology of aging*. New York: Oxford University Press.

Mitchell, D. B. (1993). Implicit and explicit memory for pictures: Multiple views across the life span. In Implicit Memory: New Directions in Cognition, Development, and Neuropsychology, P. Graf & M.E.J. Mason (Eds.). Hillsdale, NJ: Laurence Erlbaum Associates, pp. 171–190.

Morris, J., Storandt, M., McKeel, D., Rubin, E., Price, J., Grant, E., & Berg, L. (1996). Cerebral amyloid deposition and diffuse plaques in "normal" aging: Evidence from presymptomatic and very mild Alzheimer's disease. *Neurology, 46*, 707–719.

Morrison, J., & Hof, P. (1997). Life and death of neurons in the aging brain. *Science, 278*, 412–419.

Murphy, D., DeCarli, C., McIntosh, A., Daly, E., Mentis, M., Horwitz, B., & Rapoport, S. (1996). Sex differences in human brain morphometry and metabolism: An in vivo quantitative magnetic imaging and positron emission tomography study on the effect of aging. *Archives of General Psychiatry, 53*, 585–594.

O'Connor, M., Sieggreen, M., Bachna, K., Kaplan, B., Cermak, L., & Ransil, B. (2000). Long-term retention of transient news events. *Journal of International Neuropsychological Society, 6*, 44–51.

Petersen, R., Jadk, C., Xu, Y., Waring, S., O'Brien, P., Smith, G., Ivnik, R., Tangalos, E., Boeve, B., & Kokemen, E. (2000). Memory and MRI-based hippocampal volumes in aging and AD. *Neurology, 54*, 581–587.

Petersen, R., Smith, G., Waring, S., Ivnik, R., Tangalos, E., & Kokmen, E. (1999). Mild cognitive impairment: Clinical characterization and outcome. *Archives of Neurology, 56*, 303–308.

Poon, L. (1985). Differences in human memory with aging: Nature, causes, and clinical implications. In J. Birren & K. Schaie (Eds.), *Language and skills across adulthood*. New York: Van Nostrand Reinhold.

Prasad, K., Nahreini, P., & Edwards-Prasad, J. (2000). Multiple anti-oxidants in the prevention and treatment of Alzheimer's disease: Analysis of biological rationale. *Clinical Neuropharmacology, 23*(1), 2–13.

Price, J. L., & Morris, J. C. (1999). Tangles and plaques in nondemented aging and "preclinical" AD. Annals of Neurology, Vol. 45 (3), March 1999.

Rabinowitz, J., Ackerman, B., Craik, F., & Hinchey, J. (1982). Aging and metamemory: The roles of relatedness and imagery. *Journal of Gerontology, 37*, 688–695.

Raz, N., Gunning-Dixon, F., Head, D., Dupuis, J., & Acker, J. (1998). Neuroanatomical correlates of cognitive aging: Evidence from structural magnetic resonance imaging. *Neuropsychology, 12*, 95–114.

Sagar, H. J., Sullivan, E. V., Gabrieli, J. D. E., Corkin, S., & Growdon, J. H. (1988). Temporal ordering and short-term memory deficits in Parkinson's disease. *Brain, 111*, 525–539.

Salthouse, T. (1990). Working memory as a processing resource in cognitive aging. *Developmental Review, 10*, 101–124.

Salthouse, T. (1992). *Mechanisms of age-cognition in adulthood*. Hillsdale, NJ: Lawrence Erlbaum.

Salthouse, T., & Litchy, W. (1985). Tests of the neural noise hypothesis of age-related cognitive change. *Journal of Gerontology, 40*, 443–450.

Salthouse, T., Mitchell, D., & Palmon, R. (1989). Effects of adult age and working memory on reasoning and spatial abilities. *Journal of Experimental Psychology: Learning, Memory and Cognition, 15*, 507–516.

Schaie, K. (1990). The optimization of cognitive functioning in old age: Predictions based on cohort-sequential and longitudinal data. In P. Baltes & M. Baltes (Eds.), *Successful aging: Perspectives from the behavioral sciences*. New York: Cambridge University Press.

Schaie, K., & Willis, S. (1991). Adult personality and psychomotor performance: Cross-sectional and longitudinal analyses. *Journal of Gerontology, 46*, 275–281.

Schenk, D., Seubert, P., Lieberg, I., & Wallace, J. (2000). Beta-peptide immunization: A possible new treatment for Alzheimer's disease. *Archives of Neurology, 57*(7), 934–936.

Schmitt, F., Davis, D., Wekstein, D., Smith, C., Ashford, J., & Markesbery, W. (2000). Preclinical AD revisited. Neuropathology of cognitively normal adults. *Neurology, 55*, 370–376.

Shimamura, A. P. (1995). Memory and frontal lobe function. In M. Gazzaniga (Ed.), *The cognitive neurosciences* (pp. 803–813). Cambridge, MA: The MIT Press.

Shimamura, A. P., Janowski, J. S., & Squire, L. R. (1990). Memory for the temporal order of events in patients with frontal lobe lesions and amnesic patients. *Neuropsychologia, 28*, 803–813.

Sliwinski, M., Lipton, R. B., Buschke, H., & Stewart, W. (1996). The effects of preclinical dementia of estimates of normal cognitive functioning in aging. *Journal of Gerontology, 4*, 217–255.

Small, S., Stern, Y., Tang, M., & Mayeux, R. (1999). Selective decline in memory functioning among healthy elderly. *Neurology, 52*, 1392–1396.

Somberg, B., & Salthouse, T. (1982). Divided attention abilities in young and old adults. *Journal of Experimental Psychology: Human Perception and Performance, 8*, 651–663.

Sperling, R. (2000). Functional MRI studies of hippocampal activation in age and Alzheimer's disease. *Neurobiology of Aging, 21*,

Sperling, R., Cocchiarella, A., Bates, J., Rentz, D., Schacter, D., Rosen, B., & Albert, M. (2000). Functional MRI studies of face-name association in healthy elderly and mild AD. *Neurology, 54*(3), 475.

Squire, L. R. (1992). Memory and the hippocampus: A synthesis from findings with rats, monkeys and humans. *Psychological Review, 99*(2), 195–231.

Tierney, M., Szalai, J., Snow, W., & Fisher, R. (1996). The prediction of Alzheimer's disease: The role of patient and informant perceptions of congnitive deficits. *Archives of Neurology, 53*, 423–427.

Verhaeghen, P., Marcoen, A., & Goosens, L. (1993). Facts and fiction about memory and aging: A quantitative integration of research findings. *Journal of Gerontology, 48*, 157–171.

Weintraub, S., Powell, D., & Whitla, D. (1994). Successful cognitive aging: Individual differences among physicians on a computerized test of mental state. *Journal of Geriatric Psychiatry, 28*, 15–34.

Welsh, K., Butters, N., Hughes, J., Mohs, R., & Heyman, A. (1992). Detection and staging of dementia in Alzheimer's disease: Use of the neuropsychological measures developed for the Consortium to Establish a Registry for Alzheimer's disease. *Archives of Neurology, 49*, 448–452.

West, R. (1996). An application of prefrontal cortex function theory to cognitive aging. *Psychological Bulletin, 120*, 272–292.

Zelinski, E., & Burnight, K. (1997). Sixteen-year longitudinal and time lag changes in memory and cognition in older adults. *Psychology and Aging, 12*, 503–513.

CHAPTER 8

The Ontogeny of Wisdom in Its Variations

Deirdre A. Kramer

The concept of wisdom has intrigued the human mind for at least as long as recorded history. In the last two decades, interest in the topic has spread to the realm of psychological inquiry. This chapter outlines tenets of psychological models of wisdom, explores the empirical research on the construct, and elaborates on the processes contributing to its ontogeny. At least two variations of wisdom, each with its own developmental path, emerge in research. One is characterized by "postformal" cognition and is rare, while the other is characterized by cognitive processes engaged in deducing "truths" from experience. At the heart of both variations are the ability to find meaning in life's often adverse experience, openness to experience, generativity, and ego integrity, although these tasks might be enacted at different levels of complexity.

CONCEPTUALIZATION OF WISDOM

Wisdom is generally thought of as excellent judgment about human affairs. It is hypothesized to involve distinct cognitive processes and is usually conceived of in multidimensional terms as involving an integration of cognitive, emotional, and behavioral processes.

Cognitive Processes in Wisdom

Two kinds of cognitive process emerge in psychological models of wisdom: (1) insight and (2) awareness of the relativistic, uncertain, and paradoxical nature of human problems. They are not mutually exclusive. With respect to *insight*, Brent and Watson (1980) argued that wisdom involves both breadth and depth of

DEIRDRE A. KRAMER • Department of Psychology, Rutgers, The State University of New Jersey, Piscataway, New Jersey, 08854-8040.

Handbook of Adult Development, edited by J. Demick and C. Andreoletti. Plenum Press, New York, 2002.

insight, which allows the individual to reflect on the immediate situation in order to extract its more abiding meanings. Others describe a transcendence of self that allows for a construction of meaning on a more abstract or universal level (Achenbach & Orwoll, 1991) or the formulation of societal solutions to problems (Clayton, 1982). Still others have proposed that wisdom involves the ability to individuate oneself from the conventional norms that govern adult behavior, and to integrate private, subjective experience with externally defined conventional reality (Labouvie-Vief, 1990; Pascual-Leone, 1990). Each model includes a process of reflecting on the particulars of contextually embedded experience to gain insight into deeper or more encompassing human truths.

The other set of cognitive processes inherent in wisdom is awareness of the relativistic, uncertain, and paradoxical nature of reality. Real-life problems often necessitate a recognition of the inherent limitations of abstract logic; the subjectivity inherent in what constitutes a problem and what would be an adequate solution; the inherently dynamic, contradictory nature of human experience; and the necessity for using a "logic" based on paradox. This form of thinking is described in the philosophical literature on wisdom (Clayton, 1982; Moody, 1983; Taranto, 1989) and has been conceptualized in modern day models of wisdom both as "postformal operations" (e.g., Kramer, 1990a; Labouvie-Vief, 1990; Pascual-Leone, 1983, 1990) and as a mode of information processing (e.g., Dittmann-Kohli & Baltes, 1990).

Why is thinking predicated on uncertainty, change, and contradiction believed to be important to wise judgment? One reason is that human dilemmas are complex, subjective, and constantly changing. The most important emotional and existential dilemmas in life may not lend themselves to linear, rational modes of thinking, but require alternative modes of representation, such as imagery, art, metaphor, and nonlinear "logic." Another reason has to do with the increasing complexity of evolving social structures in a global world. Staudinger (1996) writes:

> According to cultural-historical analyses, knowledge related to the conduct, interpretation, and meaning of life is one of the first borders of knowledge to develop in any human community ... Especially in the early phases of cultural evolution, this collective knowledge (i.e., knowledge shared by members of a human community) becomes manifest in sayings, proverbs, and tales .... With cultural evolution proceeding—that is, with the increasing size and complexity of the human community—the number of proverbs increases and proverbs subsequently become more and more detached from the concrete situations in which they were originally coined. At this later 'stage' of cultural evolution, the key question then becomes when to apply a particular piece of that body of knowledge ... Wisdom then is contained not in the sayings and proverbs themselves, but rather in their insightful application to a given problem situation. (Staudinger, 1996, pp. 282–283)

Thus, wisdom involves exceptional breadth and depth of knowledge about the conditions of life and human affairs and reflective judgment about the application of this knowledge. To exert judgment about when knowledge is applicable in a complex, dynamic human sphere, it is important to reflect on one's subjective standpoint to consider alternative frameworks and to be receptive to alternative modes of representation.

Multidimensionality

A number of models position wisdom at the intersection of multiple psychological dimensions. Research on lay conceptions of wisdom has yielded anywhere from three (Clayton & Birren, 1980) to five or six (Holliday & Chandler, 1986; Sternberg, 1985) dimensions, including such qualities as reflectiveness, emotional understanding, social unobtrusiveness, judiciousness, communication skills, and concern for others. Psychological theory-driven models of wisdom are multidimensional in positing wisdom at the intersection of cognitive, affective, and behavioral dimensions (Birren & Fisher, 1990; Kramer, 1990a; Labouvie-Vief, 1990; Orwoll & Perlmutter, 1990; Pascual-Leone, 1990).

Using an object relations perspective, Kramer (1990a) and Orwoll and Perlmutter (1990) propose that mature development entails the integration of previously repressed emotional and conflictual experience. This makes integration of conflicting representations of both self and other possible. Kramer (1990a) argued that, as a result, the individual is less likely to project unwanted conflictual aspects of the self onto others, allowing for a more genuine, empathic bond with them. The more integrated person is less judgmental, more tolerant, and more accepting of opposing perspectives and of human limitations. In Pascual-Leone's (1990) theory, affect, reason, and action schemas become increasingly integrated over time through a series of structural reorganizations fueled by active exertion of will. The result is an increased capacity to reflect on one's experiences and choices; disembed onself from ingrained ways of thinking, feeling, and acting; connect empathically with the experience of others; and eventually transcend the self to be receptive to "unwilled" experience. Like Pascual-Leone (1990), Achenbach and Orwoll (1991) and Labouvie-Vief (1990) also propose that the often long and arduous process of coming to know oneself and grappling with the inner reaches of emotion eventually leads to some form of *transcendence*, a detached, but encompassing, concern with life itself, or an ability to separate, experience, and own one's emotions as apart from social conventions.

Wisdom is reflected not only in the private realm of thought or affect, but also manifests itself in constructive action. Kramer (1990a) proposes that the wise person is capable of interacting with others in a way that does not put those others on the defensive. Orwoll and Perlmutter (1990) emphasize compassionate action in a fully developed wise person. Pascual-Leone (1990) argues for the importance of active exertion of will, which he defines in part as the ability to counter and change automatic processes in order to achieve greater self-realization and greater concern for core human experience. Each of these involves a behavioral component.

Wisdom as a Cognitive Domain or Collection of Person Attributes?

Wisdom as a Cognitive Domain.   Wisdom has been conceptualized in at least two distinct, but not mutually exclusive, ways: as a cognitive domain or as a conglomerate of personal attributes (that may develop with maturity). Paul Baltes, Freya Dittmann-Kohli, Jacqueline Smith, Ursula Staudinger, and others at the Max Planck Institute in Berlin, Germany have developed a model of wisdom that conceptualizes it as a form of cognitive expertise about the domain of human affairs. By studying wisdom as a form of expertise rather than characterological attributes of people, they allow the possibility that wisdom presents itself in a wide variety of forms, such as the written word, common-sense maxims, religious teachings, and historical documents. Humans represent but one conduit of wisdom (Baltes & Smith, 1990).

A fundamental assumption of the Max Planck model is that wisdom is a manifestation of a pragmatic, crystallized form of intelligence that is selectively maintained in the later years. In contrast to fluid forms of intelligence proposed to be directly dependent on biological structures of the brain, and that show more age-related decline, experiential knowledge is ostensibly not as sensitive to changes in physiological structure (Dittmann-Kohli & Baltes, 1990). With age and experience may come the ability to use the extensive knowledge structures to offset observed losses in overall capacity. Adults who have acquired substantial experience in the pragmatics of life will optimize their functional capacity by relying on complex structures of meaning, selectively maintaining highly rehearsed systems of knowledge that compensate for other cognitive losses.

The Max Planck model characterizes wisdom as exceptional insight, good judgment, or advice about life matters. There are five specific criteria of wisdom: (1) rich factual knowledge about matters of life; (2) procedural knowledge about ways of dealing with life problems; (3) lifespan contextualism; (4) uncertainty in problem definition; and (5) relativism regarding problem solution. The first two criteria are straightforward. The third, lifespan contextualism, involves an awareness of the ways in which the context can influence problems, including sociohistorical and life-phase contexts. Uncertainty of problem definition reflects the ambiguity of human problems, which lend themselves to multiple interpretations depending on individual values, circumstances, and frames of reference. Relativism in problem solution involves the recognition of the inherent unpredictability of outcomes as well as the subjectivity inherent in what would constitute a successful resolution of the problem (Baltes & Smith, 1990; Dittmann-Kohli & Baltes, 1990).

Wisdom at the Intersection of Multiple Dimensions and Human Attributes. Achenbach and Orwoll (1991) defined wisdom as the intersection of two tripartite dimensions. The first dimension contains personality, cognition, and conation and the second the intrapersonal, interpersonal, and transpersonal domains. By cross-referencing these two tripartite dimensions, they derived nine qualities of wisdom (self-development, empathy, self-transcendence on the personality dimension; self-knowledge, understanding, and recognition of limits of knowledge and understanding on the cognition dimension; and integrity, maturity in relationships, and philosophical/spiritual commitments on the conative dimension). They characterize the three dimensions of personality, cognition, and conation as interdependent and wisdom as a quality that synergistically transcends any one or combination of the nine characteristics. In an idealized form, wisdom is characterized by the integration of all these factors. In actuality, the degree of presence of each factor will vary with the personal and historical circumstances of a given individual's life, allowing for variations in the way wisdom is manifest (Achenbach & Orwoll, 1991).

In elaborating the derivatives of this model, Orwoll and Perlmutter (1990) draw from the work of Erikson, Jung, and Kohut to propose that self-development and self-transcendence are at the heart of wisdom development. According to Orwoll and Perlmutter, "Wisdom depends on an unusually integrated personality structure that enables people to transcend personalistic perspectives and embrace collective and universal concerns. We assume, however, that these personality attributes presuppose complementary cognitive development and that wisdom is rare precisely because it entails both exceptional personality and cognitive growth" (Orwoll & Perlmutter, 1990, p. 160). Such exceptional self-development involves achievements such as integrating the ego syntonic and dystonic poles of Erikson's final, integrity vs. despair stage (i.e., acknowledging and experiencing realistic sense of despair without threatening the integrity of the self) and integrating polar opposites within the self, including the "shadow," or dark side of the personality. They propose that such development requires mature, dialectical thinking. The outcome includes empathy, mature humor, and an ability to accept the transience of life.

In conjunction with self-development in the wise individual is self-transcendence (Orwoll & Perlmutter, 1990). Collectively, the two define the wisdom construct. Orwoll and Perlmutter conceptualize self-transcendence as moving beyond a concern with the individual self to consider more collective and universal concerns. It involves recognizing the limits of self (see also Taranto, 1989 and Meacham, 1990). It also entails greater accessibility of unconscious experience as a source of insight, not only about oneself but also about the other and universal experience. Similarly, Pascual-Leone's (1990) model proposes successive stages of integrating multiple and conflicting schemes of thought, emotion, and action to promote increasing levels of active will, self-transcendence, and eventually detached receptivity to and concern for life's offerings.

EMPIRICAL INVESTIGATION OF WISDOM

The Max Planck Studies

The most systematic program of research on the construct of wisdom to date comes from the Max Planck Institute for Education and Human Development. Paul B. Baltes, his associates Jacqueline Smith, Ursula Staudinger, and others have developed a think-aloud technique for measuring wisdom in the domains of life planning and life review. Subjects, recruited both through advertisements and professional sampling strategies, are trained in this procedure and then interviewed about hypothetical dilemmas concerning important life events (related to life planning, life management, or life review). Twelve raters are drawn from carefully screened community volunteers, ten of whom are trained to rate responses on one of the five wisdom criteria (allowing for two raters per criterion) on a scale from 1 to 7. Two untrained raters provide a global wisdom judgment rating. The global wisdom judgments tend to be highly correlated with the findings of the trained raters on the five criteria, supporting the idea that there is a discernible quality "wisdom" that can be reliably identified and assessed. Subjects are also tested on selected measures of fluid and crystallized intelligence. Correlates of wisdom scores, such as amount of verbiage and psychometric intelligence, are statistically covaried in analyses. In addition, Staudinger and Baltes (1997) and Staudinger, Lopez, and Baltes (1997) administered a battery of cognitive and personality measures to assess patterns of relationship with wisdom.

The Max Planck group has tested wisdom in young, middle-aged, and older adults, including community volunteers, wise nominees, and clinical psychologists. They hypothesized that wisdom is selectively maintained in later life and that people working in human relations professions would acquire a great deal of expertise in the domain of human affairs, and thus would show the wisest performance. They also hypothesized that community-residing individuals who are nominated as exemplars of wisdom would show a higher degree of wisdom than non-nominated subjects. Collectively, their results can be summarized as follows:

1. Wise performance is a rare occurrence, evident in approximately 5% of the subjects tested, supporting their contention that wisdom is a form of expertise requiring experience, practice, or complex skills (Baltes & Smith, 1990). Among all groups tested, mean wisdom scores typically average only about 3 of a possible 7 on a given criterion.

2. Wise performance does not show a generalized age trend but rather shows stability (Staudinger, Smith, & Baltes, 1992) or slight growth (Staudinger & Baltes, 1997) on life review measures and relative stability or, occasionally, a slight decrease in average levels of performance on certain life-planning tasks (Baltes & Smith, 1990; Smith, Staudinger, & Baltes, 1994). Older adults are among the top wisdom scorers, suggesting that wisdom is a useful avenue for studying areas of selective competence in later life (Baltes & Smith, 1990).

3. People of all ages tend to reason more wisely about both life-planning and life review dilemmas most relevant to their own age groups (Baltes & Smith, 1990; Smith et al., 1994; Staudinger et al., 1992). This is most pronounced in young (Baltes & Smith, 1990) and older adults (Baltes & Smith, 1990; Staudinger et al., 1992) and on the "rich factual knowledge" criterion (Staudinger et al., 1992).

4. Professional specialization accords an advantage in wise reasoning on a life-planning task (Smith et al., 1994) and a life review task (Staudinger et al., 1992). Even so, the clinical psychologists performed only at a slightly above-average level of wise reasoning, with mean scores just below a 4 ($1-1\frac{1}{2}$ points higher than nonclinicians). Differences between clinical psychologists and controls on the life review task are most pronounced on the criteria of factual knowledge, procedural knowledge, and lifespan contextualism (Staudinger et al., 1992).

5. Among the "top performers" (top 25% of the wisdom scorers), about half were clinical psychologists. A larger proportion of the top-performing clinical psychologists were older than younger adults, whereas there was an approximately equal number of young and older nonpsychologists among the top performers. Thus, Staudinger et al. (1992) concluded that age combined with professional experience may afford people a slight advantage in wisdom when considering top performances. Analyses performed on aggregate scores, however, did not produce significant interactions between age and professional specialization.

6. Wise nominees tend to draw from human service professions or positions of leadership (56%) or have had exceptional life experiences, such as penning an autobiography (44%) or serving as Nazi resistors during the Third Reich (31%) (Staudinger, 1996).

7. Average wisdom-related performance of wise nominees exceeds that of the clinical psychologists (Staudinger, 1996).

To test the interface between the cognitive expertise of wisdom and other forms of intelligence, cognition, and personality, Staudinger et al. (1997) administered an extensive battery of tests. The personality and personality–intelligence interface (in particular, creativity and the ability to consider multiple sources of information in governing one's actions) accounted for the largest percentage of the variance in wisdom. However, wisdom also possessed a substantial amount of unique variance. In the personality domain, openness to experience and psychological mindedness were the strongest predictors of wisdom performance, with personal growth (reflective of Eriksonian development) also contributing to the predictive equation. Wisdom was not associated with social intelligence; nor was it related to cautiousness (indeed, quite the opposite, with its correlation to creativity). Staudinger et al. (1997) concluded that their measure of wisdom taps the personality, in addition to the intelligence, dimension, supporting the multidimensionality of the construct. While both fluid and crystallized intelligence measures were related to wisdom, the crystallized ("experiential") measures showed the stronger relationship.

Other Empirical Studies

There have been relatively few other large-scale studies of wisdom and none as broad-ranging in scope as the Max Planck studies. One exception is an ambitious dissertation by Tracy Lyster from University of Concordia in Montreal. Lyster (1996) measured the five criteria from the Baltes and Associates' model and added two additional wisdom criteria: affect-cognition integration and generativity. Across two to three sessions totaling about 4 to 6 hours of study, Lyster videotaped 78 wise nominees, 78 nominators, and 22 self-referred "wise" people discussing important events and a dilemma from their own lives, and reflecting on themselves, their concept of gender, and their concept of wisdom. Lyster scored these for the seven combined wisdom criteria. The interviewers also rated participants on interpersonal variables and coded the videotapes for emotions expressed. In addition, Lyster administered a wide battery of psychological tests of intelligence, personality, coping style, and paradigm beliefs, as well as a structured personal and occupational history interview.

Like the Max Planck studies, overall performances on the wisdom measure were in the low/average-to-average range, with wise nominees scoring higher than the other two groups. Also mirroring the Max Planck studies, wisdom, as assessed by the quantitative criteria, was related to global impressions of wisdom by raters, suggesting, again, that there is a distinct, perceptible quality of "wisdom," which can be reliably identified and which maps onto existing measures of wisdom. Like the findings of the Max Planck group, Lyster found no relationship between measured wisdom and age. Also mirroring the Max Planck studies, older people were among the top scorers on the wisdom criteria (despite lower scores on the fluid measure of intelligence), leading Lyster to conclude that wisdom shows a more favorable lifespan trajectory than other cognitive domains. However, a great deal of wisdom potential went untapped, even among her "wisest" participants, with the highest score on the wisdom measure being 32 of 63, or 51% of the potential score. Also consistent with the Max Planck group's findings, wise nominees were disproportionately represented in human service professions, in particular, advising professions (ministry, mental health, education).

The strongest predictor of wisdom was the personality dimension *openness to experience*, consistent with Staudinger et al.'s (1997) finding. Among the nominated subjects, "wise" individuals also showed higher dialectical and lower absolute scores on the paradigm belief measure, higher scores on the IQ subtests, greater emotional complexity, more reflection and less avoidance in coping with sadness, lower belief in internal control (despite equal *desire* for control), and less life dissatisfaction (but not greater life satisfaction, per se), despite poorer self-reported health. Wisdom was not related to educational level.

Lyster hypothesized a negative relationship between neuroticism and wisdom and a positive relationship between extraversion and wisdom. However, this hypothesis was not borne out. Sages were no more or less "neurotic" than less wise individuals and spanned the range of extraversion to introversion. Also, wiser individuals did not evince a different pattern of affective expression than less wise individuals. Some wise individuals (especially women) were highly emotionally expressive and some (especially men) showed very little emotional expressiveness. The finding of *less reported life dissatisfaction* (but no more life *satisfaction*) among the wiser individuals suggested that, despite not scoring lower on the neuroticism scale (and hence capable of experiencing anxiety and other negative emotions), wiser people did not succumb to despair. The qualitative analyses supported this contention, suggesting that the wise people did their fair share of struggling with difficult existential issues and integrating negative experiences and emotions, and seemed capable of transforming these experiences into a generally hopeful perspective. They seemed particularly astute at creating meaning and purpose in their experience. These effects held when confounding factors such as educational level were partialled out.

Lyster also asked her subjects to discuss their conceptions of wisdom and its antecedents. Wise people were able to find meaning and import in both positive and negative life experience, and used both for transformative experience. What appeared to differentiate the wisest from their less wise counterparts was the ability to transform negative experience into life-affirming and growth-affirming experiences. "A quality of affect perceived as 'serenity' in which the participants readily expressed both positive and negative emotions with an air of acceptance and tranquility was the most frequently rated emotion for all groups" (Lyster, p. 119). More than half of the wise nominees showed this quality. This finding is supported by other research reporting wise nominees to have achieved greater resolution of the integrity vs. despair crisis than non-nominees and creative nominees (Orwoll & Perlmutter, 1990) and to possess a strong sense of Eriksonian ego integrity (Lofsness, 1994; Rosel, 1988; Thomas, 1991). Acceptance, both of self and of life's imperfections, seems to be a defining feature of wisdom.

In a dissertation on dialectical reasoning, Bacelar (1998), too, found that life events in themselves do not predict dialectical reasoning (in an interview about a life dilemma), but rather the ability to meaningfully engage with and reflect on that experience. Like Lyster (1996) and Staudinger et al. (1997), Bacelar found that *openness to experience* (as assessed in a life-events interview) was the single most predictive factor in producing mature (i.e., dialectical) reasoning. Ardelt (1997) analyzed Haan's Ego Ratings and Block's California 100-item Q-sort ratings to measure the affective, cognitive, and reflective domains of wisdom in 82 women and 39 men ages 58 to 82 from the Berkeley Guidance Study. She found that wisdom, rather than objective circumstances (including physical health), impacted life satisfaction in the later years. In fact, wisdom counteracted a negative influence of age on life satisfaction (i.e., when wisdom is entered into the analysis, the negative relationship between age and life satisfaction lost its significance). As Ardelt points out, while knowledge is at the heart of wisdom, it is not *any knowledge* that defines wisdom: rather, it is the ability to interpret one's experiences in order to determine the significance of facts and events. As the aforementioned studies indicate with a wide variety of samples and measurement techniques, wise people seem adept at constructing meaning from their experiences in such a way as to counter the disappointment and despair that can easily disrupt our lives.

In addition to having achieved greater ego integrity, wise people demonstrate a greater concern with humanitarian issues, seen in the form of caring, generative concern and action (Lyster, 1996; Valdez, 1994) and global perspective (Lyster, 1996; Orwoll & Perlumutter, 1990), suggesting greater mastery of the generativity stage of Erikson's theory. In a rare longitudinal study of wisdom, Wink and Helson (1997) studied 94 women from the Mills College Longitudinal Study and 44 men who were their partners. The women were tested at ages 21, 27, 43, and 52. The male partners participated in the first and last follow-up assessments, at mean ages of 31 and 56, respectively. At all times of measurement, subjects were administered the Occupational Creativity Scale; Adjective Checklist (from which a Practical Wisdom Scale was derived); Loevinger's Sentence Completion Test of ego development; the California Personality Inventory; Q-sort clinical ratings of insight, generativity, and autonomy (Kohut's healthy self-orientation); the Myer-Briggs Scale; and ratings of life satisfaction. At the age 52 follow-up, subjects were also interviewed about wisdom for a Transcendent Wisdom Rating. Transcendent wisdom was defined as "abstract (transcending the personal), insightful (not obvious), and to express key aspects of wisdom, such as a recognition of the complexity and limits of knowledge, an integration of thought and affect, and philosophical/spiritual depth" (Wink & Helson, 1997, p. 6). Practical wisdom was defined primarily in terms of interpersonal skill and interest, along with such qualities as insight, clear thinking, reflectiveness, and tolerance.

Wink and Helson (1997) found nonsignificant correlations between the practical and transcendent measures (at age 52), suggesting that the two represent distinct domains of wisdom-type judgments. Both forms were significantly correlated with ego development, insight, autonomy (i.e., healthy self-orientation), and the psychological mindedness scale. In addition, practical wisdom was related to generativity, mentoring, dominance, and empathy, while transcendent wisdom was correlated with Jungian intuiting-sensing, occupational creativity, and flexibility. Dominance and empathy at age 21 were significantly but modestly correlated with practical wisdom at age 52, while flexibility, psychological-mindedness, and empathy were significantly but modestly correlated with transcendent wisdom at age 52, suggesting to Wink and Helson that the former assesses competency in the interpersonal domain, while the latter assesses competency in the transpersonal realm. Practical wisdom (which, unlike transcendent wisdom, was measured longitudinally) showed longitudinal increase from young to middle age in both men and women. Moreover, those women who were psychotherapists scored at a higher level on both forms of wisdom, with educational level held constant, and their wisdom scores increased more over time than those of nonpsychotherapists. This latter finding supports both the hypothesis by Baltes and associates that expertise in the domain of human affairs *promotes* the development of wisdom (rather than wiser people simply self-selecting into such professions) and Staudinger et al.'s (1992) finding on the combined importance of age and professional specialization.

ONTOGENY OF WISDOM

A growing body of research has explored the development of cognitive processes associated with wisdom. These include (1) postformal thinking and (2) the generation of abstract "truths" to formulate maxims, proverbs, and so forth. The former would tap that border of knowledge Staudinger (1996) refers to as a later evolutionary development, that is, the ability to transcend personal and contextually embedded experience to reflect on and evaluate criteria for applicability of known "truths."

"Postformal" Thinking

"Postformal" models of cognitive development typically posit a shift during adolescence and early adulthood away from dualistic, absolute, or reductive modes of thinking to relativistic modes of thinking characterized by awareness of the subjectivity of knowledge, the inability to know the world directly (unfiltered by constructed knowledge systems), and the realization that everything is in a state of flux (e.g., Basseches, 1980; Kitchener & King, 1981; Kramer, 1989; Labouvie-Vief, 1982; Pascual-Leone, 1983; Perry, 1970; Sinnott, 1984). The models diverge somewhat in their characterization of subsequent shifts in adult thinking. Despite the differences in terminology and operationalization, most allow for a reworking of the extreme subjectivity or relativistic stance that emerges in adolescence and early adulthood, to allow for value formation and commitment in adulthood. I focus on the models of Basseches (1980), Kramer (1989), and Pascual-Leone (1983), who view the developmental shift as one from absolute to relativistic to dialectical thinking.

Conceptually, postformal thinking shares features with three of the Max Planck wisdom criteria: lifespan contextualism, uncertainty in problem definition, and relativism regarding problem resolution. Empirically, Lyster (1996) found a significant correlation between scores on her wisdom interview, which is heavily loaded by the Max Planck wisdom criteria, and the Social Paradigm Belief Inventory (SPBI), which measures relativistic and dialectical beliefs. Consistent with findings by Lyster and the Max Planck group, the most advanced postformal thinking is a rare attainment.

A growing body of research suggests either a positive or a curvilinear relationship between dialectical thinking and age, although, as with any construct, the patterns are complex and vary as a function of any number of variables. Not inconsistent with both Lyster and the Max Planck findings, only a small percentage of individuals show dialectical thinking (between 0% and 5%) or transitional dialectical thinking (0% to 30% depending on the age group studied and exact measures). This cognitive component of wisdom appears to be the most difficult of wisdom-related processes to develop and may constitute just one path of wisdom, most similar to Wink and Helson's transcendent form of wisdom.

Both longitudinal (Haviland & Kramer, 1991; King, Kitchener, Davison, Parker, & Wood, 1983; Perry, 1970) and cross-sectional (Kramer & Melchior, 1990) evidence supports the hypothesis that relativistic thinking emerges during adolescence and continues to develop into early adulthood. Researchers have proposed further development toward dialectical thinking in middle age and beyond (Basseches, 1980; Kramer, 1983; Pascual-Leone, 1983). Basseches (1980) devised a comprehensive system for measuring dialectical assumptions. He interviewed a small sample of freshmen, seniors, and graduate students and found a significant positive relationship between age/education and dialectical beliefs. Despite confounds among age, educational level, and gender, his study was important in paving the way for future work.

At least three studies have systematically explored the relative effects of age and educational level by comparing traditional-aged and nontraditional-aged undergraduates and/or graduates of different levels of educational attainment. The results have been conflicting, leaving open the likelihood that either factor can serve as a facilitating condition for the execution of postformal thinking. Strange and King (1981) found evidence for an effect of education as opposed to age on the Reflective Judgment Interview (which assesses levels of postformal thinking using Perry's model) in college students ranging in age from 18 to 26. Schmidt (1985) found main effects of both age and educational level on Kitchener and King's Reflective Judgment Interview in a sample of 40 traditional-aged college freshmen (i.e., 18-year-olds), 40 traditional juniors (mean age 21 years), and 19 nontraditional-aged freshmen (mean age 21 years). In contrast, using the SPBI developed by Kramer, Kahlbaugh, and Goldston (1992) to measure absolute, relativistic, and dialectical beliefs, Soundranayagam (1996) found evidence for an effect of age but not educational level. She compared 40 traditional first-year students (mean age 18.8), 40 traditional seniors (mean age 21.9), 40 graduate students (mean age 37.6), 9 nontraditional college freshmen (mean age 38.6), and 27 nontraditional seniors (mean age 36.6) on the SPBI. Nontraditional students attained higher scores than traditional-aged students regardless of the educational level; they performed comparably to their age-peer graduate student counterparts. Soundrayanagam's study differed from Strange and King's and Schmidt's in at least three ways: (1) she employed older nontraditional-aged students, including middle-aged adults, allowing for a potentially larger impact of age due to greater heterogeneity of age and use of an age group that is expected to show the most mature form of reasoning; (2) she used a questionnaire rather than interview; and (3) her model was based on a model of dialectical reasoning that predicted development beyond early adulthood.

Kramer and Woodruff (1986) found evidence for more relativistic and dialectical thinking in a representative sample of older adults relative to middle-aged and younger adults, controlling for amount of verbiage produced. However, the content of one of her dilemmas may have differentially (in particular, adversely) affected the performance of middle-aged adults because it dealt with a middle-aged dilemma. In a follow-up, unpublished study with Jacqueline Melchior and

Cheryl Levine, Kramer manipulated the effects of age relevance of content material and explored its interaction with age of subject. We interviewed 160 adults, including 40 high-school students, 40 young adults, 40 middle-aged adults, and 40 older adults about paragraph-long dilemmas varied for age-relevance of content (i.e., an adolescent, young adult, middle-aged, or older adult version). Subjects were probed with nine questions designed to tap relativistic assumptions and four questions designed to tap dialectical assumptions (but note that all were scored on six levels, ranging from preabsolute to dialectical). The subject selection, dilemmas, procedure, and scoring are described in Kramer et al. (1992).

A three-way age (4) by gender (2) by age-relevance of dilemma (4) MANOVA was performed on responses to both the relativistic and dialectical questions. There was a significant main effect of age on the relativistic questions, $F(3,128) = 3.40$, $p = .02$), with significantly higher performance by middle-aged as opposed to older adults. There were also significant age by story interactions on responses to both the relativistic and dialectical questions. Simple effects and post hoc analyses revealed significant effects on the old-age dilemmas for both sets of questions and a near-significant effect on the young adult dilemma for dialectical questions. Older adults performed most poorly on their same-age dilemma (see Figs 8.1 and 8.2). However, they performed at a high level—attaining the highest level in response to the dialectical questions—on the young adult dilemma. Generally, the findings reveal a tendency for poorer performance on dilemmas closer to one's age group than on those further from one's age group. There were no significant differences on the other dilemmas, suggesting context-specific cognitive maintenance with age. As seen in Figs 8.1 and 8.2, however, probing about dialectical assumptions, as opposed tor relativistic ones, results in a more divergent pattern of findings across the dilemmas.

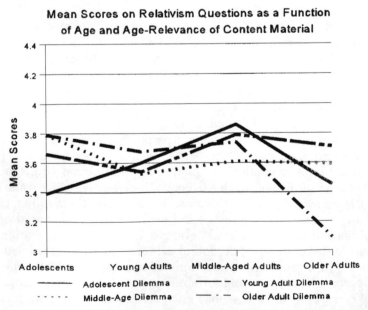

Figure 8.1. Mean scores on relativism questions as a function of age and age-relevance of content material.

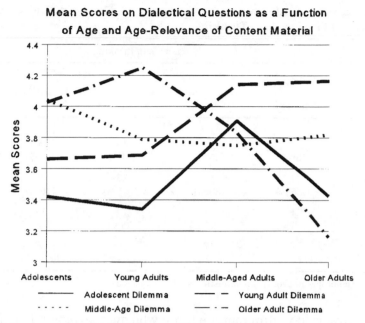

Figure 8.2. Mean scores on dialectical questions as a function of age and age-relevance of content material.

Responses on the SPBI also produced a significant main effect of age, $F(3, 152) = 3.13$, $p < .03$. Post hoc analyses revealed a significant quadratic trend, with middle-aged adults scoring the highest and older adults the lowest (mean scores for each age group were 61.2, 62.8, 63.7, and 60.5 for adolescent, young, middle-aged, and older adults, respectively).

The qualitative data tell a more complex and intriguing story. There was a significant relationship between age and modal qualitative level, $\chi^2 = 26.99$, $p < .01$ (see Table 8.1). Only three subjects in the entire study were coded as being at the highest, prototypical dialectical level, one middle-aged and two older adults; no adolescents or young adults were observed at this level. When the transitional-dialectical level is also counted, about 30% of the middle-aged and 23% of the older adults could be considered dialectical in their thinking, in contrast to 1% and 10% of the adolescents and young adults, respectively. The modal responses for the adolescent, young adult, and middle-aged groups were relativistic, while those for the older adults were transition to relativism. The youngest subjects were most heavily concentrated at the transition to relativism and relativism stages (82.5% and 87.5% of the adolescents and young adults, respectively, vs. 67.5% of the middle-aged and 60% of the older adults). In general, older adults were much more widely dispersed across the five levels observed than any of the other groups, with an equally high number of people at levels 2, 4, and 5, for example. Elsewhere, I have argued that in later life, people may reach a crossroad where, in rejecting extreme relativity, they are faced with at least three choices: (1) revert to formism; (2) remain inconsistent in their beliefs, alternating between formism and relativism (stage 3—transition to relativism); or (3) evolve a dialectical understanding (Kramer, 1990b). More sensitive coding procedures

Table 8.1. Frequency of Adolescents, Young Adults, Middle-Aged Adults, and Older Adults at Each of Five Levels of Thinking

	Levels of thinking				
	2	3	4	5	6
		Transitional		Transitional	
	Formistic	relativistic	Relativistic	dialectical	Dialectical
Age group					
Adolescent	2	14	19	5	0
	(5%)	(35%)	(47.5%)	(1.25%)	(0%)
Young adult	1	12	23	4	0
	(2.5%)	(30%)	(57.5%)	(10%)	(0%)
Middle-aged	1	13	14	11	1
	(2.5%)	(32.5%)	(35%)	(27.5%)	(2.5%)
Older	7	17	7	7	2
	(17.5%)	(42.5%)	(17.5%)	(17.5%)	(5%)

Note: There were no people at level 1, "preformistic," so it is not included here.

need to be developed to differentiate those who once favored relativism and now reject it from those who have never developed it. Longitudinal analyses of such transitions are also important for testing this hypothesis.

These findings are further supported by Bacelar (1998). She found a curvilinear relationship between age and relativistic/dialectical beliefs, using the interview about the middle-aged dilemma used in the Kramer, Melchior, and Levine study. Middle-aged adults showed higher scores than older adults. However, Bacelar found that there were no young adults among those few subjects whose responses were at a transitional dialectical or dialectical level, whereas there was a small percentage of middle-aged and older adults (10% and 6%, respectively) at one of these two levels. Older adults, generally, were more widely dispersed among the six levels of reasoning, mirroring the Kramer, Melchior, and Levine study. Bacelar also found that other variables, such as personal meaning-making and educational level, accounted for a larger share of the variance in cognitive scores than age.

In studies with college-educated mature adults using the SPBI, Kramer and her colleagues have found evidence for less endorsement of absolute beliefs and greater endorsement of dialectical beliefs among middle-aged and older adults relative to younger adults (Kramer & Kahlbaugh, 1994) or greater endorsement of dialectical as opposed to relativistic beliefs in middle-aged and older, as opposed to younger, adults (Kramer et al., 1992). In addition, Kramer and Kahlbaugh (1994) found evidence for greater dialectical processing of memory for prose by older adults as opposed to younger adults. Moreover, both younger and older adults showed poorer overall recall for a dialectical passage than a mechanistic one, despite agreeing more with the dialectical one. Kramer and Kahlbaugh (1994) argued that the later development of dialectical thinking renders it less stable than earlier developing modes of reasoning, resulting in poorer recall for dialectical ideas and the low percentage of subjects who articulate dialectical assumptions in a consistent manner (such as in the interview studies). Less developed thinkers, young or old, may recognize advantages of dialectical arguments, but it is rare for dialectical beliefs to be integrated into a coherent dialectical structure (Bacelar, 1998; Basseches, 1980; Kramer, 1990b). Irwin and Sheese (1989) found that adolescents can understand dialectical principles when explicitly prompted about them, but do not typically articulate them spontaneously.

Older adults show greater inter-individual variability in reasoning level and are overrepresented at both the highest and lowest levels. The *potential* for dialectical thinking exists in later life but is not actualized in the majority of subjects. Fischer and Bidell (1998) proposed that models of developmental growth be construed as dynamic structures that can be modeled as complex "webs." Rather than development proceeding linearly, we are capable of enacting a wide variety of self-organized, interlocking competencies that produce variable outcomes depending on contextual demands. There is a good deal of evidence for maturationally linked changes in the neurophysiological and cognitive systems, that could allow for more complex skills to evolve and be utilized under certain conditions (Fischer & Biddel, 1998). However, especially (though not exclusively) with more complex stages of cognitive processing, where less environmental press for and/or support of such cognitive modes exist, the majority of people do not perform at the most complex level in their competency range. While the *potential* for dialectical thinking may be afforded through maturational changes in neurophysiology, cognition, and personality, as well as through cognitive expertise, the potentiation and activation of such a complex mode of thought is rare. Fischer and Bidell (1998) argue:

> A person carrying out activities does not possess one fixed level of organization. The types and complexities of organizations found in dynamic skills are always changing because (a) people constantly vary their activity systems as they adjust to varying conditions and coparticipants, and (b) people are commonly in the process of reorganizing their skills to deal with new situations, people, and problems. For instance, a tennis player will play at top level one day—after a good night's rest, on an asphalt court, against a well known opponent. The same player will play at a much lower level the next day, with a bad night's sleep, on a clay court, against a new adversary. This reduction in the player's skills level is a real change in the organization of activity. It is not an illusory departure from some 'more real' underlying stage or competence. There is a change in the actual relationships among the participating systems of perception, motor anticipation, motor execution, memory ..., and so on. These relations constitute the true dynamic structure of skill. (Fischer & Bidell, 1998, p. 483)

The "postformal" cognitive processes in psychological models of wisdom may not be strongly nurtured and supported outside of very unique circumstances, such as academia and clinical settings. More mature adults might, under supportive, cued conditions, such as in recognition memory, on preference measures, or on a Likert rating scale, appear dialectical but under more rigorous, uncued production conditions, such as interview or recall memory, show little evidence of dialectical thought or understanding. Similarly, certain affective conditions, such as sadness, might set the stage for more relativistic and/or dialectical understanding, while others, such as anger, might inhibit dialectical or relativistic thinking and promote absolute thinking (Haviland & Kramer, 1991; Wilson, 2000).

Staudinger and Baltes (1997) found evidence for just such a supportive contextual effect on wisdom. They found that dialogue with others whose advice is generally valued and sought—whether in actual social interaction, followed by a brief time to reflect on the discussion, or through internal representations of such dialogues (i.e., imagining what the valued other person would say)—promotes higher wisdom performance than nondigested discussion with the valued person about a dilemma, simply thinking a problem through, or standard individual assessment. The ability to represent multiple perspectives facilitates wise judgment. This effect was most pronounced with older dyads, who tended to outperform younger dyads in one of the facilitative social interactive conditions (interaction plus time for reflection). Top wisdom performers (top 20%) benefitted disproportionately as well from both facilitative interactive conditions (Staudinger & Baltes, 1997).

Thus, the cognitive processes most often advanced in psychological models of wisdom (i.e., postformal-type cognition) appear to be rare attainments, even among wise exemplars, are highly susceptible to contextual variations, and show divergent trends with age. This leaves open the question of what cognitive processes the majority of wise exemplars exhibit. I now turn my attention to a different form of wisdom-related cognition, which may constitute a distinct path of wisdom.

Information Processing Related to Insight

In contrast to the relative rarity of postformal-type wisdom, there are perhaps more readily accessible kinds of information processing that show clearer evidence of age-related growth that are also compatible with wisdom. Such processing involves the abstraction of important truths from personal experience.

Gisela Labouvie-Vief, Cynthia Adams, and their colleagues have found evidence for more abstract, interpretive, metaphorical processing of information in middle-aged and older adults relative to young adults (Adams, 1991; Adams, Smith, Nyquist, & Perlmutter, 1997; Labouvie-Vief & Schell, 1982). Young adults engage in more verbatim, literal processing of prose material. Adams (1991) found that *narrative* texts are more suited to eliciting such advanced forms of processing than *expository* text. This is commensurate with what one would expect of wisdom, which is generally viewed less as a form of academic knowledge and more as an experiential, at times intuitive, form of knowledge that taps natural domains of experience. Bruner (1990) contends that humans engage more frequently in narrative modes of knowing and much less so in hypothetico–deductive forms of knowing or reasoning. One would expect wisdom to develop in the narrative domain.

Labouvie-Vief and her associates found that older adults could engage in a more verbatim, explicit form of encoding and retrieving prose passages if instructed to do so; however, they did not do so spontaneously (Labouvie-Vief & Schell, 1982). Adams (1991) found that younger adults could engage in a more abstract mode of reasoning if asked to summarize the main points; however, they still did not attain the levels of abstraction and metaphorical representation as the mature adults. In a follow-up study, Adams et al. (1997) found that younger and older adults processed information equally deeply but that older adults processed information more *synthetically* and *deeply*, as opposed to *analytically* and deeply. Both young and older adults could be induced to process information in a manner more congruent with the other age group's style of processing when explicitly instructed to do so; however, each group's distinctive strengths suffered somewhat under instructions to counter their usual processing style. They found those conditions less familiar and more difficult.

The researches of Labouvie-Vief and Adams in concert support a stylistic difference between young and older adults in processing contextually meaningful information, suggesting a move toward a more abstract/synthetic style of processing information by older people that synthesizes information into highly abstract, sometimes metaphorical summary statements containing general truths about the human condition. This may allow for the kind of insight described in Brent and Watson's conception of wisdom. Middle-aged and older adults appear to engage more readily in that "border" of knowledge described by Staudinger (1996) having to do with the abstraction of general truths about the human experience (e.g., formulating proverbs and common-sense maxims). Such abstraction might serve the function of aiding the individual in constructing meaning in potentially adverse life circumstances, as found in Lyster's wise exemplars. It would not require that people disembed themselves from their personal experience or contexts to reflect on the truth criteria informing their knowledge systems, as would be required of postformal thinking. The latter is a more complex developmental achievement. The abstraction of truths appears to be more readily available in the cognitive repertoire of most adults, and a particular strength of mature adults; as such, it might constitute one component (i.e., the cognitive component) of a more common variant of wisdom.

The Ontogeny of Affective Processes in Wisdom

The postformal variation of wisdom necessitates, in my view, a mature personality and cognitive structure capable of integrating positive and negative affect, and recognizing inner conflicts (see Haviland and Kramer, 1991; Kramer, 1990a). The more common form of wisdom, as seen, for example, in Wink and Helson's practical wisdom, also necessitates a high degree of emotional maturity in promoting acceptance of self and other, generativity, and ego integrity. There is evidence for greater integration of cognition and emotion with maturity. Adult maturity brings with it a capacity to reason postformally about emotionally laden material (Blanchard-Fields, 1986), to think in a more complex manner about emotions (Labouvie-Vief, Hakim-Larson, DeVoe, & Schoeberlein, 1989), and to regulate one's emotions in a more mature manner, using mature defense mechanisms (Diehl, Coyle, & Labouvie-Vief, 1996; Irion & Blanchard-Fields, 1987; Labouvie-Vief, Hakim-Larson, & Hobart, 1987; Vaillant, 1993). According to Vaillant (1993), with age the ego develops, expanding our capacity to integrate four "lodestars" of conflict: desire, conscience, external reality, and people. It enables one to recognize one's inner conflicts, outer conflicts, thoughts, and emotions—and to integrate these to better understand the source of his or her feelings.

Immature defenses, such as projection and acting out, tend to locate the source of conflict outside the self, while intermediate-level defenses are more apt to locate the source of the conflict within the self but still fail to achieve integration of the four sources of potential conflict. Mature defenses allow for a recognition of both inner conflicts and outer conflicts, along with both the thoughts and the feelings they represent. The goal of mature development, then, is not an absence of conflict or negative emotion, but the ability to recognize and tolerate one's conflicts and emotions. As one set of cognitive processes in wisdom, relativistic and dialectical thinking are both facilitated by, and in turn, facilitate the ability to overcome immature, projective defenses (Haviland & Kramer, 1991; Kramer, 1990a). The wisest of Lyster's subjects showed such integrative capacity. As Joan Erikson (1988) points out, for the virtue of wisdom to develop in later life, one must balance ego integrity with its opposite, despair. A certain amount of despair about the state of the world is realistic and essential for the survival of the species. However, the wise person is not unduly paralyzed, fragmented, or incapacitated by this despair.

One could posit that, for the type of wisdom that involves postformal thought, a high degree of integration between cognition and affect, including both conscious and unconscious processes, is required. The more extant, perhaps "practical," form of wisdom would also involve a capacity for integration, but may not involve as complex an understanding of inner, subjective, and unconscious experience. Both would engender acceptance and ego integration, but the resultant self-structure would be formulated at different levels of complexity.

CONCLUSION

Certain emotional concomitants of psychological models of wisdom, such as generativity, affect-cognition integration, mature defense mechanisms, and coping styles, consistently show a positive trend with age and/or mature development, but the cognitive processes most often proposed in our models seem restricted to a minority of individuals, young or old, generally between 5% and 10%. Empirical research on wisdom and its related processes leads me to conclude that there are at least two forms of wisdom, one characterized by developmentally advanced forms of cognition that are rarely exhibited and that require a great deal of expert and/or buttressing conditions, and one characterized by more readily accessed cognitive abstraction skills, which may characterize the majority of wise exemplars and may also develop with age. Both forms of wisdom appear to involve open-mindedness, acceptance, concern for others, and ego integrity; however, the "postformal" type also involves an ability to reflect on the contents of truth criteria. The avenues for development of particularly the postformal variety of wisdom include such factors as life experience, professional specialization, education, maturation of ego structure, and possibly, neurophysiological changes. Under ideal, supportive conditions, these could result in enhanced performance with age, or, under less supportive conditions, maintenance or loss of cognitive function with age.

One final point: whichever variation wisdom takes, it seems to serve a very significant function of rendering the human experience more coherent and meaningful. This promotes a sense of hopefulness and well-being (see also Kramer, 2000). These are not derived at the expense of emotional expression and experience, but rather encompass a wide variety of emotional experiences, making it possible to use adverse experiences as well as catalyze unwanted emotions as a source of self-reflection and growth.

REFERENCES

Achenbach, W. A., & Orwoll, L. (1991). Becoming wise: A psycho-gerontological interpretation of the Book of Job. *International Journal of Aging and Human Development, 32,* 21–39.

Adams, C. (1991). Qualitative age differences in memory for text: A life-span developmental perspective. *Psychology and Aging, 6*, 323–336.

Adams, C., Smith, M. C., Nyquist, L., & Perlmutter, M. (1997). Adult age-group differences in recall for the literal and interpretive meanings of narrative text. *Journal of Gerontology: Psychological Sciences, 52B*, P187–P195.

Ardelt, M. (1997). Wisdom and life satisfaction in old age. *Journal of Gerontology, 52B*, P15–P27.

Bacelar, W. T. (1998). Age differences in adult cognitive complexity: The role of life experiences and personality. Doctoral Dissertation, Rutgers University, New Brunswick, NJ.

Baltes, P. B., & Smith, J. (1990). Toward a psychology of wisdom and its ontogenesis. In R. J. Sternberg (Ed.), *Wisdom: Its nature, origin, and development* (pp. 87–120). Cambridge, UK: Cambridge University Press.

Basseches, M. (1980). Dialectical schemata: A framework for the empirical study of the development of dialectical thinking. *Human Development, 23*, 400–421.

Birren, J. E., & Fisher, L. M. (1990). The elements of wisdom: Overview and integration. In R. J. Sternberg (Ed.), *Wisdom: Its nature, origins, and development.* Cambridge, UK: Cambridge University Press.

Blanchard-Fields, F. (1986). Reasoning on social dilemmas varying in emotional saliency: An adult developmental perspective. *Psychology and Aging, 1*, 325–333.

Brent, S. B., & Watson, D. (November, 1980). Aging and wisdom: Individual and collective aspects. Paper presented at the 3rd Annual Meeting of the Gerontological Society, in San Diego.

Bruner, J. (1990). *Acts of meaning.* Cambridge, MA: Harvard University Press.

Clayton, F. (1982). Wisdom and intelligence: The nature and function of knowledge in the later years. *International Journal of Aging and Human Development, 15*, 315–321.

Clayton, V., & Birren, J. E. (1980). The development of wisdom across the life-span: A re-examination of an ancient topic. In P. B. Baltes & O. G. Brim, Jr. (Eds.), *Life-span development and behavior, Vol. 3* (pp. 103–135). New York: Academic Press.

Diehl, M., Coyle, N., & Labouvie-Vief, G. (1996). Age and sex differences in strategies of coping and defense across the life span. *Psychology and Aging, 11*, 127–141.

Dittmann-Kohli, F., & Baltes, P. B. (1990). Toward a neofunctionalist conception of adult intellectual development: Wisdom as a prototypical case of intellectual growth. In C. Alexander & E. Langer (Eds.), *Higher stages of human development* (pp. 54–78). New York: Oxford University Press.

Erikson, J. M. (1988). *Wisdom and the senses: The way of creativity.* New York: W. W. Norton.

Fischer, K. W., & Bidell, T. R. (1998). *Dynamic development of psychological structures in action and thought.* In W. Damon & R. M. Lerner (Eds.), *Handbook of child psychology*, Vol. 1: Theoretical Models of Human Development, 5th edit. (pp. 467–561). New York: John Wiley & Sons.

Haviland, J. M., & Kramer, D. A. (1991). Affect-cognition relations in an adolescent diary: The case of Anne Frank. *Human Development, 34*, 143–159.

Holliday, S. G., & Chandler, M. J. (1986). *Wisdom: Explorations in adult, competence.* Basel: Karger.

Irion, J. C., & Blanchard-Fields, F. (1987). A cross-sectional comparison of adaptive coping in adulthood. *Journal of Gerontology, 42*, 502–504.

Irwin, R. R., & Sheese, R. L. (1989). Problems in the proposal for a "stage" of dialectical thinking. In M. L. Commons, J. D. Sinnott, F. A. Richards, & C. Armon (Eds.), *Beyond formal operations 11: Comparisons and applications of adolescent and adult developmental models* (pp. 113–132). New York: Praeger.

King, P. M., Kitchener, K. S., Davison, M. L., Parker, C.A., & Wood, P. K. (1983). The justification of beliefs in young adults: A longitudinal study. *Human Development, 26*, 106–115.

Kitchener, K. S., & King, P. M. (1981). Reflective judgements: Concepts of justification and their relationship to age and education. *Journal of Applied Developmental Psychology, 2*, 89–116.

Kramer, D. A. (1983). Post-formal operations? A need for further conceptualization. *Human Development, 26*, 91–105.

Kramer, D. A. (1989). Development of an awareness of contradiction across the lifespan and the question of post formal operations. In M. L. Commons, J. D. Sinnott, F. A. Richards, & C. Armon (Eds.), *Beyond formal operations II: Comparisons and applications of adolescent and adult developmental models* (pp. 133–159). New York: Praeger.

Kramer, D. A. (1990a). Conceptualizing wisdom: The primacy of affect-cognition relations. In R. J. Sternberg (Ed.), *Wisdom: Its nature, origins, and development* (pp. 279–313). Cambridge, UK: Cambridge University Press.

Kramer, D. A. (1990b). A scoring manual for assessing absolute, relativistic, and dialectical thinking. Unpublished manuscript, Rutgers University, New Brunswick, NJ.

Kramer, D. A. (2000). Wisdom as a classical source of human strength: Conceptualization and empirical inquiry. *Journal of Social and Clinical Psychology, 19*, 83–101.

Kramer, D. A., & Kahlbaugh, P. E. (1994). Memory for a dialectical and a non-dialectical prose passage in young and older adults. *Journal of Adult Development, 1*, 13–26.

Kramer, D. A., Kahlbaugh, P. E., & Goldston, R. B. (1992). A measure of paradigm beliefs about the social world. *Journal of Gerontology: Psychological Sciences, 47*, P180–P189.

Kramer, D. A., & Melchior, J. (1990). Gender, role conflict, and the development of relativistic and dialectical reasoning. *Sex Roles, 23*, 553–575.

Kramer, D. A., & Woodruff, D. S. (1986). Relativistic and dialectical thought in three adult age groups. *Human Development, 29*, 280–290.

Labouvie-Vief, G. (1982). Dynamic development and mature autonomy. A theoretical prologue. *Human Development, 25*, 161–191.

Labouvie-Vief, G. (1990). Wisdom as integrated thought: Historical and developmental perspectives. In R. J. Sternberg (Ed.), *Wisdom: Its nature, origin, and development* (pp. 52–83). Cambridge, UK: Cambridge University Press.

Labouvie-Vief, G., Hakim-Larson, J., DeVoe, M., & Schoeberlein, S. (1989). Emotions and self-regulation: A life-span view. *Human Development, 32*, 279–299.

Labouvie-Vief, G., Hakim-Larson, J., & Hobart, C. J. (1987). Age, ego level, and the life-span development of coping and defense processes. *Psychology and Aging, 2*, 286–293.

Labouvie-Vief, G., & Schell, D. A. (1982). Learning and memory in later life: A developmental view. In B. Wolman & G. Stricker (Eds.), *Handbook of developmental psychology* (pp. 828–846).

Lofsness, J. K. (November, 1994). Application of a model of wisdom to one life's experience. Presented in D. A. Kramer (Organizer), *Wisdom in meaningful life contexts*, symposium presented at the 47th Annual Meetings of the Gerontological Society, in Atlanta, GA.

Lyster, T. L. (1996). A nomination approach to the study of wisdom in old age. Doctoral Dissertation, Concordia University, Montreal, Quebec, Canada.

Meacham, J. A. (1990). The loss of wisdom. In R. J. Sternberg (Ed.), *Wisdom: Its nature, origins, and development* (pp. 191–211). Cambridge, UK: Cambridge University Press.

Moody, H. R. (November, 1983). Wisdom and the search for meaning. Paper presented at the 36th Annual Meetings of the Gerontological Society of America, in San Francisco.

Orwoll, L., & Perlmutter, M. (1990). The study of wise persons: Integrating a personality perspective. In R. J. Sternberg (Ed.), *Wisdom: Its nature, origins, and development* (pp. 160–177). Cambridge, UK: Cambridge University Press.

Pascual-Leone, J. (1983). Growing into human maturity: Toward a metasubjective theory of adult stages. In P. B. Baltes & O. Brim (Eds.), *Life-span development and behavior, Vol. 5* (pp. 117–156). New York: Academic Press.

Pascual-Leone, J. (1990). An essay on wisdom: Toward organismic processes that make it possible. In R. J. Sternberg (Ed.), *Wisdom: Its nature, origins, and development* (pp. 244–278). Cambridge, UK: Cambridge University Press.

Perry, W. G. (1970). *Forms of intellectual and ethical development in the college years: A scheme.* New York: Rinehart & Winston.

Rosel, N. (1988). Clarification and application of Erikson's eighth stage of man. *International Journal of Aging and Human Development, 27*, 11–23.

Schmidt, J. A. (1985). Older and wiser? A longitudinal study of the impact of college on intellectual development. *Journal of College Student Personnel, 26*, 338–394.

Sinnott, J. D. (1984). Postformal reasoning: The relativistic stage. In M. L. Commons, F. A. Richards, & C. Armon (Eds.), *Beyond formal operations: Late adolescent and adult cognitive development* (pp. 298–325). New York: Praeger.

Smith, J., Staudinger, U. M., & Baltes, P. B. (1994). Occupational settings facilitating wisdom-related knowledge: The sample case of clinical psychologists. *Journal of Consulting and Clinical Psychology, 62*, 989–999.

Soundranayagam, L. (1996). The reorganization of sex stereotypes with adult cognitive development. Unpublished dissertation, Rutgers University, New Brunswick, NJ.

Staudinger, U. M. (1996). Wisdom and the social-interactive foundation of the mind. In P. B. Baltes & U. M. Staudinger (Eds.), *Interactive minds: Life-span perspectives on the social foundation of cognition* (pp. 276–315). New York: Cambridge University Press.

Staudinger, U. M., & Baltes, P. B. (1997). Interactive minds: A facilitative setting for wisdom-related performance? *Journal of Personality and Social Psychology, 71*, 746–762.

Staudinger, U. M., Lopez, D. F., & Baltes, P. B. (1997). The psychometric location of wisdom-related performance: Intelligence, personality, and more? *Personality and Social Psychology Bulletin, 23*, 1200–1214.

Staudinger, U. M., Smith, J., & Baltes, P. B. (1992). Wisdom-related knowledge in a life review task: Age differences and the role of professional specialization. *Psychology and Aging, 7*, 271–281.

Sternberg, R. J. (1985). Implicit theories of intelligence, creativity, and wisdom. *Journal of Personality and Social Psychology, 49*, 607–627.

Strange, C. C., & King, P. M. (1981). Intellectual development and its relationship to maturation during the college years. *Journal of Applied Developmental Psychology, 2*, 281–295.

Taranto, M. A. (1989). Facets of wisdom: A theoretical synthesis. *International Journal of Aging and Human Development, 29*, 1–21.

Thomas, E. L. (1991). Dialogues with three religious renunciates and reflections on wisdom and maturity. *International Journal of Aging and Human Development, 32*, 211–227.

Vaillant, G. E. (1993). *The wisdom of the ego.* Cambridge, MA: Harvard University Press.

Valdez, J. M. (1994). Wisdom: A hispanic perspective. *Dissertation Abstracts International, 54–12, Section B*, 6482.

Wilson, P. (2000). The effects of mood-influenced writing on cognitive complexity. Unpublished doctoral dissertation, Rutgers University.

Wink, P., & Helson, R. (1997). Practical and transcendent wisdom: Their nature and some longitudinal findings. *Journal of Adult Development, 4*, 1–15.

CHAPTER 9

Psychological Approaches to Wisdom and Its Development

DOROTHY J. SHEDLOCK AND STEVEN W. CORNELIUS

Wisdom is an age-old concept transcending Western philosophy and modern psychology. Philosophers and theologians have long discussed the topic, and laypeople from all cultures form opinions about it. Recently, however, theorists from multiple disciplines have developed a renewed interest in wisdom. In contemporary psychology, lifespan developmental psychologists have initiated empirical investigations of this phenomenon. The impressive growth in the psychological literature has progressed to multiple theoretical approaches.

Existing psychological treatments of wisdom can be organized into three general approaches: social judgment, personality, and cognitive expertise. The present chapter briefly reviews each of these approaches and discusses their strengths and limitations. It then outlines an integrative model that discusses the construct of wisdom, empirical support for the model, and the development of wisdom.

PSYCHOLOGICAL APPROACHES TO WISDOM

In the social judgment approach, the content and the structure of people's conceptions of wisdom are the central focus (Clayton & Birren, 1980; Holliday & Chandler, 1986; Maciel, Sowarka, Smith, & Baltes, 1992; Orwoll & Perlmutter, 1990; Shedlock & Cornelius, 1995; Sternberg, 1985). It may be noted that a social judgment approach is not unique to wisdom; this natural language method has been used to explore other psychological constructs such as emotions (Shaver, Schwartz, Kirson, & O'Connor), personality (Buss & Craik, 1983), intelligence, and creativity (Sternberg, 1985) as well. This approach provides a foundation for investigation because people use their implicit theories when evaluating themselves or others; its goal is to "provide ... an account that is true with respect to people's

DOROTHY J. SHEDLOCK • Department of Psychology, State University of New York at Oswego, Oswego, New York, 13126.   STEVEN W. CORNELIUS • Department of Human Development, Cornell University, Ithaca, New York, 14853-4401.

Handbook of Adult Development, edited by J. Demick and C. Andreoletti. Plenum Press, New York, 2002.

beliefs" (Sternberg, 1998, p. 348) about wisdom and wise people. Lay conceptions of wisdom derive from people's shared culture and common language. This ideal cultural model is the evaluative standard against which individuals are judged to be wise or not.

Explorations of lay ideas indicate that people largely agree that wisdom is complex, multifaceted in scope, and broad. Findings indicate that wisdom is conceived as comprising desirable and exceptional psychological functioning in many domains, incorporating (1) remarkable practical and interpersonal skills, (2) mature personality, and (3) superior cognitive expertise and intellectual ability (Berg & Sternberg, 1992; Clayton & Birren, 1980; Holliday & Chandler, 1986; Maciel et al., 1992; Shedlock & Cornelius, 1995; Sternberg & Berg, 1987). Wisdom is perceived to be distinct from other psychological constructs such as creativity and intelligence (e.g., Holliday & Chandler, 1986; Sternberg, 1985) and believed to evolve at mature ages (Heckhausen, Dixon, & Baltes, 1989; Orwoll & Perlmutter, 1990; Shedlock, Cornelius, & de Bruyn, 1994). In sum, social judgment research indicates that the construct of wisdom has psychological reality (Maciel et al., 1992) and that descriptions of the prototypical wise person are broader than conceptions offered by personality or cognitive theorists.

The second major approach to wisdom can be derived from theories of personality development. As ego theorists suggest (e.g., Erikson, 1950, 1960, 1980; Erikson, Erikson, & Kivnick, 1986), wisdom culminates a lifetime of personality growth and reflects a fully developed ego:

> [E]ach stage in the life cycle involves the individual in reintegrating, in new, age-appropriate ways, those psychosocial themes that were ascendant in earlier periods. ... in the process of bringing into balance the tension that is now focal. ... these elders are attempting to reconcile the earlier psychosocial themes (generativity and stagnation, intimacy and isolation, identity and identity confusion, and so on) and to integrate them in relation to current, old-age development. (Erikson et al., 1986, pp. 54–55)

The idea of ego maturity has influenced multiple perspective on lifespan personality development and in turn, wisdom. Wisdom is a positive aspect of aging that could function as a developmental goal and stimulate continuous growth and psychological development across the entire lifespan. Personality theorists argue that wise people have exceptionally mature, complex, and integrated personalities (Guttmann, 1978; Haan, 1985; Huyck, 1990; Jung, 1969, 1971; Kramer, 1990; Livson, 1981; Neugarten, 1964; Orwoll & Perlmutter, 1990; Perlmutter, Kaplan, & Nyquist, 1990; Wink & Helson, 1997). The essence of integration entails a wise person's ability to accommodate contradiction; this is reflected in the mediation of opposing universal psychic forces (Freud, 1964; Jung, 1969, 1971), and reconciling and resolving inner conflicts (Loevinger, 1976), or achieving a balance between knowing and doubting (Meacham, 1990).

Personality theorists (especially self-concept theorists) suggest that wisdom develops and becomes especially salient in midlife and beyond. As ego theorists suggest, it evolves from a continuously active and constructing self (Erikson et al., 1986; Pascual-Leone, 1990; Vaillant, 2000). Although potential for personal change may exist, it requires an impetus. This may arise from perceived choice and motivation (Lachman, 1986; Ryff, 1984), self-concept (Cross & Markus, 1991; Markus & Nurius, 1986), or compelling life circumstances (Fiske & Chiriboga, 1985; Ryff, 1989; Thomae, 1983).

Cognitive personality theorists (e.g., Kegan, 1982; Labouvie-Vief, 1990; Pascual-Leone, 1990) suggest that personality and cognitive growth intertwine and are manifest in dialectical or postformal modes of thinking. Some (e.g., Orwoll & Perlmutter, 1990) argue that personality development promotes the required complementary cognitive development necessary for the growth of wisdom. Finally, social cognitive theorists posit that cognitive processes underlying personality and behavior undergird wisdom (Cantor & Kihlstrom, 1985, 1987). Thus, personality theories are themselves diverse, but all view wisdom as advanced personal development. The linking of mature personality with mature intellect in wisdom gives them the same hallmark—the integration and balance of contradiction—which is commonly ascribed to wisdom.

The third approach portrays wisdom as a particular form of cognitive expertise, focusing on knowledge and cognitive style. Paul Baltes and his colleagues in the Berlin Wisdom Project (Baltes & Smith, 1990; Baltes & Staudinger, 1993, 2000; Baltes, Staudinger, Maercker, & Smith, 1995) have developed a program of research in this approach to assess people's performance on a family of criteria. They define wisdom as "expert knowledge involving good judgment and advice in the domain fundamental pragmatics of life" (Baltes & Smith, p. 95). The knowledge base of wisdom encompasses not only an extensive understanding of the life course and human development (factual knowledge), but also a broad array of knowledge about decision-making strategies (procedural knowledge). Any expert knowledge system is assumed to comprise these two types of knowledge. Besides practical life knowledge, metalevel criteria of wisdom are identified by three cognitive styles: (1) Lifespan contextualism involves understanding how life events occur in specific contexts that embody life themes and their relevant developmental relationships. (2) Relativism requires knowledge about individual and cultural differences in goals, values, and priorities; it allows one to separate one's own personal beliefs from others' problems, promotes flexible and unprejudiced values, and acknowledges numerous possible interpretations and solutions to life problems. (3) Uncertainty involves recognizing and possessing strategies to manage potential life outcomes that are relatively indeterminate and unpredictable.

The balance theory of wisdom proposed by Sternberg (1998) also includes knowledge and cognitive style. Sternberg suggests that wisdom knowledge is broader than the domain-specific expertise of practical intelligence. He suggests that the knowledge system underlying wisdom is a form of tacit knowledge. Tacit knowledge is typically acquired through informal experience with little environmental support, and is used to attain personally valued and practically useful goals (Sternberg, Wagner, Williams, & Horvath, 1995). The tacit knowledge of wisdom is used in problem solving to attain an equilibrium between self, other, and contextual interests. This balance among different interests is central to wisdom and sets it apart from social intelligence, which need not consider extrapersonal matters or the greater good. In addition, wisdom entails knowing the limits of one's context-dependent tacit knowledge: The wise know when not to give advice as well as when to give it. Sternberg (1997, 1998) identifies three generic cognitive styles (i.e., executive, legislative, and judicial) in his theory of mental self-government. The wise rely primarily on a judicial (evaluative, judgmental) style, using their knowledge to make balanced decisions about problems in their larger contexts. In short, wise problem solving or advice giving helps one to adapt to, select, or shape environments to maximally benefit all involved. Sternberg (1998) suggests that "The wise individual seeks to resolve ambiguities, whereas the traditionally intelligent person excels in problems that have few or no ambiguities" (p. 362).

STRENGTHS AND LIMITATIONS OF EXISTING APPROACHES

Existing investigations of wisdom each have their respective strengths and limitations. Obviously, no single approach can provide a full account of a psychology of wisdom. Each approach, however, has some unique contributions as well as distinct shortcomings. When considered together, commonalities emerge that point to some central themes that must be considered in any model of wisdom.

There are three major strengths of the social judgment approach. As one might anticipate, people's judgments are broad and paint a picture of a wise person encompassing interpersonal, personality, and cognitive facets of wisdom. Even so, and despite different methodological and analytic techniques, remarkable consensus exists about the attributes of a wise person. Exploring what people think about wisdom and the ideal wise person has helped to verify the psychological construct itself by identifying the scope of the domain, and informing explicit and testable theories. This approach helps to avoid the potential problem of defining the construct too narrowly. Using social judgment as a point of reference also helps to ensure the utility and applicability of empirical explorations of the construct of wisdom. People use their implicit theories of wisdom to evaluate others' and their own capabilities.

On the other hand, a few caveats about social judgment should be noted. First, numerous factors can bias people's schemata. Social psychology typically stresses how contextual factors, expectations, and situational demands help to form, maintain, and reinforce people's beliefs. It may be that the salience and accessibility of characteristics people attribute to wisdom may override their actual importance, that is, not all characteristics are necessary and sufficient for defining wisdom. Second, lay conceptions may be interesting in their own right, but they may not be universally applicable. Although wisdom is inherently social and cultural in conception, there is a dearth of cross-cultural research to determine which wisdom characteristics may be generic or unique to any particular culture. For example, cultural writings suggest that Western societies value intellect and its application while Eastern societies focus more on personal experiential, transcendental, or spiritual knowledge (Takahashi & Bordia, 2000). Third, social judgment research is based on independent (i.e., nonprogrammatic) studies of relatively small samples using data-driven methods of analysis.

Personality theorists have contributed to a psychology of wisdom in several ways. First, they identify the progression of wisdom across the life course, emphasizing wisdom as an endpoint for positive growth in the second half of life. For example, ego level increases with age (Vaillant, 2000) and is associated with social reasoning (Blanchard-Fields, 1986, 1997; Kramer & Woodruff, 1986), problem solving (King, Kitchener, & Davison, 1989), and coping strategies (Labouvie-Vief, Hakim-Larson, & Hobart, 1987). In stipulating the orderly, intrapsychic changes applicable to wisdom across the lifespan, personality approaches make a second, related contribution by pinpointing a developmental goal for individuals as well as cultures. In view of the real and perceived declines of aging, wisdom may seem to be a contradictory goal; at the level of an individual, wisdom's growth may depend on motivation, self-concept, and perceptions of personal control. In the face of declining abilities and resources, older people can age successfully by compensating or concentrating effort on a desired goal at the expense of other, less important ones (e.g., Baltes & Baltes, 1990). Third, these optimistic views of human growth and development provide a counterpoint to negative stereotypes of aging in contemporary society. Indeed, as part of a larger trend in positive psychology (Seligman & Csikszentmihalyi, 2000), multiple concepts about higher levels of human development, especially in mature adulthood, are brought together under the rubric of wisdom. Finally, at a cultural level, wisdom helps to identify the potential of older adults in particular to contribute to the well being of the larger social system.

Personality approaches are commonly criticized for relying on speculative theories and for using constructs deemed somewhat broad and fuzzy. Therefore, a general challenge for personality and self theories is to derive testable hypotheses. Although the breadth of such theories is desirable, this may occur at the expense of specificity about the critical features of wisdom per se, the factors that facilitate or hinder its acquisition, and the consequences of its development for psychological well-being. Second, growth theories of personality generally fail to acknowledge the contribution of enduring differences among individuals' traits, dispositions, and temperaments. The role of stable traits in wisdom has yet to be clarified; various studies have suggested contradictory findings, such as whether extroversion is positively (Shedlock, 1998), negatively (Maciel, Staudinger, Smith, & Baltes, 1991) or not related (Kwon, 1995) to wisdom. Finally, and not surprisingly, personality approaches also neglect to specify the processes involved in transforming experience into knowledge, and translating knowledge into behavioral action in the development of wisdom.

Given the recency of cognitive approaches to wisdom, one might anticipate they have benefited by hindsight to avoid the pitfalls and shortcomings of both social judgment and personality approaches. To some extent, this may be so. First, cognitive approaches have clarified, perhaps more than any other approach, intrinsic facets of the construct of wisdom. The Berlin Group, in particular, has presented a well-grounded rationale for multiple criteria of the construct and devised multiple methods of assessment. Their work integrates wisdom criteria with unique styles of adult thinking (i.e., dialectical, relativistic) highlighted in modern theories of adult cognitive development. In doing so, it provides important insights about developmental differences in thought between adolescents and adults, and in turn, wisdom. Second, cognitive approaches signal the roles of experience and context, and suggest that wisdom is facilitated by extensive practice and informal training in resolving enduring problematic conditions of human life (Smith & Baltes, 1990; Smith, Staudinger, & Baltes, 1994; Staudinger, 1999; Staudinger, Smith, & Baltes, 1992; Sternberg et al., 1995). Finally, cognitive approaches reinforce the "other-centered" focus of wisdom, not as an individual achievement, but rather a process fostering the well being of others and society at large.

There appear to be three chief limitations of cognitive approaches. One challenge for this area of study involves specifying cognitive processes (e.g., How do people use basic cognitive resources such as attention and memory to achieve and maintain necessary domain-specific knowledge of wisdom?) and how wisdom develops (i.e., What is the course of change from novice to expert performance in wisdom?). A second concern is the role of performance factors in efforts to assess competence. For example, a decline in basic processes, especially of memory retrieval, may mask the competence of older adults' performance in empirical assessments of wisdom. Finally, although social and personal factors are highlighted by laypersons' definitions of wisdom, these have garnered relatively little attention in cognitive approaches (however, see Staudinger & Baltes, 1996).

TOWARD AN INTEGRATIVE THEORETICAL MODEL OF WISDOM

Despite their differences, pooling the strengths of the different approaches to wisdom affords a more comprehensive view of the construct. The overlap between approaches allows some generalizations about what wisdom is and how it develops. There approaches appear to agree that the substance of wisdom involves specialized life knowledge about the conditions of human life and its variations, a flexible and unbiased cognitive style, and mature personal integration.

The Construct of Wisdom

Wisdom involves specialized life knowledge. Existing models consider the substance of wisdom to involve the conditions of human life and its variations. In short, accumulated life experience and learning appear to facilitate wisdom knowledge. Lay descriptions emphasize the breadth and content of wisdom knowledge (e.g., "knows about possible conflicts between different life domains" [Maciel et al., 1992]). Wisdom requires social knowledge and is expressed through various personal qualities that enhance human affairs and interaction (e.g., advice giving, communication skills, social competence, and practical problem-solving ability). Furthermore, the wise are assumed to have learned from personal experience. This experience includes both normative tasks at different phases of the life course as well as non-normative life challenges that provide important lessons about variations and diversity in life. Cognitive approaches focus on wisdom as a highly developed domain of knowledge acquired by some people through experience. The domain of knowledge identified by these models is extensive and wide-ranging: knowledge about the self, interpersonal relationships, and larger spheres of existence (e.g., social groups, cultures, world, spirituality, existential meaning). Wisdom knowledge can be acquired through personal experience, observation of others, as well as the contemplation of life course dilemmas that recur across generations.

Wisdom involves a flexible, unbiased, and broad cognitive style. The wise have a sophisticated ability to consider and accommodate seemingly contradictory factors in solving the problems of life that is manifest in their decision-making abilities. The wise person "can see/consider all points of view" (Holliday & Chandler, 1986), "is able to take the long view" (Sternberg, 1985), "is open-minded," and "knows when to talk and when to listen" (Shedlock & Cornelius, 1995). The wise person's abilities to resolve conflicts between competing life domains requires not only knowledge about life but also the ability to maintain a broad perspective without being overly distracted or overwhelmed by details. Personality perspectives conceive of wisdom as personal growth in affective, social, and cognitive realms. In doing so, they suggest that wisdom incorporates the development of advanced modes of emotion, motivation, and self-concept. Personality theorists suggest that contradictions arise in at least two different ways. For example, Erikson focuses on normative developmental issues across the life course that culminate in wisdom. For others, mature personal integration is achieved through the resolution of intrapsychic conflicts. Postformal modes of relativistic and contextual thinking (Kramer, 2000; Labouvie-Vief, 1990) and a balance among competing viewpoints (Sternberg, 1998) are characteristic of wisdom. Thus, it is associated with a tolerant cognitive style and attitude toward life.

A third commonality between approaches is that wisdom reflects mature personal integration and growth. People's implicit ideas underscore the wise person's continued growth and adaptation across the lifespan, enlarged perspective, and motivation to help others and to make the world a better place. Personality development and its resultant wisdom arise from advanced and mature integration of emotion and cognition, the decline of egocentrism, or the personal growth arising from the resolution of normative developmental or intrapsychic challenges. Cognitive approaches describe the personal growth and integration of wisdom as being facilitated by highly developed extensive knowledge about self, others, and the larger existence, as well as extensive practice and experience with important life problems.

Empirical Support for an Integrative Model

Empirical study supports the utility of all three approaches for a conceptual model of wisdom (Shedlock, 1998; Shedlock & Cornelius, 2000a, 2000b, 2000c). Research employing measures that assessed wisdom from each of the approaches (subjective social judgments, personality, and objective wisdom performance) yielded theoretically consistent relationships between wisdom and its hypothetical correlates (experiential factors, personality, and cognitive abilities). In our work, young, middle-aged, and older adults ($N = 105$) completed multiple measures assessing wisdom and related constructs so that their associations could be examined. To explore an empirically grounded, integrative model of wisdom, numerous microlevel indicators from separate domains were condensed to investigate their relation to wisdom. This involved creating domain scores from separate variables to represent general concepts in the model.[1]

Three scores represented the major psychological approaches to wisdom. Social judgments by others were obtained from ratings by a friend of the focal respondent on quality of advice giving and prototypic descriptors of wisdom derived from social judgment research (Shedlock & Cornelius, 1995). Self-perceived personality assessments were measured by the focal respondents' ratings of themselves on the same prototypic descriptors. A cognitive assessment of objective wisdom performance was measured by a newly developed wisdom inventory (Shedlock, 1998) assessing people's performance on the Berlin Group's five criteria: (1) factual knowledge about lifespan development; (2) procedural knowledge about life dilemmas; and (3) attitudinal assessments of contextualism, relativism, and uncertainty. Domain scores were created by summing the standardized scores of friends' ratings to represent the social judgment of others; using standardized scores of self-ratings for self-perceived wisdom, and summing participants' standardized scores on factual knowledge, procedural knowledge, contextualism, relativism, and uncertainty for objective wisdom performance.

Experiential factors are assumed to positively influence the development of wisdom. The composite score for this domain was created from several qualitative categorical variables derived from participants' written responses to a question asking them to explain any past life events or experiences they felt "played an important role in forming the person [they] are today." The life experience variables were selected to focus on a smaller set of nonredundant variables most predictive of wisdom in the present study. These included the sum of reported experiences that (1) occurred in early life (before adulthood), (2) were of an interpersonal nature (either with one other person or in a group setting such as school), (3) had a negative valence, (4) discussed significant life changes, and (5) involved the family of origin.

A domain score to represent personality correlates included the developmental dimension of Generativity and Integrity (Ryff & Heincke, 1983) and Big Five personality traits (NEO-AC Short Form, Costa & McCrae, 1992). Standardized scores for each personality indicator were combined according to their expected theoretical relationships with wisdom: Generativity and Integrity received positive weights, Neuroticism was negatively weighted, and Extraversion, Openness, Agreeableness, and Conscientiousness were given positive unit weights in creating a personality domain score.

A cognitive abilities domain score combined the standardized scores on fluid (ADEPT Letter Series Test, Blieszner, Willis, & Baltes, 1981; Matrices Relations Test, Plemons, Willis, & Baltes, 1978), and crystallized ability tests (Verbal Comprehension Test, Grimsley, Ruch, Warren, & Ford, 1957). The social and practical intelligence domain score combined standardized scores on social intelligence (selected subscales from the George Washington Social Intelligence Inventory, Moss, Hunt, Omwake, & Woodward, 1949) and practical intelligence tests (Comprehension subtest of WAIS-RC, Wechsler, 1981).

Table 9.1 shows correlations among conceptual domains and supports the utility of social judgment, personality, and cognitive approaches in the assessment of wisdom. Three main findings emerged. First, these approaches appear to complement one another: They do show overlap, but each approach also contains unique information. Social judgment is related to personality self-assessment of wisdom, which in turn is related to cognitive performance. Second, life experience and personality are positively correlated with all three approaches to wisdom. Third, both traditional cognitive abilities and social and practical intelligence are positively correlated with the performance measure of wisdom, but not with judgments by others or the self. Multiple correlations were computed between each approach to wisdom and the experiential, personality, and intellectual domain variables to summarize their interrelations: (1) for Friend's Social Judgment, $R = .41$, $F (4, 100) = 5.02$, $p < .001$; (2) for Self Perception of Wisdom,

Table 9.1. Correlations Among Conceptual Domains in the Wisdom Model

Variable domain	1	2	3	4	5	6
1. Friend's judgment						
2. Self perception	.24**					
3. Wisdom performance	.04	.29***				
4. Life experience	.20*	.17*	.35***			
5. Personality	.39***	.61***	.38***	.20*		
6. Cognitive abilities	.10	−.09	.32***	.02	.11	
7. Social–practical intelligence	−.03	−.02	.48***	.26**	.08	.60***

$*p<.10; **p<.05; ***p<.01.$

$R = .63$, F (4, 100) = 16.56, $p < .001$; and (3) for Wisdom Performance, $R = .61$, F (4, 100) = 15.19, $p < .001$. Overall, these findings are encouraging because they show that social judgment, self-perception, and objective wisdom performance are related to each other and to other conceptual domains in theoretically meaningful ways.

It is interesting that social judgments of wisdom seem largely based on individuals' personalities. Personality, because it correlates with all three approaches to wisdom, appears to bind them together. Why does personality appear to be such a strong influence? It may be that people experience difficulty trying to distinguish wisdom-related properties from perceptions of other aspects of themselves, and thus they may overestimate the contribution of personality relative to its actual manifestation. Similarly, one may question whether social judgments of wisdom rely too much on the focal person's personality. As noted earlier, one potential shortcoming of the social judgment approach is that it fails to discriminate those descriptors of a wise person that are central and necessary from more peripheral personal qualities. It should be noted, however, that personality shows about the same magnitude of relationship to wisdom performance as it does to friends' judgments. It would seem then, even from a conservative viewpoint, that personality is a key correlate of wisdom however it is assessed (see Staudinger, Lopez, & Baltes, 1997; Sternberg, 1998).

Given the preference in modern psychology for objective assessments, some might question the utility of social judgment or self-perceptions in contrast to performance. Such criticism could be answered in at least two ways. On the one hand, these phenomena are interesting in their own right (Sternberg, 1998), and certainly judgments of one's own and others' wisdom long preceded objective assessment procedures favored in contemporary psychology. On the other hand, performance measures are themselves not without limitations. Objective assessments are a relatively small sample of behavior in comparison to the information on which people's judgments are typically based and they use hypothetical vignettes as substitutes for wisdom displayed in ordinary life. This leads to the issue of whether wisdom is a trait or domain-specific knowledge, as seems to be the case for other types of expertise (Ceci, 1996). Thus, each approach to wisdom has benefits and limitations, and more accurate assessment may be revealed in the intersection of these approaches rather than in any single one. Accordingly, a single global wisdom score based on all three approaches was created. It was significantly correlated (all p's $< .001$) with scores for the domains of Life Experience

($r = .35$), Personality ($r = .67$), and Social-Practical Intelligence ($r = .21$) but not with traditional Cognitive Abilities ($r = .13$, $p > .20$).

The Development of Wisdom

One might expect to find consistent evidence for the development of wisdom in adulthood—perhaps reaching a zenith in late life—given long-established beliefs about wisdom and aging. However, despite the number of books, chapters, and articles devoted to this topic (e.g., Baltes & Staudinger, 2000; Clayton & Birren, 1980; Kramer, 2000; Labouvie-Vief, 1990; Staudinger, 1999; Sternberg, 1990) there is scant empirical evidence showing durable age-related increases in wisdom. For example, the Berlin Group has reported age-related increases in performance on life review tasks featuring an older adult target (Staudinger et al., 1992), no difference between age groups in performance (Staudinger, 1999; Staudinger & Baltes, 1996), and age-related declines in performance on life-planning tasks featuring a young adult target (Smith & Baltes, 1990). Likewise, Kwon (1995), in a study of three-generation Korean families, found age-related increases within each generation of young adult children, middle-aged parents, and their elderly grandparents but a negative trend in performance across these generations. Shedlock (1998) reported that friends' judgments of wisdom were not related to age, self-perceptions showed a slight positive trend with age, and wisdom performance showed different age relationships for different criteria. This lack of consistency suggests, as would be expected, that wisdom is a complex phenomenon sensitive to a number of contextual influences and modulating factors (e.g., see Baltes & Staudinger, 2000; Staudinger, 1999; Sternberg, 1998). Based on current theory and evidence, these factors may be grouped into experiential, personality, and cognitive domains.

In considering experience, it may be useful to distinguish between an individual's personal experience and his or her vicarious, or indirect, experiences. In a developmental context, the experience of personally undergoing normative developmental tasks would seem to be one avenue for the acquisition of knowledge about important dilemmas encountered by people across the typical life course. Such experience certainly would contribute to knowledge about the grist of lifespan development—the what, when, and how of development (cf., Baltes & Smith, 1990). In addition, non-normative life experiences may be especially salient because they provide teachable moments about life events for which there may be few norms. In this same vein, an appreciation of the common and the unique qualities not only of one's life, but also of one's cohort, would seem conducive to an understanding of the relativity of one's knowledge. Of course, personal experience is likely to be biased in a myriad of ways, but exposure to others' experiences can provide a broader basis for learning about human development. Such vicarious experiences could be gained through familial socialization, being mentored, or by specialization in professions that afford opportunities for acquiring wisdom-related knowledge (e.g., various helping professions). These, too, may contribute to expanding one's knowledge not only about the common problems of lifespan development, but also, and perhaps more importantly, about diversity and multiple life course trajectories (e.g., person and contextual factors that affect the resolution of life crises).

Personality, of course, can be linked to life experience, depending on how a person resolves normative tasks. Erikson suggested that successful resolution of current psychosocial goals ensures the stable and secure development of the ego in future tasks. Certainly one would not expect an isolated, stagnating, despairing adult to manifest wisdom. In addition to developmental facets of personality, relatively stable traits may also influence whether or not individuals seek a variety of experiences, are open to being influenced by them, and accept the inevitable "grayness" of life decisions. Some individuals may be motivated to pursue opportunities for learning that foster the development of wisdom, but others may avoid or not value them. If wisdom necessitates consideration of the common good (Baltes & Staudinger, 2000; Sternberg, 1998) we would expect that this value orientation would be most likely be acquired in cultures, societies, or families where individuals are socialized or reinforced for a collective ethic. Traits, motivation, and values probably become intertwined in the development of a judicious cognitive style that is flexible, fair-minded, and tolerant.

The role of experience and personality in the development of wisdom presumes that people actively construct their knowledge systems. From a psychological viewpoint, basic intellectual abilities are necessary for learning and memory and thus form a foundation for wisdom. This is perhaps most apparent at the early and late periods of the lifespan. In addition, it is possible that basic abilities (e.g., fluid and crystallized intelligence) are necessary but not sufficient for social and practical intelligence to be acquired. What would seem to distinguish these domains from wisdom is its novelty. For example, if we construe social intelligence as culturally constructed solutions (i.e., norms and rules) to life problems, wisdom is a kind of knowledge that has yet to be incorporated into the culture and become a target of cultural socialization. Thus, we might envision a developmental sequence in which basic abilities undergird social-practical intelligence, and are built on to become a more specialized knowledge domain of practical wisdom and perhaps at a penultimate level, philosophical wisdom.

CONCLUSION

Because wisdom is a complex multifaceted phenomenon, it is surprising the degree of consensus that has been reached about it in psychological theory and research. Instead of viewing social judgment, personality, and cognitive approaches as competing viewpoints, it seems more useful to cast them as complementary facets—showing commonalities as well as unique differences. The overlapping components of the construct entail wisdom as life knowledge, as a cognitive style, and as a mature personal integration of cognition, emotion, and motivation. At least for the foreseeable future, the combination of theses approaches is necessary to prevent premature foreclosure on a theoretical conception or mode of assessment that underrepresents the scope of wisdom. Indeed, the empirical findings reviewed in this chapter attest to the utility of all three approaches.

In the continued quest for understanding wisdom, one of the most compelling challenges will be to articulate why few achieve it, how those few do acquire it, and what hinders or prevents the majority of people from becoming wise. At the outset, we might speculate that simple additive models are insufficient to capture the interactive effects of person, process, and context (Bronfenbrenner & Morris, 1998). As Erikson proposed:

> It is through this last stage that the life cycle weaves back on itself in its entirety. ... With their special perspective on the life cycle, these [wise] individuals... serve as guides for the future of those who follow, at the same time that they must struggle to find guides on whom they can rely in considering their own futures and that of our world as a whole. (Erikson et al., 1986, pp. 55–56)

Wisdom appears to develop across the lifespan in ever-widening spheres that encompass the self, others, and the greater good. In this way, wise individuals pass on to younger generations the knowledge and values they have acquired in understanding the inevitable dilemmas, conflicts, and crises of the life course. Finally, we should acknowledge that existing work primarily focuses on practical wisdom, perhaps because it is more amenable to empirical assessment. As some (Pascual-Leone, 1990) have attempted, however, future work will need to flesh out the transcendent quality of wisdom (perhaps what is sometimes labeled philosophical wisdom) if we are to understand the full richness and essence of this age-old construct.

[1] From a statistical viewpoint, this approach had much to recommend it because the ratio of cases to variables in the present study was less than desirable to conduct inclusive analyses. Had there been enough subjects, the effects of each conceptual domain could have been estimated by entering the microlevel variables as blocks for each conceptual domain. In regression analyses, the key is to identify the direction of relationship between predictor and outcome variables (i.e., positive, negative, or neutral); models using unit weights predict about as well as those with specific regression coefficients. Thus, scores for each conceptual domain were created based on their predicted relationship to wisdom. Domain scores were determined by creating standardized scores and then combining the individual variables comprising each conceptual domain of the proposed wisdom model.

REFERENCES

Baltes, P. B., & Baltes, M. M. (1990). Psychological perspectives on successful aging: The model of selective optimization and compensation. In P. B. Baltes & M. M. Baltes (Eds.), *Successful aging: Perspectives from the behavioral sciences* (pp. 1–34). Cambridge, UK: Cambridge University Press.

Baltes, P. B., & Smith, J. (1990). Toward a psychology of wisdom and its ontogenesis. In R. J. Sternberg (Ed.), *Wisdom: Its nature, origins, and development* (pp. 87–120). New York: Cambridge University Press.

Baltes, P. B., & Staudinger, U. M. (1993). The search for a psychology of wisdom. *Current Directions in Psychological Science, 2*(3), 75–80.

Baltes, P. B., & Staudinger, U. M. (2000). Wisdom: A metaheuristic (pragmatic) to orchestrate mind and virtue toward excellence. *American Psychologist, 55*(1), 122–136.

Baltes, P. B., Staudinger, U. M., Maercker, A., & Smith, J. (1995). People nominated as wise: A comparative study of wisdom-related knowledge. *Psychology and Aging, 10*(2), 155–166.

Berg, C. A., & Sternberg, R. J. (1992). Adults' conceptions of intelligence across the adult life span. *Psychology and Aging, 7*, 221–231.

Blanchard-Fields, F. (1986). Reasoning on social dilemmas varying in emotional saliency: An adult developmental perspective. *Psychology and Aging, 1*, 325–333.

Blanchard-Fields, F. (1997). The role of emotion in social cognition across the adult lifespan. In K. W. Schaie & M. P. Lawton (Eds.), *Annual Review of Gerontology and Geriatrics, 17*, 325–352. New York: Springer.

Blieszner, R., Willis, S. L., & Baltes, P. B. (1981). Training research in aging on the fluid ability of inductive reasoning. *Journal of Applied Developmental Psychology, 2*, 247–265.

Bronfenbrenner, U., & Morris, P. A. (1998). The ecology of developmental processes. In R. M. Lerner (Ed.), *Handbook of child psychology, Vol. 1: Theory.* New York: John Wiley & Sons.

Buss, D. M., & Craik, K. H. (1983). The act frequency approach to personality. *Psychological Review, 90*, 105–126.

Cantor, N., & Kihlstrom, J. F. (1985). Social intelligence: The cognitive bases of personality. *Review of Personality and Social Psychology, 6*, 2–32.

Cantor, N., & Kihlstrom, J. F. (1987). *Personality and social intelligence.* Englewood Cliffs, NJ: Prentice Hall.

Ceci, S. J. (1996). *On intelligence: A bioecological treatise on intellectual development.* (Expanded edition). Cambridge, MA: Harvard University Press.

Clayton, V. P., & Birren, J. E. (1980). The development of wisdom across the life span: A reexamination of an old topic. In P. B. Baltes & O. G. Brim (Eds.), *Life-span development and behavior, Vol. 3* (pp. 103–135). New York: Academic Press.

Costa, P. C., & McCrae, R. R. (1992). *NEO PI-R Professional Manual.* Odessa, FL: Psychological Assessment Resources.

Cross, J. F., & Markus, H. (1991). Possible selves across the life span. *Human Development, 34*, 230–255.

Erikson, E. H. (1950). *Childhood and society.* New York: W. W. Norton.

Erikson, E. H. (1960). *Youth and the life cycle, Vol. 7* (pp. 43–49). The Children's Bureau, US Department of Health, Education, and Welfare.

Erikson, E. H. (1980). *Identity and the life cycle.* New York: W. W. Norton.

Erikson, E. H., Erikson, J. M., & Kivnick, H. G. (1986). *Vital involvement in old age.* New York: W. W. Norton.

Fiske, M., & Chiriboga, D. S. (1985). The interweave of societal and personal change in adulthood. In J. Munnichs, P. Mussen, E. Olbrich, & P. G. Coleman (Eds.), *Life-span change changes in gerontological perspective* (pp. 177–209). New York: Academic Press.

Freud, S. (1964). *Collected works, standard edition.* London: Hogarth Press.

Grimsley, G., Ruch, F. L., Warren, N. D., & Ford, J. S. (1957). *Employee aptitude survey.* Los Angeles: Psychological Services.

Guttmann, D. (1978). *Reclaimed powers: Toward a new psychology of men and women in later life.* New York: Basic Books.

Haan, N. (1985). Common personality dimensions or common organization across the lifespan? In J. Munnichs, P. Mussen, E. Olbrich, & P. G. Coleman (Eds.), *Life-span change changes in gerontological perspective.* New York: Academic Press.

Heckhausen, J., Dixon, R. A., & Baltes, P. B. (1989). Gains and losses in development throughout adulthood as perceived by different age groups. *Developmental Psychology, 25*, 109–121.

Holliday, S., & Chandler, M. (1986). *Wisdom: Explorations in adult competence.* Basel: Karger.

Huyck, M. H. (1990). Gender differences in aging. In J. E. Birren & K. W. Schaie (Eds.), *Handbook of the psychology of aging*, 3rd edit. (pp. 124–132). San Diego: Academic Press.

Jung, C. G. (1969). *The structure and dynamics of the psyche.* Princeton, NJ: Princeton University Press.

Jung, C. G. (1971). The stages of life. In J. Campbell (Ed.), *The portable Jung.* New York: Penguin Books.

Kegan, R. (1982). *The evolving self.* Cambridge, MA: Harvard University Press.

King, P. M., Kitchener, K. S., & Davison, M. L. (1989). Relationships across developmental domains: A longitudinal study of intellectual, moral, and ego development. In M. L. Commons, J. D. Sinnott, F. A. Richards, & C. Armon (Eds.), *Adult development, Vol. 1: Comparisons and applications of adolescent and adult developmental models.* New York: Praeger

Kramer, D. A. (2000). Wisdom as a classical source of human strength: Conceptualization and empirical inquiry. *Journal of Social and Clinical Psychology, 19*(1), 83–101.

Kramer, D. A., & Woodruff, D. S. (1986). Relativistic and dialectical thought in three adult age groups. *Human Development, 29*, 280–290.

Kramer, D. A. (1990). Conceptualizing wisdom: The primacy of affect-cognition relations. In R. J. Sternberg (Ed.), Wisdom: Its nature, origins, and development (pp. 279–313). New York: Cambridge University Press.

Kwon, Y. K. (1995). Wisdom in Korean families: Its development, correlates, and consequences for life adaptation. Unpublished doctoral dissertation, Cornell University, Ithaca.

Labouvie-Vief, G. (1990). Wisdom as integrated thought: Historical and developmental perspectives. In R. J. Sternberg (Ed.), *Wisdom: Its nature, origins, and development* (pp. 52–85). New York: Cambridge University Press.

Labouvie-Vief, G., Hakim-Larson, J., & Hobart, C. J. (1987). Age, ego level, and the life-span development of coping and defense processes. *Psychology and Aging, 2,* 286–293.

Lachman, M. E. (1986). Locus of control in aging research: A case for multidimensional and domain-specific assessment. *Psychology and Aging, 1,* 34–40.

Livson, F. B. (1981). Paths to psychological health in the middle years: Sex differences. In D. Eichorn, M. Honzik, & P. Mussen (Eds.), *Past and present in middle life* (pp. 183–194). New York: Academic Press.

Loevinger, J. (1976). *Ego development.* San Francisco: Jossey Bass.

Maciel, A. G., Sowarka, D., Smith, J., & Baltes, P. B. (1992). Features of wisdom: Prototypical attributes of wise people. Paper presented at the 100th Annual Meeting of the American Psychological Society, Washington, DC.

Maciel, A. G., Staudinger, U. M., Smith, J., & Baltes, P. B. (1991). Which factors contribute to wisdom: Age, intelligence, or personality? Poster presented at the 99th Annual Meeting of the American Psychological Association, San Francisco.

Markus, H., & Nurius, P. (1986). Possible selves. *American Psychologist, 41,* 954–969.

Meacham, J. A. (1990). The loss of wisdom. In R. J. Sternberg (Ed.), *Wisdom: Its nature, origins, and development* (pp. 181–211). New York: Cambridge University Press.

Moss, F. A., Hunt, T., Omwake, K. T., & Woodward, L. G. (1949). *The Social Intelligence Test, George Washington Series.* Washington, DC: Center for Psychological Service, George Washington University.

Neugarten, B. L. (1964). *Personality in middle and later life.* New York: Atherton.

Orwoll L., & Perlmutter, M. (1990). The study of wise persons: Integrating a personality perspective. In R. J. Sternberg (Ed.), *Wisdom: Its nature, origins, and development* (pp. 160–180). New York: Cambridge University Press.

Pascuale-Leone, J. (1990). An essay on wisdom: Toward organismic processes that make it possible. In R. J. Sternberg (Ed.), *Wisdom: Its nature, origins, and development* (pp. 244–278). New York: Cambridge University Press.

Perlmutter, M., Kaplan, M., & Nyquist, L. (1990). Development of adaptive competence in adulthood. *Human Development, 33,* 185–197.

Plemons, J. K., & Willis, S. L., & Baltes P. B. (1978). Modifiability of fluid intelligence in aging: A short-term longitudinal training approach. *Journal of Gerontology, 33,* 224–231.

Ryff, C. D., & Heincke, S. G. (1983). The subjective organization of personality in adulthood and aging. *Journal of Personality and Social Psychology, 44,* 807–816.

Ryff, C. D. (1984). Personality development from the inside: The subjective experience of change in adulthood and aging. In P. B. Baltes & O. G. Brim (Eds.), *Life-span development and behavior, Vol. 3* (pp. 243–279). New York: Academic Press.

Ryff, C. D. (1989). In the eye of the beholder: Views of psychological well-being among middle-aged and older adults. *Psychology and Aging, 4*(2), 195–210.

Seligman, M. E. P., & Csikszentmihalyi, M. (2000). Positive psychology: An introduction. *American Psychologist, 55*(1), 5–14.

Shaver, P., Schwartz, J., Kirson, D., & O'Connor, C. (1987). Emotion knowledge: Further exploration of a prototype approach. *Journal of Personality and Social Psychology, 52,* 1061–1086.

Shedlock, D. J., Cornelius, S. W., & de Bruyn, E. H. (1994). Implicit theories of wisdom and its development. Poster presented at the Thirteenth Biennial Meetings of the International Society for the Study of Behavioural Development, Amsterdam, The Netherlands.

Shedlock, D. J., & Cornelius, S. W. (1995). Implicit theories of wisdom: Exploration of individual differences in a prototype approach. Poster session presented at the Biennial Meeting of the Society for Research in Child Development, Indianapolis, IN.

Shedlock, D. J. (1998). Wisdom: Assessment, development, and correlates. Unpublished doctoral dissertation, Cornell University, Ithaca.

Shedlock, D. J., & Cornelius, S. W. (March, 2000a). Assessing wisdom. Poster presented at the Eastern Psychological Association Conference, Baltimore, MD.

Shedlock, D. J., & Cornelius, S. W. (April, 2000b). Wisdom: Perceptions and performance. Poster session presented at the Cognitive Aging Conference, Atlanta, GA.

Shedlock, D. J., & Cornelius, S. W. (November, 2000c). Intellectual and personality correlates of wisdom. Poster presented at the Gerontological Society of America's 53rd Annual Scientific Meeting Program, Washington, DC.

Smith J., & Baltes, P. B. (1990). Wisdom-related knowledge: Age/cohort differences in response to life-planning problems. *Developmental Psychology, 26*, 484–505.

Smith, J., Staudinger, U. M., & Baltes, P. B. (1994). Occupational settings facilitative of wisdom-related knowledge: The sample case of clinical psychologists. *Counseling and Clinical Psychology, 62*, 989–1000.

Staudinger, U. M. (1999). Older and wiser? Integrating results on the relationship between age and wisdom-related performance. *International Journal of Behavioral Development, 23*(3), 641–644.

Staudinger, U. M., & Baltes, P. B. (1996). Interactive minds: A facilitative setting for wisdom-related performance? *Journal of Personality and Social Psychology, 71*, 746–762.

Staudinger, U. M., Lopez, D. F., & Baltes, P. B. (1997). The psychometric location of wisdom-related performance: Intelligence, personality, and more? *Personality and Social Psychology Bulletin, 23*, 1200–1214.

Staudinger, U. M., Smith, J., & Baltes, P. B. (1992). Wisdom-related knowledge in a life-review task: Age differences and the role of professional specialization. *Psychology and Aging, 7*, 271–281.

Sternberg, R. J., & Berg, C. A. (1987). What are theories of adult intellectual development made of? In C. Schooler & K. W. Schaie (Eds.), *Cognitive functioning and social structure over the life course* (pp. 2–23). Norwood, NJ: Ablex.

Sternberg, R. A. (1985). Implicit theories of intelligence, creativity, and wisdom. *Journal of Personality and Social Psychology, 49*, 607–627.

Sternberg, R. J. (Ed.) (1990). *Wisdom: Its nature, origins, and development.* New York: Cambridge University Press.

Sternberg, R. J. (1997). *Thinking styles.* New York: Cambridge University Press.

Sternberg, R. J. (1998). A balance theory of wisdom. *Review of General Psychology, 2*, 347–365.

Sternberg, R. J, Wagner, R. K., Williams, W. M., & Horvath, J. A. (1995). Testing common sense. *American Psychologist, 50*, 902–927.

Takahashi, M., & Bordia, P. (2000). The concept of wisdom: A cross-cultural comparison. *International Journal of Psychology, 35*(1), 1–9.

Thomae, H. (1983). Personality and adjustment to aging. In J. E. Birren & K. W. Schaie (Eds.), *Handbook of the psychology of aging*, 3rd edit. San Diego: Academic Press.

Vaillant, G. E. (2000). Adaptive metal mechanisms: Their role in a positive psychology. *American Psychologist, 55*(1), 89–98.

Wechsler, D. (1981). *WAIS-R Manual: Wechsler Adult Intelligence Scale—Revised.* San Antonio, TX: Psychological Corporation.

Wink P., & Helson, R. (1997). Practical and transcendent wisdom: Their nature and some longitudinal findings. *Journal of Adult Development, 4*, 1–15.

CHAPTER 10

Reflective Thinking in Adulthood

Emergence, Development, and Variation

KURT W. FISCHER AND ELLEN PRUYNE

Reflective thinking is a complex form of cognition almost exclusively associated with adulthood and adult development. It was first defined by John Dewey as "active, persistent, and careful consideration of any belief or supposed form of knowledge in the light of the grounds that support it, and the further conclusions to which it tends" (1910/1991, p. 6). The key elements of Dewey's definition—the use of evidence and reasoning, the questioning of knowledge and beliefs, and the active pursuit of justifiable conclusions—constitute the basis for most contemporary theories that address, in whole or in part, the development of reflective thinking.

Reflective thinking depends on the ability to think abstractly and is therefore associated only with the more advanced stages of development. The cumulative evidence from behavior and brain research indicates that the general *capacity* for advanced abstract thinking, which serves as the foundation for reflective thinking, emerges during early adulthood. It is the emergence of this general capacity that establishes an upper limit on the level of independent functioning an individual can potentially achieve in reflective thinking or other domains involving advanced abstract thinking. This upper limit on skill development is termed the *optimal level*.

Whether reflective thinking realizes its optimal level, or even emerges at all, depends entirely on the individual's experience. Reflective thinking is not a preordained ability that automatically emerges at a particular age or stage of development; instead, it is a skill that must be painstakingly constructed like any other skill, in this case by building on the capacity for abstract systems thinking. Similar to other skills, reflective thinking appears to require a significant period of time in

KURT W. FISCHER AND ELLEN PRUYNE • Department of Human Development and Psychology, Harvard Graduate School of Education, Cambridge, Massachusetts, 02138.

Handbook of Adult Development, edited by J. Demick and C. Andreoletti. Plenum Press, New York, 2002.

which to develop, between 6 and 10 years if the necessary optimal level is present. Education has been found to be the most significant factor in this developmental process. There is also some evidence that midlife presents a second chance to construct reflective thinking skill for those who failed to do so during early adulthood.

DYNAMICS OF THE DEVELOPMENT OF REFLECTIVE THINKING IN ADULTHOOD

Development in general, and the development of reflective thinking in particular, is not linear and simply progressive in nature as commonly assumed. Instead, it is nonlinear and dynamic, demonstrating significant variability *across* individuals at any particular age or stage of development and *within* individuals in terms of both cross-domain and intra-domain skill level at any particular time (Fischer & Bidell, 1998; Thelen & Smith, 1998; van Geert, 1991). Thus, two 40-year-old adults will evidence different levels of skill in reflective thinking. Furthermore, each adult observed separately will demonstrate cross-domain differences between level of reflective thinking skill and levels of other skills, as well as intra-domain variations in reflective thinking ability over the course of a single day. This diversity in the shape of developing skills results from the interaction of multiple components, the most important of which are the dynamic properties of change and development within the growing person and the levels of support for skill mastery provided by the external environment.

Nonlinear Dynamic Development and Reflective Thinking

These dynamic and nonlinear properties suggest a constructive web, rather than the oft-cited ladder of developmental stages, as a useful metaphor for understanding and describing emergence, development, and variability in reflective thinking and other skill domains. Each person spins a unique web that accords with the pace, direction, and focus of his or her individual learning and development. Within this web, each strand of skills develops separately from other strands as illustrated in Fig. 10.1, which shows three abstract skill domains characterized by connecting strands—arithmetic concepts, self-concepts in relationships, and reflective thinking (Fischer & Bidell, 1998). Once reflective thinking emerges in the web, skill and knowledge develop along multiple independent strands that represent separate tasks and situations, such as writing essays about challenging issues in social studies class vs. encountering messy family problems at home. Each individual strand shows the developmental range in skill and knowledge that is evidenced by the individual for that particular task and situation given varying levels of contextual support.

What this means is that there is no single level of competence in reflective thinking or any other domain. Instead, in the absence of task intervention or scaffolding by others, individuals show great variation in skill levels in their everyday functioning. Optimal levels are attained only in those infrequent circumstances when environmental conditions provide strong support for high performance. Such conditions—clearly defined tasks, use of familiar materials, memory aids and cues, and the opportunity for practice (Brown & Reeve, 1987; Fischer & Lamborn,

Mathematics   Self in   Reflective Thinking
               Relationships

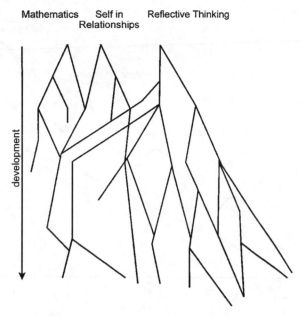

Figure 10.1. A web of development in three domains. Development across diverse contents forms a developmental web (not a unitary ladder), with strands grouping together into domains of related skills, such as the domains of arithmetic, self in relationships, and reflective thinking.

1989)—are not present in most ordinary situations. For this reason, a persistent gap exists between an individual's *functional level* under typical low-support conditions and his or her *optimal level* in the presence of high contextual support.

The gap between functional and optimal level grows with age. As Fig. 10.2 illustrates, functional level tends to be characterized by slow, gradual, and continuous growth over time, whereas optimal level exhibits stagelike spurts and plateaus within an upward trend. These two trend lines diverge because they depend on different sets of growth processes: Functional level results from the steady construction of skill in a particular domain over time, whereas optimal level—the upper limit on functioning—is achieved through strong contextual support for a skill combined with organic growth processes that reorganize behavior and brain activity in recurring growth cycles. In practical terms, *optimal level establishes the limits of the individual's independent capacity for reflective thinking or engagement in other skills with contextual support, while functional level represents the normal level of functioning the individual has attained through engagement in the activities of everyday life without contextual support.*

Research on reflective thinking and related capacities has focused primarily on functional rather than optimal levels (Basseches, 1984; Commons, Trudeau, Stein, Richards, & Krause, 1998; Cook-Greuter, 1999; Dawson, 2002; Kegan, 1982; Kitchener & Fischer, 1990; King & Kitchener, 1994; Kuhn, 1992; Kuhn, Garcia-Mila, Zohar, & Andersen, 1995; Pirttilä-Backman, 1993; Sinnott, 1998). One of the best measures of reflective judgment is the Reflective Judgment Interview (RJI), which is designed explicitly to measure reflective judgment as defined by Dewey (1910/1991, 1933) and Perry (1970)—consideration of what is known in

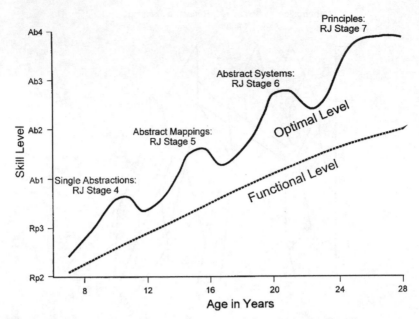

Figure 10.2. Growth curves for optimal and functional levels. Optimal level skill grows in spurts with the emergence of each new skill, while functional level grows more smoothly (or does not show systematic spurts), as shown in this idealized graph.

terms of the evidence and argument that support it as well as conflicting viewpoints and implications. The RJI is an instrument in which ill-structured problems are described from two or more opposing viewpoints, such as whether chemical additives to food are safe and helpful or dangerous and harmful. People are asked to form judgments about these problems and explain the rationale behind their judgments. Because contextual support for optimal performance is not provided consistently to the respondents in the standard interview, this instrument assesses functional level only. A more recent instrument called the Prototypic Reflective Judgment Interview (PRJ), described later in this chapter, has been devised to measure optimal levels that can be directly compared with the RJI's functional levels.

The development of the PRJ was inspired by Rest's work on moral judgment (Kitchener & Fischer, 1990), which itself was a response to the pioneering work of Lawrence Kohlberg. The Moral Judgment Interview (MJI) developed by Kohlberg and his associates poses hypothetical moral dilemmas that represent conflicts between two central values, such as the value of preserving a life vs. the value of upholding the law (Colby & Kohlberg, 1987). Each dilemma is followed by a series of questions focused on this value conflict, which are specifically designed to elucidate the respondent's moral judgment and reasoning. The MJI is a production task in that it places the onus on respondents to produce justifications for their positions and credits them only with ideas and understandings that are explicitly expressed. Hence, it tends to ignore tacit understandings and underestimate the functional level of development achieved (Rest, Narvaez, Bebeau, & Thoma, 1999).

Rest, Deemer, Barnett, and Spickelm's (1986) Defining Issues Test (DIT) was purposefully constructed to address this and other measurement issues of the MJI. The DIT poses the same moral dilemmas but does not expect respondents to independently articulate the critical issues and arguments involved; instead it offers a list of statements that respondents are asked to rate and rank. Rest et al. (1999) suggest that this recognition task may tend to overestimate the functional level of a person's development. Either the DIT or the MJI may be used to assess the functional level of an individual's moral *cognition*, taking these possible over- and underestimations into account. However, neither should be considered an effective measure of the functional level of an individual's moral *behavior*. Many other factors contribute to moral behavior, including situations, specific conceptions of virtue, and concern for others. For example, Arnold's (1993) research with adolescents suggests that conceptions of personal virtue may be better than moral judgment scores in predicting an individual's tendency to actually behave in an ethical manner. Her research with adults indicates that concern with generativity contributes along with moral stage to concern and effectiveness of value socialization of young people (Pratt, Norris, Arnold, & Filyer, 1999).

Cycles of Reconstruction in Building the Advanced Skills of Reflective Thinking

Reflective thinking develops in accordance with the recurring growth cycles that characterize human development despite the fact that it emerges at a relatively advanced "stage" in the developmental sequence. In general, development over the lifespan progresses through a series of 13 levels that begin to emerge shortly after birth and continue until the individual reaches approximately 25 to 30 years of age. The series of levels arise from the operation of two distinct growth cycles—a shorter term cycle nested within a longer term one. In the shorter term cycle, successive skill *levels* are constructed, each marked by a cluster of spurts and drops in optimal-level performance as the new capacity emerges. The levels in turn are grouped into the longer term cycle called a *tier*—single units, mappings of single units to one another, systems of coordinated mappings, and finally principles relating two or more systems (which also serve as single units forming the foundation of the next longer term growth cycle). The four skill levels in a tier together constitute a single form of action or thought, which represents one iteration in the longer term growth cycle. As Fig. 10.3 illustrates, there are (at least) four tiers that emerge over the human lifespan—reflexes in early infancy, actions in the first 2 years of life, representations in the preschool and early school years, and abstractions in late childhood and beyond (Fischer & Bidell, 1998; Fischer & Rose, 1994).

Reflective thinking does not emerge until early adulthood at the earliest, because it is a form of advanced abstract thinking in which abstract systems and principles are constructed from units and mappings of abstract information. It thus occupies a position at the tail end of the progression of skill development— the systems and principles levels of the abstract tier.

This advanced position in no way translates into an easier or faster process of skill development, but to the contrary, high-level abstract skills are slow and hard to build. Each level of reflective thinking requires building new skills by a process

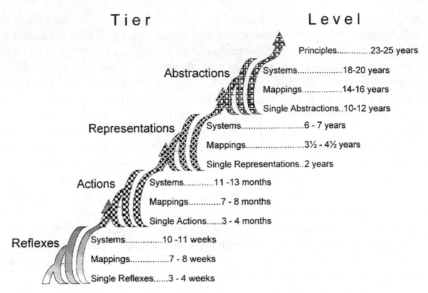

Tier   Level

Principles............23-25 years

Abstractions   Systems..................18-20 years

Mappings...............14-16 years

Single Abstractions..10-12 years

Representations   Systems.......................6 - 7 years

Mappings......................3½ - 4½ years

Single Representations..2 years

Actions   Systems............11 -13 months

Mappings.............7 - 8 months

Single Actions......3 - 4 months

Reflexes   Systems..............10 -11 weeks

Mappings...............7 - 8 weeks

Single Reflexes......3 - 4 weeks

Figure 10.3. Developmental cycles of levels and tiers of skills. Ages indicate approximate times at which a skill level emerges under optimal conditions. (Most skills remain below that level in ordinary functioning.) Tiers are listed on the left, and levels within each tier are listed on the right.

of microdevelopment—regressing back to lower levels and sequentially constructing skills through the action, representation, and abstraction tiers (Fischer, 1980; Granott, 1994, 2002; Granott, Fischer, & Parziale, 2002). A straightforward example is an architectural student regularly confronted with challenging problems in the design of public buildings. First, beginning with the action tier, the student learns to weigh the technical and physical aspects of the problem in reasoning toward a solution. Progressing to the representation tier, the student includes evidence about the symbolic meanings people ascribe to various architectural styles and how these relate to comfort, beauty, and other desired attributes of buildings. The final tier requires bringing abstract issues into consideration, such as the conflicting interests of various parties regarding the building's setting and design and the value systems that underlie these conflicts.

This skill-building process involves not regular progression to more complex skills, but instead repeatedly regressing to low levels in order to build new specific skills and progress through levels in each tier to construct a new high-level skill that can be generalized across tasks and domains. Research studies have delineated several examples of such sequences, including the development of understanding Lego robots by adults unfamiliar with them (Dunbar, 2001; Fischer & Granott, 1995; Fischer & Yan, 2002; Fischer, Yan, McGonigle & Warnett, 2000; Granott, 1994, 2002; Granott, Fischer, & Parziale, 2002; Kuhn, 2002; Kuhn et al., 1995). The process is a repetitive one that requires performing and reconstructing a particular task, typically many times, before it becomes stabilized, as illustrated in Fig. 10.4. Generalizing it to other tasks requires even more reconstruction. When people are allowed to work with this natural cyclical process of repeated reconstruction, they can gradually build broad skills and generalize them. Some

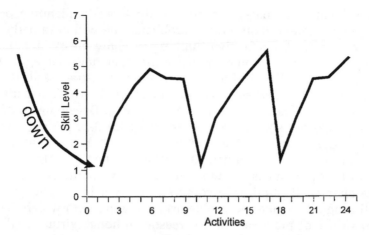

Figure 10.4. Building a new skill: repeated regression and reconstruction. In building a new skill, a person drops down to a low skill level, builds up a more complex, adequate skill, then repeats the process over and over, as shown in this smoothed graph based on the research of Granott (1994, in press).

cognitive science researchers have had difficulty producing generalizable skills in learning studies (Detterman, 1993), but natural learning produces generalization readily through regression and rebuilding (Dunbar, 2001; Fischer & Farrar, 1987; Perkins & Salomon, 1989; Sa, West, & Stanovich, 1999).

For instance, the architectural student will become skillful in thinking about difficult building problems only after working through the process several times. As the common wisdom says, "Practice makes perfect," with skills requiring multiple repetitions before consistent performance can be achieved. The learning process requires this repetition for generalization because at the start of learning, even minor changes in the task produce abrupt drops in skill level (regressions). The process of learning architectural thinking (as well as other kinds of broad skills) begins all over again with each change, as the architect's skill regresses repeatedly to lower levels and each time gradually progresses, level by level, toward advanced ability in a particular task, and eventually a broader domain.

The architectural student will find that having a high level of ability in reflective thinking with an architectural problem does not mean that she has that same ability in messy interpersonal dilemmas, ill-structured problems in a business management class, or decisions about how to vote in a national election. In these areas she will remain at a lower level of functioning unless she engages in activities or efforts that help her construct more advanced skills. If she builds reflective skills with more and more diverse tasks, her ability in reflective thinking will become increasingly stabilized and generalized until eventually she shows similar skills across several tasks and situations (King & Kitchener, 1994; Kitchener, Lynch, Fischer, & Wood, 1993). With this cross-domain consolidation, her reflective thinking skills become integrated into higher order generalized control systems of abstract systems and principles—the highest levels of general reflective thinking. In other words, if the architectural student in fact constructs reflective thinking skills for designing public buildings, and then extends those skills to interpersonal relationships, business management, citizenship, financial planning, or other tasks in which she encounters ill-structured problems, she will eventually find herself with a general ability for reflective thinking in many messy situations that she faces, especially ones with which she is familiar. The development of such reflective thinking illustrates the kind of generalization of abstract skills that is evident in the acquisition of expertise in a domain or field, which usually takes 5 to 10 years (Ericsson & Charness, 1994; Gardner, 1993). Development of new scientific theory also shows a similar process, as in Charles Darwin's development of the theory of evolution (Fischer & Yan, 2002; Gruber, 1981) as well as more ordinary scientific accomplishments (Dunbar, 1997; Klahr, 2000; Kuhn et al., 1995).

This ability is not an easy one to construct or maintain, as will be elaborated later in the chapter. It requires ongoing engagement in weighing evidence and reasoning through ill-structured problems, an activity with which many individuals have insufficient practice. For this reason, although virtually all adults have the capacity for developing reflective thinking, it appears that few attain general proficiency.

CYCLES OF DEVELOPMENT OF REFLECTIVE THINKING: COGNITION, EMOTION, AND BRAIN

Between birth and early adulthood, cognition, emotion, and brain functioning develop through a series of growth spurts that involve reorganization of capacities and neural networks marked by the emergence of new optimal levels in cognitive and emotional functioning (Fischer & Rose, 1994, 1998). In terms of reflective thinking, each reorganization introduces the *capacity* to reason and make conclusions about ill-structured problems at a more complex level, with the actual development of reflective thinking skills, and eventual functional use in everyday life, dependent on application of this capacity to complex knowledge dilemmas so as to construct reflective thinking skills over time.

Optimal Levels of Reflective Thinking and Brain Correlates

Evidence for growth spurts in the brain includes brain activity, synaptic density, and myelin formation (Dawson & Fischer, 1994; Diamond & Hopson, 1998). *Brain activity* has been measured primarily through electroencephalographic (EEG) recordings of electrical activity (brain waves) in the cerebral cortex. These recordings indicate that brain spurts follow growth cycles, as illustrated by the series of spurts in growth of EEG power shown in Fig. 10.5. The cycles seem to be modulated by the frontal cortex and to move consecutively through the cortical sections of first the right and then the left hemisphere of the brain. The cycles involve systematic change in both the power (amount of energy) in the EEG and the connections between cortical regions as measured by similarity in EEG patterns (coherence). The cortical growth cycles seem to correspond to the series of 13 skill-development levels divided into four tiers (reflexes, actions, representations, or abstractions) described in Fig. 10.3.

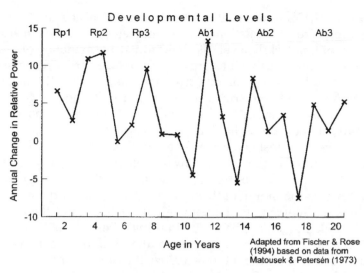

Figure 10.5. Growth spurts in EEG power corresponding with skill levels. Annual change in EEG power shows cycles of spurts and drops, which match ages of emergence of optimal skill levels in cognitive development. EEG power measures are for the alpha frequency band in the occipital–parietal area of the cortex.

Synapses are the structures that transmit electrochemical signals from one neuron to another via axons and dendrites, which are like threads that come together at synapses to send and receive messages. *Synaptic density* is the number of synapses in a specific volume of cortex. Dendrites protrude from the neuron in branchlike shapes to send signals through synapses to an axon on another neuron. In general, the denser the dendritic branches and synaptic connections, the greater the potential for learning and complex behavior, including rapid, facile development and consolidation of skills (Diamond & Hopson, 1998; Huttenlocher, 1994). Living in a richly stimulating environment produces rich growth of dendritic branches, while living in a boring, impoverished one leads to less dendritic growth.

Myelination is a process that significantly improves the transmission of neural signals via the axons that serve as transmission wires between neurons in the brain, and it also contributes to fattening the brain and tripling the brain's weight after birth. Myelin is a white fatty substance that encases and insulates the axons, improving the efficiency and speed of conduction of electrical impulses; its absence results in delayed or disrupted neural transmissions and, as a result, deficits in the speed, facility, and control normally expected in cognitive, emotional, and physical functioning. These three processes—spurts in EEG power and connectivity, dendritic branching and synapse formation, and myelination— seem to provide bases for growth of the upper limit on skill level for reflective thinking and other behaviors—that is, the optimal level of skill.

Kitchener and King's (1981) reflective judgment model effectively describes the optimal levels of reflective thinking. The model's seven stages of reflective judgment relate directly to the seven optimal skill levels that emerge between 2 and 30 years of age (Fischer & Farrar, 1987; Kitchener & Fischer, 1990; Kitchener et al., 1993;

see also Case, 1991; Commons et al., 1998; Gruber & Vonèche, 1976; Sinnott, 1998). Table 10.1 summarizes this relationship and indicates the approximate age of emergence of each optimal skill level, as determined by evidence of cognitive and emotional development, as well as growth of the three brain processes described above. The first three stages of reflective judgment, what Kitchener and King call *prereflective*, are characterized by the belief that knowledge is certain and absolute as well as the failure to use evidence to arrive at conclusions. These stages generally describe preschool and schoolage children, as well as functional (nonoptimal) levels of most adolescents and adults. A common effect of college is to stimulate growth of higher stages toward reflective thinking. In the words of one college student who was resisting the difficulties of disagreements about knowledge, "A certain amount of theory is good, but it should not be dominant in a course. I mean, theory might be convenient for them, but the facts are what's *there.* And I think that *should* be the main thing" (Perry, p. 67).[1]

Stages 4 and 5 are *quasireflective* in that individuals move beyond the absolute approach of this student to recognize the relative and uncertain nature of knowledge with regard to ill-structured problems but continue to experience difficulty in applying evidence to justify beliefs or reason toward a conclusion. These two stages generally characterize the optimal levels of adolescents and the functional levels of many adults. The final two stages are truly *reflective* in the sense that individuals recognize both the relative and constructed nature of knowledge and the need to continually evaluate their own and others' judgments and beliefs in terms of context, argument, and evidence. Stage 6 emerges at the earliest in early adulthood at around age 20 and stage 7 in the mid-20s (King & Kitchener, 1994).

What especially distinguishes reflective thinking is its requirement of active evaluation of knowledge, a process that the college student we quoted was trying to avoid (King & Kitchener, 1994). As shown in Table 10.1, the ability to engage in an active process of knowledge evaluation and construction begins at stages 4 and 5 with the achievement of abstract thinking, or what Jean Piaget (1972) called formal operations, and the further development of abstract mapping skill, the capacity to relate two or more abstract concepts (such as knowledge and evidence) to one another. The first level of reflective thinking skill (stage 6) builds further on this foundation by relating abstract mappings to one another to create an abstract system—for example, knowledge and evidence from sociology combined with knowledge and evidence from political science to form sociopolitical knowledge. The second level of reflective thinking (stage 7), currently believed to be the highest attainable skill level (but see Commons et al., 1998: Cook-Greuter, 1999: and Fischer, Stewart, & Yan, in press), consists of the ability to integrate abstract systems to arrive at general principles, such as two or more theories about human and animal evolution to formulate a general principle for the evolution of species. Thus, reflective thinkers do not take knowledge as given, but instead they evaluate it and then combine it in novel and complex ways to construct and evaluate new ways of understanding and knowing.

The emergence of reflective thinking (when the capacity for thinking about abstract systems has just been established) will normally involve the appearance of discontinuities in optimal levels for skill development in multiple domains

---

[1] For clarity, this quotation has been edited to remove conversational markers of pauses and hesitations.

Table 10.1. Skill Levels and Reflective Judgment Stages

Skill level	Skill structure	Example of behavior	Reflective judgment stage: nature of knowledge and justification	Approximate age at emergence of optimal level (Years)[a]
Level Rp1. Single representation: a concrete instance	$\left[\,A_P\,\right]$	Knowing is limited to concrete instances, such as A_P, "I know there is cereal in the box."	Stage 1. What a person believes to be true is true. No justification is necessary.	2
Level Rp2. Representational mapping: coordination of two representations	$\left[\,A_R \xrightarrow{\text{answers}} A_W\,\right]$ $\left[\,A_R \xrightarrow{\text{authority}} A_X\,\right]$	Right answers about A, A_R, are contrasted with wrong answers about A, A_W. Right answers about A, A_R, are true because of Authority X's answer about A, A_X.	Stage 2. A person can know with certainty either directly or via an authority. Justification is via an authority.	3.5 to 4.5
Level Rp3. Representational system: coordination of two aspects of two representations	$\left[\,C_S^{\,V} \xleftrightarrow{\text{uncertain}} C_T^{\,W}\,\right]$	In areas that are temporarily uncertain, Authorities V and W both know about C, C_V and C_W, and come to different conclusions about it, C_S and C_T.	Stage 3. In some areas knowledge is temporarily uncertain. Justification is based on what feels right at the moment.	6 to 7

(continued)

Table 10.1. continued

Skill level	Skill structure	Example of behavior	Reflective judgment stage: nature of knowledge and justification	Approximate age at emergence of optimal level (Years)[a]
	$\left[A_R^X \xleftrightarrow{\text{certain}} B_R^Y \right]$	In areas known for sure, Authority **X** knows about area A, A_X, and knows the right answer, A_R, and Authority **Y** knows about area B, B_Y, and knows the right answer, B_R.	In other areas, authorities actually know the truth, and knowledge is certain.	
Level Rp4/Ab1. Single abstraction: coordination of two representational systems	$\left[\begin{array}{c} \overset{\textit{uncertainty}}{A_P^X \longleftrightarrow A_Q^Y} \\ \Updownarrow \\ B_P^X \longleftrightarrow B_Q^Y \end{array} \right] \equiv \left[\begin{array}{c} \diagdown \\ \mathbb{K} \end{array} \right]$	Abstract conception that any item of knowledge is uncertain, $\mathbb{K}$: Authority **X** knows about A and B, A_X and B_X, as a function of his or her viewpoint P, A_P and B_P, while Authority **Y** comes to different conclusions about A and B, A_Y and B_Y, as a function of his or her different viewpoint Q, A_Q and B_Q.	Stage 4. Knowledge is generally uncertain because of viewpoints and situations. Simple right and wrong disappear. What we can know and how we justify beliefs is idiosyncratic.	10 to 12
Level Ab2. Abstract mapping: coordination of two abstractions	$\left[\mathbb{K}_F \xleftrightarrow{\textit{context-specific}} \mathbb{J}_F \right]$	Each (uncertain) item of knowledge in a particular context, $\mathbb{K}_F$, can be justified by an argument specific to that context, $\mathbb{J}_F$.	Stage 5. Knowledge is contextual: People know via individual conceptual filters based on viewpoints. Justification is specific to context or situation.	14 to 16

Level	Skill structure		Stage	Age[a]
Level Ab3. Abstract system coordination of two aspects of two abstractions	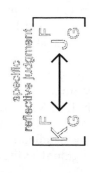 specific reflective judgment	Each item of knowledge, K, depends on several contexts or viewpoints, F and G, and can be justified by evidence from each of them, J_F and J_G.	Stage 6. Knowledge is constructed by comparing evidence and opinion on different sides of an issue or across contexts. Justification involves explaining such comparisons.	19 to 21
Level Ab4: Single principle: coordination of two abstract systems	general reflective judgment	General principle P that knowing is the outcome of the process of justifying and defending beliefs: Arguments K and M about contexts F & G and H & I, respectively, derive from justifications J and N based in those contexts, and lead to a conclusion that is probably true.	Stage 7. Knowledge is the outcome of an inquiry process that is generalizable across issues. Justification is probabilistic, involving the use of evidence and argument to present the most complete, compelling understanding of an issue.	24 to 26

Adapted from Kitchener & Fischer (1990); also Fischer & Rose (1994).

[a]These are the modal ages at which a level first emerges according to research with middle-class American and European people. They may differ across social groups.

Note: In skill structures, each letter denotes a skill component, and brackets enclose a single skill. Each large letter = a main component (a set), and each subscript or superscript = a subset of the main component. Bold letters = sensorimotor actions, italic letters = representations, and outline letters = abstractions. Lines connecting sets = relations forming a mapping, single-line arrows = relations forming a system, and double-line arrows = relations forming a system of systems. Lowercase labels above a skill structure describe the gist of the relationship embodied in the skill.

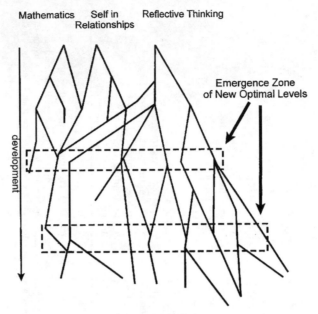

Mathematics   Self in   Reflective Thinking
              Relationships

Emergence Zone
of New Optimal Levels

development

Figure 10.6. Emergence of two skill levels across strands in a developmental web. Discontinuities (spurts, forks) in growth of optimal skill levels occur in age zones or clusters. These changes co-occur across domains that receive contextual support for high-level skills, but they occur much later for unsupported skills.

(Fischer, Kenny, & Pipp, 1990; Fischer & Rose, 1994). As Fig. 10.6 illustrates, emergence of a new skill level produces a cluster of discontinuities across many strands in the developmental web as new optimal skill levels become established across a range of domains. Discontinuities include spurts in growth, differentiation of a strand into several, and integration of several strands into one. Such discontinuities occur across domains for which the environment provides strong support for high-level skill. For example, individuals just starting to build skill in reflective thinking at stage 6 may also begin to construct an interpersonal skill for self in relationships by combining several types of honesty and kindness into a concept of constructive criticism (Fischer & Lamborn, 1989), and they may also begin to develop the mathematical skill of relating the seemingly unconnected operations of addition and division to one another (Fischer, Hand, & Russell, 1984). Likewise, individuals may show parallel changes in reflective and moral judgment, as suggested by research combining Kitchener and King's Reflective Judgment Interview with Rest's Defining Issues Test of moral judgment (King, Kitchener, Wood, & Davison, 1989; Rest et al., 1999).

These cross-domain correspondences in optimal level occur because of the combination of (1) engagement with the problems (reflective judgment, constructive criticism, addition–division relation) and (2) emergence of a new capacity/ neural network. They contrast with the pervasiveness of uneven development across tasks and domains, which Piaget (1972) called *horizontal decalage*, which is so common that it has been called the rule in cognitive development (Biggs & Collis, 1982; Fischer, 1980; Flavell, 1982). The concurrent changes mean that a new general capacity is emerging, and they occur only when high contextual support is present for learning and performance in each of the domains. Domains in which there is no such support will *not* show concurrent spurts or forks in development. Also, contrary to common assumptions, the concurrent changes do not signify any close conceptual connection between the component skills involved in reflective judgment, constructive criticism, and the addition–division relation (Fischer & Farrar, 1987).

Developmental Foundations for the Emergence of Reflective Thinking in Early Adulthood

Why doesn't reflective thinking emerge before early adulthood? The answer to this question appears to rest with the cognitive capacities and brain processes that emerge in adolescence and early adulthood, which ground the development of abstract thinking that is so essential to reflective thinking. The dynamic inter-action of cognitive/brain capacities with the presence or absence of environmental demands for serious engagement in reflective thinking are largely responsible for the timing that characterizes the emergence of reflective thinking in each individual. We will review relevant evidence regarding both brain functioning and reflective thinking.

The evidence for brain development related to the emergence of abstractions and reflective thinking involves the three characteristics described above—myelination, synapse formation including dendritic branching, and spurts in brain activity. Brain development, counter to common belief, is a long, protracted process extending well into adulthood, with myelination playing a major role, along with other changes. Myelination grows like a wave that slowly spreads, beginning in one area of the cortex and eventually covering the entire brain (Diamond & Hopson, 1998; Yakovlev & Lecours, 1967). The wave moves in a very specific pattern. During gestation and infancy it covers those regions that function the earliest, the ones involved in controlling basic survival functions such as breathing, reflexes, and sensation. It then permeates the regions involved in motor control, that is, walking, running, and object manipulation, completing this phase by approximately age 2. Last, it spreads through those areas implicated in the highest functions, such as thinking, memory, and awareness, a phase that continues at least through the second decade of life. The late myelination of these areas helps explain why reflective thinking does not emerge before early adulthood. It is precisely the coordination of one's thinking at the abstract level—the ability to integrate initially abstract mappings and then abstract systems with one another to form abstract systems and then principles—which is the hallmark of reflective thinking. Probably there are smaller patterns of change in myelin besides the slow wave, but further research is required to specify those changes.

Coincident with myelination is another brain process implicated in the development of reflective thinking—synapse formation and dendritic branching, as well as the formation of new neurons. The total number of nerve cells and synapses in the brain do not seem to change much after infancy, but dendritic branches and spines continue to spout and grow vigorously throughout life, strongly affected by experience. Contrary to beliefs a few decades ago, synapses also continue to grow, and even new neurons form every day, with new synapses and neurons replacing less useful ones, which are pruned away. Recent evidence suggests that adults grow not only new dendrites but also new synapses and new neurons, and that growth is especially prominent in the cortical regions associated with higher thinking (Gould, Reeves, Graziano, & Gross, 1999; Greenough, Wallace, Alcanta, Hawrylak, Weiler, & Withers, 1992).

Huttenlocher (1994) has found that in the frontal cortex, a brain region that plays a central role in mental organization and reasoning and thus reflective thinking, synaptic connections reach maximum density at approximately age 10, followed by a gradual decrease until late adolescence. This decline seems to occur from a process of synaptic pruning that rids the brain of excess neural connections. Chugani (1994) has produced evidence that this synaptic pruning process concludes at approximately age 16 to 18 (although we expect that it may go on at least until age 30). He found that the metabolic activity of brain cells as measured by glucose use bottoms out at adult levels at this time, signaling the attainment of a smaller and more efficient forest of dendritic connections to be maintained. This achievement—the pruning and weeding of synaptic connections to form the most efficient neural networks—probably represents a critical advance in learning ability and mental maturity in early adulthood and lays the groundwork for the emergence of reflective thinking at the optimal level.

Further, the neural connections between the frontal cortex and other regions of the brain seem to be critical to the attention, focus, mental organization, reasoning, and judgment that underlie skill in reflective thinking. These connections demonstrate systematic growth cycles for each major developmental reorganization (Fischer & Rose, 1994; Somsen, van't Klooster, van der Molen, van Leeuwen, & Licht, 1997; Thatcher, 1994). Van der Molen and Molenaar (1994) discovered that in children exercising concrete operational skills, such connections became highly active during tasks requiring speed, judgment, and control of one's response, or what is known as inhibitory control. They found that this ability increased with age and suggested that inhibitory control involving neural connections between frontal cortex and other parts of the brain is a defining characteristic of brain and cognitive development.

Thatcher's (1991, 1994) study of brain development using EEG coherence to measure connectivity offers perhaps the most compelling evidence for the role of the frontal cortex in cognitive development. He found that 90% of the coherence patterns undergoing systematic development between birth and age 20 involve the frontal cortex, with no other cortical area showing such prominence. Thatcher suggested that this area guides the formation of network connections between the cortex and other regions of the brain and in effect directs and regulates most brain development.

EEG readings have also been used to pinpoint the spurts or peaks that mark the beginnings of major reorganizations of the brain with each developmental level. Hudspeth and Pribram (1992) found three peaks during adolescence—approximately at ages 12, 15, and 19. Similarly, Fischer and Rose (1994) described brain/cognitive spurts at ages 10 to 12, 14 to 16, and 18 to 20 years, and found cognitive evidence for one more spurt in reflective judgment and other skills at approximately age 25 (Kitchener et al., 1993). These four spurts involve the emergence of the four skill levels of abstractions shown in Tables 10.1 and 10.2 and Fig. 10.3. The initial level, the capacity to use simple abstractions, first emerges in late childhood around age 10 to 12; the second level of abstract mapping, the ability to relate one abstraction to another, in mid-adolescence at age 14 to 16; abstract systems, the third level, in early adulthood at age 19 to 21; and abstract principles, the fourth level, in the mid-20s. Again this evidence indicates that the foundation for reflective thinking is established only in early adulthood, with the emergence of abstract systems and principles. The final abstract skill level and the seventh stage of reflective judgment emerge at about age 25, and EEG evidence has yet to be generated that would test for brain development changes associated with this particular discontinuity.

The Development of Functional Levels of Reflective Thinking

The new capacities that develop in early adulthood lay the groundwork for the emergence of reflective thinking, but its actual development depends on active engagement in the process of skill construction. Attaining a skill level requires microdevelopment, as explained earlier: Beginning with a low level of functioning (regression to low levels in novel tasks), people gradually and repeatedly build and rebuild skills over time, eventually consolidating and integrating components into higher order skills. In terms of reflective thinking, this means regression to concrete representations (single representations and then mappings) in reasoning with ill-structured problems, working with using specific evidence to reach particular conclusions, followed by gradual construction of complex representational systems and then abstractions, as shown in Fig. 10.4.

As an advanced skill along the developmental spectrum, reflective thinking characterizes only the final stages or levels in most theories and models of intellectual development. Like advanced moral development, it depends on environments that support high-level abstract thinking about multiple perspectives and is thus facilitated by higher education (Dawson, 2002; King & Kitchener, 1994; Rest et al., 1999; Rest & Thoma, 1985). Because reflective thinking occupies the highest levels in the developmental web, it is in a different position from skills learned earlier in life: It is not subject to developmental pressures from emergence of higher optimal skill levels, nor is it supported by most everyday environments.

Because of this absence of a general developmental impetus, substantial environmental support or pressure is necessary to spur an individual to construct reflective thinking skills. Reflective thinking, in other words, is not expected and promoted in the same way as socially valued skills emerging earlier in life, such as language, social skills, and arithmetic. Adults typically function at the levels of single abstractions and do not use the levels required for reflective thinking (Fischer, Kenny, & Pipp, 1990). They even fail most of Piaget's formal operations tasks (Martorano, 1977; Piaget, 1972; Neimark, 1981; Sinnott, 1998). For example, most adults understand and employ each of the four basic arithmetic operations—addition, subtraction, multiplication, and division, but without support, they do not relate two of these operations to one another, such as addition and multiplication. Likewise, they demonstrate an appropriate understanding of an abstract concept such as justice, but do not readily relate this concept to other abstract concepts such as democracy, law, society, intention, and responsibility.

Environmental support assumes importance for individuals both in constructing higher functional levels of reflective thinking skill and in achieving performance (albeit temporary) at their currently highest optimal level. The functional level is the highest level an individual can independently sustain, but the optimal level can be achieved in the presence of contextual support for high performance, as explained earlier. For reflective thinking, such support can be provided by teachers, coaches, therapists, supervisors, religious and spiritual advisors, or others who traffic in ill-structured problems and demand the use of evidence and abstract reasoning to arrive at well-justified conclusions and solutions. Texts and writings modeling these characteristics can also serve as contextual support.

Even in the presence of such support, however, the gap between optimal and functional levels is not erased. Once the support is removed, the individual reverts back to his or her current functional level. The optimal level serves as a long-term beacon toward which people can strive when they make arguments without contextual support (Granott, Fischer, & Parziale, 2002).

Kitchener and her colleagues (1993) devised an instrument called the Prototypic Reflective Judgment Interview (PRJ) as a means to measure optimal levels of reflective thinking that could be compared with the functional levels measured by the standard Reflective Judgment Interview (RJI). One of their primary aims was to document and further understand the gap between optimal and functional levels. They tested 104 students evenly distributed between 14 and 28 years of age; all students were highly educated—on a college track in high school, achieving well in college, or succeeding in graduate school. Each student had general practice and instruction, and as high contextual support for the PRJ each one was shown prototypic examples of good arguments for each stage. As predicted, the students scored on the RJI at the functional levels expected for their age levels, and they scored significantly higher on the PRJ. For example, with the support provided by the prototypic prompts for the PRJ, students under age 16 uniformly failed to generate any stage 6 thinking, whereas for students aged 19 to 23 approximately half of responses showed stage 6. Figure 10.7 shows the growth of stage 6 reasoning under optimal and functional conditions, as well as the growth of the overall reflective judgment score combining tasks for all stages under optimal conditions.

Cumulative research to date on reflective judgment and other complex abstract skills indicates that the emergence of optimal levels and the attainment of functional levels typically occur at the ages given in Table 10.2 (King & Kitchener, 1994). There is a lag of one to two stages or levels, or approximately 3 to 20 years, between optimal and functional levels. In other words, the slow, steady construction of reflective thinking skill over 3 to 20 years is generally required before a particular optimal level becomes fully functional and generally characteristic of an individual's behavior. As expected, college students evidence a capacity to understand stage 6 reflective thinking but do not demonstrate consistent skill in the absence of optimal contextual support (King & Kitchener, 1994; Kitchener et al., 1993). Functional stage 6 skill (without optimal support) has been found only in advanced graduate students, age 23 or older and therefore several years beyond the age of emergence of stage 6 under optimal support. Even in the late 20s students attained only about 50% stage 6 arguments without optimal support (functional stage 6 tasks in Fig. 10.7).

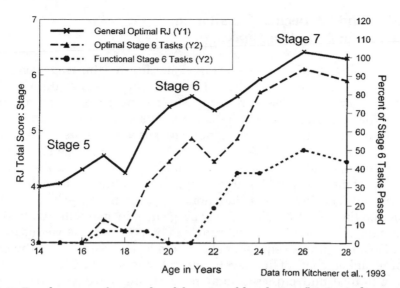

Figure 10.7. Development of optimal and functional levels in reflective judgment. Optimal level performance spurts with emergence of stages 5, 6, and 7 of reflective judgment, but functional level performance shows slower, more gradual increase. The *top line (solid)* shows the general score for reflective judgment across all tasks. The two *dotted lines* show percentage of correct performance for the specific tasks assessing stage 6, which is the beginning of true reflective thinking.

Table 10.2. Approximate Ages for Optimal and Functional Levels of Reflective Judgment

Stage of reflective judgement	Emergence of optimal level	Emergence of functional level[a]
Prereflective judgement		
Stage 3 (level Rp3)	6 to 7 years	Middle school and high school age, 12 to 17 years
Quasireflective judgement		
Stage 4 (level Ab1)	10 to 12	Late high school, college, and above, 16 to 23 years. Never for many people and domains
Stage 5 (level Ab2)	14 to 16	Early graduate school, 19 to 30 years or older. Never for many people and domains
Reflective judgement		
Stage 6 (level Ab3)	19 to 21	Advanced graduate school, 23 to 40 years or older. Never for many people and domains.
Stage 7 (level Ab4)	24 to 26	Advanced graduate school, 30 to 45 or older. Never for many people and domains.

[a]Ages for emergence of functional level vary widely, and so these estimates are coarse.

Note: This table includes only the last five of the seven stages, which are the ones that adults use most.

Reviews and research by Kitchener & Fischer (1990), Kitchener & King (1994), Kitchener et al. (1993), as well as Basseches (1984), Colby et al. (1983), Cook-Greuter (1999), Dawson (in press), Fischer et al. (1990), Perry (1970), Rest et al. (1999), and Vaillant (1977).

Education and Intellectual Stimulation:
Key Factors in Reflective Thinking

Some adults gradually become skilled at reflective thinking even in ordinary functioning without support. Others never develop reflective thinking skill, and many only learn it marginally, never mastering and generalizing it during their ifetime. What factor(s) explain why some adults achieve the functional level of reflective thinking and others never do? The answer appears to be education or comparable cognitive stimulation and enrichment. In fact, age seems to be more important for optimal levels of reflective thinking than for functional levels. The capacity for reflective thinking under optimal conditions normally emerges in early adulthood, during the 20s. Functional level, on the other hand, requires skill construction through steady, long-term engagement in an intellectually stimulating environment or milieu such as that provided by post-secondary educational institutions. Given the myriad educational and intellectual paths that individuals take, the timing and outcome of this developmental process is highly idiosyncratic.

Longitudinal and cross-sectional research on reflective judgment demonstrates that individuals engaged in higher-education programs and activities slowly and steadily increase their reflective thinking skill (King & Kitchener, 1994). As Table 10.2 illustrates, there is an increase in functional level of approximately one stage in reflective judgment for every step in educational attainment. High school students are consistently prereflective, functioning primarily at stage 3. College students, both traditional and nontraditional, and students in master's and professional programs and the first 2 years of doctoral work, are consistently quasireflective, functioning primarily at stages 4 and 5, respectively. Only advanced graduate students and beyond demonstrate consistent skill in reflective thinking, that is, stages 6 and 7.

In colleges and universities, students find it difficult to maintain the view of knowledge as absolutely right or wrong, being constantly confronted with a great diversity of approaches to knowledge, especially in the humanities and social sciences (Basseches, 1984; Kitchener & King, 1994; Perry, 1970; Rest et al., 1998). They learn, for example, that not only do eminent scholars in different fields disagree, but even eminent scholars in the same field disagree with each other on specific and important issues. They learn that a poem or novel is open to a number of interpretations. The following dialogue from Perry's (1970) study shows one student's reaction to this situation:

> *Interviewer:* You kept looking for the answer and they wouldn't give it to you ... ?
> *Freshman:* Yeah, it wasn't in the *book*. And that's what confused me a lot. *Now* I know it isn't in the book for a purpose. We're supposed to think about it and come up with the answer. (p. 78)

Students also discuss this diversity of knowledge and viewpoint with fellow students from various backgrounds and learn that their friends perceive the diversity in ways that differ from their own perceptions. They must find ways to deal with the diversity of knowledge, understand it, and develop a commitment in the face of it (Matteson, 1977).

Dealing with these matters is a high-level abstract problem, and as with all such problems people take a long time to solve it (Gruber, 1973). When few absolute right answers are forthcoming, students gradually recognize the relativity of knowledge and move into stages 4 and 5: Different viewpoints or frames of reference or contexts lead to different explanations and facts, and thus knowledge is by nature relative to a person's viewpoint and not absolutely true or false.

The argument that education is more important than age in affecting functional levels of reflective thinking rests with two kinds of evidence. First, there is no difference in reflective thinking stage for students at the same educational level who vary in age. Second, younger people who attend college have higher scores than older people who have never done so.

Furthermore, the key factor in reflective thinking does not seem to be education in general, but a certain *kind* of education—a focus on reasoning about ill-structured problems, which is central to the social sciences. Contrary to the stereotype of brilliant mathematicians and physicists, graduate students in the social sciences score significantly higher on the RJI than graduate students in the natural and mathematical sciences. The social sciences give more attention to the use of evidence and reasoning to reach conclusions about ill-structured problems from different perspectives than do the natural sciences. In this way the social sciences provide greater opportunity for students to engage in the slow, steady construction of reflective thinking skill (King & Kitchener, 1994).

The reflective judgment findings are echoed in the moral judgment domain. Formal education is the strongest predictor of scores on Rest's test of principled moral reasoning, the DIT (Rest et al., 1999), and positively correlated with scores on the Moral Judgment Interview (Colby, Kohlberg, Gibbs, & Lieberman, 1983; Dawson, 2002; Edwards, 1994). DIT levels increase while a person is in an educational setting, and after he or she leaves, they tend to plateau at the highest functional level attained (Rest et al., 1999). Colby et al. (1983) found that no individuals reach moral judgment stage 4 without attending some college or the combined stage 4/5 without finishing college. For those who get little formal education after high school, their DIT scores may actually decline over years (Rest et al., 1986), providing further evidence that education and not age is the critical factor in reflective thinking.

The kind or quality of the education is central for moral development, as it is for reflective judgment. Colby et al. (1983) found that moral judgment is related to the educational experience itself and not accounted for by intelligence test scores. In particular, cognitive/intellectual experiences are most highly correlated with moral judgment, and even when the educational effect is removed, intellectual stimulation and richness of the social environment are found to be significant variables in DIT scores (Rest et al., 1999). Rest et al. (1986) conclude that "The people who develop moral judgment are those who love to learn, who seek new challenges, who enjoy intellectually stimulating environments, ... [who] have an advantage in receiving encouragement to continue their education and their development" (p. 57), while "The critical characteristic of a college for promoting moral judgment seems to be a commitment to critical reflection" (Rest et al., 1999, p. 73).

In Perry's (1970) seminal, exploratory study of reflective judgment in college students, he found three ways that college students remove themselves from the process of intellectual and ethical development and effectively prevent further construction of reflective thinking skill. First, in *temporizing*, they make a deliberate decision to pause in their growth while they consolidate the skills they have already attained and ready themselves for movement to the next level. Second, in *retreating*, they regress in their development to a prereflective position characterized by dualistic thinking and resistance to complexity. Both Perry and Kitchener and King (1994) found that regression rare. Third, in *escaping*, students use the relativism of quasireflective thinking as an excuse to become detached and avoid intellectual commitment and responsibility, in effect to become a passive rather than active agent in their own development. Or in another form of escaping they may become opportunistic doers who avoid thinking through the implications of their actions.

Given the temporary nature of temporizing, the real stumbling block to construction of reflective thinking skill is escaping—that is, removing oneself from opportunities for continued skill development, either actively (through physical withdrawal from further education) or passively (through intellectual withdrawal from educational activities). Perry deems both retreating and escaping to represent failures in growth and maturity. Taking such a route is natural because of the great difficulty of dealing with the relativism of knowledge.

A question yet to be answered is why education and intellectual stimulation promote the development of reflective thinking. As with the age effects for optimal levels, cognitive and brain research again provide a compelling, if partial answer. Extensive classical cognitive research shows the pervasive effects of rich environments on cognition and development from an early age (e.g., Diamond & Hopson, 1998; Hebb, 1949; Hunt, 1961; Hunt, Mohandessi, Ghodssi, & Akiyama, 1976; McCall, Appelbaum, & Hogarty, 1973). In neuroscience, experiments with rats, mice, cats, monkeys, and other animals demonstrate that the degree of intellectual stimulation in the environment exerts a powerful effect on cortical thickness (especially dendritic branching and cells that support neurons in their functioning) (Diamond, 1988; Greenough et al., 1992). Parts of the cerebral cortex, particularly those implicated in learning and memory, expand and contract in response to environmental stimulation, an effect that holds for all ages from newborn to elderly. Environmental effects have produced differences in cortical thickness as great as 17%. The increase in cortical thickness resulting from an enriched environment produces smarter rats—for example, rats that run mazes better—evidence that suggests why reflective thinking is correlated with education. Equally striking may be the finding that a boring environment has a thinning effect at least as pronounced as the thickening effect obtained from a stimulating environment. In other words, people must use their skills in intellectually stimulating activity or risk losing them.

Comparable evidence has been generated from research on the human brain. Arnold Scheibel and his colleagues (Scheibel, Perdue, Tomiyasu, & Wechsler, 1990; Simonds & Scheibel, 1989) found that, in general, dendritic branching is thicker in areas that control a heavily used function. For example, the higher the level of a person's education, the greater the branching in Wernicke's area, which plays an essential role in understanding speech (Jacobs, Schall, & Scheibel, 1993). Scheibel surmises that length and level of education, vocation, hobbies, and intellectual interests—the key elements in cognitive stimulation—are significant factors in determining which areas of the brain receive heavy use and as a result continue to grow and develop over the lifespan. Here again is significant evidence that the development of skill in general and reflective thinking in particular requires consistent exercise of thinking, activities, and related brain structures.

This exercise is straightforwardly accomplished through formal education, particularly at college and graduate levels, but also appears possible through comparable forms of intellectual challenge and enrichment that are persistently and actively pursued over a sufficient span of years. Individuals who do not find themselves in such intellectually stimulating contexts will nevertheless develop skills in other domains that emerge at the same time as reflective thinking, particularly skills whose practice and mastery find sufficient support in their immediate environment, such as the complex, sophisticated skills of a blacksmith, a family leader, an automobile mechanic, or a computer repair technician. Without a supportive environment, people's development of unsupported skills will be slowed, falling far beyond the ages for the emergence of optimal levels listed in Table 10.2. In fact, unless people decide to enter an institution of higher learning or similar intellectually stimulating environment, they may find that their only remaining opportunity for the development of reflective thinking comes not by changing their environment, but simply by entering fully into the oft-dreaded and maligned experience of personal growth and development that can come in midlife.

A Second Chance at Midlife?

If individuals do not pursue formal education at the graduate level or discover a comparable vehicle for intellectual development, does that mean they have lost their only opportunity to develop a functional level of reflective thinking (stage 6 or 7)? There is tentative evidence that changes in behavior and brain associated with midlife may present a second chance to build higher level thinking skills, including reflective thinking.

The brain remains plastic and amenable to modification by experience throughout life, particularly in those regions responsible for long-term survival functions such as thinking and learning (Diamond, 1988; Gould et al., 1999). Every brain function operates according to its own timetable, and Benes (1995) has found that the process of myelination, originally believed to end in childhood, not only continues through adulthood but shows a major spurt in growth in the 40s and 50s. What is especially provocative about this finding is where the main spurt in myelin occurs in midlife—the neural pathways running from the prefrontal cortex to the limbic system. The myelination in this pathway implies that the emotional responses controlled by the limbic system become better integrated with and regulated by the higher cognitive functions of the neocortex such as reasoning and learning. Benes suggests that "The advanced phase of adult human development during adulthood might involve a more effective interplay between cognitive processing and emotional reactivity" (p. 236). A possible consequence of this connection is the usual decrease in negative affect as adults grow older (Carstensen, 2000).

Bernice Neugarten and her colleagues found midlife changes in what they called "interiority," which became evident by the mid-40s (Neugarten, 1968, 1996). In a study of 100 men and women in middle adulthood, they were "impressed with the central importance of what might be called the executive processes of personality in middle age: self-awareness, selectivity, manipulation and control of the environment, mastery, competence, the wide array of cognitive strategies." Especially notable was "reflection as a striking characteristic of the mental life of middle-aged persons: the stock-taking, the heightened introspection, and, above all, the structuring and restructuring of experience—that is, the conscious processing of new information in the light of what one has already learned and the turning of one's proficiency to the achievement of desired ends" (Neugarten, 1996, p. 148).

Given Benes' and Neugarten's findings, is there confirming evidence in the reflective thinking research? Unfortunately, there have been no longitudinal studies tracking individual development long enough to address the question. Kitchener and King's 10-year longitudinal study included 10 individuals who reached age 41 or older by the final testing, but these were all advanced graduate students, which precludes analysis of the effects of education. A cross-sectional study of reflective judgment in people 65 years of age and older found that those with only a high school education produced scores close to the college average, which is about half a stage higher than the average for high school graduates, while retired college faculty with doctorates scored close to the average for advanced doctoral students, as might be expected (Kitchener & King, 1994).

Longitudinal studies of moral judgment using either the MJI or the DIT have mostly not extended past the 30s into midlife or do not provide enough detail to determine if there is a midlife spurt in moral judgment (Colby & Kohlberg, 1987; Colby et al., 1983; Dawson, 2002; Rest et al., 1999). Longitudinal studies of personality and life accomplishments suggest that major changes in reflection and thoughtfulness commonly occur in midlife, although they are not obviously tied to a specific age within the general period (Block, 1971, 1993; Dawson, 1998; Elder, 1974; Mussen, Eichorn, Honzik, Bieber, & Meredith, 1980; Sternberg, 1990; Vaillant, 1977, 1986).

What does seem clear from Benes' and Neugarten's findings and the longitudinal studies of personality development is that midlife may prove to be a critical period in the development of reflective thinking. Future longitudinal research that controls for or goes beyond the education effect is needed to yield the evidence required to answer the question.

BASES FOR DEVELOPMENT OF REFLECTIVE THINKING IN ADULTHOOD: CONCLUSIONS

Reflective thinking is a form of cognition involving the construction of abstract systems and principles. It emerges at the earliest at the beginning of adulthood and thus is a phenomenon almost exclusively associated with adulthood and adult development. Its late appearance in development occurs because reflective thinking depends on the emergence and development of the capacity for abstract thinking at an advanced level. The capacity for understanding and using simple abstractions first emerges at age 10 to 12 and develops through a slow, gradual, and painstaking process of skill construction over time. Capacities at this and succeeding levels become established through dynamic and recurring growth cycles that reorganize and elaborate skills and neural networks, especially connections among skills and between the frontal cortex and other sections of the brain.

Capacity at the next level—the mapping of two simple abstractions to one another—is first established at age 14 to 16, and its development must follow the same process of skill construction as that for simple abstractions. Many adolescents and adults never engage in skill construction for reflective thinking at this level or subsequent levels. They develop skill in the use of simple abstractions, and progress further primarily in domains related to their everyday lives. Their capacity for advanced abstract thinking in many domains is never realized, and they never develop reflective thinking.

Whether or not individuals succeed in developing their capacity for abstract mapping in reflective thinking, they experience two additional developmental reorganizations. The first, at approximately age 20, establishes the capacity for relating two or more abstract mappings together to form abstract systems, and thus puts in place the necessary foundation for the emergence of reflective thinking. The second, at age 25, establishes the capacity for relating two or more abstract systems together to form abstract principles. This is the highest level of development for which there is clear empirical evidence, and it appears that most people attain it in only limited domains.

Dynamic changes within the brain that support an individual's capacity for thinking at a new level of abstraction are not sufficient to ensure the emergence and development of skill at that level. The other piece of the puzzle is the environmental context within which the person works, plays, and resides. The environment must support, and even demand, a person's development of that skill for it to become part of everyday functioning. It is only when the person begins to use the support and heed the demands of the environment that the skill emerges and commences its development

As the laborious and lengthy process of skill development proceeds—often spanning many years—a person demonstrates a range in skill and knowledge that varies according to the degree and quality of environmental support. In the presence of high support, he or she is capable of independent functioning at optimal level, the upper limit on supported performance. In the absence of high support, the person typically operates at functional level, the upper limit on spontaneous performance, although functioning at even lower levels will occur in many circumstances, when high-level functioning is not demanded or when the person is tired or emotionally or physically impaired.

A substantial amount of research on reflective judgment and moral judgment suggests that reflective thinking emerges and develops most commonly in environments rich in intellectual stimulation involving multiple arguments and consideration of evidence, especially post-secondary educational institutions. Facing up to the relativism of knowledge and moving beyond it is no easy task! The best support for achievement of competence in reflective thinking comes from institutions and programs that require students to consistently consider conflicting evidence and reason from evidence to feasible solutions to ill-structured problems. Functional levels of skill in reflective thinking have been found most frequently in people who are in the advanced stages of graduate programs or have moved beyond them. However, midlife seems to offer people a second chance for the emergence and development of reflective thinking, as evidenced by both an increase in cognitive interiority/reflectivity at midlife and a spurt in the formation of myelin (the insulation around neural axons). Myelin growth connecting prefrontal cortex to limbic system spurts in the 40s and 50s, a change that probably integrates the emotional responses of the limbic system more effectively into the neocortical functions involving reasoning and learning. This kind of integration will naturally facilitate interiority and reflection, which are characteristic of many people in midlife. These findings suggest the provocative idea that midlife presents an important opportunity to rediscover and develop ability in reflective thinking.

REFERENCES

Arnold, M. L. (1993). The place of morality in the adolescent self. Unpublished doctoral dissertation, Harvard University.

Basseches, M. (1984). *Dialectical thinking and adult development*. Norwood, NJ: Ablex.

Benes, F. M. (1995). A neurodevelopmental approach to the understanding of schizophrenia and other mental disorders. In D. Cicchetti & D. J. Cohen (Eds.), *Developmental psychology, Vol. 1: Theory and Methods*, (pp. 227–253). New York: John Wiley & Sons.

Biggs, J., & Collis, K. (1982). *Evaluating the quality of learning: The SOLO taxonomy (structure of the observed learning outcome)*. New York: Academic Press.

Block, J. (1971). *Lives through time*. Berkeley, CA: Bancroft Books.

Block, J. (1993). Studying personality the long way. In D. C. Funder, R. D. Parke, C. Tomlinson-Keasey, & K. Widaman (Eds.), *Studying lives through time: Personality and development* (pp. 9–41). Washington, DC: Amerian Psychological Association.

Brown, A. L., & Reeve, R. (1987). Bandwidths of competence: The role of supportive contexts in learning and development. In L. S. Liben (Ed.), *Development and learning: Conflict or congruence?* Hillsdale, NJ: Lawrence Erlbaum.

Carstensen, L. L., Pasupathi, M., Mayr, U., & Nesselroade, J. R. (2000). Emotional experience in everyday life across the adult life span. *Journal of Personality and Social Psychology, 79*, 644–655.

Case, R. (1991). *The mind's staircase: Exploring the conceptual underpinnings of children's thought and knowledge*. Hillsdale, NJ: Lawrence Erlbaum.

Chugani, H. T. (1994). Development of regional brain glucose metabolism in relation to behavior and plasticity. In G. and K. W. Fischer (Eds.), *Human behavior and the developing brain*. New York: Guilford Press.

Colby, A., & Kohlberg, L. (1987). *The measurement of moral judgment, Vol. 1: Theoretical foundations and research validation*. Cambridge, UK: Cambridge University Press.

Colby, A., Kohlberg, L., Gibbs, J., & Lieberman, M. (1983). A longitudinal study of moral judgement. *Monographs of the Society for Research in Child Development, 48*(1, Serial no. 200).

Commons, M. L., Trudeau, E. J., Stein, S. A., Richards, F. A., & Krause, S. R. (1998). Hierarchical complexity of tasks shows the existence of developmental stages. *Developmental Review, 18*, 237–278.

Cook-Greuter, S. R. (1999). Postautonomous ego development: A study of its nature and development. Unpublished doctoral dissertation, Harvard Graduate School of Education, Human Development & Psychology, Cambridge, MA.

Dawson, G. & Fischer, K. W. (Eds.) (1994). *Human behavior and the developing brain*. New York: Guilford Press.

Dawson, T. L. (2002). New tools, new insights: Kohlberg's moral reasoning stages revisited. *International Journal of Behavior Development, 25*, 154–166.

Dawson, T. L. (1998). "A good education is ... " A life-span investigation of developmental and conceptual features of evaluative reasoning about education. Unpublished doctoral dissertation, University of California, Berkeley, CA.

Detterman, D. K. (1993). The case for the prosecution: Transfer as an epiphenomenon. In D. K. Detterman & R. J. Sternberg (Eds.), *Transfer on trial: Intelligence, cognition, and instruction* (pp. 1–24). Norwood, NJ: Ablex.

Dewey, J. (1910/1991). *How we think*. Amherst, NY: Prometheus Books.

Dewey, J. (1933). *How we think: A restatement of the relation of reflective thinking to the educative process*. Lexington, MA: Heath.

Diamond, M. (1988). *Enriching heredity: The impact of the environment on the anatomy of the brain.* New York: Free Press.

Diamond, M., & Hopson, J. (1998). *Magic trees of the mind.* New York: Dutton.

Dunbar, K. (1997). How scientists think: On-line creativity and conceptual change in science. In T. Ward, S. Smith, & S. Vaid (Eds.), *Conceptual structures and processes: Emergence, discovery, and change* (pp. 461–493). Washington, DC: American Psychological Assocation.

Dunbar, K. (2001). The analogical paradox: Why analogy is so easy in naturalistic settings, yet so difficult in the psychological laboratory. In D. Gentner, K. J. Holyoak, & B. Kokinov (Eds.), *Analogy: Perspectives from cognitive science.* Cambridge, MA: The MIT Press.

Edwards, C. P. (1994). Cross-cultural research on Kollberg's stages: The basis for consensus. In B. Puka (Ed.), *New research in moral development, Vol. 5* (pp. 373–384). New York: Garland.

Elder, G. H., Jr. (1974). *Children of the Great Depression: Social change in life experience.* Chicago: University of Chicago Press.

Ericsson, K. A., & Charness, N. (1994). Expert performance: Its structure and acquisition. *American Psychologist, 49,* 725–747.

Fischer, K. W. (1980). A theory of cognitive development: The control and construction of hierarchies of skills. *Psychological Review, 87,* 477–531.

Fischer, K. W., & Bidell, T. R. (1998). Dynamic development of psychological structures in action and thought. In R. M. Lerner (Ed.) & W. Damon (Series Ed.), *Handbook of child psychology: Vol. 1. Theoretical models of human development,* 5th edit. (pp. 467–561). New York: John Wiley & Sons.

Fischer, K. W., & Farrar, M. J. (1987). Generalizations about generalization: How a theory of skill development explains both generality and specificity. *International Journal of Psychology, 22,* 643–677.

Fischer, K. W., & Granott, N. (1995). Beyond one-dimensional change: Parallel, concurrent, socially distributed processes in learning and development. *Human Development, 38,* 302–314.

Fischer, K. W., Hand, H. H., & Russell, S. L. (1984). The development of abstractions in adolescence and adulthood. In M. Commons, F. A. Richards, & C. Armon (Eds.), *Beyond formal operations* (pp. 43–73). New York: Praeger.

Fischer, K. W., Kenny, S. L., & Pipp, S. L. (1990). How cognitive processes and environmental conditions organize discontinuities in the development of abstractions. In C. N. Alexander & E. J. Langer (Eds.), *Higher stages of human development: Perspectives on adult growth.* New York: Oxford University Press.

Fischer, K. W., & Lamborn, S. D. (1989). Mechanisms of variation in developmental levels: cognitive and emotional transitions during adolescence. In A. de Ribaupierre (Ed.). *Transition mechanisms in child development: The longitudinal perspective.* Cambridge, UK: Cambridge University Press.

Fischer, K. W., & Rose, S. P. (1994). Dynamic development of coordination of components of brain and behavior. In G. Dawson & K. W. Fischer (Eds.), *Human behavior and the developing brain.* New York: Guilford Press.

Fischer, K. W., & Rose, S. P. (1998). Growth cycles of brain and mind. *Educational Leadership, 56*(3), 56–60.

Fischer, K. W., Stewart, J., & Yan, Z. (in press). Cognitive development in adulthood: Dynamics of variation and consolidation. In J. Valsiner & K. Connolly (Eds.), *Handbook of developmental psychology.* Thousand Oaks, CA: Sage.

Fischer, K. W., & Yan, Z. (2002). Darwin's construction of the theory of evolution: Microdevelopment of explanations of species' variation and change. In N. Granott & J. Parziale (Eds.), *Microdevelopment: Transition processes in development and learning.* Cambridge, UK: Cambridge University Press.

Fischer, K. W., Yan, Z., McGonigle, B., & Warnett, L. (2000). Learning and developing together: Dynamic construction of human and robot knowledge. In J. Weng & I. Stockman (Eds.), *Workshop on Development and learning: Proceedings of an NSF/DARPA workshop* (pp. 50–59). East Lansing, MI: Michigan State University.

Flavell, J. (1982). On cognitive development. *Child Development, 53,* 1–10.

Gardner, H. (1993). *Creating minds.* New York: Basic Books.

Gould, E., Reeves, A. J., Graziano, M. S. A., & Gross, C. G. (1999). Neurogenesis in the neocortex of adult primates. *Science, 286,* 548–552.

Granott, N. (1994). Microdevelopment of co-construction of knowledge during problem-solving: Puzzled minds, weird creatures, and wuggles. *Dissertation Abstracts International, 54*(10B), 5409.

Granott, N. (2002). How microdevelopment creates macrodevelopment: Reiterated sequences, backward transitions, and the zone of current development. In N. Granott & J. Parziale (Eds.), *Microdevelopment: Transition processes in development and learning*. Cambridge, UK: Cambridge University Press.

Granott, N., Fischer, K. W., & Parziale, J. (2002). Bridging to the unknown: A fundamental mechanism in learning and problem-solving. In N. Granott & J. Parziale (Eds.), *Microdevelopment: Transition processes in development and learning*. Cambridge, UK: Cambridge University Press.

Greenough, W., Wallace, C., Alcanta, B., Hawrylak, A., Weiler, I., & Withers, G. (1992). Development of the brain: Experience affects the structure of neurons, glia, and blood vessels. In N. Anastasiow & S. Harel (Eds.), *The at-risk infant: Vol. 3. Intervention, families, and research*. Baltimore: Paul H. Brookes.

Gruber, H. E. (1981). *Darwin on man*, 2nd edit. Chicago: University of Chicago Press.

Gruber, H. E. (1973). *Courage and cognitive growth in children and scientists*. In M. Schwebel & J. Raph (Eds.), Piaget in the classroom. New York: Basic Books.

Gruber, H. E., & Vonèche, J. J. (1976). Reflexions sur les opérations formelles de la pensée. *Archives de Psychologie, 44*, 45–55.

Hebb, D. O. (1949). *The organization of behavior*. New York: John Wiley & Sons.

Hudspeth, W., & Pribram, K. (1992). Psychophysiological indices of cerebral maturation. *International Journal of Psychophysiology, 12*, 19–9.

Hunt, J. M. (1961). *Intelligence and experience*. New York: Ronald Press.

Hunt, J. M., Mohandessi, K., Ghodssi, M., & Akiyama, M. (1976). The psychological development of orphanage-reared infants: Interventions with outcomes (Tehran). *Genetic Psychology Monographs, 94*, 177–226.

Huttenlocher, P. (1994). *Synaptogenesis in human cerebral cortex*. In G. Dawson & K. W. Fischer (Eds.), Human behavior and the developing brain (pp. 137–152). New York: Guilford.

Jacobs, B., Schall, M., & Scheibel, A. B. (1993). A quantitative dendritic analysis of Wernicke's area in humans. II. Gender, hemispheric, and environmental factor. *Journal of Comparative Neurology, 327*, 97–111.

Kegan, R. (1982). *The evolving self: Problem and process in human development*. Cambridge, MA: Harvard University Press.

King, P. M., & Kitchener, K. S. (1994). *Developing reflective judgment: Understanding and promoting intellectual growth and critical thinking in adolescents and adults*. San Francisco: Jossey-Bass.

King, P. M., Kitchener, K. S., Wood, P. K., & Davison, M. L. (1989). Relationships across developmental domains: A longitudinal study of intellectual, moral, and ego development. In M. L. Commons, J. D. Sinott, F. A. Richards, & C. Armon (Eds.), *Adult development, Vol. 1* (pp. 57–72). New York: Praeger.

Kitchener, K. S., & Fischer, K. W. (1990). A skill approach to the development of reflective thinking. In D. Kuhn (Ed.), *Developmental perspectives on teaching and learning thinking skills, Vol. 21* (pp. 48–62). Basel: Karger.

Kitchener, K. S., & King, P. M. (1981). Reflective judgement: Concepts of justification and their relation to age and education. *Journal of Applied Developmental Psychology, 2*, 89–116.

Kitchener, K. S., Lynch, C. L., Fischer, K. W., & Wood, P. K. (1993). Developmental range of reflective judgment: The effect of contextual support and practice on developmental stage. *Developmental Psychology, 29*, 893–906.

Klahr, D., with Dunbar, K., Fay, A. L., Penner, D., & Schunn, C. D. (2000). *Exploring science: The cognition and development of discovery processes*. Cambridge, MA: The MIT Press/Bradford.

Kuhn, D. (1992). Thinking as argument. *Harvard Educational Review, 62*, 155–178.

Kuhn, D. (2002). A multi-component system that constructs knowledge: Insights from microgenetic study. In N. Granott & J. Parziale (Eds.), *Microdevelopment:Transition processes in development and learning*. Cambridge, UK: Cambridge University Press.

Kuhn, D., Garcia-Mila, M., Zohar, A., & Andersen, C. (1995). Strategies of knowledge acquisition. *Monographs of the Society for Research in Child Development, 60*(4, Serial No. 245).

Martorano, S. C. (1977). A developmental analysis of performance on Piaget's formal operations tasks. *Developmental Psychology, 13*, 666–672.

Matteson, D. R. (1977). Exploration and commitment: Sex differences and methodological problems in the use of identity status categories. *Journal of Youth and Adolescence, 6*, 353–374.

McCall, R. B., Appelbaum, M. I., & Hogarty, P. S. (1973). Developmental changes in mental performance. *Monographs of the Society for Research in Child Development, 38*(3, Serial no. 150).

Mussen, P. H., Eichorn, D. H., Honzik, M. P., Bieber, S. L., & Meredith, W. H. (1980). Continuity and change in women's characteristics over four decades. *International Journal of Behavioral Development, 3*, 333–348.

Neimark, E. D. (1981). Confounding with cognitive style factors: An artifact explanation for the apparent nonuniversal incidence of formal operations. In I. Sigel, D. Brodzinsky, & R. Golinkoff (Eds.), *New directions in Piagetian theory and practice.* Hillsdale, NJ: Lawrence Erlbaum.

Neugarten, B. L. (Ed.) (1968). *Middle age and aging.* Chicago: University of Chicago Press.

Neugarten, D. A. (Ed.) (1996). *The Meanings of age: Selected papers of Bernice L. Neugarten.* Chicago: University of Chicago Press.

Perkins, D. N., & Salomon, G. (1989). Are cognitive skills context-bound? *Educational Researcher, 18*, 16–25.

Perry, W. G. (1970). *Forms of intellectual and ethical development in the college years.* New York: Holt, Rinehart & Winston.

Piaget, J. (1972). Intellectual evolution from adolescence to adulthood. *Human Development, 15*, 1–12.

Pirttilä-Backman, A. M. (1993). The social psychology of knowledge reassessed: toward a new delineation of the field with empirical substantiation. Unpublished Doctoral dissertation, University of Helsinki, Helsinki.

Pratt, M. W., Norris, J. E., Arnold, M. L., & Filyer, R. (1999). Generativity, moral development as predictors of value-socialization narratives for young persons across the adult life span: From lessons learned to stories shared. *Psychology & Aging, 14*, 414–426.

Rest, J. R., Deemer, D., Barnett, R., Spickelmier, J., & Volker, J. (1986). Life experiences and developmental pathways. In J. R. Rest (Ed.), *Moral development: Advances in theory and research.* New York: Praeger.

Rest, J., Narvaez, D., Bebeau, M. J., & Thoma, S. J. (1999). *Postconventional moral thinking.* Mahwah, NJ: Lawrence Erlbaum.

Rest, J. R., & Thoma, S. J. (1985). Relation of moral judgment development to formal education. *Developmental Psychology, 21*, 709–714.

Sa, W. C., West, R. F., & Stanovich, K. E. (1999). The domain specificity and generality of belief bias: Searching for a generalizable critical thinking skill. *Journal of Educational Psychology, 91*, 497–510.

Scheibel, A. B., Perdue, S., Tomiyasu, U., & Wechsler, A. (1990). A quantitative study of dendritic complexity in selected areas of the human cerebral cortex. *Brain & Cognition, 12*, 85–101.

Simonds, R. J., & Scheibel, A. B. (1989). The postnatal development of the motor speech area: A preliminary study. *Brain & Language, 37*, 42–58.

Sinnott, J. D. (1998). *The development of logic in adulthood: Postformal thought and its applications.* New York: Plenum Press.

Somsen, R. J. M., van 't Klooster, B. J., van der Molen, M. W., van Leeuwen, H. M. P., & Licht, R. (1997). Growth spurts in brain maturation during middle childhood as indexed by EEG power spectra. *Biological Psychology, 44*, 187–209.

Sternberg, R. J. (Ed.) (1990). *Wisdom: Its nature, origins, and development.* Cambridge, UK: Cambridge University Press.

Thatcher, R. W. (1991). Maturation of the human frontal lobes: Physiological evidence for staging. *Developmental Neuropsychology, 7*, 397–419.

Thatcher, R. W. (1994). Cyclic cortical reorganization: Origins of human cognitive development. In G. Dawson & K. W. Fischer (Eds.), *Human behavior and the developing brain* (pp. 232–266). New York: Guilford Press.

Thelen, E., & Smith, L. B. (1998). Dynamic systems theories. In R. M. Lerner (Ed.) & W. Damon (Series Ed.), *Handbook of child psychology: Vol. 1. Theoretical models of human development*, 5th edit. (pp. 563–634). New York: John Wiley & Sons.

Vaillant, G. E. (1977). *Adaptation to life.* Boston: Little, Brown.

Vaillant, G. E. (Ed.) (1986). *Empirical studies of ego mechanisms of defense.* Washington, DC: American Psychological Association.

Van der Molen, M. W., & Molenaar, P. C. M. (1994). Cognitive psycho-physiology: A window to cognitive development and brain maturation. In G. Dawson & K. W. Fischer (Eds.) *Human behavior and the developing brain.* New York: Guilford Press.

van Geert, P. (1991). A dynamic systems model of cognitive and language growth. *Psychological Review, 98*, 3–53.

Yakovlev, P. I., & Lecours, A. R. (1967). The myelogenetic cycles of regional maturation of the brain. In A. Minkowsky (Ed.), *Regional development of the brain in early life* (pp. 3–70). Oxford: Blackwell.

CHAPTER 11

Four Postformal Stages

Michael Lamport Commons and Francis A. Richards

The term "postformal" has come to refer to various stage characterizations of behavior that are more complex than those behaviors found in Piaget's last stage—formal operations—and generally seen only in adults. Commons and Richards (1984a, 1894b) and Fischer (1980), among others, posited that such behaviors follow a single sequence, no matter the domain of the task, for example, social, interpersonal, moral, political, scientific, and so on.

Most postformal research was originally directed toward an understanding of development in one domain. The common approach to much of the work on postformal stages has been to specify a performance on tasks that develop *out of* those described by Piaget (1950, 1952) as *formal-operational* or out of tasks in related domains (e.g., moral reasoning). The assumption has been made that the predecessor task performances (formal operations) are in some way necessary to the development of their successor performances and proclivities (postformal operations). Unlike many of the other theories, the Model of Hierarchical Complexity (MHC), presented here (Commons, Trudeau, Stein, Richards, & Krause, 1998), generates one sequence that addresses all tasks in all domains and is based on a contentless, axiomatic theory.

In any case, the collection of theories form part of the field that is known as *Positive Adult Development* (Commons, 1999; Sinnott, 1987, personal communication). This field examines ways in which development continues in a positive direction during adulthood.

This chapter first reviews evidence supporting the general idea that postformal operations exist. One form of evidence is a discussion of ways in which Piaget's formal operational stage of development cannot adequately describe more complex forms of thought (Piaget, 1954; Piaget & Inhelder with Sinclair-de Zwart, 1973).

Michael Lamport Commons • Department of Psychiatry, Harvard Medical School, Boston, Massachusetts, 02115-6196.   Francis A. Richards • Office of School Improvement, Rhode Island Department of Education, Providence, Rhode Island, 02903.

Handbook of Adult Development, edited by J. Demick and C. Andreoletti. Plenum Press, New York, 2002.

Specific proposals for how postformal thinking might differ from formal thinking are then briefly reviewed. The history of postformal research and writing indicates convergence between different theories in the types of reasoning described. We will argue that the Model of Hierarchical Complexity provides a framework within which the diversity of other theories can be placed, and illustrate a possible consensus view as to what the postformal stages might look like.

MODEL OF FORMAL OPERATIONS

Piaget (1954, 1976) used propositional logic as a model of formal operations. Here, he followed a direction developed earlier by Frege (1950) and Peano (1894), who attempted to generalize propositional logic in order to represent other forms of reasoning, most notably arithmetic. Because propositional logic was developed to reason about other forms of reasoning, it can be used to reason about the logics of classes and relationships. It can represent logical statements about both relationships (e.g., p is heavier than q) and classes (e.g., x is a member of the class A). However, in this form, propositions can be treated as objects in themselves, rather than as the direct reflection of some concrete reality. Hence, propositions have formal properties that are different from the properties of objects and the actions that organize propositions are different from the actions that organize concrete reality.

Propositions are also statements to which the truth values *true* (T) or *false* (F) can be assigned. Or the propositions may be built out of variables other than truth values. The actions that organize propositions work under the basic constriction of this bivalent system of truth values. For example, the truth values impose a restriction on the operation of negation: a proposition and its negation cannot both be true. To negate a proposition is to change its truth value.

Piaget ties the logic of propositions to the Boole's (1854) system of combination, a system in which *nuclear* propositions are combined into larger *molecular* propositions using the of connectives: 'not', – 'and', $\wedge$; 'or', '$\vee$' (*vel* = 'and/or'); 'if', $\rightarrow$; 'if and only if', $\leftrightarrow$. All these connectives are operations. The connective *not* was just discussed in connection with the operation of negation. The connectives *and* and *or* join nuclear propositions into molecular propositions of unlimited length. If one of the propositions in a string constructed only with the connective *and* is false, then the entire molecular proposition is also false. In contrast, only one member of a string connected by *or* need be true for the molecular proposition to be true.

OBJECTIONS TO FORMAL OPERATIONS

For all its formal and empirical power, objections have been made to Piaget's model of formal operations and the mechanics it implies. One argument, advanced and expanded by Broughton (1977, 1984), has to do with the nature of the integration of formal and empirical reasoning at this stage. Simply put, Broughton argues that there is no integration; rather there is domination of the latter by the former. That is, formal operations do not integrate the Boolean logical operations with the empirical manipulations used by participants to answer questions. The formal structure of reasoning becomes preeminent at the formal-operations stage and the empirical structure can no longer function in its negating, and ultimately dialectical, role. As a result, formal operations cannot be used to reason about certain types of phenomena, notably those of a non-Boolean nature. Perhaps they cannot even be used to observe such phenomena. Broughton concludes that, if Piaget's developmental sequence leads to such a barren endpoint, it ought to be abandoned altogether.

Piaget (1971) seems already to have countered Broughton's contention about the *closure* inherent in formal operations. Specifically, he discusses the activity of negating key axioms in formal structures as a method of transforming and keeping open hypothetico–deductive structures. He provides an example of this activity in the negation of axioms in Euclidean geometry that led, during the 19th century, to the development of several non-Euclidean geometries. In addition, he discusses the *decomposition* of systems into more basic systems, and the subsequent *recombination* of basic systems into new systems, as activities that keep the formal-operational structure from ossifying.

The argument advanced in the body of research to be reviewed here, however, is that these latter activities are postformal rather than formal. Their appearance can then be explained as a development necessitated by limitations of a linear formal system. Generally, the argument runs that formal operations, instantiated in propositional operations and employing the system of Boolean connectives mentioned, are adequate to formulate and analyze linear logical and causal relations. The latter are particularly useful for reasoning about situations in which dependent and independent variables are postulated to exist. Together, they create a kind of reasoning that will be referred to here as *functional analysis*.

In fact, almost all theories, including developmental theories, require, at a minimum, the use of systems to adequately explain the phenomena they address. Systems are characterized by relations that are not only functional, but transformational as well. Such transformations require nonlinear conceptions of causality. At this order of complexity, formal operations are not sufficient. For a more detailed discussion of this point, see Commons and Richards (1978); Commons, Richards, and Kuhn (1982); Richards and Commons (1984); and Commons and Richards (1984a).

PIAGET AND POSTFORMAL THEORY

It is important to distinguish between Piaget's theoretical framework and the products of that framework, in particular, formal operations. While the argument has just been made that formal operations are inadequate as a model for the activities that have most likely been operating in Western thinking for approximately a century, the same claim is not made about Piaget's theoretical framework. Piaget advanced a complex theory of processes of assimilation, accommodation, and autoregulation that could be formulated only by using postformal operations. If Piaget's own explanatory system is explicable in terms of a higher level logic than that of formal operations, then it is the former but not the latter logic that is of use to psychologists attempting to understand adult stages of development.

A REVIEW OF SOME POSTFORMAL RESEARCH

Reasoning such as that shown in the preceding, which is characteristic of the reasoning of some adults, is more complex than formal operational reasoning, as defined by Piaget (e.g., Inhelder & Piaget, 1958; Piaget & Inhelder, 1969). In earlier work (Commons & Richards, 1984a, 1984b; Commons et al., 1998) we have argued that this kind of reasoning represents one of several proposed new adult stages. Both empirical and analytic evidence for these stages has been presented. The existence of such reasoning demonstrates that development continues beyond adolescence and into adulthood, into the postformal realm.

A number of different postformal reasoning theories have been described, including those of Arlin (1975, 1984); Armon (1984); Basseches (1980, 1984) following Riegel (1973); Benack (1984); Commons and Richards (1978, 1984a; Commons et al., 1982); Demetriou (1990; Demetriou & Efklides, 1985); Fischer, Hand, and Russell (1984); Kohlberg (1990); Koplowitz (1984); Labouvie-Vief (1980, 1984); Pascual-Leone (1984); Powell (1984); Sternberg (1984); and Sinnott (1984). All argue in common that postformal behavior involves one or more of the following: perceiving, reasoning, knowing, judging, caring, feeling, or communicating in ways that are more complex or more all-encompassing than formal operations. How the theories have generated their particular form of postformal operations, however, differs.

The most common method of extending stage theory into the postformal area is to locate limitations in formal operations, then describe a kind of thinking, often drawing from other traditions, that enables the individual to transcend these limitations. Examples of this include: Arlin (1975, 1977, 1984), who draws on creative reflection; Basseches (1980, 1984), who draws on the dialectical tradition; Linn and Siegal (1984), who draw on philosophy of science; Koplowitz (1984), who draws on General Systems Theory and Buddhism; Sinnott (1981, 1984), who draws on relativity theory; or Armon (1984), who draws on moral philosophy. What do some of these conceptions of postformal reasoning look like?

Arlin had the first explicit notion of a stage beyond formal operations. Arlin's (1975, 1977, 1984) concept of postformal operations is based on the hypothesis that a radical change occurs in the way formal operations are used. While accepting the idea of formal-operational structure, she proposes that the whole function of that structure changes. Her argument is that a replacement process takes place whereby problem-solving operations disappear and problem-finding operations appear. To find a problem requires reflection on what a problem is. Part of discerning what a problem is requires reflection on how problems are solved. The form of a problem is partially determined by the possibility of some form of solvability.

Basseches (1980, 1984), for one, argues that postformal thinkers use the idea of *form* rather than the idea of *thing*. Forms are structures whose fundamental function is to change. As such, forms have systemlike properties. Things are structures whose fundamental function is to maintain their stability or identity. They have the properties of simple, linear, causal models seen in formal operations. In postformal thinkers, structure can never be temporally crystallized, but it can still be used to interpret society, nature, and the self as organizations in constant transformation.

Sinnott (1981, 1984), using the concept of relativity (Einstein, 1950), proposes relativistic frameworks that contain and coordinate more particular frameworks. Each framework can be thought of as a system of relationships among elements. A relativistic framework would then be a more general system for relating systems. Sinnott uses the concept of system metaphorically, so a system need not be attached to concepts of energy, mass, speed, and so on. It could be a system of relations that coordinates people. While a person who thinks in a formal-operational manner could reason within one such system, a person who thinks in a postformal manner deals with the problem of integrating local systems into a framework, and deals successfully with the relativity of the systems.

A variant of this argument appears in Koplowitz's (1984) description of unitary operations. He argues that, as thinking becomes more developed, the perceived boundaries between people become less useful. A child, for example, cannot be understood outside of its family. In a real sense, a child is part of a larger whole, from which that child cannot be disassociated. Koplowitz's *unitary operations* are used to comprehend wholes that have internal parts. Consequently, they organize and bring those parts into relation to one another.

The next group of researchers maintains that postformal cognition attempts to accomplish the same functions as formal cognition, but that the complexity of the patterns of thought, and the complexity posited in the objects of thought, is at a new level. These researchers analyze the nature of the developmental process, rather than the limitations inherent in formal operations. Instead of concentrating on a demonstration that change does occur, this approach attempts to show *how* it could and *must* occur. Piaget (1970) had proposed a general process of *equilibrium* and a somewhat more specific process of *reflective abstraction* to account for stages of development. Commons et al. (1982), Sternberg and Downing (1982), Commons and Richards (1984a, 1984b), Fischer et al. (1984), Pascual-Leone (1984), and Sternberg (1984) all focus explicitly on proposing a variety of mechanisms of intellectual development. They attempt not only to clarify proposed mechanisms of development, but also to show that the continued operation of these mechanisms should result in postformal thinking.

One approach, found in Fischer (1980), Fischer et al. (1984), and Sternberg (1984), to describing this new level of complexity is to use the analogy of *unfolding dimensionality*. The concept of unfolding dimensionality uses dimensions in space to convey the idea of the new size and complexity of postformal cognition. Although size may be thought of as quantitative, dimensional increase in size generates complexities that must be thought of as qualitative. Importantly, different arithmetics, geometries, and algebras are variously possible and impossible in different dimensions. For instance, adding a dimension to two-dimensional space makes it possible for the angles of a triangle to sum to more than 180 degrees and for parallel lines to intersect. Intuitively, the complexity of geometric systems increases as the size of the space containing them increases.

In a related approach, found in Commons and Richards (1984a, 1984b), Labouvie-Vief (1984), Powell (1984), and Richards and Commons (1984), sets of axioms, or other system properties, are used to describe the increased complexity of postformal reasoning. Labouvie-Vief (1980, 1984), for instance, uses the properties of different systems of logics. She describes the limitations of different logics and asserts that these limitations are due to their *strength*. A *strong* logical system is one that has several limiting assumptions. When a logic contains many restricting assumptions, it seems clear, but causes confusion when applied in areas that do not conform to those assumptions. Postformal reasoning arrives at an understanding of the inflexibility involved in thinking "overlogically." It locates the limitations of excessive assumptions and formulates a more flexible, *weaker* logic containing fewer assumptions. Although this logic is weaker than the logics it replaces, it retains these assumptions because, with further restrictions, it directs their use in appropriate situations. By releasing formal thinking from overly restrictively strong logics, a weaker logic allows the development of new kinds of thinking.

Richards and Commons (1984) likewise describe the new complexity of postformal thinking in terms of systems, but attempt to describe systems formally. Their argument for the qualitative nature of change is consequently less tied to the particular nature of either logic or physics. This argument is based on the notion that higher-stage thinking is irreducible to lower-stage thinking. This means that in the process of stage transformation, new objects of thought appear that cannot be successfully thought about at a lower stage. Considerable attention is devoted to defining irreducibility in Bickhard (1978); Campbell and Richie (1983); Commons and Richards (1978, 1984b); Commons et al. (1998); and Commons, Goodheart, and Bresette (1995).

A different perspective on this argument appears in Powell's (1984) description of *category operations*. Category operations have been developed in mathematics, partly in response to the Bourbaki (1939) program. One of the concerns of the Bourbaki program has been to place the various branches of mathematics in relation to one another. Their approach has been to locate mathematical *mother structures* that can be transformed and combined to produce the various mathematical disciplines (discussed in Piaget, 1970). Category operations were invented to reach the same goal, but to do so by examining the nature of mathematical operations rather than mathematical structures. Since category operations characterize the nature of mathematical activity, they model postformal thinking as an understanding of the ways activity can be related.

Both of these general approaches led to the claim that adult thinking contains the formal-operational framework, but employs at least one more encompassing framework as well. Some of the more complex adult behavior is characterized by multiple-system models (e.g., Kallio, 1991, 1995). Some adults are said to develop alternatives to, and perspectives on, formal operations. They use formal operations within a "higher" system of operations and transcend the limitations of formal operations. In any case, these are all ways in which these theories argue and present converging evidence that adults are using forms of reasoning that are more complex than formal operations. There are also at least two sets of differences between the theories. One is that different theories have different stopping points. Some posit only one postformal stage; others posit up to four.

A second difference is that many suggest that there is a "transcendental" stage after the "regular" stages. They also suggest a sequence for the development of "consciousness." The current chapter will not deal further with proposals for transcendental stages or levels of consciousness. The tasks that transcendental or consciousness levels address are not claimed to have material of substantive base and are therefore not addressed by task analysis.

THE MODEL OF HIERARCHICAL COMPLEXITY

The Model of Hierarchical Complexity (MHC) (Commons & Richards, 1984a, 1984b; Commons et al., 1998) is an across-domain or universal system that classifies the task-required hierarchical organization of responses. Every task contains a multitude of subtasks (Overton, 1990). When the subtasks are completed in a required order, they complete the task in question. Tasks vary in complexity in two ways, which are defined next: either as *horizontal* (involving classical information); or as *vertical* (involving hierarchical information).

Horizontal (Classical Information) Complexity

Classical information describes the number of "yes–no" questions it takes to do a task. For example, if one asked a person across the room whether a penny came up heads when he or she flipped it, his or her saying "heads" would transmit one bit of "horizontal" information. If there were two pennies, one would have to ask at least two questions, one about each penny. Hence, each additional one-bit question would add another bit. Let us say he had a four-faced top with the faces numbered 1, 2, 3, or 4. Instead of spinning it, he tossed it against a backboard as one does with dice in a game. Again, there would be two bits. One could ask him whether the face had an even number. If it did, one would then ask if it were a 2. *Horizontal complexity*, then, is the sum of bits required by just such tasks as this.

Vertical (Hierarchical) Complexity

Specifically, *hierarchical complexity* refers to the number of recursions that the co-ordinating actions must perform on a set of primary elements. Actions at a *higher order of hierarchical complexity:* (1) are *defined* in terms of actions at the *next lower* order of hierarchical complexity; (2) *organize* and *transform* the lower-order actions; and (3) produce organizations of lower-order actions that are new and *not arbitrary*, and cannot be accomplished by those lower-order actions alone. Once these conditions have been met, we say the higher order action *co-ordinates* the actions of the next lower order. *Stage of performance* is defined as the highest-order hierarchical complexity of the task solved. Commons, Goodheart, and Dawson, 1997; Commons, Richards, Trudeau, Goodheart, & Dawson, 1997) found, using Rasch (1980) analysis, that hierarchical complexity of a given task predicts stage of a performance, the correlation being $r = .92$ (hierarchical complexity of the task that is completed).

Formulating the Postformal Orders of Hierarchical Complexity

Commons et al. (Commons & Richards, 1978; Commons et al., 1982, 1998) showed that the postformal stages were true hard stages in the Kohlberg and Armon (1984) sense, but with some small modification. They used a mathematical axiomatic system derived from Duncan Luce's (e.g., Krantz, Atkinson, Luce, & Suppes, 1974; Krantz, Luce, Suppes, & Tversky, 1971) work on measurement. Each proposed stage was checked with the main three axioms. Again, these axioms state that any given higher-stage action has to be defined in terms of an associated lower one and organize those lower-stage actions in an nonarbitrary way.

Commons and Richards' concerns lay with the general specification of any empirical task that possibly could be used to demonstrate either the presence of, or the development into, a postformal stage. They deemphasize the reconstruction of the "reality" of a person "at a given stage." Instead, they attempt to develop a general way to specify the organization of tasks in any domain that a person "at a given stage" can do. Other attempts to specify what it means to be at a postformal stage can be found throughout the work reviewed here.

Postformal Orders of Complexity

Four postformal orders of hierarchical complexity have been proposed (Commons & Richards, 1984a, 1984b), beginning with systematic thinking and developing through metasystematic to paradigmatic and cross-paradigmatic thinking. The four postformal orders, according to the MHC, are displayed in Table 11.1.

Systematic Order. This stage was introduced by Herb Koplowitz (personal communication,1982).[1] At the systematic order, ideal task completers discriminate

Table 11.1. Postformal Stages, as Described in the General Model of Hierarchical Complexity

	What they do	How they do it	End result
11 *Systematic operations*	Constructs multivariate systems and matrices.	Coordinates more than one variable as input.	Events and ideas can be situated in a larger context. Systems are formed out of formal-operational relationships.
12 *Metasystematic operations*	Constructs multisystems and metasystems out of disparate systems.	Compares systems and perspectives in a systematic way. Reflects on systems, and creates supersystems of systems.	Supersystems and metasystems are formed out of multiple systems.
13 *Paradigmatic*	Fits metasystems together to form new paradigms.	Synthesis	Paradigms are formed out of multiple metasystems.
14 *Cross-paradigmatic*	Fits paradigms together to form new fields.	Forms new fields by crossing paradigms.	Fields are formed out of multiple paradigms.

[1] He suggested that metasystems and general system must operate on systems while we were working on his chapter at Dare Institute.

the frameworks for relationships between variables within an integrated system of tendencies and relationships. The objects of the systematic actions are formal-operational relationships between variables. The actions include determining possible multivariate causes—outcomes that may be determined by many causes; the building of matrix representations of information in the form of tables or matrices; and the multidimensional ordering of possibilities, including the acts of preference and prioritization. The actions generate systems. Views of systems generated have a single "true" unifying structure. Other systems of explanation or even other sets of data collected by adherents of other explanatory systems tend to be rejected. Most standard science operates at this order. At this order, science is seen as an interlocking set of relationships, with the truth of each relationship in interaction with embedded, testable relationships. Researchers carry out variations of previous experiments. Behavior of events is seen as governed by multivariate causality. Our estimates are that only 20% of the US population can now function at the systematic order without support.

Metasystematic Order.   At the metasystematic order, ideal task completers act on systems; that is, systems are the objects of metasystematic actions. The systems are made up of formal-operational relationships. Metasystematic actions compare, contrast, transform, and synthesize systems. The products of metasystematic actions are metasystems or supersystems. For example, consider treating systems of causal relationships as the objects. This allows one to compare and contrast systems in terms of their properties. The focus is placed on the similarities and differences in each system's form, as well as constituent causal relationships and actors within them. Philosophers, scientists, and others examine the logical consistency of sets of rules in their respective disciplines. Doctrinal lines are replaced by a more formal understanding of assumptions and methods used by investigators.

As an example, we would suggest that almost all professors at top research universities function at this stage in their line of work.

Paradigmatic Order.   At the paradigmatic order, people create new fields out of multiple metasystems. The objects of paradigmatic acts are metasystems. When there are metasystems that are incomplete and adding to them would create inconsistences, quite often a new paradigm is developed. Usually, the paradigm develops out of a recognition of a poorly understood phenomenon. The actions in paradigmatic thought form new paradigms from supersystems (metasystems).

Paradigmatic actions often affect fields of knowledge that appear unrelated to the original field of the thinkers. Individuals reasoning at the paradigmatic order have to see the relationship between very large and often disparate bodies of knowledge, and coordinate the metasystematic supersystems. Paradigmatic action requires a tremendous degree of decentration. One has to transcend tradition and recognize one's actions as distinct and possibly troubling to those in one's environment. But at the same time one has to understand that the laws of nature operate both on oneself and one's environment—a unity. This suggests that learning in one realm can be generalized to others.

Examples of paradigmatic order thinkers are perhaps best drawn from the history of science. For example, the 19th-century physicist Clark Maxwell constructed a fields paradigm from the existing metasystems of electricity and magnetism of Faraday, Ohm, Volta, Ampere, and Oersted using the mathematics of fields and waves. Maxwell's (1871) equations, showing that electricity and magnetism are united, formed a new paradigm. The wave fields can be easily seen as the rings that form when a rock is dropped in the water or a magnet is placed under paper that holds iron filings. This paradigm made it possible for Einstein to use notions of curved space to describe space–time to replace Euclidean geometry. The waves were bent by the mass of objects so that the rings no longer fit in a flat plane. From there modern particle theory has been able to add two more forces to the electro-magnetic forces.

Cross-Paradigmatic Order. The fourth postformal order is the cross-paradigmatic. The objects of cross-paradigmatic actions are paradigms. Cross-paradigmatic actions integrate paradigms into a new field or profoundly transform an old one. A field contains more than one paradigm and cannot be reduced to a single paradigm. One might ask whether all interdisciplinary studies are therefore cross-paradigmatic? Is psychobiology cross-paradigmatic? The answer to both questions is "no." Such interdisciplinary studies might create new paradigms, such as psychophysics, but not new fields.

This order has not been examined in much detail because there are very few people who can solve tasks of this complexity. It may also take a certain amount of time and perspective to realize that behavior or findings were cross-paradigmatic. All that can be done at this time is to identify and analyze historical examples.

Copernicus (1543/1992) coordinated geometry of ellipses that represented the geometric paradigm and the sun-centered perspectives. This coordination formed the new field of celestial mechanics. The creation of this field transformed society—a scientific revolution that spread throughout world and totally altered our understanding of people's place in the cosmos. It directly led to what many would now call true empirical science with its mathematical exposition. This in turn paved the way for Isaac Newton (1687/1999) to coordinate mathematics and physics, forming the new field of classic mathematical physics. The field was formed out of the new mathematical paradigm of the calculus (independent of Leibniz, 1768, 1875) and the paradigm of physics, which consisted of disjointed physical laws.

Rene Descartes (1637/1954) first created the paradigm of analysis and used it to coordinate the paradigms of geometry, proof theory, algebra, and teleology. He thereby created the field of analytical geometry and analytic proofs. Charles Darwin (1855, 1873) coordinated paleontology, geology, biology, and ecology to form the field of evolution which, in its turn, paved the way for chaos theory, evolutionary biology, and evolutionary psychology. Albert Einstein (1950) coordinated the paradigm of non-Euclidian geometry with the paradigms of classical physics to form the field of relativity. This gave rise to modern cosmology. He also co-invented quantum mechanics. Max Planck (1922) coordinated the paradigm of wave theory (energy with probability) to form the field of quantum mechanics. This led to modern particle physics. Lastly, Gödel (1931) coordinated epistemology and mathematics into the field of limits on knowing. Along with Darwin, Einstein, and Planck, he founded modern science and epistemology.

Bringing Simplicity to Multiplicity

Reviewing his career, Piaget (1952) remarked:

My one idea...has been that intellectual operations proceed in terms of structures-of-the-whole. These structures denote the kinds of equilibrium toward which evolution in its entirety is striving; at once organic, psychological and social. (p. 256)

In part, the work of all the researchers mentioned here is a response to this one idea. Their work represents part of a broad attempt to grow out of the form Piaget gave to a wide variety of 19th century thought. The question remains whether the growth of postformal theories is itself proceeding in terms of some sort of structure-of-the-whole. Broughton (1984) argues that this is not the case and suggests abandonment rather than revision. Another approach to this question is to assume that postformal research does not talk about many different stage sequences, but about many different manifestations of the same stage sequence.

The problem of specifying what is meant by a stage and by a stage sequence remains a critical issue in developmental theory. Elsewhere, Piaget (1972); Flavell and Wohlwill (1969); Kohlberg (1969, 1981, 1984); Flavell (1971, 1972, 1977, 1982); Bickhard (1978, 1979); and Campbell and Richie (1983) have devoted considerable attention to it. This specification is centrally important in Kohlberg and Armon (1984), and Commons and Richards (1984a, 1984b). Kohlberg and Armon's concern is to distinguish *functional*, *soft*, and *hard stages*. *Functional stage* refers to the Eriksonian (1959, 1978) model in which each stage develops to perform a new task or function. *Soft stage* refers to development that is conditioned by particular experiences. These experiences could arise from differences in personality characteristics, education, class, age, and so on. *Hard stage* refers to developmental sequences that occur universally, arising out of the overall reorganization of an underlying intellectual framework.

Table 11.2 presents one way that the stage sequences presented here can be aligned across a common *developmental space*. The harmony in the alignment shown in Table 11.2 suggests a possible part of a reconciliation of Kohlberg and Armon's (1984) hard- and soft-stage distinction. Hard stages were to have some logical basis whereas soft stages might be based on some functional ordering based solely on empirical findings. Although each of these stages may be soft stages individually, taken as a whole, they indicate the development of some hard stages. The Model of Hierarchical Complexity provides a proposal for what the logic may be underlying the sequences reported in Table 11.2.

The nature of these various postformal stages or levels cannot be determined from Table 11.2. Their extent may range beyond the developmental areas so far described. Their empirical nature has been emerging with a clearer understanding of the similarities and differences of the various stage conceptions. For this to have happened, the nature of elements and operations has to have been communicated among the researchers studying the various developmental sequences that appear in Table 11.2. Part of what this suggests is that the proposed postformal stages have been adequately formalized in a way that has facilitated comparison. Similarly, theories of stage transition must be formalized.

Table 11.2. Comparative Table of Concorded Theories of Formal Stage

Researchers	Abstract	Formal	Systematic	Metasystematic	Paradigmatic	Cross-Paradigmatic	Transcendental
Commons & Richards (1984)	9 (=4a)	10 (=4b)	11 (=5a)	12 (=5b)	13 (=6a)	14 (=6b)	
Sonnert & Commons (1994)	Group	Bureaucratic	Institutional	Universal	Dialogical		
Inhelder & Piaget (1958)	Formal III-A	Formal III-B	Postformal	Polyvalent logic; systems of systems			
Fischer, Hand, & Russell (1984)	7	8	9	10			
Sternberg (1984)		First-order relational reasoning		Second-order relational reasoning			
Kohlberg (1981)	3 Mutuality	3/4	4 Social system	5 Prior rights/social contract 6 Universal ethical principles			
Benack (1994)	4	5	6	7			
Pascual-Leone (1983)	Late concrete	Formal and late concrete	Predialectical	Dialectical			Transcendental
Armon (1984)	3 Affective mutuality	3/4	4 Individuality	5 Autonomy		6 Universal categories	
Powell (1984)	Early formal	Formal	Stage 4a/interactive empathy	Category operations [?]			
Labouvie-Vief (1984)		Intrasystematic	Intersystematic	Autonomous			
Arlin (1984)	3a Low formal (problem-solving)	3b High formal	4a Postformal (problem-finding)	4b Relativism of thought 4c Overgeneralization 4d Displacement	4e Late postformal (dialectical)		

Sinnott (1984)		Formal	Relativistic/relativize systems, metalevel rules	of concepts Unified theory: interpretation of contradictory levels		
Bassesches (1984)	Phase 1b: Formal early foundations	Phase 2: Intermediate dialectical schemes	Phase 3: 2 out of 3 Clusters of advanced dialectical schemes	4 Advanced dialectical thinking		
Koplowitz (1984)		Formal	Systems	General systems		Unitary concepts
King & Kitchener (2002)	4	5	6	7		
Torbert (1994)	Diplomat	Technician	Achiever	Ironist		
Kegan (1994)	3 Interpersonal	3/4	4 Institutional	5		
Loevinger (1998)	Conformist-conscientious	Conscientious	Individualistic	Autonomous integrated[2]		
Cook-Greuter (1990)	3/4	4	4/5	5	5/6	6
William Gray (personal communication, 1999)						
Trevor Bond (personal communication, 1999)						
Dawson (1998)	9	10	11	12	13	14
Kallio (1991, 1995)	Formal 1	Formal 2	Formal 3 generalized formal	Postformal		
Demetriou (1990, 1995)						
Broughton (1977, 1984)	3 Person vs. inner self	4 Dualist or positivist; cynical, mechanistic	5 Inner observer differentiated from ego	6 Mind & body experiences of an integrated self		

[2]Loevinger table contents as modified by Commons (present volume).

Table 11.3. Empirical Comparison of Individuals' Scores on
Different Measures of Postformal Reasoning

	Ms	GL	MJ	Loev	Structural loading
Multisystems (**Ms**)		.44**	.31**	.13	.75
Good Life (**GL**)			.41**	.00	.85
Moral Judgment (**MJ**)				.22	.64
Sentence Completion (**Loev**)					.26

**indicates an alpha level at or of less than .01

An Empirical Comparison of Some Measures of Postformal Behavior. One way to test whether there might be an necessary ordinal sequence underlying all of the separate postformal theories is to empirically compare performances across instruments developed within each theory. Although this has not yet been done for all of the theories, some preliminary work has been done comparing a few of the instruments. In Commons et al. (1998) scores on the Multisystems Task, developed to assess the Model of Hierarchical Complexity, were correlated to tasks from a few postformal theories, including Armon's Good Life Interview, Kohlberg's Moral Judgment Interview, and Loevinger's Washington University Sentence Completion Task. As Table 11.3 shows, only Loevinger's scores were not related to the other measures of postformal stages. Scores from the instruments were also factor-analyzed, using principal-components analysis. All of the instruments, except for Loevinger's Sentence Completion Task (SCT), showed significant positive loadings on the first factor, which was termed the structural factor. Sentence Completion Task is related to the inverse of education. (-.11 ns). King, Kitchener, Wood, and Davison (1989; also see Kitchener & King, 1990) have obtained similar results for the SCT.

Although these results provide some beginning empirical support that there might be a common variable underlying these different postformal conceptions, they do not suggest what that variable might be.

Because the model of hierarchical complexity is based on sequencing the order of hierarchical complexity of tasks, the model logically contains all the stage systems. The sequence of stages and levels generally lines up across theorists. At this date, we would be surprised by an error in the sequence after infancy. If there were such it would be in infancy. The Model of Hierarchical Complexity does not address interest, as Carol Gilligan (1982) does. Nor does it address domain, content, or the conditions under which performance is obtained.

Commons et al. (1998) present Theorem 3: A linear order of development may exist only within a single domain, on single sequences of tasks. This theorem shows inconsistences in development across tasks and domains. The corresponding Model of Hierarchical Complexity Scoring Scheme discusses many systematic ways of producing variation in performance.

Attaining Postformal Stage Performance. Commons and Miller (1998) and Commons and Richards (2002) have described both stage transition and reasons why transition takes place or fails to take place. The first three steps (deconstruction) start with initially high loss of perceived reinforcement opportunity. But, during the advance through these initial steps, more reinforcement is obtained. Psychologically, the results are consistent with Jesus Rosales and Donald Baer's

(1997) work on *behavioral cusps*. They posit more cusps than there are stages, however. Most of the proposed psychological mechanism of transition seems to be consistent with these theories. Despite this, most theories do not operationalize clearly the steps in transitions or the empirical basis for transition.

Both within many neo-Piagetian accounts (e.g., Case, 1974, 1978, 1982, 1985) and Precision Teaching (e.g. Binder, 1995) accounts, automatization of previous stage behavior is predicted to improve the rate of obtaining next-stage performance. From the data from Precision Teaching, fluency in lower-stage tasks greatly enhances the acquisition of the new stage tasks. As a task is completed near the maximum rate and errors almost disappear, the actions are said to become automatized. Hence, such over-learning leads to automatization. The task stimuli are said to become "chunked." That is, each individual stimulus in the task has to be discriminated individually, but still as a whole.

Although all tasks must have an order of hierarchal complexity, performance on such tasks depends on many other task characteristics. They include: level of support (Commons & Richards, 1995; Fischer et al., 1984), horizontal complexity, fluency of performance on the component tasks, "talent," interest, and so forth. Hence, one expects complex interrelationships between measures of performance on tasks and conditions of measurement. The stage of performance should be curvilinear when plotted against subject's chronological age (Armon & Dawson, 1997; Dawson, 1998) and linear when plotted against log age. Any variability should increase with age, and it does. Yet, there is some evidence that at the higher stages there is less spread. The proclivity to integrate relationships and systems and even paradigm from many domains probably increases with postformal stage.

Implications. There are a number of reasons that postformal stages are important. Without our giving order of importance by order of presentation, here are some of the reasons. Postformal stages may account for part of unequal accomplishment. They might account for part of the differences among individuals within a social group as to things such as income and academic performance. Stage of performance might be used to evaluate the effect of a culture on social, political, and educational development (Bowman, 1996; Commons & Goodheart, 1999; Commons, Krause, Fayer, & Meaney, 1993; Commons & Rodriguez, 1990, 1993). They might make clear the evolutionary implications of attaining postformal performance (Commons & Bresette, 2000). Because education is a strong predictor of stage of performance, we might be better at finding out why, if we understand the postformal stages and their measurement better. Society might be organized to produce and pay off the acquisition of postformal performance. Individually, the benefits and costs of postformal development in the wide domains investigated could be understood.

There are clear social benefits, including improved moral and ethical atmosphere, to the attainment of increased postformal development. As social perspective skills increase with these stages, the plight of the underclasses will improve. Dialogical means are used in order to conduct real discussion in the process of making policy. The abuse of power is decreased.

Another and important social benefit is that postformal development is associated with increased innovation. Innovators functioning at each of the four stages do tasks of different hierarchical complexity that do not overlap with one another. They do the different tasks using skills that are increasingly rare. The end results are entirely different for society. People have been known to buy the expertise of people functioning at the systematic and metasystematic stages; however, we posit that a person must function in the area of innovation at least at the metasystematic order of hierarchical complexity or higher to produce truly creative innovations. That means that at least two multivariate systems must be coordinated.

The results of innovation become much more expensive at the paradigmatic and cross-paradigmatic stages. In fact, at the cross-paradigmatic stage, so few people exist that societies have no mechanisms to encourage such activity as far as we know. Yet it is the cross-paradigmatic skills that change the course of civilization.

The development of complexity in human societies depends on innovations by single individuals. The innovator has the tendency to discern and discriminate relationships among elements that are extremely complicated. Making an innovation is much more difficult than learning about it after it is made. Major cultural innovations require paradigmatic complexity (Commons & Richards, 1995) at least because there is no support whatever from within the cultures themselves. The difficulty of an action depends on level of support in addition to the horizontal information demanded in bits, and the order of hierarchical complexity. The *level of support* (Fischer et al., 1984) represents the degree of independence of the performing person's action and thinking from environmental control provided by others in the situation.

There is little support for major innovations in culture because the history of the necessary hierarchical complexity surrounding the task is absent. Nor is there a history of reinforcement that would induce the subject to detect new phenomena. "Finding" a given question increases complexity demand by one order of complexity over solving a posed problem with no assistance. Finding the question allows for finding a problem to address that question, which increases the complexity demanded by one further order (Arlin, 1975, 1977, 1984). Finding and identifying the underlying phenomenon requires still a third additional order of complexity.

Lastly, there are interpersonal and personal benefits to moving through the postformal stages. Relationships are seen in more equitable terms. The struggle for independence and dependence is integrated into a more functional interdependence in which a contribution to the needs and preferences of others is part of nonstrategic interaction. Unresolved conflicts are dealt with within a larger framework of co-constructing a workable dialog.

ACKNOWLEDGMENT.   Some of this material comes from three sources: (1) Commons, M. L., & Goodheart, E. A. (1999). The philosophical origins of behavior analysis. In B. A. Thyer (Ed.), *The philosophical foundations of behaviorism*. New York: Kluwer. (2) Commons, M. L., & Bresette, L. M. (2000). Major creative innovators as viewed through the lens of the general model of hierarchical complexity and evolution. In M. E. Miller & S. Cook-Greuter (Eds.), *Creativity, spirituality, and transcendence: Paths to integrity and wisdom in the mature self*. Stamford, CT: Ablex. (3) Richards, F. A., & Commons, M. L. (1990). Postformal cognitive-developmental research: Some of its historical antecedents and a review of its current status. In C. N. Alexander & E. J. Langer (Eds.), *Higher stages of human development:*

Perspectives on adult growth. New York: Oxford University Press. Patrice Marie Miller and others have edited the manuscript and made major suggestions for its improvement.

The standard stage sequence was constructed by Commons and Richards working with many people. Most importantly, from 1981 on, Pascual-Leone discussed half stages with Commons. Fischer and Commons also talked about them. Richards and Commons had most of the stages by 1978, although they missed the systematic and paradigmatic stages and were unsure of how to define the abstract stages as mentioned above. By 1983, they had them all. Most important to this process/accomplishment was the input of Fischer. He supplied the arguments for abstract stage, as well as his four levels per tier (but did not have empirical evidence for performance at the higher two levels). Koplowitz suggested the systematic stage while in conference with Commons during the editing of his chapter. He said if there is to be a general system or metasystem, it must be about some system, that is, it must consist of simpler systems. Elena Jomar (personal communication) suggested the paradigmatic stage at a lecture held at Harvard School of Education in 1984 by Carol Gilligan. She said that if there were a cross-paradigmatic stage there had also to be a paradigmatic stage to cross.

Sonnert and Commons (1994), after scoring a number of protocols, using both the Colby & Kohlberg (1987a, 1987b) manual and the Model of Hierarchical Complexity, came to see Kohlberg's moral judgment stage 4/5 as a transition. At Dorothy Danaher's wedding party in 1984, Higgins, Miller, and Danaher discussed with Kohlberg why he did not see his stage 4 as systematic and stage 5 as metasystematic postformal stages. He said that, although his stages were postformal, he had not thought to relate them to the post-Piagetian postformal work. Later, Kohlberg (1990) wrote a chapter relating them.

REFERENCES

Arlin, P. K. (1975). Cognitive development in adulthood: A fifth stage? *Developmental Psychology, 11,* 602–606.

Arlin, P. K. (1977). Piagetian operations in problem finding. *Developmental Psychology, 13,* 247–298.

Arlin, P. K. (1984). Adolescent and adult thought: A structural interpretation. In M. L. Commons, F. A. Richards, & C. Armon (Eds.), *Beyond formal operations, Vol. 1: Late adolescent and adult cognitive development* (pp. 258–271). New York: Praeger.

Armon, C. (1984). Ideals of the good life and moral judgment: Ethical reasoning across the life span. In M. L. Commons, F. A. Richards, & C. Armon (Eds.), *Beyond formal operations, Vol. 1: Late adolescent and adult cognitive development* (pp. 357–380). New York: Praeger.

Armon, C., & Dawson, T. L. (1997). Developmental trajectories in moral reasoning across the Lifespan. *Journal of Moral Education, 26*(4), 433–453.

Baer, D. M., & Rosales-Ruiz, J. (1998). In the analysis of behavior, what does "development" mean? *Revista Mexicana de Analisis de la Conducta, 24*(2), 127–136.

Basseches, M. A. (1980). Dialectical schemata: A framework for the empirical study of the development of dialectical thinking. *Human Development, 23,* 400–421.

Basseches, M. A. (1984). Dialectical thinking as a metasystematic form of cognitive organization. In M. L. Commons, F. A. Richards, & C. Armon (Eds.), *Beyond formal operations: Late adolescent and adult cognitive development* (pp. 216–257). New York: Praeger.

Benack, S. (1984). Postformal epistemologist and the growth of empathy. In M. L. Commons, F. A. Richards, & C. Armon (Eds.), *Beyond formal operations, Vol. 1: Late adolescent and adult cognitive development* (pp. 340–356). New York: Praeger.

Bickhard, M. H. (1978). The nature of developmental stages. *Human Development, 21,* 217–233.

Bickhard, M. H. (1979). On necessary and specific capabilities in evolution and development. *Human Development*, *22*, 217–224.

Binder, C. (1995). Promoting Human Precision Teaching (HPT) innovation: A return to our natural science roots. *Performance Improvement Quarterly*, *8*(2), 95–113.

Boole, G. (1854). *Laws of thought*. London: Walton and Maberly.

Bourbaki, N. (1939). *Elements de mathematique*. Paris: Hermann.

Bowman, A. K. (1996). The relationship between organizational work practices and employee performance: Through the lens of adult development. In partial fulfillment of the requirements for the degree of Doctor of Philosophy, Human and Organization Development, The Fielding Institute, Santa Barbara.

Broughton, J. M. (1977). Beyond formal operations: Theoretical thought in adolescence. *Teachers College Record*, *79*(1), 87–98.

Broughton, J. M. (1984). Not beyond formal operations, but Beyond Piaget. In M. L. Commons, F. A. Richards, & C. Armon (Eds.), *Beyond formal operations: Late adolescent and adult cognitive development* (pp. 395–411). New York: Praeger.

Campbell, R. L., & Richie, D. M. (1983). Problems in the theory of developmental sequences: Prerequisites and precursors. *Human Development*, *26*, 156–172.

Case, R. (1974). Structures and strictures: Some functional limitations in the course of cognitive growth. *Cognitive Psychology*, *6*, 544–573.

Case, R. (1978). Intellectual development from birth to adulthood: A Neo-Piagetian interpretation. In R. Siegler (Ed.), *Children's thinking: What develops?* Hillsdale, NJ: Lawrence Erlbaum.

Case, R. (1982). The search for horizontal structures in children's development. *The Genetic Epistemologist*, *11*(3), 1–12.

Case, R. (1985). *Intellectual development: Birth to adulthood*. Orlando, FL: Academic Press.

Colby, A., & Kohlberg, L. (1987a). *The measurement of moral judgment: Vol. 1. Theoretical foundations and research validation*. New York: Cambridge University Press.

Colby, A., & Kohlberg, L. (Eds.) (1987b). *The measurement of moral judgment: Standard form scoring manuals*. New York: Cambridge University Press.

Commons, M. L. (1999). Threads of adult development. *Adult Developments*, 1–2.

Commons, M. L., & Bresette, L. M. (2000). Major creative innovators as viewed through the lens of the general model of hierarchical complexity and evolution. In M. E. Miller & S. Cook-Greuter (Eds.), *Creativity, spirituality, and transcendence: Paths to integrity and wisdom in the mature self* (pp. 167–187). Stamford, CT: Ablex.

Commons, M. L., & Goodheart, E. A. (1999). The philosophical origins of behavior analysis. In B. A. Thyer (Ed.), *The philosophical legacy of behaviorism* (pp. 9–40). London: Kluwer Academic.

Commons, M. L., Goodheart, E. A., & Bresette, L. M. with N. F. Bauer, E. W. Farrell, K. G. McCarthy, D. L. Danaher, F. A.. Richards, J. B. Ellis, A. M. O'Brien, J. A. Rodriguez, & D. Schrader (1995). Formal, systematic, and metasystematic operations with a balance-beam task series: A reply to Kallio's claim of no distinct systematic stage. *Adult Development*, *2*(3), 193–199.

Commons, M. L., Goodheart, E. A., & Dawson, T. L. (1997). A Saltus analysis of developmental data from the balance beam task series. Presented at The Ninth International Objective Measurement Workshop, Chicago.

Commons, M. L., Krause, S. R., Fayer, G. A., & Meaney, M. (1993). Atmosphere and stage development in the workplace. In J. Demick & P. M. Miller (Eds.), *Development in the workplace* (pp. 199–218). Hillsdale, NJ: Lawrence Erlbaum.

Commons, M. L., & Miller, P. M. (1998). A Quantitative behavior-analytic theory of development. *Mexican Journal of Experimental Analysis of Behavior*, *24*(2), 153–180.

Commons, M. L., & Richards, F. A. (1978). The structural analytic stage of development: A Piagetian postformal stage. Presented at Western Psychological Association, San Francisco.

Commons, M. L., & Richards, F. A. (1984a). A general model of stage theory. In M. L. Commons, F. A. Richards, & C. Armon (Eds.), *Beyond formal operations, Vol. 1: Late adolescent and adult cognitive development* (pp. 120–140). New York: Praeger.

Commons, M. L., & Richards, F. A. (1984b). Applying the general stage model. In M. L. Commons, F. A. Richards, & C. Armon (Eds.), *Beyond formal operations: Late adolescent and adult cognitive development* (pp. 141–157). New York: Praeger.

Commons, M. L., & Richards, F. A. (1995). Behavior analytic approach to dialectics of stage performance and stage change. *Behavioral Development Bulletin*, *5*(2), 7–9.

Commons, M. L., Richards, F. A., & Kuhn, D. (1982). Systematic and metasystematic reasoning: A case for levels of reasoning beyond Piaget's stage of formal operations. *Child Development, 53*, 1058–1068.

Commons, M. L., Richards, F. A., Trudeau, E., Goodheart, A. E., &. Dawson, T. L. (1997). Psychophysics of stage: Task complexity and statistical models. Presented at The Ninth International Objective Measurement Workshop, Chicago.

Commons, M. L., & Rodriguez, J. A. (1990). "Equal access" without "establishing" religion: The necessity for assessing social perspective-taking skills and institutional atmosphere. *Developmental Review, 10*, 323–340.

Commons, M. L., & Rodriguez, J. A. (1993). The development of hierarchically complex equivalence classes. *Psychological Record, 43*, 667–697.

Commons, M. L., Trudeau, E. J., Stein, S. A., Richards, F. A., & Krause, S. R. (1998). The existence of developmental stages as shown by the hierarchical complexity of tasks. *Developmental Review, 8*(3), 237–278.

Copernicus, N. (1992/1543). [De revolutionibus orbium caelestium.] On the revolutions/Nicholas Copernicus; translation and commentary by Edward Rosen. Baltimore, MD: Johns Hopkins University Press.

Darwin, C. (1855). *On the origin of the species.* London: Murray.

Darwin, C. (1873). *Expressions of the emotions in man and animals.* New York: D. Appleton.

Dawson, T. L. (1998). "A good education is … " A life-span investigation of developmental and conceptual features of evaluative reasoning about education. Unpublished doctoral dissertation, University of California at Berkeley, Berkeley, CA.

Demetriou, A. (1990). Structural and developmental relations between formal and postformal capacities: Towards a comprehensive theory of adolescent and adult cognitive development. In M. L. Commons, C. Armon, L. Kohlberg, F. A. Richards, T. A. Grotzer, & J. D. Sinnott (Eds.), *Adult development, Vol. 2: Models and methods in the study of adolescent and adult thought* (pp. 147–173). New York: Praeger.

Demetriou, A., & Efklides, A. (1985). Structure and sequence of formal and postformal thought: General patterns and individual differences. *Child Development, 56*, 1062–1091.

Descartes, R. (1637/1954). The geometry of Rene Descartes: With a facsimile of the first edition, 1637/translated from the French and Latin by David Eugene Smith and Marcia L. Latham. Garden City NY: Dover.

Einstein, A. (1950). *The Meaning of Relativity.* Princeton, NJ: Princeton University Press.

Erickson, E. H. (1959). Identity and the life cycle. *Psychological Issues Monograph, I.*

Erickson, E. H. (1978). *Adulthood.* New York: W. W. Norton.

Fischer, K. W. (1980). A theory of cognitive development: The control and construction of hierarchies of skills. *Psychological Review, 87*, 477–531.

Fischer, K. W., Hand, H. H., & Russell, S. (1984). The development of abstractions in adolescents and adulthood. In M. L. Commons, F. A. Richards, & C. Armon (Eds.), *Beyond formal operations: Late adolescent and adult cognitive development* (pp. 43–73). New York: Praeger.

Flavell, J. H. (1971). Stage-related properties of cognitive development. *Cognitive Psychology, 2*, 421–453.

Flavell, J. H. (1972). An analysis of cognitive-developmental sequences. *Genetic Psychology Monographs, 86*, 279–350.

Flavell, J. H. (1977). *Cognitive development.* Englewood Cliffs, NJ: Prentice-Hall.

Flavell, J. H. (1982). Structures, stages, and sequences in cognitive development. In W. Andrew Collins (Ed.), *The concept of development.* Hillsdale, NJ: Lawrence Erlbaum.

Flavell, J. H. & Wohlwill, J. F. (1969). Formal and functional aspects of cognitive development. In D. Elkind & J. H. Flavell (Eds.), *Studies in cognitive development: Essays in honor of Jean Piaget.* New York: Oxford University Press.

Frege, G. (1950). *The Foundations of Arithmetic* (J. L. Austin, Trans.). Oxford: Oxford University Press.

Gilligan, C. (1982). *In a different voice: Psychological theory and women's development.* Cambridge, MA: Harvard University Press.

Gödel, K. (1977). Some metamathematical results on completeness and consistency; On formal undefinable propositions of *Principal Mathematica* and related systems I; On completeness and consistency. In J. Heijehoort (Ed.), *From Frege to Gödel: A source book in mathematical logic 1879–1931.* Cambridge, MA: Harvard University Press, 1977. (Originally published 1930, 1931, 1931, respectively.)

Inhelder, B., & Piaget, J. (1958). *The growth of logical thinking from childhood to adolescence: an essay on the development of formal operational structures* (A. Parsons & S. Milgram, Trans.). New York: Basic Books (originally published, 1955).

Kallio, E. (1991). Formal operations and postformal reasoning: A replication. *Scandinavian Journal of Psychology, 32*(1), 18–21.

Kallio, E. (1995). Systematic reasoning: Formal or postformal cognition? *Journal of Adult Development, 2*, 187–192.

King, P. M., Kitchener, K. S., Wood, P. K., & Davison, M. L. (1989). Relationships across developmental domains: A longitudinal study of intellectual, moral, and ego development. In M. L. Commons, J. D. Sinnott, F. A. Richards, & C. Armon, (Eds.). *Adult Development, Vol. 1: Comparisons and applications of adolescent and adult developmental models* (pp. 57–72). New York: Praeger.

Kitchener, K. S., & King, P. M. (1990). Reflective judgement: Ten years of research. In M. L. Commons, C. Armon, L. Kohlberg, F. A. Richards, T. A. Grotzer, & J. D. Sinnott (Eds.), *Beyond formal operations, Vol. 2: Models and methods in the study of adolescent and adult thought* (pp. 63–78). New York: Praeger.

Kohlberg, L. (1969). Stage and sequence: The cognitive-developmental approach to socialization. In D. A. Goslin (Ed.), *Handbook of socialization theory and research.* Chicago: Rand McNally.

Kohlberg, L. (1981). The meaning and measurement of moral development. *The Heinz Werner Lecture Series (1979), Vol. XIII.* Worcester, MA: Clark University Press.

Kohlberg, L. (1984). *Essays on moral development, Vol. 2: The psychology of moral development: Moral stages, their nature and validity.* San Francisco: Harper & Row.

Kohlberg, L. (1984). *Essays on moral development, Vol. 2: The psychology of moral development: Moral stages, their nature and validity.* San Francisco: Harper & Row.

Kohlberg, L. (1990). Which postformal levels are stages? In M. L. Commons, C. Armon, L. Kohlberg, F. A. Richards, T. A. Grotzer, & J. D. Sinnott (Eds.), *Adult Development, Vol. 2: Models and methods in the study of adolescent and adult thought* (pp. 263–268). New York: Praeger.

Kohlberg, L., & Armon, C. (1984). Three types of stage models used in the study of adult development. In M. L. Commons, F. A. Richards, & C. Armon (Eds.), *Beyond formal operations: Late adolescent and adult cognitive development* (pp. 383–394). New York: Praeger.

Koplowitz, H. (1984). A projection beyond Piaget's formal operational stage: A general system stage and a unitary stage. In M. L. Commons, F. A. Richards, & C. Armon (Eds.), *Beyond formal operations: Late adolescent and adult cognitive development.* (pp. 272–295). New York: Praeger.

Krantz, D. H., Atkinson, R. C., Luce, R. Duncan, & Suppes, P. (1974). *Contemporary Developments in Mathematical Psychology, Vol. 2: Measurement, psychology, and neural information processing.* San Francisco: W. H. Freeman.

Krantz, D. H., Luce, R. D., Suppes, P., & Tversky, A. (1971). *Foundations of measurement, Vol. 1: Additive and polynomial representations.* New York: Academic Press.

Labouvie-Vief, G. (1980). Beyond formal operations: Uses and limits of pure logic in life-span development. *Human Development, 23*, 141–161.

Labouvie-Vief, G. (1984). Logic and self-regulation from youth to maturity: A model. In M. L. Commons, F. A. Richards, & C. Armon (Eds.), *Beyond formal operations: Late adolescent and adult cognitive development* (pp. 158–179). New York: Praeger.

Leibniz, G. W. (1768). L. Deutens (Ed.), *Opera Omnia.* Geneva: Deutens.

Leibniz, G. W. (1875). *Die Philosophische Schriften.* Berlin: Gerhardt.

Linn, M. C., & Siegal, H. (1984). Postformal reasoning: A philosophical model. In M. L. Commons, F. A. Richards, & C. Armon (Eds.), *Beyond formal operations: Late adolescent and adult cognitive development* (pp. 239–257). New York: Praeger.

Maxwell, J. C. (1871). *Theory of heat.* London: Longmans and Green & Co., (2nd ed.).

Newton, I. (1687/1999). *The Principia: mathematical principles of natural philosophy.* I. Bernard Cohen and Anne Whitman, assisted by Julia Budenz (trans.); preceded by a guide to *Newton's Principia* by I. Bernard Cohen. Berkeley: University of California Press.

Overton, W. F. (1990). Competence and procedures: Constraints on the development off logical reasoning. In W. F. Overton (Ed.), *Reasoning, necessity and logic: Developmental perspectives* (pp. 1–32). Hillsdale, NJ: Lawrence Erlbaum.

Pascual-Leone, J. (1984). Attention, dialectic, and mental effort: Towards an organismic theory of life stages. In M. L. Commons, F. A. Richards, & C. Armon (Eds.), *Beyond formal operations: Late adolescent and adult cognitive development* (pp. 182–215). New York: Praeger.

Peano, G. (1894). *Notations de Logique mathematique.* Turin: G. Guadagin.

Piaget, J. (1950). *The psychology of intelligence.* (M. Piercey and D. E. Berlyne, Trans.). London: Routledge & Kegan Paul.

Piaget, J. (1952). Autobiography. In E. G. Boring, H. S. Langfeld, H. Warner, & R. M. Yerkes (Eds.), *A history of psychology in autobiography, Vol. IV.* Worcester, MA: Clark University Press.

Piaget, J. (1954). *The construction of reality in the child.* (M. Cook, Trans.). New York: Ballantine Books.

Piaget, J. (1970). *Structuralism.* (C. Maschler, Trans.). New York: Harper and Row.

Piaget, J. (1971). *Biology and knowledge: An essay on the relations between organic regulations and cognitive processes.* Chicago: The University of Chicago Press.

Piaget, J. (1972). Intellectual evolution from adolescence to adulthood. *Human Development, 15(1),* 1–12.

Piaget, J. (1976). *The grasp of consciousness: Action and concept in the young child.* Cambridge, MA: Harvard University Press.

Piaget, J., & Inhelder, B. (1969). Intellectual operations and their development. In P. Fraisse & J. Piaget (Eds.), *Experimental psychology: Its scope and method, Vol. VII.* (T. Surridge, Trans.). New York: Basic Books.

Piaget, J. & Inhelder, B., with Sinclair de Zwart, H. (1973). *Memory and intelligence.* (A. Pomerans, Trans.). New York: Basic Books.

Planck, M. (1922). *The origin and development of the quantum theory,* by Max Planck, trans. by H. T. Clarke & L. Silberstein; being the Nobel prize address delivered before the Royal Swedish Academy of Sciences at Stockholm, 2 June, 1920. Oxford: The Clarendon Press.

Powell, P. M. (1984). Stage 4A: Category operations and interactive empathy. In M. L. Commons, F. A. Richards, & C. Armon (Eds.), *Beyond formal operations, Vol. 1: Late adolescent and adult cognitive development* (pp. 326–339). New York: Praeger.

Rasch, G. (1980). *Probabilistic model for some intelligence and attainment tests.* Chicago: University of Chicago Press.

Richards, F. A., & Commons, M. L. (1984). Systematic, metasystematic, and cross-paradigmatic reasoning: A case for stages of reasoning Beyond formal operations. In M. L. Commons, F. A. Richards, & C. Armon (Eds.), *Beyond formal operations, Vol. 1: Late adolescent and adult cognitive development* (pp. 92–119). New York: Praeger.

Riegel, K. F. (1973). Dialectic operations: The final phase of cognitive development. *Human Development, 16,* 346–370.

Rosales-Ruiz, J., & Baer, D. M. (1997). Behavioral cusps: A developmental and pragmatic concept for behavioral analysis. *Journal of Applied Behavioral Analysis, 30,* 533–544.

Sinnott, J. D. (1981). The theory of relativity: A metatheory for development? *Human Development, 24(5),* 293–311.

Sinnott, J. D. (1984). Post-formal reasoning: The relativistic stage. In M. L. Commons, F. A. Richards, & C. Armon (Eds.), *Beyond formal operations: Late adolescent and adult cognitive development* (pp. 298–325). New York: Praeger.

Sonnert, G., & Commons, M. L. (1994). Society and the highest stages of moral development. *The individual and society, 4(1),* 31–55.

Sternberg, R. J. (1984). Higher-order reasoning in postformal operational thought. In M. L. Commons, F. A. Richards, & C. Armon (Eds.), *Beyond formal operations: Late adolescent and adult cognitive development* (pp. 74–91). New York: Praeger.

Sternberg, R. J., & Downing, C. J. (1982). The development of higher-order reasoning in adolescence. *Child Development, 53,* 209–221.

CHAPTER 12

Postformal Thought and Adult Development

Living in Balance

JAN D. SINNOTT

OVERVIEW

Adulthood may be described as a stage of life in which we wrestle with the mystery of existence, in the midst of life's chaos, in the here-and-now rather than in some perfect potential future. The unique qualities of adulthood can be difficult to describe developmentally. Adults are neither changing *toward* some defined endpoint, nor changing *away* from some past perfection. Instead, healthy adults keep a dynamic homeostasis; they balance. They change as if in a dance, or as if practicing an Eastern art such as tai chi, moving through forms and paces. In dancing, what matters and what is enjoyable—the whole purpose in fact—is movement in the present moment, and, of course, not falling over. In dancing, the process itself *is* the goal. To choose to dance is to pick a form and simply do it; the walk within the dance does not "get" anywhere! But for those of us in Western cultures, especially those of us who study human behavior and are enculturated to seek goals and linear causes, the metaphor of the "dance" of development might become an *irritating* metaphor, too, because Westerners want to know *where* the dance of adult development is going. If we do demand to know the destination (in terms of life trajectory), the dance seems to be going to a destination no grander than death. Aware of this, the conscious development of adults seems to include the ability to take part enthusiastically in the dance of life for its own sake, with an awareness of mortality.

JAN D. SINNOTT • Department of Psychology, Towson University, Towson, Maryland, 21252.

Handbook of Adult Development, edited by J. Demick and C. Andreoletti. Plenum Press, New York, 2002.

Students of adult psychosocial development also encounter challenging qualitative, nonlinear processes in adults, processes with terms that are difficult to quantify. Such processes come from studies of conscious adults who say that they strive to live "meaningfully," who report qualitative differences stemming from events in the stage of adulthood, and who say they more and more live "in the present" and "for the sake of others" because they know their futures are finite. "Success" or "successful development" in such individuals means that a unique dance goes on for each individual, with each dancer in *balanced movement*, until the ending of individual dances; it does not necessarily mean something so concrete as higher scores on tests or fewer doctor visits. Such processes are more difficult to describe because they resemble the emergent processes of self-regulating systems.

In this chapter my goal is to address this confusing psychosocial developmental dance movement and to study it more deeply. I chose to approach it from the direction of cognitive development, knowing that cognitive development is linked to emotion, spirituality, community, and existential meaning. I'll describe the relationship between complex cognition and the experiences and behavior of adults, as they develop in adulthood. I chose this *cognitive* quality for two reasons. First, I am interested in how adults develop their thinking processes as they develop overall. Second, my studies have shown that adults use their thinking to cocreate the reality of their adult lives. Thus the complexity of their thinking has huge implications for the quality of their lives and their relationships. It organizes the relationships among the subpersonalities within the person; it organizes relationships among persons; and it organizes the relationships between the adult and the transcendent or spiritual. Their cognition links up with their perception, their social and community milieu, their emotions, their life stage demands, their spirituality, and any other of their behavioral or experiential domains. The type of cognition that I have found to be associated with the greatest balance in adult development is postformal thought, which I describe briefly in the language of my own Theory of Postformal Logical Thought (a term that is capitalized when it refers to my own specific theory). Logic is one way of describing relationships or world views on which one's "truth" is built. Postformal logic describes the cognitive dance that in turn gives a language to the behavioral and experiential dance of balanced adult development. So the theory of postformal thought is a kind of shorthand that describes the way adults can epistemologically interact with the "objective" phenomena of their lives, cocreating balanced developmental dances.

To make this complex argument, this chapter is divided into three sections. First I show how the level of complex cognition is key in organizing the experience and behavior of adults as they develop a balanced dance of adulthood. Second I briefly describe the nature of Postformal Thought as described in my Theory of Postformal Thought (Sinnott, 1998b). Finally I show how having this complex ability allows one to develop with greater balance as he or she dances through some issues of adult life, for example, facing mortality, building intimate relationships, creating community, and experiencing spirituality (Erikson, 1950, 1982). Intimate relationships and Postformal Thought are analyzed in greater depth. I argue that each adult person uses Postformal Thought to create the consciously known balance, identity, meaning, and goals of his or her uniquely changing dance of existence.

HOW COGNITION ORDERS CONSCIOUS ADULT DEVELOPMENTAL EXPERIENCES

The process of adult psychological and psychosocial development might be described as an integrated series of conscious or unconscious actions within the person, actions to construct and balance the self, fueled by internal or external energy and events.

Here is a simple set of terms to illustrate what I have in mind. This abstract behavioral sequence emphasizes cognition. The process is a circular one with five steps; the process is nested within three variables; and the process further influences three variables outside of it. Nested within the *physical self*, within *society*, and within *historical and transcendental time*, (1) *prior cognition* influences (2) *perception*, which influences (3) *behavior*, which provides (4) *feedback*, which alters (5) *cognition*. The accumulated effects of cycles through this loop change variables outside the loop: effects change *identity*, and have an impact on the *physical self* and *society*.

As a concrete example, here is a story. Nested within the facts that I am a woman in my 50s (physical self) among my family and friends in Washington, DC (society) at the end of the millennium (historical time) and connected in spirit to all existence (transcendental time), I see my child (perception) about to graduate high school. I find coming to mind all the memories and awareness of what it meant to me to graduate and how I felt when other children of mine have done so and left for college (prior cognition). I code this as "a bittersweet, potentially sad occasion" and begin to take numerous pictures (behavior), trying to hold on to the moment to keep her from leaving the nest. During the graduation ceremony my parents turn to me and say (feedback) "You may have the pictures, but when they're gone the pictures aren't enough. It's not the same. But it's just the beginning of lots of happy *adult* times you'll share." I begin to realize that this is true, and holding on to the past is no fun (altered cognition). Through new eyes (new mental constructs) I begin to see myself (identity) as the successful mother of a new adult who will make my life interesting over the years. I begin to relax, lessening stress on my body (physical change), and begin to see others in the group "catch" my more upbeat view. "We did a good job!" I say, congratulating my partner, reconceptualizing the sad event as a triumph we can share to make us closer (family/social change). I am now ready to assimilate the next cognitive event through a slightly changed cognitive filter, ready to go through the circle again. But this next round will start off differently with an upbeat belief about what it means to have a child graduate.

Keeping this example in mind, here are some further descriptions of the abstract terms used in the preceding. Notice that some of these descriptions may deviate from classical definitions of the same terms. The *physical self* is the physical body including the bodily changes associated with the years between adolescence and death. *Society* is the set of relationships with others that each adult experiences. These may include intimate relationships with partners, family relationships, work and friendship relationships, cultural experiences, and any other relationships that form the interpersonal community in which the individual takes part. *Historical or transcendental time* means the particular historical era in which the person is developing and the spiritual background against which each adult sees life unfolding. In a sense each person is in a relationship with that more lasting, larger entity (perhaps "the Universe," "God," "Spirit," "History") that gives that person's life existential meaning from a transcendent perspective. *Prior cognition* is based on Piagetian theory and indicates the cumulative mental constructs, constructed by the knower up to this point in life, lenses by which the knower now grasps "objective" reality. *Perception* means the filtered new material received for processing. *Behavior* refers to any activity that a human performs consciously or unconsciously. *Feedback* is whatever follows from the behavior, in this case, that which alters the cognition or identity of the person. *Cognition* is framed from a Piagetian perspective and includes the person's current ways of constructively knowing the objective world. Finally, *identity* is the person's cognitive concept of self. I realize that these terms could be defined in very precise alternative ways, but have selected these definitions to facilitate a dialogue that takes the whole person's known experience into account.

Notice that the circular process described here involves in*ter*personal, in*tra*personal, and *trans*personal relationships. He or she does not develop alone. The first and most obvious kind of relationships are those that link the developing adult with other persons in a couple, a family, a society, or history. The second type of relationship is the one *within* the person, one relating the various sides of the self or various aspects of the personality. The third relationship, one that is especially pertinent to developing adults as they become aware of their mortality, is the relationship between the adult and the transcendent, that is, that which goes beyond historical time. This is the adult's relationship with God, spirit, or "background of existence" (to name just a few terms). The three types of relationships may be differentially important to an adult, and relationships of any sort may matter in different degrees to any specific adult. These relationships provide stimulation for further adult development by (among other things) challenging the present cognitions that organize action and identity. The adult must successfully bridge across the differing "logics" of the three types of relationships, an act that appears to require Postformal Thought. Notice that this shifting of logics, or any other part of the process, may be conscious or unconscious.

So how does cognition order conscious adult developmental experience? And what does it matter? Recall that I am attempting to show the connection between thinking in a complex way and successfully developing during the stage of life termed "adulthood." To do this I need to show the connection between cognition and adult development in general. The circular process described in the preceding links what the developing adult *thinks* to what he or she *does* or holds as a self definition. It links how reality is *filtered or received* by that person with how that person *interacts with the reality of the three types of relationships* of which he or she is part. Within the circle model, the adult will not be able to balance the existential, physical, identity, social, emotional, or other demands and disturbing feedback unless the cognitive part of the circle can handle and organize this complex information load, shifting among logics. Handling the load is exactly what complex Postformal Thought seems to help the adult do. In that way cognition (Postformal Thought) helps him or her live in balance and develop in adulthood.

WHAT IS COMPLEX POSTFORMAL THOUGHT?

Over the past few years we lifespan developmentalists have been faced with several new challenges to our cognitive epistemological theories: the challenge of the "new physics," that is, relativity theory and quantum physics; and the challenge of systems theory, chaos theory, and complexity theories. The new physics and related new theories demand a paradigm shift in the understanding of "objective" reality, as, for example, new physics defines objective large-scale physical reality as incorporating paradoxical logical contradictions and necessarily subjective choices about the nature of that physical reality (e.g., Sinnott, 1981, 1998b; Wolf, 1981; Zukav, 1979). At the same time these developments were unfolding, we developmentalists were faced with the task of accounting for the *positive* intellectual development that we saw in some adults as they matured and aged, positive development in spite of documented losses such as slower nerve conductance velocity, poorer vision, poorer memory, fewer choices about jobs, family changes such as kids leaving home, and so forth. Earlier research on adult development had focused heavily and almost exclusively on the negative changes that come in every domain at the end of the lifespan. Now we were collecting data that indicated numerous positive changes after adolescence, including cognitive ones. It seemed to be astounding and radical in certain circles to even suggest that experience and development might make at least some people wiser, permit them to function at higher levels, and let them experience more personal happiness, but it seemed true nonetheless.

Responding to these challenges instigated by new paradigms led me to create a theory of *Post*formal Complex Thought (e.g., Sinnott, 1981, 1984b, 1994a, 1994b, 1994c, 1994d, 1994e, 1996a, 1996b, 1998b; Sinnott & Cavanaugh, 1991). In my theory Postformal Thought is the last step in logical development during the lifespan. My Postformal Thought goes beyond Piaget's traditional stages of logical development (i.e., sensorimotor operations, preoperations, concrete operations, and formal operations). Knowers who are postformal sometimes find themselves working with multiple contradictory formal logical systems. Postformal Thought helps to organize these logically conflicting formal operational systems and demands. This logical superorganization is done within a specific context at a specific time. At that point Postformal Thought allows a choice of one formal logical reality (namely, one formal system), a choice that is adaptive for that time and context. That choice is then incorporated into ongoing thought and behavior, made "real" through subsequent living done within its framework (Sinnott, 1996b, 1998b).

Postformal Thought as I define it develops especially through the logical contradictions about "reality" encountered during social interactions, when one person's formal operational logical "truth" about the interaction contradicts that of another person or contradicts the shared truth of the culture. At that point of conflicting logics, a necessarily somewhat subjective (albeit passionate) choice is made of which reality to consider "real" in this case. The knower decides the rules of the game as part of playing the game, and goes on to *live* the reality selected, therefore cocreating reality, so to speak. This Postformal Thought could describe the form of the way that an Einstein could know the strange logical reality of the new physics. It could also describe the way individuals know on a deeply adaptive level the coconstructed realities of their relationships with one another and the coconstructed realities of a culture or a spiritual awareness.

For example, when I begin teaching any college class, the class and I begin to structure the reality of our class relationship by first mutually deciding the reality of our relationship; second, behaving based on that decision; and finally mutually creating our class relationship based on those decisions in the days that follow.

One student may see me as a surrogate parent and act within the Piagetian formal logic appropriate to that vision, to which I might respond by being parental. Another student may logically construct me as a buddy and act within that logical frame, to which I might respond by thinking and acting in the buddy logical frame (or by thinking and being more parental to compensate). The result over the time of a semester will be a unique relationship with this class that is created based on the logics we have chosen. These choices and decisions might be conscious.

Based on my 20 years of earlier research, Postformal Logical Thought seemed to develop later in life, after a certain amount of intellectual and interpersonal experience (Sinnott, 1998b). For example, only after experiencing intimate relationships, with the mutually constructed logics about the reality of intimate life together that come with those relationships, can a person be experienced enough to know that, "If I think of you as an untrustworthy partner, then treat you that way, you are likely to become an untrustworthy partner."

One of the most interesting aspects of Postformal thinking is the way in which it interlaces with the demands of life periods we customarily call middle age and old age. For example, a thinker in middle age optimally needs to consciously synthesize and balance work and family demands, not just choose one or the other of the logical structures of family *or* work. Likewise, an aging individual needs to *integrate* the formal logic of identity that says "I *am* partly my body" with the formal logic of identity that says "I *am not* entirely defined by my body."

Another interesting aspect of Postformal thinking is the way it can help describe and clarify familiar life-stage conflict themes such as those articulated by Erikson, for example, intimacy vs. isolation, generativity vs. stagnation, and integrity vs. despair. For example, a thinker hoping to be generative or reach integrity, in Erikson's terms, needs to be able to handle the logics of self vs. others and the logics of the many possible interpretations of the meaning of his or her life. Postformal thinking seemed to describe the unique ways that an individual expands and makes sense of the creations we call his or her "identity" and "life."

After the challenges of new physics paradigms and positive development in aging led me to construct my theory of Postformal Thought, I began to discover how useful a tool it was for describing many kinds of complex thought in adult life settings. For example, I gradually came to realize that this Postformal Self-Referential Thought helped describe what went on in the minds of mature and older adults who were addressing the issue of their changes, or even the issues raised by deeply mystical experiences and near-death experiences. The choosing of realities (by some of these mature respondents) was akin to that of individuals in therapy or of spiritual seekers who consciously move among different realities or different constructions of reality and are aware of their part in that process. Thus it appears that similar wise intellectual processes seemed to be serving several kinds of knowers processing many kinds of issues, some of them developmental (Sinnott, 1992, 1994a, 1994b, 1997, 1998a).

Postformal Thought Is Composed of Thinking Operations

To give a richer description of Postformal thinking it is important to look more closely at the details of this sort of thinking. The examination of operations that together make up the stage will make it easier for us to operationalize the concept.

The operations were taken from responses within open-ended dialogues, responses made by individuals who seemed to exemplify the wise, complex, generative, mature adult. These adults happened to cross my path while I was busy with other projects. But they caught my attention as adaptive and interesting people who not only survived the onslaughts of adulthood, but actually thrived there in that developmental period. I wondered in particular what aspect of their cognition made them so good at life.

When I looked at the articulated thoughts of these adaptive people I sensed they had a special way of describing processes of solving problems, problems in both the narrower sense of structured logical problems and problems in the broader sense of difficult situations encountered in the course of a life (or the course of a day!). I began to categorize the ways these people interacted with, constructed, and knew reality as they thought about it. That set of categories coalesced around 11 key operations.

These operations were the main ones, although others were present. The operations reflected some of the key thought patterns that distinguish the new physics from Newtonian physics. They also overlapped nicely with some variables in cognitive problem solving literature. Mapping their points of impact in the problem-solving process seemed possible if I used tools such as the thinking aloud approach from cognitive studies and tools such as artificial intelligence models from the computer-based AI literature (e.g., see Ericsson & Simon, 1984; Newell & Simon, 1972).

A description of the key thinking steps or operations on reality that appear to be present in Postformal Thought is given in the following section. It is possible to consider respondents to be Postformal thinkers even if they lack some operations in their answers. The more operations they show, however, the more certain the analyst is that they truly have attained Postformal Thought. Like so many other psychological qualities, Postformal cognitive ability seems to be analog rather than digital in quality.

It has proved useful to score some of the operations using a simple Present/Absent; others seemed to call for a count to be made of the times the operation appeared. This dual approach is based on practical concerns: certain operations seem to occur once at most in response to a given problem or issue while others occur often within one given issue or problem. For the latter, counting frequency can be used to do further analyses related to creativity, productivity, divergent thinking, and Postformal Thought. Within the list of operations, the following have generally been scored as simply Present/Absent: metatheory shift, process–product shift, pragmatism, paradox, and self-referential thought. The rest are scored for frequency, producing ratio-scale data, which can be reduced to nominal data as needed.

Operations and Rationale

Metatheory shift indicates that the respondents are able to think in at least two logic systems because they have shifted between an idealized interpretation of the problem and a practical interpretation of the problem and solved within those constraints. This is important for Postformal Thought because one essential element of Postformal Thought is the ability to order several formal logical operational systems.

Problem definition is a second way to get at the respondent's ability to move within those two or more formal operational systems. If the respondent overtly labels the problem to be a class of logical problems (as is required to receive a point for this operation) the respondent is, by definition, excluding classes of problems that it is not. Therefore the respondent is ordering more than one system. Problem definitions are counted because the more there are, the better the assurance that logics are truly being shifted.

Process–product shift occurs when the respondent indicates that a problem is solved with either a *process* that is a logical system that would work in many cases like this problem case or a *product* which is a concrete solution, for example, a correct numerical answer. Again, two logical systems are coordinated within the thinking of the respondent.

Parameter setting involves the respondent's limiting or organizing the problem space. This ability relates to Postformal Thought as defining the space of the problem opens or limits the logical structure(s) of the problem. Again, to define the problem presumes that it could be otherwise defined, potentially having a different logic. Here the number of defining acts becomes somewhat relevant. If the respondent gives only one parameter of the problem he or she is less likely to be holding at least two logics about the problem.

Pragmatism, defined as being able to select one of several solutions as "better," is included in the operations set because the Postformal thinker needs to be able to choose a single logic among more than one of them and make a commitment to go forward with that logic, as opposed to one of the other logics in play.

Multiple solutions is another of the counted operations. It is included because if one problem is posed but several solutions that are considered correct are generated, experience and probes of answers have led to the conclusion that more than one logic exists. This suggests Postformal Thought. *Multiple goals* and *multiple methods* and *multiple causality* are counted operations with the same rationale as multiple solutions has.

Paradox is a device in literature and humor that takes advantage of the intellect's ability to find the weird aspect of the overlap of two logics. Therefore it indicates the presence of ability to order logical systems. Paradox is not generated as frequently in structured and abstract testing situations as in everyday sorts of testing situations. The interesting people I noticed at first, when I was starting this series of studies, used paradox spontaneously as they spoke casually, and used it often.

Self-referential thought is the articulation of the respondent's awareness that he or she must be the ultimate judge of the logic to commit to. Here the respondent is conscious of using Postformal Thought.

Measures of Postformal Thought

Measures that have been developed so far include a standardized interview form; a thinking aloud form, with and without probe questions; a paper-and-pencil form specific to selected job contexts; and a computerized version. All of these versions are useful to obtain information on a respondent's use of postformal operations. All forms but the paper-and-pencil form (which asks directly about the use of operations at work) use between 6 and 12 problems that are based on formal combinatorial and proportionality reasoning in various contexts. The various forms are reliable and have face validity, predictive validity, and construct validity. Transcripts can be reliably coded using the scoring methods outlined earlier. Postformal thinking operations reliably appear in the thought of adults in every subsample tested to date.

Having described what complex Postformal Thought is, let's turn to a description of how this cognitive quality might influence the life of an adult. Our focus is on the areas of mortality, relationships, learning, and community.

EXAMPLES OF THE INFLUENCES OF POSTFORMAL THOUGHT ON ADULT DEVELOPMENT: INTIMATE RELATIONSHIPS

To illustrate how an adult's *cognitive* development (allowing him or her to use Postformal Thought) both changes and promotes the adult's *overall* development, I'll discuss cognitive development in the context of intimate relationships. Further discussion of these topics and of the relationship between Postformal Thought and existential questions, spirituality and mysticism, community, work, models of women's midlife development, adult learning at the university level, creativity, identity formation, healing, psychopathology, teaching, and other issue areas can be found in my earlier work (e.g., Sinnott, 1993a, 1994a, 1994b, 1997, 1998a, 1998b). Readers are invited to go to original sources in the reference list, as the discussion in this chapter is necessarily brief.

Intimate Relationships: Couple and Family Relationships

The ability to use Postformal Thought can change the quality of intimate relationships, the formation of which is a key task for the developing adult. The more complex nature of relationships between intimates capable of *postformal* cognition can challenge development further, stimulating further growth.

Individuals must coordinate multiple realities postformally, at least some of the time, if they are to succeed in couples and in families. Intimate relationships are intense interactions, by definition, with emotions weaving through each interaction and often contributing heat to any light that cognition may shed. Framing actions of family members in one cognitive context rather than another has implications for daily encounters and decisions. For a couple or a family that remain such for a longer period of time, the entire enterprise is colored by the history of past cognition, emotion, and action.

Styles of dealing with this intensity are defined by the emotional defense patterns, the shared cultural reality, the history, and the cognitive development of each person in the relationship. The presence or absence of postformal thinking can skew not only defense styles, but also responses to shared cultural reality and the history of that couple or family. Similarly, the emotionally based defense mechanisms, the shared culture, and the relationship history can distort the development or use of Postformal Thought by individuals or intimate groups at later points in their history. Further, these patterns are taking shape within the psychodynamics of each individual, within the individual's ongoing dialogue between the ideal and real self-in-relationship, between members of the couple, in the *couple* as a unique living system in *its own* right, between any of these "selves" and society, *and* in the family as a living system. Notice that even the relationship itself begins to take on a life of its own, going on with a history somewhat separate from the histories of the individuals within it. It is as if additional layers of complexity were overlaid on triangular theories of intimate relationships such as Sternberg's (Sternberg, 1986) or Marks' (Marks, 1986), and on relational "stage" theories such as Campbell's (Campbell, 1980) such that each triangle of relationship features or each stage becomes four dimensional and transforms over time! And each of these realities has its own "logic."

We start our discussion by examining the interplay and mutual causality of each of the main relational elements (which interact with Postformal Thought) just mentioned: defense patterns, cultural reality, and relationship history. How might postformal thinking help couples and families make it through life? How might their relational life stimulate the development of Postformal Thought? We examine how these connections relate to marital and family harmony, distress, and potential therapeutic strategies. Then we summarize a sample of some research on couple relationships, a project initiated by Rogers (1989; Rogers, Sinnott & Van Dusen, 1991), in which Postformal Thought is taken into account as a factor.

Interplay of Postformal Thought and Other Elements in Intimate Behavior

The adaptive value of Postformal Thought is that it can help bridge logical realities so that partners in a relationship can reorder logically conflicting realities in more complex ways. This skill can let the knower(s) handle more information, live in a state of multiple realities that logically conflict, and become committed enough to a chosen reality to go forward and act, thereby reifying the chosen (potential) reality.

In an intimate relationship individuals attempt to join together to have one life, to some degree. As the Apache wedding blessing says, "Now there is one life before you ... (so) ... enter into the time of your togetherness." For a couple, three "individuals" begin to exist: partner 1, partner 2, and the relationship that begins to take on a life of its own. For a family, of course, there are even more "individuals" present, as family therapist Virginia Satir (1967) noted when she worked with not only the real humans in her office but also with the remembered aspects of other absent relatives with whom the real humans psychologically interact. To have that one life together, to whatever extent they wish to have it together, the logics of the individuals must be bridged effectively. Those logics might be *about* any number of things, some of which do not sound especially "logical," including concepts, roles, perceptions, physical presence, emotions, and shared history.

Effects of the Individual's Cognitive Postformal Skills on Intimate Relationships

In some cases the individual has access to a cognitive bridge across realities, but not a postformal one. After all, realities *are* bridged, though poorly, if one person in a relationship dominates another and that one's reality becomes the other's, too. But this domination does not require a synthesis of logics since one logic is simply discarded.

There is a variation, too, on this theme. Two members of an intimate union, because of religious beliefs or personal emotional needs, also may drop their own logics to give preference to that of the new "individual," namely "the relationship," letting its role-related reality dominate both of their own individual ones. This, too, is not a postformal synthesis but a capitulation.

A converse scenario might find the logic of that third individual, "the relationship," dominated and discarded by one or both partners' logics. In all these cases one individual has lost part of the self, in order to maintain the relationship.

Members of the intimate partnership may find this winner/loser solution to conflicts large and small to be the best fit for them; they may not be capable of any higher-level solution, may not be motivated to find it, or may not be emotionally ready for one. But *Postformal* Thought is not involved in this sort of resolution. As Maslow (1968) might suggest, this less skilled move simply might be the best move they can think of right now to deal with their situation, even if it raises new problems (in relationship terms) down the line. Perhaps this less skilled behavior is so incorporated into their shared history that it would be a challenge to their relational identity for them to have a relationship without it. But it is not postformal thinking, and it is less adaptive overall than postformal thinking would be.

Predictable Couple Problems When One Logic Dominates

When one relational logic simply dominates the other a chance for growth is lost. There also come to be emotional overtones that begin to color the relationship processes and relationship history. The partner or family member whose logic was dominated usually exacts a price, whether consciously or unconsciously, expecting a payback to allow the balance of power to return to the relationship. And less information can be taken in, integrated, and acted on by an individual using one simpler logic as opposed to two logics, or as opposed to a more complex single one such as postformal logic.

There are several predictable outcomes that are less than optimal when one logic dominates another. Lack of movement forward toward Postformal Thought slows the individual's growth. Having a single dominant logic begins or continues a story in the relationship history that is a story about winners and losers and simplistic cognitions about complex life events. It slows the individuals' movement toward understanding and learning how to work with the shared cultural reality [or the shared cultural trance, as Ferguson (1980) puts it]. It does prevent any challenges to whatever emotional defenses the individuals may have used in the past, but keeps the peace at a cost.

A special case occurs if the intimate group is a family with minor children, none of whose members happen to use Postformal Thought. The predictable difficulties mentioned in the last section are multiplied in a situation with more individuals. It is harder for children to gradually learn a more skilled cognitive approach where there are no daily role models. It is harder for the children to grow up, create their own personal view of the world, and leave that family when power struggles centering on control of the family reality have been going on for so long in their own family history.

Benefits of Using a Complex Postformal Logic

Alternatively, the bridge across realities may be a postformal one. The incompatible several logical realities within the relationship might be orchestrated more easily and orchestrated at a higher cognitive level to permit a more complex logic of the relationship to emerge. Postformal thinkers can adapt to the challenges of intimate relationships better than those without Postformal Thought because no one's logic needs to be discarded for the relationship to go on. For the postformal couple power and control are not the same level of threat looming on the relational horizon as they are for the individuals without Postformal Thought who must worry about cognitive survival in their relationships. Shared history for the postformal intimates reflects the synthesis of cognitive lives rather than alternating dominance of one reality over another. Each logical difference or disagreement ends up being another piece of evidence that the relationship remains a win–win situation for individuals within it. This enhances the relationship's value and tends to stabilize it even further.

In the situation where one individual in the relationship is the *only* postformal one, we see a different opportunity and challenge. Several resolutions are possible each time an interpersonal logical conflict occurs. Perhaps the less cognitively skilled individual will use this chance to grow cognitively, with predictable benefits. This is the best outcome and one of the desirable features of having intense intimate relationships.

Alternatively, perhaps the increasingly aggravated postformal individual will let emotions overtake him or her and will temporarily resolve the situation by regressing cognitively and acting out against or withdrawing from others in the situation. This will lead to the previously mentioned predictable problems for couples using lower level logics.

Perhaps the more cognitively skilled individual will decide to wait and hope that the other will come around to a more skilled view of the situation. This tactic is easier for that postformal person to tolerate because the postformal individual sees the bigger picture and does not have to take the power struggle quite as seriously as other members of the intimate group. But predictable conflicts will still occur and growth may be stalled.

Once again a special situation occurs when this intimate group is a family with minor children. The postformal parent has reason to believe that the children will develop further and possibly become postformal thinkers themselves. The postformal parent might consciously try to encourage the cognitive development of those children in the direction that the postformal parent has mastered.

The mismatch in cognitive levels will not be a cause for frustration, in this case, but rather for challenge and hope for the future development of the children.

The Other Side of the Coin: Effects of Intimate Relationship Factors on Postformal Thought Development

The factors in ongoing intimate relationships that we have been discussing, mainly emotional defense mechanisms, shared relationship history, and shared cultural reality, potentially can *influence* the development and use of Postformal Thought, not just be influenced by it. For example, if an individual is emotionally damaged and is responding to all situations out of need (Maslow, 1968), that person is less likely to take a risk in a relational situation and let go of his or her own cognitive verities long enough to be willing to bridge to someone else's realities. Even the postformal thinker would not be likely to use that level of logical thought in such an interpersonal contest in which he or she is emotionally needy or "one down." Just as negative emotions often dampen the higher-level creative spirit, emotional damage means the individual tries to regain safety before meeting the higher level needs of the relationship. Children in a relationship where the parents are damaged emotionally, or children who themselves are emotionally damaged, will find it harder to learn postformal responses to family dynamics. Such families function at the lower levels of unproductive patterns on the Beavers Scale, for example (Beavers & Hampson, 1990), described so well by Scarf (1995) in her book on the intimate worlds of families. The life-and-death emotional struggles that occupy such families prevent those children from having the emotional space or energy to bridge realities.

Shared relationship history also is a strong force influencing postformal thinking. The habits of relating that individuals have developed in earlier years tend to perpetuate themselves over the lifetime. If those habits do not include postformal cognitive processes for relating at the time that a given relationship begins, and if many years are spent bridging the related individuals' realities in a nonpostformal way, then it will be increasingly difficult for anyone (child or adult) in the relationship to move on to a postformal way of relating, violating earlier habit.

An exception to this is the family situation in which parents may be postformal but young children are not. Postformal parents may find this cognitive discrepancy easier to bear than less cognitively skilled parents, but still will be influenced by living in relationships in which they are always using a logical level that is beneath their own. One's tendency to permanently distort perceptions about an intimate's logical skills, based on the cognitive skills they have shown during the history of our relationship with them, is very strong. We see just how strong when we see parents relating to their adult children as if those adults are *still* very young children. Parents must make significant efforts to overcome history, or at least reconsider whether historical patterns need to be revised before using them to predict today's behavior. Intimates influenced by their history face an equally daunting task if they want to relate on a new (to them), more skilled, logical level.

Shared cultural reality (or social forces and roles) also influences the ways that postformal thinking can be used in intimate relationships. This is a domain in which social roles often interfere with the choice of possible processes of relating. The shared reality of the social roles "appropriate" for various intimate relationships must be a "lowest common denominator" reality that the vast majority in a society can achieve, or pretend to achieve. The shared social role reality for couples and family relationships is often structured enough and at a low enough skill level that no bridging of conflicting logical realities is necessary at all. All that is necessary is to act out the appropriate roles in a convincing way and to make the socially appropriate comments about feelings connected to those roles. Tradition does save cognitive energy!

Notice that the first time in this discussion that we need to consider whether a couple or family is heterosexual or homosexual, legally married or not, childless or not, a May/December union, divorced or previously married, with or without stepchildren or other relatives, and so forth, is in the context of this "shared social cognition" element. Other than here in this paragraph, the processes discussed in this chapter apply to *all* of these differing roles and family configurations. Only the shared social cognition element discriminates among the various family configurations. Persons in all the various family configurations *can* use the same cognitive relational processes and can experience the same styles of relating.

A conflict may occur when the views of any knower (e.g., one member of the couple) about the reality of their relationship come into conflict with the views of society about the same intimate relationship. This conflict might be the stimulus for the growth of a postformal way of seeing their relationship and seeing the world. For example, it has not been many years since the existence of a childless marriage was considered an ongoing tragedy, for everyone, and, if intentional, a sign of problems in one's personality and maturity level. Imagine a member of a couple who feels very happy in the relationship, even secretly happy to have evaded the encumbrance of children, coming face to face with this tragic and pathologized view of the "selfishness and immaturity" of childless life together. Knowing that such a view does not square with personal knowledge may be the impetus for this person to develop postformal elements of knowing, perhaps for the first time. The motivation in this case is social.

When the reality of one person in the relationship clashes with another's view, the intensity of the bond is what motivates them to seek a resolution. This is a push toward development of Postformal Thought, or perfection of it, as lower level logics will leave the conflict unresolved. Since framing such a situation postformally can help keep blame and anger at bay, postformal skills are often welcome conflict resolution devices!

The recent past has been a time of social change, especially in regard to the forms of intimate relationships. While there are inherently limited possibilities for intimate relationship behavior in the human behavioral repertoire, some of those possibilities are more fashionable than others at a given time in history. Living at a time of social change means that the individual and even the relationship is challenged to cross logical realities about self without losing self, all the while under shifting shared social reality pressures. Access to Postformal Thought makes it easier for the social change shift to occur while an identity is maintained by a person or a relationship.

Gender Roles

Let's look at one more specific practical variable in intimate relationships, one in which Postformal Thought may make a difference: the roles related to gender (sex roles, sex role stereotypes) and behavior related to those masculine/feminine roles. I have written about this topic rather extensively (e.g., Cavanaugh, Kramer, Sinnott, Camp, & Markley, 1985; Sinnott, 1977, 1982, 1984a, 1986a, 1986b, 1987, 1993b; Sinnott, Block, Grambs, Gaddy, & Davidson, 1980; Windle & Sinnott, 1985) and have included it in my research efforts, because gender roles and the cocreation of social roles have been a central aspect of historically recent social changes in the United States. "Gender role" is a different concept than sexual identity, sexuality, or masculine/feminine behavior. Gender roles may at various times be ambiguous, polarized into opposites, synthesized into an androgenous larger versions, reversed, or transcended entirely. The general age-related progression of gender role development is from polarizing masculine/feminine roles to transcending roles entirely in favor of giving energy to other parts of identity. Gender-related roles enter discussions of intimate relationships because couples tend to divide the work of living together, and gender has often been used by society to define roles. So couples enter relationships, even homosexual ones, with ideas of what proper socially dictated masculine and feminine behavior is. Sometimes identity is being challenged when there is conflict over role-related behavior, making an apparently simple negotiation over something concrete such as housework into a complex, full-blown struggle over identity and worth. If a couple is struggling about gender role related behavior, Postformal Thought makes it easier to sort things out. A postformal partner can readily understand that, if he or she gets beyond emotional or habitual reactions, the roles can be validly coconstructed by them in any number of ways, as logical systems to which they commit themselves and weave into their lives. That partner can also understand that a gender role and its related behavior is only a minor part of his or her constantly transforming identity and is a poor index of personal worth. For the postformal partner(s) the negotiation then moves back to the domain of, for example, "What job do I want?", rather than remaining in the domain of identity and worth "I'm a terrible person if you make more money (less money) than I do, and my identity is in danger."

A STUDY OF COUPLE RELATIONSHIPS AND POSTFORMAL THOUGHT

If the availability of Postformal Thought is related to the quality of intimate relationships, we should be able to see an empirical connection between those two variables. Rogers (1989; Rogers et al., 1991) set out to investigate the joint cognition of two persons trying to solve the postformal problems together. These two persons might be longer term married adults or strangers in a dyad, which might influence their cognition. Rogers also wanted to examine marital adjustment and social behaviors evident during problem solving. She expected that well adjusted married dyads would demonstrate more postformal problem solving and more socially facilitative behaviors than the poorly adjusted married dyads.

Forty heterosexual couples between the ages of 35 and 50 were recruited. They were mainly Caucasian, married for an average of 15 years, 75% for the first time, 25% having had a long-term previous marriage also. Forty-one percent had a Bachelor's, Master's, and/or Ph.D. degrees. After individuals were prescreened for intelligence they were tested for marital adjustment using Spanier's (1976) Dyadic Adjustment Scale, a widely used self-report instrument, which tests for, among other things, diadic cohesion, consensus, and satisfaction. The individuals were randomly assigned to work in one of the following contexts: well-adjusted couple, working as a couple; poorly adjusted couple, working as a couple; well-adjusted-couple individuals, working with someone not their spouse; and poorly adjusted couple individuals working with someone not their spouse. Then each "couple" (real or artificial) was videotaped solving the postformal logical problems. Tapes were scored according to the coding schemes of Pruitt and Rubin (1986) and Sillars (1986) to obtain counts of the social behavior factors of avoidance, competition/contention, accommodation/yielding, and cooperation/collaboration.

While there were no marital adjustment or dyadic context relationships with using *formal* logical operations, there were relationships with using *postformal operations*. Eighty percent of the maritally well adjusted dyads, both real couples and strangers paired together, gave evidence of significantly more postformal thinking operations than the poorly adjusted did. This was especially true for responses to the problems with an interpersonal element, just as was true in research referenced earlier in this chapter. Analyzing facilitative social behaviors from the videotape, Rogers once again found that ability to use formal operations did not relate to the social behaviors while use of postformal operations did. For example, dyads without Postformal Thought demonstrated more contentious and competitive behaviors while problem solving. The social behaviors demonstrated by respondents during testing were also strongly related to marital adjustment. For example, well adjusted respondents demonstrated fewer avoidance behaviors.

Rogers' results support the theory described earlier in this chapter. Postformal thinking and adjustment in intimate relationships are positively related. Some generalized ability seemed to be present that operated whether or not the spouses were working with each other or with strangers. It may have operated by means of facilitating positive types of social behaviors and interactions, as evidenced by the fact that postformal thinkers produced more cooperative and fewer avoidant behaviors. Postformal thinkers seemed to explore and create to a greater degree, tolerate others' ways of seeing reality, and ultimately be able to commit to one solution. When working with strangers, they also took more pains to communicate "where they were coming from" in their views of a problem's many realities. Rogers' work suggests that postformal thinking is useful in intimate relationships.

SUMMARY

In this chapter I have argued that possessing a cognitive ability, namely Postformal Thought, colors and stimulates adult development. The nature of

Postformal Thought was described. Finally examples of the relationship between Postformal Thought and adult development were described using the issue area of intimate relationships.

REFERENCES

Beavers, W., & Hampson, R. (1990). *Successful families: Assessment and intervention.* New York: W. W. Norton.

Campbell, S. (1980). *The couple's journey: Intimacy as a path to wholeness.* San Luis Obispo, CA: Impact.

Cavanaugh, J., Kramer, D., Sinnott, J. D., Camp, C., & Markley, R. P. (1985). On missing links and such: Interfaces between cognitive research and everyday problem solving. *Human Development, 28,* 146–168.

Ericsson, K., & Simon, H. (1984). *Protocol analysis.* Cambridge, MA: The MIT Press.

Erikson, E. (1950). *Childhood and society.* New York: W. W. Norton.

Erikson, E. (1982). *The lifecycle completed.* New York: W. W. Norton.

Ferguson, M. (1980). *The Aquarian conspiracy: Personal and social transformation in the 1980s.* Los Angeles: Tarcher.

Marks, S. (1986). *Three corners: Exploring marriage and the self.* Lexington, MA: D.C. Heath.

Maslow, A. H. (1968). *Toward a psychology of being.* New York: Van Nostrand Reinhold.

Newell, A., & Simon, H. (1972). *Human problem solving.* Englewood Cliffs, NJ: Prentice-Hall.

Pruitt, D., & Rubin, J. (1986). *Social conflict.* New York: Random House.

Rogers, D. R. B. (1989). The effect of dyad interaction and marital adjustment on cognitive performance in everyday logical problem solving. Doctoral dissertation, Utah State University, Logan, Utah.

Rogers, D., Sinnott, J., & Van Dusen, L. (1991, July). Marital adjustment and social cognitive performance in everyday logical problem solving. Paper presented at the 6th Adult Development Conference, Boston, MA.

Satir, V. (Ed.) (1967). *Conjoint family therapy.* Palo Alto, CA: Science and Behavior Books.

Scarf, M. (1995). *Intimate worlds: Life inside the family.* New York: Random House.

Sillars, A. (1986). Manual for coding interpersonal conflict. Unpublished manuscript, University of Montana, Department of Interpersonal Communications.

Sinnott, J. D. (1977). Sex-role inconstancy, biology, and successful aging: A dialectical model. *The Gerontologist, 17,* 459–463.

Sinnott, J. D. (1981). The theory of relativity: A metatheory for development? *Human Development, 24,* 293–311.

Sinnott, J. D. (1982). Correlates of sex roles in older adults. *Journal of Gerontology, 37,* 587–594.

Sinnott, J. D. (1984a). Older men, older women: Are their perceived sex roles similar? *Sex Roles, 10,* 847–856.

Sinnott, J. D. (1984b). Postformal reasoning: The relativistic stage. In M. Commons, F. Richards, & C. Armon (Eds.), *Beyond formal operations* (pp. 298–325). New York: Praeger.

Sinnott, J. D. (1986a). *Sex roles and aging: Theory and research from a systems perspective.* New York: S. Karger.

Sinnott, J. D. (1986b). Social cognition: The construction of self-referential truth? *Educational Gerontology, 12,* 337–340.

Sinnott, J. D. (1987). Sex roles in adulthood and old age. In D. B. Carter (Ed.), *Current conceptions of sex roles and sex typing* (pp. 155–180). New York: Praeger.

Sinnott, J. D. (1992). Development and yearning: Cognitive aspects of spiritual development. Paper presented at the American Psychological Association Conference, Washington, D.C.

Sinnott, J. D. (1993a). Teaching in a chaotic new physics world: Teaching as a dialogue with reality. In P. Kahaney, J. Janangelo, & L. Perry (Eds.), *Theoretical and critical perspectives on teacher change* (pp. 91–108). Norwood, NJ: Ablex.

Sinnott, J. D. (1993b). Sex roles. In V. S. Ramachandran (Ed.), *Encyclopedia of human behavior, 4,* 151–158.

Sinnott, J. D. (1994a). Development and yearning: Cognitive aspects of spiritual development. *Journal of Adult Development, 1,* 91–99.

Sinnott, J. D. (1994b). *Interdisciplinary handbook of adult lifespan learning.* Westport, CT: Greenwood.

Sinnott, J. D. (1994c). New science models for teaching adults: Teaching as a dialogue with reality. In J. D. Sinnott (Ed.), *Interdisciplinary handbook of adult lifespan learning* (pp. 90–104). Westport, CT: Greenwood.

Sinnott, J. D. (1994d). The future of adult lifespan learning. In J. D. Sinnott (Ed.), *Interdisciplinary handbook of adult lifespan learning* (pp. 449–466). Westport, CT: Greenwood.

Sinnott, J. D. (1994e). The relationship of postformal learning and lifespan development. In J. D. Sinnott (Ed.), *Interdisciplinary handbook of adult lifespan learning* (pp. 105–119). Westport, CT: Greenwood.

Sinnott, J. D. (1996a). Postformal thought and mysticism: How might the mind know the unknowable? *Aging and Spirituality: Newsletter of American Sociological Association Forum on Religion, Spirituality and Aging, 8,* 7–8.

Sinnott, J. D. (1996b). The developmental approach: Postformal thought as adaptive intelligence. In F. Blanchard-Fields & T. Hess (Eds.), *Perspectives on cognitive change in adulthood and aging* (pp. 358–383). New York: McGraw-Hill.

Sinnott, J. D. (1997). Developmental models of midlife and aging in women: Metaphors for transcendence and for individuality in community. In J. M. Coyle (Ed.), *Women and aging: A research guide* (pp. 149–163). CT: Greenwood.

Sinnott, J. D. (1998a). Creativity and postformal thought. In C. Adams-Price (Ed.), *Creativity and aging: Theoretical and empirical approaches* (pp. 43–72). New York: Springer.

Sinnott, J. D. (1998b). *The development of logic in adulthood: Postformal thought and its applications.* New York: Plenum Press.

Sinnott, J. D., Block, M., Grambs, J., Gaddy, C., & Davidson, J. (1980). *Sex roles in mature adults: Antecedents amd correlates.* College Park, MD: Center on Aging, University of Maryland College Park.

Sinnott, J. D., & Cavanaugh, J. (Eds.) (1991). *Bridging paradigms: Positive development in adulthood and cognitive aging.* New York: Praeger.

Spanier, G. B. (1976). Measuring dyadic adjustment: New scales for assessing the quality of marriage and similar dyads. *Journal of Marriage and the Family, 38,* 15–28.

Sternberg, R. J. (1986). A triangular theory of love. *Psychological Review, 93,* 119–135.

Windle, M., & Sinnott, J. D. (1985). A psychometric study of the Bem Sex Role Inventory with an older adult sample. *Journal of Gerontology, 40,* 336–343.

Wolf, F. A. (1981). *Taking the quantum leap.* New York: Harper & Row.

Zukav, G. (1979). *The dancing wu li masters: An overview of the new physics.* New York: Bantam.

CHAPTER 13

Developmental Trajectories and Creative Work in Late Life

Laura Tahir and Howard E. Gruber

THE EVOLVING SYSTEMS APPROACH TO
CREATIVE WORK ACROSS THE LIFESPAN

The evolving systems approach to creativity (see Gruber, 1980; Gruber & Davis, 1988; Wallace & Gruber, 1989) focuses on facets of the development of the creator, his or her work, and the contextual frame in which this work occurs. From a developmental perspective we understand that creativity is a process that changes over the lifespan. Keegan (1995) explored the connection between creativity in childhood and creativity in adulthood. In another line of work, several researchers focused on early life creative work, or "starting out" (Brower, 1996; Bruchez-Hall, 1996; Gruber, 1996; Keegan, 1996; Tahir, 1996). Several questions are raised in the present chapter: What happens to this process as the person ages? Is late life creative work different from that produced earlier? Are there any general trajectories of creative work?

Case studies of creative people suggest that creative work indeed can continue well into late life. Furthermore, while there are continuities over the lifetime of a creative person, there may also be "late blooming" creative achievements. The evolving systems approach to adult development and to creative work in particular does not posit a single model of the creative personality from which to theorize about creative people in general, nor does it posit general stages through which the creative process moves. As Gruber (1980) writes, "Since every creative achievement is unique in precisely the ways that draw our attention to it, we may well find that the appropriate relevant list of attributes of the creative person varies from task to task and from individual to individual" (p. 286). Thus, the evolving systems approach asks: How does creative work work? What is it that the creative

Laura Tahir • Director of Psychology, Garden State Youth Correctional Facility, Highbridge Road, Yardville, New Jersey, 08620.   Howard E. Gruber • Columbia Teacher's College, Bronx, New York 10461.

Handbook of Adult Development, edited by J. Demick and C. Andreoletti. Plenum Press, New York, 2002.

person is doing when he or she is being creative? The model is ideographic and tracks the developing creative individual over time.

The evolving systems approach focuses on three interrelated subsystems that interact to make creative work happen. These subsystems are knowledge, purpose, and affect. Organization of knowledge refers to the structures of the creative individual's thought. Creative thought does not take place with "one great moment of insight," nor is it "one monotonic gradual change" (Gruber, 1977, p. 6). Rather, it is a process characterized by repetition and perseverance. Organization of purpose refers to the way a person orchestrates his or her activities to achieve optimal work. The creative person pursues a network of enterprises that is crucial to his or her productive life. Organization of affect refers to the feelings that occur when a person is being creative, or functioning optimally. A paramount requirement for steady creative work is the construction and maintenance by the creator of an appropriate affective system. Its function is not to make the creator happy or to relieve her or him of the burden of neurosis. Rather, it is to nurture the creative process, a function that can be carried out in diverse ways, ways that evolve developmentally throughout the life history. Positive emotions such as curiosity, hope, joy of discovery, and love of truth are motivators crucial to the maintenance of an ongoing creative system.

The study of late life creativity is described here not as the study of adaptation to the deficiencies of old age, but rather as the study of the process of an evolving system as it continues to be productive. This chapter is intended to cast light and doubt on the stereotype of elderly people as enfeebled and unproductive and, as the stereotype is applied to creative people, suffering from a loss of creative vigor and zest for innovation. Our method is simply to examine some instances of late life creative work to highlight the diversity of developmental patterns to be found. We steer clear of statistical analyses, opting instead for more detailed examination of a few cases.

CASE STUDIES OF LATE LIFE CREATIVITY

Creative Work in the Arts

One of the most famous examples of late-life creativity in the United States is that of Georgia O'Keeffe (1887–1986), a pioneer in American abstract art. O'Keeffe was in her 30s when she first began to show her work at the New York City gallery of her patron (and later husband), Alfred Stieglitz. In addition to her abstract art, some of her work was representational. Her work is characterized by sexual symbolism, particularly her flower paintings (although the artist herself denied the work was sexual). O'Keeffe was labeled a "new woman" for her independent lifestyle in the early years of the 20th century, and the FBI reportedly described her as an "ultraliberal security risk." Shortly after Stieglitz died in 1946, O'Keeffe moved from New York to New Mexico and spent the rest of her long life there, frequently using southwestern motifs of desert and mountain in her art. She wrote and supervised the publication of her memoirs when she was 89 years old. In spite of her failing eyesight, in her 90s O'Keeffe began to work in pottery. Describing her work in this new art form, she said, "It takes more than talent.

It takes a kind of nerve ... a kind of nerve and a lot of hard, hard work" (quoted in Strickland, 1992, p. 141). O'Keeffe's work at this time was inspired in part by Juan Hamilton, a man nearly 60 years younger than she whom she met in 1973 when she was 86. Hamilton was an amateur artist struggling financially when he was offered a job at O'Keeffe's ranch in Abiquiu, New Mexico. O'Keeffe's eyesight was poor and worsening at this time and she came to depend on Hamilton, at first as a handyman, and then as a personal assistant and close friend. Hamilton accompanied O'Keeffe on her daily walks and helped her with other activities of daily living. He provided intellectual stimulation for her and served as her muse. She became his mentor, and her need for generativity was in part satisfied through her relationship with him. Unfortunately, the relationship became complicated and strained over the next 14 years, and it is widely believed that Hamilton took advantage of O'Keeffe financially. Nevertheless, it was O'Keeffe's wish that her estate be left to Hamilton. O'Keeffe died in Sante Fe in the spring of 1986 at the age of 98.

Anna Mary Robertson Moses (1860–1961), also known as Grandma Moses, was an American folk artist most noted for her longevity and for the fact that she got quite a late start in her artistic career. Grandma Moses grew up on a farm in upstate New York in a family of ten children. She married at age 27 and moved with her husband to Virginia, where they worked for several years as tenant farmers. The couple had ten children, only five of whom survived childhood. They returned to upstate New York because of homesickness, and Grandma Moses lived there, as well as in Bennington, Vermont, for the rest of her life. Although untrained in art, she liked doing needlework. When she was in her 70s, she began to paint with oils depicting country landscapes and farm life. She reports that she "painted for pleasure, to keep busy and to pass the time away" (Kallir, 1948, p. 129). After a few of her paintings were exhibited at county fairs and the neighborhood drug store in Hoosick Falls, her work was recognized by Louis J. Caldor, an engineer and art collector who provided the artist with the necessary support to launch her career. In October 1940, when Grandma Moses was 80 years old, she had her first New York City exhibit at the Galerie St. Etienne on West 57th Street. She continued to paint and her fame continued to grow. In 1949 she was invited to the White House to receive an award from President Truman. Her financial success puzzled her, as she saw art as primarily a way to supplement her farm income. While she traveled when she needed to promote her career, she preferred to stay at home. It was characteristic of Grandma Moses not to mention suffering or hardship in her memoirs, and her paintings similarly portray simple farm scenes of hard work and contentment. Describing her life's work in an autobiography she wrote during her late 80s and early 90s, she wrote, "I look back on my life like a good day's work, it was done and I feel satisfied with it. I was happy and contented. I knew nothing better and made the best out of what life offered. And life is what we make it, always has been, always will be" (Kallir, 1948, p. 140). Grandma Moses continued to paint for the rest of her life. In 1956, at the age of 96, she was commissioned by the presidential cabinet to portray President Eisenhower's Pennsylvania farm retreat. She died at the age of 101.

The Irish-English writer George Bernard Shaw (1856–1950) was born in Dublin, the third child and only son of George Carr, a barely successful corn merchant, and Lucinda Elizabeth Gurly Shaw, an accomplished musician and teacher. Never a good student, Shaw preferred self-education, visiting the Dublin National Gallery and the Royal Theatre and playing and listening to music with his mother's associates (most notably George John Lee (1831–1886), his mother's singing teacher and a close friend of the family). Shaw quit school at age 15 and became a junior clerk, but in 1876, 20-year-old Shaw became restless with the cultural limitations in Dublin and emigrated to London to live with his mother and sister, who had gone there earlier to work with Lee. In London, he began to ghostwrite music and art criticism. In his own name he wrote a total of five novels, most of which were only marginally successful. He was greatly influenced by the work of Marx and other socialist writers. During the 1880s he was active in the Fabian Society and the Land Reform Union, middle-class socialist groups. In his early 30s, Shaw gave up the novel as a genre, preferring the theatre, where he thought he would be able to influence a broader audience. He saw the stage as a means of social criticism on which he could broadcast his belief in the evils of capitalism, romanticism, and traditional morality. In all, he wrote nearly 50 plays. He was 42 years old when he married Charlotte Payne Townshend, and their marriage, which both agreed would not be consummated, lasted until Charlotte's death in 1943. Although they had no children of their own, Shaw is said to have been fond of children, and he and Charlotte had a very close relationship with T. E. Lawrence (1888–1935), who adopted the Shaw name in 1927. Shaw wrote plays and prefaces, thousands of letters, and other works of fiction and nonfiction, into his 90s. *The Intelligent Woman's Guide to Socialism and Capitalism* (1928) echoed many of the feminist themes of his earlier plays. The novella *The Adventures of the Black Girl in Her Search for God* (1932) describes Shaw's creative evolution and the idea of the "life force," or energy of progress. According to Shaw, the life force was a metaphor for a universal will that each individual has and that strives to improve the human race. Shaw recorded his memoirs in *Sixteen Self-Sketches*, first published in 1939, and revised in 1947 when he was 91. Shaw died on November 2, 1950, from complications due to a fractured left thigh he sustained when he fell while pruning a branch in his garden 2 months prior to that. He was 94 years old.

Dorothy West (1907–1998) was the last surviving member of the Harlem Renaissance, a group of writers that included Langston Hughes, Zora Neale Hurston, Countee Cullen, and Wallace Thurman. West was born in Boston in 1907, the only child of an elderly freed slave who later became a successful wholesale fruit seller. She wrote her first story at age 7, and by age 14 she won several writing competitions sponsored by *The Boston Post*. West traveled to New York City as a teenager when she was awarded a prize in a national writing contest. Soon after that she moved to New York, where she and her African-American colleagues made a valuable contribution to American literature. In 1932 she went to Moscow to appear in a Communist film about race relations in the United States, but the project was derailed by a white American. West's father died and she returned to New York. She founded the magazine *Challenge*, which became a forum for writers such as Richard Wright, Margaret Walker, and Ralph Ellison. The Great Depression drove most of the writers from New York because they could no longer survive financially. West stayed in New York and worked as a welfare relief investigator with the Works Progress Administration Writers Project. She also wrote short stories for the *New York Daily News*. In 1943 West moved to Martha's Vineyard where her mother lived, and it was there that she published her first novel, *The Living is Easy*, in 1948. Although critics praised the book, the *Ladies' Home Journal* refused to run excerpts of it for fear that racist readers would cancel their subscriptions. West seemed to have been thwarted by the magazine's rejection and didn't publish another novel for nearly 50 years. She also received disapproval from those who felt that she should have used her writing to express anger about interracial conflict. West's work was not directly political. Rather, she sought to describe complex subtleties of feeling. West remained in Martha's Vineyard, where for the next 20 years she wrote a column for the "Vineyard Gazette," a community newspaper. In 1995, when West was in her late 80s and with the encouragement of her friend Jacqueline Kennedy Onassis, she published her second novel, *The Wedding*. This novel was well received and she wrote *The Richer, the Poorer*, a collection of short stories and reminiscences, the same year. West died in 1998 at the age of 91.

Italy's greatest composer, Guiseppe Verdi (1813–1901), had a long and productive career in spite of the setback and tragedy that often characterized his life. He was born into a poor family of farmers, and owing to his lack of money and status, he got a late start in music. Verdi's first opera, *Oberto*, was performed in 1839 when he was in his mid-20s and it was a great success. However, soon after he composed a comic work that turned out to be a disaster, and his wife and two children died in a 22-month span (1838–1840). Verdi became so depressed that he wanted to give up his operatic career, but with the help of the impressario Bartolomeo Merelli, he undertook a third opera, *Nabucco*, which was first performed in 1842 and was a great success. Verdi went on to compose many more operas, such as *Rigoletto* (1851), *Il Trovatore* and *La Traviata* (1853), *Simon Boccanegra* (1857), *La Forza del Destino* (1862), and *Aida* (1871). At the age of 60, he memorialized the great Italian writer Alessandro Manzoni with the famous *Requiem*. Soon after that, he was nominated for a seat in the Italian parliament, and for several years he spent more time in politics and farming than in music. He then met the composer and author Arrigo Boito, and the two collaborated on *Othello* (1886), Verdi's first opera in 16 years and a great success. Verdi began his last masterpiece, *Falstaff*, in his mid-70s. The composition of this opera took considerable time because Verdi experienced the deaths of several close friends and again became depressed and even feared that he would not live long enough to complete the opera. *Falstaff* premiered in 1893 when Verdi was 80 years old. It was his only comic opera (other than the early failed attempt), and in it Verdi seems to be poking fun at the Romantic era in general and at the foibles and follies of aging in particular. The final ensemble is a fugue in which Falstaff sings "Tutto nel mondo e burla" ("Everything in the world is a joke"). Four years later, Verdi's second wife died of pneumonia. To deal with his desolation, he founded the Rest Home for Aged Musicians in Milan, which still bears his name and is supported by his royalties. In 1900, while visiting his adopted daughter in Milan at Christmas time, Verdi sustained a stroke and died soon after in January 1901 at age 87.

Marian Anderson (1897–1993) is considered by many the greatest American contralto of the 20th century. She was born in Philadelphia of poor African-American parents. Her talent became apparent when she sang in the choir of the Union Baptist Church in Philadelphia. In her 20s she appeared with the New York Philharmonic and made her debut at Carnegie Hall. At the same time critics praised her, she had to travel in segregated railway cars. In Europe she received overwhelming admiration. Toscanini introduced her at a recital in Salzburg by saying that "a voice like yours is heard only once in a hundred years" (quoted in Bogle, 1980, p. 107). However, Anderson had to deal with incessant racial discrimination. Even after having achieved enormous success in Europe, Japan, and South and Central America, in 1939 she was denied the use of Constitution Hall in Washington, DC, by its owners, the Daughters of the American Revolution. First Lady Eleanor Roosevelt resigned her membership in the DAR in protest. An alternate concert was scheduled at the Lincoln Memorial on Easter Sunday, which 75,000 wildly enthusiastic people attended. Juxtaposed with this overwhelmingly fervent reception was the fact that Anderson had not been able to find a hotel room in Washington that would accommodate blacks. While acknowledging bitterness, she was stoic and tried to understand racism. Her temperament was gentle, and she wrote, "I would be fooling myself to think that I was meant to be a fearless fighter" (Anderson, 1956, p. 188). Anderson married architect Orpheus Fisher in 1943. In 1955, Anderson was the first black to become a permanent member of the Metropolitan Opera. She was a goodwill ambassador for the United States and in 1958 she was made a delegate to the United Nations. Anderson made a four-continent farewell concert tour in 1964–1965. After her formal retirement from singing, she continued to perform as narrator with symphony orchestra performances of Copland's *Lincoln Portrait* until 1977, when she was 80 years old. She also gave concerts on special occasions, made a few recordings, and lectured as part of Sol Hurok's speakers' bureau. Anderson was a member of the Board of Directors and a trustee for many organizations concerned with music, education, adoption, and world peace. She continued to accumulate international accolades, as well as honorary doctorates from more than 20 American schools. She spent the last years of her life on a farm in Connecticut, caring for her husband, who was partially disabled by a stroke. After he died in 1992, she remained at her home with some assistance until 1992 when she moved to Portland, Oregon to live with her nephew James De Priest, music director of the Oregon Symphony. Anderson died in 1993 at the age of 96.

Martha Graham (1894–1991) broke new ground in twentieth-century American dance during a time when that art was characterized predominantly by the frivolity of vaudeville and lighthearted musicals. She was born in Allegheny, Pennsylvania. Her family moved to California when Graham was 14, and it was there that at age 17 she saw Ruth St. Denis perform on stage and decided that she wanted to become a dancer. Graham's father, a physician, wanted his oldest daughter to pursue an academic career. In spite of his protests, Graham enrolled in the Denishawn dance school, founded by her idol Ruth St. Denis and her husband, Ted Shawn. Graham was a hard worker, and as the years went on she became a prolific performer and choreographer. Her early works were imitative of the Denishawn style, but with time she created her own dance vocabulary, characterized by angular movement, use of torso, and dramatic falls. Her career spanned over 70 years and she danced all over the United States and in Europe. She spent many years with collaborators at Bennington College, which became a Mecca for modern dance. She developed professional associations with dancers, writers, musicians, and designers. Graham had a long-term relationship with Denishawn's music director, Louis Horst, who wrote numerous new scores for her dances. After a 20-year relationship/collaboration that has been described as having "changed the face of modern dance" (Lee, p. 432), the two separated, and Graham married Eric Hawkins, a young dancer she had secretly been living with for years. Their marriage was short-lived, apparently because Hawkins felt he could not be an equal in the relationship. Graham went into a deep depression after the marriage broke up. She immersed herself in a period of self-isolation and did not dance for 2 years. She began to drink heavily and suffered from alcohol-related liver damage. Even so, Graham continued to dance. At age 64, she premiered her full-length dance drama *Clytemnestra*, which was recognized as one of the great choreographic contributions to modern dance. Her final performance was at age 74. She had been slowed by arthritis; however, she had been living with the illusion that she was immortal. In her later years she was director of her troupe, the Martha Graham Dance Company. She continued to choreograph and perform lecture-demonstrations, and she agreed to allow the re-creation of some of her earlier works, a practice she had previously forbidden. It seems that her realization that she was not immortal enabled her to preserve her work for future generations. When her Dance Company was rejected a matching funds grant from the National Endowment for the Arts in 1983, Graham publicly protested, accusing the NEA of age discrimination. In subsequent years the Company received funding. Graham was in the process of choreographing *Maple Leaf Rag* when she died in April 1991 at the age of 97.

Russian-American composer Igor Stravinsky (1882–1971) grew up in czarist Russia. His father was an opera singer and lawyer. Young Stravinsky started piano lessons at the age of 9, but he was not considered a musical prodigy. He entered law school at his father's wishes, but he soon gave up law to study music. Stravinsky composed his first major work, *Fireworks*, in 1908, the year in which his first teacher, Rimsky-Korsakov, died. *Fireworks* was based on Russian folk tunes and was fairly successful. Stravinsky then met the impressario Diagalev, the founder and director of the Russian Ballet, who had a large following of musicians and dancers and other artists whom he supported. Through this first collaboration, Stravinsky began to compose the ballet music for which he is most noted. *The Fire Bird* (1910) was his first ballet, also based on the Russian nationalist tradition and characterized by the orchestration he had learned from Rimsky-Korsakov. *The Rite of Spring* (1913), Stravinsky's third ballet, has been called "undoubtedly the most famous composition of the early twentieth century" (Grout, 1960, p. 631). This huge work was received with outrage by the audience at the Paris premiere in May 1913, who perceived it to be discordant and unconventional. Meanwhile, because of the turn of events in Russia, Stravinsky and his wife and their four children were political exiles in western Europe for many years. They settled in France in 1920. Stravinsky's music in the 1920s and 1930s was neoclassical and characteristic of the French avant-garde rather than Russian nationalism. Stravinsky became a French citizen in 1934. Several years later, however, both his daughter and wife died of consumption, and he himself became very ill. He moved to the United States in 1939 and held the Chair of Poetry at Harvard University for the academic year 1939–1940. He married Vera de Bosset and they became US citizens and settled in Hollywood. In 1948, Stravinsky began a 3-year collaboration with Chester Kallman and W. H. Auden on the opera *The Rake's Progress*, based on a series of paintings by Hogarth.

This work is considered to mark the end of Stravinsky's neoclassical period. During this time Stravinsky met the young American conductor Robert Craft. Stravinsky's turn to serialism is often attributed to his association with Craft, who had introduced the composer to new recordings of Schönberg, Berg, and Webern. Stravinsky created his first completely serial work, *Threni* (1958), a large-scale work for soloists, chorus, and orchestra, when he was in his 70s. He was 84 when he completed his last major work, *Requiem Canticles* (1966). Lillian Libman, Stravinsky's personal manager, wrote about the composer's vast energy and hard work, noting that his itinerary during his 80th year "would have hospitalized a dozen people I can think of who were two decades his junior" (Libman, 1972, p. 108). However, ill health began to slow his activities. The household moved to NYC in 1969 and Stravinsky died there in April 1971 at age 88.

Creative Work in the Physical, Social and Political Sciences

The English scientist who established the theory of evolution, Charles Robert Darwin (1809–1882), was born in Shrewsbury, England. His grandfather, Erasmus Darwin (1731–1802), was a physician and poet. His father was a country doctor and his mother was the daughter of Josiah Wedgwood, the potter and industrialist. Both parents were in their 40s when young Darwin was born, and his mother died when he was 8. He was not a particularly good student and certainly not a child prodigy, but he developed a keen interest in rocks, insects, flowers, and natural history. His father sent him to Edinburgh University to study medicine, but young Darwin had little interest in this. He then began religious studies at Christ's College at Cambridge University from which he was graduated in 1831. That same year, at age 22, he accepted a position as naturalist aboard the British naval ship HMS Beagle, which left England for South America. Later he wrote that the 5-year voyage had been "by far the most important event in my life and has determined my whole career" (Barlow, 1969, p. 76). On his return from the Beagle voyage he wrote a journal of the trip, *A Naturalist's Voyage Around the World*, which was widely acclaimed. There was no mention of evolution in this work, but Darwin had begun thinking about it. Two years later he read the work of Malthus and assimilated those ideas into his own thoughts about natural selection. Darwin married his first cousin, Emma Wedgewood, in 1839. Her fortune enabled them and their seven children to live comfortably. Darwin continued to accumulate and assimilate data that would support his theory of evolution. His greatest work, *On the Origin of Species*, was published in 1859 after a gestation of some 25 years. Almost equally important was *The Descent of Man and Selection in Relation to Sex*, published in 1871, with a companion volume, *The Expression of Emotions in Man and Animals* (1872). During the 1860s and 1870s he continued botanical work begun earlier on plant physiology, morphology, taxonomy, and plant behavior. Although all of this work bore some relation to his evolutionary concerns, it nevertheless formed a separate enterprise. In the 1860s he published about 25 papers on botanical subjects, and about 15 in his remaining years. In his final year, 6 months before his death at age 73 in 1882, he published *The Formation of Vegetable Mould Through the Action of Worms, with Observations on Their Habits*. This last work continued an enterprise he had begun in the

1830s, the study of how living organisms transform their environment. In 1882 he added something new, unheralded in his or anyone else's previous work: his observations on the behavior of these invertebrate creatures.

Lise Meitner (1878–1968) was an Austrian-Swedish physicist and mathematician who discovered the protactinium-231 isotope and contributed to the development of the atomic bomb. She was the third of eight children whose father was a lawyer and mother was a piano teacher. Although the family were Jewish, Meitner's father was agnostic and wanted his children to learn about science. Even so, he insisted that Lise first complete education to become a teacher before agreeing to pay for a tutor to help her gain entrance into the university. (Ordinarily Viennese girls at that time ended their education by age 14, or went on to study teaching.) Lise Meitner was 23 when she and only three other women were accepted into the University of Vienna. In 1905 she became only the second woman to earn a physics degree there. After teaching briefly at a girls' school, she was accepted to work with Nobel Prize winner Max Planck—for no pay—at the University of Berlin. There she met the young chemist Otto Hahn, and the two of them started what would be a lifelong collaboration. Hahn had the title of professor. Meitner was an unpaid volunteer who was confined to the basement of the chemistry building because women were not allowed upstairs, and because there were no restrooms in the building for women, Meitner had to use the facilities at a nearby hotel. Meitner lost her financial support when her father died in 1911, but in 1912 Max Planck offered her a small stipend to work for him grading papers and helping him organize his seminars. She continued to work with Hahn, but after their collaboration on protactinium, which required both physics and chemistry (and for which Hahn took most of the credit even though Meitner did most of the work), Meitner began to focus on her own interests in physics. After Germany was defeated in World War I and the Kaiser was overthrown, the role of women improved somewhat and Meitner was for the first time allowed to lecture at the university in 1922. In 1926, at the age of 48, she became Germany's first woman physics professor. She again teamed up with Hahn and three other scientists to work on uranium physics, but at the peak of her career in 1934, Hitler began to fire all non-Aryans. Meitner was considered the leader of her team, but she was forced to escape from Berlin in July 1938. At the invitation of Neils Bohr, she fled to the Copenhagen Institute, where her nephew Otto Robert Frisch was a physicist, and from there she went to the Institute of Physics in Stockholm. She continued her work and shared her results with her team as well as with her nephew. It has been documented that it was Meitner who initiated the experiment and that Frisch explained the process of the nucleus of an atom splitting and releasing an enormous amount of energy. However, Hahn received the Nobel Prize in chemistry in 1944 and did not acknowledge Meitner's contribution to his work. While Meitner never publicly or privately complained about not having received a Nobel Prize, she openly criticized Hahn for his passivity in abiding by the rules of the Nazis. In 1947 Meitner refused an invitation to return to Berlin to teach. She stayed in Sweden, was physically active, climbed mountains, and visited the United States several times to work and see family. Meitner spent the last 8 years of her life in Cambridge with her nephew, and died a few days before her 90th birthday.

Annie Wood Besant (1847–1933) was a British social reformer and theosophist of Irish descent. Her father died when she was 5 years old, leaving her mother with little financial resources to care for her three children, one of whom soon died in infancy. Mrs. Wood was an enterprising woman and started a boarding house for students. She was able to earn enough money to send her son through Cambridge and to save enough money to buy a house and retire comfortably. Young Annie, however, was not provided with a formal education because it was not the custom to educate girls for anything other than marriage. Rather, she was sent to live with Miss Marryat, a pious religious fanatic. Miss Marryat was very kind to Annie, who was struggling with religious issues. At age 20 Annie Wood married Frank Besant, an Anglican vicar, with whom she had two children. She soon lost her religious faith and then left her husband. She joined forces with Charles Bradlaugh, a crusading atheist. A fiery and effective orator, as well as a patient teacher, Annie Besant worked on behalf of whatever cause captured her enthusiasm. In 1877 she and Bradlaugh were found guilty of moral depravity for their role in disseminating literature on birth control and were sentenced to a fine and 6 months' imprisonment. They won an appeal and never had to serve the sentence, but as a consequence Besant lost custody of her children. In 1885 Besant joined the Fabian Society and worked closely with her friend Bernard Shaw, who shared her socialist views. She was one of the contributing authors to *Fabian Essays in Socialism*, which Shaw edited in 1889. During that same year she sought out and met Madame Blavatsky and converted to a powerful religious movement, Theosophy. Besant served as president of the Theosophical Society from 1907 until her death. The Society's headquarters were in Adyar, India, and Besant eventually moved there. She learned Sanskrit and translated the *Bhagavad Gita*. In her 70s she toured Western nations with Krishnamurti, whom she regarded as the new messiah, preaching mysticism. In addition to her spiritual pursuits, she did political work for Indian independence and became President of the Indian National Congress in 1917. Besant was a prolific writer and one of the most influential women of her era. Although increasingly frail, she was active in the pursuits that had long absorbed her until almost the very end—a span of nearly 60 years of service to her causes. She died in 1933, shortly before her 86th birthday.

American reformer and feminist Elizabeth Cady Stanton (1815–1902) was one of 11 children born to a conventional, wealthy judge, Daniel Cady, and Margaret Livingston Cady, a progressive, vivacious, independent woman. Four of the five Cady sons died in childhood, and the fifth died at age 20. Elizabeth Cady recalls her grieving father wishing that she had been a boy. Her strong desire for her father's approval motivated her to achieve a sense of mastery. She attended the Willard School in Troy, New York. At age 25, she married a prominent lawyer and abolitionist 10 years her senior, Henry Stanton. They attended the first World Anti-Slavery Convention in London in 1840, although women were denied seats because of gender. Although Henry Stanton initially supported women's activity in antislavery activity, he later changed his view because he felt the presence of women would bring ridicule to the abolitionists. The Stantons continued to go their separate ways politically and their marriage became one that Elizabeth tolerated for the sake of their seven children. Eventually, on friendly terms, they took up separate households, Henry in NYC, and Elizabeth in Tenafly, NJ. Elizabeth and several other women organized the famous Seneca Falls Women's Rights Convention in 1848. In March 1851, Elizabeth Cady Stanton and Susan B. Anthony met for the first time at an antislavery meeting and formed a lasting friendship.

In 1869, they organized the National Woman Suffrage Association (NWSA) in New York City to reconcile the feminists and abolitionists. The American Woman Suffrage Association (AWSA), which was willing to defer woman's suffrage in favor of racial issues, was founded that same year. In 1890 the suffragists united under the banner of the National American Woman Suffrage Association, with 75-year-old Elizabeth Cady Stanton its first president, and 70-year-old Susan B. Anthony its first vice president. In 1876 Stanton and several other women began to write a history of the women's movement. It was not published until 1922, 2 years after the passage of the Nineteenth Amendment. In later life, Elizabeth Cady Stanton endured physical ailments, death of friends, and financial insecurity. Henry Stanton died in 1887. By 1894 she needed two canes to walk and a nurse to bathe and dress her. She was close to her children and grandchildren, but they were scattered in all parts of the world. Obesity (she loved to eat and weighed 240 pounds) and failing eyesight confined her to her New York City apartment, where she lived with two of her children during the last years of her life. At a lavish 80th birthday celebration at the Metropolitan Opera House, filled with spectators and their gifts, Stanton didn't have the energy to complete her speech, which had to be read for her. She used the forum to proclaim her antipathy toward all that oppresses women. Stanton wrote two major works in her later years. The first was *The Woman's Bible*, a best seller coauthored with five other women, which was completed in 1895 when Stanton was 80. In this work she attacked the literal mindedness of the clergy and their conservative impact on society. Stanton published her last major work in 1898 at age 83, an autobiography, *Eighty Years and More*. Stanton died in her NYC apartment on October 26, 1902. Her funeral was small and private, with family and a few close friends, including Anthony. Her casket was covered with flowers and a photo of Anthony.

NOMOTHETIC APPROACH AND STEREOTYPES ABOUT AGING

The prevailing work on aging and creativity from the nomothetic approach tends to focus on statistical "age curves" demonstrating correlations between productivity and age. Early work by Lehman (1953), for instance, led him to conclude that most creative efforts occur during an individual's third and fourth decades of life. There have been many large-scale quantitative studies since that time (see Simonton, 1988) that similarly demonstrate averages for creative peaks across different domains.

The nomothetic approach often presumes a fixed set of dimensions characteristic of stages through which people move. Jaques (1970), for instance, suggests that there is a period of "young creativity" during which the person in his or her 20s and 30s is "hot from the fire" and is spontaneous and excited about new ideas. He referred to creativity of the late 30s and beyond as "sculpted creativity," which includes intermediary processing between the initial and final stages of a creative product.

The nomothetic approach has created oversimplifications about late life creativity, however, which are not always generalizeable to the unique cases of highly creative people. For instance, Lubart and Sternberg (1998) describe an "old age style" consisting of four main characteristics. The first is subjectivity. A prevalent view is that the older artist is more subjective, or inner-oriented. Thus, Cohen-Shalev (1989) reports that as creative writers age they are more likely to use introspective approaches and to focus on inner experiences. Cumming and Henry (1961) even suggest a stage of "disengagement" in which older people withdraw from others. The second characteristic is an emphasis on unity and harmony. Simonton's (1989) investigations support this notion by demonstrating increased melodic simplicity and shorter duration in composers' late works, what he calls their "swan songs." Similarly, Kemper (1990) found that older subjects tend to use fewer complex sentence structures in speech and writing than do younger subjects. Third, the creative works of older people are supposedly more a summing up of ideas they formed earlier in life. That is, older people are said to be more integrative rather than original in their products (J. Erikson, 1988; Lehman, 1953). Finally, a fourth component of the "old age style" is that content is assumed to emphasize aging itself, focusing on deficits and impending death. For example, Verdi's late life opera *Falstaff* is a story about the loves and mishaps of an aging knight.

THE UNIQUENESS OF LATE LIFE
CREATIVE WORK

Research on aging and creativity from both the nomothetic perspective (e.g., Simonton, 1989) and the evolving systems approach (e.g., Wallace & Gruber, 1989) is optimistic, revealing that creative work continues well into old age. The average age at time of death of the cases in this chapter is 90. The vitality and longevity described here are consistent with the results of the recent MacArthur Foundation Study of Aging in which several myths of aging are debunked (Rowe & Kahn, 1998). But are there any generalities about late life creative work? Rather than suggest that creative work follows a particular trajectory or that there is an old age style, the cases in this chapter reflect uniqueness of trajectory as well as uniqueness of style.

Trajectories Are Unique. Peaks Are Different

Each person described in this chapter traveled a different route in achieving her or his creative work, which is expected from our knowledge of the increasingly diverging developmental pathways in adulthood. There is no average age or "peak" at which each did the most or best work. Some were prodigies and received recognition for their work early in life. Georgia O'Keeffe declared her intention to become an artist when she was in the eighth grade and her talents were soon noticed by her teachers. Dorothy West was an award-winning author in her teens. Marian Anderson joined her church choir at the age of 6. Elizabeth Cady Stanton, motivated in part by a need for her father's approval, excelled in her studies as a young girl. Eight-year-old Lise Meitner kept a math book under her pillow at night and would question her elders about such phenomena as the colors of an oil slick. Others got relatively late starts. Grandma Moses began painting in her 70s.

Martha Graham did not begin classes at Denishawn until she was 22, and even then her teachers were not particularly impressed with her. Igor Stravinsky demonstrated an early interest in painting, theater, and music, but he was not considered a musical prodigy. Bernard Shaw was never a good student and quit school when he was 15. Neither Guiseppe Verdi nor Charles Darwin was considered to be remarkable in youth. While Annie Besant was always recognized as an intelligent child and young woman, it was only after her marriage failed that she started her serious work in social reform.

After the initial steps, creative work is achieved at different rates. Sometimes the work is punctuated by debilitating silences, and at other times the work appears to progress steadily or rapidly. Although Meitner was a precocious student as a girl, her entrance to the University was delayed because her father wanted her first to complete a "feminine" course of study. Meitner's career peaked when she was in her 50s, as is often the case for retired scientists who are dependent on academia for their source of support. Darwin continued to write his major works until the last year of his life. Verdi got a late start because his family didn't have the money to send him to school and he had to wait for a kind benefactor's aid. The trajectories of O'Keeffe, Verdi, and Graham were often characterized by periods of depression that interrupted their work. West didn't write her second novel for almost 50 years after the publication of her first, although she continued to write articles during the intervening years. The creative trajectories of Anderson and Graham were influenced in part by the physical integrity of voice and body. While both had to eventually stop performing owing to the natural effects of aging, they were still able to participate in their respective arts until very late in life. Anderson taught voice and supported the arts through charitable organizations, and Graham choreographed and administered a dance company.

Enterprises in a network each have their own trajectory. Work along the path of one enterprise may move at a faster rate than another. For instance, Shaw undertook several other careers prior to becoming a playwright in his late 30s. At times he could work on several enterprises at once, or he could choose to abandon one for another. Darwin's work as a taxonomist formed an enterprise that developed more quickly than his theoretical work of constructing a theory of evolution, a more risky enterprise given the hostility he anticipated from his critics. Verdi composed very little during the relatively long period in which he was devoted mainly to farming and politics.

Style Is Unique

The notion of an "old age style" of late life work is not reflected in the present cases. While it is true that almost all wrote autobiographical work in their old age, the perspectives were not necessarily more subjective than earlier accounts. Although Besant and Stanton both dealt with religious themes in their late life, their styles were not esoteric or subjective. Stanton's *Woman's Bible* was her most outspoken work. In her diary, Stanton described herself as becoming "more radical" with age (see Griffith, 1984, p. 195). O'Keeffe was able to work in a new medium, pottery, when her failing eyesight interfered with the precision she needed for painting. Stravinsky's serialism certainly does not reflect melodic simplicity. He himself described his late works as "the new poetry." Verdi's *Falstaff* has been described as "freed from the harmonic constrictions of his earlier period" (Osborne, 1977, p. 360). While Verdi's last great work dealt with the topic of aging, West returned to themes from her youth in her last novel. Some of Shaw's later works, such as *Back to Methuselah* (1921) and *The Simpleton of the Unexpected Isles* (1934), were innovative and futuristic. Graham's late life focus was not on her past, but rather, she stated, "I'm very bad about time. I don't think about the past time very much. I think about what I've got to do and what I want to do in the future" (quoted in Lee, 1998, p. 441).

COLLABORATIONS AND GENERATIVE ENGAGEMENT IN LATE LIFE CREATIVE WORK

One generalization we can make from the cases presented here is that collaborations are important in creative work. This is especially true of late life work. Research in the evolving systems approach consistently reveals that creative people are not at all isolated, but rather, they are driven to create a milieu in which they can engage with others, in which they can be nurtured by feedback from others. Brower (1995), in his careful documentation of the lives of creative people, provides numerous examples of collaborations in which great artists and thinkers benefited from their interaction with others. Similarly, in answer to a question about how Piaget was different from other people, Gruber replied that the creative person's life is organized differently.

> The "common" man has his social world for his team. [However], Piaget builds his team, and what he does is considerably more than any one person could do by himself. He works closely with his team. For other creative men, it's the same thing, even if the team isn't present in fact; bonds are developed with other scholars, and that's a kind of team, too. Contrary to the popular notion, the creative man has close links with the world; he needs the world in order to correct himself and to find new ideas. I think he's less alone than the average man. (quoted in Bringuier, p. 80)

The collaboration between Elizabeth Cady Stanton and Susan B. Anthony was particularly crucial to the history of the women's movement. Their friendship was mutually beneficial in that each shored up the other. Assuming the role of older mentor, Stanton advised Anthony on how to improve her public speaking and how to handle hecklers. When Stanton was overwhelmed with public and private demands, Anthony would move in and watch Stanton's children while her friend wrote without interruption at the dining room table. Anthony was unmarried and free of the domestic burdens that often interfered in Stanton's work. The relationship was not competitive, and each woman acknowledged the cooperation of the other. According to Anthony, Stanton wrote all her speeches. According to Stanton, "Our speeches may be considered the united product of two brains" (quoted in Griffith, 1984, p. 183).

The musical career of Guiseppe Verdi was greatly influenced by his collaboration with Arrigo Boito (1842–1918). Nearly 30 years his junior, Boito wrote two libretti based on Shakespeare's *Othello* and *Falstaff* and was able to convince Verdi to provide the music. This was at a time in Verdi's late life when he had settled into a comfortable retirement to farm life and had refused requests from publishers to continue to compose. Boito and Verdi developed a deep and lasting friendship. Of Verdi's death, Boito wrote, "Verdi is dead; he has carried away with him an enormous quantity of light and vital warmth. We had all basked in the sunshine of his Olympian old age" (quoted in Osborne, 1977, p. 62).

Stravinsky's late work is rarely discussed without some reference to his young conductor friend, Robert Craft. Craft not only assisted the composer personally and helped him record his complete works, but he also published a series of books of personal reminiscences and musical discussions. After Stravinsky's death, Craft organized the construction of *A Reliquary for Igor Stravinsky (1974)*, an orchestral piece by Charles Wuorinen based on a number of incomplete musical sketches by the late Stravinsky.

Integral to Shaw's work were the relationships and friendships he developed in the theater with countless actors, writers, musicians, and others. Jackie Kennedy Onassis was able to encourage her older friend Dorothy West to write, and Juan Hamilton similarly served as muse for the greatly admired and elderly Georgia O'Keeffe. Graham received sustenance from her troupe of young dancers. When she isolated herself during her depression and alcoholic binges, she was unable to dance. While Meitner spent hours by herself experimenting and thinking about physics, she also led teams of other scientists to assimilate their work and accommodate her own structures of thought.

Distinctly different from the idea of disengagement in old age is the idea of generative engagement. Generative engagement refers to the kind of collaborations that older people pursue to ensure their work will survive them. Erik Erikson wrote that the crucial realization of middle age is "I am what survives me" (1968, p. 141). Generativity can take the form of leaving one's genes behind. Grandma Moses, Charles Darwin, and Elizabeth Cady Stanton all had many children and grandchildren, many of whom directly carried on the work of their parents and grandparents. Stanton saw the experience of creating and rearing children as a microcosm of a larger sense of generativity when she wrote, "As I sit here beside Hattie with the baby in my arms, and realize that three generations of us are together, I appreciate more than ever what each generation can do for the next one, by making the most of itself" (quoted in Griffith, 1984, p. 182).

While creating and rearing children can enhance the sense of generativity, it requires time and energy that can interfere with other creative work. More than half the cases here are of men and women who remained childless. Greenacre (1960) postulated that creativity is the correlate of childlessness. This generalization perhaps applies more to women than to men since women are more likely to be affected by the competing demands of childbearing and childrearing which can interfere with creative achievement outside this role. Indeed, Gedo (1963) notes that the great majority of women "geniuses" have been childless. (In addition to the freedom to create that childlessness affords many women, the presence of a strong relationship has similar positive effects. In her book, *Nobel Prize Women in Science* (1993), McGrayne notes that all but one of the husbands of the women she studied supported their wives' science, sometimes at considerable sacrifice.)

Engaging in creative work, especially when one collaborates, fosters a sense that there is something beyond the self. People doing creative work are often characterized at some time in their lives as behaving selfishly or as autocrats. Rather than mere selfishness, this behavior can be explained as a need to give the work precedence over everything else. O'Keeffe's friends, for instance, remarked that as she grew older she was often inflexible in her need to have things go her way. Her friend Virginia Robertson, who assisted the 90-year-old artist in writing her autobiography, commented on the laborious task of working with O'Keeffe, who was as meticulous about her writing as she was with her painting. But the work of O'Keeffe, as much as the person, was magnetic and inspiring, and Robertson persevered.

In generative engagement, the creative product is more important than the person making it, and the person is interested in giving the product to future generations. Anderson had no children, but she believed that education was a key to social and racial equality and she often spoke at elementary schools. She spent the last years of her life active in charitable organizations providing education for needy and talented students. Verdi died a wealthy man and had no surviving children. The chief beneficiary of his will was the rest home for aged musicians he started in Milan, which still exists and in whose chapel he and his second wife are buried. West never married and never had children, but her legacy was her writing, which paved the way for other African-American women such as Rita Dove, Ntozake Shange, Alice Walker, and Toni Morrison. The work of the women discussed in this chapter gives permission to the following generations of women to give work priority in their lives.

Meitner expressed her concern with future generations in a famous transatlantic interview and broadcast with Eleanor Roosevelt in 1945. Roosevelt praised her work in atomic physics, but Meitner was uncomfortable being associated with the powerful destruction of the atomic bomb. She commented:

> Women have a great responsibility, and they are obliged to try, so far as they can, to prevent another war. I hope that the construction of the atom bomb not only will help to finish this awful war, but that we will be able, too, to use this great energy that has been released for peaceful work. (quoted in Greene, 1964, p. 397)

In sum, creative work can continue throughout the lifespan. While the creative work and developmental trajectories of the cases in this chapter differ from person to person, a common concern with those involved in late life creative work is the need to share the work with future generations. In late life, with increasing awareness of one's immortality, collaborations become more important. An emphasis is placed on creating an environment in which the creative work can continue to grow.

REFERENCES

Anderson, M. (1956). *My Lord, what a morning.* New York: Viking Press.

Barlow, N. (Ed.). (1969). *The autobiography of Charles Darwin, 1809–1882.* New York: Norton.

Bogle, D. (1980). *Brown sugar.* New York: DaCapo Press.

Bringuier, J.-C. (1980). *Conversations with Jean Piaget.* Chicago: University of Chicago Press.

Brower, R. (1995, August). Creativity: Correlates and development. Paper presented at the Annual Meeting of the American Psychological Association, New York.

Brower, R. (1996). Vincent van Gogh's early years as an artist. *Journal of Adult Development, 3*(1), 21–32.

Bruchez-Hall, C. (1996). Freud's early metaphors and network of enterprise: Insight into a journey of scientific self-discovery. *Journal of Adult Development, 3*(1), 43–57.

Cohen-Shalev, A. (1989). Old age style: Developmental changes in creative production from a life-span perspective. *Journal of Aging Studies, 3*(1), 21–37.

Cumming, E., & Henry, W. E. (1961). *Growing old.* New York: Basic Books.

Erikson, E. H. (1968). *Identity: Youth and crisis.* New York: W. W. Norton.

Erikson, J. (1988). *Wisdom and the senses.* New York: W. W. Norton.

Greenacre, P. (1960). Woman as artist. In *Emotional Growth* (Vol. 2). New York: International Universities Press.

Greene, J. E. (1964). *100 great scientists.* New York: Washington Square Press.

Griffith, E. (1984). *In her own right, the life of Elizabeth Cady Stanton.* New York: Oxford University Press.

Grout, D. J. (1960). *A history of western music.* New York: W. W. Norton.

Gruber, H. E. (1977). The study of individual creativity: A report on the growth of a paradigm with some excerpts from a little known document by the young Piaget. Paper presented at the Seventh Annual Symposium of the Jean Piaget Society, Philadelphia.

Gruber, H. E. (1980). The evolving systems approach to creativity. In S. & C. Modgil (Eds.), *Toward a theory of psychological development.* Windsor, UK: NFER.

Gruber, H. E. (1996). Starting out: The early phases of four creative careers—Darwin, van Gogh, Freud, and Shaw. *Journal of Adult Development, 3*(1), 1–6.

Gruber, H. D., & Davis, S. N. (1988). Inching our way up Mount Olympus: The evolving systems approach to creative thinking. In R. J. Sternberg (Ed.), *The nature of creativity.* Cambridge, UK: Cambridge University Press.

Jaques, E. (1970). *Work, creativity, and social justice.* London: Heinemann.

Kallir, O. (Ed.) (1948). *Grandma Moses, my life's history.* New York: Harper & Brothers.

Keegan, J. (1995, August). Creativity from childhood to adulthood: A difference of degree and not of kind. Paper presented at the Annual Meeting of the American Psychological Association, New York City.

Keegan, J. (1996). Getting started: Charles Darwin's early steps toward a creative life in science. *Journal of Adult Development, 3*(1), 7–20.

Kemper, S. (1990). Adults' diaries: Changes made to written narratives across the lifespan. *Discourse Processes, 13*, 207–224.

Lee, S. A. (1998). Generativity and the life course of Martha Graham. In D. P. McAdams & E. de St. Aubin (Eds.), *Generativity and adult development.* Washington, DC: American Psychological Association.

Lehman, H. C. (1953). *Age and achievement.* Princeton, NJ: Princeton University Press.

Libman, L. (1972). *And music at the close: Stravinsky's last years.* New York: W. W. Norton.

Lubart, T. I., & Sternberg, R. J. (1998). Life span creativity: An investment theory approach. In Adams-Price (Ed.), *Creativity & successful aging.* New York: Springer.

McGrayne, S. B. (1993). *Nobel Prize women in science.* New York: Carol.

Osborne, C. (1977). *The dictionary of composers.* New York: Taplinger.

Rowe, J. W., & Kahn, R. L. (1998). *Successful aging.* New York: Pantheon Books.

Shaw, B. (1949). *Sixteen self-sketches.* New York: Dodd, Mead.

Simonton, D. K. (1988). Age and outstanding achievement: What do we know after a century of research? *Psychological Bulletin, 104*, 251–267.

Simonton, D. K. (1989). The swan-song phenomenon: Last-works effects for 172 classical composers. *Psychology and Aging, 4*(1), 42–47.

Strickland, C. (1992). *The Annotated Mona Lisa.* Kansas City: Andrews and McMeel.

Tahir, L. (1996). Their initial sketch and the growth of a creative network of enterprises in early adulthood: George Bernard Shaw. *Journal of Adult Development, 3*(1), 33–41.

Wallace, D. B., & Gruber, H. E. (Eds.). (1989). *Creative people at work: Twelve cognitive case studies.* New York: Oxford University Press.

CHAPTER 14

The Development of Possible Selves During Adulthood

Jane Allin Bybee and Yvonne V. Wells

Possible selves show dramatic change in form and function over the lifespan. In the present review, the ideal self-image (the self as one would like to be) is examined first. Developmental changes in the discrepancy between the real (or current) self and the ideal self are reviewed. Attention is given next to whether or not a high ideal self is adaptive for individuals at different life stages. Age-related changes in the content of the ideal self-image, as revealed in participants' own ideal self-descriptions, are then examined. The nightmare self (the self as one does not want to be) is also considered. Marked changes with age in the content of this possible self are reviewed. Other possible selves such as the moral (or ought) self and the fantasy self (the self as one would like to be if anything were possible) are briefly considered and the dearth of research on developmental changes in these selves is noted.

DEVELOPMENTAL CHANGES IN THE IDEAL SELF-IMAGE AND SELF-IMAGE DISCREPANCY

Striking changes appear in the ideal self-image across the lifespan. The ideal self-image is set progressively higher with development during childhood, reaching its zenith during adolescence and young adulthood (e.g., Bybee & Zigler, 1991; Katz & Zigler, 1967; Katz, Zigler, & Zalk, 1975; Phillips & Zigler, 1980). This developmental increase appears in cross-cultural studies as well (Velasco-Barraza & Muller, 1982). Similarly, the disparity between the real and ideal self-image becomes more pronounced with age over this period. These trends halt during young adulthood and the patterns thereafter reverse. The ideal self-image is set

Jane Allin Bybee and Yvonne V. Wells • Department of Psychology, Suffolk University, Boston, Massachusetts, 02114.

Handbook of Adult Development, edited by J. Demick and C. Andreoletti. Plenum Press, New York, 2002.

ever more modestly with advances in age and self-image discrepancy becomes less pronounced (Bybee, Piastunovich, & Glick, 1995).

Increases in the ideal self-image with development prior to adulthood appear to reflect improvements in cognitive reasoning (Achenbach & Zigler, 1963). Consistent with the movement toward cognitive differentiation with development during childhood, older individuals employ more categories and finer distinctions in the real self-image (e.g., Harter, 1981; Mullener & Laird, 1971). Similarly, with maturity, children are better able to distinguish the real from ideal self-image. Children set the ideal self-image progressively higher relative to the real self-image with age as they conceptualize what they want to be as something different than who they currently are (Phillips & Zigler, 1980). In addition, with the advent of abstract and hypothetical reasoning, young people are better able to construe a world of possibilities and may raise their sights accordingly.

The rise in the ideal self-image prior to adulthood may also reflect the effects of socialization (Bybee & Zigler, 1991). As societal mores and standards are incorporated with age, heightened expectations may be reflected in a higher ideal self-image (and lowered self-esteem over failure to meet these standards). Moreover, parents, teachers, coaches, and the like may expect more responsible and disciplined behavior from more mature students. So too may personal and idiosyncratic demands for self-centered ideals such as material possessions, physical attractiveness, or academic excellence increase with development. As the bar is raised, the ideal self may become more stringent. Evidence that the ideal self-image becomes progressively more intertwined with social strictures and demands of conscience is provided by Bybee and Zigler (1991), who report the emergence of stronger links between the ideal self-image and guilt with age among young people.

Further evidence that both level of intellectual abilities and socialization play a role in how high or low the ideal self-image is set is provided by Bybee and Zigler (1991). Adolescents in higher compared to lower academic tracks in this study had higher ideal self-images. Indeed, the bright group had self-image disparity scores over three times as great as the less intellectually gifted group. Brighter young people may be more facile with respect to abstract reasoning and may have a greater capacity for hypothetical thought. Hence, they may be better able to imagine the self as being different than it is presently and may, accordingly, set the ideal self-image higher. Moreover, educators, employers, and parents may have heightened expectations for, and give greater responsibilities to, brighter students. Greater expectations may translate into a higher ideal self-image. In addition, more intellectually gifted students may have different future prospects than intellectually average students. They may have more opportunity to gain entry to college. They may have a better chance of continuing onward to graduate or professional school. They may have more room for career advancement. In short, brighter students may have more doors open to them. A high ideal self-image among the intellectually gifted group may reflect the broader range of opportunities open to them as well as the great personal discipline needed to reach higher rungs of educational and career achievement.

Around young adulthood, age-related movement toward an increased ideal self-image and greater self-discrepancy ceases. Several reasons for this halt seem evident. Age-related improvements in cognitive differentiation abilities and in abstract, hypothetical reasoning do not continue ad infinitum. By early adulthood, individuals may be fully capable of distinguishing the real from the ideal self-image and more than able to construct possibilities for themselves in the future. Further advances in cognitive reasoning abilities may not result in an increased ideal self-image with adult development (although the effects of intellect henceforth may exert influence as an individual difference variable as discussed later). The work of socialization also may be largely completed around the beginning of young adulthood. Society awards the full range of rights and responsibilities (i.e., voting, purchasing alcohol, being tried for crimes as an adult, marrying, and living independently) to individuals by or before age 21. Responsibilities (and expectations from parents, employers, and others) may increase somewhat into young adulthood and then level off as young adults move away from home, complete their education, attain financial independence, and enter the workforce.

Ryff (1991) reports that during adult development, the ideal self-image declines as self-image disparity decreases. Specifically, the ideal self-image declines in the domains of self-acceptance, autonomy, positive relationships with others, purpose in life, environmental mastery (competence in managing day-to-day affairs), and personal growth. Decreases are almost always linear, with declines in the ideal self-image from young and middle adulthood to late adulthood always proving significant. Present (real) self-ratings generally do not vary with age according to Ryff. The few developmental differences that appear indicate an increase in the real self with adult development. The only exception is in the personal growth domain where the real self is set lower among old-aged compared to young adults. Ryff (1991) reports that the effects of age on ideal and present self-ratings are not qualified by sex.

Ryff (1991) explains her results in the following way. Young adults, who have their whole lives before them, foresee big changes afoot in the future. Expectations are not well grounded in experience and may be overly naïve. Middle-aged adults see continued growth and progress in life goals, although expectations are more tempered. Older adults rein in their ambitions for the future. They may also become more accustomed to and accepting of themselves, faults and all. Hence, with development, the once high ideal self-image declines as the real and ideal self-image are brought into line with one another.

The age-related declines in the ideal self-image and reductions in self-image discrepancy noted by Ryff (1991) are found in a predominantly middle-class sample who, on average, had some college education. Similarly, Bybee et al. (1995) report that, among middle status adults, global ideal self-image scores inch downward with age (although the decline did not reach significance) and self-image disparity becomes progressively smaller at higher age levels. Not surprisingly, developmental effects are more pronounced in the Ryff than in the Bybee et al. study. Most age-related effects in the Ryff study appear after middle age. The Bybee et al. study does not include adults past middle age (participants ranged in age from 23 to 67).

Bybee et al. report that the effects of adult development on the real and ideal self-image vary by socioeconomic status. Changes in the ideal self-image appear among middle but not high socioeconomic status adults. For middle status adults, the real and ideal self-image may more closely approximate each other because occupational, familial, and financial goals may be attained with age and individuals may come to accept aspects of their personality that are not amenable to change. Among high status adults (those who had completed an advanced degree or held a professional position), effects of development on the ideal self-image are virtually identical for individuals at various stages of young and middle adulthood, showing no signs of age-related decline. High status adults have more avenues for career and personal advancement available to them than do middle status adults. Education and financial resources open doors and possibilities. A professional at midlife may imagine moving into senior or administrative positions, pulling in a major grant or large client, or founding their own company. In their personal life, high status adults at midlife may envision building their dream home, traveling the world, or vacationing in the Mediterranean. High compared to middle status individuals are more likely to have the resources to attain numerous and lofty ideals and, hence, may be more likely to envision change as part of their ideal self. In addition, high compared to middle status individuals may not only have a greater capacity for abstract, hypothetical thought, but may choose to spend more time thinking at that level. The ideal self-image, being in the realm of the imagination, would presumably be set higher among those more inclined to think abstractly. In addition, for high status adults, a high ideal self-image may be adaptive over much of adulthood and may, as a result, remain constant rather than decline with development. Evidence that a high ideal self-image is more adaptive for high compared with middle status individuals is reviewed in the next section.

DEVELOPMENTAL CHANGES IN THE
ADAPTIVENESS OF THE IDEAL SELF-IMAGE

A large disparity between the real and ideal self-images has traditionally been viewed as ominous and a marker of psychopathology (Rogers, 1961). Clinically depressed individuals compared to nondepressed individuals score higher on real–ideal discrepancy (Strauman, 1989). Indeed, self-image disparity items are included in depression inventories as diagnostic criterion (Blatt, 1979; Blatt, D'Afflitti, & Quinlan, 1976). Not only is a low real self-image seen as maladaptive, but a high ideal self-image is also viewed as problematic. Having a perfect idealized vision of how one might be may make the present self-image seem woefully lacking in comparison and may evoke self-loathing, guilt, and self-recrimination (Rosenberg, 1979). Lofty goals, expectations, and hopes embodied in a high ideal self-image may exert pressure, stress, and strain. When the ideal self is out-of-sync with actual attributes or accomplishments, as reflected in the real self-image, unpleasant internal conflict may result. Failing to reach the ideal self-image may cause the individual to suffer from disappointment and self-disparagement (Higgins, 1987).

The ought self represents the self one should be and contains elements of conscience, role demands, and duties to others (Bybee, Luthar, Zigler, & Merisca, 1997). Just as a high ideal self-image has been put forth as a source of psychiatric symptomatology, Higgins (1987) advances a theoretical conceptualization that casts the moral or ought self as a source of mental anguish and disorder. Higgins (1987) hypothesizes that because violating rules and deserting responsibilities may lead to punishment or sanction, actual–ought disparity may be associated with symptoms of agitation and with anxiety disorders. Consistent with this position, disparity between the actual (real) and ought selves is greater among individuals with than without social phobia, an anxiety disorder (Strauman, 1989). (Note that Tangney, Niedenthal, Covert, & Barlow [1998] report that the ideal and ought self do not show distinctive relationships with symptoms of dejection and agitation.)

The fantasy self, the self one would be if anything were possible, contains: playful daydreams; desires for perfectionism; and idealistic, grandiose strivings (Bybee et al., 1997). Individuals who are more preoccupied with their fantasy self are more antisocial and more maladjusted on measures of academic performance and social competence (Bybee et al., 1997). One interpretation of these findings is that just as individuals who daydream have trouble concentrating on work (Golding & Singer, 1983), those who ruminate over the fantasy self overlong may have problems concentrating at school and in social relationships. Alternatively, fantasies of power, imperialism, and dominance may lead to rough or ruthless treatment of others. Or, perhaps, being socially isolated and rejected or failing in school leads to more daydreaming and preoccupation with a less threatening fantasy world.

In contrast to work that explores the darker aspects of possible selves, Bybee and Zigler (1991) (see also Glick & Zigler, 1985; Markus & Nurius, 1986) emphasize possible adaptive components of a high ideal self-image. A high ideal self-image may serve as a source of motivation and goal direction, providing an arena for testing out new roles, setting goals, and planning ways of reaching them (Markus & Nurius, 1986). Acting consistently with one's standards or attaining an aspect of the ideal self-image may engender feelings of pride, satisfaction, and pleasure, feelings that might then serve as incentives for the attainment of future goals (Bybee & Zigler, 1991). Consistent with this position, Bybee (1987) reports that school-aged students with higher compared to lower ideal self-images have better scholastic performance, less aggressive behavior, and greater frustration tolerance. A complex ideal (or possible) self may also serve as a psychological buffer. Indeed, individuals with a more complex possible self show less extreme affective reactions when undergoing evaluation (Niedenthal, Setterlund, & Wherry, 1992). Moreover, individuals with a high ideal self-image show better social adjustment and better recovery after a life crisis (Bybee, 1987; Markus & Nurius, 1986).

The ought or moral self-image may also serve positive functions. A high moral self-image may reflect the incorporation of a strong sense of duty and responsibility and a strict ethical code. Discrepancy between the ought and real self-image is related to more active participation in religious activities (Lilliston & Klein, 1991). Individuals preoccupied with their moral self-image are more warm, altruistic, trusting, and straightforward (Bybee et al., 1997).

The fantasy self-image might also play a positive role in adjustment. Daydreams are useful in rehearsing difficult situations and maintaining stable mood states (Singer & Switzer, 1980). Similarly, thoughts of the fantasy self might be expected to play a role in mood regulation or frustration tolerance. Adaptive correlates of the fantasy self, although anticipated, do not appear in a study by Bybee et al. (1997).

Whether possible selves are adaptive or not may vary with age. To young people who have their whole adult lives before them, a high ideal self-image may represent wanting to move forward into complex and challenging adult roles. Having a high ideal self-image during childhood may indicate a healthy welcoming of changes that will inevitably occur at the beginning of adulthood.

Moreover, with development, young people do typically change and mature, showing growth in the type of traits found on self-image inventories. Parents and teachers anticipate and even demand changes in personality among children. In contrast, adults who show dramatic personality changes and whose personalities are difficult for others to gauge are viewed by others as unstable or mentally unhealthy (Colvin, 1993). Dramatic changes in physical appearance are normative during child and adolescent development. For adults, radical change is not the norm and others often look askance at those who attempt it. High ideals that are socially acceptable and anticipated by others may be more adaptive than those that are not welcomed.

Further, for a young person, wanting to be different, even radically different, may be realistic. For the young, many data points are not yet set. Even extremely difficult goals may be attainable. Who is to say whether or not a child who wants to earn a Ph.D., become a millionaire, have three children, and spend half the year in the tropics can attain these ideals? Individuals often can through dint of effort move up (or down) the social and economic ladder (although this varies widely across culture and so on). Desire and hard work can effect change. A high ideal self-image may serve as a particular source of motivation and goal direction for young people. With age, many individuals may attain long awaited ideals such as finishing college or getting married. Achievement of goals over time may result in a lowering of the ideal self-image (as components that are realized no longer appear as ideals). Hence, a high ideal self-image might be correlated with better adjustment among adults.

As the years roll by, just as some goals are attained, other doors begin to close. The biological clock for childbearing ticks ever louder and runs out. Commitments and fiduciary responsibilities may make it ever more difficult to return to school or change occupations. Many life decisions, once made, cannot easily be undone. Radical change may no longer be attainable at midlife. A high ideal self-image in later life may reflect goals that can no longer be attained. As such, it may become a source of bitterness and serve to engender regret, frustration, and disappointment over dreams unrealized.

The pattern of results in the literature is quite striking and easily summarized. A high ideal self-image appears to show the most beneficial relationships among those individuals who are in positions because of their age, educational or occupational group, or gender that permit the greatest upward socioeconomic mobility. The ideal self-image is highest and most adaptive among those individuals who are young, highly educated or in professional positions, or male (Bybee, Piastunovich, & Glick, 1994). Among children and adolescents, as mentioned earlier, a high ideal self-image is related to lowered aggression, a high real self-image, academic achievement, and frustration tolerance (Bybee, 1987). Similarly, for high status adults under 30, a high ideal self-image is related to less hostility, higher self-esteem, and a less negative worldview. For adults aged 30 to 44, relationships are more moderate in strength. For adults over 45, the ideal self-image is unrelated to the measures of adjustment.

A high ideal self-image, then, is seemingly most beneficial for individuals who have room for career and financial advancement. For middle status professions where seniority may play a larger role than individual initiative in determining responsibilities, status, and salary, and where upward mobility is more restricted, a high ideal self-image generally does not show relationships—positive or negative—with indicators of mental well-being. The results are likewise strikingly different across various sex and status groupings. For high status males, a higher ideal self-image is correlated with better adjustment. For females and middle status males, the ideal self-image shows virtually no relationships with adjustment. In no age, status, or gender grouping in the aforementioned studies are maladaptive correlates of a high ideal self-image found.

Most studies on the moral and fantasy self appear in the social/personality literature and include college undergraduates as participants. The literature is too sparse to address whether the relationship of the moral and fantasy selves to adjustment differs across age.

CHANGES ACROSS THE LIFESPAN IN
THE CONTENT OF THE IDEAL SELF-IMAGE

Possible selves provide an arena for imagining oneself in new roles, constructing goals, and planning methods of attaining aspirations (Glick & Zigler, 1985; Markus & Nurius, 1986). Changes in roles and life tasks across the lifespan are reflected in the ideal self-image. Repeatedly, studies show an increase in mentions of a particular ideal as the time for attaining that ideal approaches and finally arrives. Thereafter, the ideal is less frequently and eventually never mentioned. One of the main changes in the content of the ideal self-image over the lifespan involves references to adult roles. Early adulthood is a formative period when major and often intractable life decisions are made. Much of childhood and adolescence is spent anticipating, imagining, and preparing for adult life when roles such as college student, spouse, parent, and employee are assumed for the first time. Conversely, when older adults reminisce about consequential experiences, memories are most typically drawn from the formative early adult years (Mackavey, Malley, & Stewart, 1991). The main pattern of change in the ideal self over the lifespan may be simply stated. Mentions of wanting to attain or succeed in adult roles rise as young adulthood approaches and subsequently decline in later young adulthood and beyond as roles are either attained or the time for achieving them passes.

Many of the most common ideal self-descriptions of children, adolescents, and young adults involve the life tasks of young adulthood. Themes of completing educational degrees, getting married, having children, and selecting or establishing a career are prevalent among children and become more pronounced with age (Bybee, Glick, & Zigler, 1990; Bybee et al., 1997). The percentage of students mentioning desires to go to college, for instance, quadruples and then triples with development through puberty (Bybee et al., 1990). During adult development, themes surrounding educational attainment decline (Bybee & Merisca, 1995). The percentage of adults mentioning desires to complete an educational degree in their ideal self-descriptions declines from early young adulthood into later young adulthood and middle age.

This rise and then fall occurs in the family domain as well. With development from the 5th to the 11th grade, students are three times as likely to mention desires to get married and have children (Bybee et al., 1990). By later young adulthood and middle age, mentions wane (Bybee & Merisca, 1995). Declines occur some-what later in qualitative ideals such as wanting to be a good parent or have a better marriage, with a drop-off becoming apparent during middle age (Bybee & Merisca, 1995).

In the career domain, life tasks of early adulthood again loom large during childhood and adolescence. Occupation is, by far and away, the most commonly mentioned component of the ideal self-image among children and adolescents (Bybee et al., 1990). Nearly one in five students mentions wanting career success. Equally common are desires for certain job characteristics (such as good hours and benefits). Career aspirations are widely mentioned even at the youngest age level and continue in force into post-puberty. With adult development, career is less commonly mentioned as a component of the ideal self-image. Although desires for career success, for instance, are the second most common constituent of the ideal self-image among adults under 30 (Bybee & Merisca, 1995), the per-centage of individuals mentioning career success is halved by later young adult-hood and halved again by middle age. Mentions of desires for career change show a similar decline. References to occupation likewise become less prevalent with age level in a sample of adults aged 19 to 86 (Cross & Markus, 1991).

Rather than examining changes in the ideal self across a broad section of the lifespan, several studies have taken as their focus those changes that occur during a life event or transitional period. Before assuming a new role, individuals may imagine themselves in that position and may construct hoped-for selves as a guide to action. As the time for assuming a role approaches, aspects of the role would be expected to become more predominant in the ideal self-image. Just such an effect is found in a study of the transition to fatherhood (Strauss & Goldberg, 1999). The role of parent is more predominant in the ideal self-image of first-time fathers of neonates than among fathers-to-be (drawn from Lamaze courses). After the role is firmly established, presumably there is less need for imaginings as ele-ments of the ideal self have been translated into reality. Such an effect is found in a study by Hooker, Fiese, Jenkins, Morfei, and Schwagler (1996). Mothers and fathers of infants are more likely to have hoped-for parenting selves than are par-ents of older children. Moreover, the prevalence of parenting in the ideal self is markedly higher among parents of infants (60%; Hooker et al., 1996) than among other groups (comparable percentages are 43%, 11.1%, 18.3%, 19%, and 10%, in order, for high schoolers [Bybee et al., 1990]; college students [includes wanting to get married] [Bybee et al., 1997]; adults aged 23–30 [Bybee & Merisca, 1995]; and adults aged 18–24 and 25–39 [includes wanting to get married] [Cross & Markus, 1991]). The large role that life events and transitions play in the develop-ment of the ideal self-image, then, is quite evident in lifespan studies as well as studies targeting the time before, during, and after a transitional period.

Not all changes involve life transitions. Changes in references to the body in ideal self-descriptions across the lifespan show considerable change across the adult years as detailed in Bybee and Wells's (in press) review. The nature of changes depends on whether a more global and sweeping vs. specific approach to assessing body ideals is taken. Cross and Markus (1991) report that the physical domain becomes more prevalent with adult development. In contrast, studies adopting a more fine-tuned approach indicate that constituent domains (appear-ance, fitness, health) follow radically different developmental trajectories. Desires for changes in physical appearance reach their peak among the young. Turning to specific studies, Bybee et al. (1991) report that one in three prepubescent students expressed desires to be more physically attractive. With development through and past puberty, mentions of physical appearance become less frequent. The authors suggest that, with maturation, changes in physical appearance (e.g., becoming taller or having longer legs) may become less possible and hence the domain may be less frequently mentioned. In addition, the decline in mentions of physical traits may reflect the broader movement during child development away from physical and observable self-descriptors and toward more abstract traits (Damon & Hart, 1982).

Despite the apparent decline in mentions of physical appearance with child development, Bybee et al. (1997) found that one in three college students mentions desires for increased physical attractiveness as a part of their ideal self-image. A spike in mentions of physical appearance around early young adulthood, if replicable, would be consistent with movement into the intimacy vs. isolation life stage (see Erikson, 1963). Desires for enhanced physical and sexual appeal may reflect individuals' wish to attract romantic partners.

In later stages of adulthood, desires for physical attractiveness are resounding in their absence. Fewer than 5% of adults in the Bybee et al. (1995) sample mention physical attractiveness in their ideal self-descriptions. Why would one of the most prevalent ideal self-descriptors among the young all but disappear by mid-young adulthood? The answer seems to lie, at least in part, in the fact that allusions to the body continue in force but take on another form, that of concern with health and physical fitness. Indeed, the only constituent of the physical domain that becomes more prevalent with development through middle age is health. Desires for good health are twice as prevalent among adults over 45 as among those aged 23 to 30 (Bybee & Merisca, 1995).

The percentage of individuals mentioning desires for improved physical fitness and abilities as a part of the ideal self has an apparent bimodal distribution, with a major mode at puberty and a minor mode in young adulthood. The developmental pattern must be considered in light of the very strong sex differences in mentions of the category. Among the children and adolescents in the Bybee et al. (1990) sample, males are three times as likely as females to mention desires for improved athletic abilities. The huge increase in the number of students mentioning wanting to be more muscular and powerful around puberty in the Bybee et al. study coincides with the increase in musculature and build that accompanies sexual maturation for males. Males may anticipate, hope for, and welcome the changes that puberty brings to their physique as maturation brings them closer to the male ideal body type (Striegel-Moore, Silberstein, & Rodin, 1986). Mentions of physical abilities appear to recede after puberty, then spring up once more (although only to half the earlier level) in young adulthood only to decline again at middle age (Bybee et al., 1990, 1995, 1997). This rise may reflect desires to ward off the effects of aging and maintain a youthful appearance.

Ideal self-descriptions of adults include yearnings that virtually never appear prior to adulthood such wanting to travel and undergo spiritual growth (Bybee et al., 1990, 1995, 1997). Cross and Markus (1991) report that adults over, compared to under, age 60 are more likely to mention lifestyle (e.g., traveling, developing hobbies, living in the tropics) in their ideal self-image. In addition, personal characteristics become more predominant and diverse in ideal self-descriptions.

Adults, for instance mention desires for balance, open-mindedness, and productivity as a part of the ideal self-image, traits that are not widely mentioned prior to adulthood (Bybee et al., 1990, 1995, 1997).

Several constituents of the ideal self-image appear in similar prevalence across the lifespan. Desires for self-assurance, for considerateness toward others, and for social acceptance and respect appear to be universally included as a part of the ideal self.

DEVELOPMENTAL CHANGES IN THE CONTENT OF THE NIGHTMARE SELF-IMAGE

The ideal self-image stands in contrast to the nightmare self. The nightmare, feared, or undesired self may serve as a repository for unwanted social identities, insecurities, negative possibilities, and personal worst case scenarios (Bybee, Wells, & Merisca, 2000; Cross & Markus, 1991; Ogilvie & Clark, 1992). When asked to describe what they do not want to be like in the future, individuals commonly express wishes not to assume a socially deviant persona such as drug addict, criminal, bad parent, or bag lady (Bybee et al., 2000). Undesired selves may also reflect past selves that were once true and could potentially become real again. Individuals who were once overweight, for instance, may constantly fear that they will become fat in the future.

Negative possible selves may be formed by experiences of death or loss, by observing travails of family members, or by learning of tragedies in the lives of others. Many individuals on hearing that Magic Johnson had AIDS, for example, incorporated desires not to contract this disease in the feared self. Other experiences may be more idiosyncratic. An adolescent watching a father with a drinking problem might, for instance, fear becoming an alcoholic. Lapses in willpower, defeats, and losses may all engender or activate feared selves.

Nightmare selves may provide insight into individuals' behaviors and thought processes. These undesired selves imbue meaning and lend significance to events. College students who go on a drinking binge and black out may be much more terrified and alarmed if they have a feared possible self of becoming an alcoholic like their father than if they do not.

Undesired selves may serve as deterrents or disincentives. Individuals harboring fears of becoming fat may be more likely to pass on dessert. Feared possible selves may arouse considerable anxiety, stress, and dread, resulting in poor mental health. Yet individuals such as defensive pessimists who plan out worst case scenarios and strategize ways to prevent feared from becoming actual selves may be better adjusted (Norem & Illingworth, 1993). Feared selves, similar to possible selves in general, may be most vulnerable and responsive to change. Long before real or imagined adversities are reflected in actual self-descriptors, they may be voiced as possible selves.

We were unable to locate any studies examining the development of the nightmare self prior to adulthood although several have examined aspects of the nightmare self not related to development among children (e.g., Bybee & Wells, in press; Knox, Funk, Elliott, & Bush, 2000). Two studies contain reports of the effects of adult development on the content of the nightmare self: Cross and

Markus (1991) and Bybee et al. (2000). Important differences across these studies affect interpretation of the results having to do with the nightmare self. Compared to the Bybee et al. study, the Cross and Markus (1991) study examines changes in broad band descriptors rather than specific constituent categories. Constituents do not always show the same pattern as the overarching domain. Cross and Markus (1991) report that from the college to retirement years, for instance, adults were less likely to mention feared self-descriptors in the family domain. Bybee et al., adopting a more fine-tuned approach, report fears of not having a family decline with age whereas fears of not being appreciated by family members increase. Age ranges examined also differed across studies. The Cross and Markus (1991) study contains a broad age range that includes college and retirement aged adults. Most age-related changes (outside the family and material domain) involve differences between retired adults and those in the workforce. Adults in their retirement years compared with younger groups, for instance, are less likely to mention the occupational and abilities/education categories and are more likely to mention physical and lifestyle fears. The Bybee et al. sample is limited to young and middle-aged adults. A final difference has to do with sex composition. The Cross and Markus sample is predominantly female. The more even gender distribution in the Bybee et al. permits examination of main effects and interactions of gender.

With increased age from college to retirement years, individuals are less likely to mention family in the feared self (Cross & Markus, 1991). Examination of specific constituents indicates that at higher age levels, adults are less likely to mention fears of being without a family of their own (Bybee et al., 2000). Young adults preoccupied with issues of intimacy and isolation (Erikson, 1963) may be especially fearful of not establishing meaningful relationships with a life partner. Older adults are more likely to be married and to have children, thus reducing the fear of being alone. In addition, adults who remain single and/or childless in midlife may become reconciled to the situation and may be less likely to mention it as a fear. With age, adults are less likely to mention fears of having an unhappy family life. Younger compared to older adults place an especially high premium on intimacy and may be particularly worried that spousal relationships will be overly superficial or emotionally unsatisfying. Alternatively, younger adults may be fearful of having an unhappy family because they are involved in unstable early marriages or have had children before they were ready to assume this responsibility.

In contrast, the percentage of adults mentioning fears of being unappreciated by their family increases with age, doubling from the youngest to the oldest age level (Bybee et al., 2000). As adults approach midlife, concerns with generativity such as giving back to the community and nurturing the next generation move to the forefront. Older adults may be especially worried that their efforts will not be recognized and appreciated by their family. A second, more dramatic increase with age is found in mentions of becoming dependent on family members. Indeed, along with poor health, this is the number one fear of the 45 and over groups. For adults in their financial prime who are concerned with giving and helping others, thoughts of one day being in need of assistance themselves may be an affront to their pride and an area of tremendous concern.

Mentions of material fears, such as being poor or financially insecure, increase with age from young (18–39) to middle adulthood (40–59) (Cross & Markus, 1991).

As mentioned earlier, the Cross and Markus nonstudent sample was predominantly female. In the Bybee et al. study, fears of becoming destitute, homeless, and financially insecure increase with age among women, whereas fears of becoming homeless and financially insecure ease with age among men. These sex differences may reflect the greater wage-earning potential of men in this society. In addition, because of their greater life expectancy, women do face greater odds they will end up in a costly nursing home or that they will run out of money set aside for retirement. This may account as well for the gender difference. Fears of becoming overly materialistic were less frequent among the over 45 compared to the younger age groups. Youthful idealism may give way to financial realism with age.

Cross and Markus (1991) report an age-related decline in mentions of occupational fears. Specific constituents of the career category show different developmental patterns, however, according to Bybee and Wells. They find a marked decline in fears of being unsatisfied with one's job. Proportionately fewer adults in the older groups express worries that they would feel unfulfilled or unaccomplished in their careers. Studies of occupational development indicate that workers in midlife have either attained positions of power and influence or become increasingly detached from their jobs as they move toward retirement. Members of the older group may, then, either feel fulfilled or come to devalue career accomplishment as other life tasks assume greater importance. In contrast, Bybee and Wells find those over 45 are more likely than the younger groups to mention fears of becoming unproductive. Perhaps this increase reflects worries that job skills will become obsolete or fears that the older adults are merely "punching the time-card," while waiting for retirement.

Cross and Markus (1991) report age-related increases in fears stemming from the physical domain. By examining constituent categories, Bybee and Wells find age-related increases in fears of becoming physically incapacitated, mentally incapacitated, and unhealthy, but marginal declines in fears of being unattractive and physically unfit. Consistent with the view that possible selves may reflect the approach of undesirable as well as positive life events (Bybee et al., 1990; Markus & Nurius, 1986), adults may increasingly mention fears of losing their physical facilities, mental acumen, and good health as they approach old age. On the other hand, concern with physical appearance and might seems to lessen.

Neither study contains reports of age-related changes in personal traits. With age, adults are less likely to mention fears of substance abuse such as becoming or remaining addicted to alcohol, cigarettes, or drugs (the three substances show parallel declines). Habits that are not broken may become increasingly engrained with age.

REFERENCES

Achenbach, T., & Zigler, E. (1963). Social competence and self-image disparity in psychiatric and nonpsychiatric patients. *Journal of Abnormal and Social Psychology, 67*, 197–205.

Blatt, S. J. (1979). Depressive experience questionnaire. Unpublished manuscript. Yale University.

Blatt, S. J., D'Afflitti, J. P., & Quinlan, D. M. (1976). Experiences of depression in normal young adults. *Journal of Abnormal Psychology, 85*, 383–389.

Bybee, J. A. (1987). Domains of the real and ideal self-image: Relationships to gender, depression, and adjustment. Paper presented at the New England Social Psychological Association, New Haven, CT.

Bybee, J., Glick, M., & Zigler, E. (1990). Differences across gender, grade level, and academic track in the content of the ideal self-image. *Sex Roles, 22*, 349–358.

Bybee, J. A., Luthar, S., Zigler, E., & Merisca, R. (1997). The fantasy, ideal, and ought selves: Content, relationships to mental health, and functions. *Social Cognition, 15*, 37–53.

Bybee, J., & Merisca, R. (April, 1995). Age-related changes in the content and adaptiveness of future selves. Paper presented at the meeting of the Eastern Psychological Association, Boston, MA.

Bybee, J., Piastunovich, A., & Glick, M. (August, 1994). Higher ideal self-images are adapative for young, high status males. Poster presented at the Annual Convention of the American Psychological Association, Los Angeles, CA.

Bybee, J. A., Piastunovich, A., & Glick, M. (April, 1995). The development and adaptiveness of self-image discrepancies in adulthood. Poster presented at the Biennial Meeting of the Society for Research in Child Development, Indianapolis, IN.

Bybee, J. A., & Wells, Y. (in press). Possible selves: Diverse perspectives across the life span. *Journal of Adult Development.*

Bybee, J. A., Wells, Y., & Merisca, R. (2000). The nightmare self: Changes in adulthood. Manuscript in preparation.

Bybee, J. A., & Zigler, E. (1991). The self-image and guilt: A further test of the cognitive–developmental formulation. *Journal of Personality*, 733–745.

Colvin, C. R. (1993). "Judgable" people: Personality, behavior, and competing explanations. *Journal of Personality and Social Psychology, 64*, 861–873.

Cross, S., & Markus, H. (1991). Possible selves across the life span. *Human Development, 34*, 230–255.

Damon, W., & Hart, D. (1982). The development of self-understanding from infancy through adolescence. *Child Development, 53*, 841–864.

Erikson, E. H. (1963). Childhood and society, 2nd edit. New York: W. W. Norton.

Glick, M., & Zigler, E. (1985). Self-image: A cognitive-developmental approach. In R. Leahy (Ed.), *The development of the self.* New York: Academic Press.

Golding, J. M., & Singer, J. L. (1983). Patterns of inner experience: Daydreaming styles, depressive moods, and sex roles. *Journal of Personality and Social Psychology, 45*, 663–675.

Harter, S. (1981). A model of mastery motivation in children: Individual differences and developmental change. In W. A. Collins (Ed.), *Aspects of the development of competence*, Vol. 14. Minnesota: The Minnesota Symposium on Child Psychology.

Higgins, E. T. (1987). Self-discrepancy: A theory relating self and affect. *Psychological Review, 94*, 319–340.

Hooker, K., Fiese, B. H., Jenkins, L., Morfei, M. Z., & Schwagler, J. (1996). Possible selves among parents of infants and preschoolers. *Developmental Psychology, 32*, 542–550.

Katz, P., & Zigler, E. (1967). Self-image disparity: A developmental approach. *Journal of Personality and Social Psychology, 5*, 186–195.

Katz, P., Zigler, E., & Zalk, S. R. (1975). Children's self-image disparity: The effects of age, maladjustment and action–thought orientation. *Journal of Personality and Social Psychology, 11*, 546–550.

Knox, M., Funk, J., Elliott, R., & Bush, E. G. (2000). Gender differences in adolescents' possible selves. *Youth and Society, 31*, 287–310.

Lilliston, L., & Klein, D. G. (1991). A self-discrepancy reduction model of religious coping. *Journal of Clinical Psychology, 47*, 854–860.

MacKavey, W. R., Malley, J. E., & Stewart, A. J. (1991). Remembering autobiographically consequential experiences: Content analysis of psychologists' accounts of their lives. *Psychology and Aging, 6*(1), 50–59.

Markus, H., & Nurius, P. (1986). Possible selves. *American Psychologist, 41*, 954–969.

Mullener, N., & Laird, J. D. (1971). Some developmental changes in the organization of self-evaluations. *Developmental Psychology, 5*, 233–236.

Niedenthal, P. M., Setterlund, M. B., & Wherry, M. B. (1992). Possible self-complexity and affective reactions to goal-relevant evaluation. *Journal of Personality and Social Psychology, 63*, 5–16.

Norem, J. K., & Illingworth, K. S. S. (1993). Strategy-dependent effects of reflecting on self and tasks: Some implications of optimism and defensive pessimism. *Journal of Personality and Social Psychology, 65*, 822–835.

Ogilvie, D. M., & Clark, M. D. (1992). The best and worst of it: Age and sex differences in self-discrepancy research. In R. Lipka & T. M. Brinthaupt (Eds.), *Self-perspectives across the life span* (pp. 186–222). Albany, NY: State University of New York Press.

Phillips, D., & Zigler, E. (1980). Children's self-image disparity: Effects of age, socioeconomic status, ethnicity, and gender. *Journal of Personality and Social Psychology, 39*, 689–700.

Rogers, C. R. (1961). *On becoming a person*. Boston: Houghton Mifflin.

Rosenberg, M. (1979). *Conceiving the self*. New York: Basic Books.

Ryff, C. D. (1991). Possible selves in adulthood and old age: A tale of shifting horizons. *Psychology and Aging, 6*, 286–295.

Singer, J. L., & Switzer, E. (1980). Mindplay: The creative uses of daydreaming. Englewood Cliffs, NJ: Prentice-Hall.

Strauman, T. J. (1989). Self-discrepancies in clinical depression and social phobia: Cognitive structures that underlie emotional disorders? *Journal of Abnormal Psychology, 98*, 14–22.

Strauss, R., & Goldberg, W. A. (1999). Self and possible selves during the transition to fatherhood. *Journal of Family Psychology, 13*, 244–259.

Striegel-Moore, R., Silberstein, L., & Rodin, J. (1986). Toward an understanding of risk factors for bulimia. *American Psychologist, 41*, 246–263.

Tangney, J. P., Niedenthal, P. M., Covert, M. V., & Barlow, D. H. (1998). Are shame and guilt related to distinct self-discrepancies? A test of Higgins's (1987) hypotheses. *Journal of Personality and Social Psychology, 75*, 256–268.

Velasco-Barraza, C. R., & Muller, D. (1982). Development of self-concept in Chilean, Mexican, and United States school children. *Journal of Psychology, 110*, 21–30.

CHAPTER 15

The Good Life

A Longitudinal Study of Adult Value Reasoning

CHERYL ARMON AND THEO LINDA DAWSON

Philosophers and social critics have promoted different conceptions of the good human life for some 2000 years. Such philosophical conceptions always included, or relied entirely on models of good psychological functioning or mental health. In contrast, psychologists have only recently entered the debate about the Good Life. It was not until the 19th and 20th centuries that theorists such as Baldwin (1906), James (1890), Freud (1961), Horney (1937), and Erikson (1963) began to articulate models of mental and psychological health. These models can be understood as attempts to define (in part) a good human life. Many contemporary philosophers believe that there are as many conceptions of the Good Life as there are persons who seek it (e.g., Nozick, 1974; Rawls, 1971). The findings of this study, however, indicate that although the *sources* of our conceptions differ widely across time and culture, the number of actual views of the Good Life may be finite. Moreover, despite the 2000 years that passed between ancient philosophers and early psychologists, these groups, too, produced some strikingly similar ideas about the Good Life. Further, there is dramatic similarity between the Good Life concepts of many philosophers and the work of contemporary developmental psychologists. Finally, the results presented here demonstrate that educated adults, many of whom have studied neither philosophy nor psychology, also construct similiar good life concepts.

This study provides a general, developmental model of value reasoning about the Good Life and presents empirical findings from a 13-year study of young and older adults. It has been difficult to appreciate many of the substantial theoretical and empirical commonalities in this area between philosophical and psychological studies, as well as among developmental studies themselves. Researchers

CHERYL ARMON • Education Programs, Antioch University Southern California at Los Angeles, Marina del Rey, California, 90292.   THEO LINDA DAWSON • Graduate School of Education, University of California at Berkeley, Berkeley, California, 94720.

Handbook of Adult Development, edited by J. Demick and C. Andreoletti. Plenum Press, New York, 2002.

working in different disciplines and subdisciplines are often unfamiliar with one another's work. There are so many models, findings, and assertions about the development of reasoning about values, it is often difficult to see the forest for the trees. The empirical work and philosphical justification for the analysis of ethical judgments provided here represents an advance in understanding some core commonalities. A general, developmental model of value reasoning about the Good Life can incorporate many of these typically separate findings.

Although ethical philosophers describe the Good Life for adults, not children, there has been little societal interest in adult value reasoning or adult morality. While it is adults who make social policy, establish a community's quality of life, decide to wage war, and determine methods of parenting and educating the young, little attention is given to how adults reason about values and morality, and how such reasoning develops. In a democratic society in which adult citizens are expected to participate intelligently, it is particularly bewildering that interest in adults' cognitive, philosophical, and psychological capacities for this sort of intellectual work is so lacking. This phenomenon may be partly a result of psychology's longstanding rejection of the notion of adult development, that is, the idea that adults are capable of changing their reasoning after their early 20s. Some contributions to this volume begin to fill in this barren landscape (e.g., Brabeck and Schrader, this volume).

This chapter provides a partial report on the four assessments of a longitudinal study of developing conceptions of the Good Life begun in 1977, which included both children and adults. After the first follow-up in 1981, a developmental model of reasoning about the Good Life was constructed. The model relied on both structural–developmental and ethical theory for its theoretical framework. At that time, the Good Life Scoring Manual was completed, and it was proposed that developing conceptions of the Good Life could be effectively modeled with an invariant sequence of five stages (Armon, 1984b). There was also an emphasis on adulthood, particularly the investigation of structural–developmental change during the second half of the lifespan. The purpose of this chapter is to explore the nature of this reasoning and development of the adult subjects.

REVIEW OF THE LITERATURE

The Developmental Psychology of the Good Life: A Theoretical Framework for the Stage Model

Structural–Developmental Approaches. The structural–developmental approach (Colby & Kohlberg, 1987; Kohlberg, 1969, 1984; Piaget, 1960, 1968) has been used to investigate the structural organization of concept and value development in many different domains. The focus of the approach is on the hierarchical, structural organization of reasoning, and development is typically modeled in stages or levels. For example, Loevinger (1976) describes stages of ego development, Selman (1980) defines stages of good friendship concepts, Damon and Hart (1990) delineate stages of self-understanding, Gilligan (1981) describes the development of care and responsibility, Fowler (1981) studies faith development,

Rosenberg (1988) describes stages of political reasoning, and Piaget (1932) and Kohlberg (1981, 1984) construct developmental models of justice and fairness. From a philosophical viewpoint, each of these areas fall within the ethics of human experience, and most could be said to fall within the Good Life. My own work has focused explicitly on the development of value reasoning about the Good Life, generally, and about good work, good relationships, and the good person, in particular (Armon, 1984a, 1984b, 1988, 1993, 1998; Armon & Dawson, 1997).

While philosophical and methodological differences among these models exist, commonalties are more striking, and they reside within the evaluative dimension. The models all describe similar paths of value development within one or more domains. At the earliest, or lowest, stage of development, subjects tend toward egoistic, impulsive, and undifferentiated value concerns. Next develops instrumental, controlled gratification and a focus on power. This is followed by group or interpersonal norms and affective emphasis, which is succeeded by societal or systematically constructed, self-authored values. The final level is always some form of autonomy, interdependence, or dialectical construction of values. (Table 15.1 presents developmental stage theories for comparisons.)

Where significant differences between developmental models do exist, they appear at the highest stages. It is unclear whether these differences reflect actual empirical differences, theoretical or ideological differences between researchers (e.g., liberal bias), or simply an inability to produce definitive constructions owing to the rarity of performances at these stages. The paucity of higher stage performances may be due, in part, to the fact that most developmental studies do not include older adults who are more likely to produce them.

Hard and Soft Stage Theories. The good life stage model is general and can be seen to incorporate some of the more domain-specific stage models within it. Generalizability is dependent, in part, on the extent to which the theory can withstand a "structural analysis." Theorists regularly debate the structural integrity and/or exclusivity of various stage models. Notable is Kohlberg and Armon's (1984) discussion of "hard" and "soft" stage models.

For some time, Kohlberg argued that "hard" stage theories meet *Piagetian criteria* for a stage,[1] and are more generalizable (perhaps universal). In contrast, "soft" stage models may meet some of these criteria in a general way, but no serious attempt is made to meet them all. Kohlberg claimed that his stage model of moral judgment and Piaget's stage model of cognitive development represented the only stage models that have met the necessary criteria. Yet, despite his more than 20-year effort, Kohlberg fell short of adequately demonstrating all four Piagetian criteria for the moral judgment stages. This is particularly evident in the area of defining the logic (hierarchical integration) of the stage sequence (Habermas, 1979; L. Kohlberg,

---

[1] Piagetian criteria for a hard stage theory are: (1) A qualitative difference in structures, or modes of thinking, that serve the same basic function at various points in development; (2) the different structures form an invariant sequence in individual development; (3) each of these different and sequential modes of thinking forms a "structured whole," in which structures appear as a consistent cluster of responses in development; and (4) stages form an order of increasingly differentiated and integrated *structures* to fulfill a common function. Accordingly, higher stages integrate the structures found at lower stages (Kohlberg & Armon, 1984).

Table 15.1. Comparison of Developmental Sequences

Piaget (1958) Cognitive	Kohlberg (1983) Justice reasoning	Selman (1980) Armon (1984) Perspective-taking	Loevinger (1970) Ego	Damon & Hart (1988) Self-understanding	Kegan (1982) Self	Gilligan[a] (1982) Caring	Fowler (1981) Faith
Pre-Operational	Stage 1: Heteronomous morality	Stage 1: Differentiated and subjective	1: Symbolic 2: Impulsive	Level 1: Categorical identifications	Stage 1: Impulsive		Stage 1: Intuitive-projective
Concrete Operational	Stage 2: Individualism, instrumental purpose, and exchange	Stage 2: Self-reflective	Δ: Self-protective	Level 2: Comparative assessments	Stage 2: Imperial	Survival	Stage 2: Mythic-literal
Early Formal	Stage 3: Mutual interpersonal expectations; conformity	Stage 3: Third-person mutuality	3: Conformist 3/4: Conscientious/Conformist	Level 3: Interpersonal implications	Stage 3: Interpersonal	Goodness	Stage 3: Synthetic-conventional
Consolidated Formal	Stage 4: Social system maintenance; conscience	Stage 4: In depth and societal; Symbolic	4: Conscientious	Level 4: Systematic beliefs and plans	Stage 4: Institutional	Truth	Stage 4: Individuative-reflective
	Stage 4/5: Subjective relativism		4/5: Individualistic				
	Stage 5: Prior rights; Social contract or utility	Stage 5: Second-order reciprocity	5: Autonomous		Stage 5: Interindividual	Non-violence	Stage 5: Paradoxical-consolidative
	Stage 6: Universal ethical principles	Stage 6: Second-order mutuality	6: Integrated				Stage 6: Universalizing

[a] Gilligan may not agree with this characterization of her stages.
From Damon, W., & Hart, D. (1988). Self-understanding in childhood and adolescence. New York: Cambridge University Press; Fowler, J. (1981). Stages of faith: The psychology of human development and the quest for meaning. San Francisco: Harper & Row; Gilligan, C. (1982). In a different voice: Psychological theory and women's development. Cambridge, MA: Harvard University Press; Inhelder, B., & Piaget, J. (1958). The growth of logical thinking from childhood to adolescence. New York: Basic Books; Kegan, R. (1982). The evolving self: Problem and process in human development. Cambridge, MA: Harvard University Press; Kohlberg, L., Levine, C., & Hewer, A. (1983). Moral stages: A current formulation and a response to critics. Contributions to Human Development, 10, 174; Loevinger (1970). A Theory of Ego Development. San Francisco: Jossey-Bass. Selman, R. L. (1980). The growth of interpersonal understanding: developmental and clinical analyses. New York: Academic Press; Armon, C. (1984). Ideals of the good life: Evaluative reasoning in children and adults. Unpublished doctoral dissertation. Harvard University, Cambridge, MA.

personal communication, June 3, 1986; Sonnert & Commons, 1994; cf. Kohlberg, Levine, & Hewer, 1983). Nor is it clear that Piaget's model met his own criteria for generalizabilty.[2] Piaget's stage of formal operations, for example, has been criticized repeatedly for not providing a full account of its transformational laws, and many have questioned the logic of the stages and substages (e.g., Broughton, 1984; Ennis, 1978; Feldman, 1980).

It is unlikely that this form of structural analysis will ever be completed or even attempted for the many stage models currently in use, at least to the degree demanded by Piaget or Kohlberg. Nevertheless, clearly articulated analytic models, applied in longitudinal studies, using valid and reliable scoring schemes, and including cross-cultural tests can help immensely in distinguishing between content-oriented, phase-type theories ("soft-stage models," Erikson, 1963; Levinson, Darrow, Klein, Levinson, & McKee, 1978) and sequential, structural theories.

Newer forms of analyses, however, continue to support fundamental structural–developmental constructs (e.g., the General Stage Model, Commons & Richards, 1984; Lam, 1994; Rasch Analyses, Dawson, 1997).

Structural Development in Adulthood.   During the 1960s and 1970s, when humanistic psychologists were celebrating adults' self-realization and capacities for change (e.g., Erikson, 1963; Maslow, 1971; Rokeach, 1973), most developmental psychologists continued to accept Piaget's (1972) assertions that (1) formal operations developed in early adolescence and (2) formal operations was the final stage of cognitive development. In the last two decades, however, many researchers have posed alternatives to these claims. These alternatives focus on the idea that formal operations is not necessarily the last stage in cognitive development, or that current definitions of formal operations do not include all of the cognitive functions of that period, or that the endpoint of cognitive development does not necessarily occur during adolescence, but rather in adulthood, or combinations of these ideas (e.g., Alexander, Drucker, & Langer, 1990; Richards, Armon, & Commons, 1984). During the late 1980s and early 1990s, an array of books, textbooks, articles, and a professional journal devoted exclusively to theory and research about progressive development in adulthood appeared (e.g., Alexander & Langer, 1990; Commons, Armon, Kohlberg, Richards, Grotzer, & Sinnott, 1990; Commons, Richards, & Armon, 1984; Commons, Sinnott, Richards, & Armon, 1989; Demick, 1994; Rybash, Roodin, & Santrock, 1991; Stevens-Long & Commons, 1990). It should no longer be necessary to defend the study of development during adulthood. Today, progressive research programs are less likely to investigate whether there is adult development but, rather, its prevalence, form, and nature.

Lifespan Developmental Approaches.   While not focused so narrowly on structural development, lifespan developmental psychology supported early studies of progressive adult development and aging (Baltes, 1983; Baltes, Hayne, &

---

[2] For Piaget (1970), "stages" represent both organizations of structures and the processes and contingent phenomena involved in structural development. Thus, it was primarily the proposed *structures* that must meet criteria, not the stages, which are more general. His criteria for structures concern: (1) wholeness; (2) transformations; and (3) self-regulation. A structure is a system of transformations, or a system of laws of transformations. Thus, a structural analysis will identify the transformations, the transformational system, and the laws that govern them, of each structural organization or "stage."

Lipsitt, 1980). In its formative years, lifespan psychology was defined both in terms of universals (Birren, 1964) and in terms of the particular historical and personal events that influence an individual life (Baltes, Reese, & Lipsitt, 1980). Researchers in this area have by now amassed a large body of work on adult development and aging. Although acquired and discussed differently than those of the structural developmental approach, these results also help describe ideals of the Good Life in adulthood. For example, there is extensive literature on psychological well-being in the second half of life that provides theoretical models of positive functioning, life satisfaction, happiness, adjustment, and the like, which can be seen as different ways of talking about the Good Life in adulthood (e.g., Cutler, 1979; Lawton, 1984; Stock, Okun, & Benin, 1986).

More recent lifespan work on "successful aging" even more closely parallels attempts to describe the Good Life. This work has gone beyond theoretical models and investigated "how middle-aged and older adults themselves define positive functioning" (Ryff, 1989, p. 195). Performing a content analysis on adults' responses to questions such as "What is most important in life?" and "How would you define an ideal person?" Ryff (1989) reports a greater incidence of the type of responses commonly found at the higher, adulthood stages in structural–developmental models. For example, while structural–developmental researchers report *work-related accomplishments* as central to the conventional level of development in adulthood (e.g., Armon, 1993; Kegan, 1982, 1994), Ryff (1989) reports job-related experiences as the second most important activity of midlife. Similarly, while structural–developmental researchers, such as Loevinger (1976), Armon (1984a), Colby and Kohlberg (1987), and others report the *maintenance of interpersonal relationships* as central to the earlier (and more commonly found) stage of the conventional level in adulthood, Ryff reports family relationships to be the *most* important activity of midlife. Thus, the content analyses of the lifespan researcher and the structural analyses of the structural–developmental counterpart come together to enhance our understanding of certain aspects of the Good Life in adulthood.

The Philosophy of the Good Life[3]

In addition to developmental psychology, the developmental model of value reasoning about the Good Life relies on moral philosophy, or ethics, in two ways. First, philosophy provides analyses about the possible relationships between *the good* and *the right*, which are described briefly later in this section. Second,

---

[3] There are two main reasons for the necessity of philosophy in a study of evaluative reasoning about the Good Life. The first and most obvious reason is that constructs concerning value, or the good, are ethical in nature. To say something is good is to make an ethical claim and ethical claims cannot be supported or refuted with a purely empirical model. The second reason is that the present developmental model is a normative one; that is, it is claimed that the highest stage is more adequate than those that precede it. Although the fundamental scheme of this study is psychological, and thus primarily descriptive, part of *any* developmental analysis includes an explanation of where development *leads*. Typically, such psychological work blurs the boundaries between descriptive and prescriptive work. A more complete explanation and analysis of the normative quality of the Good Life reasoning model can be found in Armon (1984b).

traditional philosophical constructions of the Good Life were used to inform the construction of the content categories used in part of the data analysis. In the initial model building of the Good Life stage model (Armon, 1984b), it was found that the material adult subjects offered when describing their ideals of the Good Life, particularly at the higher stages, was similar to professional philosophers' views. Thus, traditional philosophies of the Good Life were used to categorize adult subjects' responses. The following section briefly describes the different philosophical orientations that formed such categories for the present study. (A fuller account of these categories can be found in Armon, 1984b.)

Philosophies of the Good Life: A Theoretical Framework for Categorizing Content

Traditionally, ethics has attempted to answer value-related questions of seemingly universal interest such as "What is the Good Life?" Responses to this question can be roughly classified as either Perfectionistic or Hedonistic. Perfectionist theories define good living as the development and expression of inherent human talents and capacities. (Which capacities to perfect is oft debated.) Hedonist theories define good living as the successful acquisition and appreciation of pleasure—the ultimate intrinsic value. In Hedonism, the means to pleasure are secondary. Achieving the result, drawing pleasure from an object or activity, is key.

For the content categories, this study relied primarily on the works of three perfectionists: Aristotle, Spinoza, and Dewey. It also drew on two hedonists, Epicurus and Mill. Although these are leading theorists in their persuasions, others could have been chosen to exemplify the ethical views. Figure 15.1 illustrates these five Orientations.

Perfectionism.   For Aristotle (Ostwald trans., 1962), the distinctive human capacities, especially practical and theoretical wisdom, are what we must perfect. From this perspective, the Good Life consists of experiences that ensure the development of intellectual, ethical, and even physical abilities that are uniquely human. (This Orientation is referred to here as Perfectionism–Capacities.) For Spinoza (1949), the Good Life develops our recognition, through self-knowledge, of our interdependence with Nature or God (referred to here as the Perfectionism–Unity Orientation). Instead of a focus on particular capacities or endpoints,

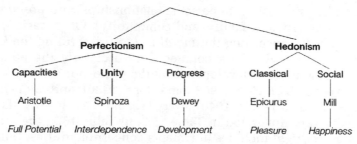

Figure 15.1. Good life philosophical orientations.

Dewey (1944, 1957, 1980) claims that experiences that contribute to continuous, progressive development are central to the Good Life. (This Orientation is called Perfectionism–Progressive in this study.)

Hedonism. In contrast to Perfectionistic views, Hedonists Epicurus and Mill claim that the experience of pleasure or happiness, and the absence of pain, should be the universal objects of human desire. Epicurus (from Laterius, 1925) constructs an essentially self-centered philosophy of the Good Life, concerning himself with individual pleasure. (Here, this Orientation is called Classical Hedonism.) Questioning the delights of sensuality and extravagance, however, he recommends pleasures resulting from peaceful equilibrium—attained through self-knowledge, a virtuous character, and the company of intellectually stimulating friends.

Another form of Hedonism is offered by Mill (1957/1861, 1978/1861). He rejects the egoistic aspect of Epicurus' thought and argues that the Good Life takes place in a *social* rather than an *individual* context in which the consideration and maximization of all persons' pleasure is paramount. (This orientation is called Social Hedonism in this study.) Like Epicurus, Mill asserts that the pleasures resulting from persons' higher faculties, including pleasure resulting from a virtuous character, are superior to "sensuous pleasures."

The good life stage model brings together the philosophy and psychology of the Good Life. There are, of course, cultural, historical, ideological, and methodological differences between these disciplines. Thus, one might expect ideals of the Good Life constructed by (some ancient) philosophers to be radically disparate from those inferred from developmental psychology. Yet, when carefully considered, particularly when focusing on adult reasoning at the higher stages of development, the philosophy and developmental psychology of the Good Life appear more similar than different, containing core elements that have remained central for thousands of years. Bringing together different developmental models and combining them with the typically separate discipline of ethical philosophy reveals commonalties that inform our understanding of a good human life from prescriptive and descriptive, theoretical and empirical perspectives.

The Good and the Right. The operational definition of the Good Life is the combined set of values that persons affirm in normative, ideal-evaluative judgments about the Good Life, in general, and about good work, good friendship or relationship, and the good person, in particular.

The domain of the Good Life is conceptualized as broad, including the *moral good* (e.g., ethical dimensions of persons, relationships) and *nonmoral good* (e.g., nonmoral aspects of work, family, and community). On occasion, subjects also produce judgments concerning the *moral right* in describing the Good Life and such judgments must also be accounted for. To classify the ethical quality of responses, categories were developed from the philosophical works of Campbell (1935), Frankena (1973), Hare (1952), and Ross (1930), and are generally consistent with the works of Lewis (1946), Perry (1926/1954), and Rawls (1971, 1980). These categories are illustrated in Table 15.2. In Table 15.2, the first category on the left, the *moral right* includes judgments concerning duties, human and legal

Table 15.2. Value Categories from Traditional Ethics

Moral right	Moral good	Moral worth (aretaic)	Nonmoral good (intrinsic)	Nonmoral good (extrinsic)
Justice; fairness; rights; obligations; duties	Care and responsiveness; welfare consequences to others	Motives; character (e.g., conscientiousness; generosity; self-sacrifice)	Ends (e.g., autonomy; knowledge; self-realization; intelligence; nature; freedom; consciousness)	Means (e.g., paintings; cars; travel; education)

rights, and absolute principles. The second category contains judgment of the moral good. Included in this category are emotional care and responsiveness, and concerns for the welfare consequences of others. The third category is moral worth, the focus of which are *aretaic* judgments. Moral worth resides in the person, such as traits of character.

The fourth category is intrinsic, nonmoral good, and contains judgments about generally accepted human values, for example, knowledge, sociality, or artistic expression, which in themselves are nonmoral. These are sometimes referred to as "end-values" (Rokeach, 1973). The final category is extrinsic, nonmoral good, which contains judgments about "goods" that people value because of what they bring or do, not because of what they are in themselves. This category would include cars, pencils, or houses. These are sometimes referred to as "means values" or "instrumental values."

Distinctions Between the Good and the Right. As described, judgments about the Good Life often concern the moral good and, sometimes, the moral right. Some developmental psychologists, notably Kohlberg (1981), have gone to great lengths to keep these categories of judgments distinguishable. Others, for example, Gilligan (1981) and Selman (1980), have ignored these distinctions.

Ethical philosophers typically expend great effort in defining the relationship between the good and the right. For example, Rawls' (1971, 1980) deontic perspective holds that conceptions of the good are pluralistic. He focuses on the primacy of universal principles of right action (moral right, justice) for the distribution of the conditions for the attainment of any good life. Thus, conceptions of the good are always subordinate to conceptions of the right. In contrast, Mill's (1957/1861, 1978/1861) utilitarian view requires that the Good first be defined since the moral obligation, or first principle, is to maximize the Good.

For the present study, the debate concerning the primacy of the Right or the Good need not be resolved. It is proposed here only that they can be distinguished and that a consistent and generalizable psychological theory of reasoning about the Good Life may be possible.

This report of the longitudinal findings briefly covers the first and second assessments (described in detail in Armon, 1984b) and provides more information on the the results of the third and fourth assessments of the study of good life reasoning of a group of adults over a period of 13 years. The study examines stage development during the lifespan in four domains: Good Life, Good Work, Good Friend, and Good Person. Findings from the broad good life domain are reported here. The main purpose of this follow-up report is to validate the sequentiality of the stage model and to continue investigating adult structural development. The following hypotheses were specified for the analyses of follow-up data: (1) Conceptions of the Good Life can be represented by a developmental, sequential stage model, ordered by qualitative increase in cognitive complexity (inclusiveness) and social perspective-taking capacities; (2) structural development (stage change) occurs in adulthood; and (3) the higher stages of the model will appear only in adulthood, even with a "privileged" sample.

METHOD

Study Design

The design of the original cross-sectional study, which was later followed up longitudinally, was determined by a number of theoretical concerns. There was a direct attempt to create a design that would be most likely to identify development throughout the lifespan, particularly in adulthood. Three variables were most important in the design: age, interest, and higher education. In accordance with Dewey's (1944) notion that an individual's interests are key factors in his or her learning and development, individual interest in development or change was hypothesized to be positively associated with continued development. It was also thought that interest would affect the subjects' willingness to perform the practical duties of a long-term longitudinal subject (e.g., notifying the experimenter when relocating). Participation in higher education was also expected to be associated with continued development, as shown by other studies (e.g., Colby, Kohlberg, Gibbs, & Lieberman, 1983; Rest & Thoma, 1985; Walker, 1986). The study was first executed cross-sectionally in 1977. In 1980, most of the subjects were located and interviewed again. In total, subjects were interviewed four times—in 1977, 1981, 1985, and 1989–90.

Participants

At time 1 (1977), 50 individuals, ranging in age from 5 to 72, responded to a flyer seeking volunteers. The flyer sought "people interested in development through the lifespan, or their children," and was distributed to friends and strangers around a state university in the Los Angeles area. The first 50 individuals to respond, who ranged in age from 5 to 72, were accepted. Four were Latino, one was Asian, two were mixed race, and the rest were Caucasian. Average years of education for adult males at time 1 was 17 years; for females it was 15.5 years. This difference was not found to be significant. Four adults had completed doctorates, nine had completed some post-undergraduate work, six had completed college, and nine had completed some college. Individuals' annual incomes ranged between $18,000 and $140,000, with an average of $30,000. Average years of education and annual income rose throughout the study. Income levels were not found to be significantly related to any other variables in this study.

Only adult data are presented here. Adults were divided into two age groups, consisting of a *younger adult group* of twenty-one 23- to 45-year-olds (11 females and 10 males), and an *older adult group* of eight 50- to 72-year-olds (6 females and 2 males). Categorization into these groups was based on the view that different developmental trends may occur in adulthood and late adulthood.

Procedures

Thirty adults were interviewed at time 1, 27 were interviewed at time 2, 25 were interviewed at time 3 (1985), and 23 were interviewed at time 4 (1989–90). Five participants missed the last two interviews due to health problems or could not be contacted at either time 3 or time 4. Participants who completed only two interviews were not significantly different from those who completed three or four interviews on any of the measures used in this study.

Training of Interviewers and Scorers.   The training of two adult, advanced undergraduate students was performed by the author in 1985 and 1989. First, the two students studied the rules and procedures outlined in the Good Life Scoring Manual (Armon, 1984b). Second, they conducted practice interviews with volunteer subjects not associated with this study. The practice interviews were tape-recorded and transcribed. Third, each of the two students practice-scored the other student's interview independently, using the Good Life Scoring Manual, while the author scored both. Finally, the three of us met and reviewed the protocols and scoring to discuss errors in both the interview methods and the scoring, and differences between the three of us in the scoring. This process involved approximately 10 interviews (over 3 months) until a level of reliability and validity was reached that was acceptable to the three scorers. (See Reliability, below.)

Administration

The structural–developmental research approach (Colby & Kohlberg, 1987) was used to collect and analyze the interview data at each test time. (See Armon, 1984b, for a complete explanation of how data were collected and analyzed.) The semistructured Good Life Interview (Appendix A), administered at each test time, consists of open-ended questions in many domains beginning with "What is the Good Life?" (combined with multiple probe questions to elicit subjects' underlying reasoning, e.g., "Why is that good?"), and including questions about good work, good friendship or relationship, and good person. Demographic and life history information was also collected each time. It was not uncommon for interviewers to travel out of state, to other countries, and to federal prisons to locate and interview subjects. Thus, the environmental milieu of the interviews was variable. Interviews were always conducted, however, in privacy, with only the interviewer and the subject present in a room. The interviews were tape-recorded and transcribed. The average interview time was two and a half hours.

RESULTS

Data Analysis and Results of 1977 and 1981 Assessments

This section provides a very brief summary of the main findings from the first follow-up study. Detailed explanations can be found for all analyses and results mentioned here in Armon (1984b).

Preliminary Good Life Reasoning Stages. In 1977, working from the original cross-sectional data of 50 subjects, five preliminary stages of good life reasoning were proposed (Erdynast, Armon, & Nelson, 1978). In 1982, a construction sample of 12 follow-up cases was created by randomly selecting four cases from each of the three age groups—children, young adults, and older adults. Based on the clinical analyses of these data, which is described briefly below, the "value reasoning" construct was refined, the Philosophical Orientations were operationalized with the use of Norms and Elements (see below), and the Good Life Scoring Manual (Armon, 1984b) was constructed. The manual's scoring system converts stage scores to a continuous scale, for example, stage 1 = 100 EMS (Ethical Maturity Score).

Method of Analysis. Units of analysis were those statements subjects used to affirm something as good, worthy, or ideal and the values and reasons offered in the justification. This modeling of value judgments was adapted from Nowell-Smith (1954). The contents of the scorable units were categorized using an Issues, Norms, and Value Elements system similar to that of Colby and Kohlberg (1987), but additional Norms and Value Elements had to be added from Rokeach (1979) and Maslow (1971) to cover all the good life judgments. In a fully elaborated good life Judgment, an Issue, Norm, Modal Element, and Value Element is present.

Issues, Norms, and Elements. In assessing good life judgments, the *Issue* is merely the topic under discussion, usually identified by the question that the subject is asked (e.g., "What is a good person?") or sometimes by spontaneous response of the subject. The Issue categorizes the content that the subject is expressing an evaluation about (e.g., good person, good work). The *Norm* categorizes the general value that the subject assigns to the issue (e.g., honesty, caring, love, society). In other words, the Norm canotes the value area that the subject begins discussing in relation to the Issue (e.g., in the response, "A good person is trustworthy," "good person" is the Issue and "trustworthiness" is the Norm). There are moral and non-moral Norms. *Modal Elements* express the type of judgment; they express the "mood" or modality of ethical language, such as "having a right," or "character," or "blaming/approving." Finally, *Value Elements* are values used to support and justify Norms. Value Elements provide the final aspect of a judgment. They are terminal or end-values (Rokeach, 1973) for which the Norm serves as object.

To give an example of the Issue, Norm, and Value Element system, let us say a subject's response to "What is a good person?" and "Why?" is: "A good person works for the community." When questioned, "Why is that good?" the subject replies, "Because it shows a certain character." Questioned further: "What do you mean a certain character?" subject replies, "Well, I think a good person gets his personal satisfaction from working in community with and for others." In this judgment, the Issue is Person. The Norm is Community. The Modal Element is Character. And the Value Element is Satisfaction/Fulfillment.

Once content was categorized in this way, systems of reasoning, that is, methods of organizing the Norms and Elements, were identified in subjects' protocols independent of the particular Norms or Elements used. Similar systems were grouped and then groups were hierarchically ordered by increasing cognitive complexity, or inclusiveness, and by increasing levels of social perspective-taking (Armon, 1984a, 1984b; Selman, 1980).

Final Good Life Reasoning Stages. The resulting five-stage, hierarchical model confirmed the general construction of the initial stages derived from the cross-sectional data (Erdynast, Armon, & Nelson, 1978). It begins in early childhood with an egocentric conception of the Good Life derived primarily from pleasure-seeking fantasy (e.g., "The Good Life is having my birthday party every day") and culminates with a complex conception of the good that encompasses complex criteria, including a preeminent societal dimension ("The Good Life is the worthy life. It is the integrated life—bringing the various facets of experience into balance with my interests and talents. It is also constructed in social context. To be good, it must move the society forward in some way").

The stages are most easily observed in individuals' constructions of their evaluative criteria; that is, the criteria the subject uses to decide whether a person, idea, state, or activity is good. Table 15.3 summarizes the stages of value reasoning about the Good Life.

Reliability of the Scoring Manual for Stage Assessment. The interrater reliability of the scoring manual, including stage and Philosophical Orientation assessment, and the long- and short-term test–retest reliability of the good life reasoning construct were substantially above acceptable limits. Detailed tests were performed using the Good Life Scoring Manual in 1983 and are reported in Armon (1984b). Compared to the reliability estimates of other, similar models (e.g., the Standard Form Moral Judgment Scoring System, Colby & Kohlberg, 1987), the present model and scoring scheme is well within acceptable limits. In addition, the Good Life Scoring Manual has been used in other studies with acceptable results (e.g., Commons, Armon, Richards, & Schrader, 1989; Lam, 1994).

CURRENT METHODS OF ANALYSIS

Several analyses in this report are conducted on the pooled longitudinal and cross-sectional data. This practice has two major advantages. First, power is increased when all measurements at all test times are treated as independent observations, which is permissible when a study design involves more than two test times separated by relatively long intervals (Willet, 1989). Second, the longitudinal information incorporated into the analysis in this way adds valuable

Table 15.3. Stages of Value Reasoning about the Good Life

Stage 1 Egoistic hedonism	Stage 2 Instrumental hedonism	Stage 3 Affective mutuality	Stage 4 Individuality	Stage 5 Autonomy/community
The individual child does not possess conscious value criteria. Nor is the rational distinguished from the irrational; possible and impossible occurrences are not distinguished. Only ends are considered, not the means for their attainment, nor their possible consequences. Perceived to be good are those material objects and physical activities that provide pleasure to the self. The Good is synonymous with the desired. No distinction is made between physical pleasure, happiness, or contentment. Other persons, including family members, are rarely included.	The individual thinks instrumentally about achieving the Good Life for him- or herself. In considering the means, individuals at this stage contemplate others' interests, motives, and intentions, as well as external physical and social conditions. What characterizes these means, however, is their concrete, instrumental quality. Others are considered as separate persons with their own interests, but the focus is on how others serve the self's needs. Other people are important aspects of the good life because they are a means to the self's ends.	The Good is shared by the self and others. Reasoners see the mutual quality of interaction with others as an integral part of the Good Life. Mutuality in relationships and consensus in valuing is sought. Beyond the general distinction between happiness and pleasure, individuals attend to the form of happiness itself. Happiness has a distinct meaning, grounded in affective contentment. There is also a sense in which happiness, or the good life in general, can be defined as the absence of certain negative affective states or experiences, such as loss, crisis, loneliness, fear, anxiety, worry, and stress.	The origin of value lies within the individuated self. A central feature is a concern with individualism. This is an orientation toward self-chosen values and the freedom to go against consentual norms, if necessary, to make choices about, and to pursue, one's particular vision of the Good Life. Meaning, worthiness, and value are criteria for satisfaction, independent of others' beliefs and desires. Wide variability in individuals' values is acknowledged and tolerated. This awareness can be coupled with a form of relativism in reasoning about the good, particularly in the Classical Hedonism orientation. In the Perfectionistic Orientation, relativism is less prevalent. The conception of good tends to be generalized to other persons. The focus remains, however, on the fulfillment and realization of the self's chosen values.	Individuals attempt to construct what is of value independent of social and historical norms. Thus, value is something to be perceived and constructed by each individual. Individuals employ generalizability, universality, and/or intrinsicality, rather than individualism, as criteria of value. The emphasis is placed not on the choosing of values, but on the perception and construction of the worth of the values themselves, for both the self and others. The focus is on those traits, objects, processes, or states that possess intrinsic value and the obligation to uphold those values, once recognized. A principled, ethical view of an ideal human world is constructed in which justice is a precondition for goodness.

information about growth trends. Specifically, by including longitudinal data for individuals, it can be shown that age differences reflect actual trends at the individual level rather than artifacts of statistical averaging. In order to eliminate concerns about the possible introduction of error with this approach, all analyses were also run separately on the data for each test time. The trends found at each test time were consistent with the trends reported for the pooled sample.

Results of 1985 and 1990/91 Assessments

Manual Reliability. To continue to test the reliability of the Good Life Scoring Manual, one third of the 1985 and one third of the 1989–1990 protocols were each scored independently by the author and two trained undergraduate students using the unaltered Good Life Scoring Manual (Armon, 1984b) in 1990. There was a 100% perfect agreement rate within one stage, 93% agreement within a half stage (50 EMS points), and 88% agreement within a fourth stage (25 EMS points).

Invariant Sequence of Stages. To meet the criteria for invariant sequence, it must be demonstrated that, beyond measurement error, no subject's stage score at $T+1$ is less than their score at T (which would constitute regression) and that no subject skipped a stage while developing progressively through the sequence. These criteria were strongly supported by the follow-up data. While significant stage change did not occur in all subjects at each test time, for those whose reasoning did change, it changed toward the next, successive stage. Assuming a conservative measurement error of ±31 EMS points (just over 1/4 stage),[4] no regression was identified. In the 4 (sometimes a bit more) years between test times it might have been possible for subjects to develop beyond the next ordered stage, thereby possibly "skipping" a stage. In this sample, however, this did not occur. Figure 15.2 plots all adult subjects' scores at the four test times with age. There it can be seen that subjects' reasoning either remains at the same stage or changes toward the next, ordered stage.

Internal Consistency of Stage Scores. The internal consistency of stage scores was tested in 1981 and 1985 by measuring the proportion of reasoning at each stage in the construction protocols (12 cases, two interviews each, from three age groups, see above). The 1981 results can be found in Armon (1984b) and are almost identical to those of 1985. In 1985, on average, 74% of the scorable

---

[4] To find an upper limit for the standard error of measurement in the 77–81 analyses, the highest variation was coupled with the lowest correlation. A 95% confidence interval was computed around the long-term, test–retest correlation, .95. The SD of 100 was used with the lower limit of that interval, .90. The standard error of measure was then estimated with the following equation:

$$SE = SD\sqrt{1 - r_{xx}}$$
$$SE = 100\sqrt{1 - .90}$$
$$SE = 31.62$$

This results in a standard error of 31.62 EMS points.

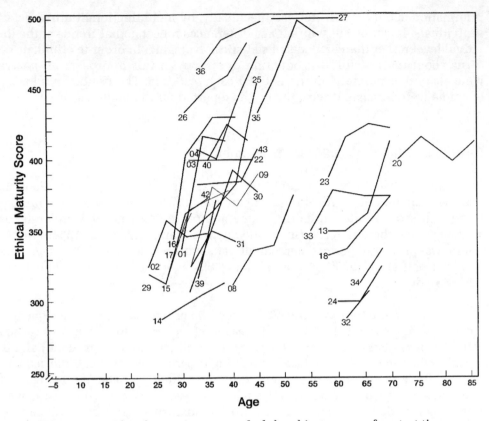

Figure 15.2. Ethical maturity scores of adult subjects across four test times.

responses in each protocol was at a single stage—the "modal stage" (Colby & Kohlberg, 1987). The mean percent of the next most often used stage, always adjacent to the modal stage, was 21%. The mean percent of the third most often used stage, also always adjacent to the modal stage, was 5%. The number of scorable responses in a given protocol was widely variable, particularly with the young children in the sample. A 6-year-old might produce as few as two scorable responses for an Issue, while an older adult might produce 20 or more. Nevertheless, in the construction sample, no subject produced a single scorable response (even a "guess score") that was assessed at a stage other than the modal or adjacent stage.[5]

Rasch Analysis

Rasch analysis was used to further examine the construct validity of the Good Life measure. Though well-known in psychometric circles, Rasch models have been employed by developmentalists only recently (Andrich, 1986; Andrich & Styles,

---

[5] Another test of internal consistency investigates the stability of stage scores across the four different Issues. Although each Issue is scored separately, independence was not guaranteed. Correlations between pairs of Issues ranged from .82 to .94 in 1977 and 1981 (see Armon, 1984b).

1994; Bock, 1991; Bond, 1994, 1995; Dawson, 1996; Demetriou, Efklides, Papadaki, Papantoniou, & Economou, 1993; Draney, 1996; Goodheart, Dawson, & Commons, 1996a, 1996b; Hautamäki, 1989; Müller, Reene, & Overton, 1994; Müller, Winn, & Overton, 1995; Noelting, Coudé, & Rousseau, 1995; Noelting, Rousseau, & Coudé, 1994; Wilson, 1985, 1989a, 1989b; Wilson & Draney, 1997). One function of these models is to examine behavior on measures intended to capture hierarchies of difficulty, which makes them highly suitable for developmental applications. The Rasch model can be employed to test the assumption that performances and items (or levels of items) form a stable, hierarchical sequence (within probabalistically determined constraints) that can be successfully modeled along a single continuum (Andrich, 1989; Fisher, 1994; Masters, 1988; Wright & Linacre, 1989).

In their raw form, little can be said about the relative distances between stage scores (Duncan, 1984; Michell, 1990; Thurstone, 1959; van der Linden, 1994). A Rasch analysis transforms these scores into interval form through a log transformation (Wright & Linacre, 1989). The result is a common metric along which both stage difficulty and respondent ability estimates are arranged. The metric is referred to as a logit scale. (For more information on logits and algorithms for estimating them, see Wright and Masters, 1982, or Ludlow and Haley, 1995.) The distance between logits has a particular probabalistic meaning. An ability estimate for a given individual, say of 1.0 logits, means that the probability of that individual performing accurately on an item at the same level is 50%. There is a 73% probability that the same individual will perform accurately on an item whose difficulty estimate is 0.0, an 88% probability that he or she will perfom accurately on an item whose difficulty estimate is −1.0, and a 95% probability that he or she will perform accurately on an item whose difficulty estimate is −2.0. (For general discussion of the properties of the Rasch model and related models, see Andrich, 1988; Fisher, 1994; Masters, 1982, 1997.)

One advantage of Rasch models is that they provide estimates of both participant abilities and item difficulties. Each of these includes an error term, which makes it possible to establish confidence intervals for all item and person estimates. Fit statistics are also provided for each estimate, so both items and persons can be examined for their conformity with the requirements of the model. Both individual items or persons and subgroups of items or persons can be examined in this way. As is demonstrated below, the results of the analysis provide a great deal of useful information about the psychometric properties of one's instrument, patterns of behavior, and the latent variable being investigated. (The software used to run this analysis was Quest [Adams & Khoo, 1993].)

To conduct the analysis on EMS scores, it was necessary to convert weighted average scores into stage scores. The translation was conducted as follows:

100–124 = Stage 1.0
125–174 = Stage 1.5
175–224 = Stage 2.0
225–274 = Stage 2.5
275–324 = Stage 3.0
325–374 = Stage 3.5
375–424 = Stage 4.0
425–474 = Stage 4.5
475–500 = Stage 5.0

It should be noted that the half-stage scores represent a mixture of reasoning at adjacent stages. The Good Life scoring system does not include scoring criteria for half-stages. Data from all test times for all three age groups was included in the analysis to maximize the sample size. The larger sample size provided more reliable item estimates. However, only case estimates for adult participants are shown in Fig. 15.3. The partial credit model (Masters, 1982, 1994), an extension of the original Rasch model designed to permit the estimation of multiple levels for each item in an instrument, was utilized for this analysis. The partial credit model is appropriate here because stage of reasoning was assessed in each of four domains (the Good Life, Good Work, Good Friendship, and the Good Person), and

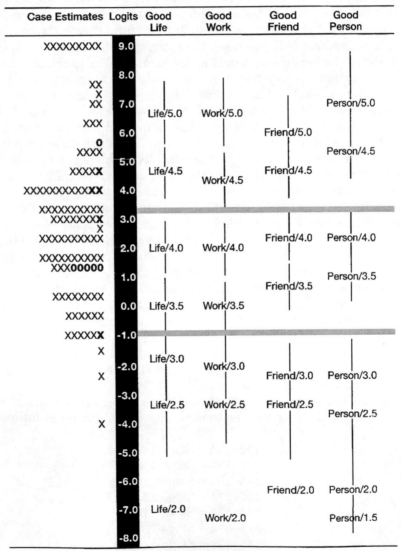

X = case with positive misfit (2.9%); O = case with negative misfit (3.4%); N =110

Figure 15.3. Rasch analysis: item and case map.

each of these domains has nine possible levels (five stages, and four half-stages). In the analysis, each domain is equivalent to an item. The proportion of the observed case estimate variance that is considered to be true (reliability of case estimates) is .93. (See Wright & Masters, 1982, for more information about reliability of case estimates.) Given the small sample size ($n = 147$) and the number of levels and items, .93 is remarkably good. Reliabilities for case estimates of .85 and up are considered acceptable in ability testing (Wilson, January 1997, personal communication). All items have acceptable infit (weighted) and outfit (unweighted) fit statistics (Wright & Masters, 1982), indicating that they all belong on the same scale or, in other words, that they measure the same dimension of ability. Infit and outfit statistics are expressed as meansquares and *t*s. Meansquares are expected to be close to 1.0 and *t*s are expected to be close to zero.

Figure15.3 shows the item and case estimates for the Good Life analysis. The case estimates are displayed on the left of the figure, and the item estimates are displayed on the right. Near the center of the figure is the logit scale. Standard errors for individual cases range from .63 to 1.08, with a mean of .78. Standard errors for the item estimates range from .44 to 1.28 logits. Two standard errors are shown in Figure 15.3 as bars around each level of each item for which a standard error could be calculated. These can be thought of as 95% confidence intervals. Standard errors are greater at the lowest and highest stages because the sample sizes on which estimates are based are small at the extremes of the range; hence, we are less confident of the locations of persons and items at these locations.

Cases that fit the model are indicated with Xs, those with negative weighted misfit are indicated with Os, and those with positive weighted misfit are indicated with Xs. Altogether, 11 performances (7.5%) do not fit the model. This is outside the expected 5% that can justifiably be attributed to measurement error. Not all of these performances, however, raise actual contradictions in terms of the theoretical model. Six cases exhibit negative misfit (4.1%). All of these are participants who performed at a single stage. If some performances at a given difficulty level are much more consistent than others at that level, they will have negative misfit. These performances are more consistent than expected, but they are congruous with the theoretical expectation that some individuals will demonstrate performances that are consolidated at a single stage. Five of the negatively misfitting cases are individuals whose performances on all items were scored at stage 3.0. Note that their estimates are located approximately 3.0 logits above the difficulty estimates for the stage 3 items. This means that they, and the other cases at their ability level have about a 88% probability of scoring at stage 3.0 and somewhat more than a 50% probability of exibiting some stage 4.0 reasoning. Only five cases exibit positive misfit (3.4%). These are the individuals whose scores span more than 1.5 stages. Although these cases pose problems for both the model and our theoretical expectations, 3.4% is well below the 5% expected error level. Overall, the reliability of case estimates, along with case and item fit statistics, provide good evidence for the construct validity of the Good Life instrument. This analysis also provides support for the invariant sequence of stage development and for the structured whole criterion. Evidence in support of invariant sequence is provided by the relatively small standard errors for most levels of most items. In Figure 15.3, the error bars clearly show gaps between each full stage and the half-stage above it (one exception occurs at the stage 2.0/2.5 transition).

Item separation of this kind is strong evidence of invariant sequence. It is also apparent from Fig. 15.3 that stages tend to cluster together. Note the horizontal gray lines between stage 3.0 and the 3.5/4.0 cluster and stage 4.0 and the 4.5/5.0 cluster. These gray lines mark the areas in which there is no overlap of error ranges between scores at one full stage and those at the next half stage (which is, as stated above, a combination of reasoning at the two adjacent stages). In contrast, note that error ranges between half-stages and the next full stage usually overlap. This pattern is supportive of the idea that, at least within a given domain, individuals tend to reason using the structures of no more than two adjacent stages, and that reasoning tends to consolidate at a given full stage before any reasoning at the next stage appears.

The combination of good case and item fit is evidence that would support the construct validity of any instrument, in the sense that it upholds the notion that a single, underlying dimension of ability is being assessed. In the present case, however, construct validity is additionally supported by evidence for two important postulates of developmental stage theory, that of invariant sequence and the structured whole criterion. These findings are consistent with those reported in other analyses of instruments based on Piagetian principles (Dawson, 1997, 1996; Dawson, Goodheart, Draney, Wilson, & Commons, 1997; Draney, 1996; Goodheart et al., 1996a, 1996b; Müller et al., 1994, 1995; Noelting et al., 1994, 1995)

Adult Development

Three questions related to adult development were addressed: (1) What is the relationship between age and stage attainment in adulthood, and are the relationships homogeneous in different age groups? (2) Are there stages that occur only in adulthood? (3) Does structural developmental change occur in adulthood and, if so, what can be said to influence it?

Adulthood Stages. The 1985 and 1989 analyses confirmed earlier outcomes about the existence of adult stages (Armon, 1984a, 1984b). No subject attained a stage 4 score prior to age 24, and no subject attained a stage 5 score prior to age 34, which confirmed hypothesis 2 that these stages occur only in adulthood, if they occur at all, even with a "privileged" sample.

Structural Developmental Change in Adulthood. The largest structural change during any of the three 4-year periods was three quarters of a stage, demonstrated by only two people (6%). Seven subjects (36%) developed one half stage, while 15 (47%) developed one fourth stage (equivalent to measurement error), and 8 subjects (27%) showed no development at any of the test times. The most change over the whole study period occurred with a single subject and was one full stage. Eight subjects demonstrated no change; 10 demonstrated one quarter stage (not beyond measurement error), 7 one half stage, and 5 three quarters of a stage.

Age. Age was not significantly related to the development of reasoning about the Good Life in adulthood when all adults were combined into one group ($r = .127$). There was a significant, linear relationship between the two in the younger adult group, however ($r = .35$, $F = 11.10$, $p < .001$).

Education. While neither income nor previous stage were found to be significantly related to adult structural change, years of education were significantly related in both adult groups $(r = .60, F = 33.36, p < .00)$. In addition, those who continued in school, or returned to school during the study period, had a significantly higher probability of an increase in their stage scores than those who terminated their education, or did not return to school during the study period $(r = .42, F = 5.82, p < .02)$. It is important to note, however, that although increased education affected many subjects' scores, it was not necessary for stage change in all cases. A few subjects developed without it, while a few others had it and did not advance in stage.

Occupational Status. Hollingshead Occupational Scale scores (Hollingshead, 1975) were significantly correlated with scores on reasoning about the Good Life $(r = .59, F\{1,100\} = 54.47, EMS = 22.24 HOL + 218.39, p < .001)$. (They also accounted for about 5% of the variance over education scores when run on the nonstandardized residuals from education X EMS: $r = .23, F\{1,100\} = 5.47$. RESID of EMS by $ED = 7.45 HOL - 54.42, p < .05$.)

Gender. The overall effect of gender on adult scores was significant, with men attaining an average of almost a half-stage above the women $(r = .41 F\{1,103\} = 21.40, EMS 45.51 MALE + 319.11, p < .001)$. In a multiple regression with education and occupational status, the salience of gender is reduced but remains significant $(r = .71, F\{3,98\} = 32.74, p < .001, EMS = \{5.96 HOL + 38.52 GEN + 146.09\}, tED = 2.86, p < .01; tHOL = 2.59, p < .05; tGEN = 4.74, p < .01)$.

DISCUSSION

The longitudinal data reported here further validate the invariant sequence stage model of value reasoning about the Good Life. Many subjects, including adults, demonstrated development and, when they did, it was always toward the next stage in the sequence. The Good Life Scoring Manual, developed in 1984, continues to demonstrate high interrater and test–retest reliability. Newer forms of analyses, for example, Rasch, also continue support for the model. The general findings of this study provide robust support for a structural–developmental model of value reasoning as well as some important findings on adult structural development.

Related Variables

Although age variables are useful in providing evidence for the developmental hypotheses of the Good Life Stage model, they provide only indirect evidence. Virtually all studies with developmental outcomes variables show high correlations between age and stage in children, particularly with middle-class samples. Indeed, at first glance, the mere passage of time appears to be salient in individual development, at least to stage 3—the stage of conventional (socially accepted) reasoning about the person, friendship, marriage, morality, and the self. (See, e.g., Colby et al., 1983; Damon & Hart, 1990; Fowler, 1981; Kegan, 1982; Selman, 1980.)

The degree of correspondence between age and stage, however, was observed to decrease to insignificance throughout adulthood.

Although adults' age and stage scores were not significantly correlated, only adults had reached the two highest stages. Hence, although advancement in age is not a sure indicator of development to higher stages, it appears to be a necessary condition. (See Adult Development, below.)

With this sample, income did not have a significant impact on development. We suspect, however, that the average income was too high to reveal differences. If sufficient income provides for a relatively comfortable lifestyle and continuing educational opportunities, income may cease to have relevance for development.

Colby et al. (1983) reported that in middle-class samples of older adults with average to high IQs, neither income, previous stage, nor age appeared to influence structural change in adulthood. With subjects beyond the age of 30, only the impact of education has been examined and has proved to be modestly significant. The findings reported here are consistent with previous studies that have consistently demonstrated the positive impact of education on stage attainment (e.g., Colby et al., 1983; DeVos, 1983; Rest & Thoma, 1985; Walker, 1986; see also Armon & Dawson, 1997). Further research is needed to investigate the relationship between specific forms of education—the actual content and activities—and structural–developmental change.

Philosophies of the Good Life. In 1984 it was demonstrated that nonphilosopher adults think much like "expert" philosophers when contemplating the Good Life. Although the Philosophical Orientations were shown to be reliable then, they proved insignificant in terms of their impact on stage development. For example, the Perfectionism Orientation was not shown to be significantly related to stage change scores as was hypothesized in 1984. These findings may be inconclusive since there were too few subjects associated with some of the Orientations, particularly the subgroups of Perfectionism, to explicitly support or reject the hypothesis.

Gender Differences. Although statistically significant gender differences were found in good life reasoning stage attainment in the adult group, they had not been found in the younger age groups. This may indicate complex relationships between historical periods and gender role socialization. Alternatively, it may indicate that women's development is truncated in the adult years. Male and female subjects were approximately equal in their age, educational attainment, occupational status, and socioeconomic status. Thus, the gender differences cannot be reduced to differences in these variables (as has been the case with moral judgment [e.g., Walker, 1984]). Only additional studies with larger samples of males and females can expand our understanding. Previously (Armon, 1984a), no gender differences were found in the distribution of Philosophical Orientations in any age group, which may indicate some interesting directions for future work in gender studies.

Adult Development

The findings presented here should encourage researchers to take structural development in adulthood seriously. More research, explicitly designed to study adult structural development needs to be done. If the findings in this study are repli-cated with different populations, there are many important implications. For exam-ple, we may need to think about including structural–developmental information in the construction of adult education, training programs, and psychological interven-tion methods in the same ways that we do for child and adolescent populations.

More germane to this study, however, is the issue of why only adults attain higher stages, and only some adults at that. We need to move beyond the question of whether adults develop, and ask the question, "How and why do adults develop?" Why is age a necessary condition for higher stage development? What are the necessary ingredients of continued change? Must one have a life of responsibility and role-taking as Kohlberg (1984) claimed? The antecedents of adult development are almost completely unknown. Even in literature where the notion of structural development in adulthood is accepted, it is conceived of as either hierarchical or absent. This dichotomy is unlikely. Other forms of develop-ment, such as decalage and consolidation, need to be investigated and distin-guished from one another.

As researchers, we need to go beyond delineating static stage models and begin investigating variables that may impact adult development. A few small studies (mostly single-case) have suggested that early religious training and/or confrontation with challenging value conflicts are salient. In Colby and Damon's (1992) case studies of moral exemplars, or Kohlberg's (1981) analysis of Andrea Simpson, it seems that some people make growth opportunities out of circum-stances they just "fall into," while others pass by such opportunities unaware.

Although only a few theorists are examining the relationships between personality or temperament and life experiences, findings are promising. For example, Blasi (1980) discusses the possibility of certain self-concept variables having influences on moral development; Kitwood (1988) argues that early psy-chological wounds inhibit development; and certain temperament variables (such as extroversion) appear to be related to successful and unsuccessful experiences in childhood (Cowen, Wyman, Work, & Parker, 1990; Masten, Best, & Garmezy, 1990). Future individual case analyses of subjects from this study may provide relevant information on some of these important issues. (See Armon, 1995, for an example.)

Limitations of this Study

A major limitation of this study results from the size and nature of the sam-ple, and its lack of generalizability to typical populations. While the sample was chosen purposefully to enhance the probability of observing adult development, it simultaneously limits how one can use the findings. Another cause for concern is the exclusive use of Western culture subjects, along with the corresponding reliance on Western philosophy for the formation of the Philosophical Orientations. The impact of this limitation cannot be tested without studying other populations. The Good Life Stage model, however, does not depend on the universalization of Western liberalism. On the contrary, one would expect if the study were to be carried out in the East, for example, other philosophies of the Good Life might be found (Zen Buddhist, for example). It has only been claimed here that there may be a finite number of Philosophical Orientations, not that this study has discovered them all. Moreover, although it is expected that different cultures might provide alternative Philosophical Orientations, it is not expected that the structural aspects of value reasoning would be significantly different.

Another shortcoming of this study is that reasoning about the Good Life says little about real-life behavior. Construct validity, however, can be maintained by demonstrating that the interview and scoring scheme provide a valid assessment of value reasoning about the Good Life, or of value reasoning *stage*, rather than a valid assessment of individuals' behavior or actual "good lives" (Colby et al., 1983). High correlations between stage attainment and Hollingshead occupational scores indicate that, on average, subjects at higher stages perform occupations with greater complexity and, possibly, social responsibility. To find out the extent to which they actually pursue the good lives they describe, however, would require in-depth case study methods.

In general, future research on structural development, particularly during adulthood, must maintain a rigorous theoretical and methodological approach so as not to fall prey to death by criticism. Moreover, the ethical dimension in social development research needs to be more clearly acknowledged. Without clearly specified constructs and a rigorous methodology to distinguish, for example, content from structure, or valuing from cognition, specific issues in human development will remain elusive. This is particularly true of the task of identifying the processes and causes of transition or development from one stage to another, which has eluded researchers for some time.

Implications for Education

The ways in which people decide what they value is of central importance to education. In the last decade, interest in value education has resurfaced. New emphasis is being placed on the role of the school and university in the formation of character, particularly since it is now recognized that all schools impact the character of its students, whether they do so planfully or not. In addition to issues of justice and fairness, character development includes reflection and action on issues of responsibility, general concern for others, and self respect. Such reasoning consists in part of value reasoning about the Good Life.

In particular, value reasoning about the Good Life in the moral domain identifies the moral good of responsibility, of persons, of the self, and even of justice (Rawls, 1971). For example, a principle of benevolence (Frankena, 1973) falls within the domain of the Good rather than the realm of the Right (justice, fairness, rights). In general, many real-life judgments concern neither physics (Piaget) nor justice (Kohlberg); rather, they are judgments of the Good Life—of values, aims, and ideals.

The developmental processes of value reasoning outlined here can inform a philosophy of education that goes beyond "values clarification" to identify reasonable and justifiable standards for values-guided education. That reasoning about the Good Life develops through a predictable sequence of stages allows educators to design curricula with developmentally appropriate methods and objectives. Moreover, our larger study, and others like it, make clear that young students have their own concepts of the Good Life that are neither an incomplete notion of adult ideals nor simply a replication of their particular environment. These students need to be encouraged to articulate their ideals of the Good Life, to learn about those of others, and to question the meaning and outcomes of various good life concepts. These practices will serve to enhance students' ethical thinking, encourage perspective-taking and mutual respect, and provide essential opportunities for reflection and critical analysis.

The findings of this study show that these issues should not be restricted to traditional elementary, high school, or college populations. It was demonstrated here not only that persons can develop their thinking about values throughout the lifespan, but also that postconventional stages of value reasoning do not emerge before adulthood. Increasing evidence supports these findings (see Kohlberg, 1986; Lam, 1994; also see reviews by Alexander et al., 1990; Richards et al., 1984). Moreover, it was shown here that development to those stages can be significantly influenced by educational experience. And, finally, today, more than one third of all university students *are* adults beyond the age of 25.

Unfortunately, typical adult-oriented education programs—whether they be pragmatic programs for continuing education, rehabilitaion, or those of prestigious graduate schools—do not see values development as central to their endeavors. Rather, because it is assumed that the character of the adult student is already set, these programs are restricted to the accumulation of skills and knowledge and, occasionally, to the advancement of critical thinking. Findings of this and other studies, however, demonstrate the inadequacy of this view. Indeed, it appears that educational experience in adulthood can be extremely important in the development of autonomy and principled ethical reasoning. Philosophers of education, psychologists, and educators need to recognize that particular forms of educational experience are probably key factors in the creation of critically reflective, ethical adults who can participate effectively in our increasingly complex and often threatened civilization.

Although a general model of value reasoning about the Good Life will tell only a part of the story of human valuing, it nevertheless goes beyond and is to be preferred to the value relativism and subjectivism so prevalent in contemporary society, generally, and in psychology in particular. From homelessness to adolescent homicide, contemporary social problems are in part, a consequence of adult value reasoning. A stage model of value reasoning about the Good Life can inform our understanding as to some of the origins of such problems and contribute to education and intervention models that attempt to address them.

ACKNOWLEDGMENTS.   Findings and analyses of the first follow-up were included as part of the first author's doctoral dissertation at Harvard University, under the supervision of Lawrence Kohlberg. The balance of the research was completed with support from Antioch University Southern California. We wish to graciously thank all of the participants for sharing their thoughts with us over thirteen years. We are indebted to our interviewers Joyce Friedman, Larry Smiley, Lura Beth Illig, and Sandra Nahan who sometimes traveled thousands of miles to unexpected places to meet with participants. We thank Michael Commons and Bill Puka for valuable input on earlier versions of this paper. We also acknowledge Paul Holland and Michael Commons for their assistance with recent statistical analyses.

REFERENCES

Adams, R. J., & Khoo, S.-T. (1993). Quest: The interactive test analysis system. Victoria, Australia: Australian Council for Educational Research.

Alexander, C. N., Druker, S. M., & Langer, E. J. (1990). Introduction: Major issues in exploration of adult growth. In C. N. Alexander & E. J. Langer (Eds.), *Higher stages of human development* (pp. 3–32). New York: Oxford University Press.

Alexander, C., & Langer, J. (1990). *The higher stages of human development.* New York: Oxford University Press.

Andrich, D. (1986, September). Intellectual development of pre-adolescent and adolescent children from a psychometric perspective. Paper presented at the International Conference on Longitudinal Methodology, Budapest, Hungary.

Andrich, D. (1988). *Rasch models for measurement.* Newbury Park, CA: Sage.

Andrich, D. (1989). Distinctions between assumptions and requirements in measurement in the social sciences. In J. A. Keats, R. Taft, R. A. Heath, & S. H. Lovibond (Eds.), *Mathematical and theoretical systems. Proceedings of the 24th International* (pp. 7–16). North-Holland, Amsterdam, Netherlands.

Andrich, D., & Styles, I. (1994). Psychometric evidence of intellectual growth spurts in early adolescence. *Journal of Early Adolescence, 14,* 328–344.

Aristotle (1962). *Nichomachean ethics* (trans. by Ostwald). Indianapolis: Bobbs-Merrill.

Armon, C. (1984a). Ideals of the good life and moral judgment: Ethical reasoning across the life span. In M. L. Commons, F. A. Richards, & C. Armon (Eds.), *Beyond formal operations: Late adolescent and adult cognitive development* (pp. 357–381). New York: Praeger.

Armon, C. (1984b). Ideals of the good life: A longitudinal/cross-sectional study of evaluative reasoning in children and adults. Unpublished dissertation, Harvard University.

Armon, C. (1988). The place of the good in a justice reasoning approach to moral education. *The Journal of Moral Education, 17,* 220–229.

Armon, C. (1993). Developmental conceptions of good work: A longitudinal study. In J. Demick & P. M. Miller (Eds.), *Development in the workplace* (pp. 21–37). Hillsdale, NJ: Lawrence Erlbaum.

Armon, C. (1995). Moral judgments and self-reported events in adulthood. *Journal of Adult Development,* Vol. 2.

Armon, C. (1998). Adult moral development experience and education. *Journal of Moral Education, 27,* 3.

Armon, C., & Dawson, T. (1997). Developmental trajectories in moral reasoning across the life span. *Journal of Moral Education, 26,* 4.

Baldwin, J. M. (1906). *Thought and things or genetic logic.* New York: Macmillan.

Baltes, P. B., Reese, H. W., & Lipsitt, L. P. (1980). Lifespan developmental psychology. *Annual Review of Psychology, 31,* 65–110.

Baltes, P. B. (1983). Life-span developmental psychology: Observations on history and theory revisited. In R. M. Lerner (Ed.), *Developmental psychology: Historical and philosophical perspectives* (pp. 79–112). Hillsdale, NJ: Lawrence Erlbaum.

Baltes, P. B., Hayne, R. W., & Lipsitt, L. P. (1980). Life-span developmental psychology. *Annual Review of Psychology, 31,* 65–110.

Birren, J. E. (1964). *The psychology of aging.* Englewood Cliffs, NJ: Prentice-Hall.

Blasi, A. (1980). Bridging moral cognition and moral action: A critical review of the literature. *Psychological Bulletin, 88* (1), 1–45.

Bock, R. D. (1991). Prediction of growth. In L. M. Collins & J. L. Horn (Eds.), *Best methods for the analysis of change: Recent advances, unanswered* (pp. 126–136). Washington, DC: American Psychological Association.

Bond, T. G. (1994). Piaget and measurement II: Empirical validation of the Piagetian model. *Archives de Psychologie, 63,* 155–185.

Bond, T. (1995). *BLOT Bond's logical operations test,* Townsville, James Cook Universary.

Broughton, J. M. (1984). Not beyond formal operations but beyond Piaget. In M. L. Commons, M. Richards, & C. Armon (Eds.). *Beyond formal operations: Late adolescent and adult cognitive development.* New York: Praeger.

Campbell, C. A. (1935). Moral and non-moral values. *Mind, 44,* 279–291.

Colby, A., & Kohlberg, L. (1987). *The measurement of moral judgment, Vols. I & II.* New York: Cambridge University Press.

Colby, A., & Damon, W. (1992). *Some do care: Contemporary lives of moral commitment.* New York: The Free Press.

Colby, A., Kohlberg, L., Gibbs, J., & Lieberman, M. (1983). Report on a 20-year longitudinal study of moral development. *Monograph of the society for research in child development, 48*(4).

Commons, M. L., Armon, C., Kohlberg, L., Richards, F. A., Grotzer, T. A., & Sinnott, J. D. (Eds.) (1990). *Adult development, Vol II: Models and methods in the study of adolescent and adult thought.* New York: Praeger.

Commons, M. L., Armon, C., Richards, F. A., & Schrader, D. E. with Farrell, E. W., Tappan, M. B., & Bauer, N. F. (1989). A multidomain study of adult development. In M. L. Commons, J. D. Sinnott, F. A. Richards, & C. Armon (Eds.), *Adult development, Vol. 1: Comparisons and applications of adolescent and adult developmental models* (pp. 33–56). New York: Praeger.

Commons, M. L., & Richards, F. A. (1984). A general model of stage theory. In M. L. Commons, F. A. Richards, & C. Armon (Eds.), *Beyond formal operations: Late adolescent and adult cognitive development* (pp. 141–157). New York: Praeger.

Commons, M., Richards, F., & Armon, C. (Eds.) (1984). *Beyond formal operations: Late adolescent and adult cognitive development.* New York: Praeger.

Commons, M. L., Sinnott, J. D., Richards, F. A., & Armon, C. (Eds.) (1989). *Adult development, Vol. I: Comparisons and applications of adolescent and adult developmental models.* New York: Praeger.

Cowen, E. M., Wyman, P. A., Work, W. C., & Parker, G. R. (1990). The Rochester child resilience project: Overview and summary of first year findings. *Development and Psychopathology, 2,* 193–212.

Cutler, N. E. (1979). Age variations in the dimensionality of life satisfaction. *Journal of Gerontology, 34,* 573–578.

Damon, W., & Hart, D. (1990). *Self understanding in childhood and adolescence.* New York: Cambridge University Press.

Dawson, T. L. (1996, September). Integration and differentiation in longitudinal moral development data. Paper presented at The Growing Mind, Geneva, Switzerland.

Dawson, T. L. (1997, June). New tools, new insights: Kohlberg's moral reasoning stages revisited. Paper presented at the Twenty-Seventh Annual Symposium of the Jean Piaget Society, Santa Monica, CA.

Dawson, T. L., Goodheart, E. A., Draney, K., Wilson, M., & Commons, M. L. (1997, March). Concrete, abstract, formal, and systematic operations as observed in a "Piagetian" balance-beam task series. Paper presented at the International Objective Measurement Conference, Chicago, IL.

Demetriou, A., Efklides, A., Papadaki, M., Papantoniou, G., & Economou, A. (1993). Structure and development of causal-experimental thought: From early adolescence to youth. *Developmental Psychology, 29,* 480–497.

Demick, J. (Ed.) (1994–). *The Journal of Adult Development.*

DeVos, E. (1983). Socioeconomic influences on moral reasoning: A structural-developmental perspective. Unpublished dissertation, Harvard University.

Dewey, J. (1944/1916). *Democracy and education.* New York: Macmillan.

Dewey, J. (1957/1922). *Human nature and conduct.* New York: Modern Library.

Dewey, J. (1980). *Theory of the moral life.* New York: Irvington.

Draney, K. L. (1996). The polytomous Saltus model: A mixture model approach to the diagnosis of developmental differences. Unpublished doctoral dissertation, University of California at Berkeley, Berkeley, CA.

Duncan, O. D. (1984). *Notes on social measurement: Historical and critical.* Beverly Hills, CA: Sage.

Ennis, R. (1978). Conceptualization of childrens' logical competence: Piaget's propositional logic and an alternate proposal. In L. S. Seigel & C. J. Brainard (Eds.), *Alternatives to Piaget: Critical essays on the theory* (pp. 201–260). New York: Academic Press.

Erdynast, A., Armon, C., & Nelson, J. (1978). Cognitive developmental conceptions of the true, the good, and the beautiful. *The eighth annual proceedings of Piaget and the helping professions.* Los Angeles: University of Southern California Press.

Erikson, E. H. (1963). *Childhood and society.* New York: W. W. Norton.

Feldman, D. H. (1980). *Beyond universals in cognitive development.* Norwood, NJ: Ablex.

Fisher, W. P., Jr. (1994). The Rasch debate: Validity and revolution in educational measurement. In M. Wilson (Ed.), *Objective measurement* (pp. 36–72). Norwood, NJ: Ablex.

Fowler, J. (1981). *Stages of faith: The psychology of human development and the quest for meaning.* New York: Harper & Row.

Frankena, W. K. (1973). *Ethics.* Englewood Cliffs, NJ: Prentice-Hall.

Freud, S. (1961). *The standard edition of the complete works of Sigmund Freud* (trans. & edited by J. Strachey). London: The Hogarth Press.

Gilligan, C. (1981). *In a different voice.* Cambridge, MA: Cambridge University Press.

Goodheart, E., Dawson, T., & Commons, M. (1996a, June). Primary, concrete, abstract, formal, systematic, and metasystematic operations as observed in a piagetian logical task series. Paper presented at the Annual Meeting of the Society for reseach in Adult Development, Boston.

Goodheart, E., Dawson, T., & Commons, M. (1996b, November). A Rasch analysis of developmental data on relations between more and less powerful persons. Paper presented at the Annual meeting of the Association for Moral Education.

Habermas, J. (1979). *Communication and the evolution of society.* Boston: Beach Press.

Hautamäki, J. (1989). The application of a Rasch model on Piagetian measures of stages of thinking. In P. Adley (Ed.), *Adolescent development and school science* (pp. 342–349). London: Falmer.

Hare, R. M. (1952). *The language of morals.* New York: Oxford University Press.

Hollingshead, A. B. (1975). Four factor index of social status. New Haven, CT: Yale University Dept. of Sociology.

Horney, K. (1936). The neurotic personality of our time. New York: Norton Press.

James, W. (1890). *Principles of psychology.* New York: Holt.

Kegan, R. (1982). *The evolving self: Problems and process in human development.* Cambridge, MA: Harvard University Press.

Kegan, R. (1994). *In over our heads: The mental demands of modern life.* Cambridge, MA: Harvard University Press.

Kitwood, T. (1988). *Concern for others: A new psychology of conscience and morality.* London: Routledge.

Kohlberg, L. (1969). Stage and sequence: The cognitive developmental approach. In D. A. Goslin (Ed.), *Handbook of socialization: Theory and practice* (pp. 347–480). Chicago: Rand McNally.

Kohlberg, L. (1981). *The philosophy of moral development.* New York: Harper & Row.

Kohlberg, L. (1984). *The psychology of moral development.* New York: Harper & Row.

Kohlberg, L. (1986). *The adult as a moral philosopher.* Unpublished manuscript, Harvard University.

Kohlberg, L., & Armon, C. (1984). Three types of stage models used in the study of adult development. In M. L. Commons, F. A. Richards, & C. Armon. (Eds.), *Beyond formal operations: Late adolescent and adult cognitive development* (pp. 383–394). New York: Praeger.

Kohlberg, L., Levine, C., & Hewer, A. (1983). Moral stages: A current formulation and a response to critics. *Contributions to human development, 10.* Basel: Karger.

Lam, M. S. (1994). Women and men's notions of the good life. Unpublished dissertation. University of Massachusetts-Amherst.

Laterius, Diogenes (1925). *Lives of eminent philosophers.* New York: Loeb Classical Library.

Lawton, M. P. (1984). The varities of wellbeing. In C. Z. Malatesta & C. E. Izard (Eds.), *Emotion in adult development* (pp. 67–84). Beverly Hills, CA: Sage.

Levinson, D. J., Darrow, C. N., Klein, E. B., Levinson, M. H., & McKee, B. (1978). *The seasons of a man's life.* New York: Knopf.

Lewis, C. I. (1946). *An analysis of knowledge and valuation.* LaSalle, Il: Open Court.

Loevinger, J. (1976). Ego development: Concepts and theories. San Francisco: Jossey-Bass.

Loevinger, J. with Blasi, A. (1970). *A theory of ego development.* San Francisco: Jossey-Bass.

Ludlow, L. H., & Haley, S. M. (1995). Rasch model logits: Interpretation, use, and transformation. *Educational & psychological measurement, 55,* 967–975.

Maslow, A. (1971). *The farther reaches of human nature.* New York: Viking Press.

Masten, A. S., Best, K. M., & Garmezy, N. (1990). Resilience and development: Contributions from the study of children who overcome diversity. *Development and psychopathology, 2,* 425–444.

Masters, G. (1997). Fundamental measurement for outcome evaluation. In R. M. Smith (Ed.), *Outcome measurement, Vol. 11* (pp. 261–288). Philadelphia: Hanley & Belfus.

Masters, G. N. (1982). A Rasch model for partial credit scoring. *Psychometrika, 47,* 149–174.

Masters, G. N. (1988). Measurement models for ordered response categories. In R. Langeheine & J. Rost (Eds.), *Latent trait and latent class models* (pp. 11–29). New York: Plenum Press.

Masters, G. N. (1994). Partial credit model. In T. Husén & T. N. Postlethwaite (Eds.), *The international encylcopedia of education* (pp. 4302–4307). London: Pergammon.

Michell, J. (1990). *An introduction to the logic of psychological measurement.* Hillsdale, NJ: Lawrence Earlbaum.

Mill, J. S. (1957/1861). *Utilitarianism*. Indianapolis: Bobbs-Merrill.

Mill, J. S. (1978/1861). On Liberty. In R. Wolheim (Ed.), *John Stuart Mill: Three Essays*. Oxford: Oxford University Press.

Müller, U., Reene, K., & Overton, W. F. (1994, June). Rasch scaling of a deductive reasoning task. Paper presented at the Annual Symposium of the Jean Piaget Society, Philadelphia, PA.

Müller, U., Winn, M., & Overton, W. (1995, June). Rasch analysis of two recursive thinking tasks. Paper presented at the Annual Symposium of the Jean Piaget Society, Berkeley, CA.

Noelting, G., Coudé, G., & Rousseau, J. P. (1995, June). Rasch analysis applied to multi-domain tasks. Paper presented at the Twenty-Fifth Annual Symposium of the Jean Piaget Society, Berkeley, CA.

Noelting, G., Rousseau, J. P., & Coudé, G. (1994, June). Rasch analysis applied to Piagetian-type problems. Paper presented at the Annual Symposium of the Jean Piaget Society, Chicago, Il.

Nowell-Smith, P. (1954). *Ethics*. Baltimore, MD: Pelican Books.

Nozick, R. (1974). Anarchy, state, and utopia. New York: Basic Books.

Perry, R. B. (1954). *General theory of value*. Cambridge, UK: Cambridge University Press.

Piaget, J. (1932). *The moral judgment of the child*. London: Routledge & Kegan Paul.

Piaget, J. (1960). The general problems of the psychological developement of the child. In J. M. Tanner & B. Inhelder (Eds.), *Discussions in child development: Proceedings of the world health organization study group on the psycholobiological development of the child, Vol. 4*. New York: International Universities Press.

Piaget, J. (1968). *Six psychological studies*. New York: Random House.

Piaget, J. (1970). *Structuralism*. New York: Basic Books.

Piaget, J. (1972). Intellectual evolution from adolescence to adulthood. *Human Development, 15*, 1–12.

Rawls, J. (1971). *A theory of justice*. Cambridge, MA: Belknap Press.

Rawls, J. (1980). Kantian constructivism in moral theory: Rational and full autonomy. *The Journal of Philosophy, 77*(9), pp. 515–572.

Rest, J. R., & Thoma, S. J. (1985). Relation of moral judgment development to formal education. *Developmental Psychology, 21*(4), 709–714.

Richards, F. A., Armon, C., & Commons, M. L. (1984). Perspectives on development of thought in late adolescence and adulthood: An introduction. In M. L. Commons, F. A. Richards, & C. Armon (Eds.), *Beyond formal operations: Late adolescent and adult cognitive development* (pp. xiii–xxviii). New York: Praeger.

Rokeach, M. (1973). *The nature of human values*, 2nd edit. New York: Free Press.

Rokeach, M. (1979). *Understanding human values, individual and societal*. New York: Free Press.

Rosenberg, S. (1988). *Reason, ideology and politics*. Princeton, NJ: Princeton University Press.

Ross, W. D. (1930). *The right and the good*. Oxford: Clarendon Press.

Rybash, J. M., Roodin, P., & Santrock, J. W. (1991). *Adult development*. Dubuque, IA: W. C. Brown.

Ryff, C. (1989). In the eye of the beholder: Views of the psychological well-being among middle-age and older adult. *Psychology and Aging, 4*(2), 195–210.

Selman, R. (1980). *The growth of interpersonal understanding*. New York: Academic Press.

Sonnert, J. G., & Commons, M. L. (1994). Society and the highest stages of moral development. *The Individual and Society, 4*, 31–35.

Spinoza, B. (1949). *Ethics*. New York: Hafner Press.

Stevens-Long, J., & Commons, M. L. (1990). *Adult life: Developmental processes (4th Ed.)*. Mountain View, CA: Mayfield.

Stock, W. A., Okun, M. A., & Benin, M. (1986). Structure of subjective well-being among the elderly. *Psychology and Aging, 1*, 91–102.

Thurstone, L. L. (1959). *The measurement of values*. Chicago: University of Chicago Press, Miday Reprint Series.

van der Linden, W. (1994). Fundamental measurement and the fundamentals of Rasch measurement. In M. Wilson (Ed.), *Objective measurement: Theory into practice* (pp. 3–24). Norwood, NJ: Ablex.

Walker, L. (1984). Sex differences in the development of moral reasoning: A critical review. *Child Development, 55*, 3.

Walker, L. (1986). Experiential and cognitive sources of moral development. *Human Development, 29*, 113–124.

Willet, J. B. (1989). Some results on reliability for the longitudinal measurement of change: Implications for the design of studies of individual growth. *Educational and Psychological Measurement, 49*, 587–602.

Wilson, M. (1985). Measuring stages of growth: A psychometric model of hierarchical development. Occasional paper No. 29. Hawthorn, Victoria, Australia: Australian Council for Educational Research.

Wilson, M. (1989a). Empirical examination of a learning hierarchy using an Item Response Theory model. *Journal of Experimental Education, 57*, 357–371.

Wilson, M. (1989b). Saltus: A psychometric model of discontinuity in cognitive development. *Psychological Bulletin, 105*, 276–289.

Wilson, M., & Draney, K. (1997, March). Beyond Rasch in the measurement of stage-like development. Paper presented at the the annual meeting of the American Educational Research Association, Chicago, Il.

Wright, B. D., & Linacre, J. M. (1989). Observations are always ordinal; Measurements, however, must be interval. *Archives of Physical Medicine and Rehabilitation, 70*, 857–860.

Wright, B. D., & Masters, G. N. (1982). *Rating scale analysis.* Chicago: Mesa Press.

CHAPTER 16

Moral Metacognition in Adolescence and Adulthood

DAWN E. SCHRADER

Metacognition, or thinking about thinking, has been studied extensively in the domains of memory, reading comprehension, and knowledge acquisition (e.g., Baker & Brown, 1984; Brown, 1980; Flavell, 1978, 1985; Markman, 1979). Yet little empirical research on metacognitive thinking exists in the social domain. How do we think about our thinking when problems arise between two people or between people and institutions? Between our social–moral obligations and relationships and our moral ideals?

The question of metacognition is of significant import, however, especially with regard to how our thinking becomes increasingly self-reflective and inclusive of increasingly complex skills required for social–moral problem solving. As Wood (1983) demonstrated, social conflicts and problems are ill-structured problems, and with ill-structured problems come ill-defined solutions. This makes the social domain a particularly fertile ground for exploring complex cognitive processes. But with such an exploration, one must turn toward an examination of the ways in which people think about their own thinking about such problems, and broaden our outlook beyond a focus solely on whether and how individuals solve such problems.

Thinking about thinking involves many types of cognitive processes. Some researchers argue that metacognition appears in young children—as monitoring, detecting, and correcting errors in their thinking (e.g., Brown, 1980); while others such as Piaget contend that thinking about one's thinking is, at least, a formal operational task or a task that appears with greater ability in adulthood (Dewey, 1933). Through the burgeoning array of postformal operations studies, we have seen that attainment of formal operations does not explain all cognitive development that occurs in adulthood (Commons, Armon, Kohlberg, Richards, Grotzer, & Sinnott, 1990; Commons, Richards, & Kuhn, 1982). Indeed, changes in form of reflective thought may intersect with attainments beyond formal operations. According to

DAWN E. SCHRADER • Department of Education, Cornell University, Ithaca, New York, 14853-4203.

Handbook of Adult Development, edited by J. Demick and C. Andreoletti. Plenum Press, New York, 2002.

the interview data reported later in this chapter, at least two characteristics of postformal thought coincide with developments in reflective thought: systematic operations (Commons et al., 1984, 1990; Commons, Richards, & Kuhn, 1982) and problem-finding (Arlin, 1975).

With regard to systematic operations, Demetriou (1990) examined metacognition and such postformal thinking. Demetriou broadened the definition of metacognition beyond the self-monitoring and detection of errors. He described four categories of metacognition that he thought could be used in resolving ill-structured problems. In his study, all participants were given four ill-structured problems and were asked to categorize the problems based on how people think and the methods used to arrive at solutions. The categories were: clearly similar tasks, clearly different tasks, and tasks similar in some respects and different in other. Participants were asked to explain their responses in detail. Their responses were coded using four metacognitive categories: No Reflection, Content-based reflection, Specifications of the operations involved, and Analysis and integration of operations.

Demetriou found an association between metacognition and postformal operational reasoning, particularly for students who were still in college, and attributed his findings to the fact that college students engage in problem solving daily, and that practice may be responsible for developing reflective awareness of their thinking processes. The importance of this study for the present work is that Demetriou studied metacognitive knowledge in ill-structured problems, albeit not moral ones, and found different forms of metacognition than have been discussed in the literature. Because moral dilemmas are also ill-structured problems, the basic premise of this work contends that moral metacognition has different forms in adulthood than in childhood and adolescence.

As with postformal operational thinking in the logical domain, the increasingly complex forms of metacognition involve a reflective coordination of systems of thought and thinking strategies about social relationships and moral ideals. One is thus able to not only resolve complex problems with myriad variables, but may also see patterns and generate new questions and solutions for situations and problems that recur in individuals' lives. Thus, *problem finding* is an important characteristic of both post formal and metacognitive thought.

The questions that an exploration of metacognition in the moral domain might answer are: How do we think about the ways in which we think about moral problems? Do metacognitive processes differ in the moral domain to the logical domain? These questions have no easy answers. They are moral questions that are in some sense confounded with perspectives of morality—whether from a justice or care perspective or from stage and type of moral reasoning. They extend beyond the domain of memory and knowledge, although memory and knowledge may enter into our thinking, and require further cognitive procedures than simply logical analysis. They require these kinds of thinking, but also a different form of thought—reflective thought—reflective of both our ways of thinking about problems as well as reflective of our personal and moral selves. We as individuals must think about how we will resolve these questions; mull over possible alternatives; seek out various solutions; examine the consequences of each of our proposed solutions; and weigh the personal, ethical, and moral consequences of each for ourselves and for our social relationships. We remember how similar situations turned out in the past—how we came to that past decision. Was the solution well considered? How could it have been better considered? Given that there were consequences, what have we learned from them that will affect our current ways of thinking or the alternatives of action or thought that we will hold or discard? Comprehension also plays an important role. Are we sure that we understand the situation? What do we know? What do we not yet know? And, most importantly, what do we not know that we don't know?

Substantial literature exists to demonstrate that judgment, memory, comprehension—all these are essential in looking at academic problems, and are important in looking at social ones. However, social problems, and especially ethical/moral problems, have additional components discussed in the preceding section that are not included in the literature on metacognition. For example, what about issues of care and love? Complexity of individual reactions? Individuals' conception of justice? Flavell (1985) touches on these questions when he distinguishes metacognitive knowledge from metacognitive experience. Although the two are interrelated, such interaction has not yet been described in the literature. The problem is that in logic, actions have specific and known reactions given certain parameters. When human emotions, thoughts, and memories are involved, there is a "psychologic" (Kegan, 1982). Psychologic is less predictable than logic. There is some sense of probability, sense of consistency, within individuals in certain circumstances; but individuals change, the other individuals involved in situations change, and thus the social–moral problem changes. These make the social–moral domain extraordinarily complex.

What is of interest in this chapter is the thoughts of individuals attempting to make sense of their thinking about complex moral situations, hypothetical or actual. More specifically, how aware is an individual of her or his thinking processes while resolving moral dilemmas, what thought processes are involved, and how are those thought processes related to other aspects of morality such as moral stage?

PARTICIPANTS

The three case examples in this chapter illustrate the forms and variation of metacognition in the moral domain that were developed from an exploratory study of 30 participants. Before introducing the three case examples, a brief description of all the participants follows.

Participants for the study included 30 students, 10 from high school, 10 college, and 10 doctoral level. The doctoral group was primarily white, aged 26 to 38, and nearly all had full-time work experience prior to returning to school. The undergraduate group was primarily white, aged 19 to 22, many had worked in summer jobs, and most worked part-time during the school year. The high school group was primarily white or mixed race (non-black, non-hispanic), aged 16 to 18, and nearly all worked part-time after school.

The three students used in this case analysis were selected from the larger sample based on their representativeness of various levels of metacognition, and their usefulness for providing examples of the range of metacognitive levels that appear within a protocol, and the type of changes that appear due to each task and the interview itself. The examples were selected based on clarity of their responses. Both males and females were found at all metacognitive categories.

The three students selected for this analysis were Meg, Mark, and Rob. Meg, a high school student at a selective admissions school in the eastern United

States, was 16 years old at the time of the interview, a college-bound junior, and participated in a moral education intervention in her high school. Mark was a 22-year-old psychology student at an eastern Ivy League college in the United States. Mark attended a private preparatory high school, and at the time of the interview was a student teacher for social studies in an urban high school. The third case is Rob. Rob attended a public high school and a small liberal arts college before enrolling in a doctoral program at an eastern Ivy League university in the United States. At the time of the interview, he had completed all requirements for a doctorate in Educational Psychology except the dissertation.

METHODS

All participants were interviewed by the author using an interview (referred to here as the Metacognitive Interview) which included: (1) a problem designed to elicit manifest metacognitive strategies, referred to as the "Teach Me" problem; (2) Kohlberg's Moral Judgment Interview (Colby & Kohlberg, 1987); (3) a Real Life Interview (modified from Gilligan, 1982); and (4) a series of questions designed to elicit reflective recall of metacognitive strategies used in resolving the Moral Judgment Interview and the Real Life Interview. (See Appendix for the interview schedule.) Each interview was tape-recorded and transcribed verbatim.

The system for evaluating metacognition, described elsewhere (see Schrader, 1988), provides a code for the overall interview for metacognition, as well as provides a means by which individual responses can be coded. The categories of metacognition in the moral domain are Elementary Reflection, Self-Reflective Monitoring, Identification of Processes, Explanation of Processes, and Evaluation of Processes.

In the coding manual, each of these levels[1] explicated the criteria for categorizing a response at each level via its cognitive organizing system, verbal representation, and relation to other cognitive theories/systems. A summary of the description of each level of metacognition follows, along with a brief example.

LEVELS OF METACOGNITION

Elementary Reflection (ER)[2]

This level is characterized by thinking about thinking, but the thought is based on the contents of thought, such as retelling facts of what is in their minds. Participants report that their decisions appear like "magic"—there is no apparent process of thinking known to them, and are not consciously reflectively aware of their thinking processes until prompted to reflect. You "just know." Thinking

---

[1] These are referred to as "levels" rather than categories of metacognition given the results of the analysis found in Schrader (1988).

[2] Elementary Reflection was originally called "Non-Reflection" in Schrader (1988). The change is due to the notion that there is a beginning consciousness of thought; if not immediately reflective, it is conscious and has the possibility of reflection when probed. It is considered the first step leading toward reflecting on reflective thought.

is "on line"—it is not reflected upon—no monitoring or editing of thoughts. For example, one student answered the question, "Did you consider any alternative strategies when resolving this?" by saying, "No. I don't think so. It just happens. That's what you think. Your instinct takes over."

Self-Reflective Monitoring (SRM)

At this level, people begin to notice the workings of their own mind. Their cognitive system is implicit—not completely accessible and knowable to themselves, but there is some evidence that they are self-aware of thinking about their thinking, by reference to the content factors of their decision process. A coordinated *system* of thinking is not elaborated in general terms, but individuals can talk about what procedures were used. For example, when asked, "How do you know when you've reached the best solution?" a participant answered, "Well, what you consider to be the best solution will be the solution you consider to be the correct solution, and so it's the same thing. It would be the sense of sureness that you were right and that you wouldn't have to keep *debating every time you looked at the two sides*. You would just have arrived at that and know." This quote indicates that there is some underlying cognitive strategy involved in resolving problems, namely debating two sides, but does not articulate the processes involved in such a debate.

Identification of Processes (IP)

In this level, individuals' justifications indicate knowledge of, and can identify, components of a system of cognitive processes that they use in thinking about a problem, yet the description remains on the plane of labeling. Their system of thinking becomes identifiable in conscious awareness. An example of this thinking appeared when a participant summarized their thought processes when resolving the "teach me" problem by saying, "Ok... First you have to look at what the claims are of all the people in the situation. You also know you have other issues in that, but, and then you have to find out why those things are important to them, what it is they want and why they want it and then you make a judgment... and then you have to weigh and balance the consequences to all of the parties,... and then you have to decide what is the more important, you know, whose needs are more important and why, and what these are—in this case somebody's life is more important than the drug."

The preceding example demonstrates that a cognitive process or procedure is explicit in the subject's mind, and that the steps in the process can be articulated. The factors or steps involved form a system, but are not explained in and of themselves, nor in coordination with each other.

Explanation of Processes (EXP)

People elaborate on the processes that are identified in the IP level—they explain the rules of their system of thinking, explaining how and why the components are included in their system, and how those components interact to form a system for how they think about problems. Procedures are abstracted from their thinking about the contextual characteristics of the dilemma (i.e., they go beyond the concrete task), and are explained as to how they relate to their system of resolving a problem. For example, one participant answered the first question in the metacognitive interview, "So if you were going to teach me how to go about resolving this [hypothetical dilemma], what would you say?" with the following complex answer: "I would tend to say to break it down into categories and break it down into ramifications, in that there are at least three people involved here, and given the potential that you have two choices of outcome.... Now take each one and look at the ramifications for every single person based on that decision. So you've got sort of 6 categories at that point. I guess in that way you'd almost make a spreadsheet of what are your options, and what would happen if you choose any given one. And then it would come into question as to how you feel about those ramifications and those options, and going through outcomes, deciding if you want to go by the rule of it's wrong to steal or that life is more important than money—that would choose which of the categories you're gonna look at, and then from there I'd say break it down even further to—almost like a—what's the word I want? Like a flow chart in chemistry, when you break it down into smaller and smaller components. And then once you've arrived at what you consider— you've narrowed it down to 'this is what I think'—to work backwards and see if any of your original decisions change based on what you've decided is the potential outcome, the one that you think is the correct outcome."

The person above identifies and describes the thought processes used in thinking about a problem, and then explains why those factors or procedures work together. A coordinated system is identified and explained, but is not yet taken as an object of reflection to be evaluated in terms of its overall adequacy, as in the next level.

Evaluation of Processes (EVP)

At this level executive control is used to judge the adequacy of their thinking processes in relation to the task at hand and related tasks and alternative strategies. In other words, the person takes an outside perspective on their cognitive system and evaluates its adequacy. There is an attempt to be exhaustive or comprehensive in the search for strategies—to search to include all relevant factors and processes to obtain the most satisfactory outcome to a problem. People look at the processes identified and explained, and compare and contrast alternative strategies and alternatives that they could have taken within their system of thinking, and determine the way that various components work together in defining the system or theory that they use. For example, when asked, "How would you know that you made the right choice in this case?" someone exhibiting the evaluative processes answered, "I don't know how you would know. I think that if you had an internal logic going and you didn't open up that logic to the scrutiny of others, I don't think you *can* know. ... you have self-reflection, but ultimately you are caught up in your own circle of things that you can't truly step outside of that to recognize your errors. I think that's very important, that process, and I think it's dynamic in that you change by your influence in looking."

In this quote, the person takes an outside perspective on her thinking process, is aware of some kind of system that she uses, and critiques that system as to its adequacy as well as indicating how the system might be changed.

In the following case analyses, each of these the expression of metacognition—its forms and variance, and the relation of metacognition to moral reasoning is presented. These case analyses also examine the range of levels of metacognition within each interview and the predominant level used across the tasks contained in the interview. Each of the following cases is discussed individually for each of the three parts of the interview, then compared and contrasted in terms of moral metacognition. This analysis demonstrates the changes that occur with development in age and educational experience.

CASE ANALYSES

Case 1: Meg, High School Junior

Meg's Metacognitive Interview (MCI) evidenced two of the levels of metacognition described earlier: Self-Reflective Monitoring and Elementary Reflection. She predominantly used the metacognitive level of Self-Reflective Monitoring, which was also the highest level she used in the interview. Her moral judgment stage was the transitional stage 3/4, which is common for her age group, and her weighted average score was 330. Stage 3 moral reasoning predominated Meg's Moral Judgment Interview responses, yet approximately 30% of her reasoning showed evidence of stage 4.

Teach Me Task. Meg evidenced a sense of self-discovery in her responses to the problem where she was asked to teach the interviewer the ways that she should go about reaching a solution to a hypothetical dilemma, namely, the Heinz dilemma from Colby and Kohlberg's (1987) Moral Judgment Interview. Meg began her response to the problem by saying:

> Okay. Oh boy. All right, well, I guess you just have to consider both sides of the situation. It'd be like you'd have to, I mean, you'd have to see like would, I mean, I guess it depends on who you're thinking of more. I mean, if you're thinking of about your wife, and stuff like that, 'cause I personally, I would probably just like go and break in. I just would like—and I'd deal with it later, I guess. 'Cause I mean, it's not right. I mean, the guy—I mean, the woman obviously needs the drug anyway and he tried everything possible; he was going to pay it back; it wasn't like he just went and just like broke into the store and so ...

Meg began her discourse by naming a method or procedure to begin thinking about the problem—"consider both sides of the situation." That statement indicated that some self-reflection on a cognitive process was used in resolving the dilemma, and further, Meg recognized that her solution depended on "who you're thinking of more"—another indication of self-reflective awareness. Meg then did not follow through with a discussion of the cognitive processes that she would "teach" to another person as if she was teaching them how to resolve this dilemma. Rather, Meg continued to discuss the situation itself rather than the thought processes that she might use in coming to a resolution. In other words, Meg discussed the content considerations involved in her decision making, but did not identify and explain to the learner the *way* to think about making sense of these considerations to form a conclusion to the problem. To push her thinking further, Meg was asked directly to specify strategies or methods of thinking about this problem:

> ARE THERE ANY STRATEGIES OR METHODS THAT I SHOULD USE IN SOLVING...?
>
> Ending a problem? I don't know. That sounds sneaky. You have to...I don't know. You just have to like, you know, I guess, like your pros and cons, what's going to happen, and then I guess, you know, everybody—they just have to make their own decision on like their morals. But these people who can like, "No. It's against the law; you just can't do it," but it's like, I guess it's the way you're brought up; I mean, if you love somebody, I mean, you just—I mean sometimes it just like overrides maybe common sense or what's like right, I mean, by society.

As seen in the preceding, Meg returned to her original formulation in resolving this problem—looking at the pros and cons and people involved—but her response indicated that she was uncertain of her own thinking; she faltered in her speech—a representation of the confusion with the question or of the incoherence of her understanding of her own logical system of resolving these kinds of dilemmas or in communicating that system in this particular task. To resolve this, she moved toward something more simple and familiar to her—attributing decisions to "the way you're brought up."

At the end of this task, Meg, like other participants, was asked to summarize the steps that she would tell another to use in resolving this dilemma:

> All right. I guess you just—you start with a problem; I guess you take into consideration everybody who's involved; try to make the most—I mean, what would maybe be best for not only yourself but what would, you know, for everybody who's affected—which is so difficult 'cause everyone—someone always gets messed up along the line. And then just—I guess, just do it.

Again, Meg's response began with a self-reflective awareness (SR) of her cognitive process as far as she knows them. Meg provided a very brief sense of reflective awareness of the implicit workings of her own mind (e.g., "Start with the problem, look to see what's best"). However, her processes and strategies seemed hidden from her. She reverted to what is characteristic of the elementary reflective level of metacognition—"You just do it."

Hypothetical Dilemma Metacognitive Questions. In thinking back over the way that she went about resolving the hypothetical dilemmas presented in the Moral Judgment Interview, Meg was aware of her thinking to the same extent as in the Teach Me task. She had a degree of self-reflective awareness of her thinking processes, but much of her thinking was hidden from her. She did not yet begin to identify the processes that might be involved in a coherent and explicit (or knowable) system for thinking about resolving moral problems. Her thinking about her own thinking was in the realm of speculation for her (indicated by many "I guess" statements and her disjointed speech patterns). Her responses are thus classified as Self-Reflective Monitoring:

> How did I solve them? All right. You know, I got the problem and then I guess I looked back on my morals and my feelings came into play then. I guess I put into the situation and then I maybe put myself into the other person's situation. I guess I went more with my own feelings and stuff, you know, with like Heinz or something like that. I guess I thought about the consequences of what would happen, you know, if I went this way or

if I went that way and decided which one I would have felt best with or which one my conscience would have let me feel better with. And then I just, you know, I guess I just like kind of did it, and then, you know, afterwards, I guess you just have to like accept it, you know—what happened and just deal with it.

Meg indicated self-reflective awareness of her thinking ("put myself into the other person's situation; went more with my feelings and stuff"), but her response also evidenced remnants of Elementary Reflection while responding to the dilemmas ("I just like kind of did it"). She validated the idea that she was not quite yet consolidated at the level of Self-Reflective Monitoring by stating:

WHILE YOU WERE RESOLVING [THE DILEMMAS], WERE YOU AWARE OF SOME KIND OF APPROACH THAT YOU WERE USING AT THE TIME, OR IS IT ONLY NOW THAT YOU'RE THINKING ABOUT IT ... ?

I guess now that I'm thinking about it. When I'm like trying to solve something, I don't like write it down on paper, but if I go back and think about it, it usually follows the same pattern. So whether you realize it or not, you usually follow the same thing.

Real-Life Dilemma. When discussing a moral dilemma that she actually faced, Meg recounted a recent situation of which she was currently feeling the effects. She related a situation where two of her friends did not like a third one. After describing the situation, Meg framed the conflict in the following way:

WHAT WAS THE MORAL CONFLICT FOR YOU IN THAT SITUATION?

See, I like—it would have been so much easier for me to just drop this girl and go off and be friends with them, but I mean, you know, and it would have been easy. I mean it wouldn't have been that big of a loss, because I have other friends and we never did that much anyway...[But] then last year her and I were close friends and she wasn't really good friends with anyone else, but I was pretty good friends with these two people also. I would've like left her in the dust, you know? I would've felt bad about it. I didn't want to; she was really nice, and it was like a matter of like, you know, if I went that way it would've been easier. All my problems would've been solved, because I wouldn't have to hear it from them anymore, "Oh, why are you hanging out with her?" But at the same time it was like, "Oh, come on," you know? It's like, "I'm friends with her; I can't do this." So I didn't know which way to go.

The moral conflict for Meg was a conflict between doing what she thought and felt best with, or taking the "easy way out." When asked to think about how she came to a resolution to this problem, again Meg focused on the situation itself rather than on a cognitive process, strategy, or decision making system that she could identify—either when she was in the situation itself, or in retrospect.

SO WHAT DID YOU THINK ABOUT WHEN YOU WERE DECIDING WHICH WAY TO GO?

...These—see, these two tried speaking to her because of me, but it just didn't work. You have your friends and I have mine, and we can still be friends at the same time and, you know, we talked about this and it turned out fine ... So it worked out pretty well, you know; I'm friends with both of them.

In the preceding passage, Meg did not refer to her thinking processes in this case, except by implicit reference to them in the context of the action that she took—namely, speaking with her friends. Rather, her actions and reflection on her actions implicitly inform Meg of her thinking process.

Based on Meg's answers to the metacognitive questions in her real-life dilemma, Meg's metacognitive knowledge was scored at the level of Elementary

Reflection on cognitive processes. As such, she did not indicate awareness of her thinking *processes* per se, but rather focused on the content and the consequences of her cognitive ruminations (such as talking about it, thinking a lot, etc.).

Even when asked about whether she made the best decision in her judgment, Meg sustained her stance of either not knowing or not choosing to discuss her thinking strategies, other than indicating that she "thought a lot" about the problem:

> DO YOU FEEL LIKE THE DECISION YOU MADE WAS REALLY THE BEST ONE?
>
> Yeah. 'Cause it took—that one took so much more time. It took so long. I mean it could've been resolved in, I'm sure, like a week, you know, and we would have never spoken again. But it took a lot longer, but I think it was worth it. I mean it was worth like all the trouble...
>
> DID YOU THINK ABOUT THAT WHILE YOU WERE DOING IT? LIKE WHAT KINDS OF THINGS THAT YOU WERE THINKING ABOUT AND HOW YOU WOULD WEIGH THOSE OUT?
>
> Yeah, I was. I was kind of—'cause this took place over a long time, and a lot of times you just have to sit down and think about it, you know? So yeah, you usually wind up rationalizing also. Yeah, I did. I put a lot of thought into it. I mean now that I look back on it, it was so clear-cut, it's like how could you just drop a friend so that you—just 'cause of someone else? It's like, you know, it's — It's such a pain and you're like, "Oh God," 'cause, you know, a lot of times I just can't be bothered, you know?...This went on and on; it went on throughout the summer. It was ugh! It was horrible.

Meg's responses were interpreted as Elementary Reflection. Even though Meg had a strong sense that this was a long and difficult decision to make, she did not articulate self-awareness of her cognitive capabilities. The statement Meg made toward the end of the interview highlighted her Elementary Reflection on her own thinking processes:

> HOW DID YOU FINALLY KNOW WHEN YOU'D REACHED THE DECISION?
>
> Oh, we had big arguments, you know? 'Cause we—like the 4 of us used to like hang out together, and we would just like sit each other down and just scream at each other, you know? And we'd just be—and we'd like work it out; we'd tell everything, you know. We'd just tell each other everything, you know, the way we'd feel, and then I'd be like in between them, you know? Or I'd be like, you know, I'd talk to them one day and I'd be like, "This is how I feel," and then I'd talk to like the other girl the next time. And after a while it just like finally—it just finally cleared up.

Meg was acutely aware of the outward manifestations of her thoughts and ideas about the situation, but her mastery over the decision and resolution process was not evident. The situation, she said, "just finally cleared up." In essence, Meg needed her thought processes to be "out there" in the world before she could cognitively operate on them. Thus, little self-reflective monitoring of her cognitive processes were evidenced.

In reviewing Meg's metacognitive levels as represented by the overall interview, her statements demonstrated early levels of metacognition. Meg began to self-reflect on her thinking processes, and had a strong tendency toward being elementarily reflective of her thinking processes. She articulated her thinking about problems in content terms, and indicated a vague awareness of some aspects of a cognitive system operating in her mind. For example, she was aware of examining both sides of a situation and looking at the pros and cons of solutions, but then states that she simply "decides" on a solution. Further, Meg viewed the notion of having a strategy or method as "sneaky"—indicating that her idea of "rationalizing" means something other than an awareness of thought processes or patterns of thinking. Meg did not continue to use the Self-Reflective level of metacognition when she discussed her own moral dilemma, but rather was only elementarily reflective on her cognitive processes.

Case 2: Mark, Undergraduate Senior

Mark's Metacognitive Interview exhibited the types of transformations evidenced by others during the interview. Mark's interview lasted longer than most—approximately 3 hours. Mark was a very willing interviewee, and demonstrated enthusiasm for reflecting on his thinking process in comments of self-discovery such as, "Wow—I have never thought of this before" or "I never really thought of that until I was saying it ..." These kinds of comments on self-discovery may indicate that, among other things, Mark was going through a period of transition in his metacognitive awareness; that he was beginning to think about his thinking in a way that he had not done before; or, that he is the type of person who benefits from this type of activity (an interview or discussion) to help him clarify for himself his own thinking about his thinking. A corollary to this emerging transition toward increased metacognitive awareness may be that Mark is in transition between two stages of moral reasoning. In examining his interview for both metacognition and moral reasoning stage, Mark was found to exhibit this sort of transition. His predominant metacognitive level was considered to be Self-Reflective of Processes, with a highest level of Identification of Processes, and his responses indicated vacillation between each of those levels. Further, in scoring the MJI for moral stage, Mark's reasoning exhibited a transition between stages 3 and 4. His weighted average score was 375, and his Global MJI score was at the stage 3/4 with approximately 25% of his reasoning still at stage 3.

Teach Me Task. In this problem, Mark began his response of how to teach another how to resolve the hypothetical dilemma by self-monitoring his thought processes, taking control of the thoughts going on in his mind, and establishing a procedure or a strategy that one must use to resolve this dilemma:

IF YOU WERE GOING TO TEACH ME HOW TO SOLVE THE HEINZ DILEMMA, WHAT WOULD YOU SAY?

Ok. It's easy to think about what I would do, but it's tough to think about how to approach it...Um, I guess he has to start by weighing the pros and cons of each side, what are the benefits of stealing the drug and what are the potential risks, and who is he affecting both ways. Um, by stealing the drug he is definitely affecting his wife, he's affecting himself, he's taking a risk for himself and he's partially affecting the druggist too. So you have to think about who's at a greater risk or loss, and what do the losses mean to him. Which ones are more important and less important, what can he deal with later down the road.

...HOW WOULD I GO ABOUT DECIDING WHICH TO CHOOSE IN THAT WAY?

(long pause) um, well I guess because you can't really add it up and say this person is gaining more benefit or less benefit or the benefits and costs aren't additive because he has to look at what is important to him and the people closest to him. So I guess he has to start with I think, what are the benefits that are closest to him, and how important are they. So those are the ones he should start with. I don't know if that really answers the question.

In the preceding quote, Mark indicated that there was a cognitive process that needed to be carried through in making this decision—weighing the pros and cons, looking at the risks and benefits, looking at who is affected and how. Yet Mark did not explicitly state, even when probed, how those things should be decided. He offered no **explanation** of the components of the system, although he clearly indicated that a strategy exists and can be **identified** in terms of resolving this problem. Further, Mark evidenced reflection on his thinking processes, as seen in his self-reflective statement of "I don't know if that really answers the question." He felt a sense of inadequacy of his answer, but was not sure about what further information would be needed so that someone else could be taught his method of cognitive processing. Thus, this response was coded at the Identification of Processes level of metacognition. His metacognition was beyond Self-Reflective Monitoring, but did not extend to Explanation of Processes.

When asked to summarize the steps that someone could use in thinking about this problem, Mark identified but did not explain his cognitive processes:

IF YOU COULD SUMMARIZE THE STEPS I WOULD NEED TO DO, WHAT WOULD YOU SAY?

First is to look at the problem and look at the possible solutions that you have. Um, then from the standpoint of the solutions to see how it's going to affect you, what are the possible benefits, what are the risks, what are the consequences. So examine it for yourself, then examine it for the other people that it is affecting who are close to you, then look outside that to other people. And then juggle them all.

Mark's suggested process for resolving the hypothetical dilemma was to "juggle them all," and then come up with a solution. This response is indicative of a strategy, but not a coherent process or system of cognitive operations that is explicit in his own mind—or, at least not explicit enough to describe to another. However, Mark was clear on what the important considerations are that should be used in resolving dilemmas, as well as what his implicit method for resolving such dilemmas would entail. Thus, this represented the Self-Reflective Monitoring level of metacognition, as he did not identify a process.

Hypothetical Dilemma Metacognitive Task.   In this task, Mark's responses remained consistent with his metacognitive level expressed thus far. Specifically, his thinking vacillated between the Self-Reflective Monitoring and the Identification of Processes level. For example, when first discussing his thinking processes in regard to the task, Mark replied:

WHEN YOU THINK ABOUT THE DILEMMAS, HOW DID YOU KNOW HOW TO SOLVE THEM?

Well, I just considered what my past experience has been in similar situations or, yea, I just thought of what would have been right for me in that situation, what factors I thought of and why it was important to act in a certain way, I'm asking the same question that you are asking I guess … You're asking what I did to do that. Yeah, I thought of different solutions and somehow just felt better than others. And then I explored what solutions that I thought seemed more appropriate, I thought about what the good points and bad points were of them and how they affected me and what the risks were in taking certain actions, and whether the risks mattered as much as the benefits.

In the preceding quote, Mark demonstrated consistency with the content of his remarks in the first task, and reflected on what he is doing within his own mind—"asking the same question that you are" and then indicated what else went on in his mind—asked why things are important, explored solutions, risks, benefits, etc. He discussed what he was self-reflectively aware of, rather than a coherent system or method that he followed to form a conclusion. Later he indicated self-awareness of his own lack of an explicit strategy:

WERE YOU AWARE OF THE STRATEGY OR SOME APPROACH THAT YOU WERE USING....

I wasn't I guess, until I thought about it when I guess, an answer came to me immediately, this is what I would do, and then I thought about it afterwards. So, I thought backwards in some way. I went with what, well, yea, I guess I thought of what I would do and then looked at it in comparison to other solutions. And tried to back it up.

Even at the end of this particular task, Mark did not discuss his thinking in terms of a process that he used, other than stating that he "thought about it in some way," even though he identified a process in the first part of the interview.

WHEN YOU THINK ABOUT THE DIFFERENT DILEMMAS, DID YOU USE A DIFFERENT APPROACH OR STRATEGY FOR EACH OF THE STORIES?

It seems, thinking back, it seems like it was similar... I'm thinking they were very different situations. So the factors were just really different and the person who's, Heinz is very different from Joe too, and so the perspective I'm looking from is very different too, but it seems like the considerations were, or the way I *looked* at the considerations were very much the same, even though the considerations were different. Vastly different.

OK. HOW DID YOU KNOW TO USE THE SAME APPROACH?

It kind of just happened. And I didn't know what else... I guess I never really think about how I make those decisions unless someone is you know, sit down and look at the alternatives on each side. I guess that would be an alternative strategy—sit down with someone and talking it out.

DO YOU EVER DO THAT?

Yea. Yea! I guess I do. I rely a lot on other people's opinions and judgments, experience....and then take that in.... I guess that is another consideration. I guess not, I guess it isn't a different strategy, because it is all part of the consideration in the process that I am doing. Wow!

In the preceding quote, Mark indicated that he was not explicitly aware of a strategy that he used in resolving the hypothetical dilemmas of the MJI, but that after discussing his thinking further with the interviewer, he became more explicitly aware that his *approach* to the various dilemmas is basically the same even if the content differs. Mark did not, however, explicitly state what that approach was. Thus, Mark's metacognitive level could be classified as somewhere between Self-Reflective Monitoring, that is, monitoring an approach to resolving hypothetical situations, and the beginning of identifying some of the component parts of that approach.

Another interesting aspect of this quote was his sense of surprise and discovery at realizing that he actually engaged in a way of thinking; a cognitive process or procedure of which he was heretofore unaware. He was aware that he used some way of thinking that was not knowable to him, and then brought in the notion that there was not another strategy, but "It is all part of the consideration in the process that I am doing. Wow!" This shows that his thinking was not quite at the level of Explanation of Processes, because he was unable to explain the component parts of his thinking processes in this context, and was also just barely on the fringes of the Identification of Processes level, as he was still discovering that he used a consistent strategy for resolving different situations. Thus, this passage reflects the Self-Reflective Monitoring level of metacognition, with a beginning movement toward identifying the processes in his own system.

Real-Life Dilemma. In discussing his metacognitive awareness of his thinking in an actual situation in his own life, Mark evidenced a perspective on his thinking processes that seemed less metacognitive than reflective. He demonstrated the ability to talk about his thoughts and indicated some awareness of thinking processes, but to a lesser degree than in the earlier parts of the interview.

The situation Mark described involved a moral conflict in relation to his role as a student teacher during his college years. He stated his conflict as follows:

> I teach high school kids and I'm gay and I've really had a moral problem in deciding whether to come out to my students. I guess that is a moral problem because that is something that people can see as good or bad, or something that would affect people and things like that. So that is something I thought about a lot.
>
> WHAT IS THE MORAL PROBLEM IN THAT FOR YOU?
>
> The moral problem is that, is it going to positively or negatively affect students. You know, how by my actions am I going to affect students.

In his considerations of how to go about resolving this conflict, Mark discussed the idea of being a role model for students and questioned the appropriateness of involving his personal life with his role as a teacher. At first his answers to the questions probing his metacognitive awareness in resolving this situation focused on what he did and what he needed to do to explore the possible outcomes:

> HOW DID YOU GO ABOUT WEIGHING ALL THOSE CONSIDERATIONS OR ORDERING THEM?
>
> Um, I thought, I guess I thought about it from the standpoint of the solutions. And thought about these are the possible options and solutions/options, and this is how it would turn out on both sides. I talked with other people and got their input into it too. Um, because it's hard to say, it was really difficult thinking that one solution, one consideration was really more important than another because they all seemed very important.
>
> HOW DID YOU GO ABOUT RESOLVING THIS?
>
> Um ....I tried to find a way that would satisfy myself and satisfy my students in my mind that would benefit them. Because it seemed like I started to think about the problem because I thought it would benefit kids, and that was the whole reason behind it. I wouldn't have done it if I didn't think that it would benefit kids. And then when I started to think about reasons of how it wouldn't benefit them, possibly wouldn't benefit them, I started exploring other options of approaching the situation that they are going to benefit from and feel comfortable with, and it won't go beyond the role of teacher.

Mark discussed the solutions to the dilemma, what would constitute a "satisfactory" solution both for his students and for himself, and tried to define the parameters that the solution had to meet—"the students will benefit, and it won't go beyond the role of teacher." In addition, because Mark did not know how to resolve this conflict, since the solution was not apparent to him from the outset, Mark notes that "I started exploring other options of *approaching* [emphasis added] the situation." This indicates that the novelty of the situation stimulated self-reflective awareness, but also obscured for him the implicit cognitive strategy that he espoused throughout the interview that he should weigh the risks and benefits for all involved. He intimated that those were important considerations in this dilemma, but because the outcome was not as clear as in the other dilemmas Mark became aware that he had to construct a new approach to thinking about this dilemma.

Mark made the decision to not share the information of his sexual orientation with his students. When asked, "How did you know what to do?," Mark responded:

> In the end I did what I thought was benefiting myself and benefiting my students. And was probably having very little negative effects, and it felt right for me to act as I did...Yeah, I guess the solution is more involved than just not coming out. It's much more looking at my role as a teacher, it's saying that just by being who I am and helping kids in their own way of thinking that I am letting them be more open to different ideas, and to not let kids get by, whether it's talking about black people or women, to not get by with making inappropriate remarks and things like that, and to make kids think about what they are saying, and I guess think about their own moral decisions.

In this passage Mark made his decision by considering the risks and the benefits of the people involved in the situation. Further, Mark concluded that the *process* of making the decision and the kinds of factors to be considered in the decision were more important to him than the action taken in the decision—specifically in his case, the decision to *not* tell his students about his sexual orientation. He thus resolved the situation by reconsidering the way that he constructed the problem. Instead of being a gay role model for students, he reframed the situation so that he could feel good about himself as a role model in general. In the context of the interview, however, Mark did not demonstrate metacognitive awareness of the processes or strategies he used to bring him to the point of cognitively reframing the dilemma. When asked to specify what he thought was going on in his thought process, and whether he used the same "backwards" approach that he used earlier in the interview, Mark replied:

> Yeah, well I guess it was kind of mixed because I asked some of the questions before deciding and then afterward. I first started thinking well what are the benefits and risks, and I didn't know what I was going to do before I made the decision. I don't think. ...I did a lot of thinking about it afterwards, too. I came to a decision, I thought about the two options and said that the risks...of telling my students that I am gay was just really great. I work with a lot of hispanic males, and it seems like they are the most homophobic group that I have had to work with. And that just would have set a really bad tone. So in a lot of ways I thought about, that I don't think that it would be effective to do it. And then I looked at it in terms of being a teacher and things like that, and it seemed to make a lot of sense that I wasn't going to. And I had the input of a lot of friends and colleagues.

This quote contains elements of both Elementary Reflection and Self-Reflective Monitoring levels of metacognition. It is coded Elementary Reflection in that Mark discussed the content considerations used in making the decision, told what was on his mind, but was self-reflective about how he thought about his thinking in retrospect. In that self-reflection, however, the processes or procedures that he used in his thinking, that is, his decision strategy, was not explicit.

In this quote, Mark's cognitive processes remained implicit in his thinking about his thoughts, and he tended to report or describe in a Elementary Reflection way the workings of his mind, even though the task of discussing a situation in one's own life requires reflection by it's very nature. In all, Mark's responses to the interview indicate Elementary Reflection awareness *at the time* of the situation, but in the interview itself, he self-reflects on his thinking in that situation. Mark's own report of his thinking at the time of the dilemma confirmed this:

> WERE YOU AWARE OF WHAT YOU WERE DOING AT THE TIME, OR IS IT JUST IN TALKING ABOUT IT NOW THAT YOU ARE ABLE TO RECONSTRUCT YOUR THINKING?
>
> I guess I was aware of it, more so than most decisions I make, because it seemed like a really important one. So I was pretty aware of it. I wasn't aware that I had made the final decision for a little while afterwards, you know, that I guess, when I had finally said to myself that this is what I am going to do or not going to do, then I started thinking about it and said you really knew that a little while ago.

Further, although Mark reported that he did *not* think about his thinking processes at the time of the dilemma, during the interview he noted that he was experiencing a level of reflection of which he was not previously aware:

> HAS THIS INTERVIEW AFFECTED YOUR THINKING IN ANY WAY?
>
> Yea, it's made me think more about how I think, actually, because a little while back there was something that I realized about the way that I solve problems... I can't remember what it is now. It's made me realize I didn't realize, like I was saying, about having really known when I made the decision. And that's all coming back, too, and I can see it from a little bit of perspective.

The experience of being involved in making a significant moral decision, as seen in this analysis of Mark's decision process, is not sufficient to stimulate thinking about one's cognitive strategies in resolving a conflict when in the situation itself. Rather, reflecting on that situation in retrospect, either through an interview situation such as this or through reflection in the "everyday" sense of the word, people realize that they use strategies in making the choices, and begin to identify what those strategies might be, and possibly create a system of thinking about their thinking. It is at this point that postformal operational thinking is seen. People develop systems of thought, then begin to coordinate multiple systems, compare them to one another, and evaluate them. Mark's case demonstrated the beginnings of that process. He commented that he was not explicitly aware of making a choice until it became apparent to him through some external situation. But once this was apparent, his system started becoming clear to him. Thus, Elementary Reflection is an important element in later Self-Reflective Monitoring such that without being able to at least describe the contents of one's own mind, a reflection on those contents may not occur. And without such reflection, postformal coordination of systems is not possible—even in the moral domain.

To summarize, Mark used a number of forms of metacognition within the interview. His level of metacognitive awareness depended on the task. Despite the range within tasks and the slight variation between tasks, Mark's overall metacognitive thinking is characterized predominantly by Self-Reflective Monitoring, yet some transition to Identification of Processes seems to be taking place, just as his moral stage is between stages 3 and 4.

Case 3: Rob, Doctoral Student

Rob's interview was characterized primarily by metacognitive procedures labeled Evaluation of Processes. His predominant level on the first part of the interview was Explanation of Processes, while on the other two tasks the predominant level was Evaluation of Processes. Rob's Moral Judgment Interview score was stage 5, with a Weighted Average score of 483.

In the interview Rob evidenced an extraordinarily high level of moral judgment stage and a sophisticated form of metacognition. His moral judgment score indicated that Rob was a postconventional thinker and as such, had a principled view of morality and a view of the interrelationships of people and systems.

Teach Me Task. In his response to the first task, Rob presented his moral perspective by integrating theories he learned through his educational study with his own thoughts about teaching a decision process to another person. Specifically, he uses Kant's theory as a strategy to apply to a given situation, and to inform the nature of the person's value hierarchy:

IF YOU WERE GOING TO TEACH ME TO SOLVE THIS, WHAT WOULD YOU SAY?

You are starting off with a hard one. I would ask you to consider two things. I would ask you to think about, to weigh some of the different values in the situation. By that I mean to think about the importance of upholding the law, the importance of life—the importance of the value of life, I would also ask you to think about precedent. In other words, the impact on others of Heinz breaking the law. In other words, if Heinz breaks the law, is he exonerating others who break the law. And what happens to society if people in effect take the law into their own hands—what is the *meaning* of that. I guess I think those are the factors you need to consider. And in doing that I guess I would think in more detail about what that means. If Heinz does break the law in this case, is it more important to uphold the law in this case than to save his wife's life, if by saving his wife's life, is it likely that there are others in his situation that could respond … is it inviting a social breakdown of some kind. I really *do* think about this in terms of one of Kant's Categorical Imperatives, the idea of acting according to that law which at the same time you could will to be a universal law. And could you wish this to be a universal law, and under what *terms* could you wish it to be a universal law. That's pretty much it.

OK, SO HOW WOULD I GO ABOUT WEIGHING ALL THIS? I CONSIDERED ALL THOSE FACTORS BUT HOW WOULD I GO ABOUT DECIDING…

Which is the right one? This is getting in to why *I* think it's the right one. I would think about … in terms of the law, I would think about the law and what it is intended to represent and what values its intended to represent. And I would think about whether or not … basically I think the law is a kind of convention that is important to appeal to the kind of values that the law is trying to represent and to see if it is representing them well or not well. So I would not be thinking of breaking the law on its face value. I would think of the law in terms of its underlying values. And in the course of thinking that, I would recognize that one of the things the law is intending to preserve is life. Or I think that's right. So I would give that some thought in weighing your decision of whether to break the law or not. And I would also give some thought as to whether breaking the law really does or could lead to the dissolution of an important institution or could create chaos of some kind. And to think about the degree that this is an exceptional case. I would also give some thought to thinking about … I was going to say about Heinz's relationship with his wife, but I don't think that that's the issue here. I think Heinz would be justified in doing this for a stranger or a friend, for example, in this particular case, I think that's right. Do you want more than that? Am I still too vague?

The extensive narrative above indicates that Rob was aware of employing a strategy when thinking about the way that he would resolve this dilemma—and the way that he would teach another person how to go about thinking about the dilemma. He articulated a basic process to follow: we should consider two things, weigh the values (and he explained what he *meant* by weighing the values), and examine how these processes apply to Heinz's situation. In addition, Rob examined his own cognitive processes by saying, "I guess I think those are the factors you need to consider..." and "Or, I think that's right." In this example, however, Rob did not take an outside perspective on his own thinking and evaluating his *system* of thinking and the *way* that he approaches the problem. Rather, Rob evaluated the accuracy of his identification and explanation of his cognitive processes. By answering the question of "Teach me how to go about resolving this dilemma" in this way, Rob demonstrated that he had access to a process that he uses in resolving this dilemma, articulated the main points in that process, and could explain how the process works in general and in relation to this problem in particular. Therefore, Rob exhibited metacognition in the form of Explanation of Processes. Part of this passage demonstrated that Rob evaluated his **response** to the task (e.g., "I think that is right" and "Am I still too vague"), but because he did not evaluate his general approach or **system** of thinking, it was not scored at Evaluation of Processes for this response.

Rob's narrative demonstrated an ability to identify, describe, and explain a second cognitive system that may be used to think about the dilemma:

CAN YOU THINK OF ANY OTHER WAY TO GO ABOUT MAKING THAT DECISION?

Yeah! I was thinking about this... it had to do with... I had another impulse to go this way—it was more—we were talking earlier about who was hurt or helped more in this dilemma. Does it hurt the druggist more to lose the price of this drug or does it hurt Heinz's wife more to not get the drug. I think going about the dilemma in that way, you come to the conclusion that Heinz's wife is clearly going to be hurt far more than the druggist is. And I think it is important not to hurt people. And I kind of like that way of going about the dilemma. Again I think you also have to think about it in terms of a precedent... So its not simply a matter of weighing the harm to the druggist versus the harm to the wife, but it's a matter of weighing the kinds of reverberations this kind of thing has.

SO WHICH STRATEGY DO YOU THINK IS BETTER OR BEST OUT OF THOSE TWO?

I think that... I realize that I'm more comfortable with... I think that both are kind of doing the same things, but I realize that I'm more comfortable with the second than the first... It seems like a simpler way of thinking about the problem for me.

As with the last response, Rob did not indicate metacognitive evaluation of his thinking processes. He realized that he was "more comfortable" with one process than another, yet did not *evaluate* the two ways of thinking.

Moral Judgment Metacognitive Task. Rob's score on the second metacognitive task was predominantly Evaluation of Processes, with both Explanation and Evaluation of Processes represented. He responded to the question, "Thinking back on the dilemmas that we just discussed, how did you know how to answer them?" by stating clearly that he could discuss the process that he used in answering them (Explanation), but he began to question the surety of whether he "knew" that he answered a question or addressed a problem:

IN THINKING BACK OVER THE DILEMMAS THAT WE TALKED ABOUT, HOW DID
YOU KNOW HOW TO ANSWER THEM?

I think that I didn't *know* how to answer them, I wouldn't put it that way. I can tell you
the *process* I went through in answering them, because there isn't really a set way that I
know how to do it. I'm aware of having a gut feeling about all of this, which always sur-
prises me, I am also aware there are a couple I did *not* have a gut feeling about, but
mostly I have a feeling about this is right, and I have to marshall my *reasoning* for why
this is right. And maybe because I am familiar with some of the questions that I have a
feeling of kind of reflexive response to it. And it's a process of clarifying or summoning
up reasons for why my position is right. And then occasionally, and I think this happened in one case when I was talking about the judge, and the judge's decision, I realize
that I summon up a counter-argument that may conflict with the gut feeling, and then I
have to adjust how I feel about this. So, there is some adjusting that is going on as I'm
talking about it too. And in summoning up arguments, I am also summoning up counter
arguments and imagining what the counter arguments are going to be.

This passage indicated several aspects of metacognitive thinking that is
beyond Self-Reflective Monitoring and Identification of Processes. In this
response, Rob demonstrated that he is reflective; aware of how he was thinking or
feeling about the situation—"I did not have a gut feeling about this"—and further
indicated that he has a "feeling of kind of reflexive response" to the dilemmas due
to experience with thinking about these issues previously. Rob reflected (Self-
Reflective Monitoring) on his thinking in this task and came to the realization that
he engaged in the mental process of developing counter-arguments (Identification
of Processes), explained how that fits into his usual system (Explanation of
Processes), and then indicated that, during the interview, he was beginning to
evaluate how the process fits into his current belief system by stating that "There is
some adjusting that is going on as I'm talking about it," but an actual evaluation
was not yet present. Later in the task Rob demonstrated such evaluation by stating:

...I think it is to look fairly contextually at the situation, and I think I need to push
myself sometimes to also remember that these decisions have the larger meanings that I
have been talking about and to think of them in terms of their larger meanings as well.

DID YOU CONSIDER ALTERNATIVE STRATEGIES IN SOLVING THE DILEMMAS?

At one time I did—in the Heinz dilemma. I basically, um, I basically feel like I didn't.
I sometimes get into a strict kind of utilitarian type of thing of who's right and I try to
quantify who helped and who's hurt and I sometimes think that way about moral problems. And I found myself thinking that way this time, but when I say who's helped and
who's hurt, I'm really not talking about it in a utilitarian type way, but I find myself doing
that sometimes and finding out that it really isn't faithful to the complexity of the
phenomenon, the complexity of the problem. And it's not very... it's more misleading
than guiding.

Rob indicated that he considered two factors in resolving the dilemma, and
also evaluated his thinking process. He stated, "I need to push myself sometimes
to also remember that these decisions have the larger meanings that I have been
talking about" and further evaluates the pertinence of applying his system of
thinking in certain situations: "I find myself... finding out that it really isn't faithful to the complexity of the phenomenon..." In addition to this evaluation of
his thinking process, Rob manifested a characteristic common to other postconventional level participants in the study—that is, an openness to new information

to add to either his decision or resolution of the dilemmas, or to the process or method of arriving at conclusions. For example:

HOW DO YOU KNOW WHEN A DILEMMA IS RESOLVED?

You know I don't think of, a lot of the dilemmas for me I don't think it was resolved, I think of it more as a process than as a final arriving or an ending...

SO WHEN YOU COME TO A CONCLUSION SUCH AS "HEINZ SHOULD STEAL THE DRUG," YOU DON'T THINK OF THAT AS THE END OF IT?

Well, I think of it as being, well I guess in this case in this context I think it was the end of it, but I think of a lot of solutions as being provisional in the sense that they are open to new information....Um, I think that's the answer. And I don't, I think in terms of whether I've come to the *best* solution.

The preceding passage indicated that Rob was open to new ways of thinking, that his final arriving at a solution was not the end—but rather the "process" was the important factor. He looked at solutions to the problem as "provisional"—that there were things that would change the solutions, and those things might be process-related and not simply content or factor related. The "new information" was taken in the context of the interview, referring to both a procedure gleaned from discussing the situation with others, or from added information about the dilemma itself.

Real-Life Dilemma.   When Rob was asked to discuss a dilemma that he faced that influenced his thinking, Rob relayed a situation with a lifelong friend with whom he was trying to decide whether he should share some information that Rob's friend John should know about but might hurt him. The conflict was whether Rob should tell John that John's former girlfriend was marrying another man. The situation was complicated by the fact that Rob thought that this information might be detrimental to John's newly emerging social relationship, since the breakup with his former girlfriend had a deleterious impact on John and his social life. In thinking about this problem, Rob reflected on his options:

HOW DO YOU FEEL ABOUT MAKING THAT CHOICE?

I felt like there were ways in which I was really kind of blind in the beginning of this decision. I was blind by feeling very paternalistic toward him. Initially I wasn't going to tell him, um, and I felt very protective of him and that he couldn't handle the knowledge, and I *do* feel very protective of him, paternally toward him, and he can relate to me paternally too...[and then] I realized that I wasn't really crediting him with enough...and my paternalistic feelings about him were kind of getting in the way of who he is and what he needs.

...I feel like there are a lot of really strong feelings about him that are really dominating my thinking processes a lot, crowding out other ways in which I go about making these decisions. And I think that when I finally did talk with [a friend] about it and began to think about it more...my thinking about it was just much clearer and I was doing more of the kind of things that we've been doing—which was weighing how much was this going to help and hurt him, weighing what this is going to mean in the long term with our friendship, whether this is something that I could will to be universal law—that people would do—both kinds of considerations would come in.

These passages indicate that Rob was aware of his thinking processes and reflected on the ways that he thought about the situation. In addition, Rob separated himself and his thought system from the thinking in the situation itself and evaluated his thinking processes. He was aware that his "paternalistic feelings" were getting in the way of making a decision, and that his thinking became clearer as he discussed it with others. He evaluated this strategy as "helpful," and that it enabled him to think about this situation in a way similar to how he would resolve other moral dilemmas. Thus, these passages were scored as Evaluation of Processes.

Another statement illustrating Rob's metacognitive evaluative processes follows:

> HAS THE INTERVIEW SO FAR AFFECTED YOUR THINKING IN ANY WAY, OTHER THAN WHAT YOU HAVE JUST MENTIONED IN THINKING ABOUT THE SITUATION?
>
> It has. It's made me clearer about a couple things, it's made me realize how confused I am about some things (laughs), ... also you know how you have little things ... I have unresolved pockets in my thinking and ideology about things. It's also, its made me realize that my kind of appealing to abstract values doesn't help me a lot in certain situations, it's not ... there are certain situations in which it is very informing and there are other situations in which it is not, and I really feel like where I do my best moral thinking in a way is in doing what I've just been doing in talking about my friend, which is to really challenge my own biases, to use my feelings in an informing way, to really consolidate my deepest sense of the other people in the situation, what their needs are and how their needs are going to get represented in the situation. A lot of it has to do with an ability to self-inquire, an ability to inquire into what's, you know, into what's most important here. It's pretty much of a process kind of thing.

Rob demonstrated primarily two forms of metacognition in resolving moral problems: Explanation of Processes and Evaluation of Processes. His moral stage evidenced a postconventional perspective—concerned with principles of generality (universality) and with beneficence or kindness toward others, and respect toward others in their own terms. Rob's protocol indicated that there is a relationship between moral judgment and metacognitive level. Rob used his moral judgment principles to test his decision process in the real-life dilemma situation—he questioned whether he was behaving too paternalisticly toward his friend, he questioned whether he was able to appeal to "absolute values" of who is helped and who is hurt, to address his need for respect for persons, what society or social relationships should be like, and so forth—all moral considerations at the postconventional level.

In addition to the moral concerns involved in thinking about moral problems, Rob emphasized *process* as an important element in resolving moral dilemmas. Rob saw "process" as essential to decision making, and saw the answers/resolutions as "provisional"—based on the adequacy of the process. In making his choices in the moral domain, Rob was aware of both the principles he invokes and the applicability of those principles to particular moral situations. Further, he examined the generalized implications of his choices and decisions, and whether the process or the "provisional" conclusion should be changed in order to accommodate new information. This openness to change occurs through willingness to "challenge my own biases" and through "an ability to self-inquire."

SUMMARY AND DISCUSSION

The three cases discussed here illustrated characteristics of forms of metacognition in the moral domain. They demonstrate that metacognition goes beyond the simple monitoring of cognitive process to the ability to articulate a description and explanation of such processes; then evaluate the processes in comparison to other systems or processes. Thus, metacognition may have its beginnings in childhood and early adolescence as researchers in reading comprehension and memory studies indicate, but it encompasses much more than that. Further, these case analyses demonstrate that the metacognitive levels sketched out by Demetriou (1990) in the logical domain are also seen in the moral domain. The contribution of this work is the explication of those metacognitive processes, including identifying and describing an additional level that Demetriou did not articulate, and to apply them to social–moral situations.

To review, Meg, the high school student, evidenced levels of moral reasoning from the Elementary Reflection to Self-Reflective Monitoring levels of awareness of her thinking; Mark, the undergraduate, exemplified elements of thought in the Self-Reflective Monitoring and the Identification of Processes levels; and Rob, the graduate student, demonstrated complex ways of thinking about his thinking in terms of explaining and evaluating the thought processes he used in thinking about dilemmas and problems. In their moral reasoning, Meg and Mark both scored at a conventional moral reasoning level, specifically, stage 3/4 (although Meg's moral stage was more predominantly stage 3 and Mark's was more predominantly stage 4). Meg and Mark scored at the same predominant level of metacognition, Self-Reflective Monitoring, but Meg's highest level used in the interview was Self-Reflective Monitoring, while Mark's highest level was Identification of Processes. Rob scored at the postconventional moral level, stage 5, and his metacognitive predominant and highest levels were Evaluation of Processes. From the empirical analyses conducted in a larger study (Schrader, 1988), there is a significant association between moral reasoning stage and metacognition such that the higher the moral stage, the higher the metacognitive level. This was borne out in these cases.

What accounts for the differences in metacognition between these participants in the way that they think about moral problems? Is it intelligence; experiences or practice; ability to comprehend, recall, self-monitor; age or educational level; moral stage; logical or cognitive development? It is possible that all of these have some impact on metacognition, but this study attempted to address these questions.

Intelligence

All students had equivalent intelligences based on standard achievement or IQ test. All attended selective schools. In terms of social intelligence, all had an interest in and experience with the study of moral problems. Each had taken at least one course in morality.

Experience/Practice

Of course older students have more life experiences than younger ones, and all social (as well as other) knowledge is based on experience. Each of the three cases discussed here, however, had nonsheltered life experiences and each had significant interaction with family or school interventions that discussed and practiced moral dilemma discussions.

Thus, these participants had, in other contexts, thought about the hypothetical dilemmas that were presented to them in this study. Unfortunately few people engage in active metacognitive reflection about their real life situations (Dewey, 1933). This might account for the ability to reason at a more sophisticated level of metacognition on the hypothetical problems than on their real life problems. An additional influence might be the type of schooling situations that all the participants experienced. Competitive high schools and Ivy League colleges teach their students to critically evaluate problems. Demetriou (1990) found that students did significantly better on metacognitive tasks than did adults who were no longer in school. Thus, the practice effect seemed to be of greater import than the fact of having the schooling itself.

Ability to Comprehend, Recall, Self-Monitor

The myriad researchers in metacognition tend to categorize metacognitive awareness as an all-or-nothing concept; either a child can or cannot do it well. In the case of social–moral problems, comprehension and recall are qualitatively different than in reading comprehension. However, in one sense it is taken for granted that when one is presented with or finds oneself in a social/moral situation, there is some understanding of that situation. Understanding in the moral domain has a different meaning than in reading a passage of text. It is more flexible and open to interpretation. Yet, these two components are required for learning from past experience, and therefore, for cognitive restructuring of thoughts. This is an area that warrants further investigation.

Age or Educational Level

One of the most obvious developmental variables is age. In the cases presented here, age seems to be associated with the ability to articulate more complex forms of metacognition. Just as age is positively associated with moral stage and cognitive stage, age is not sufficient to predict accurately moral, cognitive, or metacognitive level. Future studies will have to examine variations in educational level while holding age constant. Yet for these individuals, educational level was associated with metacognition. Education may be associated with metacognition for several reasons. One, as educational level increases, the amount and degree of reflectivity and critical analysis increases. One learns to examine academic issues and problems and one's relationship to those problems in any college curriculum (and in some advanced high school curriculums such as Meg's). Thus, practice with reflectivity in the logical/academic domain may relate to metacognitive thinking in the moral domain. The question remains open, however, if such encouragement is possible, especially with increasing interest in the question of generality across domains. Given that Demetriou (1990) studied metacognition in the logical domain concurrently with this present study of metacognition in the moral domain and both studies articulated similar forms of metacognition, such generality between these two domains is certainly possible.

Moral Stage

Schrader (1988) found a significant correlation between stage and metacognition, yet moral stage is also highly correlated with educational attainment and age. Thus, to determine the role of metacognition in influencing moral stage (or vice versa), one would need to analyze participants who reasoned at the same stage of moral reasoning to see if they differed in metacognitive ability.

In the case analyses presented here, Meg and Mark had similar moral stage scores; both were stage 3/4. The distinction between Meg and Mark is that the content of the protocols differ from each other despite the similar predominant moral stage and metacognitive processes. Mark produced reasons that were metacognitively more complex than Meg's. This may be due to a greater influence of higher moral stage thinking in Mark's protocol (recall that Mark's protocol was about 70% stage 4, while Meg's was about 75% stage 3). Thus, moral stage may have an influence on metacognitive level. When Rob's case is added into this comparison, the pattern is extended into postconventional reasoning. Rob's scores indicated the highest metacognitive level as well as the highest moral judgment stage. There may be a relationship such that it is necessary to articulate a process of thinking about the system that one uses resolving moral problems to move beyond conventional level reasoning—to take one's moral system as object (to borrow Kegan's [1982] terms)—and to move on to postconventional reasoning, if not postformal thinking. This is also consistent with Habermas's (1990) view of reflection on systems in postconventional thinking.

Cognitive Stage

Each of the three cases discussed were tested using a simple "formal operations" problem that was a social variation on Kuhn & Brannock's (1977) Plant Problem. All had attained levels of formal operations. Thus, formal operational ability does not account for the variation in metacognition in these cases. However, these individuals were not tested using a standard postformal operational task, yet a preliminary analysis of the responses to the formal operations problems indicated that Rob evidenced at least early postformal operational capability. Demetriou (1990) found evidence of a relationship between postformal thinking and metacognition, but did not examine the moral question. Thus, future studies will need to examine the relationship between cognitive stage and metacognition in the social–moral domain, especially since a relationship between cognitive level and moral stage has already been established (Kuhn, Langer, Kohlberg, & Haan, 1977).

In sum, these three case analyses demonstrate several aspects of this preliminary inquiry into metacognition in the moral domain. First, consistent with the work by Tappan (1990), metacognition is an important element evident in people's thinking about moral issues. Second, metacognition in the moral domain can be described in more elaborated forms that have been described in the literature in reading comprehension and memory. Third, there are differences in metacognition in adolescence and adulthood. The nature of those differences has been articulated here, but further study is needed to explain the nature of their development. What this study fails to address, even in a preliminary nature, is metacognitive experience (Flavell, 1976). Further, this work contains within it a bias toward logical processes that individuals can articulate. However, it does not exclude increasingly complex metacognitive strategies that are not as "rational" as the examples in this chapter indicate. An important empirical extension to this work would be to examine the relationship between metacognitive awareness of strategies, as described in this chapter, to epistemological perspectives as described by Belenky, Clinchy, Goldberger, and Tarule (1982) and Baxter-Magolda (1992) which illustrate gender-related patterns in "ways of knowing." Relationships to these patterns of knowing could highlight the importance of broadening the concept of metacognition as it was illustrated here which would also address the question of metacognitive experience as a metacognitive strategy.

Finding differences in metacognition in adolescence and adulthood is important for the way we examine our educational intervention programs in high schools, colleges, professional schools, and beyond. Metacognition in the moral domain is not based solely on people's logical thinking (i.e., capability of formal operational reasoning) or their moral stage. It is people's reflection on their moral reflections that can have a major impact on the kinds of social and moral decisions people make and what actions they take. Concomitantly, actions and experiences influence cognitive structuring. Thus, metacognition holds a key place in our investigation in both moral and cognitive development. Knowing the different forms of metacognition that adolescents and adults use will facilitate the development of appropriate educational programs for both moral and cognitive development.

REFERENCES

Arlin, P. (1975). Cognitive development in adulthood: A fifth stage? *Developmental Psychology, 11,* 602–606.

Baker, L. & Brown, A. L. (1984). Cognitive monitoring in reading. In J. Flood (Ed.), *Understanding reading comprehension: Cognition, language and the structure of prose* (pp. 21–44). Newark, DE: International Reading Association.

Baxter Magolda, M. B. (1992). *Knowing and reasoning in college: Gender related patterns.* San Francisco: Jossey Bass.

Belenky, M., Clinchy, B., Goldberger, N., & Tarule, N. (1982). *Women's ways of knowing.* New York: Basic Books.

Brown, A. L. (1980). Metacognitive development and reading. In R. J. Spiro, B. C. Bruce, & W. F. Brewer (Eds.), *Theoretical issues in reading comprehension* (pp. 453–481). Hillside, NJ: Lawrence Erlbaum.

Colby, A., & Kohlberg, L. (1987). *The measurement of moral judgment.* New York: Cambridge University Press.

Commons, M.L., Armon, C., Kohlberg, L., Richards, R., Grotzer, T., & Sinnott, J. (1990). *Adult Development, Vol. II* : Models & methods in the study of adolescent and adult thought. Westport, CT: Greenwood Publishing Group.

Commons, M. L., Richards, F. A., & Armon, C. (1984). *Beyond formal operations: Late adolescent and adult cognitive development.* New York: Praeger.

Commons, M. L., Richards, F. A., & Kuhn, D. (1982). Systematic and metasystematic reasoning: A case for levels of reasoning beyond Piaget's stage of formal operations. *Child Development, 53,* 1058–1069.

Demetriou, A. (1990). Structural and developmental relations between formal and postformal capacities: Towards a comprehensive theory of adolescent and adult cognitive development. In Commons, M., et al. (Eds.), *Adult development, Vol. II*. New York: Praeger.

Dewey, J. (1933). *How we think*. Lexington, MA: D. C. Heath.

Flavell, J. H. (1985). *Cognitive development*, 2nd edit. Englewood Cliffs, NJ: Prentice-Hall.

Flavell, J. H. (1978). Metacognitive development. In J. M. Scandura & C. J. Brainerd (Eds.), *Structural-process theories of complex human behavior*. Leydon: Sijthoff.

Flavell, J. H. (1976). Metacognitive aspects of problem solving. In L. B. Resnick (Ed.), *The nature of intelligence* (pp. 231–235). Hillside, NJ: Lawrence Erlbaum.

Habermas, J. (1990). *Moral consciousness and communicative action*. Cambridge, MA: The MIT Press.

Kegan, R. (1982). The evolving self: Problem and process in human development. Cambridge, MA: Harvard University Press.

Kuhn, D., & Brannock, J. (1977). Development of the isolation of variables scheme in experimental and 'natural experiment' contexts. *Developmental Psychology, 13*, 9–14.

Kuhn, D., Langer, J., Kohlberg, L., & Haan, N. (1977). The development of formal operations in logical and moral judgment. *Genetic Psychology Monographs, 95*, 97–188.

Markman, E. (1979). Realizing that you don't understand. *Child Development, 50*, 643–655.

Schrader, D. E. (1988). Exploring metacognition: A description of metacognitive levels and their relation to moral judgment. Unpublished doctoral dissertation, Harvard University.

Tappan, M. (1990). The development of justice reasoning during young adulthood: A three-dimensional model. In Commons, M., et al. (Eds.), *Adult Development, Vol. II*. New York: Praeger.

Wood, P. K. (1983). Inquiring systems and problem structure: Implications for cognitive development. *Human Development, 26*, 249–265.

APPENDIX: METACOGNITIVE AWARENESS INTERVIEW

This interview has three parts. The first two are about how to solve hypothetical problems, the third part looks at something that has happened in your own life.

Part I

I'm going to read aloud the Heinz dilemma from the Moral Judgment Interview. You may have heard this before, but we're trying a different approach to it before I ask you to resolve it for yourself. So the first question is,

1. If you were going to teach me how to answer the Heinz dilemma, what would you say?
2. What methods should I use?
3. What factors would I need to consider in deciding how to solve it?
4. What strategies or methods should I use to decide how to solve it?
5. Is this/Are these the *best* strategies? How do you know?
6. Are there alternative strategies or ways of solving this dilemma? What are they?
7. How do I know when I have reached a solution?
8. Is this solution the *best* solution? How do I know?
9. Can you summarize briefly the steps to solving the dilemma?
10. Thinking back over what you told me to do, would you change any of your instructions on how to solve this dilemma?

Administer the MJI (Colby and Kohlberg, 1987).

Part II

1. Thinking back over the dilemmas we just discussed, how did you know how to answer or solve the them?
2. What factors did you consider in deciding how to solve it?
3. How did you choose among those factors? How did you know the factors you chose were the best to consider?
4. Were you aware of a strategy or some approach that you were using to solve the dilemma? What was it?
5. Did you consider that strategy to be the best one to use in order to solve the dilemma? How did you know it was the best?
6. Did you consider alternate strategies? If so, how did you choose the one you chose?
6a. Did you use a different strategy to answer the various dilemmas? What was it? How did you know to change your approach?
6b. Do you think other people have different strategies or resolutions? What do you think they are? How valid are they (or in other words, are they just as good as the one you used)?
7. How did you know when the dilemma was resolved or when you reached an adequate solution?
8. Thinking back over the dilemma and your ways of solving it, would you change your approach to the problem or your answers? Why?
9. Were you aware of your thinking processes while you were solving the dilemma earlier, or were you able to reconstruct your process just because of this interview?

Part III

Describe a situation when you made a significant moral or ethical decision that influenced your later thinking about some other moral situation.

1. What was the situation? What was the conflict for you?
2. How did you resolve it?
3. What did you consider in deciding how to solve it?
4. Were you aware of a strategy or some approach you were using to solve it? What was it?
5. Did you consider that strategy to be the best one to use? How did you know it was?
6. Did you consider alternative strategies? (if so), How did you choose the one you chose?
7. How did you know when you reached a solution? Was that the best solution? How did you know?
8. What was the solution and the final result? How did that resolve the conflict?
9. Thinking back over the situation again and how you thought about it at the time, were you aware of your thinking about your decision process at that time, or are you able to construct what you thought about as we've been talking about it?
10. Has this interview affected your thinking in any way?

CHAPTER 17

Protest, Collaboration, and Creation of Alternative Models

Women's Health Activists Using the Internet

ALICE LOCICERO

OVERVIEW

… disease cannot be understood outside its social and cultural context
—LERNER (2001, p. 5)

We are socialized to—disease is the thing. Yeah, I slip. We all do and see patients as a disease …
—a fourth-year chief resident, quoted by Lazarus, p. 39

Many women reject the disease model of health and illness. The popular model they reject proffers biochemical, genetic, and proteomic[1] internal processes as the core, and often the sole, etiological factors causing states of health[2] or illness. Women who are dissatisfied with the disease model and who voice objections to the model, and to the service provisions that follow from its assumptions,[3]

---

[1] The term "proteomics" refers to the study of proteins in the body. In this context, I refer to study of protein contribution to health and illness. In reports suggesting that the Human Genome Project had yielded somewhat less information than researchers had hoped, there was a suggestion of a shifting of focus to the protein in biomedical research. This shift maintains the biological focus.

[2] I use the term health in an inclusive way, to represent well being, including what is often termed "mental" health.

[3] The assumptions of the disease model are that illness is a biogenetic process that takes place within one's body, that it is neither caused nor ameliorated by social or psychological factors (although psychological or social complications may arise, secondary to the illness). Therefore, treatment is prescribed and/or administered by an expert, is usually biochemically or technologically based. The patient, in this system, is a passive recipient of advice and services.

ALICE LOCICERO • Department of Psychology, Suffolk University, Boston, Massachusetts, 02114.

Handbook of Adult Development, edited by J. Demick and C. Andreoletti. Plenum Press, New York, 2002.

include a substantial minority of recipients, and former recipients, of mainstream services; researchers; providers of alternative services; and mainstream providers who include a nonconventional approach in their practices. Well considered attention to the essential and shared concerns of women who reject the disease model may be a necessary step in the development of complete, effective approaches to enhancement of health and prevention and treatment of illness.

Rejection of the disease model often begins with personal disappointment with services received in mainstream medical settings (Chamberlin, 1995; Cohen, 1991; Gaskin, 1997; Grobe, 1995b). The confidence and competence required for recipients to articulate objections and challenges to mainstream services reflect developmental achievements.[4] Articulating objections, however, does not guarantee being heard. Many recipients and former recipients have found affirmation and support for their critical perspective in groups of women with similar concerns. The groups have had beneficial effects. Beyond enhancing individual development of the women concerned, they have maintained a stream of discussion and debate, thus ensuring ongoing availability of critiques of mainstream services and suggestions for changes. Some local recipient groups are connected to larger organizations that convene regional or national conferences,[5] and the conventions receive some publicity. Women's health activists have had an impact on health care *service provision* in limited areas.[6] Building on, and moving beyond, critiques of service provision, some women's groups have begun to challenge the underlying *disease model* itself, and to endorse emerging alternative models of understanding health and illness. However, most critiques of service provision do not include an analysis or critique of the underlying models. Consequently, the disease model that dominates etiology and treatment in women's health has received far fewer protests than the service provision that is associated with it, and have remained intact.[7] Neither the concerns expressed by women who reject the disease model nor the emergent alternative models some of them have created and supported, have, to date, had much impact on the rapid ascent of that model as the primary explanation for health or illness, and as the basis for prevention and treatment.

Major challenges to women's health groups' attempts to develop new models include lack of resources, in comparison with the well-endowed pharmaceutical, physician, and "managed care" organizations, and challenges arising from the difficulties inherent in communication and collaboration, within and among groups, across role status, age, gender, power, and developmental differences. The World Wide Web has helped alleviate the difficulties arising from limited resources, and has allowed for the creation of a rich network of communication across differences.[8] This has not yet resulted in the formation of a powerful and

---

[4] Using terms of the theory by Belenky et al. (1986), women would, at minimum, be emerging from the second perspective, called "received," to the third, called "subjectivist," in relation to their own ability to think and know in order to make such challenges.

[5] Examples include La Leche League International and International Childbirth Education Association.

[6] See Lerner (2001) for a discussion of the impact of the women's movement on treatments for breast cancer.

[7] The disease model dominates other areas of health as well.

[8] Fox et al. (2000) report that 52 million Americans, the majority women, have looked to the World Wide Web for health information.

cohesive movement across different areas affected by the disease model, although it has contributed to collaborative efforts within those areas. Unless they form a cohesive movement, it is unlikely that women's groups will be effective in challenging the continuing ascendance and dominance of the disease model in relation to women's health.

In this chapter, I focus on two areas of women's health in which alternative *models* for service provision are available: midwifery, as an alternative to mainstream childbearing services; and egalitarian mutual support services for women with emotional, behavioral, or mental problems—an alternative to mainstream psychiatric services. They have, as common threads, not only dissent from the disease model, but, more importantly, an emphasis on connection, egalitarian decision making, attention to the social, economic, and physical environment, and individuation of care.

Challenging mainstream models and creating new ones are formidable tasks, requiring sustained effort over time. Women's groups wishing to challenge the disease model and/or to endorse emergent alternative models will have to effectively address developmental differences within and between groups. Cohesive and powerful alliances will be needed, and they will need to include groups that are limited to former recipients of services as well as those that include providers and/or researchers, along with recipients.[9] The developmental theories of Belenky et al. (1986) and Kegan (1994) suggest that direct, effective communication, with the goal of forging an alliance for collaboration that is consensus based, across developmental differences, is itself a developmentally advanced, constructivist process. However, neither patterns nor support for such collaboration is easily found (Kegan, 1994). The Internet seems to encourage communication and collaboration across considerable individual differences. As an open communication forum that is also a heavily used resource for information about health, it seems to provide women's groups greatly enhanced opportunities to garner support for significant challenges to the hegemonic dominance of the disease model of health and illness.

HEALTH AND ILLNESS

"Hard" and "Soft" Factors

> There are two fallacies to be avoided: determinism, the idea that all characteristics of the person are "hard-wired" by the genome; and reductionism, the view that with complete knowledge of the human genome sequence ... our understanding of gene functions and interactions will provide a complete causal description of human variability. (Venter et al., 2001, p. 1158)

In this passage, among the final comments in the publication of results of the initial mapping of the human genome, Venter et al. stop short of suggesting what the other sources of human variability might be. Women's health advocates who

---

[9] By contrast, most groups of mainstream providers are closed to nonproviders.

challenge the biogenetic, proteomic disease model[10] might suggest social, economic, power, and consequent environmental issues as factors that contribute to some variability. For example, differential access to nourishing food contributes to size differences. Violence and intimidation contribute to differences in the likelihood of developing depression or anxiety. Poverty and oppression limit choices, and those who must live in areas with much air pollution and other environmental toxins are more likely to experience respiratory health problems.

> In an effort to base psychiatry in "hard" science and thus raise its status ... psychiatrists have narrowly focused on the biological underpinnings of mental disorders while discounting the importance of such "soft" variables as culture and socioeconomic status. (Kleinman and Cohen, 1997, p. 87)

Focusing on "soft" economic, environmental, and/or social factors as etiological factors in human variability, health, and illness is unfashionable and unprofitable. It is unlikely to win a large share of available funds for research, owing to economic pressures mounted by financially advantaged pharmaceutical companies, mainstream medical and biological researchers, and biotechnology businesses. Perhaps it should not be surprising that, like other unprofitable, socially responsible work, the unprofitable, unfashionable work of maintaining a focus on "soft" environmental and social factors in relation to well being and variability of experience and identity has become, to a large extent, women's work.[11]

Remembering the Soft Factors: Challenging the Disease Model. In recent decades, groups of women have brought the "soft" factors into focus in discussions of health and illness. Examples of such groups include the Boston Women's Health Book Collective (The Boston Women's Health Book Collective, 1998), the International Cesarean Awareness Network (ICAN, 2000–2001), Wellesley's Stone Center Research Group (Jordan, 2000), La Leche League International (LLLI, 2001), the Midwives Alliance of North America (MANA, 2001), UNIFEM's End Violence List Serve (Members, 1999), the National Women's Law Center (NWLC, 2000), the Women's Institute for Childbearing Policy (WICP, 1994), and Unruly Women (Unruly Women List, 2000).

The work and products of these groups include critiques of the kinds of services provided and of the way they are provided. Some of them also include more

---

[10] For brevity, I refer to this model as the *disease* model. Characteristics of the model include its explicit, or implicit, but clear, assumption that the *primary* and *real* cause of illness (including mental illness) resides inside the individual afflicted. The underlying societal form supporting this perspective is individualism. The perspective is atomistic. This belief is apt to thrive in a nonreligious, capitalist society. Ironically, as the first reports of completion of the Human Genome Project surface, the somewhat less than spectacular explanatory value of genes is becoming evident. Attention is now turning to proteins as explanatory, however (thus the term "proteomic") with little mention of the effects of the physical or social environment.

[11] Support for this point may be found in multiple organizations of professionals and recipients of services. Acknowledgment that "soft" concerns are seen as women's concerns may be found on the American College of Obstetricians and Gynecologists' website (http://www.acog.org). Surprisingly, the website contains a special area for "women's issues." Since it would seem that the entire website would be dedicated to women's issues, I wondered which issues ACOG defined in this way. The categories defined as women's issues were soft issues: Adolescent Care, Underserved Women, and Violence.

radical critiques of the disease model. Some of the groups' work began with critiques of services, and evolved to critique of the model.[12] And some of the groups are involved in establishing and/or supporting service provision in models that offer alternatives to the disease model.

The focus of this chapter—groups of women who challenge the dominance of the disease model and the service provision associated with it—is not meant to imply either that all those who challenge the model are women, or that all support groups for women in relation to health and illness are composed of women who challenge the disease model. Advocates of attention to nonbiological etiological factors and treatment alternatives in relation to women's health include men as well as women.[13] And some groups of women concerned with health issues are comfortable with the disease model. (Some physicians selectively refer women to such groups, and discourage membership in groups that challenge the disease model.)[14] In this chapter, I limit my observations and evidence to women who challenge the disease model in two areas: childbearing and psychiatric diagnoses and treatment. In addition, I describe existing, functioning services using alternative models in both these areas.

WOMEN'S HEALTH ACTIVISTS' DEVELOPMENTAL ACHIEVEMENTS

Rejecting Expert Opinion: A Subjectivist Position

Rejecting or critiquing expert opinion is often a first step toward examining and challenging the disease model. The capacity to formulate objections and suggest alternatives to mainstream care—being able to hold and articulate an opinion in contrast to the ones held by professionals and experts who are ordinarily seen as holding authoritative knowledge—is a developmental achievement, a hallmark of emergence from the perspective called "received" by the authors of *Women's Ways of Knowing* (Blenky, Clinchy, Goldberger, & Tarule, 1986). The activity associated with "received" knowing, as the term implies, reflects the ability to learn knowledge from experts.[15] By contrast, the activities associated with "subjectivist" knowing involve rejecting socially designated authorities and experts as a source of knowledge and wisdom. The important developmental step in this respect is recognizing that one can choose the source of knowledge to which one will subscribe. Much like the Harvard students about whose development William Perry wrote (Perry, 1970) participants in Belenky et al.'s study have been seeking truth from experts, have confronted the limitations of that method of

---

[12] The Stone Center, at Wellesley College, for example.

[13] Indeed, a landmark work in the area of psychosocial factors in the etiology of depression is a research study by a pair of researchers, one male and the other female (Brown & Harris, 1978).

[14] Fernandez (1992), for example, suggests referral to support groups, for women with postpartum disorders, that "reinforce biological concepts of the disorder, and … increase compliance significantly." At the same time, numerous anecdotal reports suggest that physicians view some groups who advocate for alternative care as adversaries who endanger their members.

[15] The received knower's attitude toward learning would be similar to Paolo Friere's notion of education seen as parallel to "banking," where information is deposited in the learner's mind.

seeking truth, and are thus challenged to change their perspective. According to Belenky et al., subjective knowers do not trust conventional authorities or experts such as physicians or scientists. Early in the transition to this perspective (from the prior perspective, called "received"), women are more apt to turn, when seeking truth, to those with personal experience, who are close to them, such as friends or relatives. Later, as they are more comfortably in this position, they trust themselves, their "gut" or intuition.[16]

> Truth, for subjective knowers, is an intuitive reaction—something experienced, not thought out, something felt rather than actively pursued or constructed. These women do not see themselves as part of the process, as constructors of truth, but as conduits through which truth emerges. The criterion for truth they most often refer to is "satisfaction" or "what feels comfortable to me." (1986, p. 69)

This perspective clearly has limitations. For example, systematic procedures for the evaluation of evidence are not weighed strongly. When critiques of mainstream medicine are made from the subjectivist position, this limitation is often assumed to be *in contrast with* medical professionals, whose points are assumed to reflect scientific procedures. That is, the assumption is that the provider's point of view may be assumed to be scientific. However, critics of practice in mainstream medicine—or at least of some aspects of it—note that actual practice decisions are frequently based on tradition and the opinion of authority rather than the most current scientific findings (Lerner, 2001; Rodwin, 2001; Services, 2001). Such a view is stated simply, by a former recipient of psychiatric services:

> As I tell my story, and others tell theirs, you can easily dismiss us with that once picturesque and now pejorative phrase that says our accounts are "anecdotal" and therefore meaningless. We, in turn, can counter by saying that the "clinical impressions" that doctors use to assess our "progress" are also anecdotal—and both sides can, in effect, cancel each other out. (Gotkin, 1995, p. 194)

Of course, in the public discourse, former recipients seldom have their anecdotes "cancel out" the physicians' anecdotes. However, it is the step away from assuming the truthfulness of knowledge received from experts, toward the belief that one's own story has a right to exist side by side with the experts' stories, that allows recipients to provide accounts of their experiences that do not match the professional accounts typically recorded and attended to by the media and members of the community. That is, the ability and energy to critique mainstream models emerges with this developmental transition.[17]

Assuming that mainstream medical experts are good sources of scientifically correct knowledge creates, among the press and the general public, an expectation that these experts should be asked to make judgments about critiques of their work. This makes it harder for those who critique the mainstream view to receive a fair hearing. (Exceptions seem to occur when either there is a critical mass of

---

[16] Michael Basseches pointed out (personal communication, April 2000) that, in terms of adult development, the stage or position is a capacity of the person. The "stage" (probably mislabeled as such) is part of the person; the person should not be seen as being in one stage or another, but as having the ability to function at any stage up to the most one most recently achieved. Thus, a woman who has achieved the ability to engage in subjectivist processes does not lose her ability to receive and believe knowledge, for example.

[17] Mezirow's "transformative learning" concept is relevant here.

recipients making similar critiques—enough so that it becomes news [as when in 1957 many women responded to an article about birth practices in a popular magazine] [Wertz & Wertz, 1977] or the critiques made by recipients find their way to a provider who agrees with the recipients, and has enough power to be heard [as did women who questioned the need for radical mastectomies, in the 1970s] [Lerner, 2001].)

In most instances, women's critiques are not taken seriously, because the training, practice, and professional (guild) pressures toward endorsement of the current paradigm limit the vision of mainstream practitioners. The resulting tendency to endorse the current paradigms makes it hard for them to seek or even pay attention to critiques of their fields, and thus keeps their blind spots blind. It seems clear that assessment of services should not be left to the providers of the services (Chamberlin, 1978).

Unheard Protests: Provider Paradigms, Beliefs, and Evidence. It is not surprising that contemporary mainstream practitioners often dismiss critiques of the dominant (disease) model and the practices associated with it. The histories of science, religion, psychology, and medicine are replete with stories of truths unseen in spite of evidence, simply because the truths did not fit with the dominant paradigmatic belief system. While a given paradigm prevails, evidence against it is dismissed. After a paradigm shift, the evidence, there all along, suddenly becomes plain (Kuhn, 1996). An example of truth unseen, even when the evidence was plain, *and even after it had been empirically demonstrated*, in the area of women's health, is puerperal fever. A largely iatrogenic illness, puerperal fever is caused by infection suffered by women following childbirth. During epidemics in the 19th century, puerperal fever was far more common following births in hospitals than after home births and was even more common in births that took place in the hospital than in births that took place on the hospital steps, when the hospital wards were full or the laboring women did not reach the hospital in time to be admitted. The germ theory had not been developed, and simple methods to avoid contagion were unknown at the time, but Philip Ignaz Semmelweis, in Hungary, and Oliver Wendell Holmes, in Boston, came to understand, by observation and data collection, in the 1840s, that doctors were carrying the disease from one woman to another and both recommended hygienic practices, such as hand-washing, among doctors attending women giving birth. Semmelweis was able to demonstrate the success of hand washing practice, but both Semmelweis and Holmes were ridiculed, and the practices and theories rejected by their peers (Wertz & Wertz, 1977). The idea that doctors were involved in the spreading of puerperal fever was rejected as "a vile, demoralizing superstition" (p. 122). Semmelweis died before his contribution was recognized.

Many initially assess the story of Semmelweis as a strange, unfortunate, and unique story in the history of medicine. But eerily similar stories of physicians' practices in attending childbirth that continue after the preponderance of evidence finds them useless or even harmful continue into modern times (Goer, 1995; Tew, 1993). For example, routine shaving of the pubic area before birth at early and mid-century in the United States was done for reasons that seemed, to some, rational: prevention of infection. However, it turned out that it was, at best, ineffective, caused a great deal of discomfort, and was associated with increased, rather than decreased, "maternal morbidity" (Enkin, Keirse, Renfrew, & Neilson, 1995, pp. 200–201). Despite the fact that a controlled study showed no beneficial effects in 1922, the practice continued to be routine in much of the United States into the 1960s.

Contemporary childbirth practices that research has not supported include routine use of episiotomy.[18] (Enkin et al., 1995; Goer, 1995; Cochrane Collaborative Consumer Network, 2001; The Boston Women's Health Book Collective, 1998). Indeed research does not support routine involvement of obstetricians in the care of all pregnant and childbearing women (Durand, 1992; Enkin et al., 1995; Rooks et al., 1989). Midwifery services for women without serious risk factors are "likely to be beneficial" (Enkin et al., 1995, p. 394; WICP, 1994). Furthermore, services provided by midwives or general practitioners are associated with greater satisfaction, and similar outcomes, to care by obstetricians (Cochrane Collaborative Consumer Network, 2001). Despite these findings, the vast majority of pregnant women in the United States, including those with no known risk, seek and receive childbearing services from obstetricians, and the majority of those who do not have cesarean sections receive episiotomies.

Reflections on Witch Hunts: Lessons from the Past, for the Present. The history of witch hunts and witch trials in Europe and the United States presents another opportunity to reflect on the fact that learned and well-respected people of a given historical time, ignoring logic and evidence, have done grave injustice and, in this case, seriously decreased the female population (Ussher, 1991) while claiming a righteous motive. It is not surprising to most people that we can look back over centuries and see errors of superstition and prescientific use of rumor as evidence. It is not hard to lament the devastation that resulted. Most people looking at witchcraft, however, see one lesson and miss another. The lesson seen is that ignorance and an unscientific attitude can lead to terrible atrocities. Unfortunately, the assumption that ignorance is a thing of the past disallows the second lesson: that well-respected, educated people in roles of authority in any time period can do terrible harm when their vision is narrow (Fausto-Sterling, 1992; Nechas & Foley, 1994). A text written in 1584 (Scot, 1972) demonstrates that it was possible, using 16th century knowledge, logic, and observation, to conclude that witches did not, and could not, exist.

There is little doubt that contemporary women's health service providers are vulnerable to being blind to evidence that does not fit the paradigm in which they were trained and in which they practice (Lerner, 2001). What are they missing?

Blind Spots: Poorly Served Recipients. Most contemporary medical providers are missing, or not taking seriously, the stories of numerous, perfectly lucid women who feel they have been mistreated, sometimes brutally, in contemporary, conventional systems of childbearing and mental health service provision in the United States (The Boston Women's Health Book Collective, 1998; Chamberlin, 1978; Cohen, 1991; Davis-Floyd, 1992; Grobe, 1995a; McInnes, 2000; Oakley, 1986;

---

[18] "Routine episiotomy is the incision through skin and muscles in the perineum—the area between the vagina and the anus—to enlarge the opening through which the baby will pass" (The Boston Women's Health Book Collective, 1998, p. 489).

Shimrat, 1997; Ussher, 1991). Moreover, numerous researchers have documented systematic failures and impersonal, disrespectful services in both contemporary mental health and childbearing care systems. (See, e.g., Davis-Floyd, 1992; Lazarus, 1988; Ross & Pam, 1995; Szasz, 1994). There are, then, a large number of women for whom none of the existing alternatives in mainstream American mental health service provision is a viable choice. There is an even larger group of women for whom none of the existing alternatives in mainstream childbearing service provision is a viable choice. Many scientists and practitioners who do not focus on women's health issues actively advocate for a similar expansion of etiological and treatment models (Brown & Harris, 1978; Cohen & Kleinman, 1996; Karon & Vandenbos, 1995; Kleinman & Cohen, 1997; Lewontin, 2000; Schulman, 2001; Venter, 2001). However, women benefit least and suffer most as a result of the predominance of the disease model. This is because the majority of patients (in the United States) making health care office visits,[19] emergency room visits, and/or hospital outpatient visits, receiving surgical procedures, and experiencing hospital admissions are women (National Center for Health Statistics, 2001). In addition, women take more prescription medicines than men (Office of Women's Health, Food and Drug Administration, 1999), and are more often the care providers for ill relatives. Equally important, the adverse impact on women's health caused by gender inequity—such as the disproportionate number of women who are poor, single parents, who work in lower-paying service positions, who are providing unpaid care for ailing relatives, and/or who are victims of sexual or domestic violence—disappears from the discussion of women's health when the disease model alone is the focus of etiology and treatment.

It is clear, then, that current conventional childbearing services and current conventional policies in treatment of women diagnosed with psychiatric illness are doing harm to a large subgroup of recipients. Providers are not intentionally hurting the women who receive their services.[20] Like the colleagues of Semmelweis and Holmes, they would find it very hard to imagine, much less to believe, that they, who have devoted their careers to providing services for others, and who are practicing what they learned from mentors whom they respected highly, could be the source of pain and suffering, rather than relief. But they are unaware of their own blind spots.

Further Development

From Bad Services to Bad Models

So my own [baby's] birth was a big learning experience. I wasn't afraid and was prepared to go in and have the baby without anesthesia—but I had the great surprise that I wasn't allowed to do so. I would have had to get violent and slug the nurse in the delivery room if I didn't want to have spinal anesthesia ...

So when I started to attend births I treated women the opposite way that I'd been treated. ... (Gaskin, 1997, p. 127)

---

[19] Women dominate overall, and the difference in number of physician visits is statistically significant at every age group between 24 and 65 years, including, but not limited to, childbearing years.

[20] There are some exceptions, for example, women receiving psychiatric services—often involuntarily—have been raped and abused.

The author of the preceding passage, Ina May Gaskin, is now the director of The Farm Midwifery Center, where over 1800 women have given birth in the midwifery model. She also wrote *Spiritual Midwifery*, a midwifery classic. (Penfield Chester, 1997.)[21]

> I had entered the hospital in an attempt to find relief from the pain of living, and found, instead, that any attempt to show this pain to my "healers," so they could understand and help me, was swiftly met with repression. A good patient was one who swallowed the medication, followed the hospital routine, and indicated a willingness to "cope" with the life circumstances that had brought about hospitalization in the first place. (Chamberlin, 1995, pp. 60–61)

The author of this passage, Judi Chamberlin, is a founding member of the Ruby Rogers Advocacy and Drop-In Center, "a self-help center run by and for people who have received psychiatric services" and author of *On Our Own: Patient-Controlled Alternatives to the Mental Health System* (National Association for Rights Protection and Advocacy, 2000).[22]

Many women who reject the disease model, and the types of services that follow from its assumptions, and who either design or support alternative models, generally begin with a feeling that something in their own experience, or something they observed, wasn't quite right.[23] Gradually, they begin to see a pattern of things that don't seem right. Eventually, they come to believe that the system of service provision is wrong-minded, that it often creates the very difficulties it is supposed to prevent, and intensifies the difficulties it is supposed to treat. At some point in this process, some of the women who started by critiquing the mainstream services become dissatisfied with the fact that the disease model dominates both understanding and treatment in women's health. Some believe that the disease model's continued success as the core and sole basis for diagnosis and treatment, despite its failings, is due to efforts by those with economic interests, including physicians and stakeholders in pharmaceutical companies (McInnes, 2000). Some try alternative services, or even create new services, based on a different model.

The sorts of experiences and critiques made by Judi Chamberlin and Ina May Gaskin are not unusual. Since it is from experiences and critiques that new models will emerge, I am including a few more examples of what seem to be typical comments from women who have felt they received bad services—they were poorly treated as recipients of mainstream services. The first set of quotes is from women who describe experiences in relation to mainstream psychiatric services. The second set of quotes is from women who describe experiences in relation to mainstream childbearing services.

---

[21] Recently mainstream researchers, using rigorous matched control methods, concluded that, for the low risk women served by Gaskin's service, home birth was equally safe, and led to fewer interventions, than births in hospitals, attended by obstetricians.

[22] Ms. Chamberlin is currently engaged in research evaluating self-help programs such as the one she co-founded.

[23] Belenky et al. note that such experiences are often the challenge that leads to transition from the received to the subjectivist position. It seems likely that some critiques of mainstream services are generated by individuals who have recently become able to express such critiques, owing to their accomplishment of transition to the subjectivist position.

Descriptions of the experience of receiving psychiatric services follow:

> During the three years I spent in institutions, I saw and experienced a lot of abuse. Some of it was violent, some of it was more subtle, but it was almost always present in one form or another, like a polluted river, running through our lives. ... (Dundas, 1995, p. 35) They lead me, bruising my arms, to the seclusion room, so I can think about the lie they have given me to live, and change my mind. ... I left that hospital, having washed thousands of dishes, and having lost my soul. (Unzicker, 1995)

Descriptions of the experience of receiving childbearing services follow:

> They'd checked my urine and I was dehydrated and starved. No duh! (sic) after nothing but ice chips and 12 hours of walking and labor.[24] ... The next insult—Pitocin. I felt like a wounded deer slowly dying in the forest ... The epidural didn't take the first time so they gave another. I was numb from the neck down and completely paralyzed. I couldn't even feel my breathing and thought I was suffocating during repair. ... (Daryl, 2000)

> As his hairy little head crowned, I was given an episiotomy [see footnote 18] without my consent. (Woods, 2000–2001)

> Oh yeah, the general (anesthesia)'s the other thing from hell. ... I certainly didn't choose the risk and loss of bonding that come with the stuff. Can any of you wonderful women figure out why I had general anesthesia for an "emergency section" that took 80 minutes to complete from time we were told we were having it to my baby's birth??? What's that I hear? To shut me up? Oh surely not! :) (sic) (Griebenow, 2000)

Common themes—often stated explicitly—in these and many other stories of recipients who are not happy with their experiences include not being heard or understood (Lauer-Williams, 2001) and either being pressured to accept unwanted treatment, or having one's perception of needing help minimized. Unwanted treatment is administered in what appears to be an insensitive manner, at best, and brutally in some cases. Related to pressures to accept unwanted treatment are difficulties such as lack of communication, respect, and connection between provider and recipient. The sense that the provider believes he or she can decide what the recipient needs without consulting the recipient is consistent with, and perhaps a hallmark of, a disease model of health and illness.

The Treatment Pair: Physician and "Patient" Roles. In both of this chapter's areas of focus—childbearing and psychiatric diagnoses and treatment—the treatment pair for which services were designed, *and from which pathological conditions were categorized and questions regarding the causes of these conditions were framed*, was a woman in the role of "patient" treated by a man in the role of "doctor." Although the number of physicians and medical students who are women has increased substantially in recent years, the fields of psychiatry and obstetrics–gynecology were both established by men. Most of the architects of the practice models, definitions of pathology, and inferences regarding causality that dominate both fields were men, while women were, and are, recipients of all the services in one field, and the majority of the services in the other.[25]

It seems likely that gender stereotypes have had a strong impact on the treatment paradigms in psychiatry and obstetrics.[26] That is, the design of treatment

---

[24] Depriving women in labor of food and water is another routine practice that the preponderance of evidence would suggest should be dropped. (Enkin et al., 1995)

[25] Perhaps this in itself should lead to some reservations about the beliefs and conclusions drawn.

[26] Lerner (2001) makes it clear that gender stereotypes were highly salient in treatment for breast cancer.

was very likely influenced by the expectation that women would be dependent and passive, while men were active and instrumentally effective (LoCicero, 1993; Oakley, 1986; Ussher, 1991). It must have seemed obvious at the time the service models were developed that the (mostly male) physicians would actively make decisions and administer interventions while the (female, or mostly females) recipients would be trusting, dependent, and passive. Indeed, many women experience their providers, men or women, as being offended if they do not assume a trusting, dependent role. A participant in a study of postpartum depression relayed her experience in seeking help from her obstetrician regarding her feelings.

> She made some sort of off hand comment like I could have predicted it in someone like you. People like you often have this happen. I asked her what that meant, and she said, sort of casually, almost jokingly, when you came in with your little date book. ... I had my questions in there, every appointment. ... But the kind of person that I looked like, a list maker, I felt like I was being judged. ... (sic) (Lauer-Williams, 2001)

The majority of contemporary leaders in fields of psychiatry and obstetrics–gynecology are still men.[27] A reflection of the ongoing influence of men in shaping the practice of obstetrics and gynecology, for example, is a current debate about "patients,"—the term used for perfectly healthy women who are pregnant—rights to choose to give birth by Cesarean section, whether or not there are medical indications for it. The widely publicized participants in the debate are all men (e.g., DeMott, 2000; Wagner, 2000; Walters, 1999).[28] To a large extent, then, groups of women who resist and reject the disease model as a basis for treatment in these areas are rejecting and resisting models of service provision designed and supported by men, to be provided to women.

Beyond Guild Pressures. To sum up, then, mainstream services to women in psychiatry and obstetrics have been established in keeping with the disease model and with gender stereotypes. Authoritative knowledge is attributed to mainstream scientists and providers, who are given responsibility to assess the importance of critiques of their models (Jordan, 1997). It is a developmental achievement for women to be able to critique these services; however, women who critique the services typically do not feel heard. Professionals are subject to guild pressures to accept the dominant paradigm, and to discredit critiques of that paradigm. Indeed these pressures often lead to continuation of practices long after they have been shown to be ineffective.

In addition to guild pressures and gender stereotypes, contemporary providers are challenged in any efforts to take critiques seriously by two other sources. The

---

[27] See websites for the American Psychiatric Association, http://www.psych.org, and the American College of Obstetrics and Gynecology, http://www.acog.org.

[28] The relevance of the debaters being men is suggested by a recent survey taken among attendees at a session of the American College of Obstetricians and Gynecologists annual conference. Of those surveyed, 56.5% of the male obstetricians and 32.6% of the female obstetricians said they would prefer to have a cesarean section, rather than give birth vaginally (Gabbe & Holzman, 2001). The high percentages are somewhat astonishing, and if the difference in response between the male and female obstetricians reflects a more general difference among obstetricians in practice, the gender of leaders in the field may well make a difference in practice standards. That is, if leaders in obstetrics are very likely to see cesarean birth as preferable to vaginal birth, it would not be surprising if that affected the percentage of women who actually do end up with a cesarean section.

first source is that of the legal system in the United States, which dictates that physicians provide services similar to those offered by their peers. Common practice, or "standards of care" (Walters & Quillinan, 1999) (sometimes referred to as "community standards of care") is a mechanism of judgment whereby a physician who practices more or less like colleagues is more or less safe from malpractice judgments and/or prosecution. This serves to protect consumers from providers who might practice creatively in ways that are risky; it also serves to prevent consumers (and providers) from having a wide range of choices in terms of available services.

The third pressure on providers is associated with risk assessment, and it is a cognitive challenge. This cognitive challenge is posed by the pressure of vivid memories of past catastrophic outcomes, and the impact of those memories on present decision making. Recent developments in cognitive theory suggest that past experiences are likely to result in caution that is inconsistent with the actual probability of the kind of event feared or dreaded (Loweenstein, Weber, Hsee, & Welch, 2001). This might be assumed to adversely affect providers of psychiatric and childbearing services as follows: Fairly typical service provision experiences in obstetrics and psychiatry would not be expected to highly impact those who provide them.[29] But the terrible and unexpected—a recipient of psychiatric services who commits suicide or homicide, or who attacks a provider, or a baby or mother who dies while receiving obstetrical services—might be expected to make a strong emotional impact on the provider, and thus might be expected to interfere with optimal decision making in future situations.[30] A psychiatrist, remembering a recipient of psychiatric services who committed suicide, is likely to be more restrictive with future recipients than warranted. An obstetrician, remembering a stillbirth, may perform more tests and interventions than optimal in future labors. Still more distressing, those assessing risk are, at times, unaware that their gut reactions reflects vivid memories and dread, rather than realistic probabilistic assessment of the risk–benefit ratio. Thus, decision makers may be convinced of the rightness of their judgment of risk. Clearly, if authoritative knowledge is ascribed to someone who also believes in his/her own incorrect assessment, some less than optimal decisions are apt to be made, and they might become routine, being passed on from trusted mentor to resident. This perspective on risk assessment does cast some doubt on the value of "clinical impressions" and standards of care. (This may explain the persistence of interventions long past the time they are shown to be unjustified.)

Development Toward a "Procedural" Approach

Fixing the Disease Model.   The disease model, then, has a bias toward the passive recipient being treated by the more knowledgeable provider. But

---

[29] Indeed, one of the concerns expressed by midwives and others critiquing mainstream obstetric service is that the providers do not recognize or celebrate the intensely moving aspects of birth—that birth becomes a series of technical problems, rather than a positive, awe-inspiring event.

[30] Since, in the United States, the rate of infant mortality is about 7 or 8 per thousand, and by the end of residency, obstetricians have likely attended more than 1000 births, one would expect that they have seen some infant deaths. Psychiatrists who practice in hospital settings likewise may be assumed to have seen patients who have attempted suicide.

providers are apt to be pressured to practice in ways that reflect a mix of evidence, guild pressures, personal experience, common practice, and vividly imagined outcomes, with inaccurate probabilistic assessment of risk. Further, they are not apt to be open to critiques (or sometimes even questions) about their practices.

Serious attempts to counter these problems by advocating for evidence-based medical practices are underway (The Boston Women's Health Book Collective, 1998; Rodwin, 2001; Services, 2001). This kind of practice would require standards of service provision to reflect preponderance of evidence from results of controlled scientific studies.[31] This would seem to be a major improvement over the current system of decision making.

Earlier, I referred to activities of some who critique the mainstream providers as reflecting a subjectivist perspective, with respect to the theory developed by Belenky et al. The limits of that perspective include a lack of emphasis on logical procedures or systematic data gathering as a basis for knowledge. I noted that many internal and external critics of professional decision making point to a process that is not entirely dissimilar from the subjectivist process. The activities associated with advocating for evidence-based practice are likely to reflect another developmental transition, for critics and providers, from subjectivist to what Belenky et al. call "procedural." Women who understand the world from this more "objective" position tend to be "absorbed in the business of acquiring and applying procedures for obtaining and communicating knowledge ... the emphasis on procedures, skills, and techniques was common to all." (1986, p. 95)[32] Transition to the procedural approach may take place through mentoring, formal or informal education, or role models. Or it may occur through independent seeking of knowledge.

Although the practice of evidence-based medicine does appear to be a clear improvement over the current mainstream care provision system, it is not likely to be ideal. Political pressures may be brought to bear on (or by) service providers, to interpret studies in one way or another, for example (Rodwin, 2001). And there is likely to be disagreement about what the preponderance of evidence would support. Studies may be designed in a way that highlights one outcome while ignoring others that are equally salient. Last, many providers will not want to follow the results of studies that do not seem to fit with their own experience.[33] It is possible that evidence-based medicine will eventually be required by law, if it is not accepted widely by providers, but procedures for deciding what the evidence dictates will have to be established. There are two problematic aspects of evidence-based practice, however. First, it will, at least initially, tend to require uniformity even more stringent than the current standard of "standards of care," thus paying less attention to individual differences and to "soft" factors. Second, it will leave the recipient, as well as the provider, out of the decision-making process.

---

[31] The Cochrane Collaborative (Cochrane Collaborative Consumer Network, 2001) provides reviews that may be effectively used for this purpose.

[32] Women in this position tended to use one of two similar ways of understanding the world, which the WWK authors refer to as "connected" and "separate" ways of knowing (1986, p. 100). However, the distinction does not seem highly salient in this discussion.

[33] Indeed, a recent popular work by an obstetrician (Walters, 1999) details his own process of dismissal of scientific findings, concluding with the cliché that there are "lies, damn lies, and statistics."

Challenges to Collaboration: Subjectivist and Procedural Visions

Given how difficult it is to be heard and taken seriously when criticizing mainstream service provision, women who are engaged in critiques of providers' activities from the subjectivist perspective and those engaged in critiques from the procedural perspective must collaborate to make an impact on service provision itself or on the model it reflects. In preparing this chapter, I followed several Internet-based women's groups, all of which challenged the dominant disease model and its impact on women. Anyone wishing to explore the availability of groups whose activities reflect a variety of approaches to these challenges can easily do so by use of a search engine.[34] In the groups I followed, there were some messages that reflected a subjectivist perspective as well as some that reflected a procedural perspective. In each group, there were also participants whose messages reflected a more advanced perspective, called "constructivist" by Belenky et al.

Toward a Constructivist Perspective

The most highly integrated, the most "advanced" perspective described by Belenky et al. is that of "constructed knowledge." The authors describe the basic insights of constructivist thought as follows: "*All knowledge is constructed*, and *the knower is an intimate part of the known*" (p. 137; italics in original).

> Constructivists seek to stretch the outer boundaries of their consciousness—by making the unconscious conscious, by consulting and listening to the self, by voicing the unsaid, by listening to others and staying alert to all the currents and undercurrents of life about them, by imagining themselves inside the new poem or person or idea that they want to come to know and understand. (p. 141)

It is not hard to see, from the preceding quote, how the constructivist perspective might be helpful in analyzing with clarity what one is experiencing, and in also seeing connections among aspects of the experience. The constructivist perspective would enable one to analyze the model underlying a system, and to imagine what might have to be changed for the system to change. But the engagement in constructivist activities is also crucial for collaboration among activists with different perspectives.

> Compared to other positions, there is a capacity at the position of constructed knowledge to attend to another person and to feel related to that person in spite of what may be enormous differences. (p. 143)[35]

There are "enormous differences" among those who would challenge the disease model and the services associated with it. A few kinds of differences that I noted were differences in race, ethnicity, religion, degree of health, sexual orientation, professional status, education, income, commitment to career, country of residence,

---

[34] Entering the term "disease model" in a search engine in May, 2001, the engine (Google.com) found 932,000 sites. "Disease model challenge" found 326,000 sites. "Women challenge disease model" netted 160,000 sites. One will readily find list serves and information from sources around the world with search terms such as "psychiatric survivors" or "midwifery and home birth" or "cesarean section."

[35] The constructivist position is in many respects similar to Kegan's Interindividual position.

region of the world, degree of anger at the difficulties experienced, degree of traumatization, and level of hope. Perhaps most salient to this discussion, there were differences in the developmental level of the messages contributed. An example of an Internet-based women's health discussion among diverse participants was one sponsored by UNIFEM, a United Nations organization, during 1998 and 1999. The topic was ending violence against women worldwide (UNIFEM, 1999).

It seems likely to me that the members of groups such as the UNIFEM group who continue to communicate and offer support to one another across enormous differences are engaging in a constructivist process, although it is extremely unlikely that many of them would have scored at a "constructivist" level, if they were to be assessed according to adult development interview scoring procedures.[36] That is, it is unlikely that many of them commonly engaged in constructivist activities across various dimensions of their lives. Further, it seems unlikely that many of the participants would have described what they were doing in constructivist terms. It is possible for someone who usually engages in less developed processes to participate in a higher level process, as long as that process has been designed or facilitated by someone who is commonly engaged in and understands higher level activities (Basseches, personal communication, May 2000). Clearly, then, the presence of individuals whose contributions and participation in the groups reflected a constructivist perspective were instrumental in allowing for collaboration within groups in terms of provision of mutual support, information, and affirmation across differences. In addition to the presence of individuals whose contributions reflected constructivism, communication via the Internet also probably allowed for greater collaboration within groups. Developmental theorists (Belenky et al., 1986; Kegan, 1994) have not seen widespread support for the sort of activity that I observed in the online communications described above. And, since the communication is virtual rather than face-to-face, it probably lacks some of the elements expected in a fully constructivist process. Nevertheless, it appears to me that in at least some of the Internet-based groups I observed, constructivist processes were occurring, at least in relation to the limited area of shared concern. Messages that the senders called "rants" were posted alongside reports of controlled scientific research. Threads of conversation about specific concerns eventually picked up both the subjectivist, experiential and the procedural, scientific aspects of the issue, and often even wove them together into new, shared knowledge, in a co-constructivist process.[37]

Communication via the World Wide Web: Co-Construction

Indeed, the World Wide Web seems to be a highly valuable resource for women rejecting the disease model. Not only have traumatized and distressed women found willing "listeners," a requisite step in healing from trauma (Herman, 1997), but web-based information and support outlets may have contributed to making some women more confident in their efforts to negotiate with

---

[36] Procedures reflect theories of Kegan, Belenky et al., Commons, or Loevinger.

[37] In this respect, the conversations were much like those advocated by educators concerned with adult learning, such as Mezirow, Tarule, and with advocates such as Belenky.

providers regarding the care they receive. Furthermore, scientists and professionals working in contexts in which their dissenting views are not acknowledged
have found affirmation, both in the stories of women who have been patients and
in collaborative dialogue with other like-minded practitioners.

The potential long-term effectiveness of the World Wide Web as a tool of collaboration and dissent are not yet clear, although there are promising reports of effective grassroots organizing in using the Internet in areas other than women's health
(Wittig & Schmitz, 1996). Efforts at collaboration and dissent regarding the disease
model as explanation for women's health issues prior to the Web's emergence as a
powerful communication tool have been unsuccessful at slowing the ascendance of
the disease model. Yet current communications sequences appear promising.

How Does the Web Influence Communication Across Differences?   Numerous
differences exist with respect to speaking out on the Web vs. speaking up in person, at meetings, and even by phone or mail. When posting a message to a list
serve, one may reasonably expect to be responded to in terms of the content of the
message, rather than in terms of one's appearance, identity (which can be dis-
guised if desired), rank, gender, ability/disability, social class, race, ethnicity,
attractiveness, or age. In other words, there is a diminution of the impact of preju-
dice based on differences and on attributions based on rank or status when one is
communicating online as compared with communicating in person or at meetings
(McKenna & Bargh, 1998). Even where people are known to one another, status differences and expectations based on gender, status, or other observable attributes
have diminished impact when people are communicating via the Internet
(L. Moorehead, personal communication, November 2000). In almost all settings,
status differences lead to less attention and recognition being given to women.
Perhaps this is why new scholarship suggests that women assess Web-based team
projects and communications more favorably than men (Lind, 1999, 2001).

Other advantages of communicating via the World Wide Web in contrast with
telephone or in person communication include (1) the possibility of sending a
message whenever one is moved to do so, without worry about whether it is a
good time for the recipient to give their attention (or, for the recipient, the ability
to attend to messages at your convenience); (2) having time to edit what one is
about to say; (3) it is free or inexpensive; and (4) you can choose to be anonymous.
In contrast with postal service communication, e-mail is also quick and simple,
and does not require having a stamp or going to a mailbox or post office. (This last
advantage is true only for some women. Indeed, the majority of the world's
women do not have any easier access to the World Wide Web than to the postal
service. Despite optimism by some in the business sector [Kanter, 2001], I do not
think it likely that the digital divide will disappear very soon.)

EMERGING SUPPORT FOR
ALTERNATIVE MODELS

The Internet is an important resource for those seeking health information
(Berland et al., 2001; Fox et al., 2000) as well as for those who are seeking

affirmation and support in expressing their dissatisfaction, disappointment, and even the traumatization resulting from the services they have received from mainstream providers. It would seem likely that the Internet resources would allow for a more integrated and quicker process, then, for those disappointed with mainstream services to proceed through the steps beginning with dissatisfaction and moving toward consideration of establishing or utilizing existing alternative services. There are many examples of links between sites for those disappointed with services and those providing or advocating services based on alternative models—for example, a link between a site for women concerned with cesarean births (many of whom have been traumatized by their birth experiences) and a site that advocates midwifery and home birth. Or a link between a site for psychiatric survivors and a site for consumers' self-help resources.

Alternative Models in Women's Health: Common Themes

Various groups of women who resist the disease model prefer, and some are working toward, the development of alternative models such as midwifery or self-help for recipients of psychiatric diagnoses and/or services. These models pose egalitarian social connection, social well being, and a healthy environment as central forces of health (Chamberlin, 1978; Lewis & Bernstein, 1996; LoCicero, 2001; Miller & Stiver, 1997). Similarly, social disconnection, violence, inequity, and abuse of the environment are seen as core factors in the etiology of disease (K. A. Greene, personal communication, January 2001; Jordan, 2000). In this sort of model, the relationship between provider and recipient is a crucial element in treatment. That is, since, in this model, egalitarian social connection is seen as health promoting, providers and recipients must practice such connection in service-provision encounters.

Many women's health advocates believe in individual differences in response to treatment, and seek to protect the *recipient's* right to choose from among a number of approaches to health and treatment. Advice and perspective from providers should include knowledge of the results of controlled studies, the provider's point of view, and alternative perspectives. This more inclusive stance puts standard medical treatments on the same "menu" with alternative treatments such as meditation, home-based care, mind–body approaches, herbs, nutrition, spirituality, and self-hypnosis (K. A. Greene, personal communication, January 2001). Knowledge about care standards and about outcomes of controlled studies informs, but does not dictate, treatment. The recipient ultimately uses the provider as one of her consultants in choosing one or more from among various treatment options. If information and advice are not initially understood, continued efforts are expected to be made until the information is effectively delivered in the recipient's preferred learning style. This newer approach also demands that recipients be informed participants and decision makers, and often includes former recipients, or recipients farther along in their treatments, as mentors to newer patients.

> People need choices…Informed consent means people have access to information to make choices based on what's helpful to them. (Frado, 2000)

CONCLUSION

It seems likely that both the model of care and the processes by which coordination of women's challenges to the dominant model might take place will require ongoing constructivist activities—activities that coordinate and integrate the intuitive, experience-based knowledge processes of the subjectivist perspective with the logical, rational processes of the procedural perspective. While it may seem counterintuitive, since communication via the Internet is limited in several ways (such as by excluding nonverbal aspects of communication), the Internet seems, from my analysis, to be providing a forum that facilitates an advanced kind of communication, at least in areas of shared concern. Further, it seems likely that some of those engaged in Internet-facilitated co-constructivist processes will establish or participate in services offered in alternative service models. With the speed of communication, and the numbers of people seeking health information on the Internet, it seems likely that the numbers of recipients of services who are exposed to critiques of mainstream services will increase quickly. In addition, those who are dissatisfied with the services they receive will find support and affirmation, those who might be inclined to seek alternatives will have easy access to knowledge about them, and those who wish to establish alternatives will easily find supporters. While the pharmaceutical companies, insurance companies, medical associations, and others invested in the disease model also use the Internet, the medium itself serves a somewhat democratizing function. "Soft" factors are just as readily found by search engines as are "hard" factors. To some extent, for some period of time, the Internet appears to be leveling the field.

REFERENCES

Belenky, M. F., Clinchy, B. M., Goldberger, N. R., & Tarule, J. M. (1986). *Women's ways of knowing.* New York: Basic Books.

Berland, G. K., et al. (2001). Health information on the internet. *Journal of the American Medical Association, 285*(20), 2612–2621.

The Boston Women's Health Book Collective (Ed.) (1998). *Our bodies, ourselves for the new century.* New York: Simon & Schuster.

Brown, G. W., & Harris, T. (1978). *Social origins of depression.* New York: Free Press.

Chamberlin, J. (1978). *On our own.* New York: McGraw-Hill.

Chamberlin, J. (1995). Struggling to be born. In J. Grobe (Ed.), *Beyond Bedlam* (pp. 59–64). Chicago: Third Side Press.

Cochrane Collaborative Consumer Network (2001). *Database of Abstracts of Reviews of Effectiveness (DARE).* The University of York NHS Centre for Reviews and Dissemination. Available: http://www.cochraneconsumer.com/index.asp

Cohen, A., & Kleinman, A. (1996). Untold casualties. *Harvard International Review, 18*(4), 12–17.

Cohen, N. W. (1991). *Open season.* Westport, CT: Bergin & Garvey.

Daryl (2000). *Greta.* Childbirth.org. Available: http://www.childbirth.org/articles/stories/2000/00greta.html.

Davis-Floyd, R. E. (1992). *Birth as an American rite of passage.* Berkeley, CA: University of California Press.

DeMott, R. K. (2000). Commentary: A blatant misuse of power? *Birth, 27*(4).

Dundas, D. W. (1995). The shocking truth. In J. Grove (Ed.), *Beyond Bedlam* (pp. 33–36). Chicago, IL: Third Side Press.

Durand, M. (1992). The safety of home birth: The farm study. *American Journal of Public Health, 82*(3), 450–452.

Enkin, M., Keirse, M. J. N. C., Renfrew, M., & Neilson, J. (1995). *A guide to effective care in pregnancy and childbirth,* 2nd edit. New York: Oxford University Press.

Fausto-Sterling, A. (1992). *Myths of gender: Biological theories about women and men.* New York: Basic Books.

Fernandez, R. J. (1992). Recent clinical management experience. In J. A. Hamilton & P. N. Harberger (Eds.), *Postpartum Psychiatric Illness: A Picture Puzzle* (pp. 78–89). Philadelphia, PA: University of Pennsylvania Press.

Fox, S., et al. (2000). *The online health care revolution: How the Web helps Americans take better care of themselves.* Washington, DC: Per Charitable Trust.

Frado, L. (2000, January 2, 1000). Point of view: What I know about craziness. *Toronto Star.* Jan. 2, 2000. http://www.thestar.com/static/archives/search.html.

Gabbe, S. G., & Holzman, G. B. (2001). Obstetricians' choice of delivery. *The Lancet, 357,* 722.

Gaskin, I. M. (1997). Ina May Gaskin. In P. Chester (Ed.), *Sisters on a journey* (pp. 125–134). New Brunswick, NJ: Rutgers University Press.

Goer, H. (1995). *Obstetric myths versus research realities.* Westport, CT: Bergin & Garvey.

Gotlein, J. (1995). Bearing witness. In J. Grobe (Ed.), *Beyond Bedlam* (pp. 111–122). Chicago, Third Side Press.

Griebenow, J. (2000). Jenny's tale—saga of a birth gone wrong. GentleBirth.org. Available: http://www.gentlebirth.org/archives/jennytal.html.

Grobe, J. (1995a). *Beyond Bedlam: Contemporary women psychiatric survivors speak out.* Chicago: Third Side Press.

Grobe, J. (1995b). Hospital records. In J. Gobe (Ed.), *Beyond Bedlam* (pp. 65–72). Chicago: Third Side Press.

Herman, J. (1997). *Trauma and recovery.* New York: Basic Books.

ICAN (2000–2001). *International Cesarean awareness network.* Available: http://www.ican-online.org/.

Jordan, B. (1997). Authoritative knowledge and its construction. In R. E. Davis-Floyd & C. F. Sargent (Eds.), *Childbirth and authoritative knowledge: Cross-cultural perspectives* (pp. 55–79). Berkeley, CA: University of California Press.

Jordan, J. (2000). A model of connection for a disconnected world. In J. J. Shaw & J. Whellis (Eds.), *Odysseys in psychotherapy* (pp. 147–165). New York: Ardent Media.

Kanter, R. M. (2001). *Evolve!* Boston, MA: Harvard Business School Press.

Karon, B., & Vandenbos, G. R. (1995). *Psychotherapy of schizophrenia: The treatment of choice.* New York: Jason Aronson.

Kegan, R. (1994). *In over our heads.* Cambridge, MA: Harvard University Press.

Kleinman, A., & Cohen, A. (1997). Psychiatry's global challenge. *Scientific American, 276*(3), 86–89.

Kuhn, T. S. (1996). *The structure of scientific revolutions.* Chicago: University of Chicago Press.

Lauer-Williams, J. (2001). *Postpartum depression: A phenomenological exploration of the woman's experience.* Massachusetts School of Professional Psychology, Boston, MA.

Lazarus, E. S. (1988). Theoretical considerations for the study of the doctor–patient relationship: Implications of a perinatal study. *Medical Anthropology Quarterly, 2*(1), 35–60.

Lerner, B. (2001). *Breast cancer wars.* New York: Oxford University Press.

Lewis, J. A., & Bernstein, J. (1996). *Women's health: A relational perspective across the life cycle.* Sudbury, MA: Jones and Bartlett.

Lewontin, R. C. (2000). *The triple helix: Gene, organism, and environment.* Cambridge, MA: Harvard University Press.

Lind, M. R. (1999). The gender impact of temporary virtual work groups. *IEEE Transactions on Professional Communication, 42*(4), 276–285.

Lind, M. (2001). An exploration of communication channel usage by gender. Work Study, *50*(6), 234–240.

LLLI (2001). *La Leche League International.* Available: http://www.lalecheleague.org/.

LoCicero, A. (2001). Women's health and the world wide web. Paper presented at the New England Sociology Association Meeting, April 28, 2001, Fairfield, CT.

LoCicero, A. K. (1993). Explaining excessive rates of Cesareans and other childbirth interventions: Contributions from contemporary theories of gender and psychosocial development. *Social Science and Medicine, 37*(10), 1261–1269.

Loweenstein, G. F., Weber, E. U., Hsee, C. K., & Welch, N. (2001). Risk as feelings. *Psychological Bulletin, 127*(2), 267–286.

MANA (2001). *Midwives Alliance of North America.* Available: http://www.mana.org/.

McInnes, S. C. (2000). *I am you.* Available: http://rondak.org/sasha1.htm.

McKenna, K. Y. A., & Bargh, J. A. (1998). Coming out in the age of the Internet: Identity "demargnialization" through virtual group participation. *Journal of Personality and Social Psychology, 75*(3), 681–694.

Members, L. S. (1999). *End violence list serv* [List Serve]. Reviewed in UNIFEM (1999) Women @ work against violence: Voices in cyberspace. New York: United Nations Development Fund for Women.

Miller, J. B., & Stiver, I. P. (1997). *The healing connection.* Boston, MA: Beacon Press.

National Association for Rights Protection and Advocacy (2000) *Judi Chamberlin. National Association for Rights Protection and Advocacy*-. Available: http://www.connix.com/~narpa/chamberlin.htm.

National Center for Health Statistics (2001). *Fasts Stats A to Z.* Available: http://www.cdc.gov/nchs/fastats/men.htm.

National Women's Law Center, FOCUS, Lewin Group. (2000). *Making the grade on women's health: Executive summary* (Executive Summary). Washington, DC: National Women's Law Center, FOCUS, The Lewin Group.

Nechas, E., & Foley, D. (1994). *Unequal treatment.* New York: Simon & Schuster.

Oakley, A. (1986). *The captured womb.* New York: Basil Blackwell.

Office of Women's Health, Food and Drug Administration (1999). *Women's health: Take time to care.* US Food and Drug Administration. Available: http://www.fda.gov/womens/tttc.html.

Penfield Chester, E. (1997). *Sisters on a journey.* New Brunswick, NJ: Rutgers University Press.

Perry, W. G. (1970). *Forms of intellectual and ethical development in the college years.* New York: Holt, Rinehart, & Winston.

Rodwin, M. A. (2001). The politics of evidence-based medicine. *Journal of Health Politics, Policy, and Law, 26*(2), 439–446.

Rooks, J. P., Weatherby, N. L., Ernst, E. K. M., Stapleton, S., Rosen, D., & Rosenfield, A. (1989). Outcomes of care in birth centers: The National Birth Center Study. *New England Journal of Medicine, 321*(26), 1804–1811.

Ross, C. A., & Pam, A. (1995). *Pseudoscience in biological psychiatry.* New York: John Wiley & Sons.

Schulman, R. (2001). *Volunteers in psychotherapy,* Colloquium presented at Suffolk University, Boston, MA, Feb. 7, 2001.

Scot, R. (1972). *The discovery of Witchcraft.* New York: Dover.

Services, Learning and Information Services (2001). *Evidence-based medicine.* Available: http://www.herts.ac.uk/lis/subjects/ health/ebm.htm.

Shimrat, I. (1997). *Call me crazy: Stories from the mad movement.* Vancouver, BC: Press Gang.

Szasz, T. (1994). *Cruel compassion.* Syracuse, NY: Syracuse University Press.

Tew, M. (1993). Do obstetric interventions make birth safer? *British Journal of Obstetrics and Gynecology, 93*(7), 659–674.

UNIFEM (1999). *Women at work to end violence.* New York: United Nations Development Fund for Women.

Unruly Women (2000). Unruly Women List. Available at http://www.topica.com/lists/unrulywomen/

Unzicker, R. E. (1995). From the inside. In J. Grobe (Ed.), *Beyond Bedlam* (pp. 13–18). Chicago, IL: Third Side Press.

Ussher, J. (1991). *Women's madness: Misogyny or mental illness.* Amherst, MA: University of Massachusetts Press.

Venter, C. (2001). *Genome research* (Weekend Edition). NPR Online. Available: http://search.npr.org/cf/cmn/cmnpd01fm.cfm?PrgDate=02%2F11%2F2001&PrgID=10.

Venter, C., et al. (2001). The sequence of the human genome. *Science, 291*(5507), 1153–1158.

Wagner, M. (2000). *Technology in birth: First do no harm.* Midwifery Today. Available: http://www.midwiferytoday.com/articles/technologyinbirth.asp.

Walters, D. C. & Quillinan, E. (1999). *Just take it out.* Mt. Vernon, IL: Topiary.

Wertz, R. W., & Wertz, D. C. (1977). *Lying-in: A history of childbirth in America.* New York: Free Press.

Wittig, M. A., & Schmitz, J. (1996). Electronic grassroots organizing. *Journal of Social Issues, 52*(1), 53–70.

Woods, A. (2000–2001). The birth of my first 3 children[sic]. International Caesarian Awareness Network. http://www.ican-online.org/info/bs-anitaz.htm

Women's Institute for Childbearing Policy (1994). *Childbearing policy within a national health program: An evolving consensus for new directions.* Boston, MA: Women's Institute for Childbearing Policy.

CHAPTER 18

Gender Differences in Intellectual and Moral Development?

The Evidence that Refutes the Claim

MARY M. BRABECK AND ERIKA L. SHORE

When Samuel Johnson was asked, "Who is smarter, men or women?" he is reported to have answered, "Which man? Which woman?" (As quoted in Nicholson, 1984, p. 77). With this seemingly simple response, Johnson moved out of the traditionally dichotomous paradigm that has guided much of psychology's exploration of gender differences. For decades, philosophers, historians, and social science researchers have focused on the differences between men and women, often entering their investigations with the implicit assumption that differences exist, and then attempting to explain these differences with anecdotal and/or empirical "evidence." Psychological theory and research on intellectual and moral development have been no exception.

When gender differences in morality are asserted, such claims are frequently linked with gender differences in reasoning. Historically, assertions of gender differences have disadvantaged women in comparison to men, characterizing their moral thinking and reasoning as less developed. In this chapter, we review the theoretical claims and the empirical support for gender differences in moral and intellectual development. We focus on two of the most often cited (Hurd & Brabeck, 1997) developmental theories that assert gender differences in moral orientation (Gilligan, 1982; Lyons, 1983; Noddings, 1984) and ways of knowing/epistemology (Belenky, Clinchy, Goldberger, & Tarule, 1986). We trace the historical context

---

MARY M. BRABECK AND ERIKA L. SHORE • Lynch School of Education, Boston College, Chestnut Hill, Massachusetts, 02467.

Handbook of Adult Development, edited by J. Demick and C. Andreoletti. Plenum Press, New York, 2002.

from which these theories were developed and popularized. We then describe and critique the theoretical claims made by gender theorists and examine the evidence supporting their claims. We conclude with a discussion of recent research in the area of gender and moral development and the impact of this work on the field.

HISTORICAL CLAIMS OF GENDER DIFFERENCE IN MORAL REASONING AND EPISTEMOLOGICAL DEVELOPMENT

That there continues to be prolific writing arguing for or against gender differences in men and women's moral reasoning (Brabeck, 1983, 1989; Jaffee & Hyde, 2000) is not surprising given the strong historical roots of these arguments. Historically, the fields of psychology and philosophy observed the distinctly divided roles of men and women and sought to explain, expand on, and ultimately support the validity of these respective positions and roles in society.

Arguing from a Darwinian paradigm, discussions of gender differences in moral reasoning initially were attributed to "natural," biologically destined differences between the sexes. In his classic essay, "Psychology of the Sexes," Herbert Spencer (1875) stated,

> The love of the helpless, which in her maternal capacity woman displays in a more special form than man, inevitably affects all her thoughts and sentiments; and, this being joined in her with a less developed sentiment of abstract justice, she responds more readily when appeals to pity are made than when appeals are made to equity. (p. 36)

While ultimately suggesting that these gender differences may change as society evolves, Spencer attributes woman's unique capacity for caring to her role and identity as a mother, which he then intertwines with his assertion that women are also less developed intellectually. The conclusion is that women are not "fit" to reason abstractly, and, by extension, less able to apply "abstract justice" to moral dilemmas. Here, we clearly see a mutually exclusive construction of care and justice moral perspectives that are linked to one's intellectual capacity and gender. Spencer supports his assertion of woman's care-based morality by pointing to her more concrete, proximal, or experientially grounded intellect. He writes:

> A further tendency, having the same general direction, results from the aptitude which the feminine intellect has to dwell on the concrete and proximate rather than on the abstract and remote. The representative faculty in women deals quickly and clearly with the personal, the special, and the immediate; but less readily grasps the general and the impersonal. (p. 36)

In the words of Spencer, woman's morality is limited by her intellectual capacity. Furthermore, we hear Spencer's strong essentialist and dichotomous construction of moral reasoning along gender lines.

Extending Spencer's argument, Freud (1925/1961) claimed that as a result of their inferior superego women possessed less sense of justice than men, and therefore had a greater propensity to be influenced by feelings. He stated, "For women the level of what is ethically normal is different from what it is in men, ... [women] show less sense of justice than men, ... are less ready to submit to the great exigencies of life, ... they are more often influenced in their judgments by feelings of affection or hostility" (Freud, 1925/1961, pp. 257–258). Like Spencer, Freud equates mature morality with rationality and the ability to think abstractly, free from the influence of one's emotions. Both view women's tendency toward affective responses as limiting the development of a superior, rational morality, and both support a dichotomization and the separation of thoughts and feelings, a common construction in research literature today.

Many prominent figures in psychology, such as Albert Bandura (1973), Erik Erikson (1968), and Jean Piaget (1932), have contributed to our thinking about morality and moral development. However, Lawrence Kohlberg's (1969) theory of the development of principled moral thinking has dominated the field. Building on the work of Piaget, Kohlberg developed a structural model of cognitive moral development that posited a sequential, hierarchical, and universal pattern of moral development across the lifespan. Kohlberg's (1969, 1981, 1984) theory was based on a notion of morality in which the ability to think abstractly and apply universal principles of justice was considered the ultimate goal of moral development. In a longitudinal study of his six-stage model, Kohlberg and Kramer (1969) reported that women were more likely to score in stage 3 ("Interpersonal Concordance") and men were more likely to score in stage 4 ("Law and Order"). Based on the proposed hierarchical ordering of stages, this result led to the conclusion that men were morally superior to women.

Kohlberg claimed that moral development, like cognitive development, resulted from an interaction between maturation (age) and experience. Given that the male and female participants in his original study were similar in age, Kohlberg attributed the observed gender differences in moral development to differential role assignments in society. He claimed that because men had greater access to education and were more likely to be in high status professional roles, they had greater opportunity to acquire and apply abstract principles of justice to moral issues that arose while in these roles. Men's more advanced moral development was therefore due to greater experience with roles that demanded abstract, justice-oriented moral reasoning.

Influenced by Kohlberg's stage model, William Perry (1970) charted a theory of epistemological development among youth and young adults. Perry's theory posited nine levels of intellectual development proceeding through three broad stages: dualistic, multiplistic/relativistic, and committed reflective judgments. Building on the work of Perry (1970), Kitchener and King (1981) developed their theory of Reflective Judgment (Kitchener & King, 1981). Kitchener and King sought to separate identity development (the commitment part of Perry's theory) from reasoning (the epistemological part of Perry's theory) so as to focus only on intellectual development. To test their theory, they developed the Reflective Judgment Interview (Kitchener & King, 1981). This instrument has been used in a large number of studies of adolescent and adult development (King & Kitchener, 1994). Although Perry never claimed to study gender and intellectual development, his original study and subsequent theory building were based on an all-male sample from Harvard University.

WOMEN RESPOND TO CLAIMS OF GENDER
DIFFERENCES IN MORAL AND
EPISTEMOLOGICAL DEVELOPMENT

Beginning in the 1960s and 1970s, feminist psychologists began arguing against claims suggesting women's psychological inferiority to men. They articulated how research on women had been largely ignored, that psychological theories were developed on male participants, and that these theories were then inappropriately generalized to women.

In response to the bias against women in psychological theories (e.g., Freud), the field of androgyny burgeoned. Androgyny theorists attempted to show that gender differences in personality and cognitive abilities were negligible, if they existed at all. Those differences that were found were attributed to differences in socialization (Maccoby & Jacklin, 1974). Arguing against gender differences (e.g., Bem, 1974), they suggested that women and men were more similar than different; that women could be just as assertive, competitive, analytic, dominant, ambitious, and aggressive as men. However, feminist psychologists in the 1980s (e.g., Morawski, 1985) began to point out that, by suggesting that these qualities could and should be adopted by women, androgyny theories maintained an androcentric view, privileging traditional male characteristics. Post-androgyny researchers (Morawski, 1985; Ruddick, 1989) argued that the corrective contained its own problems. Recognizing the inadvertent elevation of these traditionally "masculine" qualities and the implicit denigration of traditional characteristics associated with women, other feminist thinkers and researchers pointed to the inadequacy of this model (Gilligan, 1977), and began creating and advocating for a new norm that celebrated alternative, "feminine" virtues (Ruddick, 1989).

Responding in particular to the work of Lawrence Kohlberg (1969) and William Perry (1970), Gilligan (1982) and Belenky et al. (1986) put forth theories of moral and epistemological development, respectively, that were grounded on research with women, and informed by a feminist lens that viewed women's experiences and attributes as strengths rather than deficits. Characteristics associated with women, they argued, were not better or more superior to men's moral and epistemological development, just "different."

Nevertheless, in an attempt to ascribe more value to traditionally "feminine" characteristics or qualities and give "voice" to perspectives not highlighted by men's experience, Gilligan (1977, 1982) makes a difference argument similar to that of her predecessors (e.g., Freud, 1925; Spencer, 1875). She states, "The psychology of women that has consistently been described as distinctive in its greater orientation toward relationships and interdependence implies a more contextual mode of judgment and a different moral understanding" (Gilligan, 1982, p. 22). In later work, Brown, Gilligan, and Tappan (1995) trace the roots of the two respective moral orientations back to infancy. They posed that the dynamics of inequality and attachment in infancy lay the foundation for two moral visions, a justice and a care orientation. These two different visions lead to different ways in which one assesses moral thoughts, feelings, and actions. Justice and care constitute distinct organizing frameworks (Gilligan & Attanucci, 1988), with developmental origins in early childhood. Individuals "prefer" one voice over another even though children as well as adults are capable of using both.

Brown et al. (1995) claim that it is possible to "differentiate transformations that pertain to equality (and justice) from transformations that pertain to attachment (and care)" (p. 315). Ultimately, they link these problems with inequality (equality) and detachment (attachment) to gender, positing that they are organized differently in male and female development. Gilligan and colleagues claim these differences persist throughout adulthood (Gilligan & Wiggins, 1987).

Belenky et al. (1986) took a similar approach, attempting to map the terrain of women's (as distinct from men's) ways of knowing. They conducted interviews with women and concluded that as they "examined their [women's] accounts of what they learned and failed to learn, of how they liked to learn, some common themes emerged, themes that are distinctively, although surely not exclusively feminine" (Belenky et al., 1986, p. 191). Despite the fact that they included only women in their original studies, Belenky et al. and Gilligan both claim their findings identify a different way of knowing and a distinct moral voice. While not claiming that these differences are exclusive to women, the oft repeated association of women with Gilligan's "different [from men] voice" and Belenky et al.'s "women's ways of knowing" has led many to believe otherwise (Brabeck & Larned, 1997; Hurd & Brabeck, 1997).

EMPIRICAL INVESTIGATIONS OF GILLIGAN'S THEORY

The work of Gilligan (1979, 1982) and Belenky, Clinchy, Goldberger, and Tarule (1986) sought to bring women's perspectives, experiences, and "voices" into focus in psychological investigations. Importantly, their work filled in what had traditionally been left out of psychological theories and reframed those "feminine" aspects that had traditionally been devalued, pathologized, or defined as "less than" when compared to a male norm. A great deal of thoughtful and critical debate and research has resulted.

As indicated in her title, *In a Different Voice*, Gilligan (1982) emphasized gender differences in moral orientation. She argued that the ethics of justice and of care are equally important moral orientations, linked to gender. Although men and women understand and can use both ethical orientations, Gilligan claims that men favor an orientation of justice and women favor an orientation of care. Explicit in her work is the notion that women are more connected to others, and value and are more concerned with maintaining relationships than men.

Gilligan's original claim that women are guided by a different moral orientation or moral "voice" than men was based on three studies she conducted (Gilligan, 1982). In her first study, she interviewed 29 women faced with the decision about whether or not to have an abortion. The women, ages 15 to 23, were interviewed before and 2 years after making this decision. Twenty-one of the women were re-interviewed. Using a similar methodology, Gilligan interviewed 25 students in their senior year of college and then again 5 years later. Finally, in her third study, referred to as the "Rights and Responsibility" study, Gilligan intensively interviewed 2 males and females between the ages of 6 and 60 years for a total of 36 interviews.

Based on her in-depth interviews with women facing a decision about abortion, Gilligan described three levels and two transition periods in the development of a care orientation. Level I is called "Orientation to Individual Survival." Individuals at this level are concerned with their own well being. This was followed by a transition "from Selfishness to Responsibility." Level II is referred to as "Goodness in Self Sacrifice." At this level, an individual's moral concerns center around care of others. The second transition is "From Goodness to Truth," followed by the third and final level of the theory of care, "The Morality of Nonviolence." Gilligan did not report evidence for the validity of these levels, and attention to the developmental levels was dropped in favor of a more global understanding of the ethic of care as evidenced in attention to relationships, harmony, and responsibilities to others (Lyons, 1983; Rogers, Brown, & Tappan, 1994). As reviewed later, some researchers have begun to develop measures (e.g., The Ethic of Care Interview) to assess the developmental aspects of Gilligan's theory (Skoe & Marcia, 1991; Skoe & Diessner, 1994).

Research examining gender and moral reasoning development as theorized by Kohlberg and Gilligan has taken various approaches. Some studies have examined whether or not there is bias in Kohlberg's theory, and asked whether there are gender differences in moral reasoning as measured by Kohlberg and his colleagues. Other studies have examined whether there are indeed two distinct moral orientations (care and justice) and if use of these orientations is linked to gender. These studies ask if there are gender differences in moral orientation. We take up each of these questions next.

Are There Gender Differences in Justice Reasoning?

A great deal of research has examined whether there are gender differences (and possible bias) in Kolberg's model of moral development. Several meta-analytic studies have been conducted to examine this question. In his meta-analysis of 152 samples, Walker (1984) found no significant gender differences in moral reasoning. He reported similar findings in a later meta-analysis, with gender only accounting for 1/20th of 1% of the variance in moral reasoning scores (Walker, 1986). The longitudinal work of Colby, Kohlberg, Gibbs, and Lieberman (1983) supports this result as well, finding that females are as likely as males to advance in the sequential order of development as predicted by Kohlberg's theory.

Building on the results of previous studies of gender and moral development and making several methodological improvements, Thoma (1986) conducted meta-analyses and secondary analyses on 56 samples of more than 6000 participants. All samples included participants who had responded to the *Defining Issues Test* (DIT), Rest's (1979) objective measure of moral reasoning of concepts of justice. Thoma (1986) reported that across age/educational level, females scored significantly higher than males in moral reasoning. Yet, consistent with Walker (1986), gender was a weak predictor of moral reasoning. Gender explained less than half of 1% of the variance in moral reasoning on the DIT. Notably, age/education was 250 times more powerful than gender in explaining the variance in moral reasoning. Thoma's samples were in college and high school.

Bebeau and Brabeck (1989) conducted another meta-analysis of Rest's (1979) DIT scores, for seven groups of adult dental students. Their findings were consistent with Thoma's (1986), reporting a negligible difference in moral reasoning between males and females favoring women, and a small effect size for gender.

More recent empirical research continues to support the claim that men and women do not differ in their moral reasoning (Conley, Jadack, & Hyde, 1997; Garmon, Basinger, Gregg, & Gibbs, 1996; Jadack, Hyde, Moore, & Keller, 1995; Skoe & Diessner, 1994; Snarey, 1998; Walker, 1984, 1989; Walker, Pitts, Hennig, & Matsuba, 1995). When gender differences in moral in reasoning are found, it is most often in samples of adolescents (Garmon et al., 1996). In a longitudinal, cross-sectional study of 44 white, well-educated, middle-class participants, Armon and Dawson (1997) reported no gender differences in moral reasoning. Interestingly, however, women's moral reasoning scores increased significantly after young adulthood, but men's moral reasoning did not change. This result points to the potentially more complex relationship between gender and moral reasoning during both adolescence and adulthood.

As researchers are developing a more complex understanding of moral development, other interesting findings regarding gender and moral development are emerging. In a study examining the relationship between ego development and moral development, Snarey (1998) found no gender differences in moral reasoning, as measured by Kohlberg's Moral Judgment Interview (MJI). However, when comparing moral development and ego development scores, men were more likely to have higher moral development scores than ego development. For females the relationship was reversed, with females more likely to have higher ego development scores than moral development scores (Snarey, 1998).

Researchers also have examined whether one's moral stage is a trait, independent of context, or rather a state, dependent on context. Researchers have examined this question by exploring whether individuals are consistent in their use of moral orientation and their stage of moral development across various moral dilemmas. The findings have been mixed. Walker et al. (1995) found consistency in moral judgment stage scores, with 80% of male and female participants evidencing the same or adjacent level of moral reasoning across three dilemmas (a "recent" dilemma, a "most difficult" dilemma, and a "prototypic" dilemma). However, Wark and Krebs (1996) reported that most participants in their study did not base their moral judgment on one stage-structure or one moral orientation. For example, 85% of participants made judgments that spanned three to six substages in Kohlberg's model. This study suggests moral development scores may depend on the content of the dilemma rather than individual characteristics such as gender.

Researchers have also examined the effect of the "type" of moral dilemma on moral reasoning. Walker, deVries, and Trevethan (1987) compared participants' stage of moral reasoning when responding to hypothetical vs. real-life dilemmas. They found that participants had higher moral stage scores (justice scores) when responding to hypothetical dilemmas than when responding to real-life dilemmas. More recent empirical research has supported this finding (Wark & Krebs, 1996).

Whether one is personally involved in a moral dilemma or not affects one's moral reasoning score. When given a choice, the type of experience identified as a moral dilemma may be gender linked. In one study, participants were asked to recall two real-life incidents involving a moral dilemma, one that involved them personally and one that did not. Women had significantly higher moral maturity score than males on "personal" real-life dilemmas (Wark & Krebs, 1996). However, the gender difference found seemed to be due to the fact that men chose more justice-pulling, antisocial dilemmas (stage 2) and women chose more prosocial, care-pulling dilemmas (stage 3).

Conley et al. (1997) examined moral reasoning of men and women; half of the sample had genital herpes, a sexually transmitted disease (STD), and the other half did not. The authors examined the relationship between participants' stage of moral development, moral orientation, gender, and STD status. They presented the participants with four dilemmas. Two of the dilemmas were Kohlberg's traditional dilemmas, and the other two were created from salient issues related to the subject of sexually transmitted diseases. The authors used Walker et al.'s (1987) method to score for Kohlberg's stages of moral reasoning and Lyon's (1983; 1988) scoring method for Gilligan's moral orientation.

While the authors did not find significant gender differences in moral reasoning, they did find that participants with herpes had higher levels of moral reasoning than those without herpes. Furthermore, the authors found that participants with more frequent and more painful STD outbreaks had lower mean weighted scores on Kohlberg's stages of moral reasoning. The authors speculate that one may have a greater propensity to be self-focused and therefore reasoning at a lower stage of moral development when enduring painful symptoms. Based on their results, the authors point to the need for new moral development theories that address "domain specific experience and knowledge" (Conley et al., 1997, p. 265). These results points to the complex relationship between moral reasoning, the personal impact or experiential knowledge of the dilemma, and one's emotional and/or physical experience with the dilemma.

In conclusion, empirical research to date does not support the claim of gender differences in justice reasoning. However, the field has begun to grapple with more complex questions. For example, under what conditions may an individual employ or utilize a more or less "advanced" moral position? How might other variables such as ego development or personal experience with a particular moral dilemma affect moral reasoning and how might these factors be linked to gender? Such questions imply that men and women can hold multiple perspectives on moral issues. These kinds of questions also enable us to begin examining and understanding the complex interactions between the individual and his/her context that lead to a particular moral stance. While Gilligan's assertion that Kohlberg's model of moral development is biased against women has not withstood the test of empirical scrutiny, her work has pointed to the need to consider contextual circumstances as intertwined with an individual's moral reasoning.

Are There Gender Differences in Care Moral Orientation?

Gilligan has claimed that there are two moral orientations, and that while women may be socialized to do well in tests that measure justice reasoning women are more likely to use care reasoning when given a choice, and are more likely to score higher on care measures than are men. Research indicates that care and justice moral orientations, are distinct and can be identified, especially in real-life dilemmas (Skoe, Pratt, Matthews, & Curror, 1996). Furthermore, both men and women are aware of both and use both orientations (Daniels, D'Andrea, & Heck, 1995; Johnston, 1988; Snarey, 1998; Walker et al., 1995; White, 1994). If they do not spontaneously use both orientations, even children can produce the other orientation when asked to do so (Daniels et al., 1995; Johnston, 1988).

While initial evidence (Gilligan, 1982; Gilligan & Attanucci, 1988) suggested that gender differences in Gilligan's care orientation existed, researchers began to question the methodological confounds in these studies. They raised questions about how the "type" of dilemma employed in the study or chosen by the participants may affect the gender differences observed. In a meta-analysis, Walker et al. (1987) found that the type of dilemma discussed (e.g., a moral dilemma involving a personal issue or relationship vs. a dilemma involving conflicting claims about individual rights) was a better predictor of moral orientation than gender. They also found that when women chose their own dilemmas, they were more likely to choose personal ones, whereas men were more likely to choose impersonal dilemmas (Walker, 1991). Thus, it was unclear whether the gender differences found in moral orientation were due to gender or rather to the type of dilemma generated by the participants (Clopton & Sorell, 1993; Krebs, Vermeulen, Denton, & Carpendale, 1994; Wark & Krebs, 1996). When asked to choose a moral dilemma, women generate more personal dilemmas, which more frequently involve care reasoning, and males generate more impersonal dilemmas, which more frequently involve justice reasoning (Skoe et al., 1996; Walker et al., 1987).

Skoe et al. (1996) found that significantly more women than men reported interpersonal dilemmas and significantly more men than women reported impersonal dilemmas. Women were more likely to discuss family issues, whereas men were much more likely to discuss work, even though women and men both had previous work experience. Similarly, Wark and Krebs (1996) found women more likely to report prosocial types of dilemmas and men as more likely to report antisocial dilemmas. Sixty percent of antisocial personal dilemmas reported by men involved violations of rules, laws, and principles of fairness. However, the rate was only 25% for women. Interestingly, 75% of antisocial dilemmas reported by women involved violations of trust, social obligation, and bonds compared to only 35% for men. Nevertheless, all gender differences were eradicated when the researchers controlled for "type" of dilemma content.

In the study described above on morality among adults with genital herpes, Conley et al. (1997) did not find significant gender differences in moral reasoning (see also Jadack et al., 1995). However, they did find that moral orientation depended on the type of dilemma presented. Overall, participants were more likely to use a care orientation for the STD dilemma and more likely to use a justice orientation for Kohlberg's dilemmas. However, participants with herpes had higher levels of moral reasoning and were more likely to use a care orientation for the STD dilemma than those without herpes. Those without herpes were more likely to use a justice orientation in the STD dilemma, continuing to support the notion that one is likely to utilize different moral orientations depending on whether the dilemma is personal or hypothetical.

More recently, new measures of moral orientation and the developmental aspects of Gilligan's ethic of care have been developed and utilized in studies of gender and moral orientation (Gibbs, Basinger, & Fuller, 1992; Liddell, 1998; Skoe & Marcia, 1991; Sochting, Skoe, & Marcia, 1994). In a large study with more than 500 participants, ranging in age from 9 to 81 years old, Garmon et al. (1996) used a new measure assessing both moral stage and orientation (the Sociomoral Reflection Measure). The SRM-SF "is efficient in that it not only allows free, ecologically valid responding ... but also permits standard assessment of *both* (italics added) content and stage of moral judgment." (Garmon et al., 1996, p. 421.) Overall, this measure strikes a balance between Kohlberg's hypothetical dilemmas and Gilligan's open-ended dilemmas and appears to have relatively promising psychometric properties. The SRM-SF uses lead-in statements and general evalua- tion questions that structure or guide a participant to recall a particular type of moral dilemma. The SRM-SF presents a way to eradicate the dilemma artifact (cre- ated by asking participants to draw on their own experience) and still attain "eco- logically valid" responses. Garmon et al. (1996) report that girls and women more frequently reported "empathic role taking" than men, a construct analogous to Gilligan's care orientation. However, they were also more likely than males to evi- dence "ethical ideality," a construct similar to Gilligan's justice orientation.

In an effort to measure the development of care-based moral reasoning, Skoe and Marcia (1991) developed the Ethic of Care Interview (ECI). The ECI is a struc- tured interview format consisting of four dilemmas, with three standardized interpersonal dilemmas and one participant-generated dilemma. The interview yields a total score and a classification into one of five discrete levels (survival, responsibility, conventions of goodness, reflective care perspective, and ethic of care). Skoe and Diessner (1994) found no gender differences in responses to the ECI, MJI, or Marcia's Ego Identity measure. However, the relationship between scores on the ECI and identity was stronger for women than for men, pointing to Gilligan's assertion that a care-based identity may be particularly salient for women. This finding is supported by the previous research cited that shows that when given a choice, women are more likely to choose personal and prosocial dilemmas. This suggests that women choose dilemmas that feel congruent with their identities as women.

In a longitudinal study of adults ages 35 to 80, Skoe et al. (1996) examined the relationship between level of care reasoning, cognitive (higher levels of inte- grative complexity, level of authoritarianism, cognitive reflectiveness) and social (level of social interactional opportunities) resources, and adaptation to life tasks. Results again revealed women to have higher scores on the ECI than men. This study also provided further support for the construct validity of both the ECI and Kohlberg's MJI in that scores, as would be expected, on both these measures were negatively correlated with authoritarianism and positively correlated with inte- grative complexity of reasoning. Also, higher levels of perspective taking were predictive of more advanced care reasoning in both genders. Consultation with others was also positively correlated with care scores for both men and women, but not with justice stage scores.

In a related study of 90 college students, Sochting, Skoe, and Marcia (1994) found that sex role orientation, as measured by the Personal Attributes Ques- tionnaire (Spence, Helmreich, & Stapp, 1974), was a stronger predictor of care- oriented moral development than gender. Using the same measure, however, Wark and Krebs (1996) found no significant effects for gender role.

Liddell (1998) examined the relationship between a quantitative measure of moral orientation (MMO) developed for college students (Liddell, 1990) and a modified version of the more common semistructured interview protocol (Gilligan, 1982; Lyons, 1983). The interview consisted of three parts: self-description (whether participants describe themselves as separate or connected), moral dilemmas (in which the participant generates his or her own moral dilemma), and a discussion of the moral issues in one of the dilemmas. So as to make the dilemmas comparable, Liddell chose a dilemma in the interview that matched a dilemma presented in the MMO. Liddell found no significant relationship between the moral voice identified in the interviews and the moral voice measured by the MMO. However, she did find a significant positive correlation between participants' self-descriptions (separate vs. connected) and moral voice (as measured quantitatively and qualitatively) and a negative correlation between the Care-connected constructs and the Justice-Separate constructs. Although her findings are based on a small sample size ($n = 37$), Liddell reports that women scored significantly higher than men when comparing the care scores on the interviews.

Methodologically, what is problematic in this and many studies of moral orientation is the lack of information given about the process of defining and rating the constructs under investigation, and how different measures are related. Without a sufficient body of empirical evidence from studies that use the same valid and reliable measures, no firm conclusions about gender and moral orientation can be made.

Similar to the research on gender and moral development, the research on gender and moral orientation leads us to similar conclusions: that moral orientation is gender linked in that gender plays a role in individuals' choice of moral dilemma, experience with particular dilemmas, and aspect of identity and development.

WOMEN'S WAYS OF KNOWING: EMPIRICAL EVIDENCE ON GENDER DIFFERENCES IN EPISTEMOLOGICAL DEVELOPMENT

Based on their own experience of female students struggling with omissions in their learning and lacking confidence in their intellectual abilities together with their awareness that models of epistemological development had been constructed based on studies of men only (Perry, 1970), Belenky et al. (1986) focused their study on the ways in which women "know." In particular, they responded to Perry's (1970) study of epistemological development with men. Using an "intensive interview/case study approach," with a diverse (age, ethnicity, class, rural/urban) group of women ($n = 135$), Belenky et al. (1986) developed five "knowledge perspectives" (silence, received knowledge, subjective knowledge, procedural knowledge, and constructed knowledge) that described women's epistemologies, how they know what they know. The authors analyzed their data using a contextual analysis. However, it is unclear how the authors reached consensus about their categories and what the degree of congruence was in their ratings. The authors state that their model is not developmental. Based on retrospective reports, they "speculate" about different potential "developmental sequences or trajectories" (Belenky et al., 1986, p. 15).

Unlike work following Gilligan's original studies, there continues to be little empirical evidence to support the claims made in Belenky et al.'s (1986) original study. There appear to be no efforts in this area of study to develop more objective and randomized studies to validate their claims. In addition, the empirical research uses varying measures of "ways of knowing" and often employs small sample sizes (Brabeck, 1993). Often critical information about the scoring of the interviews and how the authors developed and rated the constructs is not provided.

Part of the problem in examining the claims regarding Women's Ways of Knowing (WWK) is that the construct is too broadly defined (Meyers, 1996). While Belenky et al. (1986) have clearly tackled a complicated aspect of women's lives and experience, they do not define women's ways of knowing in a way that enables researchers to assess the reliability and validity of their work (Brabeck, 1993; Brabeck & Larned, 1997). Belenky et al. (1986) claim:

> We examine women's ways of knowing and describe five epistemological perspectives from which women view reality and draw conclusions about truth, knowledge, and authority. We show how women's self-concepts and ways of knowing are intertwined. We describe how women struggle to claim the power of their own minds. We then examine how the two institutions primarily devoted to human development—the family and the schools—both promote and hinder women's development. (Belenky et al., 1986, pp. 3–4)

Epistemology, according to Goldberger and colleagues (1996), involves identity, one's ability to identify and analyze the relationship of gender, power, equity, gender role socialization, and self-awareness. A thorough examination of the empirical validity of the claims that women and men differ in "ways of knowing" would entail examination of gender differences in identity, self-concept, self-awareness, gender role identity, gender role socialization, reflection on gender and power, and knowledge content.

Belenky et al.'s own work is further problematic because they studied only women. Without a male comparison group, it is difficult to know whether the observations regarding ways of knowing can be attributed to gender (as the authors claim) or to some other factor. For example, most of the women in the "silence" category were from low socioeconomic and educational backgrounds; socioeconomic status or education level rather than gender may be the critical factor in development of this epistemological stance. As Unger (1989) notes, it is difficult to tell whether "silence" (level 1) may mean conformity to the press of the public context or may say something more representative of the woman in the context (perhaps this is an artificial distinction and one cannot separate self and context).

Debold, Tolman, and Brown (1996) put it this way: "Trying to hear women's voices without an explicit critique of developmental discourse as an expression of power relations, and the kind of self/subjectivity inherent in it leads to an ambiguity that can subject the women themselves to negative interpretations of their capacities and knowledge" (p. 96). They also assert that WWK posits a notion of the self as unitary. "Assuming a unitary epistemological perspective obscures power struggles that women may necessarily incorporate as splits in subjectivity and different ways of knowing... 'self' is used to convey a global sense of personness" (p. 91). Debold and colleagues argue that an individual may shift her ways of knowing depending on her context.

Some researchers have examined the intersection of race, class, and gender and the complicated constructions of epistemologies that emerge from the intersection of these identities. While not using Belenky et al.'s model, Luttrell (1989) conducted observations in the classroom and in-depth interviews with 30 white and black working class women. She examined women's shared and differential views of intelligence and common sense, and explored the conflicting interests and values promoted through these self-perceptions. She describes the hidden gender asymmetries and inequalities in working class women's ways of knowing and how these emerge differently for white vs. black women. She makes the argument that "dominant ideologies of knowledge undermine women's collective identities, claims to knowledge, and power and the consequences for the adult education of working-class women" (p. 34), giving voice to the cultural and political significance of working-class women's ways of knowing. Lutrell (1989) argues that class, not gender, is a critical determinant of ways of knowing: "Class-based concepts of intelligence and common sense pit experience against schooling and working-class people against middle class people; race-based concepts of 'ignorance' and 'real intelligence' pit whites against blacks" (Lutrell, 1989, p. 41).

In response to critiques of their original theory, Goldberger, Tarule, Clinchy, and Belenky published *Knowledge, Difference and Power* (1996). Goldberger and coauthors (1996) claim they were attempting to find "an alternative definition of reason" (Goldberger et al., 1996). Addressing the hierarchical nature of their developmental model, they reassert that constructed knowing can be considered "superior" in its flexibility. They state that constructed knowing represents a meta-perspective of knowing, a perspective enabling one to acknowledge "that different routes to knowing have their place, their logic, their usefulness" (p. 13).

Ways of knowing and intellectual development in adulthood have been the object of study for a number of researchers who have not claimed gender differences (Brabeck, 1984; King & Kitchener, 1994; Perry, 1970). Since the 1970s, researchers and theorists have developed models of adult intellectual development based on research including both males and females. Arlin's (1975) model of problem finding; Sinnott's (1981) relativistic operations; Moshman and Timmons' (1982) stages in construction of logical necessity; Labouvie-Vief's (1982) modes of mature autonomy; Fischer's (1980) cognitive skills theory; Riegel's (1979) dialectical operations; Basseches (1980) dialectical schemata; Broughton's (1978) levels of self, mind, reality, and knowledge; and Kitchener and King's (1981) reflective judgment theory were all developed with both male and female students in mind. These theories all claim that the passage from adolescence to adulthood is marked by changes in ways of thinking about thinking; the capacity to engage in more abstract reasoning; a more complex understanding of the role of authorities and evidence in knowledge; a shift from absolute, dualistic thinking to contextualized knowledge; and an increased ability to reflect on knowledge claims. Belenky et al.'s (1986) insights from the original interviews were arrived at through qualitative methods. Our discussion here addresses the existing empirical evidence, resulting mostly from quantitative studies, regarding the claim that "women's ways of knowing" are uniquely associated with women.

William Perry's theory of intellectual and ethical development (1970) and Patricia King and Karen Kitchener's theory of reflective judgment (1994) are closely tied to the WWK theory because Belenky et al. used interview questions developed by these researchers. While early work on Perry's theory did not examine gender differences (e.g., Widick & Simpson, 1978), a few researchers have recently examined gender and epistemological development as measured by the Perry model and have not found consistent patterns of gender differences (Durham, Hays, & Martinez, 1994).

Using the Measure of Epistemological Reflection (MER) Interview (Baxter Magolda & Porterfield, 1985) and a 1-hour interview to assess Perry's first five positions and Belenky et al.'s five perspectives, Baxter Magolda (1990, 1992, 1995) examined gender differences in Perry's levels of intellectual development.

Fifty men and 51 women in their first year of college were followed through each of their 4 years of schooling. Her post-college testing included 27 women and 21 men in the 7th year of the study (Baxter Magolda, 1995). During their first year, men and women came to view knowledge as more uncertain. There were no gender differences in reasoning or preferences in either peer interactive or independent activities during this year (Baxter Magolda, 1990). During their sophomore and junior years, differences between males' and females' cognitive development were negligible. Both men and women used "relational" (similar to Belenky et al.'s connected knowing) and "impersonal" knowing (similar to Perry's levels); the college women used the relational pattern of knowing slightly more often than the college men. In their post-college years, the young adults Baxter Magolda interviewed (1995) showed an integration of relational and impersonal knowing and no gender differences. Baxter Magolda's studies indicate that both relational and impersonal patterns of knowing are identifiable in qualitative studies involving (primarily white) college students, and that men and women are equally capable of using both patterns of knowing.

Kitchener and King's (1981; King & Kitchener, 1994) model assesses epistemic development, reasoning about "the kinds of problematic situations that are truly controversial" (p. 7). Reflective judgment research assesses the ability to state and defend one's beliefs, "based on the evaluation and integration of existing data and theory into a solution about the problem at hand, a solution that can be rationally defended as most plausible or reasonable, taking into account the sets of conditions under which the problem is being solved" (p. 8). King and Kitchener (1994) provide a summary of 32 studies (more than 1700 participants) that have tested the claims of the theory. Of 17 studies including both males and females, three did not examine gender differences. Seven studies found no differences between males and females. Six studies found males higher in reflective judgment scores and one study reported a class by gender interaction, with traditional age women juniors and nontraditional age freshman women (but not traditional age freshman women) scoring higher than their male counterparts. In addition, they report Wood's (1985) analysis of the Kitchener and King longitudinal data found differences in growth spurts suggesting differences in the timing of developmental changes.

Given the confounding factors of differential opportunities, maturation rates, and timing of development, King and Kitchener (1994) believe that a conclusion about gender differences in reflective judgment "remains to be seen" (p. 177). The evidence from the theories of intellectual development under investigation for over two and a half decades does not support the claim that women engage in a different way of knowing than do men.

CONCLUSIONS

Despite evidence to the contrary, Gilligan's and Belenky et al.'s theories remain persuasive to many. For example, Gilligan's theory has been used to explain gender differences in such diverse fields as children's play; the speech of children; adult conversation; women in academia; war and peace; and law, nursing, and medicine (see Hurd and Brabeck, 1997 for a review). In a similar vein, Brabeck and Larned (1997) found that Belenky et al.'s WWK theory was cited more than 450 times between 1987 and 1994 in *Social Science Citation Index*, despite the fact that only four empirical studies of men's and women's ways of knowing had been done when their study was conducted.

An influential theoretician and researcher in the field of racial identity development, Janet Helms (1994) asserts that we need to know *what about* race is related to particular the particular variable under investigation. We believe the same is true about gender. Research on gender, moral reasoning, and moral orientation has begun to examine this question.

Less research has explored what about gender is related to different epistemological stances. Crawford (1997) suggests instead that we should be asking more questions addressing the person–context interaction. What do different women know in various situations? In what roles and in what situations do women take up particular epistemological stances? What contextual factors are important in understanding the ways women develop morally and know what they know? We need to know, as Samuel Johnson asked, Which man? Which woman?

The work of those examining the moral development and orientation and its relation to ego development, gender orientation, and type of dilemma has begun to address these infinitely more complex questions about gender and moral development and orientation. The narrative analyses being conducted by Debold, Tolman, and Brown (1996) and Day and Tappan (1995) also begin to address this complexity, illustrating the context-specific and complex nature of "knowing."

Given our review of the recent literature on adult moral reasoning and moral orientation, other questions about gender differences in ways of knowing and moral development that need investigation include: How might or under what circumstances might women and men differentially employ certain epistemological or moral stances? What promotes intellectual and moral development for men and for women; do particular experiences differentially impact women and men? Are there particular contextual circumstances that make it more likely that one will employ a particular moral orientation or articulate a moral understanding reflecting a particular stage of development? Under what contextual circumstances do men and women's epistemologies and moral orientations look similar or different? For which groups of men and women does this apply?

Thus, while the evidence does not support global gender differences in moral and epistemological development, Gilligan's and Belenky et al.'s research can lead us to a more reflexive position, examining the assumptions underlying our psychological theories, developing more complex conceptualizations of morality and epistemology, and examining the impact of person–context interactions that shape this.

REFERENCES

Arlin, P. K. (1975). Cognitive development in adulthood: A fifth stage? *Developmental Psychology, 11*, 602–606.

Armon, C., & Dawson, T. L. (1997). Developmental trajectories in moral reasoning across the life span. *Journal of Moral Education, 26*(4), 433–453.

Bandura, A. (1973). *Aggression: A social learning analysis.* Englewood Cliffs, NJ: Prentice-Hall.

Basseches, M. (1980). Dialectical schemata: A framework for the empirical study of the development of dialectical thinking. *Human Development, 23*, 400–421.

Baxter Magolda, M. B. (1990). Gender differences in epistemological development. *Journal of College Student Development, 31*, 555–561.

Baxter Magolda, M. B. (1992). *Knowing and reasoning in college: Gender related patterns in students' intellectual development*. San Francisco: Jossey-Bass.

Baxter Magolda, M. B. (1995). The integration of relational and impersonal knowing in young adults' epistemological development. *Journal of College Student Development, 36*(3), 205–216.

Baxter Magolda, M. B., & Porterfield, W. D. (1985). A new approach to assess intellectual development on the Perry scheme. *Journal of College Student Personnel, 26*, 343–351.

Bebeau, M., & Brabeck, M. M. (1989). Ethical sensitivity and moral reasoning among men and women in the professions. In M. Brabeck (Ed.), *Who cares? Theory, research and educational implications of the ethic of care* (pp. 144–163). New York: Praeger.

Belenky, M. F., Clinchy, M., Goldberger, N. R., & Tarule, J. M. (1986). *Women's ways of knowing: The development of self, voice, and mind*. New York: Basic Books.

Bem, S. L. (1974). The measurement of psychological androgyny. *Journal of Consulting and Clinical Psychology, 42*, 155–162.

Brabeck, M. (1983). Moral judgment. *Developmental Review, 3*, 274–291.

Brabeck, M. (1984). Longitudinal studies of intellectual development during adulthood: Theoretical and research models. *Journal of Research and Development in Education, 17*(3), 12–27.

Brabeck, M. (1989). *Who cares? Theory, research and educational implications of the ethic of care.* New York: Praeger.

Brabeck, M. (1993). Recommendations for re-examining women's ways of knowing. *New Ideas in Psychology, 11*(2), 253–258.

Brabeck, M., & Larned, A. G. (1997). What we do not know about women's ways of knowing. In M. R. Walsh (Ed.), *Women, men and gender: Ongoing debates* (pp. 261–269). New Haven: Yale University Press.

Brown, L., Gilligan, C., & Tappan, M. (1995). Listening to different voices. In W. M. Kurtines & J. L. Gewirtz (Eds.), *Moral development: An introduction*. Boston, MA: Allyn & Bacon.

Broughton, J. (1978). Development of concepts of self, mind, reality and knowledge. *New Directions for Child Development, 1*, 75–100.

Clopton, N. A., & Sorell, G. T. (1993). Gender differences in moral reasoning: Stable or situational? *Psychology of Women Quarterly, 17*(1), 85–101.

Colby, A., Kohlberg, L., Gibbs, J., & Lieberman, M. (1983). A longitudinal study of moral judgment. *Monographs of the Society for Research in Child Development 48*(1–2), *Serial No. 200.*

Conley, T. D., Jadack, R. A., & Hyde, J. S. (1997). Moral dilemmas, moral reasoning and genital herpes. *The Journal of Sex Research, 34*(3), 256–266.

Crawford, M. (1997). Agreeing to differ: Feminist epistemologies and women'sways of knowing. In M. M. Gergen & S. N. Davis (Eds.), *Toward a new psychology of gender* (pp. 267–284). New York: Routledge.

Daniels, J., D'Andrea, M., & Heck, R. (1995). Moral development and Hawaiian youths: Does gender make a difference. *Journal of Counseling and Development, 74*, 90–93.

Day, J. M., & Tappan, M. B. (1995). Identify, voice, and the psycho/dialogical perspectives from moral psychology. *American Psychologist, 50*(1), 47–48.

Debold, E., Tolman, D., & Brown, L. M. (1996). Embodying knowledge, knowing desire: Authority and split subjectivities in girls' epistemological development. In N. R. Goldberger, J. M. Tarule, B. M. Clinchy, & M. F. Belenky (Eds.), *Knowledge difference and power: Essays inspired by women's ways of knowing* (pp. 85–124). New York: Basic Books.

Durham, R. L., Hays, J., & Martinez, R. (1994). Socio-cognitive development among Chicano and Anglo American college students. *Journal of College Student Development, 35*, 178–182.

Erikson, E. (1968). *Identity, youth and crisis*. New York: W. W. Norton.

Fischer, K. W. (1980). A theory of cognitive development: The control and construction of hierarchies of skills. *Psychological Review, 87*, 477–531.

Freud, S. (1925, 1961). Some psychological consequences of the anatomical distinction between the sexes. In J. Strachey (Ed.), *Standard edition, Vol. 19*. London: Hogarth.

Garmon, L. C., Basinger, K. S., Gregg, V. R., & Gibbs, J. (1996). Gender differences in stage and expression of moral judgment. *Merrill Palmer Quarterly, 42*(3), 418–437.

Gibbs, J. C., Basinger, K. S., & Fuller, D. (1992). *Moral maturity: Measuring the development of sociomoral reflection*. Hillsdale, NJ: Lawrence Erlbaum.

Gilligan, C. (1977). In a different voice: Women's conceptions of self and of morality. *Harvard Educational Review, 47*(4), 481–517.

Gilligan, C. (1979). Woman's place in man's life cycle. *Harvard Educational Review, 49*(4), 431–446.

Gilligan, C. (1982). *In a different voice: Psychological theory and women's development.* Cambridge, MA: Harvard University Press.

Gilligan, C., & Attanucci, J. (1988). Two moral orientations: Gender differences and similarities. *Merrill-Palmer Quarterly, 34,* 223–237.

Gilligan, C., & Wiggins, G. (1987). The origins of morality in early childhood relations. In J. Kagan and S. Lamb (Eds.), *Emergence of morality in young children* (pp. 277–306). Chicago: University Press.

Goldberger, N. R., Tarule, J. M., Clinchy, B. M., & Belenky, M. F. (1996). *Knowledge difference and power: Essays inspired by women's ways of knowing.* New York: Basic Books.

Helms, J. E. (1994). How multiculturalism obscures racial factors in the therapy process. *Journal of Counseling Psychology, 42*(2), 162–165.

Hurd, T., & Brabeck, M. (1997). Presentation of women and Gilligan's ethic of care in college textbooks, 1970–1990: An examination of bias. *Teaching of Psychology, 24*(3), 159–167.

Jadack, R. A., Hyde, J. S., Moore, C. F., & Keller, M. L. (1995). Moral reasoning about sexually transmitted diseases. *Child Development, 66,* 167–177.

Jaffee, S., & Hyde, J. (2000). Gender difference in moral orientation: A meta analysis. *Psychological Bulletin, 126*(5), 703–726.

Johnston, D. K. (1988). Adolescents' solutions to dilemmas in fables: Two moral orientations—two problem-solving strategies. In C. Gilligan, J. V. Ward, & J. M. Taylor (Eds.), *Mapping the moral domain* (pp. 49–72). Cambridge, MA: Harvard University Press.

King, P. M., & Kitchener, K. S. (1994). *Developing reflective judgment.* San Franciso: Jossey-Bass.

Kitchener, K. S., & King, P. M. (1981). Reflective judgment: Concepts of justification and their relationship to age and education. *Journal of Applied Developmental Psychology, 2,* 89–116.

Kohlberg, L. (1969). Stage and sequence: The cognitive–developmental approach to socialization. In D. A. Goslin (Ed.), *Handbook of socialization theory and research* (pp. 347–480). Chicago: Rand McNally.

Kohlberg, L. (1981). *The philosophy of moral development.* San Francisco: Harper & Row.

Kohlberg, L. (1984). *Essays on moral development. The Psychology of Moral Development, Vol. 2.* New York: Harper & Row.

Kohlberg, L., & Kramer, R. (1969). Continuities and discontinuities in childhood and adult moral development. *Human Development, 12,* 93–120.

Krebs, D. L., Vermeulen, S. C., Denton, K. L., & Carpendale, J. I. (1994). Gender and perspective differences in moral judgement and moral orientation. *Journal of Moral Education, 23*(1), 17–26.

Labouvie-Vief, G. (1982). Dynamic development and mature autonomy. A theoretical prologue. *Human Development, 25,* 161–191.

Liddell, D. L. (1990). The measure of moral orientation. (Available from author, University of Iowa, Division of Counseling, Rehabilitation and Student Development, Iowa City, IA 52242).

Liddell, D. L. (1998). Comparison of semi-structured interviews with a quantitative measure of moral orientation. *Journal of College Student Development, 39*(2), 169–178.

Luttrell, W. (1989). Working-class women's ways of knowing: Effects of gender, race and class. *Sociology of Education, 62,* 33–46.

Lyons, N. (1983). Two perspectives: On self, relationship and morality. *Harvard Educational Review, 53,* 125–145.

Lyons, N. (1988). Two perspectives: On self, relationships, and morality. In C. Gilligan, J. Ward, & J. Taylor (Eds.), *Mapping the moral domain* (pp. 21–48). Cambridge, MA: Harvard University Press.

Maccoby, E., & Jacklin, C. (1974). *The psychology of sex differences.* Stanford: Stanford University Press.

Meyers, D. T. (1996). Emotion and heterodox moral perception: An essay in moral psychology. In D. T. Meyers (Ed.), *Feminists rethink the self.* Boulder, CO: Westview Press.

Morawski, J. G. (1985). The measurement of masculinity and feminity: Engendering categorical realities. *Journal of Personality, 53*(2), 196–223.

Moshman, D., & Timmons, M. (1982). The construction of logical necessity. *Human Development, 25,* 309–323.

Nicholson, J. (1984). *Men and women.* Oxford: Oxford University Press.

Noddings, S. N. (1984). *Caring: A feminine approach to ethics and moral education.* Berkeley: University of California Press.

Perry, W. G. (1970). *Forms of intellectual and ethical development in the college years: A scheme.* New York: Holt, Rinehart and Winston.

Piaget, J. (1932). *The moral judgment of the child.* New York: Free Press.

Rest, J. R. (1979). *Development in judging moral issues.* Minneapolis, MN: University of Minnesota Press.

Riegel, K. F. (1979). *Foundations of dialectical psychology.* New York: Academic Press.

Rogers, A., Brown, L., & Tappan, M. (1994). Interpreting loss in ego development in girls: Regression or resistance? In A. Lieblich & R. Josselson (Eds.), *Exploring identity and gender: The narrative study of lives, Vol. 2* (pp. 1–36). Thousand Oaks, CA: Sage.

Ruddick, S. (1989). *Maternal thinking: Toward a politics of peace.* Boston: Beacon Press.

Sinnott, J. (1981). The theory of relativity: A metatheory for development? *Human Development, 25,* 293–311.

Skoe, E. E., & Diessner, R. (1994). Ethic of care, justice, identity and gender: An extension and replication. *Merrill Palmer Quarterly, 40*(2), 272–289.

Skoe, E. E., & Marcia, J. E. (1991). A care-based measure of morality and its relation to ego identity. *Merrill-Palmer Quarterly, 37,* 289–304.

Skoe, E. E., Pratt, M. W., Matthews, M., & Curror, S. E. (1996). The ethic of care: Stability over time, gender differences, and correlates in mid- to late adulthood. *Psychology of Aging, 11*(2), 280–292.

Snarey, J. (1998). Ego development and the ethical voices of justice and care: An Eriksonian interpretation. In P. M. Westenberg, A. Blasi, & L. D. Cohn (Eds.), *Personality development: Theoretical, empirical, and clinical investigations of Loevinger's conception of ego development.* (pp. 163–180). Hillsdale, NJ: Lawrence Erlbaum.

Sochting, I., Skoe, E. E., & Marcia, J. E. (1994). Care-oriented moral reasoning and prosocial behavior: A question of gender or sex role orientation. *Sex Roles, 31*(3/4), 131–147.

Spence, J. T., Helmreich, R., & Strapp, J. (1974). The personal attributes questionnaire: A measure of sex role stereotypes and masculinity-femininity. *JSAS Catalog of Selected Documents in Psychology, 4,* 42 (No. 617).

Spencer, H. (1875). The psychology of the sexes. *Popular Science Monthly, 4,* 30–38.

Sutton, R. E., Cafarelli, A., Lund, R., Schurdell, D., & Bichsel, S. (1996). A developmental constructivist approach to pre-service teachers' ways of knowing. *Teaching and Teacher Education, 12*(4), 413–427.

Thoma, S. (1986). Estimating gender differences in the comprehension and preference of moral issues. *Developmental Review, 62,* 165–180.

Unger, R. (1989). *Representations: Social construction of gender.* Amityville, NY: Baywood.

Walker, L. J. (1984). Sex differences in the development of moral reasoning: A critical review. *Child Development, 55,* 677–691.

Walker, L. J. (1986). Sex differences in the development of moral reasoning: A rejoinder to Baumrind. *Child Development, 57,* 522–526.

Walker, L. J. (1989). A longitudinal study of moral reasoning. *Child Development, 60,* 157–166.

Walker, L. J. (1991). Sex differences in moral reasoning. In W. M. Kurtines & J. L. Gewirtz (Eds.), *Handbook of moral behavior and development: Vol. 2.* Hillsdale, NJ: Lawrence Erlbaum.

Walker, L. J., deVries, B., & Trevethan, S. D. (1987). Moral stages and moral orientations in real-life and hypothetical dilemmas. *Child Development, 58,* 842–858.

Walker, L. J., Pitts, R. C., Hennig, K. H., & Matsuba, M. K. (1995). Reasoning about morality and real-life moral problems. In M. Killen & D. Hart (Eds.), *Morality in everyday life: Developmental perspectives* (pp. 371–405). New York: Cambridge University Press.

Wark, G. P., & Krebs, D. L. (1996). Gender and dilemma differences in real-life moral judgment. *Developmental Psychology, 32*(2), 220–230.

Widick, C., & Simpson, D. (1978). Development concepts in college instruction. In C. A. Parker (Ed.), *Encouraging development in college students* (pp. 27–59). Minneapolis, MN: University of Minnesota Press.

White, J. (1994). Individual characteristics and social knowledge in ethical reasoning. *Psychological Reports, 75,* 627–649.

Wood, P. (1985). A statistical examination of necessary but not sufficient antecedants of problem solving behavior (Doctoral dissertation, University of Minnesota, 1985). *Dissertation Abstracts International, 46,* 2055.

PART III

Social Development in Adulthood

CHAPTER 19

Attachment Theory and Research

Contributions for Understanding Late Adolescent and Young Adult Development

MAUREEN E. KENNY AND CATHERINE E. BARTON

Over the last 20 years, attachment theory has fostered considerable theoretical writing and research, with the vast majority of studies concentrating initially on infancy and early childhood. Both Bowlby (1980) and Ainsworth (1989) maintain, however, that attachment processes are relevant to personality functioning across the lifespan, including the successful attainment of autonomy and the establishment of close intimate relationships beyond the family. Bowlby (1969/82) claims that attachment behaviors are characteristic of humans from "cradle to grave" (p. 127). According to the attachment perspective, parental attachment does not cease with the attainment of desirable levels of autonomy in adulthood and the formation of long-term romantic attachments. In adults, the emotional response generated by the death of a parent is just one illustration of the persistence of parental attachment bonds over the life course (Ainsworth, 1989).

Research interest in attachment relationships among adults began in the 1970s, with studies of adult bereavement (Bowlby & Parkes, 1970) and marital separation (Weiss, 1982). The 1980s (Armsden & Greenberg, 1987; Kenny, 1987; Kobak & Sceery, 1988) and 1990s (Allen, Hauser, Bell, & O'Connor, 1994; Kenny & Donaldson, 1991; Kenny, Moilanen, Lomax, & Brabeck, 1993; Kenny & Perez, 1996; Kobak, Sudler, & Gamble, 1992; Rice, 1990; Rice & Whaley, 1994) witnessed increasing theoretical and research attention to the importance of attachments

MAUREEN E. KENNY • Lynch School of Education, Program in Counseling Psychology, Boston College, Chestnut Hill, Massachusetts, 02146.   CATHERINE E. BARTON • Boston College, Chestnut Hill, Massachusetts, 02467-3813.

Handbook of Adult Development, edited by J. Demick and C. Andreoletti. Plenum Press, New York, 2002.

throughout the adolescent years. The development of the Adult Attachment Interview (Main & Goldwyn, 1985) in the 1980s made possible the assessment of internal/representational aspects of attachment among adults and contributed to a large body of research on adult attachment representations (Bretherton, 1992). The study of adult attachment styles, including strategies that individuals use to maintain feelings of security, feelings about the self in the context of relationships with others, and internal representations that guide interpersonal behavior, began in the late 1980s and proliferated in the 1990s (Bartholomew, 1994; Hazan & Shaver, 1987; Shaver & Hazen, 1988; Simpson & Rholes, 1998).

This chapter first presents a brief overview of selected tenets of attachment theory and their theoretical relevance for understanding development in late adolescence and adulthood. We describe three lines of research, including parent attachment among late adolescents and young adults, adult attachment representations, and adult attachment styles, and then examine the conceptual and methodological status of this work. We conclude with an assessment of the current contributions of attachment theory to the understanding of adult development and discussion of the conceptual and empirical work that needs to be done to further this knowledge.

SELECTED ATTACHMENT CONCEPTS

The origins of attachment theory can be traced to the 1930s as an outgrowth of Bowlby's interest in the links between maternal loss or deprivation and adult personality (Bretherton, 1992). Bowlby introduced the tenets of attachment theory in three classic papers presented to the British Psychoanalytic Society from 1958 through 1960, and extended these ideas in his three seminal volumes on Attachment and Loss (1969/82, 1973, 1980). Mary Ainsworth's development of the Strange Situation for assessing the infant–caretaker attachment in a laboratory playroom made it possible to empirically assess questions that Bowlby had considered difficult to study and laid the methodological foundation for a burst of empirical work concerning mother–infant attachments (Bretherton, 1992).

In Bowlby's framework (1988), the attachment system is one of five behaviorally rooted systems of motivation; others include exploration, parenting, sex, and eating—each serving a distinct biological function. The innate predilection to form attachments is believed to have evolved because of the vital protective function served by these relationships. The young infant emits signals to the environment and quickly establishes an attachment with the caretaker, who responds to the infant's cries and engages the infant in social interaction. Proximity to the caretaker is sought to reduce feelings of anxiety. Attachment behavior of the infant and parenting behavior of the caretaker interact reciprocally in the development and maintenance of attachment patterns. The availability and sensitivity of the caretaker in responding to the infant's signals is believed to foster feelings of security, and hence, the use of the attachment figure as a "secure base" of support for exploration of the environment and as a "safe haven" to which the infant can return for reassurance when feeling threatened.

Using the Strange Situation procedure, designed to examine the balance of attachment and exploratory behavior under situations of high and low stress,

Ainsworth, Blehar, Waters, and Wall (1978) observed infants' reactions to separations and reunions with caretakers. Three categories of caretaker–infant attachment patterns and their corresponding parenting behavior were identified, with a fourth category later observed by Main and Solomon (1990). The "secure" pattern of attachment is characterized by available and sensitive responses to the infant's signals, infant reassurance as a result of comfort provided by the caretaker, and infant confidence in exploring the environment (Ainsworth et al., 1978). When presented with a stressful situation, the infant who has formed an "anxious resistant (ambivalent)" attachment either rejects the affection of the caretaker or clings to the caregiver, who is inconsistently responsive to the infant. The "anxious avoidant" pattern of attachment is characterized by a caretaker who ignores the child's signals and the infant who avoids contact with the caretaker and displays no overt signs of distress (Ainsworth et al., 1978; Main & Goldwyn, 1985). The fourth category, "disorganized attachment," is characterized by infants who meet the caretaker on reunion with a dazed appearance, or a freezing or startled reaction. Parental histories of abuse, psychosis, and unresolved loss are associated with this pattern (Main & Solomon, 1990).

Secure attachments are theorized to contribute to a number of positive emotional, social, and cognitive development outcomes. Secure attachments should enable the child to tolerate anxiety-arousing situations, and should foster environmental exploration, and hence, the development of instrumental competence (Kobak & Sceery, 1988). Moreover, characteristics of the infant–caretaker relationship are hypothesized to become internalized over time to form "internal working models" of self, others, and the physical environment. Infants, for example, who experience caretakers as reliable and responsive are believed to develop internal working models of the self as worthy of consistent response, thus contributing to a generally positive view of self. Secure attachments are similarly hypothesized to contribute to expectations of others and the environment as predictable and trustworthy, providing a foundation of basic trust, enhancing environmental exploration, and the willingness to turn to others as a source of help (Bowlby, 1973).

Extensions to Adolescent and Adult Development

Attachment theory emphasizes the importance of affectional bonds throughout the life course, although the essential biological function of attachment in securing a source of safety and protection is less obvious beyond childhood. Elaborating on Bowlby's work, however, Kobak and Duemmler (1994) maintain that fear-provoking, challenging, and conflictual situations normatively activate attachment behavior among adults. Elderly adults, for example, may demonstrate heightened attachment behavior as they become more frail and less able to care for themselves (Main, 1999). Bowlby (1988) emphasizes the security-enhancing functions of the attachment relationship and maintains that an important goal for human development across the lifespan should be the maintenance of a secure base that supports both autonomy and independence.

The writings of Bowlby and Ainsworth have inspired several independent lines of research among developmental, clinical, and counseling psychologists interested in identifying nuclear family characteristics that contribute to individual differences in adaptive and maladaptive development across the lifespan, and among personality/social psychologists interested in the association of attachment patterns with relationship satisfaction and personal adjustment in adult life. The appeal of the theory across diverse fields of psychology likely stems from its compatibility with revised understandings of optimal adolescent–parent relationships (Kenny & Donaldson, 1991); critiques of traditional developmental theories that emphasize separation (Guisinger & Blatt, 1994; Josselson, 1988); its capacity to integrate (through the concept of the internal working model) psychoanalytic, social–cognitive, and ethological constructs (Rothbard & Shaver, 1994); and its relevance for the burgeoning field of developmental psychopathology (Cicchetti, 1984).

The understanding of attachment and exploration as complementary, rather than as opposing constructs, is consistent with contemporary views of optimal parent–child relationships throughout adolescent and young adult development. Traditional theories of adolescent development (Blos, 1967; Freud, 1969), emphasizing the importance of rebellion against parents in fostering psychological separation and growth, are challenged by normative studies of adolescent development (Douvan & Adelson, 1966), which reveal that adolescence is not necessarily a period of inevitable turmoil. Subsequent theory and research (Allen, Moore, & Kuperminc, 1997; Grotevant & Cooper, 1986; Hill & Holmbeck, 1986) suggest that individuation occurs most adaptively within a caretaker relationship that is transformed rather than broken during adolescence. From an attachment perspective, Ainsworth (1989) similarly maintains that the optimal outcome for adolescence is an attainment of autonomy, while continuing to value attachment.

Consistent with this revised understanding of optimal adolescent–caretaker relationships, several researchers (Allen & Hauser, 1996; Armsden & Greeenberg, 1987; Hill & Holmbeck, 1986; Kenny, 1987; Kobak & Sceery, 1988; Lapsley, Rice, & Fitzgerald, 1990) have examined the relevance of attachment theory in understanding parental relationships throughout the adolescent years. Kenny, Moilanen, Lomax, and Brabeck (1993), for example, suggested that secure parent–child attachments can provide a protective base of security for the early adolescent who is coping with numerous biological, social, and psychological changes. In addition, for the late adolescent, leaving home for college has been conceptualized as a naturally occurring "Strange Situation" (Ainsworth et al., 1978) for which the secure base offered by parental attachment can promote exploration and mastery of the college environment (Kenny, 1987). This work assumes that although parent–child relationships change during the adolescent years, the affective bond remains, and parents can continue to provide a secure base by supporting exploration and offering a safe haven of advice and comfort when needed. From this perspective, a call home to parents from a college student may be understood as an adaptive use of a secure base, rather than as a sign of dependency.

The developmental complementary of attachment and independence has been highlighted in other theoretical work and research over the past two decades, which have likely enhanced interest in applications of attachment theory for adolescent and adult development. Relational theorists (Gilligan, 1982; Miller, 1976; Surrey, 1991), for example, have been critical of development models that emphasize the attainment of autonomy and independence, and neglect the ongoing importance of relationships. Shaver, Papella, Clark, Koski,

Tidwell, & Nalbone (1996) suggest that attachment security in adults, derived presumably from a history of secure attachment to caretakers, represents an optimal model of personality for both men and women, combining desirable levels of self-confidence and autonomy with a healthy capacity for establishing intimate, emotionally expressive relationships.

The concept of the internal working model has been central in the application of attachment theory to adolescent and adult development. Attachments established earlier in the life course may continue to be adaptive because of the positive internal working models they provide, rather than as a source of actual assistance. The mental image of the attachment figure, or recollection of what that individual would say or do, may provide a source of comfort or security for the adolescent or adult. From a developmental psychopathology perspective, internal working models may serve as risk or protective factors (Kobak & Cole, 1994). Secure internal working models, for example, provide a basis for maturation along positive developmental pathways (Bowlby, 1980; Sroufe, 1989), with a positive view of self, others, and the environment contributing to psychological resilience by enabling the individual to adapt positively despite stressful life circumstances. Internal working models may impact subsequent development by providing cognitive filters that influence how new experiences are interpreted (Bretherton, 1985). The individual, for example, who has a positive and trusting view of others (associated with a secure internal working model), may be able to adaptively use available social support, buffering the negative impact of life stress.

Although attachment theory suggests that the effects of early attachments are often observable in adult life, internal working models are not assumed to be stable throughout the life course (Bartholomew & Shaver, 1998). Bowlby (1969/1982) suggests that internal working models are most flexible early in life, becoming more resistant to change over time as a result of repeated and similar interactions with the caretaker. Changes in internal working models, nevertheless, should be possible as a result of new relationships and the development of abstract cognitive reasoning, which allow the individual to reinterpret the meaning of past relationships (Bowlby 1969/82). Kobak, Sudler, and Gamble (1992) suggest that internal working models of parents are often reorganized during adolescence, as the increased cognitive capacities of the adolescent create new opportunities for updating and revising models of self and parents. The cognitive capacity of the adolescent to understand the perspective of self and parent, for example, makes it possible to develop a more mature understanding of parents based upon mutuality, rather than as obstacles to autonomy.

The work of Mary Main and her colleagues (Main, Kaplan, & Cassidy, 1985) has had a large impact in moving adult attachment research beyond behavioral observations to a focus on internal working models. Building on Bowlby's premise that adult parenting impacts infant attachment, Main et al. proposed that adults' current internal working models of their childhood attachments are the mechanisms that influence parenting behavior, which in turn influence the attachment patterns of infants and young children. In support of this premise, Main et al. (1985) report findings that parents' attachment representations are associated with their infants' behavior in the Strange Situation procedure, which had been assessed 5 years earlier. These researchers also report that some adults, despite adverse early family experiences, articulate coherent accounts of their childhood experiences and are parents of securely attached infants. These findings suggest that the intergenerational transmission of insecure attachment can be interrupted when adults' cognitive and emotional capacities permit them to reflect on and coherently integrate their attachment experiences. The identification of parents who experienced insecure attachment histories, but nevertheless demonstrate competent parenting as adults, has been of considerable interest theoretically. These parents highlight the limits of early experience in predicting later attachment and parenting and provide a basis for exploring factors that contribute to continuity and discontinuity in attachment across the lifespan and across generations (Pearson, Cohn, Cowan, & Cowan, 1994; Phelps, Belsky, & Crnic, 1998).

The concept of the internal working model of others has also been instrumental to research on adult attachment styles advanced by personality and social psychologists. The large body of work on adult attachment styles assumes that adult relationships are characterized by styles of interaction that are prototypic of the attachment types observed among infants/children and their caretakers, and that these styles are to some extent the product of internal working models of earlier attachment relationships (Hazan & Zeifman, 1999). For example, although parental attachments generally remain important as a source of support throughout the adolescent and young adult years, attachment needs and behaviors are gradually transferred from parents to peers (Allen & Land, 1999). In turn, peer attachments are believed to serve as a mechanism for forming primary attachment relationships with adult romantic partners. Romantic relationships, developing from a convergence of the sexual and attachment systems, are thus viewed as taking on functions of parental attachments over time as shared interests, strong affection, and a capacity to provide and obtain support and comfort in the romantic relationship increases. In support of this premise, Fraley and Davis (1997) report that college undergraduates use peers for proximity seeking, but are still in the process of transferring safe haven functions from parents to peers, especially to romantic partners.

The Measurement of Attachment Among Late Adolescents and Adults

Efforts to study attachment among late adolescents and adults have been complicated by difficulties in identifying and measuring attachment beyond early childhood. The attachment behavioral system is described as a motivational/control system, which has the goal of promoting safety and feelings of security (Bowlby, 1969/82). Behaviors, such as searching, clinging, and crying are observable elements of the system among infants, but age-relevant reflections of the attachment behavioral system are more difficult to specify and observe in adult relationships.

Sroufe and Waters (1977) suggest that attachment should be conceptualized as an "organizational construct," rather than as a series of discrete behaviors. As such, attachment can be understood as a system that regulates the dynamic balance between attachment and exploration across all phases of development (Hazan & Zeifman, 1999). Although theoretically relevant across the lifespan, this balance changes as a function of maturation and developmental shifts in the importance of connection and separation (Cicchetti, Cummings, Greenberg, & Marvin, 1990). With maturation, substantial increases are expected in the amount of time and distance spent apart from attachment figures before anxiety is experienced.

Moreover, over the course of development, new and multiple attachment figures may be acquired, and some attachment figures may decrease in importance (Ainsworth, 1989; Kenny & Rice, 1995). Other new relationships are formed, including friendships and professional acquaintances, that do not qualify as attachment relationships. Ainsworth (1989) distinguishes between attachments and other significant relationships, based on the capacity of attachment relationships to provide feelings of security and reduce anxiety in the presence of threat. Attachment relationships among adults may also serve varied non-attachment functions, including companionship, shared experience, and sexual bonds (Ainsworth, 1989). Relationships also become less hierarchical and more complementary over time, so that by adulthood, attachment relationships are often based on mutual offerings of comfort and security (Hazan & Zeifman, 1999). Because of the complexities of assessing attachment behavior among adults and the salience of internal working models, researchers have tended to focus on individual assessment, rather than the assessment of relationships, and have tried to capture the cognitive and emotional underpinnings of the attachment behavior system by assessing perceptions through self-report and interview methods (Crowell & Treboux, 1995).

In the domain of adolescent attachment, several self-report measures have been developed to assess current attachment relationships with parents. These measures assess levels of attachment security, as opposed to categorical classification of attachment style. The Inventory of Parent and Peer Attachment (IPPA; Armsden & Greenberg, 1987) seeks to assess the internal working model of parent and peers as attachment figures by focusing on affective and cognitive experiences associated with trust in parents' and peers' accessibility and responsiveness. The Parental Attachment Questionnaire (PAQ; Kenny, 1987) was designed to adapt Ainsworth et al.'s conceptualization of attachment for use with adolescents and young adults in a self-report format and assesses current use of parents as a secure base, as well as the security of an individual's internal working model. Researchers have also used the Parental Bonding Instrument (PBI; Parker, Tupling, & Brown, 1979) and the Mother–Father–Peer Scale (Epstein, 1983) for assessing memories (prior to age 16) of perceived parental attachments, as recalled by adolescents and adults (see Lopez and Gover [1993] for a review).

Adult attachment representations are assessed by the Adult Attachment Interview (AAI; Main & Goldwyn, 1985), which classifies adults into a category for overall "state of mind" or mental representation concerning attachment. Like the Strange Situation and adolescent self-report measures, the AAI focuses on attachments within the family of origin (Simpson & Rholes, 1998), and consists of questions and follow-up queries concerning the individual's recollection of early attachment relationships. The three major classification groups are Autonomous–Secure, Insecure–Dismissing, Insecure–Preoccupied or Entangled, with some interviewees being placed into a fourth category, Insecure–Unresolved, corresponding to the fourth childhood category. The AAI is retrospective in terms of the questions asked, but is presumed to assess adult attachment security, given that current reports of past experiences are filtered through current internal working models. The AAI does not assess attachment to any specific individual, but evaluates an overall state of mind, that has likely been influenced by a variety of relationships (Hesse, 1999). Security is believed to be characterized by the ability to think and speak coherently about attachment relationships (Main & Goldwyn, 1985). Thus, a purported advantage of the AAI is its ability to distinguish, based on the evaluation of coherence within the interview, those individuals whose internal working models are secure from those who report an idealized attachment relationship because of a highly defensive state of mind (Steele & Steele, 1994). Self-report measures, including the PBI, IPPA, and PAQ, have no means for detecting idealized defensive responses.

Measures of adult attachment style focus on attachments to peers and romantic partners, rather than to members of the nuclear family or family of origin. These are mostly brief self-report measures, which have also been criticized for limitations in detecting defensiveness (Simpson & Rholes, 1998). Hazan and Shaver (1987) developed the first self-report measure of adult attachment style, comprised of three brief paragraphs that describe secure, anxious, and avoidant relationships styles among adults, paralleling the behaviors, feelings, and dynamics displayed by infants and caregivers. Individuals choose which paragraph best describes their feelings. Among adults, the "ambivalent style" is characterized by fear of abandonment, and clingy, suspicious, dependent, jealous, controlling, and sometimes domineering behavior in relationships (Hazan & Shaver, 1987, 1990, 1994). In contrast, the person with an "avoidant style" strives to avoid emotional dependence and is uncomfortable with intimacy. The "secure" adult is comfortable depending on others and finds it easy to establish close relationships. Hazan and Shaver's (1987) categorical measure was transformed into a continuous scale format, the Adult Attachment Scale (AAS) by Collins and Read (1990), and another continuous version was developed by Simpson (1990).

Bartholomew (Bartholomew, 1990; Bartholomew & Horowitz, 1991) expanded Hazen and Shaver's original conceptualization by identifying two categories of avoidant individuals: fearful and dismissing. The resulting four-group taxonomy classifies individuals according to positive and negative models of self and other, with the fearful group (negative view of self and negative view of others) corresponding roughly to the disorganized childhood group. Two questionnaires, the Relationship Questionnaire (RQ) and Relationship Scales Questionnaire (RSQ), in addition to coded interviews, such as the Peer and Family Attachment Interviews, were developed to assess the four attachment categories and the two dimensions of view of self and others (Bartholomew & Horowitz, 1991; Griffin & Bartholomew, 1994a, 1994b). Griffin and Bartholomew (1994b) maintain that these measures allow for the classification by attachment types, as well as assessment of the individual's similarity with the prototypical representative of an attachment category, thus combining positive features of both categorical and continuous/dimensional approaches to the assessment of adult attachment.

Relatively few studies have examined the extent to which similar phenomenon are assessed by varied attachment measures. Initial efforts to assess correspondence between the the AAI and measures of adult romantic attachment, such as the AAS and RQ, and between the AAI and measures of young adult parental attachment reveal little convergence (Crowell, Treboux, & Waters, 1999; Hamilton, 1995; Kobak & Hazan, 1991). Correspondence between different measures of adult attachment style, such as the AAS and RQ, however, is strong (Crowell & Treboux,

1995). Adult attachment styles, as assessed by the RQ and AAS, have also been associated in theoretically expected ways with self-reports of parental attachment assessed by the PAQ and IPPA (Brennan, Clark, & Shaver, 1998; Crowell et al., 1999; Rothbard & Shaver, 1994).

Overall, evidence suggests that self-report measures of attachment style and parental attachment are assessing somewhat different phenomenon than what is assessed by the AAI. Although method variance, which is related to issues of self-awareness and defensiveness, may in part explain these differences, the measures also seem to be assessing different notions of security. For the AAI, security is related to one's understanding of the meaning and significance of attachment experiences, whereas the self-report measures are related to the quality of experiences with parents and partners (Crowell et al., 1999). Clearly, continued efforts to operationalize the adult attachment system and to clarify what domains and processes are assessed by different measures remain important areas for further research. The task of carefully delineating and assessing attachment markers across development periods remains a principle challenge for further advancements in examining attachment theory across the lifespan.

SELECTED RESEARCH IN ADOLESCENT AND ADULT ATTACHMENT

Attachment has not attained the status of a complete theory, but remains a limited set of propositions to be tested empirically (Berman & Sperling, 1994). The process of evaluating the theory is being accomplished through incremental efforts in testing specific postulates. In our selective and brief review of the attachment research, we seek to assess current evidence, as provided by the research on late adolescent and young adult parental attachment, adult attachment styles, and adult attachment representations, as it pertains to the existence and function of attachment bonds and patterns among late adolescence and young adults. The review serves to elucidate directions for further research.

Attachment Bonds Among Late Adolescents and Adults

Evidence to support the existence of attachment bonds among late adolescents and young adults is derived primarily from self-descriptions of parental relationships and attachment styles that are consistent with Bowlby's descriptions of attachment bonds and with the infant attachment patterns identified by Ainsworth et al. (1978) (Crowell & Treboux, 1995). Among late adolescents and young adults, for example, descriptions of parental relationships as affectively positive, facilitating autonomy, and being available as a source of support when needed are consistent with the secure attachment pattern (Kenny, 1987, 1994). Hazan & Shaver (1987) found that adults were able to classify themselves into three prototypic attachment styles: 56% secure, 25% avoidant, and 19% anxious/ambivalent, which was consistent with the percentage of infants in each of the three infant groups identified by Ainsworth et al. (1978). The three classifications identified through the Adult Attachment Interview also corresponded in roughly similar percentages to the infant attachment groups (50% autonomous/ secure; 25–30% dismissing; 10–15% preoccupied) (Steele & Steele, 1994). Because the construct of attachment is not directly observed, but inferred from self-description, critics (Kirkpatrick, 1998) continue to question whether individual differences in romantic relationships are truly differences in attachment patterns and internal working models. Because attachment bonds and internal working models are not directly observable, researchers have sought to provide validation by examining theoretically relevant correlates related to the importance and function of attachment relationships and by examining the stability of attachment patterns over time.

The Importance and Function of Attachment in Late Adolescence and Adulthood

Characteristics of current parental attachment as reported by adolescents and young adults, measures of adult attachment representations, and adult reports of attachment styles patterns have been associated with a variety of social, emotional, and cognitive characteristics that affirm the importance and function of attachment in late adolescence and adulthood. The vast majority of studies, however, have assessed contemporaneous correlates of attachment and have relied on white, middle-class samples, especially college students.

Research assessing current reports of parental attachment among late adolescents and young adults, with the Parental Attachment Questionnaire and the Inventory of Peer and Parental Attachment, reveal numerous correlations with indicators of emotional, social, and cognitive functioning. Among undergraduates, for example, secure parental attachments are positively related to social, academic, and emotional adjustment to college (Kenny, 1987; Kenny & Donaldson, 1991; Larose & Boivin, 1997) and enhanced coping resources for responding to stress (Brack, Gay, & Matheny, 1993), as well as negatively associated with psychological distress (Bradford & Lyddon, 1993) and gender role stress and conflict among males (Fischer & Good, 1998). Attachment research among racially diverse individuals is limited. Taub (1995), however, reports that the quality of parental attachment is related positively with autonomy development among African-American, Asian American, white, and Hispanic college students. Hinderlie and Kenny (1999) found a positive association between parental attachment and college adjustment among black students.

Studies assessing retrospective reports of parental attachment with the Parental Bonding Instrument or Mother–Father–Peer scale among late adolescents and young adults have similarly found theoretically expected associations with varied emotional, social, and cognitive indicators, such as self-esteem (McCormack & Kennedy, 1994; Rice & Cummins, 1996), loneliness (Kerns & Stevens, 1996), constructive thinking (Lopez, 1996), and perceived available support from peers and parents (Sarason & Pierce, 1991). Furthermore, among black and white undergraduates, retrospective reports of attachment to fathers, but not mothers, is a significant correlate of social competence (Rice, Cunningham, & Young, 1997). Carnelley, Pietromonaco, and Jaffe (1994) report, among undergraduate women, but not among married women, that retrospective reports of negative childhood attachments are associated with symptoms of depression. Among lesbian adults, perceived parental attachment is predictive of parental support in the coming out experience (Mohr & Fassinger, 1997). Despite these sparse studies, research addressing race, gender, and sexuality as they relate to attachment remains limited.

Current reports of attachment styles have been associated in theoretically expected ways with stress experienced at work and school among late adolescents (Burge et al., 1997), attachment to peers and levels of mutual caring/support and trust/intimacy (McCutcheon, 1998), and emotional expressivity among undergraduates (Searle & Meara, 1999). In addition, insecure adult romantic styles have been associated with loneliness (Hazan & Shaver, 1987), decreased life satisfaction (marital satisfaction, financial strain, and stress) (Hobdy & Hayslip, 1997), negative affectivity (Simpson, 1990), neuroticism (Shaver & Brennan, 1992), and low self-esteem (Brennan & Morris, 1997; Collins & Read, 1990).

Attachment representations assessed through the Adult Attachment Interview have also been associated with numerous emotional, social, and cognitive indicators. For example, Allen, Hauser, and Borman-Spurell (1996) compared adults who had been psychiatrically hospitalized 11 years previously to adults with no history of psychiatric hospitalization and found that adults with a history of hospitalization had a reduced ability to recall attachment experiences, were far less coherent, and showed higher levels of derogation of attachment relationships and lack of resolution of previous trauma. Related to cognitive functioning, Bernier and Larose (1999) found that among undergraduates, adult attachment classification was related to willingness to seek academic counseling and with adjustment to college and academic achievement. In addition, attachment representations among undergraduates were positively related to reasoning skills (van IJzendoorn & Zwart-Woudstra, 1995).

Although contemporaneous correlations have provided tentative theoretical support for the importance of attachment in adulthood, these findings have not established causal relationships, and therefore contribute little to our understanding of how attachment functions developmentally as a protective or risk factor. Although a substantive body of research has followed young infants into the schoolage years, longitudinal research assessing attachment outcomes among adolescents is only beginning to emerge. Examples of existing longitudinal research include Grossman and Grossman (1991), who report that aspects of social development, such as the establishment of close friendships and reliance on others as a strategy for coping with stress, are predictable at age 10 from attachments in infancy. Similarly, Elicker, Englung, and Sroufe (1992) found that children classified as secure infants were rated at ages 10 to 11 by camp counselors as self-assured, competent, and emotionally healthy, and were rated by peers as more popular, sociable, and displaying more prosocial interaction than children who had been classified as insecure. Carlson (1998) reported that the classification of disorganized attachment assigned in infancy was predictive of behavioral problems as reported by teachers throughout preschool, elementary, and high school, and with self-reported depressive symptoms and disassociation at ages 17 and 19.

Stability of Attachment Patterns

Although theory suggests that internal working models are both modifiable beyond early childhood and resistant to change, studies assessing stability and change in attachment status/type by following the same individuals from infancy to adulthood are also limited. While several studies found no significant relationship between infant attachment and later adolescent attachment (Zimmermann, Fremmer-Bombik, Spangler, & Grossmann, 1995), other research has yielded significant results. For example, Waters, Crowell, Treboux, Merrick, and Albersheim (1995) found that attachment security assessed in infancy was related to that individual's attachment representation 20 years later. More than 70% of adults retained the same secure vs. anxious classifications that they had received as infants. Those who experienced a change in attachment security were likely to have also experienced a negative life event during the period from infancy to early adulthood. Researchers focusing on adolescence (Allen & Hauser, 1996; Becker-Stoll & Fremmer-Bombik, 1997) report no direct relationship between infant attachment classification and adolescent attachment representation as assessed by the AAI, yet report evidence that observations of autonomy and relatedness in interactions with parents during adolescence are predictive of AAI classification in young adulthood.

Research using the AAI has provided evidence to link parent attachment representations, and their infants' patterns of attachment. In more than 18 different studies, infants' behavior in the Strange Situation has been shown to correspond at a rate of 75% to 80% with attachment representation of their caretakers assessed with the AAI, whether the interviews are collected before, after, or concurrent with the Strange Situation (Hesse, 1999; Steele & Steele, 1994; van IJzendoorn, 1995). In one prospective study, a 70% correspondence for classification type was found among AAI attachment classification of high-risk unmarried mothers assessed prior to the birth of the child and the Strange Situation attachment classification of the infant at 15 months (Ward & Carlson, 1995). Longitudinal research has not yet been completed, however, to assess the relationship of parents' early attachment and their parenting as adults, and hence data linking early attachment and later parenting remain inconclusive (Berlin & Cassidy, 1999).

Thompson (1999) concludes that existing research has yet to establish substantive and reliable, long-term consequences of infant attachment. This may be the case because the prediction of later outcomes is contingent on many factors, including outcome domain, the time span between attachment and later behavior, stability and change in caregiver influence, and limitations of measurement, as well as numerous developmental challenges, historical changes, and other intervening and mitigating factors (Thompson, 1999). Because the outcome variables are multidetermined, the predictive power of any single variable, such as attachment, is perhaps difficult to isolate.

FURTHER EFFORTS: TOWARD CLARITY AND INTEGRATION

Research on late adolescent and adult attachment has proliferated over the past 15 years based on work generated from several independent research traditions.

Through this work, using different methods and samples, awareness of the nature and importance of attachment relationships across the lifespan has increased. Findings from the varied traditions of developmental and social/personality psychology have often been complementary, yet at other times have been at odds, contributing to controversies about the definition and measurement of attachment. Current knowledge is limited by an abundance of correlational research, using diverse measures with limited correspondence.

Further research is needed, first of all, to address methodological limitations of existing work. A growing body of longitudinal data, made possible by the increasing maturity of individuals involved in infant attachment studies, offers new opportunities to study the sequelae of early attachment in adult life. Longitudinal research is needed to examine the mechanisms through which internal working models change and remain the same (Collins & Read, 1994), and to identify the processes through which internal working models contribute to adaptive and maladaptive outcomes (Main, 1999). Further research is also needed to assess the applicability of attachment theory for ethnically and racially diverse samples and to determine whether the mechanisms that contribute to adaptive and maladaptive outcomes are similar or different across individuals and contexts (Berlin & Cassidy, 1999). The current literature has assessed the association between attachment and a wide variety of behaviors and characteristics. To more carefully delineate the importance of attachment across the lifespan, specific theoretical claims need to be examined by testing competing models (Main, 1999), and outcomes need to be carefully selected to test what should and should not be impacted by attachment characteristics (Thompson, 1999). Measures need to be selected with an awareness of the different theoretical assumptions underlying each measure and knowledge that the choice of measure will determine what aspects of the theory are assessed (Griffin & Bartholomew, 1994a).

Much work remains in refining the conceptualization and measurement of attachment beyond childhood. Although minimal correspondence has been found between the AAI and attachment style measures, these measures have independently predicted similar outcomes in theoretically expected ways. This has been interpreted as evidence that the AAI and Attachment style measures assess common underlying constructs (Bretherton & Mulholland, 1999). Main (1999) calls for increased efforts to integrate the field of attachment as it currently stands, combining interview methods with self-report, probing the connections and lack of connections between early experiences, current state of mind, and relationships with romantic partners and children.

Leaders in the adult attachment field have identified directions for further research that recognize the value of integrating work across the field of attachment and with other areas of psychology, biology, and public policy. As theory and research move towards identifying linkages between attachment and other cognitive (attention and memory), emotional, and physiological (neurochemical, brain organization) systems, the mechanisms that link these processes will need to be explored (Main, 1999). Conceptual models can be tested that combine measures of multiple attachment and non-attachment relationships, and include important individual and contextual variables, such as temperament, physical health, cultural norms, and neighborhood safety (Berlin & Cassidy, 1999). Such integrative research should help to illuminate the relative contributions of early and later experiences to emotional and social adjustment in adult life, with important clinical implications for how and when change is possible (Berlin & Cassidy, 1999). This knowledge offers promise for devising methods of prevention and intervention that support positive development across the lifespan and for informing public policies that promote the development of security in families and in other significant relationships.

An agenda of attachment research that tests complex multivariate models, as described in the preceding, offers potential for increasing understanding of factors that enhance human development across the lifespan and for specifying the types of environments and policies that contribute to healthy human relationships. The usefulness of applying attachment theory across the lifespan will depend on future advances in measurement and research, and the translation of findings to inform public policies that foster positive outcomes across cultures and socioeconomic status.

REFERENCES

Ainsworth, M. D. S. (1989). Attachments beyond infancy. *American Psychologist, 44*, 709–716.

Ainsworth, M. D. S., Blehar, M. C., Waters, E., & Wall, S. (1978). *Patterns of attachment: A psychological study of the Strange Situation*. Hillsdale, NJ: Lawrence Erlbaum.

Allen, J. P., & Hauser, S. T. (1996). Autonomy and relatedness in adolescent–family interactions as predictors of young adults' states of mind regarding attachment. *Development and Psychopathology, 8*, 793–809.

Allen, J. P., Hauser, S. T., Bell, K. L., & O'Connor, T. G. (1994). Longitudinal assessment of autonomy and relatedness in adolescent–family interactions as predictors of adolescent ego development and self-esteem. *Child Development, 65*, 179–194.

Allen, J. P., Hauser, S. T., & Borman-Spurrell, E. (1996). Attachment theory as a framework for understanding sequelae of severe adolescent psychopathology: An 11-year follow-up study. *Journal of Consulting and Clinical Psychology, 64*, 254–263.

Allen, J. P., & Land, D. J. (1999). Attachment in adolescence. In J. Cassidy & P. Shaver (Eds.), *Handbook of attachment: Theory, research, and clinical applications*. (pp. 319–335). New York: Guilford Press.

Allen, J. P., Moore, C., & Kuperminc, G. P. (1997). Developmental approaches to understanding adolescent deviance. In S. S. Luthar, J. A. Burack, D. Cichetti, & J. Weisz (Eds.), *Developmental psychopathology: Perspectives on adjustment, risk, and disorder* (pp. 548–567). Cambridge, Cambridge University Press.

Armsden, G. C., & Greenberg, M. T. (1987). The inventory of parent and peer attachment: Individual differences and their relationship to psychological well-being in adolescence. *Journal of Youth and Adolescence, 16*, 427–454.

Bartholomew, K. (1990). Avoidance of intimacy: An attachment perspective. *Journal of Social and Personal Relationships, 7*, 147–178.

Bartholomew, K. (1994). Assessment of individual differences in adult attachment. *Psychological Inquiry, 5*, 23–67.

Bartholomew, K., & Horowitz, L. M. (1991). Attachment styles among young adults: A test of a four-category model. *Journal of Personality and Social Psychology, 61*, 226–244.

Bartholomew, K., & Shaver, P. R. (1998). Methods of assessing adult attachment: Do they converge? In J. Simpson & W. S. Rholes (Eds.), *Attachment theory and close relationships* (pp. 25–45). New York: Guilford Press.

Becker-Stoll, F., & Fremmer-Bombik, E. (1997, April). Adolescent-mother interaction and attachment: A longitudinal study. Paper presented at the Biennial Meeting of the Society for Research in Child Development, Washington, DC.

Berlin, L. J., & Cassidy, J. (1999). Relations among relationships: Contributions from attachment theory and research. In J. Cassidy, & P. R. Shaver (Eds.), *Handbook of attachment: Theory, research, and clinical applications* (pp. 688–712). New York: Guilford Press.

Berman, W. H., & Sperling, M. B. (1994). The structure and function of adult attachment. In M. Sperling & W. H. Berman (Eds.), *Attachment in adults: Clinical and developmental perspectives* (pp. 3–28). New York: Guilford Press.

Bernier, A., & Larose, S. (1999, April). Attachment status and help-seeking behaviors in academic counseling. Paper presented at the Biennial Meeting of the Society for Research in Child Development, Albuquerque, NM.

Blos, P. (1967). The second individuation process of adolescence. *Psychoanalytic Study of the Child, 72*, 162–186.

Bowlby, J. (1988). *A secure base: Parent–child attachment and healthy human development.* New York: Basic Books.

Bowlby, J. (1969/82). *Attachment and loss. Vol. I: Attachment.* New York: Basic Books.

Bowlby, J. (1973). *Attachment and loss. Vol. II: Separation.* New York: Basic Books.

Bowlby, J. (1980). *Loss, sadness and depression. Vol. III.* New York: Basic Books.

Bowlby, J., & Parkes, C. M. (1970). Separation and loss within the family. In E. J. Anthony & C. Koupernik (Eds.), *The child in his family: International yearbook of child psychiatry and allied professions* (pp. 197–216). New York: John Wiley & Sons.

Brack, G., Gay, M. F., & Matheny, K. B. (1993). Relationships between attachment and coping resources among late adolescents. *Journal of College Student Development, 34*, 212–215.

Bradford, E., & Lyddon, W. J. (1993). Current parental attachment: Its relation to perceived psychological distress and relationship satisfaction in college students. *Journal of College Student Development, 34*, 256–260.

Brennan, K. A., Clark, C. L., & Shaver, P. R. (1998). Self-report measurement of adult attachment. In J. Simpson & W. S. Rholes (Eds.), *Attachment theory and close relationships* (pp. 46–76). New York: Guilford Press.

Brennan, K. A., & Morris, K. A. (1997). Attachment styles, self-esteem, and patterns of seeking feedback from romantic partners. *Personality and Social Psychology Bulletin, 23*, 23–31.

Bretherton, I. (1985). Attachment theory: Retrospect and prospect. In I. Bretherton & E. Waters (Eds.), *Growing points of attachment theory and research* (pp. 3–35). *Monographs of the Society for Research in Child Development, 50* (1–2, Serial No. 209).

Bretherton, I. (1992). The origins of attachment theory: John Bowlby and Mary Ainsworth. *Developmental Psychology, 28*, 759–775.

Bretherton, I., & Munholland, K. A. (1999). Internal working models in attachment relationships: A construct revisited. In J. Cassidy & P. R. Shaver (Eds.), *Handbook of attachment: Theory, research, and clinical applications* (pp. 89–114). New York: Guilford Press.

Burge, D., Hammen, C., Davila, J., Daley, S. E., Paley, B., Herzberg, D., & Lindberg, N. (1997). Attachment cognition and college and work functioning two years later in late adolescent women. *Journal of Youth and Adolescence, 26*(3), 285–301.

Carlson, E. A. (1998). A prospective longitudinal study of attachment disorganization/disorientation. *Child Development, 69*(4), 1107–1128.

Carnelley, K. B., Pietromonaco, P. R., & Jaffe, K. (1994). Depression, working models of others, and relationships functioning. *Journal of Personality and Social Psychology, 66*(1), 127–140.

Cicchetti, D. (1984). The emergence of developmental psychopathology. *Child Development, 55*, 1–7.

Cicchetti, D., Cummings, E. M., Greenberg, M. T., & Marvin, R. S. (1990). An organizational perspective on attachment beyond infancy: Implications for theory, measurement, and research. In M. T. Greenberg, D. Cicchetti, & E. M. Cummings (Eds.), *Attachment in the preschool years: Theory, research, and intervention* (pp. 3–49). Chicago: University of Chicago Press.

Collins, N. L., & Read, S. J. (1990). Adult attachment, working models, and relationship quality in dating couples. *Journal of Personality and Social Psychology, 58*, 644–663.

Collins, N. L., & Read, S. J. (1994). Cognitive representations of attachment: The structure and function of working models. *Advances in Personal Relationships, 5*, 53–90.

Crowell, J. A., & Treboux, D. (1995). A review of adult attachment measures: Implications for theory and research. *Social Development, 4*, 294–327.

Crowell, J. A., Treboux, D., & Waters, E. (1999). The adult attachment interview and the relationship questionnaire: Relations to reports of mothers and partners. *Personal Relationships, 6*, 1–18.

Douvan, E., & Adelson, J. (1966). *The adolescent experience.* New York: John Wiley & Sons.

Elicker, J., Englund, M., & Sroufe, L. A. (1992). Predicting peer competence and peer relationships in childhood from early parent-child relationships. In R. D. Parke & G. W. Ladd (Eds.), *Family–peer relationships: Models of linkage* (pp. 77–106). Hillsdale, NJ: Lawrence Erlbaum.

Epstein, S. (1983). The Mother–Father–Peer Scale. Unpublished manuscript, University of Massachusetts, Amherst.

Fischer, A. R., & Good, G. E. (1998). Perceptions of parent–child relationships and masculine role conflicts of college men. *Journal of Counseling Psychology, 45*(3), 346–352.

Fraley, R. C., & Davis, K. E. (1997). Attachment formation and transfer in young adults' close friendships and romantic relationships. *Personal Relationships, 4,* 131–144.

Freud, A. (1969). Adolescence as a developmental disturbance. In G. Kaplan & S. Lebovici (Eds.), *Adolescence: Psychosocial perspectives.* New York: Basic Books.

Gilligan, C. (1982) *In a different voice.* Cambridge, MA: Harvard University Press.

Griffin, D. W., & Bartholomew, K. (1994a). The metaphysics of measurement: The case of adult attachment. *Advances in Personal Relationships, 5,* 17–52.

Griffin, D. W., & Bartholomew, K. (1994b). Models of the self and other: Fundamental dimensions underlying measures of adult attachment. *Journal of Personality and Social Psychology, 67,* 430–445.

Grossman, K. E., & Grossman, K. (1991). Attachment quality as an organizer of emotional and behavioral responses in a longitudinal perspective. In C. M. Parkes, J. Stevenson-Hinde, & P. Marris (Eds.), *Attachment across the life cycle* (pp. 93–114). London: Routledge.

Grotevant, H. D., & Cooper, C. R. (1986). Individuation in family relationships. *Human Development, 29,* 82–100.

Guisinger, S., & Blatt, S. (1994). Individuality and relatedness: Evolution of a fundamental dialectic. *American Psychologist, 49,* 104–111.

Hamilton, C. E. (1995, April). Continuity and discontinuity of attachment from infancy through adolescence. Paper presentation at the Biennial Meeting of the Society for Research in Child Development, Indianapolis, IN.

Hazan, C., & Shaver, P. R. (1987). Romantic love conceptualized as an attachment process. *Journal of Personality and Social Psychology, 52,* 51–524.

Hazan, C., & Shaver, P. R. (1990). Love and work: An attachment theoretical perspective. *Journal of Personality and Social Psychology, 59,* 270–280.

Hazan, C., & Shaver, P. R. (1994). Attachment as an organizational framework for research on close relationships. *Psychological Inquiry, 5*(1), 1–22.

Hazan, C., & Zeifman, D. (1999). Pair bonds as attachments: Evaluating the evidence. In J. Cassidy & P. Shaver (Eds.), *Handbook of attachment: Theory, research, and clinical applications* (pp. 336–354). New York: Guilford Press.

Hesse, K. (1999). The adult attachment interview: Historical and current perspectives. In J. Cassidy & P. R. Shaver (Eds.), *Handbook of attachment: Theory, research, and clinical applications* (pp. 395–433). New York: Guilford Press.

Hill, J. P., & Holmbeck, G. N. (1986). Attachment and autonomy during adolescence. *Annals of Child Development, 3,* 145–189.

Hinderlie, H., & Kenny, M. E. (August, 1999). Parental attachment, social support and college adjustment among Black college students on White campuses. Presented at the 107th annual convention of the American Psychological Association, Boston, MA.

Hobdy, J., & Hayslip, B. (1997). Attachment style and adjustment to life event in adulthood. Presented at the Annual Meeting of the American Psychological Association, Chicago, IL.

Josselson, R. (1988). The embedded self: I and thou revisited. In D. K. Lapsley & F. C. Power (Eds.), *Self, ego, and identity: Integrative approaches* (pp. 91–108). New York: Springer.

Kenny, M. (1987). The extent and function of parental attachment among first-year college students. *Journal of Youth and Adolescence, 16,* 17–27.

Kenny, M., & Donaldson, G. (1991). Contributions of parental attachment and family structure to the social and psychological functioning of first-year college students. *Journal of Counseling Psychology, 38,* 479–486.

Kenny, M., Moilanen, D., Lomax, R., & Brabeck, M. M. (1993). Contributions of parental attachments to depressogenic cognitions and depressive symptoms among early adolescents. *The Journal of Early Adolescence, 13,* 408–430.

Kenny, M. E. (1994). Quality and correlates of parental attachment among late adolescents. *Journal of Counseling and Development, 72,* 399–403.

Kenny, M. E., & Perez, V. (1996). Applications of attachment theory to adjustment and identity among ethnically diverse first-year college students. *Journal of College Student Development, 37*(5), 527–535.

Kenny, M. E., & Rice, K. G. (1995). Attachment to parents and adjustment in late adolescence: Current status, applications, and future considerations. *The Counseling Psychologist*, *23*, 433–456.

Kerns, K. A., & Stevens, A. C. (1996). Parent–child attachment in late adolescence: Links to social relations and personality. *Journal of Youth and Adolescence*, *25*(6), 323–342.

Kirkpatrick, L. A. (1998). Evolution, pair-bonding, and reproductive strategies: A reconceptualiation of adult attachment. In J. A. Simpson & W. S. Rholes (Eds.), *Attachment theory and close relationships* (pp. 353–393). New York: Guilford Press.

Kobak, R., & Cole, H. (1994). Attachment and meta-monitoring: Implications for adolescent autonomy and psychopathology. In D. Cicchiette (Ed.), *Rochester symposium on development and psychopathology: Vol. 5. Disorders of the self* (pp. 267–297). Rochester: University of Rochester Press.

Kobak, R. R., & Duemmler, S. (1994). Attachment and conversation: A discourse analysis of goal-corrected partnerships. In K. Bartholomew & D. Perlman (Eds.), *Advances in personal relationships: Vol. 5. Attachment processes in adulthood* (pp. 121–149). London: Jessica Kingsley.

Kobak, R., & Hazan, C. (1991). Attachment in marriage: The effects of security and accuracy of working models. *Journal of Personality and Social Psychology*, *60*, 861–869.

Kobak, R. R., & Sceery, A. (1988). Attachment in late adolescence: Working models, affect regulation, and representation of self and others. *Child Development*, *59*, 135–146.

Kobak, R. R., Sudler, N., & Gamble, W. (1992). Attachment and depressive symptoms during adolescence: A developmental pathways analysis. *Development and Psychopathology*, *3*, 461–474.

Lapsley, D. K., Rice, K. G., & Fitzgerald, D. P. (1990). Adolescent attachment, identity, and adjustment to college: Implications for the continuity of adaptation hypothesis. *Journal of Counseling and Development*, *68*, 561–565.

Larose, S., & Boivin, M. (1997). Structural relations among attachment working models of parents, general and specific support expectations, and personal adjustment in late adolescent. *Journal of Social and Personal Relationships*, *14*, 579–601.

Lopez, F. G. (1996). Attachment-related predictors of constructive thinking among college students. *Journal of Counseling and Development*, *75*, 58–63.

Lopez, F. G., & Gover, M. R. (1993). Self-report measures of parent–adolescent attachment and separation-individuation: A selective review. *Journal of Counseling and Development*, *71*, 560–569.

Main, M. (1999). Epilogue: Attachment theory: Eighteen points with suggestions for future studies. In J. Cassidy, & P. R. Shaver (Eds.), *Handbook of attachment: Theory, research, and clinical applications* (pp. 845–887). New York: Guilford Press.

Main, M., & Goldwyn, R. (1985). Adult attachment rating and classification system. Unpublished manuscript, Department of Psychology, University of California at Berkeley.

Main, M., & Hesse, E. (1990). Parents' unresolved traumatic experiences are related to infant disorganized attachment status: Is frightened and/or frightening parental behavior the linking behavior? In M. T. Greenberg, D. Cicchetti, & E. M. Cummings (Eds.), *Attachment in the preschool years: Theory, research, and intervention* (pp. 161–182). Chicago: University of Chicago Press.

Main, M., Kaplan, N., & Cassidy, J. (1985). Security in infancy, childhood and adulthood: A move to the level of representation. In I. Bretherton & E. Waters (Eds.), *Growing points of attachment theory and research* (pp. 66–106). *Monographs of the Society for Research in Child Development*, *50* (1–2, Serial No. 209).

Main, M., & Solomon, J. (1990). Procedures for identifying insecure–disorganized/disoriented infants: Procedures, findings, and implications for the classification of behavior. In M. Greenberg, D. Cicchetti, & M. Cummings (Eds.), *Attachment in the preschool years: Theory, research, and intervention* (pp. 21–160). Chicago: University of Chicago Press.

McCormick, C. B., & Kennedy, J. H. (1994). Parent–child attachment working models and self-esteem in adolescence. *Journal of Youth and Adolescence*, *23*(1), 1–17.

McCutcheon, L. E. (1998). Self-defeating personality and attachment revisited. *Psychological Reports*, *83*, 1153–1154.

Miller, J. B. (1976). *Toward a new psychology of women*. Boston: Beacon Press.

Mohr, J., & Fassinger, R. (1997). *Romantic attachment, parental attachment, and lesbian identity development*. Chicago: American Psychological Association.

Parker, G., Tupling, H., & Brown, L. B. (1979). A parental bonding instrument. *British Journal of Medical Psychology*, *52*, 1–10.

Pearson, J., Cohn, D. A., Cowan, P. A., & Cowan, C. P. (1994). Earned- and continuous-security in adult attachment: Relation to depressive symptomatology and parenting style. *Development and Psychopathology*, *6*, 359–373.

Phelps, J. L., Belsky, J., & Crnic, K. (1998). Earned-security, daily stress, and parenting: A comparison of five alternative models. *Development and Psychopathology, 10*, 21–38.

Rice, K. G. (1990). Attachment in adolescence: A narrative and meta-analytic review. *Journal of Youth and Adolescence, 19*, 511–538.

Rice, K. G., & Cummins, P. N. (1996). Late adolescent and parent perceptions of attachment: An exploratory study of personal and social well-being. *Journal of Counseling and Development, 75*, 50–57.

Rice, K. G., Cunningham, T. J., & Young, M. B. (1997). Attachment to parents, social competence, and emotional well-being: A comparison of black and white late adolescents. *Journal of Counseling Psychology, 44*(1), 89–101.

Rice, K. G., & Whaley, T. J. (1994). A short-term longitudinal study of within-semester stability and change in attachment and college student adjustment. *Journal of College Student Development, 35*, 324–330.

Rothbard, J. C., & Shaver, J. C. (1994), Continuity of attachment across the life span. In M. Sperling & W. H. Berman (Eds.), *Attachment in adults: Clinical and developmental perspectives*. New York: Guilford Press.

Sarason, B. R., & Pierce, G. R. (1991). Perceived social support and working models of self and actual others. *Journal of Personality and Social Psychology, 60*(2), 273–287.

Searle, B., & Meara, N. M. (1999). Affective dimensions of attachment styles: Exploring self-reported attachment style, gender, and emotional experience among college students. *Journal of Counseling Psychology, 46*(2), 147–158.

Shaver, P. R., & Brennan, K. A. (1992). Attachment styles and the "big five" personality traits: Their connections with each other and with romantic relationship outcomes. *Personality and Social Psychology Bulletin, 18*, 36–545.

Shaver, P. R., & Hazen, C. (1988). A biased overview of the study of love. *Journal of Social and Personality Relationships, 5*, 473–501.

Shaver, P. R., Papella, D., Clark, C., Koski, L., Tidwell, M., & Nalbone, D. (1996). Androgyny and attachment security: Two related models of optimal personality. *Personality and Social Psychology Bulletin, 22*, 582–597.

Simpson, J. A. (1990). Influence of attachment styles on romantic relationships. *Journal of Personality and Social Psychology, 59*, 971–980.

Simpson, J. A., & Rholes, W. S. (1998). Attachment in adulthood. In J. Simpson & W.S. Rholes (Eds.), *Attachment theory and close relationships* (pp. 3–21). New York: Guilford Press.

Steele, H., & Steele, M. (1994). Intergenerational patterns of attachment. *Advances in Personal Relationships, 5*, 93–120.

Sroufe, L. A. (1989, May). Pathways to adaptation and maladaptation: Psychopathology as developmental deviation. Paper presented at the Clinical Developmental Institute, Development and Psychopathology: Clinical Developmental Perspectives, Cambridge, MA.

Sroufe, L. A., & Waters, E. (1977). Attachment as an organizational perspective. *Child Development, 48*, 1184–1199.

Surrey, J. L. (1991). The self-in relation: A theory of women's development. In J. Jordon, A. Kaplan, J. B. Miller, I. Stiver, & J. L. Surrey (Eds.), *Women's growth in connection: Writing from the Stone Center*. New York: Guilford Press.

Taub, D. J. (1995). Relationship of selected factors to traditional-age undergraduate women's development of autonomy. *Journal of College Student Development, 36*, 141–151.

Thompson, R. A. (1999). Early attachment and later development. In J. Cassidy & P. R. Shaver (Eds.), *Handbook of attachment: Theory, research, and clinical applications* (pp. 265–286). New York: Guilford Press.

van IJzendoor, M. H. (1995). Adult attachment representation, parental responsiveness, and infant attachment: A meta-analysis on the predictive validity of the Adult Attachment Interview. *Psychological Bulletin, 177*, 387–403.

van IJzendoorn, M. H., & Zwart-Woudstra, H. A. (1995). Adolescents' attachment representations and moral reasoning. *Journal of Genetic Psychology, 156*, 359–372.

Ward, M. J., & Carlson, E. A. (1995). Associations among adult attachment representations, maternal sensitivity, and infant–mother attachment in a sample of adolescent mothers. *Child Development, 66*, 69–79.

Waters, E., Crowell, J., Treboux, D., Merrick, S., & Albersheim, L. (1995, April). Attachment security from infancy to early adulthood: A 20-year longitudinal study. Poster presentation, Biennial Meeting of the Society for Research in Child Development, Indianapolis, IN.

Weiss, R. (1982). Attachment in adult life. In C. M. Parkes & Stevenson-Hinde (Eds.), *The place of attachment in human behavior* (pp. 171–184). New York: Basic Books.

Zimmermann, P., Fremmer-Bombik, E., Spangler, G., & Grossmann, K. (1995, April). Attachment in adolescence: A longitudinal perspective. Poster presentation at the Biennial Meeting of the Society for Research in Child Development, Indianapolis, IN.

CHAPTER 20

Adult Development and Parenthood

A Social–Cognitive Perspective

Sandra T. Azar

CRISIS OR TRANSITION: IS PARENTING A "NECESSARY" AND UNIQUE STAGE OF ADULT DEVELOPMENT?

Before discussing this chapter's topic, the question needs to be raised as to whether it merits a separate chapter in this volume. There is some debate as to the universality of parenthood as a stage of adult development. On the one hand, parenting has been seen as providing a necessary and unique context for the development of psychological maturity. In contrast, it has also been seen as just one of many stressors that adults encounter that can lead to either personal growth or maladaptation. This dichotomy reflects more generally tensions within our field between the idea of a "universal" structure to the timing of psychological changes in adulthood vs. a more "open-ended" view, with more "randomness" in such "transforming" experiences (Brim & Ryff, 1980; Gergen, 1980; Palus, 1993). On one side, theorists such as Erikson (1963), Gould (1978), and Levinson (1978, 1986) have argued for a progression of stages through which most adults move and that are demarcated by age and specified periods, with parenting occurring in one's 20s or early 30s and accompanied by the negotiation of specific and unique tasks. On the other side are views that have argued that an adult's life course is more flexible, random, and more driven by context (i.e., "aleatoric" view). In this view, parenthood is one *potential* pathway to "maturity," but there may be many others producing similar changes.

Sandra T. Azar • Department of Psychology, The Pennsylvania State University, University Park, Pennsylvania, 16802-3104.

Handbook of Adult Development, edited by J. Demick and C. Andreoletti. Plenum Press, New York, 2002.

Research has reflected one side of this debate more than the other. The stage perspective has attempted to document the developmental capacities (e.g., psychological integration, secure attachment) that are required to carry out the new changes inherent in parenthood, and the successful reaching of this stage is typically assessed through parental well being, smooth family transactions (e.g., marital satisfaction), and positive child outcomes. Studies have also examined major disturbances in the parents (psychiatric problems), their adult life (marriage, work), or parenting "disorders" (e.g., postpartum depression, child abuse). Difficulties in this role are attributed to prior vulnerabilities (failures in some earlier stage), rather than to the task itself "creating" vulnerability or to failures in contextual supports. As will be seen later, this view has implications for expectations about the self and may lead to greater self-blame when "failures" are perceived. Societal blaming may also occur that may have a further impact on the self (Farber & Azar, 1999).

Less research has focused on the alternative view—that there are a multitude of transformative experiences in adulthood, of which parenting may be one. Parenting from this perspective can be studied within more general developmental models or stress and coping models, examining what the individual brings to this role and the nature of what it requires of him or her. Studies from this view focus on parenting among an array of other stressors. If they focus specifically on parenting alone, they usually explore extreme parenting stressors (e.g., birth of a disabled child), and parental, marital, and contextual factors leading to better coping are studied. Expectations about the self based on this view may be more flexible, allowing a wider range of enactment of the role of parent, and thus as a "personal theory" may lead to greater well being and more positive growth.

Although disparate, both views have one common thread, that parenthood may lead to change that may be labeled as development and thus is worthy of study, although the latter sees it as less unique than the former. While not being able to resolve this debate, this chapter attempts to argue that parenthood is basically a relational task that has some unique features, but generally involves capacities that are required for many domains of adult adjustment. It is also argued that the most central of these are social–cognitive ones. Development in parenthood, it is argued, requires flexible and appropriate expectancies about the role and active capacities to learn "in the moment" and modify one's thinking, affect, and behavior. The relational tasks require adjusting one's interpretations and responses to a moving target (i.e., the developing capacities of children) and balancing the child's needs with those of the adult, their other relationships (e.g., marital needs), and the context (e.g., life in poverty). It also requires balancing long- vs. short-term goals (e.g., balancing responses that will resolve a momentary crisis such as "giving in" to a tantrum in a grocery store vs. engaging in responses that will build long-term interpersonal regulation in interaction with the child). Thus, it is adults' capacities as strategists that may allow for both successful outcomes (that provide them with a source of efficacy essential for further actions aimed at reaching new levels of functioning) and to keep tension within the role at a level in which new capacities may evolve and older ones be refined.

Furthermore, if parenthood is a stressor (like others), this chapter sees it more like a chronic rather than an acute one (i.e., the challenges begin long before the birth of a child and continue throughout the life course[1]). Given it is seen as a commitment that cannot be broken, it is a stressor that the parent knows will not go away,[2] which has implications for responses to its challenges. Further, as the role evolves, different parts of adult's social–cognitive and behavioral repertoire may be taxed as the "job" description changes (e.g, producing regulation in interaction with infants requires different capacities than doing so with a teenager).[3]

Further, it is argued that the role is socially constructed, and thus need not be "universal" in nature across ethnic, racial, or class lines. "The images of motherhood and fatherhood reveal our shared ideals, standards, beliefs, and expectations regarding men and women as parents" (Thompson & Walker, 1989, p. 859). Cusinato (1994) has described these images as both *enduring* and *emerging*. For example, parents who were raised during the Great Depression, may have unrealistically high expectancies for raising children who are economically stable, and thus may react more strongly to their children's job loss or failure to attain an education.

Finally, the requirements of parenting vary with context (e.g., parenting in a high-risk inner city neighborhood may challenge the adult differently than in a low-risk suburban one; parenting a special needs child is different than parenting a well functioning one; adopting an older child may mean parenting differently than parenting a biological child from infancy). Even individuals' reasons for becoming a parent vary by context (i.e., Kagitcibasi, 1982). Thus, adult development may be challenged and evolve in different ways in different contexts.

A developmental–functional contextual perspective is therefore taken. It is argued that there are individual differences in parental expectancies of themselves, their children "in relation" to themselves, and what they perceive is required of them by others. As contextual factors change (children's development, family context, country of residence), these expectancies and the behavioral expression of parenting that follow from them must be flexible. If not, energy that should be directed to development will be directed toward coping with the crisis of violated expectancies and reductions will occur in self-efficacy. Maladaptive responses may result, and ultimately this creates situations that further decrease developmental potential.

This chapter begins with a functional analysis of the tasks involved in parenting that highlights briefly the manner in which the expectations regarding the role have varied over historical eras, culturally, and contextually, as well as the extreme heterogeneity in the actual role requirements. Present models of parenting are faulted for their lack of attention to this heterogeneity. This chapter then outlines a global set of social–cognitive capacities required across the tasks of adulthood and how they are challenged and evolve within the role of parent.

---

[1] The transition is most often studied because it is here that "development" may be most apparent. However, the narrow focus on this period may mislead students of development into believing that it is the only place where development occurs. The author sees development as occurring in small increments and that there is potential for it to occur continuously in a stream of actions and reactions to one's environment. Thus, the potential for development in this role can occur throughout the life course. Examples from the literature on aging parents will be presented to illustrate this.

[2] It is discussed later how this may be a social construction.

[3] Clearly, there is parenthood in which greater continuity tasks occur (e.g., with mentally retarded children) and such parents may "do" development differently (Tobin, 1987).

Such a model, it is argued, better describes the tasks involved for the full life course of parenting and across the full spectrum of parents. Because this author has come at this topic from programmatic research focusing on at-risk parents (e.g., abusive and neglectful parents, intellectually limited parents), data on atypical parents, along with developmental literature, are discussed. In this discussion, particular attention is also given to sociocultural issues.

IS PARENTING A DOABLE DEVELOPMENTAL LIFE TASK? WHAT ARE THE DEMANDS?

Tasks Are Context Driven

Caregiving in general, and parenting in particular, challenges many interpersonal, emotional, and cognitive capacities. Identifying a "universal" set of these capacities is difficult in that they are defined by children's developing needs and the availability of contextual supports. For example, in societies in which extended family are not involved in the raising of children and in situations in which resources are limited (e.g., poverty), becoming a parent dramatically increases the burdens adults have and carries with it more risk (tasks that may overwhelm adults' capacities), as well as opportunities for growth (tasks on the edge of the individual's "developmental reach" [Vygotsky, 1934], which can pull them to higher levels of maturity). This aspect of parenting is underplayed in practice (e.g., parenting is said to "come naturally"). Pressure to assume this role is great and the failure to do so, either for voluntary or involuntary reasons (e.g., infertility), carries with it a stigma. Indeed, areas where this burden exceeds adults' capacities have been viewed as anomalies (e.g., child abuse, postpartum depression), whereas individuals being equal to the task is seen as "normative" and evidence of "psychological integration" (Vondra & Belsky, 1993).

Assuming or even contemplating the role of parent may thus be viewed with trepidation because it is seen as having a narrow set of socially constructed tasks that even under the best of circumstances may be overwhelming to adults' sense of autonomy. Indeed, in recent years, pressure in the role has increased further with a growing emphasis on being "child centered" in one's family organization and a push for fathers to be more involved. While this may have been within acceptable limits in earlier eras where at least one member of the parenting team was reared with an emphasis on connectedness and deferring her interests and needs to those of others, it may be incongruent with current definitions of young adulthood that emphasize having diversified interests, free exploration, and a focus on the self (Arnett, 2000; Rossi, 1968).

This contrast, however, may not be as sharp as it is typically portrayed. The goals of parenting that are inherent in the role may be more oriented toward *parents*' needs than has been acknowledged (e.g., socialization of children for later old age care; Kagitcibasi, 1982) and the enhancements to the self that come with increases in connectedness have begun to be documented. Conceptions of the role may also vary with children's development ("The parents of teenagers should do...."), children's gender ("I expect my daughter will be the care provider in my old age"), and by other characteristics of the child and the parent's relationship with them ("My son needs a firm hand"; "I was working so much when my son was young that I did not have time for him. I want to make up for it now."). Generational differences in conceptions of the role may also provide alternative models that may facilitate adjustment and/or produce conflict. How doable the task is therefore reflects the nature and flexibility of the expectancies one has constructed and the ones demanded by society. More is said about this in the following.

The Defining Mandates of the Parenting Role

In her philosophical discussion of parenthood,[4] Ruddick (1989) argues that parents are labeled as such just because of and to the degree that they are committed to meeting the demands that define parental work. In the primary social group with which a parent is identified, whether by force, kinship, or choice, there are demands that parents raise their children in a manner acceptable to them. Thus, parenting tasks are in part socially defined (e.g., when parents violate community standards, laws allow the state to remove children). She has identified three areas of demands: *preservation, growth*, and *social acceptance*. A similar set of three basic tasks has been outlined by others. LeVine (1974), for example, highlighted three basic goals: (1) promoting the physical survival and health of children, ensuring that the child will live long enough to have children of his or her own (*the survival goal*); (2) fostering the skills and behavioral capacities that the child will need for economic self-maintenance as an adult (*the economic goal*); and (3) fostering behavioral capabilities for maximizing other cultural values (e.g., morality, achievement) (*the self-actualization goal*). Similarly, Epstein et al. (1982) discuss *basic* (providing food, shelter), *emergency* (responding appropriately to hazardous situations), and *developmental tasks* (provision of activities, responses that allow for developmental needs). Cultures differ in the emphasis on each of these goals, and at times these goals may be in conflict with each other (Azar & Cote, in press).

Ruddick (1989) views parents as meeting these demands through preservative love, nurturance, and training. When children are seen as demanding care, the reality of their vulnerability and the necessity of a caring response seem "unshakable" in her words. However, the *perception* of vulnerability, she argues, is optional as is the responding with care. Benedek (1959), making the same point, noted that children's needs are *absolute* and parents' need to respond *relative*. Thus, both posit that cognition mediates the tasks involved.

The demands of the role are set by which group parents identify with to judge "acceptable" social behavior. If parents are ambivalent about their social group or feel harassed by the demands they place on them, there may be difficulties. Furthermore, as noted earlier, these demands are often contradictory—mothers in urban neighborhoods must choose to allow their children to walk to the store to do errands to encourage their autonomy, and at the same time to do so leaves them vulnerable to risks in the environment. This requires parents to be strategists.

---

[4] Ruddick's work dealt with motherhood specifically, but her discussion can easily be extended to all parenthood. It might be argued, however, that the demands she outlines are felt differently by each gender.

Moreover, cultures differ in what they value and this may change with changes in context. For example, work by Okagaki, Sternberg, and Divecha (1990, cited in Okagaki & Divecha, 1993) has suggested that immigrants to this country differ by culture in how they think parents socialize various behaviors, but immigrant status also affects the prioritizing of behaviors (independence vs. conformity). The latter may have more to do with changes in context for these parents than universal conceptions of roles.

Parenting is therefore not divorced from more general class/cultural models (about how the world works) that are derived from social conditions and that drive a host of social behaviors (Azar & Benjet, 1994; Goodnow & Collins, 1990; Kohn, 1963). Other contextual factors (work, marriage, media, the characteristics of the child) also influence the nature of the role. Finally, the *status* of the social group with which the parent identifies may also influence the nature of tasks (e.g., minority parents may see as their role preparing children for exposure to discrimination). Thus, there may be individual differences in construction of parenthood and these differing constructions and their interaction with context and societal demands can challenge or facilitate adult development.

The Realities of Parenthood: The Lack of Homogeneity

As this discussion suggests, the realities of parenthood are not as homogeneous as the literature or our societal icons would lead us to believe (Azar, 1996). Parenthood can be deconstructed in our society as meaning two adults, a mother and a father, both of whom are biologically the child's parents and who are living together in a household, typically alone with the child or children. The typical picture is of a middle-class, Caucasian family of Western European ethnic background. These parents are believed to assume this role in their 20s and see the child through to adulthood, which occurs when the child leaves home to marry or go to college. It is also assumed that *most adults* will take on this role. In fact, the *who, how, when, where,* and *how long* of parenting are more diverse than the preceding description. Even *whether* adults assume this role has changed over time. As will be seen, this discrepancy between images and realities is a source of internal conflict for adults.

Further, this role transition is assumed, along with other transitions, to be synonymous with "adulthood." A recent survey by Arnett (1997), in fact, however, disputes this link. Young adults see "adulthood" in less tangible and in more gradual, psychological, and individualistic terms than the occurrence of a role transition such as parenthood.[5] This finding and the discussion regarding the constructed nature of the role suggest an important anchor in this chapter's discussion of parenthood and adult development. It focuses our attention on the idea that cognitive processes (i.e., our knowledge structures about the role—our beliefs/assumptions/expectancies) play an important part in determining our responses) and whether development is facilitated or not.

---

[5] Those who had become parents still saw it as one of the most important "markers" of adulthood (Arnett, 2000), which is not the same as the experience that *produced* adulthood.

Whether? The typical life course—a sequence of leaving home, marriage, child rearing, launching, and survival at age 50 with the first marriage still intact, unless broken by divorce—had not been the dominant pattern of family timing before the early 20th century. Prior to 1900, only about 40% of females in this country experienced this life pattern (Uhlenberg, 1974 cited in Harevan, 1982). The remainder either never married, never reached marriageable age, died before child-birth, or were widowed while their children were still young. Thus, current social constructions of adult life as including parenthood is a fairly recent phenomenon.

Further, even our current ideas of parenthood are open to question. Rates of women marrying each year have declined over the last century from a high of first marriage of 143 per 1000 in the mid-1940s to 76 per 1000 in 1988. Those who do marry are doing so later (e.g., in the United States, the median age for women rose from 21 to 25 and for men from 23 to 27 from 1970 to 1997; US Bureau of the Census, 1997). (The age at first childbirth has followed a similar pattern.) These rates vary by culture. African-American women are much less likely to marry. In 1990, fewer than 75% of them married compared with 90% of white women. Hispanic women have similar patterns.

Fertility rates have also decline significantly over time (from a high of 3.0 children to current rates of 1.5). There is also a rising group of women who will not bear children. It has been estimated, for example, that there are 5.3 million infertile couples (Leiblum, 1997). The expectancies of being a parent and the potential for experiencing such infertility as "loss" and having to "settle" for or "choose" alternatives to childbearing (e.g., adoption) may have unique effects on how parenthood is experienced. Groups have also emerged for whom parenting might have not been possible in the past (e.g., gay and lesbian parents [Anderson, 1999]). For such parents, children may lead to a new phase of "coming out" to extended family that may produce new adult challenges.

Who? Besides traditional configurations of two biological parents, today's parents include adoptive parents, stepparents, gay and lesbian parents, and grandparents raising their children's offspring. With various custody arrange-ments, individuals can go in and out of the parenting role (e.g., shared custody, foster parenting, loss of parental rights, and adoption disruptions). About 26% of all children (17 million) live with a divorced, separated, or stepparent. In some cases, divorce may mean no or little contact with one's children. In 1979, one sur-vey found that 36% of children of divorce had not seen their fathers in the pre-ceding 5 years or did not know where they were living (Furstenburg, Nord, Peterson, & Zill, 1983). Although the transition to parenting has been seen as unique compared to marriage in that it was not reversible (e.g, it was said you can have an ex-husband, not "ex-children" [Rossi, 1968]), individuals can and do come and go in the role.

The image of children living in two-parent homes may also not provide an accurate picture (in 1990, only 57.6% of children lived with both their biological parents, with another 11.3% living in married stepfamilies). In 1988, 28.2% of children lived with a never married mother and for African-American children, this figure is 52%. This is a marked change from earlier eras (e.g., these figures in 1960 were 3.4% for all mothers and 4% for African-American mothers). Although the percentage is small, some children also live with only their fathers. Single parenthood may foster adult growth in different ways than does dual parenting (e.g., requires more resourcefulness).

The ethnic, racial, and social class portrayed in "typical" parenthood also defies the facts. One of every five US children is either an immigrant or has immigrant parents, with most being Hispanic or Asian (National Research Council, 1998). In 1990, about 30% of children were members of minority groups. By 2030, this figure will be 50% (National Research Council, 1998). One in five children live in poverty. Despite these facts, in developmental journals from 1979 to 1993, only 3% to 8% of articles had a significant focus on minority samples (Cauce, Ryan, & Grove, 1998; Graham, 1992). Parenting studies have also mostly focused on mothers, not fathers. Thus, the database from which our models of parenthood have been developed is quite narrow, raising issues about the relevance of constructs identified as crucial to this role (Azar & Cote, in press). Moreover, samples studied that vary from this "norm" are typically "at-risk" ones (teenage mothers[6]).

The images of parenthood for men and women and for different cultural/ racial/social class groups within our society vary when it comes to parenting. For example, Furstenberg (1988) has argued that cultural images of fatherhood have been shaped by a dichotomy for good dad/bad dad fathers—with opposing portrayals of nurturing fathers with deadbeat dads. Social class and race interact with such portrayals. Lawson and Thompson (1999) have argued that scholars have perpetuated negative stereotypes of black fathers based on who is studied and the characteristics about them that are studied. Similarly, this author has argued that we hold images of lower socioeconomic status mothers as being helpless, incompetent, hopeless, and even despicable (Azar, 1996). Not surprisingly, these groups are not oblivious to such views and even may share some of them— coming to perceive themselves in a similarly negative way and becoming discouraged (Polakow, 1993). The author has heard such views expressed in her own work with mothers. For instance, a young teenage mother who was on AFDC angrily reported how people looked at her in the grocery store—"You could see it in their eyes .... like I'd made a big mistake. Like I screwed up my life and that there was no way I could be a good mother." This adds burden to an already difficult role.

These negative images can be part of the schema that adults from these groups carry with them as they assume this role and they may affect their views of themselves. For example, one African-American young teen who became pregnant told me she did not want to go back to school because "Everyone will say, 'There she is, another black girl who got herself pregnant!'" These images can thus influence developmental trajectories.

Similar bias occurs in considering alternative family structures. Alternative family forms (e.g., single parenthood, living with extended family) are seen as *inherently* less "ideal" and a source of "risk." That is, studies are not constructed with an eye toward examining strengths of these alternative family forms or contexts (i.e., ways adult development may be enhanced). In some cultures, the functions of parents are carried out by extended family members and in such cultures,

---

[6] It needs to be noted that teenage parenting produces risk mainly because of the lower education and economic status of such individuals, suggesting in part contextual factors, not person-based ones, are at fault.

single parenthood may have a different influence on adults than ones in which two people share all the functions. We have clung to studying parent–child dyadic interactions alone to gauge outcomes in parenthood, which may be myopic in some groups.

How? Biology alone does not determine parenthood. Technology has created new types of parenthood. *In vitro* fertilization; artificial insemination; and implanting of eggs, fertilization, and adoption "creates" parenting by various short or long routes that color the individual's experience of this role transformation. The implications for adult development are only just beginning to be explored (Leiblum, 1997). Adoptive parents who assume the role after years of infertility treatments may come to the role either "scarred" or stronger for their experience.

Other new "choices" are possible. For example, in international adoptions, parents are sent pictures of potential children and are allowed to "pick" or "reject" children. Public agencies trying to facilitate adoptions have created new recruitment approaches (e.g., adoption parties where children are put "on display" and potential parents come and "take a look"). The emotional challenges these practices present have not been examined.

Thus, the transition to parenthood may be a more gradual process and have many more choice points and a multitude of "transitions" rather than one abrupt one. These experiences may influence the self-schema of the parents and how they feel others view them. This also influences their expectations about their role and that of children in their lives. Such images may complicate every parenting decision. As one adoptive parent described her experiences parenting: "Everyone was looking at me ... I had to be a better mother than others—because I had been given this wonderful child." Parents' expectations about how children "turn out" may also affect their development in the role (e.g., expectations of adopting the "perfect" child or of adoption "saving" a child may be shattered if problems arise). The processes involved in parenthood may be very different for this diverse group of parents (e.g., introducing a child who is obviously of a different racial background to others may feel much more complex and stressful). Identifying processes that are common to all such groups is a particular challenge to the study of adult development.

When? The norms for when parenthood occurs have changed over history. Teenage parenting was common in the past, whereas, with lengthening lifespans, it is now viewed as a "risk".[7] Further, with birth control, couples postponing childrearing, and new technology increasing the age at which biological parenting can occur, older and older adults are having children. They, however, may encounter biases in society as to when parenthood *should* occur.

Self-perception may be affected by this bias, as well as how children respond to their "out of sync" parents. Yarrow (1991), in a discussion of "latecomer" parents, noted their children reported being embarrassed by them and that the parents experienced "guilt" in exposing their children to parents who had less energy and

---

[7] It should be pointed out that the relationship between parenting in the teenage years may not carry the risk commonly perceived in the literature. It is more risky because it is associated with less educational attainment and lower economic resources.

were out of the norm. These authors described examples of parents lying about their age, or age being a forbidden topic. Assuming this role after encountering many other life events that can have a "maturational" effect or increase vulnerability can have implications for adult development. Some adults may have had opportunities to explore multiple arenas in the work world and may not experience the losses from this area. The demands of parenting may be less stressful, whereas assuming the role before there is achievement of success may make the necessary curtailment of efforts to achieve work success more stressful. The impact may be felt throughout the life course. Fathers who have passed their prime in terms of work and who have not been "successful" in their own eyes may view their son's successes with jealousy. This may make the tasks of aging harder.

Further, the age at which one initially becomes a parent will have a cascading effect across the lifespan. Early parenting, for example, may mean unique aspects to one's relationship with offspring (e.g., young single mothers may be dating alongside their teenage daughters and thus may experience feelings of "competition"). It can mean changes in timing of other transitions, such as earlier grandparenthood which may produce more closeness with one's children as a certain amount of co-parenting can occur (Obeidallah & Burton, 1999). More tension is also possible as offspring enter their reproductive years as mothers are leaving theirs (La Sorsa & Fodor, 1990).

Where? Development in parenthood cannot be divorced from its context. Contexts determine the functionality of one's capacities, and in addition can foster or be an obstacle to new learning. The goals of socialization are different if one is raising a child to work in an agrarian society compared to raising one to become a stockbroker or lawyer.

For the most part, early models implicitly assumed that all parents have the basic physical resources to parent, and thus all parenting can take place in the same way. *Universalism* has reigned in this realm. Such models did not consider how context and child characteristics may influence what "adequate" parenting should look like (e.g., they assumed that the single parent living in a high crime neighborhood should engage in the same set of parenting behaviors as the middle-class mother without a partner who lives in a suburban setting, and that those same sets of parenting practices are "best" in either setting). Current research has dispelled such views. For example, more restrictive parenting may actually result in better outcomes in high-risk parenting environments and a more authoritative style result in worse ones (Baumrind, 1993; Baldwin, Baldwin, & Cole, 1990). Society, however, may label the restrictive parent as "bad" at parenting and may institute interventions to "train away" this behavior. Again, this message of not "measuring up" despite doing what in fact is better for one's children can result in much tension for the parent.

Early models also assumed that all families should prioritize the same socialization goals (e.g., all value self-reliance as an outcome), and in cases where they found deviations from these "norms" the group was considered more "primitive" or "less developed" or deviant. Even a cursory look at other ethnic and social class groups finds these weak assumptions. Urban parenting may require much higher parenting skills and demand more from a parent than suburban living. Poverty may mean parenting in an "unforgiving" environment. Errors have stronger consequences. Because the tasks of raising children in high-risk environments often have contradictory elements (e.g., parents must be overprotective of their children in high-risk environments, but not thwart their self-reliance [Collins, 1990]), they may result in mastering more complex cognitive problems, and thus promote adults' capacities. Add to this perhaps the preparation one's offspring may need to enter a society that may discriminate against them (because of their ethnic or racial background) and the role of parent may have a more complex job description.

Why? The reasons for becoming a parent have not received much attention, except in cross-cultural research. Kagitcibasi (1982), for example, found that in highly industrialized societies, the "old age security" provided by children is a less important reason for having children or for wanting another child, whereas in less developed countries where formal systems of support for old age are not present and in lower socioeconomic status groups where these supports are less assured, it is still a value. (Interestingly, it is more important for women in these situations, perhaps because they are less economically independent.) This fact would influence parenting practices (e.g., sibling caregiving) and thus challenges to adults. The "developed" parent within such cultures would have sets of practices that might not be considered developed in another culture. Parents who immigrate between such cultures must have more flexibility and engage in more refined decision making, recognizing the changing parameters of their new environments. This ability to transform socialization goals or hold cognitively two goals as ideals simultaneously (i.e., biculturalism [Parke & Buriel, 1998]) is only beginning to be studied.

Only a few efforts have taken place to examine the universality of expectancies regarding children. Differences in expectations regarding children taking on adult role responsibilities have been linked to parenting risk within the United States (e.g., child abuse and neglect [Azar et al., 1984; Azar & Rohrbeck, 1986]). Racial and cultural groups, however, may vary in these expectancies (Azar & Houser, 1993), whereas certain levels of such expectancies may still define within-group differences. Research is needed exploring ethnic and social class differences in parental reasons for assuming this role and the role definitions held for children in relation to parents in family life.

How Long? The traditional idea that the parenting role ends with children leaving the parental home is open to question as well. "Launching", as it has been called, does not take place in all cultures (e.g., older adults are incorporated into the young couple's home on marriage) or there is a period of moving in and out of the parental home (as many as 30% of young adults do so [Goldscheider & Goldscheider, 1994]). Role perceptions vary by culture. Shand (1985) reports that Japanese mothers view their childrearing as a lifelong role, whereas American mothers are described as seeing their role as more short term (until the child reaches independent living).

We also see "temporary" parenting roles in our society. Foster care is the best example of this, where a parent may take in one or more children at a time of crisis for the child and then act as "parent" for a limited time period that may end with the child returning to his or her biological parent or the foster parent or another parent adopting him or her. The impact on adult development of forming attachments to and nurturing a child for whom the adult knows he or she will not have input into the future has received little attention. Returning the child to the person who may have harmed him or her may also tax the adult's self-system and relational capacities.

In summary, the heterogeneity of the role of parent defies the narrow social construction presented at the beginning of this section. Our knowledge base and the theories of adult development and parenting have little to say regarding many groups. Most of our models of parenting that grew out of such images were not developed with this heterogeneity in mind. Consequentially, our psychological concepts of parenthood (its tasks, the role it plays in adults' lives) too may need to be adjusted.

At the same time, parents in our culture are influenced by and respond to this social construction. Foster parents may feel the same pressure as that of a biological parent to remain calm with a child's tantrum in a grocery store and perceive the same social censure if their interventions go badly. Our current models (evolutionary ones, attachment theories) are too narrow in their focus. They require biological relationships with children, or that parenting begin within some critical period. They are insensitive to different family structures and ignore context. They do not provide an understanding of the flexibility that is required for adults to adjust to the ways in which they diverge from the "expected" patterns of this role (e.g., personality theorists did not envision tasks such as negotiating with the doctors who may be biased against lesbian parenting or dealing with home studies that are required in adoption). A model that focuses more on processes/capacities across relationships more generally and across other types of adult tasks and that at the same time highlights unique elements within the parenting realm may provide more flexibility to encompass the developmental work of the variety of individuals who undertake the role of parent and the variety of contexts in which it is undertaken. Such a model that is based in social–cognitive theorizing is presented below.

SOCIAL COGNITION AND ADULT DEVELOPMENT

Lifespan developmentalists and family theorists have posited many domains of adult functioning that are stimulated by the tasks of parenthood including everything from sex typing to dialectical thinking (Palkovitz, 1996). While promising, this discussion has been either very general or focused on single narrow aspects of development. A more comprehensive model of the processes involved in such development is lacking. Social–cognitive theorizing may provide such a model for understanding how parenthood might act as a generator of adult development. Such a theory is very sophisticated in explaining "learning" in the context of social relationships, as well as the disturbances in social and emotional functioning that might interfere with it. Recent empirical work in parenting has also supported the role that such social–cognitive factors play in adult competence, well being, self-efficacy, and life satisfaction.

A basic assumption to this approach is that for development to occur the task before the individual has to be within his or her *zone of proximal development*.

This means it has to be within their developmental reach such that with minimal scaffolding by others and effort on their own part, adults can move to a higher level of functioning (Vygotsky, 1934). As discussed earlier, parenting (no matter what form it takes) places demands on adults and it is these challenges that can act to stimulate development. These challenges are not entirely new. Some may have been faced and mastered already or mastered to a certain level. If those that are new challenges are within the adult's developmental "comfort" zone and are met well, they provide opportunities for higher levels of functioning (e.g., a greater sense of competency in the domain challenged, higher self-efficacy, and continued persistence in learning in that domain).

A number of factors keep tasks within an adult's "zone." The individual must possess a certain level of foundational capacities. These include certain kinds of social knowledge about relationships, child care in particular, and a repertoire of cognitive and behavioral skills. Unlike in other theories of the origins of parenting (e.g., personality, attachment theories), however, these foundational capacities are not static, but evolving. Individual differences, however, exist. There are some individuals whose capacities are such that learning is more difficult (e.g., parents with certain type of cognitive limitations or whose connection to reality is tenuous). Even with trial-and-error learning, with specific training, or with help the help of others, such individuals cannot achieve adequate enough competency in enough of the areas relevant to parenting or consistently display it enough to meet the needs of children (similarly, not everyone can be an excellent computer programmer). That is, there are individuals for whom the role acts very little as a generator of development. The burdens experienced are just too overwhelming.

Along with a certain level of foundational capacities, there must also be adequate contextual supports and/or the capacity to acquire them. From a social–cognitive perspective, a social support network (e.g., feedback/modeling from one's partner, extended family, and the responses from one's offspring; information from the media) and other environmental supports (e.g., formal interventions that provide scaffolding, adequate resources) can help the adult fill in his or her gaps, provide a relaxed learning environment, and allow development to progress despite earlier inadequacies. Again, rather than being stable over the lifespan, the schema and scripts required to perform the role requirements of parenting are thus viewed as malleable throughout the life cycle. That is, under the best of circumstances, these schema and scripts are open to new information (e.g., children's evolving capacities; situational needs), and thus constantly changing. For example, if a parent has inadequate perspective taking, a partner or friend may narrate the meaning of a child's behavior for the parent or he or she may model better skills (e.g., "He's only three. They can't hold things well at that age. I usually hold the glass for him when he is pouring milk."). Such narration also helps keep parents calm (i.e., it allows emotional regulation—in this case less anger and/or shame) and restrains them from making a stable negative attribution (e.g., I am a bad mother or my child is a brat). Thus, this allows the parent to persist in an organized and adaptive fashion, as well as provides new information as to how to proceed (new responses). It also helps support feelings of mastery. This, in turn, keeps the adult from developing expectancies for failure in future childrearing situations, ensuring continued willingness to engage with and master the demands of the role, and thus allows further opportunities for growth (Azar, 1986, 1989a, 1998). Parents' involvement with the tasks of parenting is essential for learning to occur.

Contextual obstacles may interfere with the operation of foundational capacities enactment or the contextual needs may be such that they exceed an individual's capacity, which would have been adequate under other circumstances. Poverty, for example, may demand so much from parents to meet the preservation demands of their role that little is left over to engage in the other demands.

Implicit in this discussion is the idea that these same processes are necessary for successfully operating in all relationships (e.g., self-calming, problem solving) and in the world more generally. For example, the problem-solving capacities needed to hold a job (e.g., ability to interact appropriately with co-workers; to learn new functions in the workplace) are not entirely separate from the problem-solving capacities required to provide a child with a safe existence (e.g., anticipate dangers) or to nurture cognitive or emotional development (e.g., identify actions required for successful outcomes). The parent function, as Palkovitz (1996) has said, is not simply "on" or "off," but rather is part of the flow of all tasks in adult life. Because of the limited capacities of children and their more obvious vulnerability, the parenting context, however, may provide one place for these to be refined more easily—that is, the adult may be more open to learning. Similarly, lack of social support and higher contextual stressors will limit one's capacity to learn.

This approach does not preclude some capacities that are completely unique to the role of having a more vulnerable charge in one's care (e.g., the capacity to regress in one's behavior and be playful or the capacity to recognize incapacities in others such that one must put off one's own need gratification in order to provide for theirs), but the difference may be one of degree. One may need these same capacities in other arenas (e.g., if one is a nurse, a teacher, a doctor, or in any professional training or caregiving role). Such roles require a level of emotional investment in the "other" (e.g., a teacher's pride in a child's learning to read or a physical therapist's sense of efficacy in seeing a patient who broke her foot being able to walk again). This allows for a contiuum of "parent-like relationships" and the heterogeneity in parents discussed earlier. Similar to demands being placed on parents, societal expectancies of one's functioning in these other roles may also pressure individuals to persist and develop at a higher rate. For example, societal images of "ideal" teachers may produce tension for them. Similar to parental demands, there are enduring role demands as well as emerging ones for teachers. We have idealized them and demonized them, much as we have parents (Farber & Azar, 1999).

With adequate foundational capacities and contextual supports, a final requirement for learning is motivation to engage with the task. According to social–cognitive theories of motivation (Heider, 1958), when an individual encounters an action of another, an interpretation of the meaning of that action must be made for the selection of a response to occur (e.g., play with my child, restrain his or her action, provide food). That is, cognitive processes are seen as mediating interpersonal emotional and behavioral responses. The appropriateness of the response made is seen as being dependent on the accuracy (or at the very least the adaptiveness) of the interpretation made (which is dependent on one's perceptual and other capacities). Adaptive responses are ones that facilitate continued contact with others, produce success in interpersonal transactions and with the environment, and maintain as much as possible a positive mood state. For example, it is adaptive for parents to focus on strengths in their children, from a mastery perspective (e.g., "I have produced a competent child").

A number of social–cognitive elements are essential to this construction of meaning in the context of parenthood and how it might power development in adults. To keep this discussion managable, special emphasis is placed on one element, expectancies (relational schema) in parenting and their role in adult development (a theme that was highlighted in the early part of this chapter). Along the way, examples are provided of research documenting the impact of social–cognitive factors on adult outcomes in parenting, as well as their interaction with contextual factors.

Elements of social–cognitive theorizing are typically described as being of four types: (1) structural, (2) content, (3) process, and (4) products. Structural and content elements are difficult to describe separately. These elements have been discussed in many ways in the literature (including prototypes, appraisals, working models, relational schema, lay theories, personal constructs, and mental models) (Baldwin, 1992; Fletcher & Thomas, 1996). They cut across three overlapping targets: the self, relationships, and the partner (Baldwin, 1992). There are *general* theories about roles and ones that are linked to *specific* relationships. *General* theories include ones about how adults function as parents (e.g., parents should always have control over their children), how parent–child relationships ideally should work (e.g., children always listen to their mothers), and children (e.g., preschoolers need lots of guidance). (There are also expectancies regarding partners: "Once they become fathers, husband should…".) Similarly, we have schema as to how we *specifically* function as a parent (e.g., I am bad at discipline), our parent–child relationship (e.g., my child and I don't get along), our child (e.g., she is a brat), and our partners ("He is a good father."). The overlapping part might look like: "Most children listen to their mothers, but mine do not." or "Most husbands don't help out with the kids, but my husband does.").

Clashes among general and specific theories can be jarring and cause tensions. Small amounts of tension can motivate learning, whereas larger ones are more problematic. In some relationships (e.g., romantic, friendships), when such clashes cause too much tension, there may be a break (e.g., leaving the relationship). In the parenting one, at least in the younger years, such a physical break is not typically possible (except in cases where parents decide to abandon the child). Lower levels of "role abandonment," however, may occur (e.g., avoidance of providing for the child—neglectful parenting) or engaging in active rejection of the child (e.g., emotional or physical abuse). This cuts off avenues for parenting as a context for adult development.

The structural aspects of relational schema include their complexity, flexibility, integration, and differentiation. For example, self-schema must be differentiated enough to prevent challenges to one aspect of the self from spilling over to another. General child schemas must be (or become) flexible and complex enough to allow for having a child who is developmentally disabled (e.g., one who does not follow the same trajectory as other children) or who is oppositional (e.g., one for whom one's interventions will be *by definition* less effective and one's self-efficacy will be less challenged).

As these examples illustrate, the content of one's schema must also be appropriate. Thus, schema that see young children as equivalent to adults and able to take care of parents' needs may lead to misjudgments regarding children's behavior and our own efficacy. Indeed, rigid, inflexible, and unrealistic expectations of children have been associated with disturbances in parenting (e.g., child abuse [Azar, 1989b; Azar et al., 1996b; Barnes & Azar, 1990]). Schema must also be positively toned. Positive biases help parents deal with the aversive moments in parenting (e.g., children's running across the street being seen as independence rather than willfulness). On the other hand, they must not be so positive that they ignore places where improvements are needed (e.g., they must be open to negative feedback that creates moderate tension). Otherwise the parents will not identify errors in their expectancies and/or responses and make adjustments.

The adaptiveness of ones' general role expectations about parenting requires both exposure to adequate life experiences and role models that have elaborated repertoires of parenting behaviors, that are marked by adaptive and accurate elements and ones who have flexiblity in these expectancies, such that new learning can occur. For example, Russell (1974) found that men who had more preliminary preparation for parenthood experienced more gratification (e.g., pride in the baby's development, feelings of fulfillment, and purpose in life). Role expectations also may be subject to the effects of training given adequate cognitive ability (Azar, 1989a, 1998). The manner in which role discrepancies are handled may also be modeled (e.g., if role conflicts in one's family were handled in an avoidant or rejecting way then this may occur in one's own parenting).

The foundation for these basic structural and content elements has been shown to emerge early. It can be seen in very young children as they "play house." They practice the role, first being put in charge of younger children for brief moments, and then typically during adolescence babysitting for others' children. In our culture, these early assignments may be quite limited (i.e., parenting is seen as an adult activity), while in other cultures, more aspects of the role may be expected of children very early (e.g., as early as between the ages of 3 and 6 [Whiting & Whiting, 1975; Weisner & Gallimore, 1977]). The development of nurturance skills has been studied (Fogel & Melson, 1986). It has been argued that it is the intent to caretake that is modeled first, while the actual skills are not yet available until later (Bryant, 1982). Thus, the mandate to caretake precedes the capacities to do so. This is gendered. Girls are expected to take on these roles earlier than boys (Barry, Brown, & Child, 1957). This may set the stage for the later genderedness that emerges more strongly in couples as they make the transition to parenting. Caretaking by children is less competent than that of adults, thus supporting the idea that a learning curve in the role exists. The role of contextual factors in this learning has been documented for individuals who take on the role suddenly or when they are less mature (e.g., social support is predictive of adoptive parents and teenage parents functioning in the role [Levy-Shiff, Goldshmidt, & Har-Even, 1991; Azar, 1989b]).

The other half of the script (being cared for) is also rehearsed in childhood. Siblings must accept being taken care of (Bryant, 1982). This is important. As parents age, they must give up elements of the expectancy to care for their offspring and allow themselves to be cared for by them (i.e., the task of achieving "generational maturity"). Thus, there is even rehearsal of a major role transformation that occurs later in life. This later stage of parenthood may be marked by much tension regarding such expectancies. For example, Silver (1993) in her qualitative study of aged parents found some women had given up and were angry that they were not getting the care from their children that they felt their culture had promised and became demanding. Others were able to still see themselves as carrying out the tasks of motherhood and balancing the need to be cared for. For example, women described themselves as coming to a day health center in order that adult children could have time to themselves or hoping their deaths would be quick so as not to burden their children. Such responses were not always related to physical health and capacities. This researcher argues there may even be a group of elderly who refuse care needed, and thus their lives may be threatened (their expectancies take precedent over their needs). Findings from other quantitative studies also point to the importance of expectations. The greater the expectations that aged parents will receive help from their offspring, the less effective communication with the child is perceived (Quinn & Keller, 1983). Conversely, the meeting of role expectations has been associated with the quality of the relationships elderly mothers perceive with their daughters (Johnson, 1978).

There are cultural differences in parental expectancies regarding children's support during aging. For example, for African-American elderly, it is not uncommon that the caregiving role might be filled by a stranger who acts in the role of adopted daughter. This replicates what happens in parenting, in that maternal caregiving within this group is carried out by multiple adult female figures, some of whom may not be biologically related (Greene, 1990). This capacity to "flex" role definitions and to give and receive care from someone who is not biologically related may pose unique cognitive and affective opportunities and challenges to the aged adult in this group.

Expectancies regarding parenting can include expectancies regarding the self (e.g, how competent one will be, how much control one will have in interaction with one's child), the role of parent (e.g., how much time/effort it will take, how central the role of parenting is to the self), one's partner in relation to parenting (e.g., how parenting will affect your marital relationship, how much help one's partner will provide), extended family members and other supports in relation to parenting (e.g., how much assistance they will provide, how much parenting will affect activities with friends), and the child (e.g., what children can do at specific points in development, how much children should listen to parents, children's perspective taking capacities). Expectations of control in parenting have been linked to adult outcomes (Bugenthal, Mantyla, & Lewis, 1989).

The effects of violated expectancies are felt strongly at the transition to parenthood. Discrepancies between prebirth expectations of roles within the family and postbirth realities have been shown to be a source of tension for parents in this stage, especially within the marital system and especially for women (Belsky, Ward, & Rovine, 1984; Garrett, 1983; Kach & McGee, 1982). Those with more flexible expectancies show less such distress. Self-schema that involve the work role as central to identity, more self critical thought patterns, and an emphasis on autonomy predicted adjustments to the transition to parenting for women (Hayworth, 1980; Levy-Schiff, 1994), whereas for men, a self that had elements of affiliation predicted adjustment. Kalmuss, Davidson, and Cushman (1992) in a prospective study of expectancies of a large group of married middle-class, white women pregnant for the first time found all experienced discrepancies between their anticipated adjustment during pregnancy and the realities 1 year postbirth (adjustments being more difficult than expected). The level of discrepancies in specific domains, however, predicted poorer adjustment, including expectations about their relationship with their spouse, physical well being, maternal competence, and maternal satisfaction. High expectations regarding child care assistance from the spouse and support from extended family were associated with more difficulty (controlling for actual amount of assistance received). Thus violated expectancies can be problematic for adults' development during this period.

As hinted at in the earlier part of this chapter, different types of parents, however, may experience this transition differently. For example, adoptive parents in one study had more positive expectations regarding parenthood prior to assuming the role, that were maintained after becoming parents than did biological parents (Levy-Shiff, Goldschmidt, & Har-Even, 1991).

Consensus about family functioning across domains has also been found to facilitate marital adjustment (Goldberg, Michaels, & Lamb, 1985), suggesting that parents having the same subjective set or script for their family and its functioning may process incoming information in the same way, and thus work as a team more easily. Marital adjustment also appears to be a predictor of men's adjustment to the transition to parenthood. Thus, expectancies may be important for the adjustment of both men and women in this role.

Not only the content of expectancies, but how complex they are and how rigidly they are held is also important. For example, parents who have more rigid expectancies have been shown to be more directive and critical with their children (Hoover & Milich, 1994). Parents who have unrealistic expectancies regarding what is appropriate behavior to expect from children make more attributions of negative intent to even unintentional child behavior and also are more coercive in their responses in naturally occurring discipline situations and less adaptive in responses (e.g., use of explanation) with their preschool children (Azar, 1989b, 1991; Barnes & Azar, 1990). Interestingly, these relationships hold even before these individuals become parents. That is, in one study by the author, teenagers who held greater unrealistic expectations of children made more negative intent attributions to hypothetical aversive child behavior, and assigned it greater punishment (Azar et al., 1996). Rigid and inappropriate relationship beliefs have been shown to predict other types of relationship difficulties (e.g., marital distress, Eidelson & Epstein, 1982; anger and aggression in teenagers and young adults [Azar, Plante, Brehm, & Ferraro, 1994]). Thus, the structure and context of expectancies affect a myriad of adult relationships.

The depth of one's knowledge structures regarding child care may also be crucial. Lack of knowledge may exacerbate interactional problems with one's children, increasing burdens and the aversiveness of childrearing. For example, supplying infants with cereal too early can result in intestinal discomfort and greater irritability, and thus infants' general stressfulness. Infants whose parents provide them with inadequate amounts of stimulation may become less responsive and less able to self-soothe and regulate affect, making them less engaging and more aversive.

While these relational schema may sound similar to the internal working models posited by attachment theory, they are distinct in that they can emerge at any point in the life cycle (e.g., immigrant parents may need to develop a grossly different idea of what children are allowed to do and how much say they will have in controlling them in the new culture) and will be refined throughout the life cyle. As discussed previously, in aging, the identity of oneself as a giver of care to one's child may need to be replaced by, or coexist with, one of someone who is cared for by one's child. Thus, the structure and content of relational schema have utility for understanding the potential for development in adulthood and in the role of parent.

Hinted at in the preceding discussion is the third type of cognitive element—capacities for active processing of information (executive functions). For example, problem-solving capacities (e.g., the ability to notice a problem, identify alternative solutions, and select and enact the most effective ones) are crucial to childrearing. Even if one's schema regarding children are distorted in some way (e.g., one mislabels incapacity as intentional negative willful behavior), the ability to generate alternative perspectives about a childrearing event may allow for further "search" for and/or acceptance of new information offered by others (e.g., being open to one's partner's more positive interpretation) and "better" ultimate solutions (products) that ultimately may be more adaptive (e.g., deciding the child's negative behavior was not intentional) and readjusting one's affective and behavioral reponse. Other information processing skills include perspective taking (an ability to internally represent events and examine them from the perspective of another) and tolerance for ambiguity. These capacities may be especially crucial in parenting in that young children cannot provide the parent with much information (attention to small cues and extensive search processes are required) and children are constantly changing, and thus presenting "problems" for parents to solve (i.e., old learning about the child—knowledge structures—assists the parent only to a point).

Further, there may be individual differences in how "relational" one's thinking is. For example, women do more more monitoring of relationships. Men, on the other hand, although they can do such monitoring, are more likely to do so when the relationship is having problems (see Acitelli & Young, 1996 for a discussion of gender and thought in relationships). In parenting, this emphasis on monitoring relational cues is crucial. Indeed, parents see more meaning in infants' and toddlers' behavior than do nonparents (Adamson, Bakeman, Smith, & Walters, 1987). Crittenden (1993) has posited that neglectful parents fail to perceive child cues. Consistent with this view is the fact that parents who have difficulty in their role have been shown to be less discriminant in their responses to infants and toddlers and to respond in an insensitive and noncontingent manner and to be randomly intrusive and aversive (Bee, Disbrow, Johnson-Crowley, & Barnard, 1981; Crittenden, 1982). This may result from a combination of schema regarding children that are not differentiated enough to pick up on fine nuances in their behavior, inadequate knowledge structures as to how to respond, poor problem-solving capacities in generating alternative strategies, and negative attributions to what may be normal child behaviors. Indeed, one study with cognitively limited parents (IQs less than 79) showed that levels of unrealistic expectations, and poor problem solving were associated with less parent–child synchrony (Azar, Povilaitis, Johnson, Ferraro, & Soysa, 1999). Thus, transactions between the parent and child become disordant when these social–cognitive factors show disturbances. Such difficulties are a major predictor of multiple forms of adult distress.

The final cognitive element is a product of cognitive activity. For example, when an individual does not behave in a way consonant with his or her own "parent" schema or his or her child's behavior does not meet expectancies, he or she must engage in generating attributions for the causes of the discrepancy. These products, the explanations generated, can influence other thoughts about what has occurred (views of self, their children, the situation), their emotions, and behavioral responses. Take the example of a tantruming child in a grocery store who attracts the attention of other shoppers. If this event is filtered through an expectation that children *should not* embarrass their parents, it may result in attributions to the self regarding incompetence as a parent and/or attributions of negative intent to children (e.g., "He's trying to embarrass me.") and generate experiences of shame and anger. Parental attributions have been shown to be related to their affective and behavioral responses to children's behavior (Dix, Ruble, & Zambarano, 1989). Attributions have been shown to be sensitive to developmental changes in children and where they are not to bode poorly for smooth parent–child interactions (Azar, 1989a, 1998). Attributional biases also have been associated with greater parental distress (e.g., mothers who overestimate their control in childrearing show greater "depression-proneness" [Donovan & Leavitt, 1989]). These biases may be more general. For example, Geller and Johnston (1995) found a signficant relationship between mothers' attributions for their own and their children's behavior. Miller and Azar (1996) found child abuse risk was associated with biases in attributions to self and others.

Such products of cognitive activity impact on adult's well being throughout the life cycle, not just when their offspring are young. For middle-aged parents, interpretations of their role in how their children "turned out" and adult well being have been linked. Children's attainment has been shown to influence a myriad of later life adult outcomes. That is, parents' self-evaluations (e.g., self-acceptance, purpose in life, environmental mastery, and depression) have been shown to be closely tied to their offsprings' adjustment (more than their attainment) (Ryff, Schmutte, & Lee, 1996). Attributional processes come into play in moderating this effect. If parents see themselves as responsible for their children's success, their well being has been found to suffer more. Parents who see themselves as having been less involved in their children's lives may feel like they violated role expectations if the child did not "turn out well" (i.e., they should been more involved and the child might have done better if they had been). Findings of one qualitative study provide support for this. The well being of fathers was greatly decreased when their offspring was abusing substances. These fathers lamented their inability to fix their situations (Greenberg, 1991). Interestingly, mothers who perceive their children as doing *better* than they themselves had done in young adulthood (as opposed to just doing as well) have been shown to have lower well being (Ryff et al., 1996), suggesting social comparison processes observed with strangers have similar impact on adults with their children during this period.

Facilitating children's social development requires identifying and labeling children's social acts and providing feedback when they exceed bounds set up by societal expectations. In the face of the complexity of these tasks and the burdens of child care, adults must be able to self-nurture and self-regulate to stay calm and maintain their own well being. If they encounter childrearing situations that they are confused by, where they make mistakes, or find themselves initially unsuccessful (e.g., a crying infant who cannot be consoled, a child who is having trouble being toilet trained; an adolescent who has stayed out past curfew, a daughter who stays in an abusive marriage), they must be able to soothe the self and protect themselves against sharp decreases in self-efficacy or making global attributions to their own abilities or something malevolent about their child. This allows them to solicit new input (collect more information, seek out help, distract themselves) and persist in the interaction with the child. Such processes are crucial if parents are to view parenting as an ego-enhancing task and to be able to "grow." That is, these processes keep the tasks of parenting within the adult's "zone of proximal development" such that they can feel a sense of mastery and control and such that they can achieve new levels of cognitive and emotional development (e.g., if they find their very young child provides continuous evidence of their incompetence, they may experience shame and withdraw from parenting. They may not engage in the play with the child that is important to children's development, but also provides a vehicle for adults to regress and enjoy themselves). These four elements act in concert with each other and where one element fails, another acts as a safety net. Thus, they provide a "space" in which development can occur.

SUMMARY

It is clear from this brief description that parenting is a lifelong task and that it comes in a myriad of forms. With it come many challenges, as well as opportunities for using one's already acquired skills, refining them, and acquiring new ones that failed to develop earlier. A social–cognitive framework provides a framework to conceive of this adult development and to track the changes that take place over time. Furthermore, this approach has enough breadth to account for the multiple types of parenting that occur and the variety of contexts in which it is done. It provides a way to think about how the tasks of parenting outlined earlier for preservation, growth, and social acceptance in ones' offspring may be construed differently by different parents and in different societal contexts, as well as how the descrepancy between the expected role requirements and the support one will receive (from one's partner, extended family, and society) may create tension. When that tension is time limited or within a narrow range, then potential for growth occurs. However, when it is chronic or the discrepancy is too great, then new information cannot be assimilated and/or new knowedge structures that are more complex cannot emerge.

REFERENCES

Acitelli, L. K., & Young, A. M. (1996). Gender and thought in relationship. In G. J. Fletcher & J. Fitness (Eds.), *Knowledge structures in close relationships* (pp. 147–168). Mahweh, NJ: Lawrence Erlbaum.

Adamson, L. B., Bakeman, R., Smith, C. B., & Walters, A. S. (1987). Adults' interpretation of infants' acts. *Developmental Psychology, 23,* 383–387.

Anderson, C. M. (1999). Single-parent families: Strengths, vulnerabilities, and interventions. In B. Carter & M. McGoldrick (Eds.), *The expanded family life cycle*, 3rd edit. (pp. 399–416). Boston, MA: Allyn and Bacon.

Arnett, J. J. (1997). Young people's conception of the transition to adulthood. *Youth & Society, 29*, 1–23.

Arnett, J. J. (2000). Emerging adulthood. A theory of development from the late teens through the twenties. *American Psychologist, 55*, 469–480.

Azar, S. T. (1986). A framework for understanding child maltreatment: An integration of cognitive behavioral and developmental perspectives. *Canadian Journal of Behavioral Science, 18*, 340–355.

Azar, S. T. (1989a). Training parents of abused children. In C. E. Schaefer & J. M. Briesmeister (Eds.), *Handbook of parent training* (pp. 414–444). New York: John Wiley & Sons.

Azar, S. T. (1989b, November). Unrealistic expectations and attributions of negative intent among teenage mothers at risk for child maltreatment: The validity of a cognitive view of parenting. Paper presented at the annual meeting of the Association for Advancement of Behavior Therapy, Washington, DC.

Azar, S. T. (1991, November). Is the cognitively low functioning mother at risk for child maltreatment? Presented at the annual meeting of the Association for Advancement of Behavior Therapy, New York.

Azar, S. T. (1996). Cognitive restructuring of professionals' schema regarding women parenting in poverty. *Women & Therapy, 18*, 147–161.

Azar, S. T. (1998). A cognitive behavioral approach to understanding and treating parents who physically abuse their children. In D. Wolfe & R. McMahon (Eds.), *Child abuse: New directions in prevention and treatment across the life span* (pp. 78–100). Berkeley, CA: Sage.

Azar, S. T., & Benjet, C. L. (1994). A cognitive perspective on ethnicity, race, and termination of parental rights. *Law & Human Behavior, 18*, 249–268.

Azar, S. T., & Cote, L. (in press). Sociocultural issues in the evaluation of the needs of children in custody decision-making: What do our current frameworks for evaluating parenting practices have to offer? *Internation Journal of Law & Psychiatry.*

Azar, S. T., & Houser, A. (1993, November). Unrealistic expectations, negative attributions, and parenting responses: Further validation of a social cognitive model with African Amercian and Puerto Rican mothers. Paper presented at the annual meeting of the Association for the Advancement of Behavior Therapy, Atlanta.

Azar, S. T., Plante, W., Brehm, K., & Ferraro, M. (1994, November). Unrealistic expectations about relationships and aggression in women: A test of a cognitive behavioral model with mothers and young adults. Paper presented at the annual meeting of the Association for Advancement of Behavior Therapy, San Diego.

Azar, S. T., Povilaitis, T., Johnson, E., Ferraro, M. H., & Soysa, C. (1999, April). Maternal expectations, problem solving and disicipine and synchrony: A test of a social–cognitive model. Presented at the biennual meeting of the Society for Research in Child Development, Albuquerque, NM.

Azar, S. T., Robinson, D. R., Hekimian, E., & Twentyman, C. T. (1984). Unrealistic expectations and problem solving ability in maltreating and comparison mothers. *Journal of Consulting and Clinical Psychology, 52*, 687–691.

Azar, S. T., & Rohrbeck, C. A. (1986). Child abuse and unrealistic expectations: Further validation of the Parent Opinion Questionnaire. *Journal of Consulting and Clinical Psychology, 54*, 867–868.

Azar, S. T., Spence, N., Breton, S. J., Lauretti, A. F., Povilaitis, T. Y., Pouquette, C. L., Glenn, L., & Gehl, K. (1996, August). Social cognition and child maltreatment: The need for more comprehensive theory to address the relational, developmental, behavioral, and systemic problems faced by children. Paper presented at the annual meeting of the American Psychological Association, Toronto.

Baldwin, A., Baldwin, C., & Cole, R. E. (1990). Stress-resistant families and stress-resistant children. In J. E. Rolf, A. S. Masten, D. Cicchetti, K. N. Wechterlein, & S. Weintraub (Eds.), *Risk and protective factors in the development of psychopathology* (pp. 257–280). New York: Cambridge University Press.

Baldwin, M. W. (1992). Relational schema and the processing of social information. *Psychological Bulletin, 112*, 461–486.

Barnes, K. T., & Azar, S. T. (1990, August). Maternal expectations and attributions in discipline situations: A test of a cognitive model of parenting. Paper presented at the annual meeting of the American Psychological Association, Boston.

Barry, H. H., Brown, M. K., Child, I. L. (1957). A cross-cultural survey of soem sex differences in socialization. *Journal of Abnormal and Social Psychology, 55*, 327–332.

Baumrind, D. (1993). The average expectable environment is not good enough: A response to Scarr. *Child Development, 64*, 1299–1317.

Bee, H. L., Disbrow, M. A., Johnson-Crowley, N., & Barnard, K. (1981). Parent–child interactions during teaching in abusing and non-abusing families. Paper presented at the biannual convention of the Society for Research in Child Development, Boston, April.

Belsky, J., Ward, M. J., & Rovine, M. (1984). Prenatal expectations, postnatal experiences and the transition to parenthood. In R. Ashmore, & D. Brodzinsky (Eds.), *Perspectives on the family* (pp. 110–145). Hillsdale, NJ: Lawrence Erlbaum.

Brim, O. G., & Ryff, C. D. (1980). On the properties of life events. In P. B. Baltes & O. G. Brim (Eds.), *Life span developmental psychology, Vol. 3* (pp. 367–388). New York: Academic Press.

Bryant, B. K. (1982). Sibling relationship in middle childhood. In M. Lamb & B. Sutton-Smith (Eds.), *Sibling relationships: Their nature and signficance across the life span*. Hillsdale, NJ: Lawrence Erlbaum.

Bugenthal, D. B., Mantyla, S. M., & Lewis, J. (1989) Parental attributions as moderators of affective communication to children at risk for physical abuse. In D. Cicchetti & V. Carlson (Eds.), *Child maltreatment* (pp. 254–279). New York: Cambridge University Press.

Cauce, A. M., Ryan, K. D., & Grove, K. (1998). Children and adolescents of color, where are you? Participation, selection, recruitment, and retention in developmental research. In V. C. McLoyd & L. Steinberg (Eds.), *Studying minority adolescents: Conceptual, methodological, and theoretical issues* (pp. 147–166). Mahwah, NJ: Lawrence Erlbaum.

Collins, P. H. (1990). *Black feminist thought. Knowledge, consciousness, and the politics of empowerment*. Cambridge, UK: Harper Collins Academic.

Crittenden, P. M. (1982). Abusing, neglecting, problematic, and adequate dyads: Differentiating by patterns of interaction. *Merrill-Palmer Quarterly, 27*, 201–218.

Cusinato, M. (1994). Parenting over the family life cycle. In L'Abate, L. (Ed.), *Handbook of developmental family psychology and psychopathology* (pp. 83–115). New York: John Wiley & Sons.

Dix, T. H., Ruble, D. N., & Zambarano, R. J. (1989). Mothers' implicit theories of discipline: Child effects, parent effects, and the attribution process. *Child Development, 60*, 1373–1391.

Donovan, W. L., & Leavitt, L. A. (1989). Maternal self efficacy and infant attachment: Integrating physiology, perception, and behavior. *Child Development, 60*, 460–472.

Eidelson, R. J., & Epstein, N. (1982). Cognition and relationship maladjustment: Development of a measure of dysfunctional relationship beliefs. *Journal of Consulting and Clinical Psychology, 50*, 715–720.

Epstein, N. B., Bishop, D. S., & Baldwin, L. M. (1982). McMasters model of family functioning. In F. Walsh (Ed.), *Normal family processes* (pp. 115–141). New York: Guilford Press.

Erikson, E. H. (1963). *Childhood and society*, 2nd edit. New York: W. W. Norton.

Farber, B., & Azar, S. T. (1999). Blaming the helper. The marginalization of teachers and parents of the urban poor. *Journal of Orthopsychiatry, 69*, 515–528.

Fletcher, G. J., & Thomas, G. (1996). Close relationship lay theories: Their structure and function. In G. O. Fletcher & J. Fitness (Eds.), *Knowledge structures in close relationships*, (pp. 3–24). Mahweh, NJ: Erlbaum.

Fogel, A., & Melson, G. F. (1986) *Origins of nurturance. Developmental, biological and cultural perspectives on caregiving*. Hillsdale, NJ: Lawrence Erlbaum.

Furnstenberg, F. (1988). Good dads-bad dads: Two faces of fatherhood. In A. J. Cherlin (Ed.), *The changing nature of American family and public policy* (pp. 193–213). Washington, DC: US Urban Institute Press.

Furstenburg, F. F., Nord, C. W., Peterson, J. I., & Zill, N. (1983). The life course of children of divorce: Marital disruption and parental contact. *American Sociological Review, 48*, 656–668.

Garrett, E. (1983, August). Women's expectations of early parenthood: Expectations of early parenthood: Expectations versus reality. Paper presented at the meeting of the American Psychological Association, Anaheim, CA.

Geller, J. & Johnston, C. (1995). Predictors of mothers' responses to child noncompliance: Attributions and attitudes. *Journal of Clinical Child Psychology, 24*, 272–278.

Gergen, K. J. (1980). The emerging crisis in life-span developmental theory. In P. B. Baltes & O. G. Brim (Eds.), *Life span developmental psychology, Vol. 3* (pp. 32–65). New York: Academic Press.

Goldberg, W. A., Michaels, G. Y., & Lamb, M. E. (1985). Husband's and wives adjustment to pregnancy and first parenthood. *Journal of Family Issues, 6*, 483–503.

Goldscheider, F., & Goldscheider, C. (1994). Leaving and returning home in the 20th century America. *Population Bulletin, 48*, 1–35.

Goodnow, J. J., & Collins, W. A. (1990). Development according to parents: The nature, source, and consequences of parents' ideas. Hillsdale, NJ: Lawrence Erlbaum.

Gould, R. L. (1978). *Transformations: Growth and change in adult life.* New York: Simon & Schuster.

Graham, S. (1992). Most of the subjects were white and middle class. *American Psychologist, 47,* 629–639.

Greenberg, J. R. (1991). Problems in the lives of adult children: Their impact on aging parents. *Journal of Gerontological Social Work, 16,* 149–161.

Greene, B. (1990). Sturdy bridges: Role of African-American mothers in the socialization of African-American children. In J.P. Knowles & E. Cole (Eds.), *Women-defined motherhood* (pp. 205–225). New York: Harrington Park Press.

Hareven, T. (1982). American families in transition. In F. Walsh (Ed.), *Normal family processes* (pp. 446–466). New York: Guilford.

Hayworth, J. (1980). A predictive study of post-partum depression: Some predisposing characteristics. *British Journal of Medical Psychology, 53,* 161–167.

Heider, F. (1958). *The psychology of interpersonal relations.* New York: Wiley.

Hoover, D. W., & Milich, R. (1994). Effects of sugar ingestion expectancies on mother-child interaction. *Journal of Abnormal Psychology, 22,* 501–515.

Johnson, E. S. (1978). 'Good" relationships between elderly and their adult children. *The Gerontologist, 18,* 301–306.

Kach, J. A., & McGhee, P. E. (1982). Adjustment of early parenthood. *Journal of Family Issues, 3,* 374–388.

Kagitcibasi, C. (1982). Old-age security value of children. Cross-national socioeconomic factors. *Journal of Cross Cultural Psychology, 13,* 29–42.

Kalmus D., Davidson, A., & Cushmn, L. (1992). Parenting expectations & adjustment to parenthood: A test of the violated expectations framework. *Journal of Marriage and the Family, 54,* 516–526.

Kohn, M. L. (1963). Social class and parent–child relationships: An interpretation. *American Journal of Sociology, 108,* 471–480.

Lawson, E. J., & Thompson, A. (1999). *Black men and divorce.* Thousand Oaks, NJ: Sage.

La Rosa, V. A. & Fodor, I. G. (1990). Adolescent daughter-midlife mother dyad: A new look at separation and self definition. *Psychology of Women Quarterly, 14,* 593–606.

Leiblum, S. R. (1997). *Infertility: Psychological issues and counseling strategies.* New York: John Wiley & Sons.

LeVine, R. A. (1974). Parental goals: A cross-cultural view. *Teachers College Record, 76,* 226–239.

Levinson, D. J. (1978). *The seasons of man's life.* New York: Knopf.

Levinson, D. J. (1986). A conception of adult development. *American Psychologist, 41,* 3–13.

Levy-Shiff, R. (1994). Individual and contextual correlates of marital change across the transition to parenthood. *Developmental Psychology, 30,* 591–601.

Levy-Shiff, R., Goldschmidt, L., & Har-Even, O. (1991). Transition to parenthood in adoptive families. *Developmental psychology, 27,* 131–140.

Miller, L. R., & Azar, S. T. (1996). The pervasiveness of maladaptive attributions in mothers at-risk for child abuse. *Family Violence & Sexual Assault Bulletin, 12,* 31–37.

National Research Council (1998) *From generation to generation: The health and well-being of children in immigrant families.* Washington, DC: National Academy Press.

Obeidallah, D. A., & Burton, L. M. (1999). Affective ties between mothers and daughters in adolescent childbearing families. In M. J. Cox & J. Brooks-Gunn (Eds.), *Conflict and cohesion in families. Causes and consequences* (pp. 37–49). Mahweh, NJ: Erlbaum.

Okagaki, L., & Divecha, D. J. (1993). Development of parental beliefs. In T. Luster & L. Okagaki (Eds.), *Parenting: An ecological perspective* (pp. 35–67). Hillsdale, NJ: Lawrence Erlbaum.

Palkovitz, R. (1996). Parenting as a generator of adult development: Conceptual issues and implications. *Journal of Social and Personal Relationships, 13,* 571–592.

Palus, C. J. (1993). Transformative experiences of adulthood: A new look at the seasons of life. In J. Demick, K. Bursik, & DiBiase (Eds.), *Parental development* (pp. 39–58). Hillsdale, NJ: Lawrence Erlbaum.

Parke, R. D., & Buriel, R. (1998). Socialization in the family: Ethnic and ecological perspectives. In W. Damon (Series Ed.) & N. Eisenberg (Vol. Ed.), *Handbook of child psychology, Vol. 3. Social, emotional, and personality development,* 5th edit. (pp. 463–552).

Polakow, V. (1993). *Lives on the edge. Single mothers and their children in the other America.* Chicago: University of Chicago Press.

Quinn, W. H., & Keller, J. F. (1983). Older generations of the family: Relational dimensions and quality. *The American Journal of Family Therapy, 11,* 27–34.

Rossi, A. S. (1968). Transiton to parenthood. *Journal of Marriage and the Family, 30,* 263–275.

Ruddick, S. (1989). *Maternal thinking. Toward a politics of peace.* New York: Beacon.

Russell, C. S. (1974). Tansition to parenthood: Problems and gratification. *Journal of Marriage and Family, 36,* 294–302.

Ryff, C. D., Schmutte, P. S., & Lee, U. H. (1996). How children turn out: Implications for parental self evaluation. In C. S. Ryff & M. M. Seltzer (Eds.), *The parental experience in midlife* (pp. 383–422). Chicago: University of Chicago Press.

Shand, N. (1985). Culture's influence in Japanese and American maternal role perception and confidence. *Journal for the Study of Interpersonal Processes, 48,* 52–67.

Silver, M. (1993) Balancing "Caring and being cared for" in old age: The development of mutual parenting. In J. Demick, K. Bursik, & DiBiase (Eds.), *Parental development* (pp. 225–239). Hillsdale, NJ: Lawrence Erlbaum.

Thompson, L., & Walker, A. J. (1989). Gender in families: Women and men in marriage, work, and parenthood. *Journal of Marriage and the Family, 51,* 845–871.

Tobin, S. S. (1987). A non-normative old age contrast: Elderly parents caring for offspring with mental retardation. In D. G. Gutmann (Ed.), *Reclaimed powers: Toward a new psychology of men and women in later life* (pp. 124–142). New York: Basic Books.

US Bureau of the Census (1997). *Households, famiiles, and children. A 30 year perspective. Current population reports,* Series P. 23, No 181. Washington, DC: US Government Printing Office.

Vondra, J., & Belsky, J. (1993). Developmental origins of parenting: Personality and relationship factors. In T. Luster & L. Okagaki (Eds.), *Parenting: An ecological perspective* (pp. 1–34). Hillsdale, NJ: Lawrence Erlbaum.

Vygotsky, L. S. (1934). *Thought and language.* Cambridge, MA: The MIT Press.

Weisner, T., & Gallimore, R. (1977). My brother's keeper: Child and sibling caretaking. *Current Anthropology, 18,* 169–180.

Whiting, B., & Whiting, J. W. (1975). Children of six cultures. Cambridge, MA: Harvard University Press.

Yarrow, A. L. (1991). *Latecomers. Children of parents over 35.* New York: Free Press.

CHAPTER 21

Superwomen Raising Superdaughters

Whatever Happened to Black Girlhood?

Yvonne V. Wells

When one first encounters the work of Vanzant (1998, 1999) a new-age black writer focused on the self-development of women, it seems to contradict what is commonly known about black women and girls. One may think, "This is all wrong. This does not represent black women and girls in their true light as tough, competent, mature, and sage." Pink sheets, pink pajamas, and Barbie dolls with extra sets of doll clothes are mentioned in one book as important for girlhood survival. Vanzant's (1999) simple self-help phrases conflict with all that feminist and black womanist thinking would endorse. Should a serious scholar of black women's experience toss this casually aside? Yet, African-American women and men are reading this woman. At a recent book signing at a major bookstore, black men and women professionals were present.

Why is Vanzant so popular? She raises questions about the presumed need for independence and self-reliance in girls. Vanzant (1999) raises questions about long held assumptions that self-reliance and maturity in black girls is common at an early age. Her work, to some, seems to dilute values that popular thinking and scholarship suggest African-American black mothers would instill in their daughters. African-American girls have never had as much time for frills as their mainstream counterparts. Even in privileged black families girls have lots of work to do, higher education to achieve, and tired mothers to help. Vanzant's (1999) point, however, is clear. Some indulgence in frivolities, daydreaming, playing, indeed, girlhood is a necessity for healthy female development. The necessity of girlhood has been traditionally ignored for black women in American society.

Yvonne V. Wells • Department of Psychology, Suffolk University, Boston, Massachusetts, 02114.

Handbook of Adult Development, edited by J. Demick and C. Andreoletti. Plenum Press, New York, 2002.

Michelle Wallace (1979) in her book, *Black Macho and the Myth of the Superwoman*, defined the myth of the black superwoman. Wallace recalled mothers, aunts, and peers in the black community who appeared to be super and invulnerable. They were symbolic "goddesses." Yet, in reality, they were powerless and misunderstood. Black men, families, and communities referred to historical stereotypes and media images for definitions of black women. The black woman was rarely given an individual voice in her own description. She was given a powerful image that contrasted with a lowly and difficult reality. This image of the black "superwoman" has empowered African-American women throughout the history of American society. At times, however, the image has been out of the reach of black women trying to attain optimal self-development. Sometimes it has led their daughters on a collision course with an expectation of a speedy metamorphosis into a woman of unusual power and character. In the eyes of the popular media it has become a mysterious legacy, or perhaps a consolation rather than a model of what black women could become through the process of lifespan development.

For Collins (1991) popular images of African-American women seem positive in their portrayal of black women as stronger or tougher than women in other American cultures. But these images often slip into the shadows of the negative "matriarch" or "welfare queen." Negative images can be transmitted via institutions such as schools and government agencies that administer to African-Americans families. Positive myths that in some instances have had a healing effect come to objectify and alienate African-American women and their developing daughters from the mainstream of healthy social–cognitive and biological development.

When parents and communities appropriately administer the available images of black womanhood to black girls, they function as enhancements to their developmental processes. African-American boys and girls can improve their chances for healthy educational, emotional, and social development by focusing on reasonable models from their own communities and mastering the steps to achieving what ancestors, parents, and significant adults have achieved. But history and current political reality finds many black families and communities slipping farther and farther behind the mainstream in terms of economic development, cultural gains, and social development. In short, the black family is experiencing more stress and fewer resources, which means less time and ability to shield and explain the extreme images that black girls are likely to see of themselves.

Black people may be finding it increasingly difficult to take the necessary active role in interpreting society's continually unrealistic images of black people to their boys and girls. (See Wilson, 1996 for a full report on the current economic and social status of the black family.) Black families under continued stress may conclude that they do not need to pay much detailed attention to black girls' psychological development. After all, "Aren't these powerful superhuman, black mothers somehow raising them"?

Collins (1991) believes that the idea of the black superwoman can become an alienating concept for black women. Black men may accept the idea that black women are more important to their children than they are. They may view the single-parent families headed by poor young black women as acceptable, relegating them to increasing stress and difficulties, and depriving their sons and daughters of important male role models. When black women buy into the myth of superhuman power they may ignore the need for emotional support. Based on a myth, black women may forego the development of Erikson's (1980) stage of Intimacy vs. Isolation. Black women and women who exclude Intimacy development from their life struggle may never reach optimal self-development. Some black girls may move from no girlhood to raising children alone, into middle age and eldercare. The failure to develop and nurture nuclear family relationships may hasten the approach of an old age spent alone, poor, and in physical need. Overburdened young black women collapse at an alarming rate in post-modern America, leaving an inordinate number of black grandmothers wearily attempting kinship care (see McLean & Thomas, 1996) instead of peacefully contemplating wisdom and sharing it comfortably with their families.

A lack of cognitive and emotional tools resulting from a lost connection with masculine elements in family and community cannot help but seriously limit the African-American woman's resources toward complete self-understanding. African-American women are becoming increasingly deprived of examples of working out issues with the male and female aspects of themselves in the way that psychoanalytic perspectives suggest are necessary for the completion of the self. Violence, risk-taking behavior, and despair among young black males may have something to do with the disconnection from extreme images of black womanhood, which are not analyzed and explained to developing black boys.

Where the emotional spirit of the black woman is broken, her developmental and mothering difficulties are not understood by society. Society lashes back with impunity. Black men and boys are often demonized by the media and victimized by the justice system. Black women and girls face increasing ignorance of who real black women living their day-to-day lives actually are. Harsh projections of vulgar and sexualized media images may be a result of a kind of invisibility born of a desire to turn away from the difficulties of black womanhood that everyone sees in negative media projections, but does not encounter through interaction with a real black woman.

Deep injuries to the black woman's psyche and spirit are evident in life stories about the self-demise of black women whom everyone thought had "made it." Edwards and Polite (1992) discuss the life stories of successful black women such as Lenita McClaine who rose to fame as a reporter for the *Chicago Tribune* only to commit suicide at the age of 32. Developmental conflicts between a "superwoman self" and an unlived childhood may have been a part of her story.

This chapter examines the conflicts between the black superwoman and the notion of black girlhood in the lifespan development of black women. Historical and political images may inform psychological theory and practice as it pertains to black women's development. Selections of these images and their parallels in the behavioral research are considered. Even as scholars from various fields praise the black mother and daughter unit as a model of social competence and dignified survival in the face of crises, popular society, and the black woman herself, may be abandoning slowly nurtured self-development. Historical images, the media, literature, and research are further examined. Applied Afrocentric models of psychology are explored for possible strategies to protect black childhood and integrate black women's self-development with traditional developmental models.

MIXED MEDIA: HISTORICAL, LITERARY, AND MEDIA IMAGES OF BLACK WOMEN IN AMERICAN SOCIETY

When one thinks of African-American women, the fierce protection of children; the raising of daughters toward independence; and special bonds between mother and daughter readily come to mind. This kind of image is found in the book *Incidents in the Life of a Slave Girl*, an authentic narrative of the life of Linda Brent (1818–1896), resurrected from her own voice, edited by Maria Child.

Child (1973) gives an account of her life as a black mother during slavery, with emphasis on the idea of the superpower of motherhood as necessary to save daughters from the fate of slavery. Child (1973) hides in an attic for several years, watching over her enslaved daughters. She waits, sneaks into the house for food, lives at times in the swamp, and sometimes hears though she cannot directly see the abuse of her children. Her final escape involves months of hiding in swamps and a long legal struggle to finally "buy" her children into freedom.

This tenacious power of black motherhood does not surprise the reader, especially the African-American female reader of writers such as Angelou (1974), Head (1974), Morrison (1982, 1994), Walker (1982), and Washington (1987). All of these writers have recorded and designed real and fictional images of powerful black women. Forced to envision African-American life against a background of extreme negativity and in the absence of an honest historical account, such writers have consistently pointed out the amazing strengths of black women and mothers. The powerful images available for the black community reflect that where African-American culture has survived amazing feats of determination by black mothers have been the norm.

The African-American community has never completely controlled the distribution of such images to the American society, nor has it always controlled the effects of these images on individual black women. Fordham (1993) described the assertiveness of black girls. Great in their achievements, strong in their political fortitude, they owe much of their success to their black mothers who are often not educated themselves. This aspect of black girlhood is often translated in a negative way by the academic institutions in which the best black female students often find themselves. Yet the very strength that brought them to top level college environments places them in very ambivalent positions. Seen as aggressive, loud, and problematic in some mainstream educational and professional settings, they may wonder if they have to give up some of the attachments to the attitudes of the black mothers who have raised them. They were not prepared to meet this "flip side" of the strong black woman image that they had come to believe everyone would admire.

What allows the American psyche to hold a romantic view of the black woman as invulnerable, even while raising her children, but, at the same time, maintains a view of the black woman as not at all expert at raising successful daughters? Historical imagery and current media reflect and maintain the conundrum of black motherhood as powerful, ancient, and superhuman, yet inadequate.

Pieterse (1992) analyzed the 3000-year history of imagery of African people. Sometimes Pieterse (1992) finds evidence of romance, beauty, awe, and power in

African images from the Middle East, Mediterranean, and Eastern Europe. In contrast, he finds that images of enslavement, exploitation, and comic bestiality dominated the Western European and American imagination for the past 500 years. One image he found is of Africa itself depicted as a woman willingly giving "gold and her sons in chains." Pieterse (1992) analyzed portrayals of black women enslaved and auctioned as well as being violently snatched away from black men. Images were seen of black women with numerous naked children presumably fathered by many men. Daughters, sons, and the men in the lives of the black woman appear constantly as brutalized, unprotected creatures surrounding her as she kneels in servitude. These kinds of images set the stage for the developmental hegemony of today which still maintains an image of the black mother as always in distress.

The Western minds from the 15th century on saw little in praise of African parenthood. There were few, if any, African generated images to counter these views. Any notion of black motherhood and her daughter's innocence protected by that motherhood was also nonexistent. These images had to have had a lasting devastating effect on the African people who would eventually become citizens rather than slaves in Western societies. Projections about the sexual availability of black women, caricatures of their African features, and the general absence of any contrasting dignified public images combined with a typically grueling life of either slavery or the most undesirable work.

Giddings (1984) discusses the fact that dignified images of black woman and innocent images of black children were sought after by free blacks at the turn of the century. Journalists such as Ida B. Wells, black thinkers, orators, and black professionals in the period following the emancipation of slaves promoted black media. Ida B. Wells and her husband Ferdinand Barnett, a well known black lawyer at the time, owned Chicago's first black newspaper, *The Conservator*. By seizing their own corner of the media scene at the turn of the 20th century, black Americans such as these understood that they must improve the image of black men as upstanding citizens, but most of all, they had to reinvent images of black women as moral and dignified beings. It is thought that their use of a uniquely black medium helped to counter the brutal system of lynching and terrorism against black men and women by sensitizing ordinary Americans and black people themselves to the humanity and dignity of black people.

Psychologically, writings by and about blacks such as Booker T. Washington, Fredrick Douglass, Mary Church Terrell, and W. E. B. Dubois must have had a healing effect on the black mind. Images of black women as dignified members of "ladies' societies," wives, and mothers no doubt gave guidance to those still struggling with the after-effects of slavery. Images of powerful black people, the presence of black women as writers and speakers, but mostly at the time as wives and mothers, all created a model for the direction of black self-identity. At this time in history the developmental path to a good life as a black woman must have been so much easier to define. Stories in black newspapers were about successes, marriages, and children being born along with complaints about injustices.

Today, a different trend in journalism and popular media has developed with regard to black people. Black people who are great are still portrayed as public icons, yet the paths to their success are not spelled out, and their types of success are extremely limited. While anyone can call to mind numerous famous black women divas, comediennes, superstars, and superwomen, black women as connected to black men in healthy family contexts are not consistently promoted. Furthermore, the pedagogical steps to the state of black greatness are not emphasized. The image of the African-American woman in three black films popular today illustrates this trend.

The Color Purple (Spielberg and Jones, 1985) is one film that contains many images of black women, and of black women particularly as powerful sisters and mothers. In the film, a villain becomes the tormentor of a young wife who has been "bartered" to him by an evil stepfather. The wife grows up under the domination of her husband with the help of women in the film. They fight the evil men in the film and eventually bond with each other. All the women survive with their children. Some go into business. The main character even reunites with her children and sister sent to Africa with missionaries. The imagery with respect to warring between the sexes in the already brutalized black community received a great deal of critical comment.

In the end, the film prevailed as a great black epic of the 20th century, and, in many ways, it was great. It celebrated the coming of age of a great black female writer. It featured an all-star black cast. Young black girls no doubt watched the film despite its restricted rating and its controversial subject matter because, "Where else would one see so many black people and so many real snippets of black lives that all could identify with?

The film also depicted an absence of sexual boundaries between young girls and old men. Husbands disrespected wives. A lover moved in abruptly and ignored the space of the tormented married girl. All of these images were presented as spicy aspects of the film. But a theme of chaotic and disordered women who because of their powerlessness would always be preyed upon was also subtly apparent. The conclusion with regard to the image of black womanhood was one of survival. Still, many of the women did not emerge from the film as adequate family members who could keep respectful and dignified homes in connection with good black men where daughters could grow, protected into women. While the novel by Walker (1982) would have given mothers reading it space between incidents to qualify and delete inappropriate material for their black girls, the film blared like a shutter that one could not close.

Thus the film capitalized on the sensationalism of stressful black lives and may have left some tender young viewers with the impression that black women are not meant to thrive in a cooperative and complete family, which includes both men and women, but are meant to fight and be ultimately defeated, or to achieve, but to live only with their children and other women.

Popular black films such as this one can exert an important influence on those who identify with the salient black images of generations of female characters. Because films such as The Color Purple are presented against a relative absence of ordinary and commonplace images of black life, they emphasize extreme and striking images of black women. Such portrayals of black women (black men and black families) as either pathological or superhuman deconstruct the faulty relationships between black women and black men, but they do not put them back together very well.

A second film, Eve's Bayou, directed by Kasi Lemmons (1997) exploits the images of black women and girls together, yet tantalizes the viewer with a rare and glamorous all-black cast full of actors familiar to black Americans. The heroines again are generations of black women and their daughters. The film alludes to some interesting history of blacks in New Orleans, but the usual themes of slavery; sex with little girls nearby children; comical appearances; and black men dying violently overwhelm the viewer. So many racial and sexual buttons are pushed that it is nearly impossible to see clearly the danger of these images to the developmental process of real African-American children, and in particular girls. Black children will again probably watch the film, as children are the main characters.

The black father in the film offends the goddess-like mother and so he dies in a twisted misunderstanding that appears to involve his youngest daughter. The viewer is left to somehow believe that the image of the mother alone with daughters who are intimately connected with her husband's death is a triumphant one. It appears as if a rich and powerful black woman is actually better off widowed. Yet, a closer look reveals a woman who is in reality so powerless that her youngest daughter must remove her offending husband. The film unwittingly enforces the cognitive availability of but one popular image of the black woman. That is the image of the superwoman looking lovely on the outside, but in actuality being quite powerless.

The third film, *Just Another Girl on the IRT*, is an independent film by black producer and director Leslie Harris II (1993). This film is one example of a troubling yet more useful and relevant image of black female development. The film is not a Hollywood production, and thus has probably not been as widely watched and discussed in the larger black community. But even without the appropriate commentary by black mothers or role models it is a useful story for the developing black women and her daughters. Harris (1992) chronicles with blatant honesty how both the black woman and the black daughter are robbed of innocence and childhood.

The black mother in this film lacks any true authority at all in her daughter's life. She and the father work day and night. Both parents believe the daughter can handle anything because she seems so tough. They rarely follow up on their demands that she come home after school. They say she cannot go to parties, yet she always defies them. The horror of this parental leniency is made clear to the viewer when the daughter succeeds in hiding a pregnancy destined for miscarriage from them for 7 months.

The life of the black girl in the film tells the viewer very clearly that the image of black women as powerful superwomen is an unreasonable fantasy for the developing female. It further shows that an absence of a childhood occurs when black people, like the mom and dad in this film (and the daughter and her boyfriend as well) become parents too soon, as the film shows they did. They rush into a stage of generativity and work beyond the home, without being able to finish the childrearing process involved in the development of Intimacy vs. Isolation. One wonders if they too were snatched away from a slowly nurtured childhood.

The main character sees herself as an invincible powerful goddess who can fool her mother, have wild sex with her boyfriend, and somehow after all the fun go off to college. We see at the end, not a superwoman, but a teenager with a child and young boyfriend to support. Harris's (1993) theme of the dangers of the early entrance into a super womanhood is different. It does not strike the viewer as a triumphant or normative theme. Another black woman has lost her childhood but we see it definitely as a tragedy.

Just Another Girl on the IRT informs black women in their process of development of a same need for balance. The extreme superwoman images alone is not
a part of healthy ego development in black girls. The powerful black women does
not get through to the black girl and does not reflect real black womanhood. The
developmental effectiveness of the superwoman image requires a developmental
path. Harris's (1993) work showed clearly that the black girl did not understand
the steps for getting to that powerful black womanhood that she appeared to
believe was hers by virtue of her birth as a black female. Harris also makes very
clear that those steps ought to contain a dialogue with black boys, black men,
black families, and black communities before new generations of black children
are thrust into early adulthood.

The children of black parents who perform the family functions of both father
and the mother may have much more time alone to search for images of themselves
on TV and in films. The sexually provocative, physically oriented black superwoman and superman images that are big commodities in the media industry today
will be very salient to black children. As the media looms ever larger in American
life in general, black mothers and fathers must become more critical about which
black cultural variables are communicated to their black sons and daughters. Those
who serve black families (as clinicians, researchers, teachers, and service providers)
must also keep a steady awareness of history, societal bias, and ways to explain and
mediate what is in the world of images of black men and women.

EMPIRICAL VOICES FOR THE SUPERWOMAN

It is difficult from Parham, White, and Ajamu's (1999) perspective to find
research that demonstrates a positive or healthy aspect of black life or experience.
Some research however, has reflected a search for more positive perspectives.
Such research into the lives of black women and their children is prevalent across
disciplines such as antropology, sociology, behavioral medicine, and psychology.

Tanner (1974) is one example of this kind of positive attempt at research about
black women. Tanner (1974) discussed the "matrifocal" nature of kinship development among African-American women. Women in the communities studied by
Tanner formed intense attachments with women and men other than their children's fathers to get the job of childrearing done. Tanner recognized that the breach
in black male–female cultural bonds was a function of racism in society. She saw
black women as experts at forming extended friend and family systems to help
cope with such losses. Tanner (1974) provided some of the empirical evidence
supporting the notion supermothering capabilities of African-American women.

Themes of intense, extended black female; black mother–child; and black
mother–daughter attachment are favored in current research about African-
Americans today. Hill, Hawkins, Raposo, and Carr (1995); Blake (1994); Jacobs
(1994); Jarrett (1994); McClean and Thomas (1996); Taylor and Roberts (1995);
Gonzales, Cauce, and Mason (1996); and Hurd (1998) are good examples of those
who have done research into positive aspects of black womanhood and black
motherhood.

Hill et al. (1995) and Jarrett (1994) studied black and Latina mothers' use of
(respectively) close maternal contact and closeness to grandmothers as positive

ways of coping with the difficulties of growing up outside of mainstream middle-class America. Hill et al. (1995) found that while African-American mothers were keeping their children "under their wing" to cope with violence in their communities they also used the opportunity increased mother–child interaction afforded them to work on extra studies. They played with their children more; took them into work and community settings; and involved them more in spiritual activities such as praying or going to church. By playing roles of tutor and advisor, these mothers made positive strides even in their difficult situations.

Jarrett (1994) found that the three generations of African-American and Latina women in her study deliberately delayed complete separation of daughters, mothers, and grandmothers for the protection and the development of their children and for the positive enhancement of their own lives.

Rucker and Cash (1992), Harris (1994), and others have even found that black women's sense of invulnerability and powerful self-esteem, also appears in their sense of the physical self. According to these researchers, African-American women seem to have higher bodily esteem related to a more relaxed acceptance of their body size. Others such as Flynn and Fitzgibbon (1996) and Usmiani and Daniluk (1997) find that this acceptance of the physical self is passed down to black girls directly from their mothers.

It would seem that close mother–daughter attachments and expertise at forming kinship bonds enhance every aspect of black women and girl's development. Would this positive research convince Parham, White, and Ajamu (1990) to relax their vigilance around the issue of deficiency oriented work? As it turns out, the empirical work confirms the usual enticing images of black women in struggle. This research consistently focuses on poor black women who have the primary life task of scraping by with their daughters. Thus, the research demonstrates that the black mother is still the powerful black goddess, but as usual, she is still in trouble.

Even positive physical self-esteem does not lead to positive outcomes for black women and their daughters. This positive body esteem may lead African-Americans to believe that it is acceptable for young black girls to be overweight relative to their white counterparts. Indeed many, such as Stone et al. (2000), report that African-American women have nearly twice the rate of obesity as Caucasian women. Furthermore, the prevalence of obesity in black girls begins as early as age 9. There is no dearth of research into the relationships between cardiovascular disease, diabetes, breast cancer, ovarian cancer and increased rates of obesity in African-Americans.

Jacobs (1994) and Hurd (1998) bring so much compassion for the black woman into their work that they may end up blinded by their own acceptance of nonoptimal behaviors. Jacobs (1994), for example, establishes adolescent child-bearing among her African-American participants as a normative developmental "career choice." There is a nice "ring" to this notion. Yet it is not difficult to imagine the self-fulfilling potential of such notions. A too easy acceptance of this positive view of parenting at an early age might unwittingly direct black girls toward the physically, socially, and economically dangerous choice of early motherhood.

Wilson (1996) would remind us that accepting single black teen parenting as normative behavior is not reasonable in the present American economic system. Professionals who accept untenable family structures as normal options for black women and men can only continue to promote suboptimal development for black mothers and their children. Such notions can only hasten the eventual destruction of the black family in America, the extinction of the black male, and the subsequent spiritual and emotional destruction of black women and girls.

Parham, White, and Ajamu (1999) would probably appreciate that much of the aforementioned research about black women and their daughters is qualitative and involved the participation rather than the measurement of black women. Empirical approaches, such as the ones reviewed here, do not reach the average black woman as readily as media images, but, because they are systematic and typically community based, they are open to more adjustment by the researchers and the women who participate in the research.

Qualitative methodology may include the use of actual experience of research participants in the development of new theories. This use of qualitative methodology is valuable to the Afrocentric perspectives of Asante (1987), Myers (1993), and Parham, White, and Ajamu (1999). These theorists and others search for a Afrocentric psychological perspectives in the development of research to be applied to a worldwide transformation for all people of African descent. The particular value of such a new theoretical perspective for black women and their daughters is discussed in the final section.

A MODERN AMERICAN GIRLHOOD FOR THE AUTHENTIC SUPERWOMAN

Afrocentric theories such as those outlined by Akbar (1984), Asante (1987), Diop (1984), Mbiti (1970), Myers (1993), Nobles (1972), and Jones (1983), Parham, White, and Ajamu (1999) have been applied to the real lives of black people in developmental struggles for more than 30 years. With the evolution of *The Journal of Black Psychology* in the past 10 years into a mainly empirical journal, these interesting perspectives are beginning to be taken very seriously among the very few African-Americans who actually practice clinical work mainly with African-American populations. The major ideas taking shape in the Afrocentric realm are: (1) the search for historical information relevant to black people themselves as an aspect of black lifespan development; (2) the study of black families, with fathers, mothers, and extended generations as optimal systems; (3) the study of the spiritual lives of African-Americans; and (4) the search for connections or "echos" of the family and social structures of continental Africans today in the lifestyles and values of African-Americans and Africans living throughout the world.

Benefits for African-American women and girls of the development and application of Afrocentric theories and research are many. The systematic presentation of history as a regular part of the raising of black children provides a means for them to separate fact from myth. Black girls buffered with truth can be protected from harsh or embarrassing images of themselves via a media owned by people who care nothing about them. They can move to ensure that they do not act inappropriately as they may see black women in films acting, but can also know that reasons behind preconscious behavior in actual black lives may be very complex. Boys who learn about images of black men in successful families will know what they have to gain economically, socially, and historically and also developmentally by working out their knowledge of who black girls are. Boys raised by single mothers who teach them the importance of male connections may learn why their parents separated so they can eventually master skills that will protect their future relationships and thereby continue the process of nurturing and protecting authentic black female development.

History from the viewpoint of your own culture is always a story told as a teaching tool. No group is eternally oppressed and every person will have his or her own time in history. The African continent has never recovered from the loss of so many African people during the era of slavery. However, many important historical and material resources can be salvaged and developed, even today. Black women may well find that they do not wish to keep from the past all outdated definitions of womanhood, even if they can be shown to be African in origin. But they will need to learn from that history how to further develop their own stories of progress. African folklore also contains powerful images of the black mother, as a goddess or as the ocean or as the mother of all in the universe. But on the African continent, just as in American, this may be in contrast to a stark reality of lower survival rates , lower literacy rates, deplorable health conditions, and little real economic or political power. The development of a holistic psychological perspective for African-American women may find them better equipped to enter a useful process of transforming the world image of the African women by applying a common spiritual or folkloric image of womanhood in their own way. Just as Ida B. Wells found that imparting greater dignity to black women humanized and improved their living conditions in the "turn of the century south," progressive American black women may find that the promotion of ideals arising from a unique black history reflects positively on the concerns of struggling women and families in Africa today.

From the standpoint of a holistic model, all the positive and negative images, myths and histories, and research and theories present possibilities and challenges, and are ingredients in the direction of positive reconstruction of black reality in America. Black women cannot help but be healed and vindicated by research such as that by Burnett and Lewis (1993); Brody, Stoneman, and Zelinda (1994); and Hollinsworth (1999), who (respectively) establish the value of extended families, two-parent families, religion, and the unique symbolism of black familyhood. In these strengthened and spiritualized contexts, black girls will surely find their lost childhood and black mothers will regain a place of authentic strength and wisdom as just one of two or more complementary heads of their households. Little girls can relax and carry on for a while with their play and black woman can get back to their special developmental work, with their families, with their brothers, men, and uncles at their sides and their families "at their backs."

REFERENCES

Akbar (1984). Light from Ancient Africa. Tallahassee: Light Productions.

Angelou, M. (1974). I Know Wy the Caged Bird Sings. New York: Bantam Books.

Asante, M. K. (1987). The afrocentric idea. Philadelphia: Temple University Press.

Blake, I. K. (1994). Language development and socialization in young African-American children. In P. M. Greenfield & R. R. Cocking (Eds.), Cross-cultural roots of minority child development (pp. 167–196). Hillsdale, NJ: Lawrence Erlbaum.

Brody, G., Stoneman, C., and Zelinda, A. (1994). Religion's role in organizing family relationships. *Journal of Marriage and the Family.*

Burnett, M. C., & Lewis, E. A. (1993). Use of African-American family structures and functioning to address the challenges of European-American post divorce families. *Family Relations, 42*(3), 243–249.

Child, L. Maria (1973). *Incidents in the life of a slave girl.* New York: Harcourt.

Collins, P. H. (1991). *Black feminist thought: Knowledge, consciousness, and the politics of empowerment.* New York: Routeledge, Chapman and Hall.

Diop, C. A. (1984). *The African origin of civilization: A myth or reality?* Westport, CT: Lawrence Hill Books.

Edwards, A., & Polite, C. (1992). *Children of the dream: The psychology of black success.* New York: Anchor Books.

Erikson, E. (1980). *Identity and the life cycle: A reissue.* New York: W. W. Norton.

Flynn, K., & Fitzgibbon, M. (1996). Body image ideals of low-income African American mothers and their preadolescent daughters. *Journal of Youth and Adolescence, 25*(5), 615–629.

Fordham, S. (1993). "Those loud black girls": (Black) women, silence, and gender "passing" in the academy. *Anthropology and Education Quarterly, 24*(1), 3–32.

Giddings, P. (1984). *When and where I enter: The impact of black women on race and sex in America.* New York: Bantam Books.

Gonzales, N. A., Cauce, A. M., & Mason, C. A. (1996). Interobserver agreement in the assessment of parental behavior and parent–adolescent conflict: African American mothers, daughters, and indepedent observers. *Child Development, 67,* 1483–1498.

Harris, L. II. (1993). *Just Another Girl on the IRT.* (Los Angles: Artisan/Fox Video.) Film.

Harris, S. M. (1994). Racial differences in predictors of college women's body image attitudes. *Women & Health, 21*(4), 89–104.

Head, B. (1974). *A question of power.* Portsmouth, NH: Heinemann.

Hill, H. M., Hawkins, S. R., Raposo, M., & Carr, P. (1995). Relationship between multiple exposures to violence and coping strategies among African-American mothers. *Violence and Victims, 10*(1), 55–69.

Hollinsworth, D. (1999). Symbolic Interactionism, African American families, and the transracial adoption controversy. *Social Work, 44*(5), 443–445.

Hurd, T. L. (1998). Process, content, and feminist reflexivity: One researcher's exploration. *Journal of Adult Development, 5*(3), 195–203.

Jacobs, J. (1994). "Gender, Race, Class and the Trend Toward Early Motherhood; A Feminist Analysis of Teen Mothers in Contemporary Society". *Journal of Contemporary Ethnography, 22*(4), 442–462.

Jarrett, R. L. (1994). Living poor: Family life among single parent, African-American women. *Social Problems, 41*(1), 30–49.

Lemmons, K. (1997). *Eve's Bayou.* (Burbank, CA: Trimark/Vidmark.) Film.

McLean, B., & Thomas, R. (1996). Informal kinship care populations: A study in contrasts. *Child Welfare, 77*(5), 489–505.

Morrison, T. (1982). *Sula.* New York: Penguin Books.

Morrison, T. (1994). *The bluest eye.* New York: Penguin Books.

Mbiti, J. S. (1970). *African religious philosophies.* New York: Anchor Books.

Myers, L. J. (1993). *Understanding an afrocentric world view: Introduction to an optimal psychology,* 2nd edit. Dubuque, IA: Kendall/Hunt.

Nobles, W. (1972). Africa's philosophy: Foundations for black psychology. In R. Jones (Ed.), *Black Psychology.* New York: Harper & Row.

Parham, T. A., White, J. L., & Ajamu, A. (1999). *The psychology of blacks: An African-American perspective,* 3rd edit. Englewood Cliffs, NJ: Prentice-Hall.

Pieterse, J. N. (1992). *White on black: Images of Africa and Blacks in Western popular cultures.* New Haven: Yale University Press.

Rucker, C., & Cash, T. (1992). "Body Images, Body-Size Perceptions, and Eating Behaviors Among African-American and White College Women." *International Journal of Eating Disorders. 12*(3) 291–299.

Spielberg, S., & Jones, Q. (1985). *The Color Purple.* (New York: Warner Studios.) Film.

Stone, E. J., Dwyer, J. T., Yang, M., & Parcel, G. S. (2000). "Prevalence of Marked Overweight and Obesity in a Multiethnic Pediatric Population". *Journal of the American Dietetic Association. 100*(10), 1149–1156.

Tanner, N. (1974). "Matrifocality in Indonesia and Africa and Among Black Americans". In Eds. Rosaldo, M.Z. and Lamphere, L. Women Culture and Society. Stanford: Stanford University Press.

Taylor, R. D., & Roberts, D. (1995). Kinship support and maternal and adolescent well-being in economically disadvantaged African-American families. *Child Development, 66*, 1585–1597.

Usmiani, S., & Daniluk, J. (1997). Mothers and their adolescent daughters: Relationship between self-esteem, gender role identity, and body image. *Journal of Youth and Adolescence, 26*(1), 45–62.

Vanzant, I. (1998). *Faith in the valley.* New York: Simon & Shuster.

Vanzant, I. (1999). *Don't give it away.* New York: Fireside Books.

Walker, A. (1982). *The color purple.* San Diego, CA: Harcourt, Brace.

Wallace, M. (1979). *Black macho and the myth of the superwoman.*

Washington, M. H. (Ed.) (1987). *Invented lives: Narratives of black women 1860–1960.* New York: Doubleday.

Wilson, W. J. (1996). *When work disappears.* New York: Knopf.

CHAPTER 22

Revisions

Processes of Development in Midlife Women

RUTHELLEN JOSSELSON

> *What one must do to bring her to life was to think poetically*
> *and prosaically at one and the same moment, thus keeping in*
> *touch with fact—that she is Mrs. Martin, aged thirty-six,*
> *dressed in blue, wearing a black hat and brown shoes; but*
> *not losing sight of fiction either—that she is a vessel in which*
> *all sorts of spirits and forces are coursing and*
> *flashing perpetually.*
>
> —VIRGINIA WOOLF, *A Room*
> *of One's Own*, p. 66

The cataclysmic change in the social construction of women's roles in the last quarter of the 20th century erased the old definitions of the "social clock" for women in midlife. Historically, midlife women have been represented in the psychological literature primarily in regard to their reproductive role and their marital status (Gergen, 1990). More recently, this literature has swung in the other direction and has been focused on women as career achievers. But women in midlife are to be found in a panoply of life roles, sometimes sequentially, sometimes simultaneously. Women in the middle of their life course might be grandmothers or new mothers, career women in positions of authority or women undertaking new or first careers, divorced several times or never married, formerly heterosexual and now lesbian, disenchanted with political commitment or newly engaged, or people with various other life projects central in their lives. The absence of universal markers creates a challenge to a developmental psychology that would try to conceptualize a generalized process of midlife development in women.

RUTHELLEN JOSSELSON • School of Psychology, The Fielding Graduate Institute, Santa Barbara, California 93105 and Department of Psychology, The Hebrew University of Jerusalem, Jerusalem, Israel.

Handbook of Adult Development, edited by J. Demick and C. Andreoletti. Plenum Press, New York, 2002.

To construct a theory of midlife development in women, one must be able to explain psychological patterns that underlie the timing, duration, spacing, and order of life events and/or changes in internal psychological structure such as ego strength, time perspective, or achievement motivation (Rossi, 1980). But the variations among women due to social class, race, ethnic identity, educational level, health, marital status, motherhood status, and sexual orientation have defied the search for any such patterning, as has the generally greater fluidity of the life cycle that is today marked by an increase in role transitions and the disappearance of traditional timetables (Neugarten, 1968).

The influx of women into the labor force and onto career paths captured the attention of psychological researchers during much of the 1970s, 1980s and 1990s. Although it is well documented that women's careers are much influenced by social expectations and norms (Stewart & Healy, 1989), some women nevertheless chart their own journeys and create life structures quite different from those in currency in their times (Helson, Mitchell, & Moane, 1984; Tangri & Jenkins, 1986). While there is now a large literature comparing women who have different patterns of work and family life, these studies are often contradictory, and in any case shed little light on the *psychological* realities of midlife.

The only developmental model of sufficient generality that has had continuity of use across generations of psychological researchers is that of Erik Erikson. Using Erikson's stage model, women at midlife should be concerned with matters of generativity, but a careful study of women in this developmental period suggests that issues of identity and intimacy merge with generativity in such a way that the boundaries of generativity blur (Josselson, 1996). Who a woman is and for whom she focuses her efforts are intertwined. Thus, it becomes nearly impossible to differentiate with conceptual accuracy whether a woman's investments in midlife serve her sense of identity, her need for intimacy, or her wish to extend herself into the future—or all three simultaneously.

In contrast to previous conceptualizations that stressed the melancholy of the "empty nest" and the physical agonies of menopause as hallmarks of women's middle years, recent studies show that midlife is a sanguine moment in the life cycle for women (Barnett & Baruch, 1981; Mitchell & Helson, 1990).[1] Longitudinal studies of women's development using personality inventory scales show that as women reach midlife, they show increases in self-discipline, commitment to duties, independence, confidence, coping skills, and ego development (Helson & Moane, 1987). As women move toward midlife, they feel more confident, more energetic, and less vulnerable in contrast to their younger selves (Schuster, Langland, & Smith, 1993) but are not likely to equate their success in life with income, social status, or professional accomplishment—the usual benchmarks for male success. Instead the most content among them are those who feel they have met more personal standards for success that involve balancing among relationships, family, work, and learning (Josselson, 1996). Some researchers discover a "postmenopausal zest" in which women speak more assertively and assess themselves quite deliberately without reference to others' expectations and desires

---

[1] While most longitudinal studies of women have been conducted on white women, Carlson & Videka-Sherman (1990) have shown that increases in life satisfaction in midlife are particularly marked for blacks.

(although not without reference to others' needs) and experience increased energy to pursue new goals (Apter, 1995).

There is no single trajectory of women's lives. Longitudinal studies show that a range of different life patterns can be satisfying and that no particular pattern is a formula for mental health or happiness (Stewart & Vandewater, 1993; Barokas, 1992; Baruch & Barnett, 1986; Lachman & James, 1997). Midlife is a period in which women may expand, reshape, and recreate themselves, but this is a gradual process. No clear evidence for crisis has been found (Lieblich, 1986).

A woman in midlife might best be conceptualized as being in search of balance—balance among aspects of herself and aspects of her life. Daniel Levinson defines "life structure" as the underlying pattern or design of a person's life, and this life structure must evolve if only because all the elements of any individual's life cannot be fit together simultaneously. Midlife is a developmental period of such evolution, a period of picking up threads of what has gone before, integrating them in a new way or attending to aspects of experience that have been in the background in earlier years as well as taking advantage of new social, occupational, and relational opportunities that may arise. The Mid-Life Transition, which Levinson (1996) places at around age 40, involves an effort to make new choices that better fit a self that is better known and understood. But the process of such integration, of putting the disparate parts of lives and selves together, is a very difficult task (McAdams, 1993).

The most visible of the revisions that women make as they mature are shifts in their expression of competence and connection (Josselson, 1996). These two arenas of experience both reflect and herald rearrangements in the deeper strata of the self. Changes in jobs, ambition, projects, partners, friendships, and the advent and challenges of children or the decision not to have them—all these appear on the surface of the life tableau. But more deeply, women experience subtle but profound changes in their experience of authority for their lives, in their understanding of their relationships and in their needs and venues for self-expression. I don't believe that there are definable "stages." Each woman fits these pieces in place in a sequence and pattern different from those of another. Each learns different things at different times. Some women take up their competence first, then later focus more on their connection to others. Other women need to feel their connections well established before exploring the limits of what they can *do* in the world. Still others attempt to work on both simultaneously. The lessons of life that one woman masters in her 20s become clear to another only in her 40s. For some women, major losses or disappointments usher in a period of conscious efforts at revision. For others, arriving where they intended to go sparks a reenvisioning of identity for the next phase of life. Those who are able to make use of their "regrets" to motivate changes in their lives fare better than those who regret their choices but do not make changes (Stewart & Vandewater, 1999); the capacity to make change, however, is not necessarily related only to external barriers to change.

Some writers and researchers have theorized, following Jung, that women at midlife reawaken and reclaim the masculine sides of their nature in developing more the assertive, ambitious sides of their nature (Gutmann, 1987; Labouvie-Vief, 1994; Young-Eisendrath & Wiedemann, 1987). But intensive studies of women (Apter, 1995; Josselson, 1996) show that while women at midlife *do* develop aspects of their nature that might have lain fallow, these are not always the "masculine." For some contemporary women for whom career opportunities were available in earlier adulthood, midlife may herald an emphasis on their more communal inclinations as they develop or deepen relationships that may have been slighted in the rush to career success.

Our current construction of women, within both the media and feminist theory, assumes that women know what they want but are kept from it by an oppressive society. What we don't depict is the anguish of trying to know desire, the false starts, the endless discussions a woman might have with herself about what might be right for her, about what parts of herself she most wants to express in the context of an ever-changing social world that may or may not allow her such expression.

Revision of the self during midlife is most often revision of desire, the recognition that what one seemed to want at an earlier stage of life was a false desire, a wish that felt at the time like it was one's own but that now seems to have belonged to someone else. Or what seemed like mission at an earlier phase of life is later revealed to be caprice. Women themselves are often not sure if their choices are their own—whether they have "fallen into" a way of being or fully and consciously chosen to do so. As life progresses, they may see that what seemed like a choice at the time was in fact determined by fear, passivity, impulsivity, unconscious wishes, or external pressure. And sometimes, the attainment of desire only gives rise to a new, more compelling longing. Recognition of desire is what impels a woman outward into the world. The course of revision for women, then, is an internal process of more clearly understanding the nature of her desire—and more firmly grasping the reins.

Women at midlife discover, if they haven't discovered it before, that they are in fact comprised by many "selves." Rather than prizing consistency, growth at midlife involves legitimizing and containing conflicting needs and desires and finding flexible responses to each new environment in which she experiences herself (Young-Eisendrath & Wiedeman, 1987). Indeed, research demonstrates that the most contented midlife women are women with multiple investments (Baruch, Barnett, & Rivers, 1983). The earliest identity choices often favor certain parts of the self and relegate others to the shadows. As women grow, they struggle to make space for these disused or disavowed parts of the self, widening the expanse of identity to encompass what was left behind. Many women, for example, spending their 20s and 30s learning to adapt to work roles, to organizations in which they are employed, to the demands of motherhood or marriage, find they have shelved their creativity or their playfulness, and refinding these becomes a quest of their 40s and/or 50s. Others forfeit their spirituality in their passionate effort at worldly success, only to reawaken in this need later on. Some women, in search of themselves, leave behind their heritage and the part of themselves that belonged to it, only to rediscover their roots at midlife.

Marjorie Fiske (1980), in her study of adult transition, concludes that adult development is best understood as a process of shifting priorities among what she conceptualizes as four basic commitments: interpersonal, altruistic (including ethics and religion), mastery/competence, and self-protection. I would group the first two priorities under "connection" and the latter two as forms of "competence." The capacity for realigning the emphases of life makes it possible for women to move among their multiple roles, roles that highlight competence and roles that highlight connection, both contemporaneously and at different times in their lives. For many women who are mothers, for example, devotion to children feels paramount when their children are small, but they keep their investment in their careers alive, recognizing that the structure of priorities of interest and purpose may shift. Others begin to take their need for leisure and their avocational interests more seriously and create time to develop in these areas. Needs change as they develop over the life course, but not necessarily in predictable or generalizable ways. What offers exercise for faculties at one stage of life may become routine and lifeless at another, as different aspects of the self emerge to speak their part in desire. And as each new element emerges, a new balance—between agency and communion, competence and connection—must be achieved to ensure well being (Stewart & Malley, 1989).

Variable-based research aggregates women and tends to focus on such issues as career involvement, relationship satisfaction, and various measures of well being. From these studies, we gain a picture of midlife women who are generally happier in careers than in traditional roles, renegotiating primary relationships as children leave home (if they have had children), and more able to cope with midlife changes if they are more ego resilient (Klohnen, Vandewater, & Young, 1996). Multiple roles, although they involve stress and planning, seem to provide positive psychological effects (Crosby, 1991; Pietromonaco, Manis, & Frohardt-Lane, 1986).

Phenomenological studies paint a richer picture of the complexities of women's lives and their efforts to make new meaning of their lives in their middle years. There has been a good deal of intensive qualitative work on midlife women, but most of it is in doctoral dissertations that remain unpublished. All of these studies document complex and varied transitions in both agentic pursuits and the experience of relationships. Ravenna Helson's intensive longitudinal study of Mills College graduates of 1958 and 1960 documents that in their early 40s (upper middle class) women underwent a period of taking stock, questioning what they had done, trying to determine what was missing and charting a new direction. Helson and McCabe (1994) understood this to be an effort to find themselves anew, an effort to "become one's own person" by achieving a perch in life in which they had something valuable to give to others. Thus, the revision of self at midlife involved the merging of the quest for identity and generativity.

My own intensive study of midlife women focuses on a group of (white) women born in 1950 who graduated from college in 1970, women I studied as they graduated from college, and re-interviewed when they were 32 (Josselson, 1987) and again when they were 43 (Josselson, 1996). Knowing these women deeply gave me a view of the way in which external changes in relationships and careers were signs of the more pervasive internal psychic rearrangements that were taking place.

INTERNAL REVISION

Regina, a married college professor with two children, at age 43, told me the following dream: "I dream that I'm in a big house where I've lived for quite some time. And all of a sudden I learn that in the center of the house—it's almost like between floors—there is a whole suite of rooms that I never knew was there—it's hidden from sight and the only way that you can have access to this suite of rooms is through a cobwebbed spidery staircase that's narrow and dark, but I go up through the staircase—or maybe it's going down, I don't know—and I find myself in this suite with huge rooms. Everything is suffused with this golden light and my reaction to all of this is delight. I think 'there's room for everyone here' and I never knew it was here. I've thought about that dream a lot lately. I don't yet know what these rooms are going to be like to actually live in them, but I know that they are there."

This dream captures the essence of the process of revision, as I have understood it, in women's lives. Revision is like discovering rooms that have always been there, hidden inside, accessible through the previously darkened and unused regions of the self, but taking care to make room for everyone who matters.

Work and family commitments are the most visible parts of a woman's identity. This is the way we, as a culture, identify someone. But the most profound revisions of midlife are not visible at all. They are internal—the house with hidden rooms in the metaphor of the dream. The most important revisions involve making meaning and thinking about the self in new ways, effecting fundamental transformations in experiencing life. Consistent with my own findings, Jane Kroger and Kathy Green (1996), in a retrospective study of adult identity, investigated events associated with identity change between the ages of 40 and 63. They found that people were more likely to attribute identity shifts to "internal change"—new awareness or shifts in perspective—than to external influences. Women describe these moments of change as "awakenings," knowing in a new way what they have in some ways known all along.

As Donna, age 43, reflected on the unfolding of her life, dividing her life into chapters, she noted, "From chapter to chapter, there is an increasing consciousness." Going through the process of thinking aloud about their lives, most women I interviewed found themselves lacking the words to tell me about the deepest sense of their revision, but all spoke in one way or another about "increasing consciousness."

Clara, struggling for words, reflecting on her experience of life, said, "[These experiences] have opened a window into what other people feel. All these experiences just make you *understand so much more*." Understanding so much more, opening up more, having more confidence—these are the ineffable revisions these women tried to describe to me. The memorable turnings of their lives are ones that allowed them to look upon the world with expanding vision and to know better their own connection to it. They articulated in experience Robert Kegan's (1982) general statement that "The evolution of the activity of meaning [is] the fundamental motion in personality" (p. 15). For some women, this involved learning about the compromises and inevitabilities of life and finding a place for herself within that—"an angle of repose" in Wallace Stegner's phrase. For many, this revolved around coming to terms with disappointments in the work world, the recognition that one could not do all one wished, or that work itself was often dull or unrewarding. Many had to revise their goals and dreams as they better understood what was possible to accomplish, and some saw possibilities of effectiveness that they hadn't known were there. For others, a shift in or loss of an important relationship was a catalyst to revisit their idea of themselves. Sometimes gaining a new relationship, such as in motherhood, was the force that turned the kaleidoscope.

For some, the growth of psychological awareness led them to regard more seriously the importance of their own emotional needs. Having learned very often in childhood that feelings were to be put aside, controlled, or ignored, or having learned that pleasing others took precedence over consideration of one's own desires, many women came to greater appreciation of the parts of themselves contained in their emotional responses to life. Although it arrived at different times to different women, this crisis of authenticity was an important part of identity revision. They grew to embrace their anger as well as their tenderness, their guilt, their sadness, their pain, and their joy. As Andrea tried to summarize the essence of her sense of revision, "I have grown to be more aware of myself and my feelings than ever before." (And here again, language is too flimsy to fully contain the experience.)

For other women, the increase in consciousness was more woven into their understanding of relationships. Many spoke of their quest to know better what love is, their effort to know if the feeling they had for their husbands, for example, was "really love" as it seems to be depicted by the culture. Some were aware of passionate longings that went unfulfilled, desire that they let themselves know about but kept from acting on so as not to dislodge or jeopardize other commitments they treasured. Others spoke of growing into a greater understanding of motherhood, both its rewards and its limitations—all they discovered they could feel and the perfection they could not embody.

Most of the women in my study felt that as they aged, they could understand better their own life course. Many had to gain years of distance from their childhood to be able to see how it had shaped them. For some, this understanding involved becoming critical of how their parents treated them during their childhood and seeing themselves as mistreated or oppressed in some way. Although this in part reflects the recent society-wide rush to claim victimization as well as cultural pseudopsychologizing about dysfunctional families, I think it also represents important rethinking for many of these women.

The tide of criticism of their early life also allowed many of these women to claim the energy of anger. Behind what had seemed to be an idealization of the past was often a fount of dammed-up rage that, once released, could be converted into animation to make changes in the present. Thus, insight was not a passive knowing but an enlivening vision, one in which the woman was able to resist and refuse what she had come to see had been imposed upon her.

On the other hand, insight also involved new understanding and forgiveness of their families of origin. Some were enlivened from a recognition of their mothers as women in their own right rather than idealized embodied spirits of perfection or as vilified images of imperfection. They seemed to conceptualize in a newly possible way the essential humanity and private subjectivity of their own mothers—"a woman just like me." Psychoanalysis has shown us the way that we preserve throughout life our infantile attachment to superhuman parental figures. But an important challenge of adulthood is learning to know these superbeings as real people. This doesn't supplant the inner longing for the perfect parent, but it does promote insight and adult connection to parents.

SELF AND OTHER

Nearly all researchers who have studied women (see Chodorow, 1982; Gilligan, 1982; Hulbert & Schuster, 1993; Miller, 1976; among others) have found that women are distinct from one another in creating a unique constellation of relationships with other people. Carol Franz, Elizabeth Cole, Faye Crosby, and Abigail Stewart (1994) conclude their summary of the lessons they learned from their intensive studies of women's lives by stressing that interconnection is the foundation of womens' identity. And a self that is "rooted in connectedness [is] a more variable or a more complex self because its precise nature depends on relations with diverse others" (Marcus & Oyserman, p. 120). What, then, is unique to midlife in this process?

Gisela Labouvie-Vief has conducted a number of studies that document the growth of representations of the self and of others across the life cycle. She has found empirically that, looking at people (cross-sectionally) across the lifespan, there is a peak at midlife of recognition of the unique individuality of parents as well as of more complex representations of the self (Labouvie-Vief et al., 1995). She understands these changes in a cognitive–developmental framework in which greater cognitive complexity and the ability to think in terms of transformation are associated with cognitive maturation.

As growth unfolds and life progresses, the most far-reaching modifications and revisions are in the way in which a woman positions herself in relationship to others. But changes in orientation to relationships and the location of the self within them are hard to articulate. These revisions involve a gain of knowledge about how people are together, an increase of complex awareness, and an appreciation of the nearly unfathomable variability in the human expectation of and interaction with each other. Women spoke of this to me in terms of "keeping it all in balance," or "taking account of myself *and* others." What these women were undertaking was redoing their connections with an effort to keep both self and other in focus.

In terms of actual experience, this was grounded in concern with what one could expect of others and what one could expect of oneself. What is reasonable and appropriate behavior from others, and how does one respond when someone's behavior crosses this line? And what about others' expectations of oneself? How does one say "no" and maintain the relationship? How does one let another person know what one needs? So much of life revolves around the conflicting needs of people, and although we sagely pay lip service to compromise, simple compromise is often that which neither person wants. True compromise, balance between needs of the people involved, sharing of resources, and taking account of everyone's point of view all require perspective and energy and all of this matures as women grow. As Clara put it, the challenge was "to respect their needs and my needs *at the same time*." Carol Gilligan (1982) has described this inclusion of self in the ethic of care as the highest stage of female moral development.

At the same time as the self is firming in relationship to others, relationships may themselves be transmuting into ones that affirm the self in its pursuit of creativity and productivity. Both in the lives of ordinary women (Josselson, 1996) and successful artists (Zerbe, 1992), women make increasingly differentiated use of others as self-objects who buttress their ambition.

LOSSES AND GAINS

While midlife has, within the developmental literature, traditionally been depicted as the gateway to the losses of life—losses of fecundity, physical attractiveness, health, parents, friends—it also sees the growth of new forms of empathy, creativity, humor, and wisdom (Jung, 1954; Kohut, 1977). Narcissistic and relational wounds are salved by greater emotional resources. Again, the concept of balance becomes key here.

Our culture contains conflicting myths of the midlife woman: On the one hand, she is depicted as the evil Queen in Snow White, an aging, bitter, often evil hag, envious of the youth and beauty that surrounds her; on the other hand, she is venerated (less often) as the Crone, a woman of wisdom and power (Walker, 1995). The dividing line of midlife is marked by the metaphor of menopause— and it is a metaphor because, despite its physical reality, it is culturally constructed as a "change of life"—often a change from promise and possibility to decline. Recent work, however, challenges this conception of menopause, regarding it instead as a gateway to renewed zest, spirituality, and emotional freedom (Andrews, 1993; Sheehy, 1991). For most women, menopause is a physical marker that may jolt a woman into the reassessments and reflections that are the essence of this epoch of life (Apter, 1995).

Janet Strayer (1996) argues that the parallel to Levinson's "The Dream" in women's development is "The Image." Who a woman is depends in large measure on how she is seen and who she is seen by, and it is her beauty (encoded as youth in Western society) that bequeaths her power to be seen and recognized. Given this social construction, then, the loss of youthful beauty at midlife means accepting diminished visibility, a form of neutering that brings with it decreased vitality. Women who most define themselves in terms of physical attractiveness and the ensuing masculine attention are those who have the most difficulty coping with the physical transitions of midlife (Harter, 1992). A major developmental achievement of midlife is one of better recognizing the self-images that have constricted one's life in order to enlarge the self beyond them (Strayer, 1996), resisting the invisibility that is thrust upon no-longer-young women by establishing one's own vision (Apter, 1995). Thus, healthy midlife development involves integrating the shadow of oneself, the masculine and feminine, the creative and the destructive, by relinquishing the quest for perfection or external validation and turning to an unblinkered recognition of all of one's own qualities (Jung, 1954; Neugarten, 1977). And in mourning the lost imagined immortality and perfectability of youth, midlife women simultaneously become able to be "committed as well as relativistic, to see several contexts at once, to contain opposites," (Strayer, 1996) and to maintain a sense of humor.

The very awareness of physical aging is highly variable (Troll, 1982), as is a woman's inner sense of when she is in "midlife." As 50-year-old women drop their children off at the same day care center as 22-year-old mothers, then share the college classroom with these same early adults, only to work out on the treadmill next to them after class, we might wonder what ever happened to age differentiation in society. Yet the inner psychological meanings of these activities are likely to be different. Developing the theoretical constructs that will contain these differences, however, still awaits future research.

Ravenna Helson and Laurel McCabe (1994) report that when they asked their participants at age 52 what people expected them to be doing in their lives, many of them remarked that it was an odd question. Some said they could not imagine that people cared what they ought to be doing and others said that they did not care what other people might think. Most women expected to "be there" for those who needed them, but most also expected to take more time for themselves and to set their own priorities. This is consistent with my own findings in my longitudinal study, and suggests that during midlife, there is movement toward greater self-definition, a freeing of the self from its social constraints in the context of deeper and enlarged commitments to others. It is this dialectical motion that is the hallmark of midlife development in women.

ACKNOWLEDGMENT.   I wish to thank Meg Carley and Kathleen Fitzpatrick for their assistance with the literature review for this chapter.

REFERENCES

Andrews, L. (1993). *Women at the edge of two worlds*. New York: Harper Collins.

Apter, T. (1995). *Secret Paths: Women in the new midlife*. New York: W. W. Norton.

Barnett, R. C., & Baruch, G. K. (1982). *On the psychological well-being of women in the mid years*. Working paper no. 85. Wellesley College Center for Research on Women. Wellesley, MA.

Barokas, J. (1992). Development and test of a causal model of midlife women's attainments, commitments and satisfactions (life satisfaction). Unpublished Ph.D. thesis, Virginia Polytechnic Institute and State University.

Baruch, G., Barnett, R., & Rivers, C. (1983). *Lifeprints*. New York: Signet.

Baruch, G. K., & Barnett, R. C. (1986). Role quality, multiple role involvement, and psychological well-being in midlife women. *Journal of Personality & Social Psychology, 51*(3), 578–585.

Carlson, B. E., & Videka-Sherman, L. (1990). An empirical test of androgyny in the middle years: Evidence from a national survey. *Sex Roles, 23*(5–6), 305–324.

Chodorow, N. (1978). *The Reproduction of Mothering*. Berkeley: University of California Press.

Crosby, F. (1991). *Juggling*. New York: The Free Press.

Fiske, M. (1980). Changing hierarchies of commitment in adulthood. In N. J. W. Smelser and E. H. Erikson (Eds.), *Themes of work and love in adulthood*. Cambridge, MA: Harvard U. Press.

Franz, C. E., Cole, E. R, Crosby, F. J., & Stewart, A. J. (1994). In C. E. Franz and A. J. Stewart, (Eds.), *Women creating lives: Identities, resilience, and resistance* (pp. 302–323) Boulder, CO, US: Westview Press.

Gergen, M. (1990). Finished at forty: Women's development within the patriarchy. *Psychology of Women Quarterly, 14*, 471–493.

Gilligan, C. (1982). *In a different voice*. Cambridge, MA: Harvard University Press.

Gutmann, D. (1987). *Reclaimed powers: Toward a new psychology of men and women in later life*. New York: Basic Books.

Harter, S. (1992). Visions of the self: Beyond the me in the mirror (pp. 1–40). *Nebraska Symposium on Motivation*. University of Nebraska Press.

Helson, R., & McCabe, L. (1994). The social clock project in middle age. In B. F. Turner & L. E. Troll (Eds.), *Women growing older: Psychological perspectives*. Thousand Oaks, CA: Sage.

Helson, R., Mitchell, V., & Moane, G. (1984). Personality and patterns of adherence and nonadherence to the social clock. *Journal of Personality and Social Psychology, 45*, 1079–1096.

Helson, R., & Moane, G. (1987). Personality change in women from college to midlife. *J. Personality and Social Psychology, 53*, 176–186.

Hulbert, K. D., & Schuster, D. T. (1993). *Women's Lives Through Time*. San Franciso: Jossey-Bass.

Josselson, R. (1987). Josselson, R. *Finding Herself: Pathways to identity development in women*. San Francisco: Jossey-Bass.

Josselson, R. (1996). *Revising herself: The story of women's identity from college to midlife*. New York: Oxford University Press.

Jung, C. G. (1954). *Collected works of C. G. Jung, Vol. 17*. London: Routledge.

Kegan, R. (1982). *The Evolving Self*. Cambridge, MA: Harvard University Press.

Klohnen, E. C., Vandewater, E. A., & Young, A. (1996). Negotiating the middle years: Ego-relisiency and successful midlife adjustment in women. *Personality and Aging, 11*(3), 431–442.

Kohut, H. (1977). *The restoration of the self*. New York: International Universities Press.

Kroger, J., & Green, K. (1996). Events associated with identity status change. *Journal of Adolescence, 19*(5), 477–490.

Labouvie-Vief, G. (1994). Women's creativity and images of gender. In B. F. Turner & L. E. Troll (Eds.), *Women growing older: Psychological perspectives*. Thousand Oaks, CA: Sage.

Labouvie-Vief, G., Chiodo, L. M., Goguen, L. A., Diehl, M., & Orwoll, L. (1995). Representations of self across the Life span. *Psychology and Aging, 10*, 1–12.

Lachman, M. E., & James, J. B. (Eds.). (1997). *Multiple paths of midlife development*. Chicago, IL: The University of Chicago Press.

Levinson, D. J. (1996). *The seasons of a woman's life*. New York: Knopf.

Lieblich, A. (1986) Successful career women at midlife: Crises and transitions. *International Journal of Aging & Human Development, 23*(4), 301–312.

Markus, H., & Oyserman, D. (1989). Gender and thought: The role of the self concept. In M. Crawford and M. Gentry (Eds.), *Gender and Thought*. New York: Springer-Verlag (pp. 100–127).

McAdams, D. (1993). The stories we live by. New York: W. Morrow.

Miller, J. B. (1976). Toward a New Psychology of Women. Boston: Beacon Press.

Mitchell, V., & Helson, R. (1990). Women's prime of life. Is it the 50s? *Psychology of Women Quarterly, 13*, 451–470.

Neugarten, B. (1968). *Middle age and aging: A reader in social psychology*. Chicago: University of Chicago Press.

Neugarten, B. (1977). Personality and aging. In J. E. Birren & K. W. Schaie (Eds.), *Handbook of the psychology of aging*. New York: Van Nostrand Reinhold.

Pietromonaco, P., Manis, J., & Frohardt-Lane, K. (1986). Psychological consequences of multiple social roles. *Psychology of Women Quarterly, 10*(4), 373–381.

Rossi, A. (1980). Life-span theories and women's lives. *Signs: Journal of women in culture and society, 6*(1), 4–32.

Schuster, D. T., Langland, L., & Smith, D. G. (1993). The UCLA gifted women, class of 1961: Living up to potential. In K. D. Hulbert & D. T. Schuster (Eds.), *Women's lives through time*. San Francisco: Jossey-Bass.

Sheehy, G. (1991). *The silent passage: Menopause*. New York: Random House.

Stewart, A. J., & Healy, J. M., Jr. (1989). Linking individual development and social change. *American Psychologist, 44*, 195–206.

Stewart, A. J., & Malley, J. E. Case studies of agency and communion in women's lives. In R. K. Unger (Ed.), *Representations: Social constructions of gender*. Amityville, NY: Baywood.

Stewart, A. J., & Vandewater, E. A. (1999). "If I had it to do over again...": Midlife review, midcourse corrections and women's well-being in midlife. *Journal of Personality and Social Psychology, 76*(2), 270–283.

Strayer, J. (1996). Trapped in the mirror: Psychosocial reflections on midlife and the Queen in Snow White. *Human Development, 39*(3), 155–172.

Tangri, S., & Jenkins, S. R. (1986). Stability and change in role innovation and life plans. *Sex Roles, 14*, 647–662.

Troll, L. E. (1982). *Continuations: Adult development and aging*. Monterey, CA: Brooks/Cole.

Walker, B. (1995). *The Crone: Woman of age, wisdom and power*. San Francisco: Harper & Row.

Young-Eisendrath, P., & Wiedemann, F. (1987). *Female authority*. New York: Guilford Press.

Zerbe, K. J. (1992). The phoenix rises from eros, not ashes: Creative collaboration in the lives of five Impressionist and Postimpressionist women artists. *Journal of the American Academy of Psychoanalysis, 20*(2), 295–315.

CHAPTER 23

Eldercare and Personality Development in Middle Age

ALISON H. CLIMO AND ABIGAIL J. STEWART

Social science research and theory has presented a rather bleak picture of the consequences of eldercare for middle-aged caregivers. In contrast, memoirs and autobiographical accounts of eldercare, like some fictional accounts, often portray midlife caregiving as rewarding, even transformative. In this chapter we adopt a lifespan developmental perspective and take both views seriously. We try to show how the stresses of eldercare might indeed pose challenges for middle-aged caregivers, while there might also be understudied potential for eldercare to enhance and promote their personality development.

THE NEED FOR MIDDLE-AGED CAREGIVERS FOR ELDERS

Most people who live to be elderly eventually need some kind of care. In fact, many aging people, whether cognitively impaired, physically limited, or otherwise in need, require extensive and long-term care. Eldercare can include physical care (help with ordinary activities of daily living such as bathing and dressing); other kinds of instrumental assistance (help with shopping, bills, financial, or health care planning); or emotional and relational help (providing company, advice, affection), including coming to terms with death itself.

Some individual caregivers feel a personal need to provide this care; nevertheless, care for the elderly is also undeniably a social need. Broad demographic trends, such as increasing birth rate and life expectancy, and specific demographic pressures (e.g., the population bulge known as the "baby boom") have resulted in a projected increase in the population of older adults (Rowe & Kahn, 1998). In addition, the average age of the elderly population is rising and projected to continue to

ALISON H. CLIMO • Department of Social Work, Warren Wilson College, Ashville, North Carolina, 28815. ABIGAIL J. STEWART • Institute for Research on Women and Gender, University of Michigan, Ann Arbor, Michigan, 48109.

Handbook of Adult Development, edited by J. Demick and C. Andreoletti. Plenum Press, New York, 2002.

rise. Among those over age 65, "8.8 percent were over age 85 in 1980 compared to a projected 12.8 percent over age 85 in 2020" (Himes, 1992, p. 17). As the aging population continues to grow, policy makers, researchers, practitioners, and families have become increasingly concerned about how to best provide care for the frail elderly.

A variety of institutional structures have grown up in response to the need for eldercare. We have seen an increase in community programs that provide various kinds of specific services to the elderly: meals-on-wheels, programs providing help with shopping and transportation, and in-home and adult day care services. In addition, we can expect to see continuing growth in the development of assisted living facilities, continuum-of-care retirement communities, nursing homes, and hospices for end-of-life care (Cohen, 1998).

Nonetheless, both historically and currently, a great deal of long-term care for the elderly is provided by family members, most typically women. Approximately 95% of the elderly live in their communities, not in nursing homes, even if they are disabled enough to need institutional care (Sommers & Shields, 1987). Nearly 40% of those who require care are cared for by their adult children and nearly one third of all caregivers of the elderly are adult daughters (Stone, Cafferata, & Sange, 1987). In addition, research has shown that ethnic and racial minority care recipients are less likely than white care recipients to use formal services, including long-term care; they are often more likely to be cared for by an adult child (Connell & Gibson, 1997).

Even when professional caregivers provide a great deal of the physical and instrumental assistance elders need, it is quite common for elders to depend on their adult children for emotional connection and companionship as their strength, health, and independence fade. Relationships between adult children and their aged or ill parents, even when they are not primarily oriented around physical or instrumental care, are often tinged with elements of the dependence of those with lesser mobility on those with more. Thus, relationships between adult children and their aging parents usually get reversed.

CAREGIVING AND PSYCHOLOGICAL WELL BEING

Although most elders experience some kind of dependency on their middle-aged children, this is certainly not a universal experience. Many, for example, rely on care from their spouse. Others are childless. However, even some childless elders may come to depend on middle-aged adults whose relationship in some ways resembles that of parent and child (such as mentors and mentees, including teachers and former students such as Morrie Schwartz and Mitch Albom in *Tuesdays with Morrie* [Albom, 1997]). Still other elders have outlived their own children; and some elders have experienced ruptures or distances in their relationships with their children that cannot be overcome, even in the face of the demands of great age or infirmity. In fact, some of the most poignant literature we have are tales of failures of older parents and adult children to ove come these rifts (see, e.g., Bergmann's film *Autumn Sonata* [Bergmann, 1978] about a mother and daughter, or Hugh Leonard's [1978] play, *Da*, about a father and son).

We have many other tales, however, in fiction, theater, film, and memoir, of relationships restored in the elder parents' last days (e.g., Shakespeare's King Lear and Cordelia; see Miller, 1996 for other examples). In fact, a great deal of these types of literature dealing with middle-aged adults caring for the elderly emphasizes the transformative and redemptive power eldercare can have in midlife. Emily Abel (1995) has written thoughtfully about three novelists' portrayals of eldercare; of Doris Lessing's (1984) *The Diary of a Good Neighbor* she points out, "In this novel, the ability to render care is itself the hallmark of adulthood and promotes further growth and development" (p. 53).

The social science literature on midlife that emphasizes the potential for optimal well being in this general life stage (Mitchell & Helson, 1990; Stewart & Ostrove, 1998) seems compatible with these literary accounts of the transformative potential of adult children's relations with their aged and dying parents. However, both seem contradicted by the social science literature on caregiving, which most often underscores the negative psychological, physical, social, and financial consequences borne by adult children caring for elderly parents (Aneshensel, Pearlin, Mullan, Zarit, & Whitlatch, 1995; Cantor, 1983; George & Gwyther, 1986). For example, Sommers and Shields (1987) reported that grief, loneliness, frustration, anger, and guilt are some of the emotional costs of caregiving for frail elders. Perhaps a developmental perspective on both midlife well being and the realities of eldercare can help resolve the apparent paradoxes in our social scientific and cultural images of midlife caregivers' suffering, middle age as the prime of life and eldercare as an opportunity for transcendence. We will illustrate how this might work by drawing on the research literature, as well as representations of midlife caregiving in fiction, theater, film, and memoir.

PERSONALITY DEVELOPMENT IN MIDDLE AGE

Accounts of the important developmental issues in midlife often begin with the individual's changing relationship with time, in part a function of increased awareness of personal mortality. For example, some theorists have pointed to the psychological importance of the shift from measuring one's own life in terms of "time from birth" vs. "time till death" (Neugarten, 1968). Other accounts of midlife focus on physical changes, especially losses of sexual, reproductive, and athletic capacity (Greer, 1992; Heilbrun, 1988) and increases in the rates of chronic illness and impairments (Clausen, 1986). Still others point to involvement in social roles as important components of midlife development and well being (Baruch & Barnett, 1986; Vandewater, Ostrove, & Stewart, 1997). In some cases, these social, physical, temporal, and mortality changes may precipitate a full-scale "crisis." In any case, however, they are hypothesized to present a shift in psychological focus from the external world of achievement and duty to an internal world that may include increased reflectiveness, personal growth, and spiritual development (see Chiriboga, 1989; Jacques, 1954; Jung, 1954; Levinson, Darrow, Klein, Levinson, & McKee, 1978; Neugarten, 1968).

These theories may capture certain important themes associated with midlife; however, a rather different overall picture emerges from the recent empirical literature. First, although middle age may prompt a "midlife review," Levinson (1996) and Stewart and Vandewater (1999) have shown that this review often differs from a later "life review" in that it is undertaken in the service of making important life changes while there is still time. Second, McAdams (1985; see also McAdams,

Hart, & Maruna, 1998) has suggested that awareness of mortality may motivate some of midlife personality development, but its effects are not solely internal. Instead, it often results in an active preoccupation with "generativity" in the service of producing a worthy "legacy." Thus, according to McAdams, the concern with mortality is manifest in increased efforts to fuse personal expression and self-enhancement (agency) with connections with individual others and groups (communion). (See Wakefield, 1998, for another account of how "generativity can at least transiently take some of the pain out of the prospect of death" p. 163.) Finally, Neugarten and her colleagues (Neugarten, 1968; Neugarten & Berkowitz, 1964) argued more than 30 years ago that middle age was characterized by a great increase in an individual's sense of competence and confidence. Howard and Bray (1988) found this to be true in a study of managers, and Stewart, Ostrove, and Helson (2001) did as well in longitudinal studies of college-educated women (see also Stewart & Ostrove, 1998). In his account of midlife generativity, Erikson recognized the middle-aged as the responsible pillars of society, charged with "maintenance of the world" (see Erikson, Erikson, & Kivnick, 1986, p. 50), and finally, Mitchell and Helson (1990) have suggested that the 50s might actually be "women's prime of life."

Overall, then, there is considerable evidence that middle age is a period when many individuals experience new or increased desire to make a contribution that will outlive them. Although they may experience some deepening of their introverted and spiritual interests, they are also likely to broaden their "circle of care" and their preoccupation with civic, social, and global issues. These broader interests may in part be enabled by the self-knowledge and competence accumulated as a result of earlier personality, work, and relational struggles, but they are also demanded by the large circle of individuals—both younger and older—who view the middle-aged as resources for themselves and the community.

HOW DOES "ELDERCARE" FIT IN WITH MIDLIFE PERSONALITY DEVELOPMENT?

The social science literature has demonstrated that eldercare is often extremely costly to middle-aged adults, in terms of both their physical and emotional well being. On the other hand, middle age appears to be a period in which many individuals experience new levels of confidence and competence, as well as a strong desire to "make a contribution." Why wouldn't eldercare offer an excellent outlet for the developmental needs for competence and generativity in midlife? We suspect that it can, and will draw on autobiographical, fictional, and cinematic accounts of how it has, for some individuals. However, we recognize that there are also some compelling reasons that eldercare may not always fit in well with midlife personality development.

Incompatibilities Between Midlife and Eldercare

First, precisely because middle-aged adults are viewed as the responsible "maintainers of the world," they are often swamped with many demands on their time and energy. Middle-aged men and women are likely to be engaged in demanding work lives, to be parenting children or young adults still needing advice and support, and asked to play a role in church, civic, and community affairs (Clausen, 1986). If their adult relationship with their aging parents has been quite separate and independent (as is the norm and the ideal for many groups in contemporary American life), it may be quite difficult to fit responsibilities for ill and aging parents into a life already full of other demands (Doress-Worters, 1994). Aneshensel et al. (1995) use the term *role captivity* to describe the situation of becoming an "unwilling incumbent" of the caregiving role. Moreover, if an individual's hoped-for legacy lies in the areas of work, parenting, and/or community activity, demands for eldercare may quite directly frustrate generative strivings. In a personal essay, writer Nancy Willard (1993) reflects on this when her mother rhetorically asks, "What did people do in the old days?"

> What *did* they do? Dutiful daughters struggled with lifting, feeding, and changing their aged parents. I thought of my own mother under the stress of caring for her own mother, who lived with us when I was growing up... These dutiful women—"caregivers" is the current term for them—did not go off to jobs in the morning. And they certainly were not writers. (pp. 223–224)

Second, the demands of caregiving—both physical and psychological—can be extremely difficult or trying. If middle-aged children are uncertain about what to do, or seems to have done the wrong thing, or if they are simply not physically able to complete a task (e.g., lifting a parent onto and off of the toilet), their sense of competence and command can, then, be challenged or undermined (Aneshensel et al., 1995). Phyllis Rose (1997) recounts "a fight" her brother had with their mother that might have had such an effect:

> She asked him to bring him her sleeping pills from her apartment to the hospital, and he refused. He said it was probably illegal and certainly inadvisable to add sleeping pills to her already carefully controlled hospital regime. She said, "You're so mean! I can't sleep without them! Who are you going to listen to, your own mother or some twenty-year-old intern?" She accused him of cowardice, spinelessness, cruelty, and slavish obedience to authority. She hung up on him. (p. 107)

Third, the long hours and intensive and often unpredictable nature of providing daily care to an elderly person can be extremely isolating, constraining a midlife adult's ability to develop and pursue social and personal activities (Cantor, 1983; George & Gwyther, 1986; Horowitz, 1985). For the caregiving daughter in Mary Gordon's (1978) novel *Final Payments*, Abel (1995) writes, "Caregiving forces an almost total isolation on her" (p. 48). Similarly, in Margaret Forster's (1989) novel *Have the Men had Enough?*, Abel (1995) points out how the caregiving daughter, Bridget, "becomes submerged in caregiving responsibilities, fails to carve out an adult identity, loses her sense of reality, and makes impossible demands on others" (p. 46).

Finally, there may be definitional problems. Eldercare in the social science literature is usually defined as the fairly extensive, day-to-day responsibilities for the well being of an elder; often it is measured in terms of responsibility for "activities of daily living" (dressing, bathing, toileting, shopping, etc.; Aneshensel et al., 1995). In contrast, the fictional and autobiographical literature on the transcendent benefits of eldercare for middle-aged adults often focus on relational and emotional experiences, sometimes taking place in, or perhaps even due to, the context of intermittent visits rather than ongoing care (e.g., Baker, 1982). Perhaps more crucially, memoirs are also inevitably retrospective, and more focused on the self than the caregiving. Miller (1996) suggests that "The death of parents forces us to rethink our lives, to reread ourselves" (p. xiii). Annas and Rosen (1994) point out that "The memoir tends to emphasize ... events the writer has known or witnessed" (p. 1403). There are exceptions to this picture, but overall it may well be that the stress of caring for elders documented by social scientists arises from extended responsibility for providing daily care. In contrast, the possibilities for transcendence may arise from fleeting moments of deep emotional connection and insight (whether in the context of ongoing care or outside it), recollected at some distance from the events.

Compatibilities Between Eldercare and Midlife Personality Development

Having said this, we believe there are also a number of features of middle-aged child–elderly parent caregiving that could facilitate, rather than interfere with, midlife personality development. To the extent that these features pertain, the potential for development, even transcendence, might offset the burdens of limited time and competing responsibilities. The social science literature has slowly begun to explore the ways in which middle-aged adults can benefit from caring for an aging parent (Motenko, 1989; Walker, Martin, & Jones, 1992). Numerous examples in fiction, theater, film, and memoir also illustrate various ways eldercare and midlife personality development are compatible.

For example, although the social science literature has consistently and extensively demonstrated the negative effects of providing day-to-day physical and instrumental care (Aneshensel et al., 1995), for some, the daily rituals of eldercare are a source of transcendence. In her memoir of her mother's fight against and subsequent death from cancer, Terry Tempest Williams (1992) writes,

> My days are immersed in the pragmatic details of care. And I love caring for her, we all do, even though there are times when horror splashes our skin like scalding water as we watch her writhe with nausea and pain. And the other side, always the other side, is as tender as the pain is severe—bathing her, washing her hair, rubbing her body with fine French creams, feeding her ice chips, stroking her hair, her hands, and her forehead. It is sacred time. (p. 213)

As in all relationships and all stages of life, a critical issue may be the degree to which the two partners in the relationship—adult child and elder parent—can synchronize their different developmental needs (Mutran & Reitzes, 1984). In the edited volume, *The Courage to Grow Old*, Dr. Henry Heimlich (1989), who cared for his frail father for the last year and a half of his life, captures the possibility that eldercare can be mutually beneficial in the following passage:

> I feel sorry for those who consider their elderly parents a burden, rather than a privilege. I hope one of my children will want to live with me when my physical limitations increase. Not just for my sake, but for the knowledge of aging he or she will gain, the love that is engendered by proximity when a child is older, and the gratification that comes as they later reminisce. If this desire is considered an imposition or selfishness, so be it. I know that the rewards are unlimited and can be duplicated for both parent and child in no other way. (p. 33)

Although inevitably the two trajectories will not mesh well for all parent–child pairs, it may be valuable to examine in detail the factors that make it more likely that they will mesh well, to the benefit particularly of the middle-aged partner in the dyad.

For instance, if middle-aged adults are newly and increasingly preoccupied with death and mortality, extensive interaction with those much closer to death can be comforting and liberating in a certain way (Moss, Resch, & Moss, 1997). A safe but close encounter with death may lessen middle-aged adults' fear of their own death and offer them models of acceptance while at the same time allowing the elder to face his or her death with the support of their child(ren). In his book of essays, *When Life Meets Death*, Thomas William Shane (1998) is contemplating the moment of his mother's death when he writes, "Even so, the moment is marked with wonder and, just as birth, can be a time of comparable awe and need not be terrifying" (p. 46).

On the other hand, when middle-aged adults lose parents suddenly, or must watch their own parent suffer terribly or die with great resentment or unresolved frustration, their anxiety about their own death may deepen. The exact manner of a parent's final illness or death may not always be the most important "fact," however; even when middle-aged children grieve for their parents' suffering near the end, sometimes they have gained an acceptance of death through conversation and example at earlier points. Terry Tempest Williams opens her memoir reflecting on this acceptance:

> Most of the women in my family are dead. Cancer. At thirty-four, I became the matriarch of my family. The losses I encountered at the Bear River Migratory Bird Refuge as Great Salt Lake was rising helped me to face the losses within my family. When most people had given up on the Refuge, saying the birds were gone, I was drawn further into its essence. In the same way that when someone is dying many retreat, I chose to stay. (pp. 3–4)

Similarly, midlife adults' awareness of their own failing physical powers can be given context by the much more extreme limitations and increased discomfort and pain of the elderly. This, of course, has different effects, depending on many aspects of the personalities and relationship involved. Part of Dr. Heimlich's response to caring for his father is that,

> I now play tennis three or four times a week, whereas twice a month sufficed in years past. My former one to two weeks of skiing each winter has increased to three or four weeks. I realize that I could have enjoyed sports more frequently through much of my life and sometimes wonder if I should have regrets. But I know such activity would have been boring in the past when there were more interesting goals. Succeeding in the physical challenge of winning the game—my tennis game and skiing have improved markedly—leaves me with great exaltation after each victory or new hill conquered. It is a sign that I am still young. Yet, now I am conscious of the fact that one day I will have to cut back and eventually eliminate both sports and will be taking the steps to the cane, walker, and wheelchair. (p. 33)

Mitch Albom (1997) recounts what he learned from his former college professor, Morrie, about the most extreme kinds of losses of strength and autonomy. Morrie explains that "I began to enjoy my dependency. Now I enjoy when they turn me over on my side and rub cream on my behind so I don't get sores ... I revel in it. I close my eyes and soak it up" (p. 116). He accepts his increased dependency and allows himself to enjoy his childlike status: "The truth is, when our mothers held us, rocked us, stroked our heads, none of us ever got enough of that. We all yearn in some way to return to those days when we were completely taken care of—unconditional love, unconditional attention." Mitch concludes that "At seventy-eight, he was giving as an adult [for example, with advice to Mitch] and taking as a child" (p. 116). It was, of course, a blessing not only that Morrie could find the joy in dependency again, but also that he was at the same time able to give as an adult.

Even without the capacity for such reciprocity, however, some caregivers still derive pleasure from their parents' pleasure. For example, Russell Baker (1982) begins his memoir *Growing Up*,

> At the age of eighty my mother had her last bad fall, and after that her mind wandered free through time. Some days she went to weddings and funerals that had taken place half a century earlier. On others she presided over family dinners cooked on Sunday afternoons for children who were now gray with age. Through all this she lay in bed but moved through time, traveling among the dead decades with a speed and ease beyond the gift of science. (p. 1)

Perhaps because of his mother's freedom from linear time, Baker emphasizes the new perspective his relationship with her offers him. He reflects on his relationship with his son: "Between us there was a dispute about time. He looked upon the time that had been my future in a disturbing way. My future was his past, and being young, he was indifferent to the past" (p. 8). Although framed somewhat differently from psychologists' theories, Baker is aware that his own relationship to time played a different role in his relationships with different generations:

> As I hovered over my mother's bed listening for muffled signals from her childhood, I realized that this same dispute had existed between her and me. When she was young, with life ahead of her, I had been her future and resented it. Instinctively, I wanted to break free, cease being a creature defined by her time, consign her future to the past, and create my own. Well, I had finally done that, and then with my own children I had seen my exciting future become their boring past. (p. 8)

It is precisely this dispute about time between the generations that then motivates Baker to make the memoir part of his legacy: "I thought that, when I am beyond explaining, [my own children] would want to know what the world was like when my mother was young and I was younger ... " (p. 8).

Changes in the relationship with one's parent and in both parent and child's relationship with time may lead either or both to recall and review past losses. For example, Morrie tells Mitch about his response to a letter from a teacher who taught a special class of children, all of whom had lost a parent.

> Suddenly ... Morrie stopped, bit his lip, and began to choke up. Tears fell down his nose. "I lost my mother when I was a child ... and it was quite a blow to me ... I wish I'd had a group like yours where I would have been able to talk about my sorrows." (p. 71)

Mitch learns about the durability of relationships and the pain of losses when he hears Morrie asked, "That was seventy years ago when your mother died. The pain still goes on?" and Morrie replies, "You bet" (p. 72). Mitch tells us in the book's conclusion that not long after Morrie's death, he contacted his younger brother who had insisted on fighting his battle against pancreatic cancer by himself and had moved to Europe to do so.

> I told him I respected his distance, and that all I wanted was to be in touch—in the present, not just the past—to hold him in my life as much as he could let me. "You're my only brother," I said. "I don't want to lose you. I love you." I had never said such a thing to him before. (pp. 190–191)

Conversation with elders both about what feels important now, and about regrets, is often valued by middle-aged children as they review their own lives. Morrie urges Mitch to value relationships and family more, and work less, fueling Mitch's own midlife review (see, e.g., pp. 62–66). In Bergman's (1978) painful *Autumn Sonata*, the aging pianist mother is forced by her middle-aged daughter to confront her past inadequacies as a parent. After she flees, her daughter recognizes that there is an opportunity for reconciliation; she writes to her mother:

> I met you with demands instead of affection.... There is a kind of mercy after all. I mean the enormous chance of looking after each other, of helping each other, of showing affection.... (p. 84)

The past relationship between this mother and daughter was excruciating for both; and we learn that the mother in turn bears scars of her own mother's inadequate parenting. Yet the film ends with the (albeit unrealized) possibility that in this new moment a new relationship could offer healing. Similarly, Morrie tells Mitch "It's not just other people we need to forgive ... We also need to forgive ourselves ... You can't get stuck on the regrets of what should have happened" (p. 166). He advises Mitch that "I mourn my dwindling time, but I cherish the chance it gives me to make things right" (p. 167). In short, Morrie emphasizes that regrets and life review can lead to change and forgiveness; he offers a perspective and an example. Similarly, Thomas William Shane (1998) recognizes his mother's remission from cancer as an opportunity to make things right. He writes,

> It may well be that the magic of medicine will hold off destiny. But I know the truth. Remission is different from recovery. Each day offers me an occasion to express one more truth that I never found a way to express before. Each day gives us time to make amends ... to celebrate memories ... to have a cup of coffee and to tell a story. Each day is an opportunity to try to understand more fully who this woman is who gave me life and whose history makes me the man I am. (p. 38)

Sometimes contact with elders clearly underscores middle-aged adults' sense of being in their "prime." Mitch Albom admits, "Sitting there, I felt so much stronger than he, ridiculously so, as if I could lift him and toss him over my shoulder like a sack of flour" (p. 120). The strengths are not merely physical. Phyllis Rose (1997) outlined the impact of her mother's health emergencies on her three children:

> Every time my mother goes into the hospital, my brother, my sister, and I perform an elaborate stewardship dance, whose improvised moves are worked out over the telephone. Someone has to take Mother in. Someone has to take her out. Someone has to go to the apartment and get her insurance card, her checks, her bathrobe, her cane, her eye drops, her reading glasses, her *TV Guide*, her comb and brush, her Fixodent, and clothes to come home in. Someone has to arrange for a phone and, if she's well enough, a television. Someone has to hire private duty nurses in the hospital and round-the-clock nurses for the first week or so after she gets home. (p. 88)

This "dance of stewardship" is clearly the dance of organized, competent, confident middle-aged people—people who know what needs doing and how to get it done. The complex care of the aging and dying is filled with opportunities for command and competence. Yet it also presents occasions to fail or be judged inadequate. And the echoes of past failures and disappointments can amplify new ones, while past warmth and trust can buffer them.

There is no doubt that middle-aged children can find it difficult to juggle their various responsibilities, especially when the demands for eldercare are particularly intense. However, there are some special opportunities afforded by eldercare. First, to the extent that the admittedly narrow, but so important world of the elder can be made better, middle-aged adults may exercise their power within it in ways that are deeply satisfying. Phyllis Rose (1997) describes her approach when her mother is having a medical crisis:

> I personally think that someone should be with her every possible moment in the hospital to act as her advocate. Too sick to speak up for herself, hampered, in any case, by the paralysis of a vocal chord, visually impaired, hearing impaired, she is sometimes neglected, for example, when food is left for her that she cannot even see, or when she needs painkillers she can't ask for, and when she is badgered unnecessarily by hospital routines, as when they wake her up to take her temperature. (p. 88)

The power to make things right in this limited world offers a certain satisfaction. Sometimes the power is even more basic. Rose recalls,

> When my father was dying, the only book I could tolerate was Tolstoy's *The Death of Ivan Ilyich*, which describes the dying patriarch's alienation from his bourgeois, sentimental family, his ability to stand the company only of his servants, whose treatment of his dying was matter-of-fact, oriented toward the physical. I remember the scene in which his valet bathes his feet, and he experiences a kind of peace. I have tried to provide that kind of comfort for my mother. (p. 113)

Although caregivers who are also parents often struggle to find ways to manage the demands for loving care from their children as well as their elders, sometimes the very act of caregiving is a part of parenting. Sara Lawrence Lightfoot (1988) was raising two small children during the year and a half she took weekly trips from Boston to New York to talk with her mother about her life; she worried about the toll of the separations (see, e.g., p. 14). But she points out in the preface that they "gained a lively appreciation for history and roots" (p. xx). Phyllis Rose recounts an unexpected outcome of her effort to teach her son some of the meaning of her mother's illness,

> ... in my ongoing effort to educate him in the grim realities I myself had been spared, I wanted him to witness the awfulness of the hospital, to know the smell of four unemptied bedpans, to see the shrunken face of a dying person he'd loved. We arrived to find my mother moved into a clean, bright room, with only one roommate, a lady with a nice voice, not too sick. Mother was propped up, had her glasses on and her teeth in, looked cheerful and sounded positively giggly. (p. 107)

Despite the unanticipated result of this particular event, the deeper point is that Rose, like many parents, struggled to maintain her parental project even as she was absorbed with eldercare.

Naturally, many factors—including relentless demands for care from either generation—can make it difficult for any particular middle-aged child to succeed at this balancing act. But when it does somehow work, it may confirm a sense of continuity that links the generations and enhances the caregiver's sense that a legacy is being created or confirmed through the caring relationship. Lightfoot writes near the end of her book,

> I look into my mother's eyes and see my own reflections—not mirror images but refracted, varied. I feel the stubborn lineage that has survived generations—passed down from Mom Margaret, Mary Elizabeth, Mama Lettie—but I also sense the ways in which our temperaments, our family dramas, our choices, and the historical timing have made us different from each other. (p. 311)

As their last session concludes, Lightfoot writes, "As we get ready to leave my mother's office, our eyes land on a picture of my children that sits on her desk. Their beaming faces take us backwards into the future" (p. 313).

This kind of confirmation of the continuity of generations is one of the elements of midlife generativity that may be fostered by caring for aging parents. Eldercare could also be seen as an obvious opportunity for the fusion of agency ("doing for," being effective) and communion (relationship). This is true, of course, only when the caregiver can find some way to feel effective and the interactions can be positive and warm, at least in the mind of the caregiver. Mitch's meetings with Morrie were unfailingly warm, but he struggled to find ways to be effective—genuinely to help Morrie. For a while he brought Morrie food he had once liked, without noticing he could no longer eat it. Near the end, though, he found ways to provide help:

> I sat at the far end of his chair, holding his bare feet. They were callused and curled, and his toenails were yellow. I had a small jar of lotion, and I squeezed some into my hands and began to massage his ankles.
> It was another of the things I had watched his helpers do for months, and now, in an attempt to hold onto what I could of him, I volunteered to do it myself. The disease had left Morrie without the ability even to wiggle his toes, yet he could still feel pain, and massages helped relieve it. (p. 164)

When elders are angry or out of focus, it may strain caregivers' sense of communion. Somehow, however, at least some are able to identify the character of the beloved parent even in frustrating failures to connect. Russell Baker ends his memoir with his own effort to keep his mother focused on who he is. As she drifts into and out of sleep, she asks, "Who're you?" As he tries to remind her who he and his wife are ("Russell and Mimi"), "She glared at me the way I had so often seen her glare at a dolt. 'Never heard of them,' she said, and fell asleep" (p. 278).

HOW IS ALL OF THIS GENDERED?

In this account, we have drawn freely from caregiving enacted by both women and men. Although middle-aged men and women do engage in all kinds of caregiving for their parents, it is nevertheless the case that eldercare is deeply gendered in ways that are demonstrable in the social science literature. In addition, we suspect that the literary accounts we have examined point toward a gender difference that may help resolve the paradox between the social science impression that eldercare is mainly stressful and burdensome, and our cultural impression that it offers opportunities for growth and transcendence.

First, women are more likely than men to assume caregiving roles and responsibilities (Stone et al., 1987). Women and men also tend to provide different types of eldercare, often in gender-specific ways; men are more likely to provide help with care management tasks, such as decision making, finances, home repairs, and transportation; and women tend to perform more hands-on care that includes bathing, dressing, administering medications, and household chores (Horowitz, 1985; Young & Kahana, 1989). In addition, a great number of studies have found that female caregivers fare worse than male caregivers on various measures of global and caregiving-specific well being (Cantor, 1983; Chiriboga, Weiler, & Nielsen, 1990; Young & Kahana, 1989). It may be that caregivers who are involved in more removed types of care tasks, and who thus also tend to be male, have an "aesthetic distance" that allows them to notice and reflect on the opportunity for transcendence that caregiving presents.

It should be noted that given conventional gender socialization, more women than men are also experienced with caregiving. Women may less often struggle to figure out what needs to be done, because they may have more training and experience in helping care for others (Baber & Allen, 1992). At the same time, this greater level of experience and competence may mean that women are both less likely to learn and grow from it, and more vulnerable to "caregiver burnout." Perhaps men are more likely to bring interest and enthusiasm to tasks that are new and well within their capacity (as when Mitch massaged Morrie's feet). Gender socialization may also contribute to women's sense of obligation toward caregiving, contributing to role captivity, whereas men may be more free to choose to take care of their elderly family members.

Similarly, men and women may differ in terms of the activities they value at midlife that are compromised by eldercare. In earlier midlife, women may be more likely to feel the pressure of responsibilities for parenting, while men may feel more pressure to leave a legacy of work or money (Clausen, 1986). Research has suggested that later on in midlife, "Men tend to become somewhat less preoccupied with work ... and women tend to become more preoccupied with an occupational role" (Clausen, 1986, p. 159). Women who have entered careers only in middle age may thus be especially troubled by pressure in both domains. The point here is not that men or women have it harder or easier, but that the struggle to integrate personality development with eldercare may be different, depending on the object of the middle-aged child's generative focus.

CONCLUSION

The social science literature may be focused on the stresses of caregivers because documenting the negative effects of eldercare is a crucial precondition for establishing helpful social policies and program interventions for caregivers. In contrast, our cultural images of what may be gained from the experience of caring for aging elders may be based on literary accounts that take place at some distance from the daily care. In addition, these literary accounts of growth and transcendence via eldercare are inevitably also accounts by people with substantial resources; considerable evidence shows that poverty inevitably adds to the strain of caregiving (Cantor, 1983).

The challenge, then, is to find ways to provide long-term care for our elders that both meets their physical and emotional needs and honors the developmental tasks of their middle-aged children and caregivers. In part, this challenge may require us to rethink some of our society's values surrounding care and the elderly. It will require us to value the lives and well being of the frail and ill elderly, for example, as well as the caring labor their friends and families provide. In part, this challenge may require us to develop social resources that provide some of the eldercare that is most difficult for middle-aged children to provide, thus freeing them to offer the forms of care that fit both their parents' and their own developmental needs. Morrie said to his children,

> "Do not stop your lives ... Otherwise this disease will have ruined three of us instead of one."
>
> In this way, even as he was dying, he showed respect for his children's worlds. Little wonder that when they sat with him, there was a waterfall of affection, lots of kisses and jokes and crouching by the side of the bed, holding hands. (Albom, 1997, p. 93)

In many families, this kind of death is impossible. If we are to make it possible for more families, we must complement the autobiographical and fictional literature with systematic research identifying the circumstances under which eldercare and midlife personality development are not mutually exclusive, but quite compatible.

ACKNOWLEDGMENTS.   We are grateful to Sebastian Matthews and David G. Winter for helpful comments on an early version of this manuscript and to our research group for ongoing feedback, stimulation, and support.

REFERENCES

Abel, E. K. (1995). Representations of caregiving by Margaret Forster, Mary Gordon, and Doris Lessing. *Research on Aging, 17*(1), 42–64.

Albom, M. (1997). *Tuesdays with Morrie.* New York: Doubleday.

Aneshensel, C. S., Pearlin, L. I., Mullan, J. T., Zarit, S. H., & Whitlatch, C. J. (1995). *Profiles in caregiving: The unexpected career.* San Diego: Academic Press.

Annas, P. J., & Rosen, R. C. (1994). *Literature and society.* Englewood Cliffs, NJ: Prentice-Hall.

Baber, K. M., & Allen, K. R. (1992). *Women and families: Feminist reconstructions.* New York: Guilford Press.

Baker, R. (1982). *Growing up.* New York: New American Library.

Baruch, G. K., & Barnett, R. C. (1986). Role quality, multiple role involvement, and psychological well-being in midlife women. *Journal of Personality and Social Psychology, 51,* 578–585.

Bergmann, I. (1978). *Autumn sonata.* New York: Pantheon.

Cantor, M. H. (1983). Strain among caregivers: A study of experience in the United States. *The Gerontologist, 23,* 597–604.

Chiriboga, D. A. (1989). Mental health at the midpoint: Crisis, challenge or relief? In S. Hunter & M. Sundel (Eds.), *Midlife myths: Issues, findings and practice implications* (pp. 116–144). Newbury Park, CA: Sage.

Chiriboga, D. A., Weiler, P. G., & Nielsen, K. (1990). The stress of caregivers. In D. E. Biegel & A. Blum (Eds.), *Aging and caregiving: Theory, research and policy* (pp. 121–138). Newbury Park, CA: Sage.

Clausen, J. A. (1986). *The life course: A sociological perspective.* Englewood Cliffs, NJ: Prentice-Hall.

Cohen, M. A. (1998). Emerging trends in the finance and delivery of long-term care: Public and private opportunities and challenges. *The Gerontologist, 38*(1), 80–89.

Connell, C. M., & Gibson, G. D. (1997). Racial, ethnic, and cultural differences in dementia caregiving: Review and analysis. *The Gerontologist, 37*(3), 355–364.

Doress-Worters, P. B. (1994). Adding eldercare to women's multiple roles: A critical review of the caregiver stress and multiple roles literatures. *Sex Roles, 31*(9/10), 597–616.

Erikson, E. H., Erikson, J. M. & Kivnick, H. Q. (1986). *Vital involvement in old age.* New York: W. W. Norton.

Forster, M. (1989). *Have the men had enough?* London: Chatto & Windus.

George, L. K., & Gwyther, L. P. (1986). Caregiver well-being: A multidimensional examination of family caregivers of demented adults. *The Gerontologist, 26,* 253–259.

Greer, G. (1992). *The change: Women, aging and the menopause.* New York: Springer.

Heilbrun, C. G. (1988). *Writing a woman's life.* New York: W. W. Norton.

Heimlich, H. J. (1989). Before it's too late. In P. L. Berman (Ed.), *The courage to grow old* (pp. 31–36). New York: Ballentine Books.

Himes, C. L. (1992). Future caregivers: Projected family structures of older persons. *Journals of Gerontology, 47,* 17–26.

Horowitz, A. (1985). Sons and daughters as caregivers to older parents: Differences in role performance and consequences. *The Gerontologist, 25*(6), 612–617.

Howard, A., & Bray, D. (1988). *Managerial lives in transition: Advancing age and changing times.* New York: Guilford Press.

Jacques, E. (1954). Death and the midlife crisis. *International Journal of Psychoanalysis, 46,* 502–514.

Jung, C. G. (1954). The development of personality. In W. McGuire (Ed.), *The collected works of C. G. Jung, Vol. 17* (pp. 167–186). Princeton, NJ: Princeton University Press.

Leonard, H. (1978). *Da.* New York: Atheneum.

Lessing, D. (1984). The diary of a good neighbor. In D. Lessing, *The diaries of Jane Somers* (pp. 1–253). New York: Vintage Books.

Levinson, D. J. (1996). *The seasons of a woman's life.* New York: Knopf.

Levinson, D. J., Darrow, C. M., Klein, E. B., Levinson, M. H., & McKee, B. (1978). *The seasons of a man's life.* New York: Ballantine Books.

Lightfoot, S. L. (1988). *Balm in Gilead: Journey of a healer.* Reading, MA: Addison-Wesley.

McAdams, D. P. (1985). *Power, intimacy, and the life story: Personological inquiries into identity.* Homewood, IL: The Dorsey Press.

McAdams, D. P., Hart, H. M., & Maruna, S. (1998). The anatomy of generativity. In D. P. McAdams & E. de St. Aubin (Eds.), *Generativity and adult development: How and why we care for the next generation* (pp. 7–43). Washington, DC: American Psychological Association.

Miller, N. K. (1996). *Bequest and betrayal: Memoirs of a parent's death.* New York: Oxford University Press.

Mitchell, V., & Helson, R. (1990). Women's prime of life: Is it the 50's? *Psychology of Women Quarterly, 14,* 451–470.

Moss, M. S., Resch, N., & Moss, S. Z. (1997). The role of gender in middle-age children's responses to parent death. *Omega, 35*(1), 43–65.

Motenko, A. K. (1989). The frustrations, gratifications, and well-being of dementia caregivers. *The Gerontologist, 29,* 166–172.

Mutran, E., & Reitzes, D. C. (1984). Intergenerational support activities and well-being among the elderly: A convergence of exchange and symbolic interaction perspectives. *American Sociological Review, 49,* 117–130.

Neugarten, B. L. (Ed.) (1968). *Middle age and aging: A reader in social psychology.* Chicago: University of Chicago Press.

Neugarten, B. L., & Berkowitz, H. (1964). *Personality in middle and later life: Empirical studies.* New York: Atherton.

Rose, P. (1997). *The year of reading Proust.* New York: Scribner.

Rowe, J. W., & Kahn, R. L. (1998). *Successful aging.* New York: Pantheon Books.

Seltzer, M. M., & Li, L. W. (1996). The transitions of caregiving: Subjective and objective definitions. *The Gerontologist, 36*(5), 614–626.

Shane, T. W. (1998). *When life meets death: Stories of death and dying, truth and courage.* New York: The Haworth Press.

Sommers, T., & Shields, L. (1987). *Women take care: The consequences of caregiving in today's society.* Gainesville, FL: Triad.

Stewart, A. J., Ostrove, J., & Helson, R., (2001). *Middle aging in women: Patterns of personality change from the 30s to the 50s. Journal of Adult Development, 8,* 23–37.

Stewart, A., & Vandewater, E. A. (1999). "If I had it to do over again …". Midlife review, midcourse corrections, and women's well-being in midlife. *Journal of Personality and Social Psychology, 76,* 270–283.

Stewart, A. J., & Ostrove, J. (1998). Women's personality in middle age: Gender, history and midcourse corrections. *American Psychologist, 53,* 1185–1194.

Stone, R., Cafferata, G. L., & Sange, J. (1987). Caregivers of the frail elderly: A national profile. *The Gerontologist, 27*(5), 616–626.

Vandewater, E. A., Ostrove, J. M., & Stewart, A. J. (1997). Predicting women's well-being in midlife: The importance of personality development and social role involvements. *Journal of Personality and Social Psychology, 72*(5), 1147–1160.

Wakefield, J. C. (1998). Immortality and the externalization of the self: Plato's unrecognized theory of generativity. In D. P. McAdams & Ed de St. Aubin (Eds.), *Generativity and adult development* (pp. 133–174). Washington, DC: American Psychological Association.

Walker, A. J., Martin, S. S. K., & Jones, L. L. (1992). The benefits and costs of caregiving and care receiving for daughters and mothers. *Journal of Gerontology, 47*(3), 130–139.

Willard, N. (1993). *Telling time: Angels, ancestors, and stories.* New York: Harcourt, Brace & Company.

Williams, T. T. (1992). *Refuge: An unnatural history of family and place.* New York: Vintage Books.

Young, R. F., & Kahana, E. (1989). Specifying caregiver outcomes: Gender and relationship aspects of caregiving strain. *The Gerontologist, 29*(5), 660–666.

CHAPTER 24

Grandparent–Grandchild Relationships and the Life Course Perspective

LAURA HESS BROWN AND PAUL A. ROODIN

In this chapter, we outline the importance of grandparent–grandchild relationships, review previous research, and consider the life course perspective as a theoretical framework for studying grandparent–grandchild relationships.

THE IMPORTANCE OF GRANDPARENT–GRANDCHILD RELATIONSHIPS

The relationships that develop between grandparents and grandchildren have been viewed as important in a number of ways. Hagestad (1985) pointed to the grandparent–grandchild bond as influential in family continuity, not only from the standpoint of blood lines, but also because grandparent–grandchild relationships provide indirect linkages between grandparents and parents, and parents and grandchildren. Kornhaber and Woodward (1985) described the bond as a "vital connection" for grandchildren, as grandparents serve as role models, purveyors of family histories and experiences, and caretakers. Grandparents are also negotiators between parents and grandchildren, and function as symbolic figures connecting past, present, and future for grandchildren.

Grandparents often provide continuity and a safe haven in times of family crisis or disruption, such as divorce, substance abuse, or incarceration in the parent generation (Barnhill, 1996; Cherlin & Furstenberg, 1986; Minkler & Roe, 1996; Scherman, Goodrich, Kelly, Russell, & Javidi, 1988). Grandparents give both time

LAURA HESS BROWN • Department of Psychology, State University of New York at Oswego, Oswego, New York, 13126.   PAUL A. ROODIN • Office of Experience-Based Education, State University of New York at Oswego, Oswego, New York, 13126.

Handbook of Adult Development, edited by J. Demick and C. Andreoletti. Plenum Press, New York, 2002.

and money to their children and grandchildren (Bass & Caro, 1996) and serve as models for ethnic identity development in adolescent grandchildren (Smith-Barusch & Steen, 1996). Grandchildren derive satisfaction from grandparents' love and support, as well as feelings of self-worth and pride in their achievements (Tyszkowa, 1991; Van Ranst, Verschueren, & Marcoen, 1995).

Grandchildren provide grandparents with feelings of immortality and intergenerational continuity (Hagestad, 1985), and with the exception of grandparental caregivers, derive considerable enjoyment without parental responsibility (Cherlin & Furstenberg, 1986; Jendrek, 1994). Grandparents delight in the companionship, instrumental and emotional support, and knowledge of advanced technology that grandchildren offer them (Ramirez-Barranti, 1985). Kivnick (1985) identified five simultaneous dimensions of role meanings that grandparents derive from grandparenthood: (1) centrality of the role, (2) valued elder, (3) immortality through clan, (4) reinvolvement with personal past, and (5) indulgence of grandchildren.

UNDERSTANDING
GRANDPARENT–GRANDCHILD RELATIONSHIPS

In a classic paper, Kahana and Kahana (1971) challenged researchers to consider grandparent–grandchild relationships from five levels of analysis:

1. Grandparent and grandchild *social roles*, with expectations and ascribed status
2. Individual *emotional or intrapsychic experiences* for both grandparent and grandchild
3. *Transaction* between the dyadic pair, involving interaction, reciprocity, and mutual influence
4. *Group familial process*, involving complex interaction patterns between grandparents, parents, and grandchildren (power, control, and influence are considered within kinship structures, reflecting the intergenerational nature of help patterns and family maintenance)
5. Grandparent–grandchild relationships as *symbols* of continuity, aging, and youth interdependence among generations

Despite Kahana and Kahana's plea, most previous studies have been limited to examination of: (1) the perspective of the grandparent (Thomas, 1986, 1989, 1990), (2) the adjustment to the acquisition of a new familial role (Cunningham-Burley, 1986; Fischer, 1983; McGreal, 1983; Somary & Stricker, 1998), (3) grandparenting styles (Neugarten and Weinstein, 1964), and the (4) the meaning of the grandparental role (Giarrusso, Silverstein, & Bengtson, 1996; Kivnick, 1985; Robertson, 1977; Smith-Barusch & Steen, 1996; Troll, 1983; Wood & Robertson, 1976).

Research from the perspective of grandchildren and their relationships with grandparents is scarce. In addition, most such studies have examined younger grandchildren's reactions to and/or preferences for particular grandparents (Kahana & Kahana, 1970; Kornhaber & Woodward, 1985; Ponzetti & Folkrod, 1989; Schultz, 1980). Only recently have researchers addressed the grandchild in adolescence (Clingpeel, Colyar, Brand, & Hetherington, 1992; Creasey & Koblewski, 1991; Matthews & Sprey, 1985) or in young adulthood (Eisenberg, 1988; Hartshorne & Manaster, 1982; Hodgson, 1992; Hoffman, 1980; Kennedy, 1992; Robertson, 1976).

As life expectancies for both men and women increase, the possibility of studying long-term meaningful relationships between grandparents and their young adult grandchildren increases (Crispell, 1993; Hagestad, 1988; Szinovacz, 1998; Uhlenberg, 1996). Potentially, this relationship could span four or more decades, taking the grandchild from infancy into middle age, and the grandparent from middle age to oldest-old age (85 and older) according to Hagestad (1985) and Neugarten and Neugarten (1987).

The studies published on young adult grandchildren have been exploratory in nature and limited in scope (Hoffman, 1980; Kennedy, 1992; Robertson, 1976). Only a few aspects of young adult grandchildren's relationships with grandparents have been targeted; none to date has presented a comprehensive sketch of young adult grandchildren's perspectives of their grandparents (Brown, 1998; Tomlin, 1998; Whitbeck, Hoyt, & Huck, 1993).

Despite the documented importance of the relationship for individuals and families, the grandparent–grandchild relationship is one that has been neglected in family research literature to date (Aldous, 1995). Although it is "…second only in biological linkages to the parent–child dyad…" (Kivett, 1991, p. 267), few researchers have considered the grandparent–grandchild relationship as anything but peripheral to the nuclear family and its development (Ramirez-Barranti, 1985). The parent–child relationship has been considered within a developing family system (Belsky, Rovine, & Fish, 1989), but a systemic framework has not been applied as extensively to the grandparent–parent–grandchild network of relationships (Brown, 1998; Hagestad, 1985; King, Russell, & Elder, 1998; Thompson & Walker, 1987; Whitbeck et al., 1993).

As with many areas of family research, there are few theoretical models that have been systematically applied to the study of grandparent–grandchild relationships (Cohler & Altergott, 1996; Kivett, 1991). Experts agree that the literature on this topic has lacked theoretical depth and methodological sophistication (Burton & Bengtson, 1985; Hagestad, 1985; Troll, 1985; Robertson, 1995). According to Aldous (1995):

> …The research literature…has been largely atheoretical. As a consequence, there are few overall perspectives that have been used to help integrate the findings from the large number of disparate studies…Few studies set out to test hypotheses derived from theories, conceptual frameworks, or the findings from other studies. (pp. 106–107)

Cohler and Altergott (1996) point out that in the last 50 years, little theoretical attention has been paid to extended family systems in the second half of life, when issues of grandparenting and intergenerational relationships become more salient than earlier issues surrounding mate selection and childbearing. One possible solution to the theoretical vacuum in the study of grandparent–grandchild relationships is consideration of a broader, more flexible framework. Little research has been targeted to this dimension and there has been virtually no attention directed toward developing and applying an empirical model of intergenerational relationships and bonds (Brown, 1998; Tomlin, 1998; Whitbeck et al., 1993).

THE LIFE COURSE PERSPECTIVE

The life course perspective, while not a unified, homogeneous theory, provides one useful framework for considering grandparent–grandchild relationships (Bengtson & Allen, 1993; George & Gold, 1991). It is a way of organizing and highlighting individual and family development within various temporal contexts, as well as different ecological contexts from the microsystem to the macrosystem. The life course perspective recognizes the dynamic processes at work both inside and outside the family sphere (Aldous, 1996; Bengtson & Allen, 1993; George & Gold, 1991). The life course is separate and distinct from the *lifespan*, which refers to age-related changes in individual development. The life course, in contrast, emphasizes change in social persona; "…age-related change transitions that are socially created, socially recognized and shared" (Hagestad & Neugarten, 1985, p. 35). Lives are socially defined in terms of periods of social roles, rights, privileges, and obligations, based on historically and culturally defined age norms that make up the "social clock" (Hagestad & Neugarten, 1985; Neugarten & Neugarten, 1987). In other words, the lifespan is considered within the structures and social and historical contexts of the life course (Aldous, 1996; Hagestad, 1990).

Multidisciplinary Perspectives

The contributions of many academic disciplines are utilized within the life course perspective. From developmental psychology, life course theorists have employed the concepts of age-related changes, longitudinal and cross-sequential research designs, continuity and change, historical period effects, resiliency, and social ecologies to study families over time (Baltes, 1987; Bengtson & Allen, 1993).

Anthropologists have considered the cultural origins of family systems, as well as the evolutionary functions of grandparents as protectors of young children to ensure survival of the gene pool. They also outline the tendency for maternal grandparents to be more invested in grandchildren than paternal grandparents owing to greater certainty of blood ties (Smith, 1991). Biology, especially ethological biology, has contributed greatly to our understanding of human instincts and the need to form social bonds, especially with caregivers and close kin (Bowlby, 1969; Kornhaber, 1996).

Family sociology brought a tradition of considering age stratification, resource distribution, social roles and meanings, and macrosocial societal structures (Bengtson & Allen, 1993; Hagestad & Neugarten, 1985). Demographers have charted patterns of births, deaths, marriage and divorce rates, childbearing, and generational transitions within and between cohorts to document patterns of the past, and to predict changes in grandparent–grandchild relationships in the future (Hagestad, 1986; Szinovacz, 1998; Uhlenberg, 1980). Economists have contributed aggregate patterns of affluence, poverty, work, and unemployment, as well as societal trends in saving and spending across cohorts, races, and genders (Crispell, 1993; DaVanzo & Rahman, 1993). In a multigenerational family, grandparent–grandchild relationships can be viewed as economic transactions of reciprocal give and take of goods, services, and support (Bass & Caro, 1996).

Family historians have considered the familial context of change over long periods of time, myths about the "golden days" of the extended family living in harmony under one roof (Hareven, 1994), and the impact of historical events and social policies on family forms and functions. Some examples of historical markers that have influenced family phenomena include slavery, economic depressions, wars, and the industrial revolution (Burgess, 1995; Elder, 1974; Elder & Clipp, 1988).

A Research Agenda

The development of programmatic research from a life course perspective requires adherence to the following principles: (1) family relationships are categorized by both continuity and change over time, including age, cohort, and period effects; (2) the social ecology of families requires both micro and macro levels of analysis, with role perceptions and social meanings attached to family development and transitions over time; (3) family dyadic relationships are only part of the larger family context, and should be considered within the complex familial structures and interactions between members; (4) heterogeneity exists among individuals, families, and cultures, as well as diversity across developmental periods based on gender, ethnicity, and socioeconomic status; and (5) the study of family relationships requires a multidisciplinary focus, utilizing tenets from Psychology, Sociology, Demography, Anthropology, History, Economics, and Biology (Bengtson & Allen, 1993).

An Empirical Model

Brown (1998) examined young adult grandchildren's perspectives on their relationships with their parents and grandparents to highlight four different levels of analysis: (1) symbolic role expectations grandchildren hold for grandparents and grandchildren; (2) emotional closeness grandchildren felt for parents and grandparents; (3) shared behavioral activities with grandparents; and (4) interconnecting relationships within a three-generational family system (see Fig. 24.1). The purpose of the conceptual model was to illustrate a more integrated, comprehensive study of grandparent–grandchild relationships from the perspective of the young adult grandchild.

The independent variables of family structure characteristics were used to predict different dimensions of the grandparent–grandchild relationship, with parental relationships as intervening variables. Family structure characteristics studied included gender, lineage, grandparental living arrangements, parental divorce, grandparental caregiving, geographical distance from grandparents, and race/ethnicity. Grandparent–grandchild relationship characteristics included perceived closeness, frequency of contacts, rank closeness, shared activities, and role perceptions for grandparents and grandchildren. Parent relationship variables included parent–child closeness, parent–grandparent closeness, perceived parental kinkeeping efforts, and frequency of parent–grandchild contacts. Considerable support for the model was found in the first wave of data (Brown,

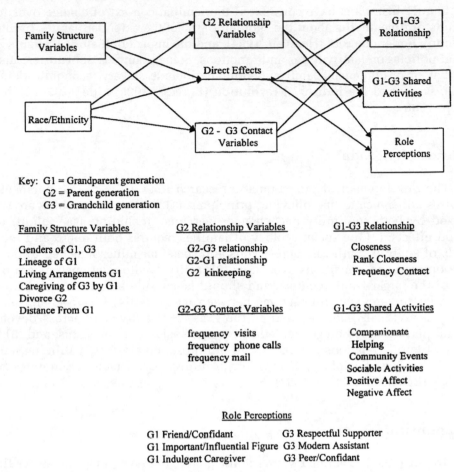

Key: G1 = Grandparent generation
      G2 = Parent generation
      G3 = Grandchild generation

Family Structure Variables   G2 Relationship Variables   G1-G3 Relationship

Genders of G1, G3   G2-G3 relationship   Closeness
Lineage of G1   G2-G1 relationship   Rank Closeness
Living Arrangements G1   G2 kinkeeping   Frequency Contact
Caregiving of G3 by G1
Divorce G2
Distance From G1   G2-G3 Contact Variables   G1-G3 Shared Activities

                                  frequency visits   Companionate
                                  frequency phone calls   Helping
                                  frequency mail   Community Events
                                                                   Sociable Activities
                                                                   Positive Affect
                                                                   Negative Affect

                        Role Perceptions

            G1 Friend/Confidant   G3 Respectful Supporter
            G1 Important/Influential Figure G3 Modern Assistant
            G1 Indulgent Caregiver   G3 Peer/Confidant

Figure 24.1. Conceptual model of young adult grandchildren's relationships with grandparents and parents.

1998). This model provides a direction for future research from the life course perspective.

Temporal Contexts

Because the individuals involved in grandparent–parent–grandchild relationships are always at very different points along the lifespan continuum, it is important to consider that each brings to the relationship developmentally unique contributions and cohort-specific interpretations (Baltes, 1987). Moreover, the relationships, like individuals, are not static, but dynamic; they change and are reconceptualized over time as grandchildren mature and grandparents age (Bengtson & Black, 1973; George & Gold, 1991; Hagestad, 1986; Silverstein & Long, 1998).

Kahana and Kahana (1970) found that children displayed different preferences for grandparental role behaviors that paralleled Piaget's stages of cognitive development. Preschoolers preferred grandparents who give them presents and treats (indicating egocentrism); in middle childhood, grandchildren stressed the importance of sharing activities such as games or outings (concrete reciprocity); and early adolescents preferred grandparents to be supportive if needed (formal symbolic thought). In a related study, Ponzetti and Folkrod (1989) analyzed children's essays about their grandparents, and found that younger children emphasized receipt of affection, nurturance, and gifts from grandparents, while older children emphasized grandparents' provision of reliable alliance, pride in family history, and guidance.

In a more recent study, Van Ranst and colleagues (1995) found that younger adolescent grandchildren in Belgium valued their grandparents more highly than did older adolescents. Young adolescents placed special emphasis on the receipt of predictable emotional support, mentoring, kinkeeping, caregiving, and reassurance from grandparents during the process of identity formation. Older adolescent grandchildren continued to view grandparents as important figures in their lives, but with less emotional dependency, and greater emphasis on links to the past. Older adolescents also differentiated more between the specific relationship characteristics they had with grandmothers vs. grandfathers, which seemed to indicate a developmental increase in cognitive classification complexity.

Young adults may renegotiate the role of grandchild with their grandparents as a part of the transition to adult maturity (Kivett, 1991). Adult grandchildren and grandparents may incorporate elements of their past relationship, such as emotional closeness and/or family-initiated contacts, but evolve more adult, reciprocal exchanges of social and instrumental aspects of the relationship (Hartshorne & Manaster, 1982; Thompson & Walker, 1987). Silverstein and Long (1998) found in a longitudinal study of grandparent–grandchild relationships that grandchildren's affection for grandparents reflected a curvilinear pattern, with affection levels high in childhood, declining during the teenage years, and rebounding moderately into adulthood.

In previous research, young adult grandchildren reported that grandparents continued to be important, valued, and influential attachment figures in their lives (Creasey & Koblewski, 1991; Kennedy, 1992; Robertson, 1976). Young adult grandchildren experience a more reciprocal exchange of social and instrumental behaviors with grandparents (Hartshorne & Manaster, 1982), and feel they give more to grandparents than they receive in return (Langer, 1990).

Demographically, young adults are much more likely to have multiple surviving grandparents than was the case a century ago. Also, as parents have fewer children, each grandparent–grandchild dyad has at least the potential for greater closeness (George & Gold, 1991; Hagestad, 1986, 1988; Uhlenberg, 1996). Most grandparents enter the role at postparental midlife; women are about 46 years old, men about 48 years (Crispell, 1993). Again there is an increasing possibility that grandparents may spend four or more decades in the role (Hagestad, 1985).

Younger grandparents tend to be more active, doing more babysitting, playing, and socializing with their grandchildren (Aldous, 1995). Older grandparents frequently have more physical limitations and consequently fewer face-to-face contacts with grandchildren. Research shows that grandparents not only provide support but also act as a "buffer" between children and their parents (Ramirez-Barranti, 1985). Of course, differences in grandparental behaviors are influenced by the relative ages of the grandchildren. Young children require more direct supervision and interaction, whereas adolescents and young adults are capable of sustaining relationships long distance (Kennedy, 1992).

Johnson (1985) identified a possible cohort effect in grandparenting styles of older and younger grandmothers responding to the divorce of their adult children. Older grandmothers (e.g., those born before 1918) raised their children within a climate of behaviorism and noninterference that were the norms of the time; after their children divorced, they were less likely to offer financial or direct support than were younger grandmothers. The younger grandmothers, born after the two world wars, were influenced in their parenting by the more humanistic teachings of Dr. Spock, and were more likely to place a greater value on familial involvement and direct social contact.

In a study of grandparent–grandchild relationships in China, Falbo (1991) demonstrated the influence of period effects on intergenerational ties. In 1979, when the Chinese government instituted the "one child" policy, it also created the 4–2–1 family structure: four grandparents and two parents influencing one grandchild. In previous cohorts, grandchildren had to share their grandparents with siblings and many cousins. In Falbo's study, only children benefited from receiving undiluted attention, affection, discipline, and socialization from parents and grandparents; the one child policy led to better academic performance when compared to children with other siblings in the family. As both family structures produced children with comparable personality outcomes, the results did not support the stereotype of the "spoiled" only child.

SOCIAL ECOLOGIES OF FAMILIES: A LIFE COURSE APPLICATION

The multiple contexts of grandparent–grandchild relationships can include microsystemic considerations of grandparents as caregivers/attachment figures (Bowlby, 1969; Jendrek, 1994; Solomon & Marx, 1995). More attention is being paid to mesosystemic connections between children's familiar daily environments— links between grandparent–grandchild relationships and schools (Solomon & Marx, 1995), religious organizations (Conroy & Fahey, 1985; Wechsler, 1985), and other social contexts (King et al., 1998). Exosystemic issues include considerations of the availability and economic feasibility of grandparents as babysitters (Presser, 1989), grandparenting in urban vs. rural environments (King et al., 1998), legal issues of visitation and custody for grandparents (Karp, 1996), and public programs and benefits available to grandparents (Mullen, 1996). Macrosystemic considerations of culture and ethnicity examine kinship patterns, culturally defined roles, and the status of the elderly as influences on intergenerational relationships (Ikels, 1998).

THE FAMILY SYSTEM

The grandparent–grandchild relationship does not exist in a familial vacuum, and it is important to consider the roles parents play in the complex interaction patterns between grandparents and grandchildren. Parents' close relationships and frequent positive contacts with their own parents and in-laws may contribute to the closeness of grandparent–grandchild bonds. Similarly tensions between grandparents and parents may hinder the development and maintenance of grandparent–grandchild bonds (Johnson, 1985; King & Elder, 1995; King et al., 1998; Matthews & Sprey, 1985).

The parent generation's efforts at maintaining kinship ties, called "kinkeeping," may provide an important bridge between grandparents and grandchildren. Hagestad (1985) interviewed three-generational families in the Chicago area, and found a remarkable amount of effort made to maintain family ties across generations. Middle-generation mothers arranged family reunions; made phone calls; and sent cards, pictures, and presents to relatives living far away. Similar evidence of parental mediation in grandparent–grandchild relationships has been found particularly following parental divorce (Gladstone, 1989; Johnson, 1985; Thompson & Walker, 1987).

DIVERSITY: GENDER, LINEAGE, RACE, AND ETHNICITY

Diversity characterizes every level of analysis when grandparent–grandchild relationships are considered. There are wide-ranging individual differences among grandparents and their extended families, as well as differences in grandparent–grandchild relationships as a function of age, gender, ethnicity, socioeconomic status, and culture (Bengtson & Allen, 1993).

Gender

Considerable evidence suggests that gender is an important variable for both grandparents and their grandchildren. Many studies indicate that same-sex ties are closer than opposite sex ties, and the strongest connections are between grandmothers and granddaughters (Creasey & Koblewski, 1991; Hagestad, 1985; Hyde & Gibbs, 1993; Kivett, 1991). Women have also been found to be more likely to monitor the "pulse" of family relationships, noting conflicts and interpersonal connectedness between members (Hagestad, 1985).

Fewer researchers have examined the roles grandfathers play in their relationships with grandchildren. Baranowski (1987) found that role meanings were closely linked to age of the grandfather. Older grandfathers far more than younger grandfathers found more meaning in the role. They found distinct pleasure in becoming reinvolved with their personal past history, accepting the centrality of grandfatherhood in their lives, and indulging their grandchildren. Health and distance from grandchildren were also salient factors for grandfather–grandchild relationships. Grandfathers in better health and those living nearby to their grandchildren had more frequent contacts and greater emotional closeness.

Kivett (1985) found that among the grandfathers she studied there was much less importance placed on the centrality of the role of grandparent in their lives than has been documented for grandmothers. Kivett also noted that grandfathers and grandchildren had relatively infrequent contacts and exchanges of assistance;

yet grandfathers had high expectations for receiving care or financial help from their older grandchildren, should such care ever be needed.

Kennedy (1992) found gender differences in the shared activities grandchildren had with grandparents. The differences fell along social vs. instrumental dimensions, with granddaughters' activities with grandparents reflecting more social involvement and grandsons' activities with grandparents reflecting a more instrumental pattern.

Lineage

Another powerful predictor for styles of relationships is the lineage of grandparents, particularly when combined with the genders of both grandparents. Grandparent–grandchild relationships have been described as having a "matrifocal tilt" (Hagestad, 1985) with maternal lineage having greater likelihood of highly maintained, close connections with grandchildren than paternal lines.

Maternal grandparents, in comparison to paternal grandparents, are consistently reported to have closer relationships with grandchildren, with closer contact theoretically maintained through the kinkeeping of middle-generation mothers and their own mothers (Hagestad, 1985). It is important to note, however, that paternal grandparents tend to be older than maternal grandparents owing to a form of hypergamy (i.e., older males marrying younger females); therefore, they are less likely to survive long enough to interact with young adult grandchildren over the long term (Uhlenberg, 1980). This differential longevity may also influence the closeness and quality of grandparent–grandchild relationships. Older paternal grandparents would also be more likely to have serious health problems which could preclude extensive travel and more active participation in their adult grandchildren's lives (Troll, 1985).

Race and Ethnicity

Researchers are becoming more aware of the importance of examining race and ethnic diversity in the study of grandparent–grandchild relationships (Hunter & Taylor, 1998; Ikels, 1998; Kamo, 1998; Williams & Torrez, 1998). Worthy of note are some recent studies including research on: (1) complexity in multiple-generation African-American family relationships (Burton, 1995; Burton & Dilworth-Anderson, 1991); (2) peer support and levels of caregiver burden in multiple-generation African-American families (Pruchno, 1999); (3) roles, intergenerational support, and language barriers among three-generational Mexican American families (Markides, Boldt, & Ray, 1986; Strom, Buki, & Strom, 1997); (4) three-generational family relationships in Japan (Strom et al., 1995) and Taiwan (Strom, Strom, Shen, Li, & Sun, 1996); and (5) adolescent grandchildren in Finland and Poland (Hurme, 1997).

Linda Burton (1995) has conducted extensive research on African-American grandmothers and their daughters and grandchildren, and found considerable complexity in the roles of black grandmothers. In longitudinal, ethnographic interviews with 48 African-American families with teenage mothers, Burton (1995) found 14 distinct patterns of intergenerational caregiving, including variations in timing of role entry, household composition, and numbers of generations involved in caregiving systems. These distinct variations indicate greater complexity of grandmothers' roles than had been suggested in previous research. Pruchno (1999) compared the black and white grandmothers who were raising their grand-children without the presence of middle-generation parents in the home. Black grandmothers were more likely than white grandmothers to have peers who were also living with their grandchildren. These black grandmothers were also more likely in their own development to have been members of multiple-generation arrangements, living together. The black custodial grandmothers in this study experienced significantly less caregiver burden and stress than did their white counterparts (Pruchno, 1999).

Researchers studying 375 Mexican-American families in San Antonio found strong intergenerational networks of assistance between grandparents and parents involving advice, health care, home maintenance, and finances (Markides et al., 1986). In contrast, there was little assistance exchanged between grandparents and grandchildren, despite the high levels of reported emotional closeness between them. Strom and colleagues (1997) suggested that a language barrier between grandparents and grandchildren might partially account for these findings. They found that Mexican-American grandparents who spoke only Spanish reported greater frustration and difficulty relating to English-speaking adolescent grandchildren than did English-speaking grandparents. However, this effect was buffered somewhat when Spanish-speaking grandparents made an effort to spend greater amounts of time with grandchildren.

Traditional Asian cultures often stress the norm of respect and honor for grandparents who derive considerable status as family elders, wise educators of the young, and keepers of cultural continuity and traditions. However, with recent changes in employment, migration, and living arrangements, many Asian grandparents report some loss of status within the family and diminution of filial piety (Kamo, 1994). Robert Strom and teams of researchers (1995, 1996) from the United States and Asia surveyed three-generational families with schoolage children in Japan and Taiwan to assess strengths and needs of grandparents in those countries. Grandparents, parents, and grandchildren were asked to rate grandparents' roles in the family on the dimensions of satisfaction, success, teaching, difficulty, frustration, and need for information. Japanese grandparents rate themselves higher on all dimensions, positive and negative, than did parents or grandchildren, although all three perceived grandparents to have respect and influence. Very similar results were found in the Taiwanese sample. In that sample 751 nonconsanguineous representatives of each generation rather than related family members were assessed using the same measures of grandparent role dimensions.

Hurme (1997) compared written essays on impressions of their grandparents from 731 adolescents from two contrasting countries: Finland, which is prosperous, secular, recently urbanized, and individualistic, and Poland, which is poorer, Catholic, collectivist, and traditionally rural. Polish adolescents described their grandparents using more positive, personal and emotional referents than did Finnish adolescents, who relied more heavily on descriptions of appearance and activities. Combined cultural, political, and religious differences between the two nations were identified as strong influences on the disparity between Polish and Finnish grandchildren's impressions of grandparents.

CONCLUSIONS AND FUTURE DIRECTIONS

This chapter highlights the importance of adopting more complex and extended developmental models to the study of grandparent–grandchild relationships. The life course perspective provides one such approach that encompasses both individual (grandparent and grandchild) and family development within historical time. This perspective, illustrated empirically by Brown (1998) (see Fig. 24.1), permits both micro and macro levels of analysis, and provides a framework to represent the importance of diversity (e.g., gender, lineage, race, and ethnicity) among grandparents, grandchildren, and their families. The life course perspective is a powerful explanatory tool that offers investigators the possibility of representing the multidimensional changes descriptive of grandparent–grandchild relationships across developmental periods. The life course perspective can also be employed to understand cultural differences and the multidisciplinary concerns of those studying grandparent–grandchild relationships.

Future researchers will want to continue to explore the longitudinal development of the grandparent–grandchild bond, extending into the later stages of the relationship as grandchildren achieve adulthood and grandparents enter extreme old age. In addition, investigators need to begin to broaden the focus of their study to examine the grandparent–grandchild relationship within familial, sociocultural, economic, and historical contexts.

With a life course perspective and a multidimensional, longitudinal model, investigators will also be better able to help professional experts in mental health, government, and law seeking guidelines and advice within their own fields of study. These experts continue to turn to social science research to help them understand grandparenting and the consequences of threats to this relationship in the family. They also seek to understand the meaning and significance of the grandparent–grandchild bond for individual development (grandparents and grandchildren) and to help them create policy and anticipate the implications of such policies. Finally, continued research from the life course perspective will help to improve the intervention strategies designed to assist families involved in grandparent–grandchild issues, including but not limited to divorce, separation, and grandparental caregiving (Barnhill, 1996; Burton, Dilworth-Anderson, & Devries, 1995; Cherlin & Furstenberg, 1986; Fuller-Thompson, Minkler, & Driver, 1997; Gladstone, 1988, 1989; Hartfield, 1996; Hirshorn, 1998; Jendrek, 1994; Johnson, 1988; Matthews & Sprey, 1984; Minkler & Roe, 1996; Solomon & Marx, 1995; Szinovacz, 1998).

REFERENCES

Aldous, J. (1995). New views of grandparents in intergenerational context. *Journal of Family Issues*, *16*, 104–122.

Aldous, J. (1996). *Family careers: rethinking the developmental perspective.* Thousand Oaks, CA: Sage.

Baltes, P. B. (1987). Theoretical propositions of life-span developmental psychology: On the dynamics between growth and decline. *Developmental Psychology, 23*, 611–626.

Baranowski, M. C. (1987). The grandfather–grandchild relationship: Meaning and exchange. *Family Perspective, 24*, 201–215.

Barnhill, S. (1996). Three generations at risk: Imprisoned women, their children, and grandmother caregivers. *Generations, 20*, 39–40.

Bass, S. A., & Caro, F. G. (1996). The economic value of grandparent assistance. *Generations, 20*, 29–33.

Belsky, J., Rovine, M., & Fish, M. (1989). The developing family system. In M. Gunnar and E. Thelen (Eds.), *Systems and development. Minnesota Symposium of Child Psychology, 22*, 119–166.

Bengtson, V. L., & Allen, K. R. (1993). The life course perspective applied to families over time. In P. G. Boss, W. J. Doherty, R. LaRossa, W. R. Schumm, & S. K. Steinmetz (Eds.), *Sourcebook of family theories and methods: A conceptual approach* (pp. 469–498). New York: Plenum Press.

Bengtson, V. L., & Black, K. D. (1973). Intergenerational relations and continuities in socialization. In P. Baltes & K. W. Schaie (Eds.), *Life-span developmental psychology* (pp. 207–234). New York: Academic Press.

Bowlby, J. (1969) *Attachment and loss, Vol. 1*. New York: Basic Books.

Brown, L. H. (1998). Young adult grandchildren's perceptions of relationships with grandparents and parents. Doctoral dissertation, Syracuse University.

Burgess, N. J. (1995). Female-headed households in sociohistorical perspective. In B. J. Dickerson (Ed.), *African American single mothers: Understanding their lives and families.* (pp. 21–36). Thousand Oaks, CA: Sage.

Burton, L. M. (1995). Intergenerational patterns of providing care in African American families with teenage childbearers: Emergent patterns in an ethnographic study. In V. L. Bengtson, K. W. Schaie, & L. M. Burton (Eds.), *Adult intergenerational relations: Effects of societal change.* (pp. 79–96). New York: Springer.

Burton, L. M., & Bengston, V. L. (1985). Black grandmothers: Issues of timing and continuity of roles. In V. L. Bengston & J. F. Robertson (Eds.), *Grandparenthood*, pp. 61–78. Beverly Hills, CA: Sage.

Burton, L. M., & Dilworth-Anderson, P. (1991). The intergenerational family roles of aged Black Americans. *Marriage and Family Review, 16*, 311–330.

Burton, L. M., Dilworth-Anderson, P., & Merriwether-Devries, C. (1995). Context and surrogate parenting among contemporary grandparents. *Marriage and Family Review, 20*, 349–366.

Cherlin, A., & Furstenberg, F. (1986). *The new American grandparent: A place in the family, a life apart*. New York: Basic Books.

Clingpeel, W. G., Colyar, J. J., Brand, E., & Hetherington, E. M. (1992). Children's relationships with maternal and paternal grandparents: A longitudinal study of family structure and pubertal status effects. *Child Development, 63*, 1404–1422.

Cohler, B. J., & Altergott, K. (1996). The family of the second half of life: Connecting theories and findings. In R. Blieszner & V. H. Bedford (Eds.), *Aging and the family* (pp. 59–94). Westport, CT: Praeger.

Conroy, D. B., & Fahey, C. J. (1985). Christian perspectives on the role of grandparents. In V. L. Bengtson & J. F. Robertson (Eds.), *Grandparenthood* (pp. 195–207). Beverly Hills, CA: Sage.

Creasey, G. L., & Koblewski, P. J. (1991). Adolescent grandchildren's relationships with maternal and paternal grandmothers and grandfathers. *Journal of Adolescence, 14*, 373–387.

Crispell, D. (1993). Grandparents galore. *American Demographics, 15*, 63.

Cunninham-Burley, S. (1986). Becoming a grandparent. *Aging and Society, 6*, 453–470.

DaVanzo, J., & Rahman, M. O. (1993). *American families: Trends and policy issues*. Santa Monica, CA: Population Index.

Eisenberg, A. R. (1988). Grandchildren's perspectives on relationships with grandparents: The influence of gender across generations. *Sex Roles, 19*, 205–217.

Elder, G. H. (1974). *Children of the great depression*. Chicago: University of Chicago Press.

Elder, G. H., & Clipp, E. (1988). War expereinces and social ties: Influences across 40 years in men's lives. In M. W. Riley (Ed.), *Social structures and human lives* (pp. 306–327). Newbury Park, CA: Sage.

Falbo, T. (1991). The impact of grandparents on children's outcomes in China. *Marriage and Family Review, 14*, 369–376.

Fischer, L. R. (1983). The transition to grandmotherhood. *International Journal of Aging and Human Development, 16*, 67–78.

Fuller-Thomson, E., Minkler, M., & Driver, D. (1997). A profile of grandparents raising grandchildren in the United States. *The Gerontologist, 37,* 406–411.

George, L. K., & Gold, D. T. (1991). Life course perspectives on intergenerational and generational connections. *Marriage and Family Review, 16,* 67–87.

Giarrusso, R., Silverstein, M., & Bengtson, V. L. (1996). Family complexity and the grandparent role. *Generations, 20,* 17–23.

Gladstone, J. W. (1988). Perceived changes in grandmother–grandchild relations following a child's separation or divorce. *The Gerontologist, 28,* 66–72.

Gladstone, J. W. (1989). Grandmother–grandchild contact: The mediating influence of the middle generation following marriage breakdown and remarriage. *Canadian Journal on Aging, 8,* 355–365.

Hagestad, G. O. (1985). Continuity and connectedness. In V. L. Bengtson & J. F. Robertson (Eds.), *Grandparenthood* (pp. 31–48). Beverly Hills, CA: Sage.

Hagestad, G. O. (1986). Dimensions of time and the family. *American Behavioral Scientist, 29,* 679–694.

Hagestad, G. O. (1988). Demographic change and the life course: Some emerging trends in the family realm. *Family Relations, 37,* 405–410.

Hagestad, G. O. (1990). Social perspectives on the life course. In R. H. Binstock & L. K. George (Eds.), *Handbook of aging and the social sciences,* 3rd edit. (pp. 151–168). San Diego: Academic Press.

Hagestad, G. O., & Neugarten, B. L. (1985). Age and the life course. In R. H. Binstock & E. Shanas (Eds.), *Handbook of aging and the social sciences,* 2nd edit. (pp. 35–61). New York: Van Rostand Reinhold.

Hareven, T. K. (1994). Aging and generational relations: A historical and life course perspective. *Annual Review in Sociology, 20,* 437–461.

Hartfield, B. W. (1996). Legal recognition of the value of intergenerational nurturance: Grandparent visitation statutes in the nineties. *Generations, 20,* 53–56.

Hartshorne, T. S., & Manaster, G. J. (1982). The relationship with grandparents: Contact, importance, role conceptions. *International Journal of Aging and Human Development, 15,* 233–245.

Hirshorn, B. A. (1998). Grandparents as caregivers. In M. Szinovacz (Ed.), *Handbook on grandparenthood* (pp. 200–216). Westport, CT: Greenwood Press.

Hodgson, L. G. (1992). Adult grandchildren and their grandparents: The enduring bond. *International Journal of Aging and Human Development, 34,* 209–225.

Hoffman, E. (1980). Young adults' relations with their grandparents: An exploratory study. *International Journal of Aging and Human Development, 10,* 299–309.

Hunter, A. G., & Taylor, R. J. (1998). Grandparenthood in African-American families. In M. Szinovacz (Ed.), *Handbook on grandparenthood* (pp. 70–86). Westport, CT: Greenwood Press.

Hurme, H. (1997). Cross-cultural differences in adolescents' perceptions of their grandparents. *International Journal of Aging and Human Development, 44,* 221–253.

Hyde, V., & Gibbs, I. (1993). A very special relationship: Granddaughters' perceptions of grandmothers. *Ageing and Society, 13,* 83–96.

Ikels, C. (1998). Grandparenthood in cross-cultural perspective. In M. Szinovacz (Ed.), *Handbook on grandparenthood* (pp. 40–52). Westport, CT: Greenwood Press.

Jendrek, M. P. (1994). Grandparents who parent their grandchildren: Circumstances and decisions. *The Gerontologist, 34,* 206–216.

Johnson, C. L. (1985). Grandparenting options in divorcing families: An anthropological perspective. In V. L. Bengtson & J. F. Robertson (Eds.), *Grandparenthood* (pp. 81–96). Beverly Hills, CA: Sage.

Johnson, C. L. (1988). Active and latent funcitons of grandparenting during the divorce process. *The Gerontologist, 28,* 185–191.

Kahana, B., & Kahana, E. (1970). Grandparenthood from the perspective of the developing grandchild. *Developmental Psychology, 3,* 98–105.

Kahana, B., & Kahana, E. (1971). Theoretical and research perspectives on grandparenthood. *Aging and Human Development, 2,* 261–268.

Kamo, Y. (1998). Asian grandparents. In M. Szinovacz (Ed.), *Handbook on grandparenthood* (pp. 97–112). Westport, CT: Greenwood Press.

Kamo, Y., & Zhou, M. (1994). Living arrangements of elderly Chinese and Japanese in the United States. *Journal of Marriage and the Family, 56,* 544–558.

Karp, N. (1996). Legal problems of grandparents and other kinship caregivers. *Generations, 20,* 57–60.

Kennedy, G. E. (1992). Shared activities of grandparents and grandchildren. *Psychological Reports, 70,* 211–227.

King, V., & Elder, G. H. (1995). American children view their grandparents: Linked lives across three rural generations. *Journal of Marriage and the Family, 57*, 165–178.

King, V., Russell, S. T., & Elder, G. H. (1998). Grandparenting in family systems: An ecological perspective. In M. Szinovacz (Ed.), *Handbook on grandparenthood* (pp. 53–69). Westport, CT: Greenwood Press.

Kivett, V. R. (1985). Grandfathers and grandchildren: Patterns of association, helping and psychological closeness. *Family Relations, 34*, 565–571.

Kivett, V. R. (1991). The grandparent–grandchild connection. *Marriage and Family Review, 16*, 267–290.

Kivnick, H. Q. (1985). Grandparenthood and mental health: Meaning, behavior and satisfaction. In V. L. Bengtson & J. F. Robertson (Eds.), *Grandparenthood* (pp. 151–158). Beverly Hills, CA: Sage.

Kornhaber, A. (1996). *Contemporary Grandparenting*. Thousand Oaks, CA: Sage.

Kornhaber, A., & Woodward, K. L. (1985). *Grandparents/grandchildren: The vital connection*. Somerset, NJ: Transaction.

Langer, N. (1990). Grandparents and adult grandchildren: What do they do for one another? *International Journal of Aging and Human Development, 31*, 101–110.

Markides, K. S., Boldt, J. S., & Ray, L. A. (1986). Sources of helping and intergenerational solidarity: A three-generations study of Mexican Americans. *Journal of Gerontology, 41*, 506–511.

Matthews, S. H., & Sprey, J. (1984). The impact of divorce on grandparenthood: An exploratory study. *The Gerontologist, 24*, 41–47.

Matthews, S. H., & Sprey, J. (1985). Adolescents' relationships with grandparents: An empirical contribution to conceptual clarification. *Journal of Gerontology, 40*, 621–626.

McGreal, C. E. (1983). Granparenthood as a symbol: Expectations for the grandparent role. Paper presented at the 36th meeting of the Gerontological Society, San Francisco, CA.

Minkler, M., & Roe, K. M. (1996). Grandparents as surrogate parents. *Generations, 20*, 34–38.

Mullen, F. (1996). Public benefits: Grandparents, grandchildren and welfare reform. *Generations, 20*, 61–64.

Neugarten, B. L., & Neugarten, D. A. (1987). The changing meanings of age. *Psychology Today, 21*, 29–33.

Neugarten, B. L., & Weinstein, K. K. (1964). The changing American grandparent. *Journal of Marriage and the Family, 26*, 199–204.

Ponzetti, J. J., & Folkrod, A. W. (1989). Grandchildren's perceptions of their relationships with their grandparents. *Child Study Journal, 19*, 41–50.

Presser, H. B. (1989). Some economic complexities of child care provided by grandmothers. *Journal of Marriage and the Family, 51*, 581–591.

Pruchno, R. (1999). Raising grandchildren: The experiences of black and white grandmothers. *Gerontoligist, 39*, 209–221.

Ramirez-Barranti, C. C. (1985). The grandparent–grandchild relationship: Family resource in an era of voluntary bonds. *Family Relations, 34*, 343–352.

Robertson, J. F. (1976). Significance of grandparents: Perceptions of young adult grandchildren. *The Gerontologist, 16*, 137–140.

Robertson, J. F. (1977). Grandmotherhood: A study of role conceptions. *Journal of Marriage and the Family, 39*, 165–174.

Robertson, J. F. (1995). Grandparenting in an era of rapid change. In R. Blieszner and V. H. Bedford (Eds.). *Handbook of aging and the family*, pp. 243–260. Westport, CT: Greenwood Press.

Scherman, A., Goodrich, C., Kelly, C., Russell, T., & Javidi, A. (1988). Grandparents as a support system for children. *Elementary School Guidance and Counseling, 23*, 16–22.

Schultz, N. W. (1980). A cognitive-developmental study of the grandparent–grandchild bond. *Child Study Journal, 10*, 7–26.

Silverstein, M., & Long, J. D. (1998). Trajectories of grandparents' perceived solidarity with adult grandchildren: A growth curve analysis over 23 years. *Journal of Marriage and the Family, 60*, 912–923.

Smith, M. S. (1991). An evolutionary perspective on grandparent–grandchild relationships. In P. K. Smith (Ed.), *The psychology of grandparenthood: An international perspective* (pp. 157–176). New York: Routledge.

Smith-Barusch, A., & Steen, P. (1996). Keepers of community in a changing world. *Generations, 20*, 49–52.

Solomon, J. C., & Marx, J. (1995). "To Grandmother's house we go": Health and school adjustments of children raised solely by grandparents. *The Gerontologist, 35*, 386–394.

Somary, K., & Stricker, G. (1998). Becoming a grandparent: A longitudinal study of expectations and early experiences as a function of sex and lineage. *The Gerontologist, 38*, 53–61.

Strom, R. D., Buki, L. P., & Strom, S. K. (1997). Intergenerational perceptions of English-speaking and Spanish-speaking Mexican American grandparents. *International Journal of Aging and Human Development, 45*, 1–21.

Strom, R., Strom, S., Collingsworth, P., Sato, S., Makino, K., Sasaki, Y., Sasaki, H., & Nishio, N. (1995). Grandparents in Japan: A three-generational study. *International Journal of Aging and Human Development, 40*, 209–226.

Strom, R., Strom, S., Shen, Y., Li, S., & Sun, H. (1996). Grandparents in Taiwan: A three-generational study. *International Journal of Aging and Human Development, 42*, 1–19.

Szinovacz, M. E. (1998). Grandparents today: A demographic profile. *The Gerontologist, 38*, 37–52.

Thomas, J. L. (1986). Age and sex differences in perceptions of grandparenting. *Journal of Gerontology, 41*, 417–423.

Thomas, J. L. (1989). Gender and perceptions of grandparenthood. *International Journal of Aging and Human Development, 29*, 269–282.

Thomas, J. L. (1990). The grandparent role: A double bind. *International Journal of Aging and Human Development, 31*, 169–177.

Thompson, L., & Walker, A. J. (1987). Mothers as mediators of intimacy between grandmothers and their young adult granddaughters. *Family Relations, 36*, 72–77.

Tomlin, A. M. (1998). Grandparents' influences on grandchildren. In M. E. Szinovacz (Ed.), *Handbook on grandparenting* (pp. 159–170). Westport, CT: Greenwood Press.

Troll, L. E. (1983). Grandparents: The family watchdogs. In T. H. Brubaker (Ed.). *Family relationships in later life*, pp. 63–74. Beverly Hills, CA: Sage.

Troll, L. E. (1985). The contingencies of grandparenting. In V. L. Bengtson & J. F. Robertson (Eds.), *Grandparenthood* (pp. 135–150). Beverly Hills, CA: Sage.

Tyszkowa, M. (1991). The role of grandparents in the development of grandchildren as perceived by adolescents and young adults in Poland. In P. K. Smith (Ed.), *The psychology of grandparenthood: An international perspective* (pp. 50–67). New York: Routledge.

Uhlenberg, P. (1980). Death and the family. *Journal of Family History, 5*, 313–320.

Uhlenberg, P. (1996). Mortality decline in the twentieth century and supply of kin over the life course. *The Gerontologist, 36*, 681–685.

Van Ranst, N., Verschueren, K., & Marcoen, A. (1995). The meaning of grandparents as viewed by adolescent grandchildren: An empirical study in Belgium. *International Journal of Aging and Human Development, 41*, 311–324.

Wechsler, H. J. (1985). Judaic perspectives on grandparenthood. In V. L. Bengtson & J. F. Robertson (Eds.), *Grandparenthood* (pp. 185–194). Beverly Hills, CA: Sage.

Whitbeck, L. B., Hoyt, D. R., & Huck, S. M. (1993). Family relationship history, contemporary parent–grandparent relationship quality, and the grandparent–grandchild relationship. *Journal of Marriage and the Family, 55*, 1025–1035.

Williams, N., & Torrez, D. J. (1998). Grandparenthood among Hispanics. In M. Szinovacz (Ed.), *Handbook on grandparenthood* (pp. 87–86). Westport, CT: Greenwood Press.

Wood, V., & Robertson, J. F. (1976). The significance of grandparenthood. In J. Gubrium (Ed.), *Time, roles and self in old age* (pp. 278–304). New York: Human Sciences Press.

CHAPTER 25

"Roots that Clutch"

What Adoption and Foster Care Can Tell Us About Adult Development

JACK DEMICK

In the United States, there are approximately 1 million children who were adopted[1] currently living in adoptive families, 4% of families with a child who was adopted, and 5 million individuals of all ages who were adopted. About 150,000 children of all races and nationalities are adopted each year.[2] At any one time, about 500,000 people are seeking to adopt with 3.3 adoption seekers for every actual adoption. As of 1998, the number of children in foster care needing adoptive families also reached approximately 150,000 in number. All of these numbers have increased and are expected to increase even more in the future.

This chapter is dedicated to my daughter, Katie Kellerman-Demick, who has taught me more about both child and adult development than she may ever know. An abbreviated version of this chapter was presented as the plenary address at the 14th Annual Adult Development Symposium of the Society for Research in Adult Development (SRAD), which took place at Salem State College on the eve of Katie's Bat Mitzvah on June 18, 1999. Thank you and I love you, Katie.

[1] Mental health professionals have discussed the need for "nonhandicapping" language in the field of adoption. Specifically, they have recommended that the words "individuals who were adopted" be substituted for "adoptees" and/or "adopted individuals." Toward maintaining the integrity of individuals as human beings, this implies that adoption is an event, not a defining characteristic or condition of persons. The use of such language parallels the guidelines to reduce bias in language recommended in the fifth edition of the *Publication Manual of the American Psychological Association* (2001).

[2] Within the adoption literature, a major distinction has been made between infant adoptions (most usually domestic adoptions within the United States) and older child adoptions (often following foster care experiences). Research on both of these populations (perhaps erroneously) has been negative, although particularly moreso for older child adoptions. The situation has become even more complex in light of a recent trend toward foreign infant adoption.

JACK DEMICK • Center for Adoption Research, University of Massachusetts Medical School, Worcester, Massachusetts, 01605.

Handbook of Adult Development, edited by J. Demick and C. Andreoletti. Plenum Press, New York, 2002.

Both federal and state legislators have recently launched adoption initiatives to increase, in a timely fashion, the number of foster children who find permanent homes. The Benchmark Adoption Survey (Donaldson Adoption Institute, 1997)—the first in-depth look at American public attitudes toward adoption based on a probability sample of 1554 adults—indicated that: (1) 6 of 10 Americans have personal experience with adoption (i.e., either a family member, close friend, or they themselves were adopted, adopted a child, or placed a child for adoption); and (2) one third have considered adopting a child at least somewhat seriously.

In light of the growing numbers of children who were adopted and/or in foster care as well as the increasing public awareness of these processes, it appears incumbent on behavioral scientists to assess the effects of these institutions on the well being of children and their families. Indeed, within the field of child development, researchers have only recently complemented the longstanding interests of behavior geneticists (i.e., comparisons of persons who were adopted with those who were not to determine the relative contributions of heredity vs. the environment on various dimensions) with the assessment of the psychological functioning per se of individuals who were adopted, their adoptive parents, and birthparents (e.g., Brodzinsky, Schechter, Braff, & Singer, 1984; Demick & Wapner, 1988b; McRoy, Grotevant, & White, 1988). This has most probably been related to changing adoption practices since the late 1970s when open adoption (communication between biological and adoptive parents) replaced traditional, closed adoption (no communication between biological and adoptive parents) as the norm.

Along with this increase in open adoption practices, the cloak of mystery that had once shrouded adoption began to be removed. Not only did more and more adoptive couples begin to meet and/or develop plans for ongoing contact with birthparents, but also more and more individuals who were adopted under the circumstances of traditional, closed adoption began to search for and find their birthparents. Soon the media were permeated with stories of open adoption, search and reunion, both disrupted private and foster care adoptions, and the like. It is argued here that as a heretofore-unexplored byproduct of these developments, psychologists now have another lens through which to examine adult development and related phenomena. That is, a heightened focus on adoption and foster care has much to contribute to our understanding of individual, familial, and societal processes involved in adult development.

As an introductory example, Melina (1989) has identified, on the basis of extensive clinical data reported by Kaplan and Silverstein,[3] the following seven issues as the core psychological issues that individuals who were adopted, birthparents, and adoptive parents must confront and deal with over their lifetimes. These issues have included: loss, rejection, guilt and shame, grief, identity, intimacy, and control. Subsequent authors (e.g., Partridge, 1991; Schecter & Bertocci, 1990) have modified and/or added to this list with the most notable addition being body image (including feelings of invisibility, etc.). It is argued here that

---

[3] It should be noted that these core issues have been based on clinical (anecdotal) rather than on empirical evidence. Thus, staff members at our center are currently assessing whether any empirical data have been accumulated to support the existence of such issues in nonclinical samples of individuals who were adopted, their adoptive parents, and birthparents.

these issues are not limited to members of the adoptive triad, but rather that they constitute a core set of issues that all of us must negotiate over the course of human development. Thus, studying individuals who were adopted, birthparents, and adoptive parents—where these core issues are highlighted and/or exacerbated— may be a useful lens through which to glean some insights about human development in general and about adult development in particular.

This idea, which is ultimately of considerable theoretical importance, is demonstrated by an analysis of previously published studies on adoption and foster care and studies that are currently underway in our laboratory at the Center for Adoption Research at the University of Massachusetts.[4] Prior to this analysis, however, a brief description of the theoretical orientation that frames our research is presented.

THEORETICAL PERSPECTIVE

For some time now, Seymour Wapner of Clark University and I (e.g., Wapner & Demick, 1990, 1998, 1999, 2000, 2002) have been working on a holistic, developmental, systems approach to person-in-environment functioning across the lifespan, which is an outgrowth and elaboration of Werner's (1940/1957) organismic–developmental theory. As is quite well known, Werner's theory was organismic insofar as he advocated that we should, as far as possible, study psychological processes as they occur within the active, striving, feeling, whole organism. We should not, for example, examine perceptual processes as if they existed in isolation, but instead should study them as they emerge from the more primitive matrices of action and feeling in which they are embedded. Werner's theory was comparative–developmental in that his view of development (based on the orthogenetic principle described below) transcended the boundaries within which the concept of development is ordinarily applied. For most psychologists, development is restricted to child growth, to ontogenesis. In contrast, Werner saw development more broadly as a mode of analysis of diverse aspects of person-in-environment functioning, encompassing not only ontogenesis, but also microgenesis (e.g., development of an idea or percept), pathogenesis (e.g., development of neuro- and psycho-pathology), phylogenesis (development of a species), and ethnogenesis (development of a culture).

Following from this, our elaboration is: *holistic* insofar as it assumes that all part-processes—biological/physical, psychological (cognitive, affective, valuative, action)—are interrelated; *developmental* insofar as it assumes, in keeping with the orthogenetic principle (Werner, 1940/1957), that development proceeds from a relative lack of differentiation toward the goal of differentiation and hierarchic

[4] The Center for Adoption Research at the University of Massachusetts, currently housed at the University of Massachusetts Medical School in Worcester, MA, is the first university-based program in the United States whose sole purpose is to study the real-world challenges that adoption and foster care present to our nation. More than 2000 individuals (from organizations such as the Open Door Society) have become "friends" of the center. In a recent mailing aimed at recruiting research participants, more than 200 individuals responded in 1 week. Whether volunteers who have have some connection to adoption (individuals who were adopted, adoptive parents, birthparents) are representative of these populations at large is an open empirical issue. At the least, data from these volunteers will provide a source of information about adoption under optimal conditions.

integration of organismic functioning; and *systems-oriented* insofar as the unit of analysis is the person-in-environment in which the *physical/biological* (e.g., health), *psychological* (e.g., self-esteem), and *sociocultural* (e.g., roles) levels of organization of the *person* are operative and interrelated with the *physical* (e.g., natural and built environment), *interpersonal* (e.g., friends, relatives), and *sociocultural* (e.g., regulations and rules of society) levels of organization of the *environment*. Further, this approach, which adopts aspects of both organismic and contextual/transactional worldviews (see Altman & Rogoff, 1987; Pepper, 1942, 1967), favors structural/organizational (drawing on Werner's [1940/1957] major theme of self–world differentiation or distancing) and dynamic (means–ends) analyses of person-in-environment functioning.

On the most general level, what is the heuristic value of these theoretical assumptions—in terms of both guiding research and interpreting empirical data—for the areas of adoption and foster care? Some time ago, Wapner and I (Demick & Wapner, 1988a) delineated the implications of this approach for the study of individuals-who-were-adopted-in-environments. A modified version can be found in Table 25.1, which lists the areas worthy of exploration in a holistic, developmental, systems analysis of problems relevant to adoption and foster care.

Such an approach is particularly useful in these problem areas, as there are clearly multiple parties involved (e.g., several sets of parents, agency workers, legal personnel), some of whom may not be present in actuality but in spirit or the collective unconscious (e.g., see Weider, 1977, on the family romance fantasies of children who were adopted). However, such an approach also highlights the usefulness of a transactionally oriented, systems orientation for the analysis of all human functioning across the lifespan and even the adult lifespan (e.g., see Elkind [1996] on the phenomena of the imaginary audience and the foundling fantasy in adolescence/young adulthood and Andersen & Berk [1998] on the transference of everyday life). Further, both structural (e.g., how does the individual who was adopted organize the various parts of his or her identity into a coherent whole?) and dynamic (e.g., how do preadoptive couples plan for, or not plan for, a pending adoption? How do individuals who were adopted plan for search and reunion with birthparents? [cf. Bacharach, 1986]), analyses have applicability not only to adoption and foster care, but also to all problem areas including adult development (e.g., How do individuals at different stages of the adult life cycle organize their self-concepts? Are there systematic changes in individuals' planning abilities over the course of adulthood?).

Our approach also has several strong implications for methodology. First, in light of the complexity inherent in our conceptualization (in line with the complex character of everyday life), we believe that holistic, ecologically oriented research is a necessary complement to more traditional laboratory work and that it should be conducted through reducing the number of focal individuals studied (persons who were adopted) rather than the number and kind of interrelationships among aspects of the person, of the environment, and of the systems to which they belong. Second, as we have reported elsewhere (e.g., Wapner & Demick, 1998), our approach has typically been concerned with describing the relations both among and within the parts (person, environment) that make up the integrated whole (person-in-environment system) as well as with specifying the conditions that make for changes in the organization of these relationships.

Table 25.1. Design of Ideal Study of Individuals-Who-Were-Adopted-in-Their-Environments

	Aspects of Focal Persons			Aspects of Environment		
	Physical/biological	Interpersonal	Sociocultural	Physical	Interpersonal	Sociocultural
Child who was adopted	Genotypic status Phenotypic status Age at placement Sex	Body experience[a] Self experience	Role as person who was adopted[a] Ethnic heritage Religious background	Objects[a] Location[a]	Family Peers[a] Neighbors[a] Teachers[a] Employers[a]	Local Community Society Legal Educational[a] Mores[a]
Adoptive parents	Genotypic status Phenotypic status Age at placement Sex	Body experience[a] Self experience	Role as adoptive parents[a] Ethnic heritage Religious background	Objects[a] Location[a]	Family Peers[a] Neighbors[a] Teachers[a] Employers[a]	Local Community Society Legal Educational[a] Mores[a]
Birthparents	Genotypic status Phenotypic status Age at placement Sex	Body experience[a] Self experience	Role as biological parent[a] Ethnic heritage Religious background	Objects[a] Location[a]	Family Peers[a] Neighbors[a] Teachers[a] Employers[a]	Local Community Society Legal Educational[a] Mores[a]

[a]Not examined in previous research.

Thus, our approach is wedded to the complementarity of explication (description) and causal explanation (conditions under which cause–effect relationships occur) rather than being restricted to one or the other. This has led us to flexible drawing from both quantitative and qualitative methodologies depending on the level of integration and nature of the problem under scrutiny (see below).

Against this backdrop, I now review studies from the areas of adoption and foster care, which have implications for the study of individual, family, and social processes within the arena of adult development. Some of these studies have been completed, while many are still on the drawing boards. Some have utilized aspects of the approach described here, while others have been generated from additional theoretical perspectives; all, however, have implications for the field of adult development.

INDIVIDUAL PROCESSES

Developmentalists have, for some time, been interested in the study of individual psychological part-processes. From our point of view, psychological part-processes inherent within the individual consist of cognitive, affective, valuative, and behavioral (action) processes. While we assume that these processes operate contemporaneously and in an interrelated fashion within the normally functioning human adult, we often separate them out for purposes of analysis. Thus, what follows is a discussion of seven individually oriented psychological processes—the first four with particular relevance for individuals who were adopted and the last three for adoptive parents—which have implications for the study of adult development.

Individuals Who Were Adopted

Attachment.   In line with the title of this chapter, attachment is probably the first process that comes to mind in thinking about the integration of adoption/foster care and adult development. This is because developmental psychologists have long documented that the infant's early attachment experience and the attachment style that results exert a continuing influence on subsequent relationships both in childhood (e.g., Ainsworth, 1989) and in adulthood (e.g., Hazan & Shaver, 1987). As Bowlby himself (1979) has emphasized: "Attachment behavior characterizes human beings from the cradle to the grave" (p. 129). Since individuals who were adopted have frequently had multiple caretakers prior to their adoption, one might speculate that they, relative to nonadopted individuals, might be prone to attachment difficulties and/or poor interpersonal relationships over the course of development.

The vast majority of research on the adjustment of individuals who were adopted has been based on studies of children, especially infants and school-aged children (e.g., Carlson, 1998; Feigelman, 1997; Howe, 1995). Fewer studies have examined adjustment in adolescents who were adopted and, surprisingly, almost no studies have assessed outcomes of adoption on American adults (with the exception of the problem of search among adults (e.g., Andersen, 1988, 1989).

Thus, as this is an open empirical issue, we are currently in the process of conducting research aimed at assessing attachment status among adolescents and adults who were adopted as infants and those who were adopted as older children. We do not expect to find differences between individuals who were and were not adopted as infants. This has been based on Singer, Brodzinsky, Ramsay, Steir, and Waters' (1985) finding of similar proportions of attachment patterns (secure, avoidant, ambivalent) in infants who were and were not adopted at birth. We do expect, however, to find differences between those who were adopted as older children and their nonadopted counterparts, partly related to a range of confounding variables often involved in older child adoptions such as special needs and/or abuse (see Rosenfeld et al.'s [1997] comprehensive update on foster care). Future research will also need to consider the variable of foreign infant adoption—from varying countries (cf. Apparo, 1997)—as preliminary data (e.g., Chisholm, 1998; Chisholm, Carter, Ames, & Morison, 1995) have suggested, for example, that children adopted after at least 8 months from a Romanian orphanage exhibit less security of attachment and more indiscriminately friendly behaviors than Romanian infants adopted before the age of 4 months and a non-adopted comparison group. Interpretations of such findings are not necessarily negative, however (see below).

Control.   Prior to the late 1980s, the vast majority of research on the experience and adaptation of individuals who were adopted was replete with references to the negative characteristics of those who were adopted, including dependency, loneliness, fearfulness, poor self-esteem, etc. (see Demick & Wapner, 1988b, for a review). However, Marquis and Detweiler (1985) reported that, relative to their nonadopted peers, adolescents who were adopted have a more positive outlook about themselves (e.g., possessing a more internal locus of control and more confidence in their own judgments) and about others (e.g., rating their parenting, for the most part, as more positive and making more favorable attributions about story characters). On the basis of these findings, they concluded that "the old stereotypic view that the adopted are 'at risk' is unfounded" (p. 1064).

This study was intriguing insofar as it represented a single attempt to document strengths rather than weaknesses within a sample of individuals who were adopted. However, in a final footnote, the authors attempted to fit their findings into the existent literature by speculating that the documented strengths might actually be a function of "... an almost Pollyannaish perspective that could inhibit a realistic view of the world in which there are sometimes negative consequences despite positive input. In a state of unpreparedness, too many doses of 'harsh reality' could cause later adjustment crises" (p. 1064). How quickly the strengths can change into weaknesses!

Nonetheless, several years later, Brodzinsky (1987) completed the job of refuting Marquis and Detweiler's positive findings. Specifically, he faulted their conclusions on the basis of: (1) a selective and biased review of the adoption literature (i.e., they failed to cite numerous methodologically sound studies); (2) subject sampling problems (i.e., they used a very small percentage of upper middle class volunteers); (3) data collection problems (i.e., participants who were adopted may have been influenced either covertly or overtly by their parents since they completed the questionnaires at home, while nonadopted participants completed them at school); (4) data analysis (i.e., multiple *t*-tests rather than multivariate analyses were employed); and (5) faulty generalization of findings (i.e., any conclusions were relevant only to infants who were adopted and not to children adopted post-infancy).

We (Aronson, Ronayne, Hayaki, & Demick, 1994) aimed to clarify some of the confusion by attempting: (1) to replicate Marquis and Detweiler's findings in light of Brodzinsky's criticisms (i.e., by sampling a range of college students without specifically recruiting those who were adopted); and (2) to rule out the possibility that those who were adopted may have a "Pollyannaish perspective" by including an additional measure of psychological defenses (Ihilevich & Gleser, 1993) that assessed the defense mechanism of repression. While our participants were significantly older than those examined by Marquis and Detweiler (young adults vs. adolescents), we found the following. Relative to their nonadopted counterparts, those young adults who were adopted: exhibited a more internal locus of control, rated protagonists on an attributional task in more positive terms, and exhibited more confidence in their judgments (replicating Marquis and Detweiler's earlier findings); the two groups did not differ, however, with respect to perceived parenting and/or any defense mechanism, including turning against object, projection, principalization, turning against self, or reversal (denial).

Taken with other more recent findings in the field of adoption (e.g., Benson, Sharma, & Roehlkepartain, 1994; Sharma, McGue, & Benson, 1996a, 1996b, 1998, have reported that, relative to their nonadopted counterparts, those adolescents who were adopted exhibited more altruistic behavior) as well as other more general findings from developmental psychology on resilient children (e.g., Anthony & Koupernik, 1974; Rosenberg, 1987; Rutter, 1984; Werner, 1993), this work has suggested that the poor adjustment of those adopted should not be overstated (as is usually the case in the clinical adoption literature); and that, for a variety of reasons (e.g., experience in problem solving, reaction to marginality), adoptive status may actually lead to psychological strength later in adult life. Such ideas are also consistent with the theme of positive adult development.

Additional studies on control and adoption/foster care, on the drawing boards, deserve brief mention. First, what is the relationship between adoptive status and different dimensions of control? Second, what if foster care children and foster parents could exert some degree of control over their placement? Ellen Langer and I are currently designing several experiments in which foster children and parents, in the context of a mixer/matching party, have some input into who is adopted by whom. The general expectation is that the placement will be more successful (lead to adoption) if the parties (particularly the children) feel some sense of control over the process (cf. Langer, 1997). Such studies have the potential to change how we currently conduct foster care, but also have clear implications for both child and adult development.

Identity and Life Stories

The study of identity—Erikson's (1968) term for a sense of who one is, what one believes in, and where one is headed in life—has also occupied a prominent position in adoption research. Specifically, investigators (e.g., Hoopes, 1990;

Kelly, Towner-Thyrum, Rigby, & Martin, 1995) have attempted to assess whether there are differences in identity formation and adjustment between those adopted and those not. While this general research strategy might be conceptualized as illustrating Werner's (1940/1957) law of *pars pro toto* (a part determines the whole), such studies fail to consider the heuristic potential of organizational analyses in the study of identity formation (i.e., the ways in which individuals organize the different aspects of their identity including sexual, political, occupational, religious, etc. over time).

In line with this, Grotevant (1997) has stated that:

> The important unanswered question is how the fact of adoption is integrated into one's overall identity; in what way is a coherent whole formed? Following Erikson's argument that an important aspect of identity is continuity across past, present, and future, coming to terms with one's identity as an adopted person should play an important role in the overall identity development process. (We need to)...understand how the different domains of identity that are assigned (such as one's adoptive status) are related to those that are more freely chosen. (p. 10)

Thus, Jennifer Kaplan and I (1999) are currently conducting a study on identity formation in three groups of black adolescent males: a group that was adopted by same-race adoptive parents, a group adopted by transracial adoptive parents, and a control group of nonadopted boys. While there has been a voluminous literature, with mixed results, on the effects of transracial adoption on children's development over time (e.g., Alexander & Curtis, 1996; Curtis, 1996; Rushton & Minnis, 1997; Vroegh, 1997), our focus has been on structural analyses of the various domains of identity (cf. Silverstein & Demick, 1994)—a problem of great relevance for adult development. Preliminary findings have suggested that adoptive status may not be the most difficult aspect of identity to integrate at least for these participants.

Further, we are working on a taxonomy for classifying the relations among levels of organization or, here, among different aspects of experience. Drawing on Wapner's (1969) analysis of the relations among cognitive processes, different aspects of experience (identity) may be *supportive* (e.g., being black and adopted are consistent), *antagonistic* (e.g., being black is incompatible with being adopted), or *substitutive/vicarious* (e.g., being black substitutes for being the most difficult aspect of identity). Again, this is an example of developmental conceptualization that has bearing on aspects of adult development more generally.

In a different yet related vein, Grotevant (1997) has suggested that a narrative approach (e.g., Polkinghorne, 1991) and, specifically, the coherence of one's life story:

> ... addresses the issue of integration and provides a window for investigating how the different domains of identity are synthesized with one another. Thus, a narrative approach to identity development in adopted adolescents permits examination of how an adolescent makes sense of being an adopted person, and the interplay between adoptive status and other aspects of identity. It can help us address questions such as, "What does it mean to be adopted? Who am I as an adopted person, and how does that relate to other aspects of my personal identity?" (p. 10)

We concur and suggest that, in keeping with our theoretical perspective, there is a place for qualitative analysis (see Eheart & Power, 1995; Kaye & Warren, 1988) as a necessary complement to quantitative inquiry (see Wapner & Demick, 1998, 1999, 2000, 2002).

Body Experience.   Anecdotal evidence in the adoption field has implicated negative body experience as a result of being adopted. Specifically, authors (e.g., Schechter & Bertocci, 1990) have speculated that, relative to nonadopted individuals, individuals who were adopted may experience problems associated with body boundaries (e.g., feelings of invisibility), body image (since the body is the only link to birthparents, the meanings of body characteristics are intensified), and sexual identity (reproductively flawed mother, sexually promiscuous birthmother). However, there are no empirical data to support these notions.

Thus, drawing on a developmental conceptualization that has proven fruitful in past investigations (e.g., d'Arrigo & Demick, in press; Demick, Ishii, & Inoue, 1997), we are currently conducting a study that assesses body experience in individuals who were vs. were not adopted. Our approach is in contrast to most past research (e.g., Lerner, Iwawaki, Chihara, & Sorrell, 1980) that has conceptualized body experience as consisting solely of the dimension of body image.[5] Instead, we assume—paralleling Werner's (1940/1957) levels of cognitive functioning—that aspects of body experience develop first within the *sensorimotor level* (body action), then within the *perceptual level* (body perception), and finally within the *conceptual level* (body and self concept). It is further assumed that there is a contingency relationship among these levels, that is, conceptual functioning depends on perceptual functioning that in turn depends on sensorimotor functioning.

Such conceptualization has the potential to discriminate among impaired, or alternatively perhaps even advanced, aspects of body experience in individuals who were adopted relative to those who were not. Such an approach has already been demonstrated to clarify aspects of body experience over the course of adult development (e.g., relative to those in late adolescence and their late 20s, those in their early 20s are more concerned about aspects of body and self experience; see d'Arrigo & Demick [in press] for a more complete discussion).

Adoptive Parents

Guilt and Shame.   An issue that has received some attention—although relevant to other members of the adoptive triad (e.g., birthparents, individuals who were adopted)—has been the feelings of guilt and shame that some infertile couples often experience prior to a following adoption. Thus, we are currently designing a program of research on depression, guilt and shame, infertility, and adoption in which we will be trying to assess whether becoming a parent through adoption increases or decreases one's positive and negative feelings as well as one's overall sense of wellbeing (cf. Abbey, Andrews, & Halman, 1994).

---

[5] For example, Lerner et al. (1980) assessed the isolated dimension of body concept (Body- and Self-Cathexis Scales) in Japanese adolescents and concluded that, relative to American youth, Japanese youth have less developed body- and self-concepts. In contrast, Demick et al. (1997)—employing a multidimensional approach—found that, relative to Americans, the Japanese had a more advanced sense of body and self experience (e.g., better body action, better body boundaries, less concern about physical appearance). Such research underscores the need for multidimensional developmental conceptualization.

Values

Perhaps the most unexplored area in developmental psychology concerns the ways in which one's value orientations (priorities) impact behavior and experience (cf. Wapner & Demick, 1998, 1999, 2000, 2002). Adoption is the perfect lens through which to study this and related problems. For example, the values held by prospective adoptive parents (e.g., see Hoksbergen, 1998) have the potential to impact their choice of adoption practice (open vs. closed). Further, the values held by society (e.g., see Cole & Donley, 1990; Miall, 1998; Soparker, Baker, Pinet, Sandberg, & Demick, 2002) concerning the nature of these processes (e.g., varying forms of adoption, foster care) represent an added dimension that potentially impacts prospective parents. This then becomes a paradigm for the study of values as well as for the examination of the relationships between the psychological and sociocultural levels of integration, a problem more generally for the field of adult development.

Parental Role

Being an adoptive parent has the potential to elucidate our ideas concerning the concept (e.g., functions) of parenting and of family, more generally. For example, we are currently conducting a study on the ways in which adoptive parents construe their roles, their families, the nature of development, and so forth. Related to this, Wapner and I are planning a comparable study on the perceptions of adoptive grandparenting. For example, do adoptive parents and grandparents—similar to biological parents and grandparents—perceive their parenting as a means to immortality (cf. Gair, 1998; Kivnick, 1985)? Do adoptive parenting and grandparenting open new avenues of adult development that may be missing in biological parenting (e.g., need to accommodate to child's interests, such as music and sports)? These and other issues have the potential to clarify many problems in the field of adult development.

Related to this is a study that we are currently conducting in collaboration with The Home for Little Wanderers' Boston Children's Institute. In another attempt to reconceptualize the practice of foster care, foster parents, potentially adoptable children, and adoptive grandparents are being integrated, in various housing complexes, as a family. Relative to those who participate in traditional foster care, does this newer arrangement lead to better outcomes for children (e.g., increased adoption) and adults (e.g., increased life satisfaction)? Such interventions have the potential to change our concept of family in a more encompassing manner, perhaps in turn leading to the notions that there are more similarities than differences in alternative parenting arrangements (e.g., see Golombok, Cook, Bish, & Murray, 1995) and that the experience of parenting in general—irrespective of how one's children are acquired—is simultaneously positive (e.g., rewarding) and negative (e.g., challenging; cf. Demick, 1999, 2002). Further, the new relationships inherent in the more recent forms of open adoption lead to the notion that adoptive relationships may serve as a paradigm for the range of new relationships that are becoming manifest in many families of the 90s (e.g., single parent families, stepfamilies, blended families).

FAMILY PROCESSES

An example of the way in which our approach shapes problem relevant to the fields of adoption and foster care—with implications for the study of family processes more generally within the context of adult development—may be seen in several of our earlier published works. Treating the problem of the effects of open vs. closed adoption (communication vs. no communication between adoptive and biological parents) on children and their families, we (Demick & Wapner, 1988b) began by theorizing about the organizational structure of the differing forms of adoption. Using four developmentally ordered self–world categories generated against the orthogenetic principle, we predicted that:

> ... adoptive families characterized by a total separation between the adopted child and his or her family of origin—as is usually the case in traditional closed adoption—may be conceptualized as dedifferentiated (all members of the family consciously or unconsciously deny that the child has been adopted), differentiated, and isolated (adoptive parents shelter the adoptee so that he or she will not learn about the biological parents from others and/or will not have to deal with the stigma of being adopted), or differentiated and in conflict (the adoptee may fantasize that the biologic parents would treat him or her differently and/or may threaten to leave the adoptive family to find the "real parents" when of age). In contrast, the adoptive family characterized by less absolute separation between the adoptee and his or her family of origin (the case in open adoption) may be conceptualized as differentiated and integrated (the adoptee may be able to integrate the various aspects of his or her dual identities, possibly mitigating potential problems with identity and self-esteem; in a similar manner, the adoptive parents may be able to integrate the different aspects of the adoptee's identity so as to avoid blaming "bad blood in the background" for any of their difficulties. (pp. 241–242)

Our major findings were as follow (Demick, 1993; cf. Demick, 1996, Demick & Wapner, 1992). First, there were no differences between adoptive and biological parents with respect to life satisfaction, control, and stress. Minimally, however, those couples who experienced an open adoption were less concerned about attachment to their child than those who experienced a traditional closed adoption. Second, there were clear selection effects: those with a *cultural and political family orientation* generally opted for open adoption, while those with a *traditional/moral/ religious family orientation* seemed to prefer closed adoption. Third, relative to other groups, adoptive mothers under open adoption experienced the lowest self-esteem (possibly related to heightened empathy from meeting the birthmother).

The implications are clear for both adoption practice and research on adult development. Theoretically, individual differences need to be considered and even integrated within developmental theory. That is, our conceptualization of developmentally based individual differences is relevant to all types of families. Further, additional work from our perspective (e.g., Kaden, 1958; Melito, 1985, 1988) has suggested that structural analyses—for example, those reflecting dominance, independence, mutuality—are useful in couple and family conceptualization (cf. Antonovsky & Sourani, 1998; Barth & Brooks, 1997). Ongoing work with Azar has also suggested that unrealistic parental expectations may lead to negative child behavior in adoptive—as well as most other—families. Methodologically, multiple dimensions and methods of assessment provide a more complete picture of the experience and action of individuals, couples, and families (cf. Finley, 1999; Howe, 1996, 1998).

SOCIETAL PROCESSES

Individuals who were adopted and their families have been reported to experience both internal and external feelings of stigmatization (e.g., March, 1995; Miall, 1987; Rosenberg & Horner, 1991). Thus, societal bias and discrimination toward members of the adoptive triad is clear. For example, the Benchmark Adoption Survey has also revealed that, even though many individuals have contact with adoption and perceive the process favorably, they also harbor some doubts about its efficacy, the (physical and mental) health of both children and parents, etc. Thus, Alex Genov and I have recently submitted a grant proposal aimed at assessing whether there are psychological correlates to marginality, a condition we believe is inherent in being adopted.

That is, Frable's (1993a, 1993b; Frable, Blackstone, & Scherbaum, 1990) work has, first, delineated marginality—or those with master status—into those whose status is: positively valued and visible (e.g., physically attractive); stigmatized and visible (e.g., overweight); positively valued and invisible (e.g., extremely wealthy); and stigmatized and invisible (e.g., bisexual); and, second, demonstrated that master status individuals are more mindful (i.e., active distinction drawing; cf. Langer, 1997) than normals in interpersonal interaction (laboratory). Based on this, we have hypothesized that individuals who were adopted in infancy and who are the same race as their adoptive parents constitute a stigmatized and invisible master status group. We also hypothesized that such individuals—in addition to being more behaviorally mindful—will also be more cognitively, affectively, and valuatively mindful. Finally, our third hypothesis is that self-perceptions of marginality are reducible through experimental manipulation and that we might be able to do this with groups of individuals who were adopted.

In a related vein, Jennifer Meyers and I (2002) have just completed a study based on Hollingshead and Redlich's (1958) now classic paradigm. Specifically, we have demonstrated that laypersons read more psychopathology into classic patient vignettes when the protagonist is additionally identified as adopted with more severe ratings related to the level of openness of the adoption. Such social processes clearly have the potential to impact the development of those who were adopted as well as to operate contemporaneously during ontogenesis and adult development (cf. Corrigan & Penn, 1999).

SUMMARY AND CONCLUSIONS

This review of our ongoing research on the problems of adoption and foster care, together with aspects of our approach, has suggested the need for those interested in adoption/foster care specifically and adult development more generally to consider their underlying theoretical assumptions; a transactional and systemic approach; developmentally based individual differences and cultural diversity; organizational analyses; dynamic analyses; multidimensional constructs and multiple methods; positive adult development; mutual value of using methodology and conceptualization from one subfield to advance the methodology and conceptualization of another and vice versa; and the roles of attachment, control, body experience, values, and so forth in adult development. In these ways, we are taking one small step toward reframing psychology both to see itself and to be seen by others as a unified science, that is, one concerned not only with the study of isolated aspects of human functioning (e.g., cognitive or affective functioning), but also with the study of problems that cut across various aspects of persons and various aspects of environments during ontogenesis whether the individual was adopted, in foster care, or not.

REFERENCES

Abbey, A., Andrews, F. M., & Halman, L. J. (1994). Infertility and parenthood: Does becoming a parent increase well-being? *Journal of Consulting and Clinical Psychology, 62*(2), 398–403.

Ainsworth, M. D. S. (1989). Attachments beyond infancy. *American Psychologist, 44,* 709–716.

Alexander, R., & Curtis, C. M. (1996). A review of empirical research involving transracial adoption of African American children. *Journal of Black Psychology, 22,* 223–235.

Altman, I., & Rogoff, B. (1987). World views in psychology: Trait, interactional, organismic and trans-actional perspectives. In D. Stokols & I. Altman (Eds.), *Handbook of environmental psychology* (pp. 7–40). New York: John Wiley & Sons.

Andersen, R. S. (1988). Why adoptees search: Motives and more. *Child Welfare, 67,* 15–19.

Andersen, R. S. (1989). The nature of adoptee search: Adventure, cure, or growth? *Child Welfare, 68,* 623–631.

Andersen, S. M., & Berk, M. S. (1998). The social–cognitive model of transference: Experiencing past relationships in the present. *Current Directions in Psychological Science, 7*(4), 109–115.

Anthony, E. J., & Koupernik, C. (Eds.) (1974). *The child in his family: Children at psychiatric risk, Vol. 3.* New York: John Wiley & Sons.

Antonovsky, A., & Sourani, T. (1998). Family sense of coherence and family adaptation. *Journal of Marriage and the Family, 50,* 79–92.

Apparo, H. (1997). International adoption of children: The Indian scene. *International Journal of Behavioral Development, 20,* 2–16.

Aronson, E., Ronayne, M., Hayaki, J., & Demick, J. (1994). *Adopted does not mean seeing the world through rose-colored glasses.* Paper presented at the annual meetings of the Eastern Psychological Association, Boston, MA.

Bacharach, C. A. (1986). Adopted plans, adopted children, and adoptive mothers. *Journal of Marriage and the Family, 48*(2), 243–253.

Barth, R. P., & Brooks, D. (1997). A longitudinal study of family structure and size and adoption outcomes. *Adoption Quarterly, 1*(1), 29–57.

Benson, P. L., Sharma, A. R., & Roehlkepartain, E. C. (1994). *Growing up adopted: A portrait of adolescents and their families.* Minneapolis, MN: Search Institute.

Bowlby, J. (1979). *The making and breaking of affectional bonds.* London: Tavistock.

Brodzinsky, D. M. (1987). Looking at adoption through rose-colored glasses: A critique of Marquis and Detweiler's "Does adopted mean different?" An attributional analysis. *Journal of Personality and Social Psychology, 52*(2), 394–398.

Brodzinsky, D. M., Schecter, D. E., Braff, A. M., & Singer, L. M. (1984). Psychological and academic adjustment in adopted children. *Journal of Consulting and Clinical Psychology, 52,* 582–590.

Carlson, E. A. (1998). A prospective longitudinal study of attachment disorganization/disorientation. *Child Development, 69*(4), 1107–1128.

Chisholm, K. (1998). A three-year follow-up of attachment and indiscriminate friendliness in children adopted from Romanian orphanages. *Child Development, 69*(4), 1092–1106.

Chisholm, K., Carter, M. C., Ames, E. W., & Morison, S. J. (1995). Attachment security and indiscriminately friendly behavior in children adopted from Romanian orphanages. *Development and Psychopathology, 7,* 283–294.

Cole, E. S., & Donley, K. S. (1990). History, values, and placement policy issues in adoption. In D. M. Brodzinsky & M. D. Schecter (Eds.), *The psychology of adoption* (pp. 273–294). New York: Oxford University Press.

Corrigan, P. W., & Penn, D. L. (1999). Lessons from social psychology on discreting psychiatric stigma. *American Psychologist, 54*(9), 765–776.

Curtis, C. M. (1996). The adoption of African American children by whites: A renewed conflict. *Families in Society: The Journal of Contemporary Human Services, 156*–165.

d'Arrigo, M., & Demick, J. (in press). Age and gender as determinants of body experience from late adolescence through young adulthood. *Journal of Adult Development.*

Demick, J. (1993). Adaptation of marital couples to open versus closed adoption: A preliminary investigation. In J. Demick, K. Bursik, & R. DiBiase (Eds.), *Parental development* (pp. 175–201). Hillsdale, NJ: Lawrence Erlbaum.

Demick, J. (1996). Life transitions as a paradigm for the study of adult development. In M. L. Commons, J. Demick, & C. Goldberg (Eds.), *Clinical approaches to adult development* (pp. 335–356). Norwood, NJ: Ablex.

Demick, J. (1999). Parental development: Theory and practice. In R. L. Mosher, D. J. Youngman, & J. M. Day (Eds.), *Human development across the life span*. Westport, CT: Praeger.

Demick, J. (2002). Stages of parental development. In M. Borstein (Ed.), Handbook of Parenting, Vol. 3, Being and becoming a parent (2nd ed., pp. 389–413). Mahwah, NJ: Lawrence Erlbaum.

Demick, J., & Andreoletti, C. (1995). Some relations between clinical and environmental psychology. *Environment and Behavior, 27*(1), 56–72.

Demick, J., Ishii, S., & Inoue, W. (1997). Body and self experience: Japan versus USA. In S. Wapner, J. Demick, T. Yamamoto, & T. Takahashi (Eds.), *Handbook of Japan–United States environment–behavior research: Toward a transactional approach* (pp. 83–99). New York: Plenum Press.

Demick, J., & Wapner, S. (1988a). Children-in-environments: Physical, interpersonal, and sociocultural aspects. *Children's Environments Quarterly, 7*(1), 28–38.

Demick, J., & Wapner, S. (1988b). Open and closed adoption: A developmental conceptualization. *Familty Process, 27*, 229–249.

Demick, J., & Wapner, S. (1992). Transition to parenthood: Developmental changes in experience and action. In T. Yamamoto & S. Wapner (Eds.), *Developmental psychology of life transitions* (pp. 243–265). Tokyo: Kyodo Shuppan.

Donaldson Adoption Institute (1997). *The benchmark adoption survey.* New York: Donaldson Adoption Institute.

Eheart, B. K., & Power, M. B. (1995). Adoption: Understanding the past, present, and future through stories. *The Sociological Quarterly, 36*(1), 197–216.

Elkind, D. (1996). Inhelder and Piaget on adolescence and adulthood: A postmodern appraisal. *Psychological Science, 7*(4), 216–220.

Erikson, E. (1968). *Identity: Youth and crisis.* New York: W. W. Norton.

Feigelman, W. (1997). Adopted adults: Comparisons with persons raised in conventional families. *Marriage & Family Review, 25*(3/4), 199–223.

Finley, G. E. (1999). Children of adoptive families. In W. K. Silverman & T. H. Ollendick (Eds.), *Developmental issues in the clinical treatment of children* (pp. 358–370). Boston, MA: Allyn & Bacon.

Frable, D. E. (1993a). Being and feeling unique: Statistical deviance and psychological marginality. *Journal of Personality, 61*(1), 85–110.

Frable, D. E. (1993b). Dimensions of marginality: Distinctions among those who are different. *Personality and Social Psychology Bulletin, 19*(4), 370–380.

Frable, D. E., Blackstone, T., & Scherbaum, C. (1990). Marginal and mindful: Deviants in social interactions. *Journal of Personality and Social Psychology, 59*(1), 140–149.

Gair, S. (1998). Coping with differences in mothering adopted children: Considering a broader model. *Adoption & Fostering, 22*(2), 16–24.

Golombok, S., Cook, R., Bish, A., & Murray, C. (1995). Families created by the new reproductive technologies: Quality of parenting and social and emotional development of the children. *Child Development, 66*, 285–298.

Grotevant, H. D. (1997). Coming to terms with adoption: The construction of identity from adolescence into adulthood. *Adoption Quarterly, 1*(1), 3–27.

Hazan, C., & Shaver, P. R. (1987). Romantic love conceptualized as an attachment process. *Journal of Personality and Social Psychology, 52*, 511–524.

Hoksbergen, R. A. C. (1998). Changes in motivation for adoption, value orientations and behavior in three generations of adoptive parents. *Adoption Quarterly, 2*(2), 37–55.

Hollingshead, A. B., & Redlich, F. C. (1958). *Social class and mental illness: A community study.* New York: John Wiley & Sons.

Hoopes, J. L. (1990). Adoption and identity formation. In D. M. Brodzinsky & M. D. Schecter (Eds.), *The psychology of adoption* (pp. 144–166). New York: Oxford University Press.

Howe, D. (1995). Adoption and attachment. *Adoption & Fostering, 19*(4), 7–15.

Howe, D. (1996). Adopters' relationships with their adopted children from adolescence to early adult-hood. *Adoption & Fostering, 20*(3), 5–13.

Howe, D. (1998). Adoption outcome research and practical judgment. *Adoption & Fostering, 22*(2), 6–15.

Ihilevich, D., & Gleser, G. C. (1993). *Defense mechanisms: Their classification, correlates, and measurement with the Defense Mechanism Inventory.* Odessa, FL: Psychological Assessment Resources, inc.

Kaden, S. E. (1958). *A formal-comparative analysis of the relationship between the structuring of marital interaction and Rorschach blot stimuli.* Unpublished doctoral dissertation, Clark University, Worcester, MA.

Kaplan, J., & Demick, J. (1999). *Identity development in black adolescents adopted by same-race versus transracial parents.* Worcester, MA: University of Massachusetts Medical School.

Kaye, K., & Warren, S. (1988). Discourse about adoption in adoptive families. *Journal of Family Psychology, 1*(4), 406–433.

Kelly, M. M., Towner-Thyrum, E., Rigby, A., & Martin, B. (1995). Adjustment and identity formation in adopted and nonadopted young adults: Contributions of family environment. *American Journal of Orthopsychiatry, 68*(3), 497–500.

Kivnick, H. Q. (1985). Grandparenthood and mental health. In V. L. Bengtson & J. F. Robertson (Eds.), *Grandparenthood* (pp. 211–224). Beverly Hills, CA: Sage.

Langer, E. J. (1997). *The power of mindful learning.* Reading, MA: Addison-Wesley.

Lerner, R. M., Iwawaki, S., Chihara, T., & Sorrell, G. T. (1980). Self-concept, self-esteem, and body attitudes among Japanese male and female adolescents. *Child Development, 51,* 847–855.

March, K. (1995). Perception of adoption as social stigma: Motivation for search and reunion. *Journal of Marriage and the Family, 57,* 653–660.

Marquis, K. S., & Detweiler, R. A. (1985). Does adopted mean different? An attributional analysis. *Journal of Personality and Social Psychology, 48*(4), 1054–1066.

McRoy, R. G., Grotevant, H. D., & White, K. L. (1988). *Openness in adoption: New practices, new issues.* New York: Praeger.

Melina, L. (1989). NACAC speakers describe seven core issues of adoption. *Adopted Child 8*(12), 2–5.

Melito, R. (1985). Adaptation in family systems: A developmental perspective. *Family Process, 24,* 89–100.

Melito, R. (1988). Combining individual psychodynamics with structural family therapy. *Journal of Marital and Family Therapy, 14,* 29–43.

Meyers, J., & Demick, J. (2002). Laypersons' judgments of psychopathology in adopted and non-adopted individuals. Worcester, MA: University of Massachusetts Medical School.

Miall, C. E. (1987). The stigma of adoptive parent status: Perceptions of community attitudes toward adoption and the experience of informal social sanctioning. *Family Relations, 36,* 34–39.

Miall, C. E. (1998). Community assessments of adoption issues: Open adoption, birth reunions, and the disclosure of confidential information. *Journal of Family Issues, 19*(5), 556–577.

Partridge, P. C. (1991). The particular challenges of being adopted. *Smith College Studies in Social Work,* 197–208.

Pepper, S. C. (1942). *World hypotheses.* Berkeley: University of California Press.

Pepper, S. C. (1967). *Concept and quality: A world hypothesis.* LaSalle, IL: Open Court.

Polkinghorne, D. E. (1991). Narrative and self-concept. *Journal of Narrative and Life History, 1*(2–3), 135–153.

Rosenberg, E. B., & Horner, T. M. (1991). Birthparent romances and identity fantasies in adopted children. *American Journal of Orthopsychiatry, 61*(1), 70–77.

Rosenberg, M. S. (1987). New directions for research on the psychological maltreatment of children. *American Psychologist, 42,* 166–171.

Rosenfeld, A. A., Pilowsky, D. J., Fine, P., Thorpe, M., Fein, E., Simms, M. D., Halfon, N., Irwin, M., Alfaro, J., Saletsky, R., & Nickman, S. (1997). Foster care: An update. *Journal of the American Academy of Child and Adolescent Psychiatry, 36*(4), 448–457.

Rushton, A., & Minnis, H. (1997). Annotation: Transracial family placements. *Journal of Child Psychology and Psychiatry, 38*(2), 147–159.

Rutter, M. (1984, March). Resilient children. *Psychology Today,* pp. 57–65.

Schecter, M. D., & Bertocci, D. (1990). The meaning of the search. In D. M. Brodzinsky & M. D. Schecter (Eds.), *The psychology of adoption* (pp. 62–90). New York: Oxford University Press.

Sharma, A. R., McGue, M. K., & Benson, P. L. (1996a). The emotional and behavioral adjustment of United States adopted adolescents: Part 1: An overview. *Children and Youth Services Review, 18*, 83–100.

Sharma, A. R., McGue, M. K., & Benson, P. L. (1996b). The emotional and behavioral adjustment of United States adopted adolescents: Part 2. Age at placement. *Children and Youth Services Review, 18*, 101–114.

Sharma, A. R., McGue, M. K., & Benson, P. L. (1998). The psychological adjustment of United States adopted adolescents and their nonadopted siblings. *Child Development, 69*(3), 791–812.

Silverstein, D., & Demick, J. (1994). Toward an organizational–relational model of open adoption. *Family Process, 33*(2), 111–124.

Singer, L. M., Brodzinsky, D. M., Ramsay, D., Steir, M., & Waters, E. (1985). Mother–infant attachment in adoptive families. *Child Development, 56*(6), 1543–1551.

Soparker, K. Baker, G., Pinet, M. Sandberg, E., & Demick, J. (2002). Community attitudes toward open and closed adoption: 1984 and 1997. Manuscript in preparation. Worcester, MA: University of Massachusetts Medical School.

Vroegh, K. S. (1997). Transracial adoptees: Developmental status after 17 years. *American Journal of Orthopsychiatry, 67*(4), 568–575.

Wapner, S. (1969). Organismic-developmental theory: Some applications to cognition. In J. Langer, P. Mussen, & N. Covington (Eds.), *Trends and issues in developmental psychology* (pp. 35–67). New York: Holt, Rinehart and Winston.

Wapner, S., & Demick, J. (1990). Development of experience and action: Levels of integration in human functioning. In G. Greenberg & E. Tobach (Eds.), *Theories of the evolution of knowing. The T. C. Schneirla conference series, Vol. 4* (pp. 47–68). Hillsdale, NJ: Lawerence Erlbaum.

Wapner, S., & Demick, J. (1998). Developmental analysis: A holistic, developmental, systems-oriented perspective. In W. Damon (Series Ed.) & R. M. Lerner (Vol. Ed.), *Handbook of child psychology: Vol. 1. Theoretical models of human development,* 5th edit, (pp. 761–805). New York: John Wiley & Sons.

Wapner, S., & Demick, J. (1999). Developmental theory and clinical practice: A holistic, developmental, systems-oriented approach. In W. K. Silverman & T. H. Ollendick (Ed.), *Developmental issues in the clinical treatment of children* (pp. 3–30). Boston, MA: Allyn & Bacon.

Wapner, S., & Demick, J. (2000). Person-in-environment psychology: A holistic, developmental, systems-oriented approach. In W. K. Silverman & T. H. Ollendick (Eds.), Developmental issues in the clinical treatment of children (pp. 3–30). Boston: Allyn & Bacon.

Wapner, S., & Demick, J. (2002). The increasing *contexts* of *context* in the study of environment-behavior relations. In R. B. Bechtel & A. Churchman (Eds.), Handbook of environmental psychology (pp. 3–14). New York: John Wiley & Sons.

Weider, H. (1977). The family romance fantasies of adopted children. *Psychoanalytic Quarterly, 46*, 185–200.

Werner, E. E. (1993). Risk and resilience in individuals with learning disabilities: Lessons learned from the Kauai longitudinal study. *Learning Disabilities Research and Practice, 8*, 28–34.

Werner, H. (1940/1957). *Comparative psychology of mental development*. New York: International Universities Press. (Originally published in German, 1926, and in English, 1940.)

CHAPTER 26

Swords into Plowshares

The Recovery Ethics of Destructive Adult Development

CARL GOLDBERG

Goodness in the greatest force in the world.

—SOMERSET MAUGHAM,
The Razor's Edge

Every day scores of men and women, by means of physical force or persuasive behavior, cause others the outrage of undeserved suffering. Many of these acts are violations of criminal statutes; still others are more appropriate to define as immoral. Nevertheless, all are evil in that they have in common treating other people with a lack of respect and consideration for the victim's humanity.

I first met Mike in my capacity as a consulting psychologist to a psychiatric facility for vicious criminals. He was 30 years of age; nearly half of his life was spent in correctional institutions. Physically imposing, he had multiple facial scars as a result of cigarette burns and brutal beatings inflicted by sadistic foster parents and correctional facility personnel. Deeply embarrassed about his appearance, Mike had been incarcerated numerous times for severely beating those who insulted him. With so many losses in his life, he found it easier to express rage at his deprivations than regret and bereavement. Nevertheless, Mike was intelligent—psychological tests administered in prison placed him in the superior range of intellectual functioning—and clever at understanding people's behavior, except in regard to his own motives. His lack of self-insight was compounded by a considerable difficulty in putting his feelings into words. He explained his actions by reference to external circumstances.

CARL GOLDBERG • Albert Einstein College of Medicine, New York, New York.

Handbook of Adult Development, edited by J. Demick and C. Andreoletti. Plenum Press, New York, 2002.

Many of the violent incidents in his life followed his substantial intake of alcohol. For example, the event that brought him back to prison was his violent attack on the club bouncer in a supper club. When Mike left the club, the bouncer had insinuated that Mike's very attractive companion could do better than Mike. Without a moment's hesitation, Mike hoisted up the man and heaved him through a glass door. The man nearly bled to death from his injuries.

OUR LACK OF KNOWLEDGE ABOUT THE REDEMPTION PROCESS IN ADULTHOOD

After each incident in which his violent rage caused him to hurt another person, Mike returned to therapy with me pleading his case: he was a victim of his upbringing; he was only doing what he had been painfully taught—to protect himself from a potentially lethal attack from others; indeed, not to defend his honor would serve to maximize that threat. Nonetheless, he stated that he wished he could live his life in a less violent way.

For the past 30 years I have treated a number of patients who have committed murder, rape, and mayhem in heinous ways, and like Mike, claimed they wished to transcend the indifference, viciousness, and cruelty that informed their daily existence. In short, they expressed the desire for a psychological and spiritual transformation. But none had any idea where to begin.

We have a tradition in our society of trying to salvage people's lives, no matter how egregious their existence has been. As a psychoanalyst concerned with moral issues, I wondered what psychological assistance I could provide Mike.

At first glance the prognosis for people such as Mike is poor. Most criminologists and forensic psychologists are dubious about a person's capacity to reverse the conditions that promote destructive behavior. They point out that there is no empirical evidence to support the efficacy of psychoanalysis for turning criminals away from crime, nor, for that matter, by any other rehabilitation approach—such as psychotherapy, group therapy, social work counseling, educational classes, or vocational training (Silberman, 1978). This view is verified by Jack Abbott (1991), a violent convict who wrote of the years he spent with legions of embittered men who leave prison more hardened and criminally proficient than when they first entered.

Pessimism about rescuing destructive lives is due to the inchoate state of our knowledge about adult moral development; we know far more about children. For example, there is a considerable literature on the early backgrounds of exemplary people—among others, studies by Levy (1946), McWilliams (1984), Midlarsky (1968), and Monroe (1991). In contrast, there are no normative data about how adult lives are transformed.

Psychoanalysis is held largely responsible for our lack of understanding of adult moral development. The theory of personality development that evolved from classical analytic notions discourages developmental exploration of adulthood. It employs clinical evidence to contend that all important aspects of personal character are inexorably set within the early years of life. This psychoanalytic bias has plagued all of the behavioral sciences (Goldberg, 1996).

The psychoanalyst concerned with recovery ethics can respond to this charge with the observation that the skillful, well-trained analysts is in the fortunate position to understand the individual best—for no other possesses as exquisite skills to listen and comprehend, and no other has the time to probe in such depth. In other words, psychoanalysts, by means of their clinical skills, have the opportunity to describe the motivations and psychodynamics of destructive people about whom they have in-depth, first-hand knowledge. As a psychologist and psychoanalysis who has spent many hours (indeed, many years!) with people who have committed crime and mayhem, my psychoanalytic practice has provided me with a perspicacious view of the conditions in the lives of people—from all walks of life—that interfered with their capacity to behave in cooperative and caring ways with others. It is this bald view of the struggle between good and evil that has enabled me to recognize that a destructive patient's overriding need in treatment is to find positive qualities about himself and his life, and to use this information to establish his personal identity in a self-enhancing way. The most difficult task, indeed, the turning point in Mike's road to psychological healing, was to gain a *trust in his own goodness* (Goldberg, 1992).

In this chapter I am concerned with two basic questions: To what extent do people have a choice and are responsible for the acts they commit or influence? Second, what are the factors that if recognized and competently acted on can change destructive lifestyles? Because we lack empirical data about changing destructive lives, I will use clinical and anecdotal evidence to examine these questions.

THE CAPACITY FOR CHOICE AND RESPONSIBILITY

To change one's destructive behavior one must first be convinced that one's way of life is dysfunctional, and, second, be able and willing to pursue constructive behavior: this is to say, capable and desirous of acting in ways in which one's own and others' best interests are respected and protected.

The crucial question of how free we are to exercise our will is a very old and controversial issue. Most theorists contend that the issue of human will is inaccessible to empirical investigation and therefore is not a proper study for psychological investigation (Lapsley, 1967). I take issue with this thesis, agreeing with Kohut (1959) that the psychoanalyst has phenomenological evidence for the freedom of will. Consider the following:

We can observe in ourselves the ability to choose and to decide. The question, then, is: Can further introspection reduce this sense into more fundamental components? In other words, can we by introspection separate the experience of making a choice into the experience of compulsion, indecision, and doubt?

My personal introspection suggests it cannot. Further exploration of the motivational basis for my experience of being able to choose leads me to a wider and more vivid sense of freedom. On the other hand, the converse psychological configuration, namely, my sense of being compelled and feeling indecision and doubt, can usually be broken down by means of introspection. As I proceed to examine these phenomenon in myself by means of resistance analysis of my underlying motives, I simultaneously experience an increasing capacity to choose and decide.

When I examine the psychological processes that follow my increased sense of an ability to choose I find that the freedom I experience is made possible by the fact that exercising my belief in psychological freedom enables me to act rather than remain passive. In short, if an individual believes that he has some freedom over his behavior, he generally acts in ways to maximize his capacity to choose. If, in contrast, he believes that he has little or no choice, then he generally behaves in ways to minimize his decisional activity. "To act, says Macmurray (1957, p. 134), is to determine, and the Agent is the determiner." To act on a belief in free will, I contend, provides a person with *agency*—the existential sense that he is capable of becoming the person he seeks to be (Goldberg, 1977).

A belief in freedom of will, then, is a *functional* position; it serves to maximize the extent of our ability to choose. But in no way does it detract from determinism as a principle governing human behavior: this belief itself is induced by the recognition that one seems to be able in some aspects of lift to have some control over one's behavior; and these actions could not take place in an indeterminate universe.

Nevertheless, it is highly unlikely that we ever will produce definitive evidence to discount free will as an epiphenomenon. The position I take here, then, is: if indeed none of us has any freedom of choice then it is pointless to continue to explore the capacity to overthrow a destructive life. Therefore, for purposes of our exploration, we must grant two assumptions: few or none of us have no choice, although some of us may be more able than others to freely choose how they behave; and the capacity for free will is not an innate attribute—consequently we would do well to learn why some people appear to have more free will than others.

FACTORS THAT IMPEDE THE ABILITY
TO DESIRE TO LIVE A DECENT LIFE

My clinical work with children suggests that there are a small number of specific factors in raising children, when found together, adequately explain why as adult they failed to develop a capacity for emotional connectedness and mutality with others, and instead have become involved in destructive behavior. The more prominent the presence of all these factors, of course, the greater is their impact on the child, and the more likely that the individual—as a child and/or and adult—will be influenced toward destructive behavior. The five I found most important (all readily present in Mike's background) are: shame, vulnerability, benign neglect, an inability to mourn, linguistic difficulty in expressing feeling, and witnessing significant people who behave as if rageful anger is a legitimate means for dealing with frustration and conflict. I refer to these as *the primary factors of destructiveness* (Goldberg, 1997b).

Shame Vulnerability

The media have given considerable attention to physical and sexual abuse in families. But no less important is the devastating effect of the shameful and humiliating ways many parents, who would be horrified to be called abusive, speak to and treat their children. Verbal shaming of a child consists of words spoken to the child in his or her presence that undermine the youngster's sense of competence and self-esteem. Humiliating actions are those that treat the child with contempt.

Each shaming experience, especially those that involve disregard and mistreatment, depletes childrens' sense of agency and personal worth; it painfully reminds them that they are inadequate. As such, these experiences undermine childrens' interpersonal relationships and their feelings of well being and security (Goldberg, 1991b).

Benign Neglect

Although the family-of-origin is not the sole agent for socializing children, it traditionally has been regarded as the major influence in personality development. Consequently, deprivation of parental attention, guidance, and compassionate regard can be as devastating to a child as physical neglect.

Not having caretakers emotionally available impedes the security and well being necessary for a child's constructive sense of personal identity. Life is not always reasonable and fair. In each child's life there inevitably will be disappointments and misunderstandings. Nevertheless, children should be protected from experiencing frequent and too prolonged emotional pain. On the other hand, if there is too little stimulation in childrens' lives, because they have been thrown on their own when they are too immature to care for themselves, they acquire an unrelenting hunger for excitement.

Children who develop a precocious restlessness, fed by a craving for stimulation, rarely develop appropriate sensitivity to the subtleties and nuances of life. They can see life only in bold black and white dimensions. These individuals have difficulties empathizing with the pain and suffering of others. Unable to read the emotions of those around them, they stumble into conflict, which more emotionally intelligent people avoid.

Inability to Mourn

Many life experiences turn out to be disheartening to the helpless and dependent child. Parents may feel ashamed in having disappointed their child. Not knowing how to set things right, they exert pressure on the child not to shed tears over hurts.

To the extent that children are unable to consciously acknowledge and give words to their inability to achieve and maintain a desired relationship with their parents, or someone else they regard as important to them, they are condemned to a world of pretense. In other words, children who cannot express their upset over their lack of control over their lives are compelled to find ways of pretending that they are not as helpless as they feel. Destructive behavior is, in an important sense, an attempt to show that the perpetrators can run their own lives—they can do whatever they please.

Linguistic Difficulties in Expressing Feelings

It is from our ability to use language intelligently that we acquire a facility in obtaining meaning for ourselves in relation with others. In short, it is meaningful words that enable us to become human rather than brute and solitary beasts.

Language disorders are frequently found in people who fail to develop a capacity for emotional closeness with others. Children who at a tender age are made to feel ashamed of having "unacceptable" feelings are discouraged from acknowledging deeply felt emotions, thereby impeding access to verbal expressions of these experiences.

These individuals are frequently pushed toward anger and rage by finding themselves in interpersonal situations in which their psychic vulnerability is painfully exposed in sensing a similar vulnerability in another person there. Unable to express a caring identification with the other, they strike out to silence their resurrected hurt.

Witnessing Significant People Who Behave as if Anger Is a Legitimate Means of Dealing with Frustration and Conflict

Parents, of course, model behavior for their children; their actions are more powerful guides than their words. A parent might tell a child that being angry is wrong, while at the same time reacting to frustration and conflict with rage.

In the absence of parental models that encourage compassion and concern for others, youngsters can be profoundly influenced by the popular media, which teaches the child that life must be exciting to be worthwhile. For these children, violence-imbued media becomes a habitual means of entertainment—easily obtained. More importantly, it serves as an untoward guide in making sense of the world in the absence of available constructive caretakers.

My claim that the primary factors taken together influence an individual toward destructive behavior is suggestive of an inevitable progression in becoming and remaining destructive. In fact, those patients with whom I worked, who to some significant extent were able to change their destructive behavioral patterns, did so by their persistent willingness and courage in struggling to recognize the adverse impact of some or all of the above destructive factors. In contrast, most behavioral scientists hold that early traumatic experiences prevent an individual from constructively modifying destructive behavior. What is the basis of their claim?

COMMON SENSE NOTIONS ABOUT
HUMAN BEHAVIOR

Most behavioral scientists seem to believe the psychological fiction that people who commit heinous behaviors suffer from a mental disorder, and this illness causes them to have less free will available to them than do "normal" people.

We can best understand this belief by examining common sense assumptions about the human mind. First, common sense tells us that human beings possess a faculty of "self-control" (which we call "free will"). Second, people assume that free will, present in all psychologically normal people, is impaired to a greater or lesser degree in psychologically abnormal people. This is because we believe that abnormal behavior—due to early trauma—has its origins in the deeper and unaware parts of the psyche, whereas more conscious and controllable factors are involved in normal behavior. Third, people believe that psychological conflicts— the presence of traumatic childhood experience—cause the person to suffer *irresistible urges* that cannot at the moment the offender is acting wrongfully be set aside and overridden by free will. Fourth, insofar as an offender is unable to control his impulses, he should not be held responsible for his psychological helplessness.

I show in the remainder of the chapter why these common sense notions are fallacious, and how by changing them, we provide the malefactor with access to constructive behavior.

THE REASONS FOR MIKE'S DESTRUCTIVE BEHAVIOR

Mike tried to impress me from the start of our clinical work together that the numerous violent altercations he had been involved in—before and during the time I was treating him—is typical of prison life, and, for that matter, of any protracted, confined area. It occurs as an automatic response for survival, predicated on the reality that in prison a person threatened cannot walk away. The prisoner has to face another inmate with whom he is in conflict every day, perhaps for many years. And at that moment when one is most vulnerable and unprepared, the dangerous other may attack with deadly force. It is far better, Mike indicated, to strike forcefully first, to intimidate the enemy into cowering timidity. In short, Mike's viciousness was based on the pernicious belief that the victim is so threatening to his well being that any destructive action he takes is justified.

Of course, Mike could have walked away from the nightclub incident that brought him back to prison. Moreover, whatever insult he perceived from the bouncer surely didn't carry a death threat. Rather it called attention to some aspect of his demeanor that a socially assured person would have likely ignored.

Mike, unsocialized from his many years in prisons and scores of foster homes and detention schools, experienced the club bouncer's words to imply that he was unwanted. Those who are raised in a subculture in which aggressive expression is the only recognized pathway to others, in perceiving a slight—even unintentional—swiftly give vent to violent reaction.

In examining Mike's life, I was shown abundant evidence that unacknowledged shame had pervasively taken over his life since he was a mistreated child, impeding him from building feelings of self-esteem. *Unacknowledged shame* refers to suppressed feelings about one's difficulty in becoming the person one seeks to be—resulting in the painful sense that one is incapable of establishing a system of shared meanings with others, and therefore one has lost and may never again regain an interpersonal bridge with significant others. In other words, Mike's persistent fear was he will be alone forever—unnoticed, uncared for, unwanted.

As such, Mike's morbid sense of shame compelled his violent rage; it was the silent, unacknowledged killer in his life, rather than death threats from dangerous others.

Choice is predicated on awareness. Because of extremely diminished intelligence, organic impairment, or chemical disposition, some people in some instances may be incapable of recognizing options and therefore have little or no freedom of choice. But this extreme incapacity, I believe, is quite rare. Mike was certainly aware of more humane options in situations in which he became violent. He told me that it was at times when he found trying to act properly uncertain that he felt a flood of *despair*: the feeling that despite recognizing what he should do, acting in an unfamiliar way was too tiring and difficult for him. It was at those moments that his rage and resentment at the injustices done to him filled his consciousness. And he made up his mind not to continue trying to do the right thing, but to get even with those who made him who he was—a vicious brute. In short, Mike was aware of options and did not follow them—because he was so resentful and enraged at other people that *he did not want to act righteously.*

Consequently, it is one matter to maintain that some offenders owing to their incapacities are unaware of moral options. It is quite another to claim that most offenders lack the capacity for responsible behavior.

Our behavior is a product not only of how we have been treated, but as importantly by the choices we make about *how to deal* with the dilemmas and ordeals in our lives. Victims of trauma and psychological conflict—such as reflected in the primary factors—become transformed into perpetrators of destructiveness *only* by their gradual and continual decisions not to consciously address their suppressed grievances and hurt.

As Carl Jung (1976) wisely claims, we deny the dark side of ourselves at extreme cost, because that which we don't bring into consciousness appears in our lives as fate. Jung's idea is that a person's past inescapably clings to him or her, and if the shadows of some of its events are too terrifying to examine, the cast of that shadow becomes one's eventual destiny.

The urges, thoughts and experiences that impel some to malevolent acts are *common to us all*. We all have a dark side: the hurts, angers and vulnerabilities that result from experiences of shame, humiliation, and the reactive feelings of self-contempt (Simon, 1996). What differentiates destructive people such as Mike from others is the unwillingness of destructive people to *self-examine* their dark side, and to come to terms with their inner forces to prevent their rage from getting displaced and acted out against anyone who has actually hurt them; those who stood by and did nothing to stop the abuse; and even those who did not know, but should have known and cared, and should have done something to stop the hurt (Goldberg, 1997a). This unwillingness to introspect keeps one a prisoner of his own toxic self-contemptuousness. Nothing in the world is as painful and unendurable as severe self-contempt. To survive the sufferer must cast these hateful feelings outward. Thus, self-contempt unexamined becomes converted into contempt against the world. The most horrific incidents that have happened in our world have resulted from intense self-hatred (Gilligan, 1996).

But cannot this unwillingness be due to the impact that early trauma had on the offender's conative capacity? Of course, but, then, there are always difficulties to self-knowledge. This is as true of people with fortunate childhoods as of those with more traumatic backgrounds. Unfortunately, common sense notions about the nature of the human mind make extreme distinctions between abnormal and normal behavior. And here lies the problem! Common sense notions about free will, based on extreme differences between the normal and abnormal mind, equate difficulty with impossibility. They are not the same. Whereas none of us has absolute free will, there are few who have no freedom of options. Consequently, there is a crucial difference in the claim that early trauma *predisposes* an individual to act in a certain way and the avowal that childhood factors determine adult behavior.

All of the destructive patients I have treated were aware of options, but chose to disregard the more humane options. Metaphorically, to act malevolently is to ignore, avert, and deny roadsigns along a hazardous highway that inform the driver that he is headed in the wrong direction. In other words, each of my malevolent patients, from similar previous events, knew the consequences of his behavior unless he took decisive action to stop the destructive pattern from continuing. But he didn't stop; his destructive behavior was an expression of informs choice.

Does this mean that common sense notions about irresistible urges are fallacious? I believe so. Let's take for instance one of the most dangerous types of destructive behavior—serial killings—and examine this type of behavior to assess my thesis. Serial killers are alleged by forensic psychologists as victims of irresistible urges: their behavior is typified by a highly compulsive set of ritualistic acts in which they select (in accord with the unwitting identification of the victim with the offender's own searing, early trauma), follow, and stalk a vulnerable victim, and then when the victim is alone and most helpless—abduct and sadistically kill.

The concept of irresistible urges was taken by psychologists from biology. As a biological concept, it does make sense: for example, beyond conscious control are attempts to breathe when suffocating and to regurgitate when digestively irritated. But as a psychological explanation the notion is highly suspect. The serial killer is not without choice. Any law enforcement officer who is knowledgeable about these killers can verify this. The officer can testify to situations in which the stalker was at some point in his ritualistic behavior apprehensive about a potential danger to himself: let's say, the unexpected arrival of other people, or a forceful fight put up by the victim.

If the serial killer's urge to commit his deadly behavior was actually irresistible, then he would be compelled to continue despite any danger to himself. But that is not what typically happens. In the usual scenario, if the stalker is seriously threatened, he halts, retreats, and comes back to pursue and kill another day. Of course, animals of prey also exhibit similar behavior. But this is probably evidence of animal volition. Like humans, animals have a history in which they develop likes and dislikes. Their cautious behavior involves choices in regard to these preferences rather than to an instinctual reflex of an organism without a brain.

The question, then, is why some people who experienced early trauma are able and willing to make humane choices, while others are not. Niccolo Machiavelli made the remarkably cynical assertion in his *Discourses* that people will always prove bad unless necessity compels them to be virtuous. In contrast, Socrates is shown in Plato's *Protagoras* contending that no one willingly acts malevolently: all virtues are forms of knowledge; the major impediment to moral insight is ignorance; the person who knows what is good, will be unable to choose wickedness.

I agree with Socrates—but I will show that it is not intellectual knowledge of good alone that promotes virtue; qualities of personal character that can be developed in adulthood are also required.

THE TASKS AND RESPONSIBILITIES OF VIRTUE

Several months into our clinical work Mike told me that he had been seeing a psychiatrist privately on the outside for 7 years, while he continued to drink heavily. Why, I asked, did he stay with the doctor if he wasn't being helped? "By continuing to tell me that I had a serious mental illness, he justified my drinking and helped me remain an alcoholic," I was told.

If ever there was an opportunity to enable Mike to reflect on his *self-deception* this was the moment. Mike had frequently claimed that he was violent because he was a victim of his upbringing, which taught him how to viciously protect himself now. By continuing to drink heavily he put himself in a vulnerable position—in which violent altercations were far more likely to occur than if he were sober.

I traced with Mike the steps involved in his habitual pattern of violence: he carried around a repository of experiences of shame and injustice that left him with considerable self-contempt; staying with these feelings was intolerable; he drank to numb these terrible feelings about himself; the drinking energized him in a search for people to hold responsible for the shames and injustices done to him; he felt a heightened excitement in finding a vulnerable candidate—the person was usually someone who unwittingly reminded him of his own hurts; he violated his victim with minimal deliberation; he felt serene and superior to the victim on to whom he had displaced his grievances.

But how, I asked Mike, does his violent displacement cycle enable him to become the person he claims he seeks to be—someone respected for his kind deeds and intelligence, rather than feared for his viciousness? For the first time Mike offered me no facile rationalizations to excuse what he readily admitted was his modus operandi. He asked if I knew how he should to change his life. I admitted that I didn't. The behavioral sciences would not be of much help. They have turned their attention more exclusively than seems warranted in examining human ills, as if we could understand virtue by studying misfortune alone (Goldberg, 1998).

It would be best, I indicated, if we sought out how Mike might redeem his life from the wisdom of those who have openly struggled with good and evil, especially those who had freed themselves from crime careers.

As I expected Mike to be equal participation in seeking how to best change his life, both of us consulted the excellent institution library for books on philosophy, theology, criminology, and classics in literature. Great literature offers us inspiration and guidance by liberating for introspection the ambiguities, paradoxes, and contradictions contained in our various views of virtue and wickedness. We also eagerly sought autobiographies of people who had successfully rehabilitated themselves. Especially helpful was a book written by Nathan Leopold (1958), who was involved in a crime said to have set the prototype for senseless murder: one of the first cases to show how victims of murder can be chosen at random because these victims as people have no importance to the murderers. Leopold spent 33 years in prison.

On the basis of written sources, my experiences working in the community with recovering drug addicts who had lived criminal lives (Goldberg, 1997a, Chapter 7), and discussions with Mike, he and I defined *virtue* as any attitude a person exhibits that enables him to be more enlightened, compassionate, and responsible both to himself and others. We arrived at the following responsibilities and tasks required to repair Mike's life: a capacity to learn the language of felt emotion, a concern with fairness and justice for others, assuming responsibility for one's own behavior, moral courage, and a willingness to self-examine. Not surprisingly, these tasks are closely related to my primary destructive factors. And while it can be reasonably questioned whether these tasks together are sufficient to restore virtue, I believe all are necessary.

A Capacity to Learn to Speak the Language of Felt Emotion

I was reading Herman Melville's *Billy Budd* at the time I treated Mike. Melville's main character explains after he felt compelled to strike and kill an abusive officer, "Could I have used my tongue, I would not have struck him."

Melville enabled me to recognize that Mike's inability to explain the reasons for his actions, ironically, revealed the very crucible in which his violent outrage was forged. Violence is a kind of language, however primitive and limited. Those who cannot communicate persuasively with their tongues strike out violently from the shame of sensing no effective alternative for verbally defining themselves to others.

Mike required linguistic training in expressing his unacknowledged feelings of shame and hurt—to erase the belief that in his essential core he was a brute creature, suffused with only a pretense of civility and sensitivity to others. In other words, Mike's language, heavily infused with aggressive, need-oriented words and concepts, reinforced his belief that he had a savage nature. So, whenever he tried to express tender or caring feelings, he generally found at his disposal only crude and emotionally shallow linguistic concepts. In working with Mike I developed Basic Emotional Communications (BEC).

BEC is designed to enhance an ongoing emotional dialogue between two (sometimes more) people involved in a significant relationship. The method demonstrates safe and effective ways of sharing hurts, fears, and desires with others, without destructive anger. As such, it disarms hurt and anger and leads to personal and relational gain (Goldberg, 1991a).

A Concern with Fairness and Justice for Others

Aristotle pointed out in *The Poetics* that since none of us lives alone, the life lived well is founded in friendship. Virtue, then, must be concerned with helping to foster the conditions that extend concern with compassion to every member of society. Unfortunately, too many people rationalize their lack of concern for others. Mike was hardly unique in telling himself that his first responsibility was to himself. Too preoccupied with his own securities, rarely was he concerned with others. Yet, in summing up one's life, wise philosophers and theologians have pointed out that there is a basic question to which each of us must respond: For the most part, has one been idle and unconcerned with the lives of others, a greedy opportunist, or someone who cared and made a difference? I confronted Mike with this question.

As a powerful presence in the institution, who was both feared and respected by other inmates and staff alike, a more verbally articulate Mike became an effective spokesman for the other patients in dealing with grievances toward staff and in settling disputes among themselves.

Learning to be of help to others is essential to the redemption of a destructive life; it is not only a social and moral responsibility of enlightened citizenship, but therapeutic, as well, for those whom we call emotionally disturbed, but who are actually emotionally impoverished. The so-called emotionally disturbed person has been deprived of equitable and balanced relationships with others. These relationships serve as a *lifeline* that sustains and maintains the individual, keeping him alive and well. When a person performs needed social and emotional functions for others and is recognized for his help, he is valued by others and by himself. Concomitantly, he experiences a greater capacity with others (Goldberg, 1998).

Assuming Responsibility for One's Own Behavior

The term responsibility covers a number of distinct but related notions. According to *The Oxford Companion to Philosophy* (Honderick, 1995), the most important uses of the term are causal responsibility, legal responsibility, and moral responsibility. To be causally (operatively) responsible for a state of affairs is to bring it about directly or indirectly. To be legally responsible is to fulfill the requirements for liability under the law. Moral responsibility has to do with assuming responsibility for our actions.

If we believe that a person has freedom of choice over his behavior, then all three of these uses are interrelated. On the other hand, many behavioral scientists have argued, if we assume that an individual's capacity to choose in some important sense is restricted, he cannot reasonably assume responsibility for his operative behavior, nor can we hold him legally responsible for the act—no more than regard him accountable for incurring cancer. The eminent psychiatrist Karl Menninger (1968) was a strong advocate of this position. I disagree.

Psychology and morality are integrally related in regard to responsibility for destructive behavior. In other words, if one has caused an action that results in harm to another, regardless of the intent or one's understanding of the behavior, morally one ought to act after the fact in such a way as to do one's best to lessen the harm one has caused.

In our present criminal justice system criminals may go to prison—for isolation and punishment—but they are not put in a position that obliges them to confront the consequences of their vicious actions. Moreover, in this impersonal administration of justice, victims and/or their loved ones are left out of the correctional system equation. Their needs, other than primitive vengeance, are not addressed.

The need to undo or lessen the harm one has caused is, I believe, the bedrock of moral wisdom, found in the teachings of the Jewish philosopher Martin Buber (1953), who stressed that the violation of a moral law is a lesser sin than the recognition that one has done wrong and yet not tried to do anything to right that wrong with those harmed. In short, from a moral perspective, lack of conscious intent does not abnegate responsibility for being the operative agent of harm. After all, if we are not responsible for our own behavior, then who is?

When I discussed these ideas with Mike he expressed an interest in my finding his victims in the community: he asked me to locate them and have them visit him in the institution. He admitted that while he was rather anxious about these encounters, he now was in touch with feelings of painful shame; he realized that he couldn't become the person he sought to be without face-to-face meetings with his victims. He recognized that only by assuming responsibility for his actions would he be able to actively investigate his own motivations in the events that had caused him trouble.

The administration of the institution would not allow me to bring in Mike's victims. But they did encourage me to set a group at the institution in which patient-victims and patient-perpetrators encountered and tried to understand what it was like to be in the skin of the other. The rationale was that the offender needs to interact responsibly with others who oppose his point of view—who have been harmed by actions fueled by his pernicious beliefs. My clinical experience with drug addicts taught me that only by gradual openness and empathy to the unfamiliar feelings and beliefs of others can the destructive person change his way of life.

Moral Courage

Our perceptions of courage, as of heroes, has shifted radically from past concepts. In ancient sagas, courage was seen as fortitude in dealing with external enemies and natural disasters. As we have become more self-aware in the contemporary age, courage has increasingly come to mean psychological bravery, the will to face our divided urges and inner terrors to create a vibrant discovered self (Goldberg, 1997a).

To act as if he were the person he sought to be rather than the despised self he experienced himself as, Mike had to actively seek out possibility. This necessitated moral courage because the effects of his possibilities could not be ascertained in advance of "turning toward possibility." Erich Fromm (1963) observed that each person may seek some ideal state of freedom, but people are not equally desirous of enduring anxiety and discomfort in pursuit of psychological freedom. Mike exhibited a courageous attitude by his asking to meet his former victims.

A Willingness to Self-Examine

Many believe that malefactors refuse to change because they *enjoy* committing wrongful acts. It is true that most destructive people justify their behavior on the basis that they are superior to the rest of us: the pernicious attitude that the victim is so weak, stupid, or incompetent that he or she can be treated as an object rather than someone who deserves decent interaction. A very different picture emerged, however, when I probed beneath the grandiose rationalizations of my malevolent patients. All had suffered deep hurt and shame from the failure to achieve emotional connectedness with their early caretakers. It was during those moments of painful shame in which they poignantly expressed their aching to be understood and cared about that my patients were willing to examine their behavior thoroughly and make amends for the evil they had done. I make no claim that this was an easy and always successful project. As Leopold's (1958) rehabilitative experience in prison indicates, it can be a long, gradual process of removing compartmentalized vulnerability and hurt, and bringing them into consciousness.

Mike met with me three or four times a week (extra sessions when emergencies arose). In these sessions Mike began to define himself in terms of his desired self and in so doing he was able to discover the "lost voice" of his deeply buried despair about his human limitations. With his more articulate and compassionate "new voice," Mike set his goal to make amends for the harm and hurt he caused. On his release from the institution he became a staff member of a halfway house program for offenders who were released into the community.

As mortals we are imperfect; we make mistakes. But no less characteristic of our humanity, we have the capacity to self-reflect, experience regret, and seek redemption for our mistakes. We should be held responsible for what we do: it is no less required to repair the harm we have caused than to receive respect and appreciation for our virtuous deeds.

REFERENCES

Abbott, J. H. (1991). *In the belly of the beast*. New York: Random House.

Buber, M. (1953). *Good and evil*. New York: Scribner.

Fromm, E. (1963). *Escape from freedom*. New York: Holt, Rinehart & Winston.

Gilligan, J. (1996). *Violence*. New York: Putnam.

Goldberg, C. (1977). The reality of human will: A concept worth reviving. *Psychiatric Annuals, 7*, 37–57.

Goldberg, C. (1991a). *On being a psychotherapist*. Northvale, NJ: Aronson.

Goldberg, C. (1991b). *Understanding shame*. Northvale, NJ: Aronson.

Goldberg, C. (1992). *The Seasoned Psychotherapist*. New York: W. W. Norton.

Goldberg, C. (1996). Introduction. The approach to positive adult development. In M. L. Commons, J. Demick, & C. Goldberg (Eds.), *Clinical approaches to adult development*. Norwood, NJ: Ablex, pp. 1–7.

Goldberg, C. (1997a). *Speaking with the devil: Exploring senseless acts of evil*. New York: Penguin.

Goldberg, C. (1997b). The Chautauqua Institution lecture: The responsibilities of virtue. *The International Journal of Psychotherapy, 2*, 179–191.

Goldberg, C. (1998). The indifference of psychoanalytic practice to social and moral responsibility. *The International Journal of Psychotherapy, 3*, 221–230.

Honderich, T. (1995). *The Oxford companion to philosophy*. Oxford, UK: Oxford University Press.

Jung, C. (1976). *The portable jung*. New York; Penguin.

Kohut, H. (1959). Introspection, empathy and psychoanalysis. *Journal of the American Journal of the Psychoanalytic Association, 7*, 459–483.

Lapsley, J. N. (1967). The concept of will. In J. N. Lapsley (Ed.), *The concept of willing*. Nashville, TN: Abingdon Press.

Leopold, N. F. (1958). *Life plus 99 years*. Garden City, NY: Doubleday.

Levy, D. M. (1946). The German anti-nazi: A case study. *American Journal of Orthopsychiatry, 6,* 507–515.

Macmurray, J. (1957). *The self as agent.* London: Farber.

McWilliams, N. (1984). The psychology of the altruist. *Psychoanalytic Psychology, 3,* 192–213.

Menninger, K. (1968). *The crime of punishment.* New York: Viking.

Midlarsky, E. (1968). Aiding responses: An analysis and review. *Merrill-Palmer Quarterly, 14,* 229–260.

Monroe, K. R. (1991). Jone Donne's people: Explaining differences between rational actors and altruists through cognitive frameworks. *Journal of Politics, 53,* 394–433.

Silberman, C. E. (1978). *Criminal violence, criminal justice.* New York: Random House.

Simon, R. I. (1996). *Bad men do what good men dream.* Washington, DC: American Psych Association.

CHAPTER 27

Discursive Practices and Their Interpretation in the Psychology of Religious Development

From Constructivist Canons to Constructionist Alternatives

James M. Day and Deborah J. Youngman

Scholars and practitioners interested in the varied dimensions of human religious conduct at any point in the lifespan owe a particular debt to those who, inspired by Jean Piaget and Lawrence Kohlberg, brought the study of religious development into the mainstream of the psychology of religion and the psychology of human development.

Models of religious development and their implications for religious education and the training of psychologists, educators, clergy, and pastoral workers have earned an important place in the developmental literature and have gained a broad readership of practitioners and members of the general public interested in their own religious development, or that of their children, parents, and peers. The dominant models in this sphere are those that have been authored by James Fowler, Fritz Oser, and Helmut Reich and their colleagues.

While researchers in the near domain of moral development—which by religious developmentalists' own admission clearly inspired and structured their own assumptions—have become increasingly aware of the metatheoretical and

JAMES M. DAY • Faculty of Psychology and Educational Sciences, Universite catholique de Louvain, Louvain-la-Neuve, Belgium.   DEBORAH J. YOUNGMAN • Department of Developmental Studies and Counseling, Boston University, Boston, Massachusetts, 02215.

Handbook of Adult Development, edited by J. Demick and C. Andreoletti. Plenum Press, New York, 2002.

methodological limitations of Kohlbergian conceptions of moral development (see, e.g., Coles, 1989; Day & Tappan, 1996; Gilligan, 1982; Packer, 1989, 1991; Schweder, 1982; Youngman, 1993), relatively little systematic attention has been devoted to similar concerns in the psychology of religious development. Given the broad variety inherent in any discussion of religious practice and experience, and the degree to which critiques of Kohlbergian models of moral development have centered on concerns for context and diversity, we find this lack of critique in the domain of religious development particularly curious. In this chapter, we endeavor to outline the major features of this cognitive–developmental paradigm of religious development and to show how research on epistemological and religious styles and in narrative approaches to the psychology of religion augur for a rereading of the cognitive–developmental paradigm and promise equally or more rewarding avenues of research and understanding.

RELIGIOUS DEVELOPMENT AS A PSYCHOLOGICAL CONSTRUCT: COGNITIVE–DEVELOPMENTAL MODELS

As recent reviews of the religious development literature make clear (see Day & Naedts, 1995, 1999; Streib, 1997; Tamminen & Nurmi, 1995; Vandenplas-Holper, 1998; Wulff, 1997), the dominant models of religious development in the psychological literature are those of James Fowler and of Fritz Oser and Helmut Reich. Fowler's work reflects his concerns with pastoral theology and his exposure, at Harvard, to the work of Kohlberg and his associates in the Laboratory for Human Development there. Oser's work has been inspired by his work among educators, and his insight early on that both Protestant and Catholic students in his seminars at Fribourg were confronted by similar pedagogical conundrums that could better be appreciated via constructivist developmental concepts than by other psychological, pedagogical, or theological concepts. Reich, Oser's colleague at Fribourg, has for some time been interested in the working out of a unified model of development that would detail the relationships among scientific and critical thinking and cognitive development in the sphere of religious constructs.

Fowler's work has focused on *faith development*, by which he means "an orientation of the total person, giving purpose and goal to one's hopes and strivings, thoughts and actions." This development involves "a dynamic pattern of personal trust in and loyalty to a center or centers of value" whose orientation can be understood in relationship to the person's trust in and loyalty to core "images and realities of power" and "to a shared master story or core story" (Fowler, 1981, 1996). Fowler's model is clearly affected by notions derived from liberal Protestant theology (Niebuhr and Tillich are both clearly represented in Fowler's notions of faith, and of faithing as a human activity) and from the field of religious studies, with the particularly phenomenological accent given the field by Wilfred Cantwell Smith and his associates in the Faculties of Arts and Sciences and of Divinity at Harvard, though it has earned popularity as a practical as well as theoretical model well beyond the confines of Cambridge, Massachusetts; of Emory University where Fowler is professor; and of liberal protestant circles. An example of his influence in work, which, furthermore, bridges the study of moral and religious development, is Youngman's analysis of lifespan retrospectives among the elderly (Youngman, 1993). Although departing from the structuralist assumptions inherent to his model, she relies on Fowler's conceptualization of the incorporated master story in her interpretation of derivative tales by which, in the degree to which dogmatic imperatives have or have not been realized, moral identities are in part revealed.

Fowler's model can best be appreciated as a multifactorial model, given that its construct of faith is so broad as to include dimensions associated with the cognitive stage notions of Piaget, Kohlberg's moral development stage formulations, Erikson's, Loevinger's, and Levinson's concepts of identity development and developmental crisis, the development of perspective-taking as articulated by Selman, and self-development as proposed by Kegan (Fowler, 1981, 1987, 1996; Tamminen & Nurmi, 1995).

Oser's work, in concert with that of Reich, has been concerned with *religious judgment* development. Essential in Oser's theory is the interpretation of the human being's relation to the Ultimate Being (God) and the action of the Ultimate Being in human life. As a person interprets the experiences of his or her life, discusses them or prays, as he or she studies religious texts and takes part in the life of a religious community, he or she actualizes the system of rules that concern his or her relationship to the Ultimate Being (Oser & Gmunder, 1991; Oser & Reich, 1990). This relationship appears in verbal form in "religious judgment" which is "some kind of cognitive pattern of religious knowing of reality" (Oser & Reich, 1990, p. 283).

Any attempt to understand Oser's perspective apart from the notion of religious "deep structure" or "mother structure" or "underlying structure"—terms that Oser borrows directly from Piaget—would be unhelpful both to students of the theory and its applications and to readers of this chapter. For this structural notion lies at the heart of Oser's theory and its adaptations by Reich, Kamminger, and Rollett, and others, and places the theory squarely in the Piagetian paradigm; Oser et al. argue that this deep structure is a *universal* feature of *religious cognition* present at every point across the lifespan, in all cultures, and irrespective of religious affiliation. Indeed both avowed atheists and agnostics are held by Oser, Gmunder, Reich, Kamminger, Rollett, and others to be concerned with fundamentally religious questions of relationship to ultimate being and purposes in their lives and in the life of the world, and to think about such questions in ways that cannot be reduced to other constructs or forms of cognition (Oser & Gmunder, 1991; Kamminger & Rollett, 1996; Oser & Reich, 1990, 1996a, 1996b).

Critical to both models are the notions of stage and sequence as they are known in the work of Piaget and Kohlberg, although Fowler's adaptation of stage constructs and theory, and the breadth of his concept of faith, offer a more fluid, and some would say, less precise, construct than does that of religious judgment in Oser's scheme. This distinction between the two theories has inclined critical some reviewers (see Power, 1991) to view his perspective as a "*soft stage*" theory and to consider Oser's model as coming closer to a "*hard stage*" model and the Piagetian criteria one would employ to assess such models—namely those criteria having to do with reversibility, invariant sequence in the stage-structured transformation of understanding, and hierarchy; Fowler's model allows greater movement across the various sequences of development envisaged in his multifactorial model than does Oser's, which is more insistent on the linearity of development and on a fixed endpoint as its most favorable, or mature, destination (see also Day 2001; Day & Naedts, 1995, 1999; Streib, 1991). Still, both theories are best understood as *stage theories* that envisage human development, and, in this case, religious development, as moving along a trajectory of mostly invariant sequences of stage transformations from less mature to more mature, and, in particular, from lesser to greater degrees and appropriations of individual autonomy.

For example, Fowler envisages a move from the heteronomous stage of "*intuitive-projective faith*," in which one is funded with long-lasting images and impressions of protective and menacing presences or powers in one's sphere of being, which are in turn given shape in the form of stories, gestures, and symbols, pregnant with emotion and dominated by the function of imagination, to "*universalizing faith*" which, dependent as it is on earlier stages of increasing individuality and autonomy and a critical capacity to stand outside one's group, tradition, and symbol system, involves movement beyond polarities inherent in any given system and is characterized by detached, yet compassionate action, of continuity with the power of being/God, in a world conceived of as moving toward fulfillment in justice and love (Fowler, 1981, 1996). As in Kohlberg's notions of moral development, and Piaget's models of the development of logical reasoning, higher stages are regarded as more adequate both because of the relative complexity they represent, and because of the adaptive value they offer, given the real nature of the world.

Such a hierarchical view to religious development is likewise apparent in the work of Oser and his associates, and perhaps more so, because it expresses a more strict concern with cognition, and both less worry about and less interest in other variables (affective and life-historical, which Fowler endeavors to take into account) in the religious experience of the subject. Oser would have us understand that development occurs when subjects move in their religious judgment from an orientation of religious heteronomy that he denotes as "*deus ex machina*" in which God is understood as all-powerful, active, and relatively capricious, acting as he (sic) would with relatively little regard for human will or action, toward the most favored endpoint of an orientation to religious *autonomy and intersubjectivity* and a feeling of *universal and unconditional religiosity*. As in Fowler's model, the sense of solidarity and continuity achieved with the ultimate is according to Oser something that can occur only after the subject has passed through a stage of *absolute autonomy* and, customarily, a rejection of religious and all other "external" authority. Like Fowler's model, Oser's assumes that higher is necessarily better on the grounds of adequacy that Piaget first asserted— higher stages are at once more sophisticated and thus represent a greater differentiation of human cognitive capacity, and they come closer to representing the world as it is; as one moves upwards in the scheme, whether in Fowler's model or in Oser's, one moves increasingly toward an approximation of the world as it really is (see Fowler, 1996; Oser & Gmunder, 1991; Piaget, 1923).

Whether in Fowler's or Oser's model, development occurs *within* a subject through a trajectory of *stages* that represent movement from *lower to higher* structures of cognition that increasingly move the subject toward greater *rational sophistication* and congruence of *approximation* with things as they really are. Researchers grasp an understanding of the subject's *level* of development through interview and/or questionnaire strategies in which the subject's language is taken to be *representative* of *private constructions* guided by *internally held deep structures* of distinctively religious cognition.

Within developmental psychology, there is general agreement that in the neo-Piagetian paradigm both models represent, Fowler and Oser et al. have moved questions concerning religious development to a place of accepted importance within the canon. Furthermore, as Tamminen and Nurmi (1995) observe, their models have provoked considerable empirical research concerning religious development in adulthood and in the entire lifespan (see also Bucher & Reich, 1989; Fowler, Nipkow, & Schweitzer, 1991). They have, in addition, detailed programs of intervention, which have served educators interested in promoting the developmental endpoints envisioned in their models.

There is some consensus as well that both Fowler's and Oser's models are flawed with considerable theoretical and methodological problems. To date, critiques of these models within the developmental literature have focused chiefly on the "relatively unsatisfactory" nature of undergirding empirical evidence (Tamminen & Nurmi, 1995, p. 302; Wulff, 1997), the lack of longitudinal data to support their inherent developmental claims, and the lack of cross-cultural evidence to support claims of universality that rely, in turn, so heavily on 20th century theological models of personhood and human relations with the "Ultimate" or "God." Despite recent and long-awaited evidence drawn from a longitudinal study (Di Loreto & Oser, 1996), recent refinements of Oser's methodology by Kamminger and Rollett at Vienna (1996), which have at long last brought more substantial numbers of subjects into the Oserian mainstream, and the development of questionnaires at Louvain for the empirical study of religious judgment (Day & Naedts, 1994, 1995, 1999), we would agree with Tamminen and Nuria (1995) in saying that "...the existing measurements and descriptions are in many cases still relatively inaccurate or artificial" (1995, p. 302).

Recently, researchers have investigated in detail the claim made by Kohlberg, and by both Fowler and Oser, that development in moral judgment would logically precede and likely trigger development in religious judgment. This research appeared to be of critical importance because of the highly conflicting data that had heretofore been obtained by researchers within Oser's and related research teams, who had found within their small samples of subjects that in some cases moral judgment levels were higher than religious judgment levels, in some cases religious judgment levels were higher than moral judgment levels, and that in other cases no significant difference between the two had been found (see Day, 2000; Day & Naedts, 1995, 1999). These studies with more than 600 Belgian adolescents and young adults at four critical age points where development is usually anticipated, from ages 12 to 24, used Gibbs, Basinger, and Fuller's (1992) Socio-Moral Reflection Measure (Short Form) to measure moral judgment levels and the Religious Reflection Questionnaire (Day & Naedts, 1995) to measure religious judgment levels. Our work led us to conclude that Oser et al.'s claims were empirically unverifiable. Indeed we found, as had previous researchers, that in some cases moral judgment levels were higher than were religious judgment levels, but that in other cases they were not, and that on the whole moral judgment was so highly correlated with religious judgment that the very notion of the independence of religious judgment as a cognitive function had to be called into question (Day, 1999; Day & Naedts, 1995, 1999).

COGNITIVE–DEVELOPMENTAL MODELS IN
QUESTION: A CRITICAL APPRAISAL

In the above-mentioned studies, our research instruments, which combined the benefits of production and recognition measures, allowed us to read the texts produced by our subjects both in terms of the structuralist criteria suggested in the Piagetian paradigm, and to reread them in concert with our attention to context-rich variables: gender, type of school, social class, and religious affiliation (Day, 1999; Day & Naedts, 1995, 1999). Rereading these texts in this light, we found, for example, that female and male subjects wrote responses emphasizing different kinds of orientations to moral and religious decision making, akin to those described by Brown and Gilligan (1991), Gilligan (1982), and Lyons (1983) in the domain of moral reasoning, and by Day and Naedts (1997); and Day (1999a, 1999b) in the domain of religious reasoning. (This despite the claim by Oser and Reich that sex differences are not pertinent to an understanding of religious judgment). In addition, we found that there were different discourse styles represented in the texts of students from elite Catholic secondary schools, and those from non-Catholic vocational and technical schools (Day, 2000). Finally, when questionnaires were completed by adolescent and young adults who also agreed to be interviewed as to their perceptions of the questionnaire we had constructed to reflect Oser's presuppositions and to accord with his clinical interview method, we found that a majority of them said that their responses to the questionnaires were ones they were *capable of making* but not necessarily *representative* of their thinking about religious questions; that many found the structure of the questions posed simply *irrelevant*; and that many saw the questions as fraught with theological assumptions to which they were indifferent or hostile.

As our interviews moved to questions of moral and religious decision making in real-life settings, we found that context mattered a great deal to the kinds of discourse produced. Girls spoke in a way vividly different from boys, women and men framed both the terms and the situations differently and structured their presentations of self using different narrative strategies, and Catholic youth talked in a way radically in contrast to how young Muslims did (Day, 2000). Thus a contextually specific set of *discursive styles* mattered as much in making sense of our participants' utterances as did any notion of cognitive deep structure, which we have already noted appeared to us an increasingly shaky notion on other empirical grounds.

Youngman's qualitative data attest to the bidirectional influence of development in the two domains—neither of which could be well accounted for by constructivist concepts, measures, or criteria. Narratives of the elderly participants in her North American study revealed that early and/or ongoing religious education and faith practices had substantially informed their capacity for moral recognition, problem solving, discourse, action, and other constitutive features of what she refers to as contextualized, intersubjective "moral presence." The majority, however, noted that moral challenges across the lifespan had frequently exceeded competence conferred. Interactive, apparently developmental changes in religious affiliation, extrareligious searches for moral authority, and transformed understandings (potentially applied) of previously held religious convictions were all consequent (Youngman, 1993).

In the light of these and other studies, we have become increasingly persuaded that although subjects are capable of producing utterances and written texts using religious language that can be read in terms of a structuralist hermeneutique, this does not mean that such utterances are representative—either of the more significant aspects of "religious" cognition, or of cognitive deep structures of the kind Kohlberg, Fowler, Oser, Reich, Kamminger, Rollett, and others would have us imagine. Such texts testify only to the efficacy of one hermeneutical strategy alongside others. They prove neither the existence nor the primacy of cognitive deep structures, and testify no more, and no less, than to one way of speaking that subjects, or research participants as we have long preferred to call them, are capable of producing (see Day, 1994).

We share Habermas' (1979, 1983) objection to Kohlbergian notions of "internal" structures of moral judgment and his advocacy of a view, which would see such structures as located in the discursive options offered and circumscribed by social settings. We would see them as interpersonally located social strategies intended to accomplish one or another end in communication or, more radically, simply as the kinds of linguistic conventions privileged in particular social circumstances, rather than as individual personal possessions or factota permitting attributions of "internal" activity. This does not mean, of course, that no valuation of such strategies, or of the social settings in which they apply, could be made. One might readily agree that some communities would be preferable to others, and thus that some verbal strategies and conventions of interpersonal exchange would be more attractive than others. We find no grounds, however, for assuming that such preferences could be explained or justified on Piagetian grounds.

In any case, that research participants are *capable* of entering into the frames ordained by Fowler and Oser et al. as properly religious does not incline us to conclude that such frames are the most interesting for understanding those participants' use of religious language. In this light, we have increasingly grave reservations as to the utility and appropriateness of the most fundamental notions of mind–language relationships informing the cognitive–developmental notion of *epistemic subject*, and the structuralist hermeneutique of interpretation with its attendant focus on *stage, structure, and sequence*, best articulated by Kohlberg in his observation that *"our hermeneutic and reconstructive stance is part of our effort to define our stages as 'hard' Piagetian stages meeting the criteria of culturally invariant longitudinal sequence, structured wholeness and hierarchical integration of a lower stage into the next stage. As a result we eliminate structurally ambiguous responses"* (Kohlberg, 1984, p. 503).

This summation of Kohlberg, near the end of his career, seems to us representative of the reductionist attitude we find objectionable in cognitive–developmental paradigms in the psychology of religious development, which by their own admission are so heavily reliant on Kohlbergian appropriations of Piagetian perspectives. As we have tried to show, we find little empirical evidence for either the validity or the utility of religious deep structure of the neo-Piagetian kind in the developmental literature of the psychology of religion, and are persuaded that evidence from both the cognitive–developmental literature and from emerging post-structuralist research paradigms augur for a rethinking of the human subject, the idea of development, and the relationship of human speech and action where religious as well as moral development is concerned.

A DISCURSIVE TURN: TEXT, STYLE,
VOICE, AND NARRATIVE

In the paragraphs that follow, we devote attention to research with a focus on *religious style* as an alternative to notions of religious stage and deep structure, and emphasize a *discursive* view of persons and mind–language relationships, arguing that *voice* and *narrative* seem to us particularly pertinent for the study of religious development in adolescence and adulthood.

We begin this excursion into alternative ways of framing religious experience in adolescence and adulthood by referring to the work of Dirk Hutsebaut and his associates at the Katholiek Universiteit Leuven (Belgium), who have for some years studied correlates among personality variables and religious attitudes among Flemish adolescents and young adults (see e.g., Hutsebaut, 1995, 1996, 1997a, 1997b). His recent appropriation (see Desimpelaere, Sulas, Duriez, & Hutsebaut, in press) of David Wulff's (1997) hypothetical categories of religious orientations and their correlates in the domain of the psychology of epistemological style seem to us worthy of note in this regard. Although it lies beyond the scope of this chapter to engage in an in-depth analysis of Hutsebaut et al.'s work, we find the following features of this most recent study to be particularly intriguing in the light of our turn toward an emphasis on *discourse*, religious *styles*, and *narrative structures and processes* in the study of adolescents and adults.

In the above-mentioned article, Hutsebaut and his colleagues offer an incisive review of the literature on epistemological style and make a case for using associated measures to plot possible correlations with Wulff's proposed orientations to religious belief, which Hutsebaut and his team (1996) have elsewhere validated and modified as empirical constructs.

Hutsebaut's team used psychoepistemological dimensions based upon the factors of Wilkinson and Migotsky (1994) completed with items from other psychoepistemological inventories, and tested correlations between such styles and three religious orientations: "Orthodoxy," "External Critique," and "Historical Relativism."

This group of researchers, following Hutsebaut's validation studies, define these terms as follows:

"Orthodoxy" means that there is a tendency to hold the opinion that on each religious question there is only one right answer, which is sustained by authority and that remains the same over time. Prototypical orthodox believers accept the answers from persons perceived as religious authorities. Moreover they are very certain about their belief and they report a positive relation to God, although this relation also includes elements of frustration, guilt, and anxiety. In addition, they are literal religious thinkers and believers, although they tend to accept any religiously colored statement. This religious style is positively correlated with anxiety in the face of new questions, with feelings of anomia, and with ethnocentricity (p. 5) (see also Hutsebaut, 1997b). (Wulff 1997 calls this orientation "Literal Affirmation.")

"External Critique" means that there is a tendency toward nonbelief, or at least the meaning and possibility of religious belief is fundamentally questioned. Prototypical persons taking the external critique position want to be sure of their belief content (which they are not) and reject literal as well as symbolic thinking about religious statements. Moreover they feel rebellious toward God and want to be autonomous, relying on their own norms instead. This religious style is positively correlated with fear of uncertainty and with feelings of anomia (p. 5) (Hutsebaut, 1997b). (Wulff's 1997 corresponding orientation is "Reductive Interpretation.")

"Historical Relativism" means that there is a tendency toward believing, but there is also a tendency to think and speak about belief in a historical way. Prototypical persons taking the historical relativism position think about religion in a symbolical way and are therefore aware of the fact that other religious meanings are also possible and that meaning can change over time. For them speaking about the absolute is a searching process, a possibility beside other possibilities. This religious style is positively correlated with openness to complex questions and negatively correlated with anxiety in the face of new questions, with feelings of anomia, and with ethnocentricity. Historical Relativism is thought to be a measure of what Wulff (1997) calls a Restorative Interpretation (Hutsebaut, 1997b).

In turn, they test the hypothesis that Wilkinson and Migotsky's (1994) epistemological dimensions will correlate with these religious ones: Orthodoxy with their dimension of "Naive Realism," External Critique with their "Logical Inquiry," and both External Critique and Historical Relativism with their "Skeptical Subjectivism."

Studying 218 subjects aged 17 to 75, divided into four age groups and four categories of educational level, and controlling for gender variables, the Leuven group analyzed the subjects' responses to six belief statements and a questionnaire containing items to which subjects responded on seven-point Likert-type scales. This study is of interest on several grounds—it contributes in an original way to the psychology of epistemological style, and pioneers an understanding of relationships of related factors to those involved in belief orientations. The study is an essay in methodological sophistication and elegance within the frame of classical personality research. For the purposes of this chapter, we want to devote particular attention to the Leuven group's remarks concerning the importance of their study for appreciating developmental notions, and in turn, offer a discursive take on their related findings.

First, despite their not having intended their study as an act of developmental research, they found that the dimensions of dualism, relativism, and commitment from Perry's (1970) developmental model of intellectual growth were more useful in describing common factors in psychological style than were those of Wilkinson and Migotsky (1994). However, and of crucial importance to the point we wish to make, the Leuven group concluded that:

> We cannot fully agree, however, with the fact that the concepts of Perry (1970) are embedded in a developmental scheme. In our investigation, for example, older subjects were found to think more dualistically, which is opposed to Perry's ideas. These results could be explained if we consider relativism as a contemporary phenomenon, which is also suggested by Perry. Anyway, we did not find any indication in our results that favors this developmental approach. (Desimpelaere et al., in press, p. 10)

We would propose that both Perry's (1970) dimensions, which he intended as descriptors of developmental transformations in intellectual functioning, and Wulff's (1997) belief orientations as validated by Hutsebaut (1997b), might most fruitfully be characterized as matters of *style*—not of personality—but of *discourse*, or *discursive frames*. Hutsebaut's observation that subjects might move across the religious styles in patterns other than linear ones, or might adopt one series of attitudes, and then another, depending on the circumstances in which they find themselves, and that even would-be developmental constructs a la Perry turn out to be descriptive of something more akin to style, or we might say, discursive strategy or stance, than to a pattern of structured movement upwards within a hierarchical scheme of stages is particularly pertinent here. Hutsebaut's insight is consistent with our observation (Day, 1991a, 1993, 1998, 1999a, 1999b; Day & Tappan, 1995, 1996; Youngman, 1993) that people use moral and religious language differently according to the audience they address, and that language is in such circumstances better described as a tool, or a convention, as a kind of performative, than as a factotum of thought lodged in the "interior" of the person. As Wittgenstein observed "It is misleading then to talk of thinking as a mental activity. We may say that thinking is essentially the activity of operating with signs."

Heinz Streib's work in the literature on religious development has roots in the dissertation work he conducted with James Fowler at Emory University, and thus in the developmental hermeneutic of Kohlberg, and its reformulations in Fowler's theory of faith development. Streib's 1991 book on faith development theory demonstrates at once his keen appreciation of Fowler's contribution, and hesitations grounded in his rereading of Fowler's approach in the light of Ricoeurian hermeneutics. As such, Streib's is one of the first publications in the developmental literature to seriously apply the concept of *narrative* derived from philosophical and literary as well as psychological literature, in rereading cognitive–developmental constructs. More recently, Streib (1997), drawing in part from Schutze (1984), has developed research strategies that have as their aim both the validation and operationalization of constructs belonging to the growing psychological literature on narrative, and comparisons of data yielded by such strategies with data coming out of the use of more traditional developmental methods. In one of the most incisive reviews of the developmental literature and its adjuncts in the psychology of religion to date, Streib (1997) draws from his research with members, leavers, and joiners of various religious groups, and in particular, of fundamentalist religious groups, to argue for the concept of *religious style* and for a rereading of the developmental literature that would redescribe Fowler's and Oser's stage descriptions as formulations of *stylistic* orientations to religious questions. Apart from Streib's hesitations with regard to these classical developmental models on empirical grounds, he argues that Fowler's and Oser's findings are *overdetermined* by their own stylistic preferences—including those for a *unified subject internally* working out questions of meaning in relatively *coherent Judeo–Christian contexts* of *religious discourse and* (as Fowler would have it) *master stories*. According to Streib, methodologies that devote serious attention to the *narrative* features of discourse about religious questions would incline us to embrace a view of the person as *multivocal*, and of religious speech as *performative and strategic* rather than representative of internally held deep structures. They would, in turn, incline us to a greater openness to the language of contemporary adolescents and adults whose speech is embedded in a life world of *religious diversity* and *overlapping frames* of discourse—religious and other—as to the meaning of life and the nature of human persons and relationships. In so doing Streib (1997) makes several references to Day (1991a, 1991b, 1993) and

Day and Tappan (1996) in developing the case for a narrative approach to questions in the field of religious development.

Ruard Ganzevoort's work presents another interesting case for moving away from a structuralist model of religious development toward an emphasis on narrative, discursive style, and religious language as a performative genre of human action that moves and develops in ways that linear and hierarchical paradigms of religious development fail to capture. Ganzevoort draws from a clinical–pastoral background, which first inclined him to study the relationship between coping strategies, religious belief, and recovery from traumatic events. From that point onwards—Ganzevoort began this work some years ago for his dissertation at Utrecht—his work has increasingly focused on *narrative* as a useful rubric for elaborating relationships among cognitive, affective, and connative dimensions of religious behavior (see, e.g., Ganzevoort, 1998a, 1998b). Studying the discourse of men from strongly religious backgrounds who have sought treatment for trauma related to histories of sexual abuse, Ganzevoort traces the evolution of concepts of self and of God in the context of therapeutic, pastoral–psychological relationships. In so doing, Ganzevoort has developed a method of *narrative analysis* that allows researchers to appreciate the emplotment of self and others, including notions of the sacred/God, in time, and in the specific patterns of relationship peculiar to the therapeutic setting. We agree with Ganzevoort in viewing both his work to date and the development of his method as promising for rethinking classical notions of religious development, and for entertaining alternative models of mind–language–action relationships in theory and research in the psychology of religion.

Our own work has moved from an early grounding in Kohlbergian moral psychology and Kegan's neo-Piagetian clinical–developmental theory during our years at Harvard to a discursive–constructionist view, a move that began with a frustration with Kohlbergian reductionism and a related concern to honor *voice, narrative,* and the *socially situated character* of developmental "structures" supposed by constructivists to reside "inside" individual subjects, as we encountered them in accounts of moral action and reports of religious experience. We have thus described our work, and it has elsewhere been described by others as defining a *narrative approach* to moral psychology and the psychology of religion (see, e.g., Day, 1991a, 1991b, 1993, 1994, 1998, 1999a, 1999b, 1999c; Day & Naedts, 1995, 1997, 1999; Day & Tappan, 1995, 1996; Ganzevoort, 1998; Gergen, 1993, 1994; Lapsley, 1996; Lourenco, 1996; Murken, 1993; Puka, 1996; Streib, 1997; Youngman, 1993).

This "narrative turn" in developmental psychology was influenced by Carol Gilligan's (1982) critique of Kohlberg's insistence on "justice" as the grounding point for understanding reasoning about moral issues. Gilligan's illustrations of alternative conceptions to "justice reasoning" reinforced our own inclination to "read back into" the text of dilemmas of moral choice and religious decision the narrative elements that had been theretofore "read out" of the text by the Kohlbergian hermeneutique. We were in turn encouraged by the parallel, and original, work of Mark Tappan and Lyn Brown who were also concerned with applying concepts from the emerging work in narrative psychology to the psychology of moral development and moral experience (see, e.g., Brown & Gilligan, 1991; Tappan, 1989, 1990, 1991a, 1991b, 1992; Tappan & Brown, 1992). This led to fruitful contacts with the work of Bruner (1986, 1990), Gergen (1993, 1994), Gergen and Gergen (1986), Sarbin (1990), and Spence (1984) in the narrative realm, and, in turn, to the conclusion that we needed to far more radically rethink mind–language–action relationships in moral and religious psychology and to reconceptualize the psychology of religious development accordingly—thus our subsequent use of Austin, Bakhtin, Foucault, Lyotard, Luria, Davidson, Harre, Rorty, Shotter, Vigotsky, Wertsch, and Wittgenstein (e.g., Day, 1993, 1994, 1998, 1999a, 1999b; Day & Naedts, 1997; Day & Tappan, 1996; Youngman, 1993).

DIALOGICAL SELVES: TOWARD A DISCURSIVE CONCEPTION OF THE DEVELOPING PERSON

Luria's point that "the chief distinguishing feature of the regulation of human conscious activity is that this regulation takes place with the close participation of *speech*" would seem to us well taken. Luria continues "whereas the relatively elementary forms of regulations of organic processes and even of the simplest forms of behavior can take place without the aid of speech, *higher mental processes are formed and take place on the basis of speech activity* (Luria, 1973, pp. 93–94). This in turn implies a move toward a different conception of self, which we have elsewhere characterized as a "dialogical self." "From a narrative perspective, self is understood not as a 'prelinguistic given' that merely employs language as a tool to express internally constituted meanings, but rather as a product of language from the start—arising out of linguistic, discursive and communicative practices" (Day & Tappan, 1996, p. 71). As Kerby (1991) observes, "Self's understanding of itself is mediated primarily through language, where language is taken to be the social medium par excellence (p. 5). The self is, therefore, neither a 'substantial entity having ontological priority over praxis,' nor an autonomous Cartesian agent "with epistemological priority, an originator of meaning' (p. 4)." Rather, it is an inhabited, decentered actor, in a theatrical world of possible stories where all action is rehearsed, justified, and reviewed according to the narrative possibilities inherent in the actual context(s) in which action occurs (Day, 1991; Day & Tappan, 1996, p. 71).

As we have elsewhere observed (Day & Tappan, 1996) we have found Bakhtin's work (1981, 1986, 1990) helpful in fleshing out part of our understanding of this dialogical self which has seemed to us so compellingly apparent in our studies. Like Vygotsky, Bakhtin argues that the human psyche is semiotically and linguistically mediated, and that it originates in the context of social relationships and social interaction (see also Wertsch, 1989, 1991). We find Bakhtin especially useful in this regard because unlike Vygotsky, who depends on a model of "internalization" to explain development in psychological processes, describing a move whereby external relationships between persons become internal relationships within the psyche, Bakhtin would appear to avoid the split between internal and external on which Vygotsky's account relies.

To more fully appreciate Bakhtin's contribution, we return to the concept of *voice*, which as we have previously noted, has been of importance to us in a variety of research projects and practices over the last decade. As Holquist and Emerson (1981) underscore, the notion of voice is for Bakhtin a dynamic relational concept that represents in effect "the speaking personality, the speaking conscious-ness." As Wertsch (1991) puts it, speaking voices are always composed of words, language, and forms of discourse, and they are always engaged in the process of social communication. We encounter a multitude of voices in the course of our lives through our embeddedness in relationships and social interactions—in this sense the social world is composed of ongoing discursive processes. The self, as Bakhtin would describe it to us, is a discursive location characterized by its partic-ular appropriations of the voices, which constitute its social world. As we have tried to demonstrate (e.g., Day 1991a, 1991b, 1998, 2000; Day & Tappan, 1995, 1996; Youngman, 1993) a serious attention to the discursive construction of the self would imply a move from the study of individual "minds" "represented" by words, to an attention to the social uses of language, the strategies of appropriation that are part of the self's differentiation and activity across varied social contexts, and to the dialogical relationship as a primary unit of analysis (see also Wertsch 1985, 1991).

DISCOURSE AND DEVELOPMENT

This in turn begs the question how we might begin to redescribe "develop-ment," and, for the purposes of this chapter, *religious* development. Here too we find Bakhtin helpful because for Bakhtin—in ways we find entirely coherent with our emerging understanding of the person—the notion of development is itself a linguistic convention, a cultural particular, which makes sense only when we understand the self as dialogical. The differentiation of this self is itself a product of linguistic processes according to Bakhtin, a process by which the person comes to view herself *as* a self insofar as she is counted on to *speak* in her own *voice*, and to *answer* to others—to become part of the dialogues of which her world is composed, to respond when spoken to, to offer moves that will incline others to speak, to par-ticipate in the making of the world through her place in the way it is linguistically constructed and shaped. For Bakhtin, what develops is *authorship*—competencies the person has to actively participate in the shaping of her world through language.

How does this emergence of *answerability* and *authorship* occur, or better, in a constructionist sense, what do we describe when we assign "responsivity" (our word) to someone, when we see her as an answerer, and author? How does a person's voice come to be constructed as her "own"?

It becomes "one's own" only when the speaker populates it with his own intention, his own accent, when he appropriates the word, adapting it to his own semantic and expressive intention. Prior to this moment of appropriation the word does not exist in a neutral and impersonal lanuage (it is not, after all, out of a dic-tionary that a speaker gets his words!), but rather it exists in other people's mouths, in other people's contexts, serving other people's intentions: it is from there that one must take the word, and make it one's own. Language is not a neutral medium that passes freely and easily into the private property of the speaker's intentions; it is populated—overpopulated—with the intentions of others. Expropriating it, forc-ing it to submit to one's own intentions and accents, is a difficult and complicated process (Bakhtin, 1981, pp. 293–294; see also Day & Tappan, 1996).

According to Bakhtin (1981), authorship emerges through a developmental process he calls ideological becoming: "The ideological becoming of a human being … is the process of selectively assimilating the words of others" (p. 341). Bakhtin continues:

> The importance of struggling with another's discourse, its influence in the history of an individual's coming to ideological consciousness, is enormous. One's own discourse and one's own voice, although born of another or dynamically stimulated by another, will sooner or later begin to liberate themselves from the authority of the other's discourse. This problem is made more complex by the fact that a variety of alien voices enter into the struggle for influence within an individual's consciousness (just as they struggle with one another in surrounding social reality). (Bakhtin, 1981, p. 348)

As Tappan (1991a,b) and we have underscored in other essays on moral and religious psychology, the goal of development is not simply a matter of speaking in one's own voice. Indeed it could be argued in a constructionist sense that romantic notions such as those of a *true* personal voice, a voice of an individual self become more *authentic*, the voice of a person *who had become more herself*, or constructivist notions of a voice become more *mature* in the sense of having *mapped* the world as it *really is* with increasing *accuracy*, *structurally* more and more adequate, would evaporate if we were to take Bakhtin seriously.

Because this person is fundamentally relational insofar as she is always part of a language that outdistances her speaking location and interweaves her with others in the flow of discourse, every point along the developmental pathway is marked by developments in the relationships of which the self is, discursively, part. Here the constructivist notion of autonomy as a necessary feature of moral and religious maturity is displaced by the notion of a self interwoven in (the) other(s) in dialogue, in which the utterances of the self are signs at the frontier of one being and another, marks at the interstice of a language mutually created by the two and more who are part of the living discursive context. As Bakhtin puts it "The word in language is half someone else" (Bakhtin, 1981, p. 293). As such the utterance of the self that marks any given point in her history is akin, as we have noted but wish again to underscore, to an act of authorship—and a such lies on the "borderline" (Bakhtin's word) between self and other(s). It is a "two-sided act … determined equally by *whose* word it is and *for whom* it is meant" (Volsinov, 1929/1986, p. 86, italics in original) (see also Day & Tappan, 1996, p. 71).

> This would, as Charles Taylor (1991) has argued, incline us to a perspective that: places dialogue at the very center of our understanding of human life, an indispensable key to its comprehension, and requires a transformed understanding of language. In order to follow up this line of thinking, we need not Mead and his like, but rather Bakhtin. Human beings are constituted in conversation; and hence what gets internalized in the mature subject is not the reaction of others, but the whole conversation, with the interanimation of its voices. Only a theory of this kind can do justice to the dialogical nature of the self. (pp. 313–314)

What would develop might then be conceived of as the *self-in-dialogue*, and the very *relationships and discursive contexts* in which the person is situated. If in the sense we wish to emphasize, selves *are* the relationships in which they are to be found, and selves *are* the narrative emplotments they are capable of persuading others to include in conversation, the very notion of development would become more radically social in character, selves as relationships would have to be conceptualized as the proper loci of psychological research, and the goals of development would thus be both a "personal" and "social" ones (we demarcate both with quotation marks because of our purpose in closing the gaps conceived in constructivist models between the two). In this vein worthwhile endpoints in development might be conceived to include the capacity to engage in dialogue, the emergence of social skills applied in such a way as to sustain conversation, to maintain relationship over time, to include others in discourse, and to create with others the kinds of discursive contexts in which increasing numbers of people would be invited and recognized as legitimate speakers. Here the definition of a person's relative "development" would include how it feels for other people to know her, work with her, interact with her, speak to and be spoken to by her.

Here "development" escapes a neat structural definition both as to person, relationship, and meaning, for it implies a participatory "as we go along" process as the developmental good, in which what develop are conversational possibilities in which the meaning is mutually constructed by the parties involved, and is always marked at once by a sense of mutual creation, discovery, falling away, and hard-to-define yet-to-be. Anderson (1997), citing several authors who have contributed in a profound way to our appreciation of this process, puts it this way: "Again, as Searle (1992) suggests, 'in a local dialogue or conversation, each speech act creates a *space of possibilities* for appropriate *response* speech acts' (emphases Anderson's) (p. 8). What emerges in this space does so on a moment-to-moment basis (Shotter, 1995) and cannot be predicted.

Meaning, for instance, according to Garfinkel (1967), is "tendentious" and cannot be stipulated in advance. *You* and *I* find out, and Garfinkel emphasizes, can *only* find out, "what it means as we go along" (Anderson, 1997, pp. 119–120). Gergen (1994, see especially pp. 264–271) describes this pro-dialogical process as a conversational frame of *supplementation*, in which the other (Shotter, 1993) feels she is a meaningful part of the conversation "only if others around them are prepared to respond to what they do and say *seriously*, that is if they are treated as a proper participant in ... 'authoring' of their reality, and not excluded from it in some way. For only then will they feel that the reality in which they live is as much theirs as anyone else's" (p. 39).

Instead, then, of "internal" "qualities" beyond, in constructionist terms, our epistemological reach, we look to applied social competence within discourse practices as the key to understanding moral and religious development. Such a shift implies a critical eye to the very social conventions that are part of the development of those social contexts and discourse practices, alert to the kinds of verbal strategies employed to make place, displace, or exclude would-be participants in the social world constituted in language. Here the "developing" person would be viewed as having increasing competence in building, maintaining, and contributing to relationships that would themselves make increasing room for other partners in conversation. She would be the kind of speaker whose voice would invite rather than exclude, value rather than denigrate, prize rather than punish, further rather than retard, the voicing and elaboration of other's words, the possibilities others would have in conversation with her. This would obviously then have a moral character—as Bakhtin himself insisted and as we elaborate further in the following paragraphs (see also Day, 1999a, 1999b, 2000; Day & Tappan 1995, 1996; Tappan 1991). We explore some of what this might mean for thinking about *religious* development in the paragraphs that follow.

DISCOURSE–ACTION: RELIGIOUS LANGUAGES
AS RELATIONAL PERFORMATIVES

We have already stressed that language is always part of an ongoing process of discursive construction of meaning, so that the individual speaker's word is always "half someone else" as Bakhtin would put it. In our work in moral psychology we have articulated a measure of how this work in speaking of the "moral audience" (e.g., Day, 1991a, 1991b)—that moral action is something that cannot be understood apart from the narrative forms in which it is imagined, explicated, played out, and that understanding this involves an appreciation of the speaker's location in relationship to those to whom she speaks, or imagines herself speaking, to whom she expects she will (have to) be accountable. These contexts inform the sense of possibilities she has for action, and indeed become part of the action, insofar as it becomes impossible to act in ways she cannot imagine describing to those who constitute the community in which she has her being as a speaker.

Thus we have argued that moral action is a function of the audience to which it is/can be played just as stories are a function of the audience to which they are/can be told. As the audience of the person shifts, so do the imaginative possibilities of intelligibility that shape the pathways further action can take. We literally talk ourselves into behavior in function of the priorities we give to those whom we imagine ourselves explaining our conduct afterward. Here time is scrambled—we plot action according to the story we can tell about it later to someone who matters to us, and as we conceive of what we imagine saying in retrospect about what we have done, we plot the action differently according to the different discursive frames in which we locate the later explanation we will give. We have held that close attention in research to narratives of moral conduct both include elements of this performative quality of moral action (Day, 1991a, 1991b; Day 1999d; Day & Tappan, 1995, 1996) and inform how we might conduct such research, from the conduct of interviews as they develop with our participants (Day, 1991a, 1991b; Youngman, 1993) to the reading and interpretation of interview texts (see also Brown & Gilligan, 1991). We have observed, for example, that shifting the imagined audience to whom the speaking participant addresses herself in the interview context shifts the relational definition that dictates what can be said, and thus produces different kinds of accounts of moral action, and that different reader stances dictate different kinds of narrative analysis on the part of researchers (Day, 1991a, 1991b; Day & Bissot, in preparation; Day & Naedts, 1995, 1999; Youngman, 1993).

In taking up a concern to better appreciate what kinds of framing might be appropriate to a discussion of *religious* development we have found it useful to explore this performative facet of language as it has been used by our research participants to describe *definitions of religion, religious experience, religious decision-making, and religious belief.* We have observed, for example, distinct discursive styles in our participants' ways of defining religion, of describing religious experience, and of framing what constitutes a religious decision, and that these styles are meaningfully related to the speakers' locations: men and women define and describe these things differently, students in elite college preparatory schools talk about them differently than do students in technical schools located in working class neighborhoods, self-identified Catholic adolescents and young adults talk about them differently than do Islamic youth, and older adults speak of them differently than do younger ones—*not* in ways traditionally conceived of by developmentalists as representing differences in developmental *level*, but instead as a function of the discursive practices that feature in the linguistically constituted worlds of which they are a part. We have also noted that as we explore shifts in the frame of the research interview, and thus create a different social world through discourse with the participant, what it becomes possible for the participant to say changes accordingly. Thus according to the rules we make with our "subjects," the "voice" of the "subject" changes—depending on these rules, for example, adolescents' talk about religion may variously represent what we have called *principled* or *relational* religious discourse, and *canonical, rebellious, or experimental* descriptions of religious belief. And we have tried to show how this implication of the researcher in the voice of the researched would necessitate new ways of reading the texts of interviews where both have been involved (Day, 1993, 1994, 1999a, 1999b, 2000; Day & Naedts, 1995, 1997, 1999).

Thus it has seemed to us fitting to say of the psychology of religion that research gives us little insight into the "interior" of the person, which, again, on the grounds we lay out, necessarily escapes our attention and ceases to be of consequence. What we have, instead, are surfaces—access to the discursive link a person has to the ever-changing relational world of which she is part through language, and what we stand to learn is not what she "believes" as an "internally held" series of attributions as to what is really out there in the world, but instead what value her discursive strategies have to her in being part of the conversations where *she* is defined and takes herself to be real in relationship with others (Day, 1993, 1994, 1998a, 1998b, 1999; Day & Naedts, 1997). This performative character of religious language thus becomes the focus of our interest—how it is that religious language as it is spoken by the person, or in a couple, or in a family, or other social group, constitutes relational possibilities, how shifts in various facets of diverse discursive frames involving religious discourse disrupt, or diminish, displace, or develop relational practices (Day, 1998a, 1998b, 1998c, 1999a, 1999b; Day & Andrews, 1999). We are then turned away from the very idea of what it would be to study what a person "believes" in which religious language was/would have been viewed as representative of internally held notions, and toward the question what religious language and other religious practice is/might be as a relational resource (see also Gergen, 1993, 1999). As we have noted elsewhere, the changing discourse of "subjects" in studies of religious psychology is particularly interesting as increasingly, these subjects themselves speak of religion with deliberate irony, coming increasingly to speak of religious language as a relational tool, and to themselves consciously reflect on how it is that they talk differently about religious belief according to the audience to whom they address their words. Whether and how religious language permits growth in the kinds, qualities, and meanings of relationship a person or couple or family or group can create becomes the focus of our concern in speaking of religious *development*, and likewise what we want to understand is how the character of *religious* language and practice contributes or presents obstacles to the same. Thus here as elsewhere in the new paradigm we want to adopt, development itself will have an as-it-happens quality marked by practices of supplementation of relationship, and will on this account more closely resemble a living work of art than a predefined plan with fixed endpoints of maturity (Day, 1993, 1998; see also other authors' use of our work in this regard, Ganzevoort, 1998b; Gergen, 1994; Streib, 1997; Wulff, 1997).

CONCLUDING/PROSPECTING REFLECTIONS ON THEORY AND RESEARCH IN THE PSYCHOLOGY OF RELIGIOUS DEVELOPMENT

As we have we have tried to show, our work is increasingly conceived of as a *postconstructivist*, and as such, *postmodern* effort to reimagine the very notion of religious development. We share what has variously been called the *constructionist* or *sociocultural* view of the *relational nature of knowledge* and the *generative nature of language*. Harlene Anderson (1997, pp. 201–210; see also Barth, 1993; Rorty, 1979, 1982, 1989) neatly outlines some of the assumptions that are to be found within the emerging community of researchers who share our perspective, to wit, that *knowledge is communal, knowledge is culture-bound, knowledge is a fluid, ongoing, process*, that *language and experience go hand in hand* that *language is active, language creates social reality*, and, we would add, *it is impossible to understand the functioning of language in development without understanding how persons construct meaning together through the conventions of narrative which permit them to define their place in time, their emplotment in meaningful action with others, and their vision of the world*. Given the traditional emphasis in various of the world's religious traditions on the performative character of religious language—that language in religious practice not only describes but recreates relationship to categories and figures of consequence—and given that religious language has traditionally been explicitly related to narratives about the meaning of life and of the world, how relationships in human communities ought to be configured and what developmental endpoints ought to be privileged, we find the psychology of religion to be a particularly interesting domain in which to work and to think from a discursive and constructionist stance regarding human development.

Curiously, this involves us in a move away from the *theological* language of constructivist developmental models, which share with a long line of thinkers from Plato to Kant (Kant's relationship is particularly well explicated by Kohlberg as central to the philosophical presuppositions inherent to constructivist developmental hermeneutics) to Descartes, "and with a host of structuralist thinkers in other domains (consider Chomsky, in linguistics, for one) the assumption that we best appreciate the manifest by unifying and grounding it in some hidden and higher level Reality. For example, people postulates grammar as a unifying, controlling and abiding intellectual structure that underlies the surface of phenomena of language. We like to see grammar hidden *a priori* standing behind language, very much as God stands behind the phenomena of the world: inescapable, always presupposed, and there whether people advert to it or not," (Cupitt, 1992, p. 9)—an elegant summation of the constructivist obsession with "deep structures" of moral, or religious functioning, where what cannot be explicated in terms of these structures evaporates from view as ceases to be of interest to the researcher, wherein, on our view, the subject comes to be seen as an instance of structure invented by researchers, and where the responsibility of such researchers in the construction of the social reality they produce in research is only little considered as being of any consequence whatsoever.

We would want to imagine religious development in a new key, once again turning to Cambridge philosopher Don Cupitt in arguing that:

> Old assumptions, distinctions and binary oppositions are being questioned. We are finally giving up the ancient idea that everything is two-leveled. Instead we are developing a new style of thinking, temporal, pragmatic, horizontal. Structure of every kind is coming to be seen as improvised and emergent within the flow of practice. Formal structures (rules, meanings, standards, and the like) are not laid up in heaven. They are not timeless, and they do not occupy a different level of being. We may come along later to abstract them from the flux and formalize them; but so far as their first appearance is concerned, grammar evolves immanently with the use of language, standards evolve immanently within practices, and so on. Nobody preplanned the *a priori*, it just grew. (Cupitt, 1992, p. 10)

It appears to us increasingly the case that constructivist accounts of religious development may be more important for how they got us here as part of the historical process of the development of developmental science itself, and what they tell us as artifacts of the philosophy of science that has dominated a particular epoch in efforts to define persons and describe, explain, and change their behavior, than about the "reality" of persons in their development. Like realist paintings in the musea of scientific art, we have come to view them as interesting ways of describing the world, but as portraying views of development that are themselves part of the reality they are meant to objectively describe—the supposedly objective researcher, like the realist painter, or the modernist abstractionist, pretends to describe things as they are or in the latter case to show what they are really made of. In the modernist idom research, like other arts, was to tell us what things really boiled down to—in painting this was color and light and so the rest was removed from view, in music it was sound and silence and so melody was deliberately fractured into its component parts, in dance it was movement and space and nothing more, in language and culture and psychology it was the deep structure of forms or grammars or structures behind the scene. Our task was to get at them, explicate the relationship between them—hidden away—and the action they guided in the world of appearances, and even more nobly, make the ignorant subject aware of them so she could better manage her life and enrich the quality of the world around her through what she would share with us; our knowledge of her deep interior. But now, we see ourselves as part of the flux, her inside escapes us, it is her surface, elaborate, languagey, relational, and prospective, that interests us. We pretend to no privileged stance as to what makes up her inside, and it is the very notion of that we allow to disappear from view. The place in between this person and ourselves is the place that interests us, and the only place we know how to speak of.

As we have tried to show, then, a critical stance regarding the assumptions that have guided the constructivist enterprise leave us suspicious about the story these developmentalists have told, and leave open alternative possibilities for redescribing the story of development, and for imagining its uses in psychological science and related research practices. In this chapter we have tried to outline what some of these alternatives might be, and how they might be of consequence for understanding some features of "religious" action, including religious "development."

REFERENCES

Anderson, H. (1997). *Conversation, language, and possibilities: A postmodern approach to therapy.* New York: Basic Books.

Austin, J. L. (1962). *How to do things with words.* Cambridge: Harvard University Press.

Bakhtin, M. (1981). In M. Holquist (Ed.), C. Emerson & M. Holquist (Trans.), *The dialogic imagination.* Austin: University of Texas Press.

Bakhtin, M. (1986). In C. Emerson & M. Holquist (Eds.), V. McGee (Trans.), *Speech genres and other late essays.* Austin: University of Texas Press.

Bakhtin, M. (1990). In M. Holquist & V. Liapunov (Eds.), *Art and answerability.* Austin: University of Texas Press.

Barth, B.-M. (1993). *Le savoir en construction: Former a une pedagogie de la comprehension.* Paris: Retz.

Brown, L., & Gilligan, C. (1991). Listening for voice in narratives of relationship. In M. Tappan & M. Packer (Eds.), *Narrative and storytelling: Implications for understanding moral development* (*New directions for child development*, No. 54). San Francisco: Jossey-Bass.

Bruner, J. (1986). *Actual minds, possible worlds.* Cambridge, MA: Harvard University Press.

Bruner, J. (1990). *Acts of meaning.* Cambridge, MA: Harvard University Press.

Bruner, J. (1991). The narrative construction of reality. *Critical Inquiry, 18,* 1–21.

Colby, A., & Kohlberg, L. (1987). *The measurement of moral judgment, Vols. 1 & 2.* New York: Cambridge University Press.

Coles, R. (1989). *The call of stories: Teaching and the moral imagination.* Boston, MA: Houghton-Mifflin.

Cupitt, D. (1991). *What is a story?* London: SCM Press.

Cupitt, D. (1992). *The time being.* London: SCM Press.

Day, J. (1991a). The moral audience: On the narrative mediation of moral 'judgment' and moral 'action'. In M. Tappan & M. Packer (Eds.), *Narrative and storytelling: Implications for understanding moral development* (*New directions for child development*, No. 54). San Francisco: Jossey-Bass.

Day, J. (1991b). Narrative, psychology, and moral education. *American Psychologist, 46,* 167–168.

Day, J. (1991c). Role-taking reconsidered: Narrative and cognitive–developmental interpretations of moral growth. *The Journal of Moral Education, 20,* 305–317.

Day, J. (1992). Narrative, moral development, and psychotherapy. Unpublished paper delivered as an invited lecture at St. John's College, University of Cambridge.

Day, J. (1993). Speaking of belief: Language, performance, and narrative in the psychology of religion. *The International Journal for the Psychology of Religion 3*(4), 213–230.

Day, J. (1994). Narratives of "belief" and "unbelief" in young adult accounts of religious experience and moral development. In D. Hutsebaut & J. Corveleyn (Eds.). *Belief and unbelief: Psychological perspectives.* Amsterdam: Rodopi.

Day, J. (1995). Sviluppo, educazione, e personalita morale. *Pedagogia et Vita, 53*(1), 31–49.

Day, J. (1998). Verhalen, identiteiten, god (en): Narratieve bemiddling en religieus discours. In R. R. Ganzevoort (Ed.), *De praxis als verhaal: Narrativiteit en praktische theologie.* Kampen: Kok Verlag.

Day, J. (1999a). Exemplary sierrans: Moral influences. In R. Mosher, D. Connor, K. Kalliel, J. Day, N. Yakota, M. Porter, & J. Whiteley (Eds.), *Moral action in young adulthood.* National Resource Center for the First-Year Experience and Students in Transition: University of South Carolina Press.

Day, J. (1999b). The primacy of relationship: A meditation on education, faith, and the dialogical self. In J. Conroy (Ed.), *Catholic education: Inside out-outside in.* Dublin: Veritas.

Day, J. (1999c). Das Gute wissen-das Gute tun: Narrationen uber Urteil und Handeln in den moralischen Entscheidungen junger Erwachsener. In D. Garz, F. Oser, & W. Althof (Eds.), *Moralisches urteil und Handeln.* Frankfurt am Main: Suhrkamp Verlag.

Day, J. (1999d). *The social construction of morality: Archives of contemporary psychology: Voices in social constructionism.* Los Angeles: Master's Work Video Productions.

Day, J., (2000). Le discours religieux en contexte: Deux études auprés d'adolescents et de jeunes adultes en Belgique francophone. In V. Sarraglov & D. Hutsebaut (Eds.). Religion et developppment humain: Questions psychologiques. Paris & Montréal: Harmattan.

Day, J., & Naedts, M. (1995). Convergence and conflict in the development of moral judgment and religious judgment. *Journal of Education, 177*(2), 1–30.

Day, J., & Naedts, M. (1997). A reader's guide for interpreting texts of religious experience: A hermeneutical approach. In J. A. Belzen (Ed.), *Hermeneutical approaches in the psychology of religion*. Amsterdam: Rodopi.

Day, J., & Naedts, M. (1999). Constructivist and postconstructivist perspectives on moral and religious judgement research. In R. Mosher, D. Youngman, & J. Day (Eds.), *Human development across the lifespan: Educational and psychological applications*. Westport, CT: Praeger.

Day, J., & Tappan, M. (1995). Identity, voice, and the psycho/dialogical: Perspectives from moral psychology. *American Psychologist, 50*(1), 47–49.

Day, J., & Tappan, M. (1996). The narrative approach to moral development: From the epistemic subject to dialogical selves. *Human Development, 39*(2), 67–82.

Desimpelaere, P., Sulas, F., Duriez, B., & Hutsebaut, D. (in press). Psycho-epistemological styles and religious beliefs. *International Journal for the Psychology of Religion*,

Dewey, J., & Tufts, J. H. (1932). *Ethics*. New York: Holt.

Di Loreto, O., & Oser, F. (1996). Entiwicklung des religiosen Urteils und religiose Selbstwirksamkeitsuberzeugung—eine Langsschnittstudie. In F. Oser & H. Reich (Eds.), *Eingebettet ins Menschsein: Biespiel Religion*. Lengerich, Germany: Pabst Science Publishers.

Dilthey, W. (1977). The understanding of other persons and their expressions of life. In W. Dilthey, *Descriptive psychology and historical understanding* (R. Zaner & K. Heiges, Trans.). The Hague: Martinus Nijhoff. (Original work published 1910).

Emerson, C. (1986). The outer word and inner speech: Bakhtin, Vygotsky, and the internalization of language. In G. S. Morson (Ed.), *Bakhtin: Essays and dialogues on his work*. Chicago: The University of Chicago Press.

Emerson, C., & Holquist, M. (1981). Glossary. In M. Bakhtin, *The dialogic imagination* (M. Holquist, Ed., C. Emerson & M. Holquist, Trans.). Austin: University of Texas Press.

Fowler, J. (1981). *Stages of faith: The psychology of human development and the quest for meaning*. San Francisco: Harper & Row.

Fowler, J. (1987). *Faith development and pastoral care*. Philadelphia: Fortress Press.

Fowler, J. (1996). *Faithful change: The personal and public challenges of postmodern life*. Nashville: Abingdon.

Fowler, J., Nipkow, K., & Schweizer, F. (Eds.) (1991). *Stages of faith and religious development*. New York: Crossroad.

Freeman, M. (1991). Rewriting the self: Development as moral practice. In M. Tappan & M. Packer (Eds.), *Narrative and storytelling: Implications for understanding moral development* (*New directions for child development*, No. 54). San Francisco: Jossey-Bass.

Gergen, K. (1982). *Toward transformation in social knowledge*. New York: Springer-Verlag.

Gergen, K. (1985). The social constructionist movement in modern psychology. *American Psychologist, 40*, 266–275.

Gergen, K. (1991). *The saturated self*. Cambridge, MA: Harvard University Press.

Gergen, K. (1993). Belief as relational resource. *International Journal for the Psychology of Religion, 3*(4), 231–235.

Gergen, K. (1994). *Realities and relationships: Soundings in social construction*. Cambridge, MA: Harvard University Press.

Gergen, K., & Gergen, M. (1986). Narrative form and the construction of psychological science. In T. Sarbin (Ed.), *Narrative psychology: The storied nature of human conduct*. New York: Praeger.

Gibbs, J., Basinger, K., & Fuller, S. (1992). *Measuring the development of socio-moral reflection: The socio-moral reflection measure—revised*. Hillsdale, NJ: Lawrence Erlbaum.

Gilligan, C. (1977). In a different voice: Women's conceptions of self and morality. *Harvard Educational Review, 47*, 481–517.

Gilligan, C. (1982). *In a different voice: Psychological theory and women's development*. Cambridge, MA: Harvard University Press.

Gilligan, C., Brown, L., & Rogers, A. (1990). Psyche embedded: A place for body, relationships, and culture in personality theory. In A. I. Rabin, R. Zucker, R. Emmons, & S. Frank (Eds.), *Studying persons and lives*. New York: Springer.

Gilligan, C., Ward, J., & Taylor, J. (Eds.) (1988). *Mapping the moral domain: A contribution of women's thinking to psychological theory and education*. Cambridge, MA: Harvard University Press.

Goffman, E. (1959). *The presentation of self in everyday life*. New York: Doubleday.

Ganzevoort, R. (Ed.) (1998a). *De praxis als verhaal: narrativiteit en praktische theologie*. Kampen, Netherlands: Uitgeverij kok.

Ganzevoort, R. (1998b). De praxis als verhaal: Introductie op een narratief perspectief. In R. Ganzevoort (Ed.), *De praxis als verhaal: narrativiteit en praktische theologie.* Kampen, Netherlands: Uitgeverij kok.

Garfinkel, H. (1967). *Studies in ethnomethodology.* Englewood Cliffs, NJ: Prentice-Hall.

Gergen, K. (1999). An invitation to social construction. London: Sage.

Habermas, J. (1979). *Communication and the evolution of society* (T. McCarthy, Trans.). Boston, MA: Beacon Press.

Habermas, J. (1983). Interpretive social science vs. hermeneuticism. In N. Haan, R. Bellah, P. Rabinow, & W. Sullivan (Eds.), *Social science as moral inquiry.* New York: Columbia University Press.

Harre, R. (1989). Language games and the texts of identity. In J. Shotter & K. Gergen (Eds.), *Texts of identity.* London: Sage.

Harre, R., & Gillet, G. (1994). *The discursive mind.* London: Sage.

Holquist, M., & Emerson, C. (1981). Glossary. In M. Bakhtin, *The diaologic imagination* (M. Holquist, Ed., C. Emerson & M. Holquist, Trans.). Austin, Texas: University of Texas Press.

Hutsebaut, D. (1995). *Een zekere onzekerheid: Jongeren en geloof.* Leuven, Belgium: Acco.

Hutsebaut, D. (1996). Post-critical belief: A new approach to the religious attitude problem. *Journal of Empirical Theology, 9*(2), 48–66.

Hutsebaut, D. (1997a). Identity statuses, ego-integration, God representation and religious cognitive styles. *Journal of Empirical Theology, 10*(1), 39–54.

Hutsebaut, D. (1997b). Structure of religious attitude in function of socialization pattern. Paper presented at the 6th European Symposium for Psychologists of Religion, Barcelona.

Jennings, W., Kilkenny, R., & Kohlberg, L. (1983). Moral development theory for youthful and adult offenders. In W. Laufer & J. Day (Eds.), *Personality theory, moral development, and criminal behavior.* Lexington, MA: DC Heath-Lexington Books.

Kamminger, G., & Rollett, B. (1996). The vienna religious judgment coding manual. Unpublished working document, University of Vienna, Austria.

Kaplan, B. (1986). Value presuppositions in theories of human development. In L. Cirillo & S. Wapner (Eds.), *Value presuppositions in theories of human development.* Hillsdale, NJ: Lawrence Erlbaum.

Kerby, A. (1991). *Narrative and the self.* Bloomington, IN: Indiana University Press.

Kohlberg, L. (1963). The development of children's orientation toward a moral order: Sequence in the development of moral thought. *Vita Humana, 6,* 11–33.

Kohlberg, L. (1969). Stage and sequence: The cognitive-developmental approach to socialization. In D. Goslin (Ed.), *Handbook of socialization theory and research.* Chicago: Rand McNally.

Kohlberg, L. (1981). *Essays on moral development, Vol. I: The philosophy of moral development.* San Francisco: Harper & Row.

Kohlberg, L. (1984). *Essays on moral development, Vol. II: The psychology of moral development.* San Francisco: Harper & Row.

Kohlberg, L., & Candee, D. (1984). The relationship of moral judgment to moral action. In W. Kurtines & J. Gewirtz (Eds.), *Morality, moral behavior, and moral development.* New York: John Wiley & Sons.

Lapsley, D. (1996). Commentary. *Human Development, 39*(2), 100–108.

Lawrence, J., & Valsiner, J. (1993). Conceptual roots of internalization: From transmission to transformation. *Human Development, 36,* 150–167.

Lourenco (1996). Reflections on narrative approaches to moral development. *Human Development, 39*(2), 83–100.

Lyons, N. (1983). Two perspectives: On self, relationships, and morality. *Harvard Educational Review, 53,* 125–145.

MacIntyre, A. (1981). *After virtue: A study in moral theory.* Notre Dame: University of Notre Dame Press.

McCarthy, T. (1982). Rationality and relativism: Habermas's "overcoming" of hermeneutics. In J. Thompson & D. Held (Eds.), *Habermas: Critical debates.* Cambridge: The MIT Press.

Mead, G. H. (1934). In C. Morris (Ed.), *Mind, self, and society.* Chicago: The University of Chicago Press.

Murken, S. (1993). *Religiositat, Kontrolluberzeugung und seelische Gesundheit bei Anonymen Alkoholikern.* Frankfurt: M. Peter Lang.

Oser, F., & Gmunder, P. (1991). *Religious judgement. A developmental approach.* Birmingham, AL: Religious Education Press.

Oser, F., & Reich, H. (1990). Moral judgement, religious judgement, world view and logical thought: A review of their relationship, Part 1. *British Journal of Religious Education, 12*(2), 94–101.

Oser, F., & Reich, H. (Eds.) (1996a). *Eingebettet ins Menschein: Beispiel Religion*. Lengerich: Pabst.

Oser, F., & Reich, H. (1996b). Psychological perspectives on religious development. *World Psychology, 2*(3–4), 365–396.

Packer, M. (1989). Tracing the hermeneutic circle: Articulating an ontical study of moral conflicts. In M. Packer & R. Addison (Eds.), *Entering the circle: Hermeneutic investigation in psychology*. Albany: State University of New York Press.

Packer, M. (1991). Interpreting stories, interpreting lives: Narrative and action in moral development research. In M. Tappan & M. Packer (Eds.), *Narrative and storytelling: Implications for understanding moral development* (*New directions for child development*, No. 54). San Francisco: Jossey-Bass.

Perry, W. (1970). *Forms of intellectual and ethical development in the college years*. New York: Holt, Rinehart & Winston.

Piaget, J. (1923). The language and thought of the child. London: Routledge, Kegan Paul.

Piaget, J. (1965). *The moral judgment of the child*. New York: The Free Press. (Original work published 1932.)

Power, C. (1991). Hard versus soft stages of faith and religious development. In J. Fowler, K. Nipkow, & F. Schweitzer (Eds.), *Stages of faith and religious development: Implications for church, education, and society*. New York: Crossroad.

Power, F., Clark, Higgins, A., & Kohlberg, L. (1989). *Lawrence Kohlberg's approach to moral education*. New York: Columbia University Press.

Puka, W. (1996). Commentary. *Human Development, 39*(2), 67–116.

Reich, H. (1997). Integrating differing theories: The case of religious development. In B. Spilka & D. McIntosh (Eds.), *The psychology of religion: Theoretical approaches*. New York: Westview/Harper Collins.

Reich, H., Oser, F., & Scarlett, W. (1996). Spiritual and religious development: Transcendence and transformations of the self. In H. Reich, F. Oser, & G. Scarlett (Eds.), *Psychological studies on spiritual and religious development: Being Human: The case of religion, Vol. 2*. Lengerich: Pabst.

Rest, J. (1979). *Development in judging moral issues*. Minneapolis, MN: University of Minnesota Press.

Rest, J. (1986). *Moral development: Advances in theory and research*. New York: Praeger.

Rorty, R. (1979). *Philosophy and the mirror of nature*. Princeton: Princeton University Press.

Rorty, R. (1982). *Consequences of pragmatism*: Minneapolis, MN: University of Minnesota Press.

Rorty, R. (1989). *Contingency, irony, solidarity*. Cambridge, UK: Cambridge University Press.

Sarbin, T. (1990). The narrative quality of action. *Theoretical and Philosophical Psychology, 10*, 49–65.

Schweder, R. (1982). Liberalism as destiny. *Contemporary Psychology, 27*, 421–424.

Searle, J. R. (1992). Searle on conversation (Compiled by H. Parret & J. Verscheuren). Amsterdam: John Benjamins.

Shotter, J. (1993). Conversational realities: Constructing life through language. London: Sage.

Spence, D. (1984). *Narrative truth and historical truth: Meaning and interpretation in psychoanalysis*. New York: W. W. Norton.

Streib, H. (1991). *Hermeneutics of metaphor, symbol, and narrative in faith development theory*. Frankfurt: M. Peter Lang.

Streib, H. (1997). Religion als Stilfrage. Zur Revision struktureller Differenzierung von Religion im Blick auf die Analyse der pluralistisch-religiosen Lage der Gegenwart. *Archiv fur Religionspsychologie 22*, 48–69.

Tamminen, K., & Nurmi, S. (1995). Developmental theory and religious experience. In R. Hood Jr. (Ed.), *Handbook of religious experience*. Birmingham, AL: Religious Education Press.

Tappan, M. (1989). Stories lived and stories told: The narrative structure of late adolescent moral development. *Human Development, 32*, 300–315.

Tappan, M. (1990). Hermeneutics and moral development: Interpreting narrative representations of moral experience. *Developmental Review, 10*, 239–265.

Tappan, M. (1991a). Narrative, authorship, and the development of moral authority. In M. Tappan & M. Packer (Eds.), *Narrative and storytelling: Implications for understanding moral development* (*New directions for child development*, No. 54). San Francisco: Jossey-Bass.

Tappan, M. (1991b). Narrative, language, and moral experience. *Journal of Moral Education, 20*, 243–256.

Tappan, M. (1992). Texts and contexts: Language, culture, and the development of moral functioning. In L. T. Winegar & J. Valsiner (Eds.), *Children's development within social contexts: Metatheoretical, theoretical, and methodological issues*. Hillsdale, NJ: Lawrence Erlbaum.

Tappan, M., & Brown, L. (1989). Stories told and lessons learned: Toward a narrative approach to moral development and moral education. *Harvard Educational Review, 59*, 182–205.

Tappan, M., & Brown, L. (1992). Hermeneutics and developmental psychology: Toward an ethic of interpretation. In W. Kurtines, M. Azmitia, & J. Gewirtz (Eds.), *The role of values in psychology and human development*. New York: John Wiley & Sons.

Taylor, C. (1991). The dialogical self. In D. Hiley, J. Bohman, & R. Shusterman (Eds.), *The interpretive turn: Philosophy, science, culture* (pp. 155–172). Cambridge: Cambridge University Press.

Valsiner, J., & Van der Veer, R. (1988). On the social nature of human cognition: An analysis of the shard intellectual roots of George Herbert Mead and Lev Vygotsky. *Journal for the Theory of Social Behaviour, 18*, 117–136.

Vandenplas-Holper, C. (1998). Le developpement Psychologique a l'age adulte et pendant la viellesse: Maturite et sagesse. Paris: Presses Universitaires de France.

Vitz, P. (1990). The use of stories in moral development: New psychological reasons for an old educational method. *American Psychologist, 45*, 709–720.

Volosinov, V. N. (1986). *Marxism and the philosophy of language* (L. Matejka & I. R. Titunik, Trans.). Cambridge, MA: Harvard University Press. (Original work published 1929.)

Volosinov, V. N. (1987). In I. R. Titunik & N. Bruss (Eds.), I. R. Titunik (Trans.) *Freudianism: A critical sketch*. Bloomington, IN: Indiana University Press. (Original work published 1927.)

Vygotsky, L. (1986). In Alex Kozulin (Ed. & Trans.), Thought and language. Cambridge: The MIT Press. (Original work published 1934.)

Vygotsky, L. (1978). In (M. Cole, V. John-Steiner, S. Scribner, & E. Souberman, Eds.), Mind in society: The development of higher psychological processes. Cambridge, MA: Harvard University Press.

Vygotsky, L. (1981). The genesis of higher mental functions. In J. Wertsch (Ed.), *The concept of activity in Soviet psychology*. Armonk, NY: M. E. Sharpe.

Wertsch, J. (1985). *Vygotsky and the social formation of mind*. Cambridge, MA: Harvard University Press.

Wertsch, J. (1989). A sociocultural approach to mind. In W. Damon (Ed.), *Child development today and tomorrow*. San Francisco: Jossey-Bass.

Wertsch, J. (1991). *Voices of the mind: A sociocultural approach to mediated action*. Cambridge, MA: Harvard University Press.

Wertsch, J., & Stone, C. A. (1985). The concept of internalization in Vygotsky's account of the genesis of higher mental functions. In J. W. Wertsch (Ed.), *Culture, communication, and cognition: Vygotskian perspectives*. Cambridge, UK: Cambridge University Press.

White, H. (1981). The value of narrativity in the representation of reality. In W. Mitchell (Ed.), *On narrative*. Chicago: University of Chicago Press.

Wilkinson, B., & Migotsky, R. (1994). A factor analytic study of epistemological style inventories. *The Journal of Psychology, 87*(3), 424–432.

Witherell, C. (1991). Narrative and the moral realm: Tales of caring and justice. *Journal of Moral Education, 20*, 237–242.

Witherell, C., & Noddings, N. (Eds.) (1991). *Stories lives tell: Narrative and dialogue in education*. New York: Teachers College Press.

Wittgenstein, L. (1958). *Philosophical investigations*. Oxford: Blackwell.

Wulff, D. (1997). *Psychology of religion: Classic and contemporary views, 2nd edit*. New York. John Wiley & Sons.

Youngman, D. (1993). Autres temps, autres mores. Unpublished doctoral dissertation, Boston University.

CHAPTER 28

Adult Development and the Practice of Psychotherapy

MICHAEL BASSECHES

This chapter explores the implications of the field of adult development for the practice of psychotherapy. The study of adult development has transformed developmental psychology from the study of child development into a lifespan developmental psychology, now capable of offering a viable alternative frame of reference for psychotherapy practice to traditional conceptual frameworks rooted in the fields of psychiatry and clinical psychology. This chapter asks: In what ways does using lifespan developmental psychology as one's *primary* conceptual frame of reference lead to differences in approaches to psychotherapy practice with adult clients, and/or to the training and supervision of adult psychotherapy practitioners?

The eight major sections that follow consider implications of thinking about the practice of psychotherapy in developmental terms for conceptualizing (1) the goals, values, and philosophical justification of psychotherapy practice in general; (2) the availability/accessibility of psychotherapy—that is, whom psychotherapy is for; (3) the nature of therapeutic processes; (4) the relationships and differences among various theories and techniques employed by therapists; (5) formulations of clients' psychological functioning, including implications for therapeutic goals and approaches for specific clients; (6) the psychological functioning of therapists, including implications for their role in the therapeutic processes; (7) why psychotherapy sometimes fails; and finally (8) the training and supervision of psychotherapy practitioners. The chapter concludes with consideration of issues regarding the integration of clinical with developmental perspectives. These issues are important in the psychotherapy practices of (1) therapists initially trained in clinical models who want to integrate insights from the field of adult development into their practice, (2) therapists initially trained in developmental models who practice psychotherapy within the broader social context of "providing health services," and (3) therapists trained in

MICHAEL BASSECHES • Bureau of Study Counsel, Harvard University, Cambridge, Massachusetts, 02138.

Handbook of Adult Development, edited by J. Demick and C. Andreoletti. Plenum Press, New York, 2002.

"clinical–developmental" psychology understood as a subdiscipline "involving the application of developmental considerations … to clinical problems and/or processes" (Demick, 1996, p. 13).

In focusing on what the field of adult development has to offer to the practice of psychotherapy, this chapter places in the background the equally important question of what the practice of psychotherapy has to offer to the field of adult development. Addressing this latter question entails comparing psychotherapy with such "natural" contexts as workplaces, families, and social networks in which adults develop, as well as with such practices as adult education, which may represent deliberate systematic efforts to foster adult development. While developmentally oriented adult education can be understood as an effort to create more optimal conditions for the development of large numbers of adults than their more "natural" environments, taken by themselves, provide, psychotherapy practices can be understood as efforts to create more optimal conditions for the development of individual clients. In effective psychotherapy, as in effective adult education, the developmental curriculum is often drawn largely from life challenges faced by adults outside the setting, but also in part created by conditions built into the structure of the setting itself. This chapter offers discussion of how psychotherapy fosters development, as well as how developmentally oriented therapists conceptualize the nature of clients' developmental curricula. However, for further explicit discussions of the relationships among therapy, other developmental contexts, and developmental curricula, the reader is referred to the work of Kegan (1982, Ch. 9; 1994, Ch. 7) and Basseches (1984, Part IV; 1989b, pp. 43–44).

IMPLICATIONS OF CONCEPTUALIZING
PSYCHOTHERAPY DEVELOPMENTALLY

The Goals, Values, and Philosophical Justification of Psychotherapy Practice

Not all lifespan developmental psychologists are genetic epistemologists in the tradition of Jean Piaget (1972). Fifteen years ago (Basseches, 1984), as the field of adult development was burgeoning, I presented research and analysis that offered a case both for the possibility of transforming genetic epistemology into an adequate basis for the study of adult development as well as child development, and for the importance of maintaining the following clear conceptual distinction within the rapidly expanding adult development field. This is the distinction between (1) studies clearly building on a conception of development that, like Piaget's, entails more complex organization and increased epistemic adequacy, and (2) studies that may employ the term "development" in their conceptualization, but that are presented in ways that the term's philosophical significance is lost (or worse, where its positive connotations are stolen, without any philosophical justification provided). This section highlights the centrality of this distinction to the question of whether work in adult development provides a *framework* with adequate philosophical underpinnings, or merely a tool, for psychotherapy practice.

The present volume is evidence of the degree to which the expansion of the field of adult development has indeed continued over the past 15 years. I hope that in the course of this expansion, this key distinction has been clarified rather than lost. I expect that this volume reveals both increased appreciation of how developmental change (in the sense of construction of more complex, epistemically adequate forms of psychological organization[1]) may occur in adulthood, as well as increased appreciation of many other psychological changes that tend to occur in adulthood, but that may or may not be developmental. I hope that my co-contributors to this volume collectively succeed in the increasingly challenging task of clarifying the distinction and relationships between these two rich bodies of work.

For the practice of psychotherapy, the broad spectrum of research on adulthood may provide useful information, but it is what I call the "dialectical–constructivist" approach that offers a significant philosophical alternative to the "clinical" justification of the practice. I refer interested readers elsewhere (Basseches, 1997a) for a fuller exposition of this approach. Here, I try to spell out how its view of the underlying goals, values, and justification of psychotherapy contrasts with the medical model in which conceptual frameworks as varied as psychoanalytic theories, cognitive–behavioral theories, and family systems theories have common roots.

Dialectical–constructivists understand development as a process of resolving conflicts which occur in experience through the construction of "novel syntheses" (Pascual-Leone, 1990) that reorganize conflicting activity schemes and representational schemes. The conflicts may be intrapsychic (among one individual's schemes) or interpersonal (actions organized by one person's schemes evoke negative, unanticipated, or disconfirming responses from others, whose activity is organized by their action and representational schemes). Novel syntheses that represent development entail increased differentiation and integration, in that they maintain previously existent capacities and meanings but have now reorganized those capacities and meanings in ways that take account of and resolve the conflict among them. They entail increased epistemic adequacy in that they facilitate adaptation to (assimilation and accommodation of) a broader range of

---

[1] I use the term "psychological organization" here, rather than "cognitive organization," although from a genetic epistemological perspective, they are fundamentally synonymous. The biological function of human "cognition" is understood as the organization of *all* human activity and experience and "representational knowledge" is understood as the extension of cognition beyond the sensorimotor level. Epistemic adequacy is a property of knowledge structures derived from their capacity for effectively performing this organizational function (ultimately, across the species, across situations, and across time). My choice of terminology is based on wanting to avoid the risk of some readers misinterpreting "cognitive organization" as excluding more emotional aspects of psychological functioning, although I bear the risk of the relationship between epistemological adequacy and psychological organization being less obvious (see Basseches, 1989a, 1997a, for further discussion of these issues of terminology and meaning in bridging the gap between developmental theory and psychotherapy practice). Piaget (1972) is most clear about the epistemological significance of his developmental research. However, I also include in this first body of work the work of neo-Piagetians (e.g., Newberger, 1980) as well as others (e.g., neo-Wernerians such as Wapner and Demick, 1998), who make clear that they equate development with increases in equilibration and organization of psychological and psychosocial systems, without explicitly specifying the epistemological significance of including knowledge systems within the purview of their work.

experience (data), and, in the case of interpersonal conflict, increase intersubjectivity and correct for egocentric and ethnocentric biases via the differentiation and integration of multiple perspectives.

As a foundation for psychotherapy, a dialectical–constructivist perspective on adult development makes development the primary goal of psychotherapy with adults. The primary value that should guide both therapist and client in the work of psychotherapy is the value on the construction of more differentiated and integrated, epistemically adequate organizations of experience, action, and meaning, which represent adaptations to the experiential challenges that bring a client to therapy and that emerge in the therapy process. Psychotherapy is justified insofar as it provides resources that supplement those clients find in themselves and their life circumstances for resolving the intrapsychic and interpersonal conflicts that they experience.

Within the medical model of psychiatry and clinical psychology, many conceptual frameworks also emphasize the importance of intrapsychic and interpersonal conflict. However, the medical justification of psychotherapy, like other medical interventions, is either pain or abnormality. Psychotherapy is justified as an effort to remove the distress and abnormality, which are usually manifest in the form of "symptoms." Thus, within a medical model, the goal of psychotherapy is the removal of symptoms of distress and abnormality. Even if symptoms are conceptualized as reflections of underlying "psychopathological" processes, and clinically oriented practitioners warn that it is the underlying disease process, rather than the specific symptoms that should be targeted, "psychopathology" can ultimately be judged only by observable experiences of distress and difference. Regardless of whether the focus of treatment is specific symptoms or the underlying processes that are presumed to generate symptoms, viewing psychotherapy as a clinical intervention implies that the values that should guide the therapist and client in their work are the ideals of freedom from suffering and conformity to "normality."

The monitoring of psychotherapy by outside managers brought by the era of "managed care" has made the aforementioned features of the medical model more vivid in the accompanying demands upon psychotherapists for symptom-focused assessments, treatment plans, and progress reports. However, in making these requests, and attempting to monitor the effectiveness of psychotherapy more scientifically, these managers are only building on assumptions of the model that were present when most practitioners were trained within a medical or clinical model of psychotherapy. Psychotherapists who don't like this intervention are forced to either confront the inadequacy of the model, or else attribute their discomfort to a reluctance to have their work systematically monitored by outsiders.

This discomfort is just one of several challenges to the clinical conception of the underlying goals, values, and justification of psychotherapy practice, which should make the developmental alternative appealing. While some of the challenges arise from the discomforts (e.g., with managed care) of psychotherapists who were themselves initially trained as clinical psychologists, psychiatrists, and clinical social workers, others have been posed from the developmental perspective. Whereas the medical model justifies psychotherapy ultimately by appeal to the value of "health," the developmental model justifies psychotherapy ultimately by appeal to the value of a human developmental process that can be viewed psychologically as ego development, biologically as intelligent adaptation, and epistemologically as truth-seeking (Kegan, 1982, p. 294). From a perspective shaped by this latter primary value, the operations of the norms of health in the "mental health" sphere appear suspect in a couple of ways.

First of all, the equation of psychological discomfort with ill health is problematic. It is epistemologically adaptive—in the service of the search for truth—for cognitive disequilibrium—a yet-to-be-resolved conflict among meaning-making structures or between expectations and experience—to be experienced psychologically as discomfort. Such discomfort motivates directing attention to the resolution of the conflict. While a developmental justification of psychotherapy affirms the developmental value of both the encounters with conflict that create discomfort and the resolutions of such conflicts that may evoke more felt satisfaction, a clinical justification of psychotherapy would only affirm the value of the resolution of conflict, and might equally affirm the avoidance of conflict or any other approach to reducing psychological discomfort (e.g., the use of medication). Now, in fact, many practicing psychotherapists would personally affirm the value of their clients facing conflict rather than avoiding conflict. However, to reconcile this value with the clinical perspective, they are required to make the case that facing the conflict will result in less discomfort for the client in the long run—a case that is often difficult to support based only on psychological theory and extant data. I suggest such therapists would be on firmer ground if they conceptualized their practice primarily using developmental models, rather than clinical ones.

The equation of mental health with "normality" is equally problematic. From the perspective of a value on truth-seeking, therapists' judgments of what conduct and experience constitutes mental health and what constitutes "deviance" or "psychopathology" are suspect as arbitrary or ethnocentric. Such judgments must ultimately appeal to some combination of what is statistically normative and to the shared norms of acceptability of those who are deemed mental health experts. The history of American clinical psychology and psychiatry's treatment of homosexuality vividly illustrates this problem. Whereas homosexuality formerly was viewed as a disorder—a form of psychopathology—to be treated (American Psychiatric Association, 1968), homosexuality now is viewed officially as a mentally healthy alternative, although "persistent and marked distress about sexual orientation" may be treated (American Psychiatric Association, 1994). Yet it is difficult to trace this change to any statistical change regarding the prevalence of homosexuality. Rather, the change appears to be the result of a sociopolitical process. Whereas formerly the "mental health" community sided with other American communities (e.g., "the religious right," "the business establishment," "major league professional sports," etc.) in excluding and devaluing members of the gay community, now members of the gay community have sufficiently effectively organized themselves and entered and influenced the mental health community to bring about a change in the latter's official platform, much as they have influenced the platform of a major political party.

If "mental health experts'" judgments are so arbitrary and subject to the sociopolitical winds blowing in their community and to the constitution of their membership, it is understandable and reasonable for clients to distrust psychotherapists who rely on these judgments in guiding their practices. This is a major problem with psychotherapists' conceptualizing their practices within a clinical frame of reference that depends on reference to what is normal and what is abnormal. Alternatively, within a developmental frame of reference, while a therapist's individual values, or the values of the communities with which the therapist identifies would be ingredients brought into the therapeutic process (as are the clients' values), only the interrelated values on epistemic adequacy and development would be appropriate as guiding values of the process. It would not be very difficult to make a convincing case that the widespread devaluation of homosexuality and social oppression of homosexuals led to systematically distorted communication (see Habermas, 1971) 30 years ago as it does today, and therefore represents an obstacle to the search for truth. This would justify therapists' most diligent efforts to avoid participating in such devaluation and oppression. In contrast, I imagine it would be much harder to make a convincing case that homosexuality is either mentally healthy or unhealthy.

In sum, to approach psychotherapy with a dialectical–constructivist lifespan developmental psychology as a primary frame of reference, is to view psychotherapy as a particular form of inquiry and experiential learning, aimed at development of more differentiated, integrated, adaptive, and epistemically adequate organizations of activity and experience. The conflicts on which the inquiry focuses are those that the client brings to the inquiry and that arise in the course of the inquiry. Such an approach is much less vulnerable to many of the internal discomforts and external challenges that therapists who approach psychotherapy primarily within a clinical/medical frame of reference encounter.

For Whom Is Psychotherapy?

Within a primarily clinical perspective, psychotherapy is for clients who suffer from some form of mental ill health. The diagnosis of a disorder (abnormality or suffering) is the justification for treatment designed to restore normality or remove suffering. Within a primarily developmental perspective, psychotherapy represents a set of additional resources which may be brought to bear, in conjunction with clients' internal and contextual resources, in constructing novel syntheses that resolve internal and external conflicts that clients face. While the next section will detail how these resources may be used, this section considers their availability and accessibility.

A developmental perspective implies greater availability and accessibility of psychotherapy. I think that it will be clear from the description of psychotherapy processes to follow, that since everyone faces internal and external conflicts, everyone could potentially benefit from psychotherapy, understood developmentally. On the other hand, to claim within a clinical perspective that everyone has a psychological disorder that could be treated by psychotherapy would be a reductio ad absurdum of the entire frame of reference, which conceives "disorder" in contrast to normal order.[2] Many therapists attempt to live, however uncomfortably, with this absurdity, and proceed to diagnose and treat any individual who

---

[2] Note, however, that if one conceives "disorder" developmentally—in contrast to "perfect order," rather than clinically—in contrast to "normal order," the concept of disorder is not very different from the concept of disequilibrium, which is, arguably, a universal experience.

enters their offices. Of course, it is not so easy for clients to live with this frame of reference. Because it is necessary to define, even stigmatize, oneself as a person with a psychological disorder to avail oneself of psychotherapeutic resources, it is understandable that a clinical frame of reference inhibits potential clients from making use of psychotherapy.

If a developmental perspective leads to the conclusion that psychotherapy is a potentially valuable resource for all people, rather than a subset of the population with mental disorders, who should pay the cost of this highly labor-intensive resource? I suspect that some psychotherapists maintain their loyalty to the clinical model in the hope of fostering accessibility to psychotherapy through the "parity" argument that health insurance should equally cover mental health costs, so that not only relatively wealthy individuals have access to treatment. (As I write this, the US Congress has recently failed to agree to legislate such parity.) But even if the parity argument were ultimately to succeed, access would still be limited to those whose conflicts are familiar, abnormal, and painful enough to be diagnosed as disorders.

Within the developmental model, the issue of subsidization of psychotherapy costs would be understood in the context of the theory of democracy. The argument that publicly supported education is crucial for democracy (see Dewey, 1916) has a much longer and much more successful history than the arguments for either universal health insurance or mental health parity. The developmental model suggests that the appropriate route to making psychotherapy more affordable through public funding would be the extension of the education-for-democracy argument to adulthood—an extension that the field of adult development now makes possible. Research in adult development that documents the importance of aspects of adult development to the realization of the ideals of democracy[3] could be conjoined with research in adult education that documents how individual, group, and/or family psychotherapy can most efficiently supplement large-group forms of adult education in successfully fostering such aspects of adult development.[4]

The Nature of Therapeutic Processes

How does the process of psychotherapy supplement a client's internal and contextual resources in fostering the construction of novel syntheses in response to intrapsychic and interpersonal conflicts?[5] It is important, first of all, to recognize that there is great diversity in the practices and procedures employed by the

---

[3] Habermas' (see McCarthy, 1978) theory of communicative competence and adult development, in my view, provides the richest extant framework for such research.

[4] As developments in technology facilitate many aspects of adult education becoming less labor-intensive, a case for the efficiency of judicious use of more labor-intensive psychotherapy to address those aspects of the developmental process to which it is best suited becomes more plausible. While the efforts to determine "judicious use" might bear some similarity to the managed care mission of cost containment through efficiency and regulation, the frame and criteria for such efforts would be more appropriate to the developmental value of psychotherapy. I propose this might mitigate those aspects of managed care which many therapists view as undermining the effectiveness of the process, while still effectively containing its cost.

[5] Significant portions of the material in this section are adapted (with permission of the publisher) from material which appeared previously in Basseches (1997a).

therapists whom adult clients consult, once the clients enter psychotherapy. This diversity is influenced by the wide range of (1) extant schools of thought or theories regarding psychotherapy; (2) extant theories or diagnostic frameworks for formulating the nature of clients' difficulties; (3) popular psychotherapy techniques; (4) idiosyncratic aspects of therapists' personalities, personal styles, and ideas affecting their practice; and (5) training experiences that have affected therapists' professional development. Psychotherapy outcome research (see *American Psychologist*, 1996: 51[10], for recent summaries and commentaries) suggests that there is indeed great diversity in the ways in which therapists may help clients achieve more successful adaptation, yet it does not demonstrate the clear superiority of any particular approach.

The language of a dialectical–constructivist model of adult development offers a frame or metatheory for understanding psychotherapy processes in general that is independent of any therapist's particular theoretical commitments or personal idiosyncracies. The section titled Developmental Perspectives on Therapists' Psychological Functioning considers therapists' formal theories within the context of the therapists' psychological organization as a whole, in assessing the impact of the therapist's development on the psychotherapy process. In this section, this model of adult development is offered to provide a frame for appreciating the contributions of a wide range of theories, practices, and therapists' personal styles to psychotherapy processes, without requiring adoption of the assumptions of particular schools of psychotherapy.

From a dialectical–constructivist perspective, the conflicts that bring clients to therapy can be understood as resulting from the interaction of limitations and distortions in the meaning-making structures and repertoire of schemes that the clients have developed in prior social contexts and the particular adaptive challenges posed by the clients' current environments. According to Pascual-Leone (1990), who has integrated adult developmental and neuropsychological perspectives in studying the construction of novel syntheses that transcend emotional conflicts, the key to creating such syntheses is the simultaneous boosting and maintenance of attention to all of the schemes, structures, and experiences that are in conflict. The prospect of help that psychotherapy offers can be understood to inhere in the various ways that psychotherapy offers additional resources to supplement the client's own procesess of attending to and making meaning of his or her experience.

After quoting from Greenberg and Pascual-Leone's (1995) overview of the role of attentional processes within psychotherapy, I will describe and illustrate three generic psychotherapeutic processes. Each of these processes involves utilizing the unique intimacy of the therapeutic relationship to offer additional resources that make adaptive novel syntheses more likely outcomes of the client's meaning-making efforts. I propose that all effective psychotherapy can be seen as involving some combination of specific forms of these three generic processes. Effective psychotherapies may differ from one another both in the specific forms and in the overall admixture of these three generic processes.

> Attentional allocation is the central processing activity determining people's awareness of themselves. What is important for therapeutic purposes is that attention is under both deliberate and automatic control. By using different types of interventions at different times, therapists can orient, direct, and monitor clients' deliberate and automatic attention

(Greenberg et al., 1993). In this way, attention provides a medium for change. People can use attention to alter their focus of awareness and to symbolize their inner experience. Personal change then can be achieved in many ways, including the following: (a) by attending to and symbolizing the internal complexity generated by automatic experience, (b) by bringing about a synthesis of new structures in therapy through coactivation of existing and newly formed schemes, (c) by generating vital explanations of currently symbolized experience, and (d) by restructuring emotional schemes by evoking them and exposing them to new input.

A dialectical constructivist perspective therefore yields a theory that recognizes the significance of the client's emotional experience as well as his or her capacity to construct meaning and develop concepts. This integration implies a view of human beings as multiple-level processors who use different types of propositional (symbolic–logical) information and affectively laden experiential (sensory, perceptual, imaginal, and representational) information. Human beings, in our view, construct representations of themselves and reality in a moment-by-moment fashion, all the while dynamically reacting to what they are attending to.

Thus, growth-promoting conscious experience derives from both deliberately controlled (often conscious, serial, and conceptual–representational) processing of information and automatic (often unconscious, parallel, and sensorimotor) processing of self-relevant information. Consequently, an adequate theory should recognize three major roots of experience: *(a) a conscious, deliberate, reflexive, and conceptual process…; (b) an automatic, direct emotional-experiential process…; and (c) the constructive, dialectical-dynamic interactions between the two (Greenberg et al., 1993).* Reflexive conceptual knowing processes provide explanations, whereas emotional schemes provide immediate reactions. The dialectical synthesis of these different sources of experience…ultimately leads the person to…psychological maturity.

In therapy, this dialectical constructive process often involves exploring differences between actual immediate experience and prior conceptually held views of how that experience should be. Contradictions between one's reflexive or acquired concepts (explanations) about how things are, or ought to be, and one's immediate experience of how things actually are constitute a great source of emotional distress, and these need to be focused on to produce new syntheses that can provide a greater sense of personal coherence. (Greenberg & Pascual-Leone, 1995, pp. 183–184, italics mine)

So some schemes and structures may function automatically, activated by elements of the situation, and some schemes and structures may be used deliberately by clients as they consciously work to make sense of their experience. The view of human experience italicized above suggests to me that therapists may influence clients' attentional, synthesis-seeking work by contributing resources to clients' "conscious, deliberative, reflexive and conceptual processes," by contributing resources to clients "automatic direct emotional-experiential processes," and/or by contributing resources to clients' efforts to bring automatically operating and more deliberately employed schemes and structures into new relationships with each other. The three generic forms of psychotherapeutic action, listed in Table 28.1, correspond to these three forms of influence. However, I begin with the latter form of influence in my descriptions, because it seems to me most essential and universal within psychotherapy.

When psychotherapy is successful, the combination of the processes of providing attentional support, offering interpretations, and participating in the enactment of novel experiences results in clients constructing more adaptive, differentiated, and integrated organizations of their own experience and action repertoires, such that the process of human development can continue to move forward. I shall now consider these three processes in greater detail.

Table 28.1. Three Generic Forms of Psychotherapeutic Action

A. ATTENTIONAL SUPPORT: Supporting the client's more effective deployment of his or her attention in organizing and reorganizing the schemes and structures that constitute his or her experience, and in addressing his or her adaptive challenges. (Contributes resources to clients' efforts to bring automatically operating and more deliberately employed schemes and structures into new relationships with each other.)

B. INTERPRETATION: Offering additional material for the client's attention in the form of novel linguistic representations of the client's experience and its social and physical contexts (i.e, offering interpretations—schemes and structures for understanding that are relevant and useful to the client's deliberate conscious efforts to make sense of his or her experience), and collaborating to varying extents in organizing these representations with those of the client. (Contributes resources to clients' "conscious, deliberative, reflexive and conceptual processes.")

C. ENACTMENT: Creating additional material for the client's attention by participating with the client in the enactment of novel experiences, thereby broadening the client's repertoire of immediate experience, while collaborating in efforts to organize and make sense of that experience. (Contributes resources to clients "automatic direct emotional-experiential processes.")

Forms of Attentional Support.   The forms of attentional support that therapists offer clients fall into two overarching categories. One category of attentional support includes those ways in which the therapist directly assists the client in attending to the full range of the client's experiences. This facilitates the reorganization of those experiences in ways that help to resolve conflicts and meet adaptive challenges so that the client's development may continue. The second category includes ways in which the therapist shares or alleviates the client's burden of managing extremely painful, threatening, or disorganizing emotions, including the panic, anxiety, shame, or sense of isolation, which may be brought on by the subjective experience of failure in the effective maintenance of a coherent, adequate organization of activity and experience. Successful establishment of a therapeutic alliance often entails the client feeling less alone in the experience of disorganization, more hopeful regarding a successful resolution of the conflict, and assured that someone else can recognize and respect the integrity of one's personhood, even when one can't recognize it oneself. This frees the client's attention from the sole management of the distress and the crisis of meaning-making failure, and makes more attention available for renewed efforts to meet the adaptive challenges that precipitated the crisis.

The first category of forms of attentional support includes such varied therapist activities as asking clients questions, empathically acknowledging what one has understood clients to be expressing, drawing attention to aspects of clients' behavior and experience, and reminding clients of experiences that the clients previously described. Such activities represent some of the ways in which therapists from a wide range of therapeutic traditions guide and support their clients' attention to their own experience. Through such activities, therapists boost their clients' awareness of aspects of the clients' own experience which they might on their own tend to ignore. Therapists also assist their clients in simultaneously holding together in consciousness aspects of their own experience, behavior, and sense-making that they might on their own tend to "split" (i.e., only experience sequentially in different times and/or contexts).[6] Consider the following interchange:[7]

> T: How are you today?
> C: Sick and tired of being told what to do by my mother.
> T: Really had it with her meddling in your life?
> C: Yeah, can you believe she called again today to check on whether I was doing my schoolwork?
> T: She's just always right there. … Kinda like you really don't even get the chance to try facing it on your own?
> C: Well, last semester she went about a month without calling me.
> T: Is that sadness I hear in your voice?
> C: (starting to cry) Yes, dammit. I felt completely awful and ended up with two Ds.
> T: So you felt badly *and* you were disappointed with your grades?
> C: I fell flat on my face! I just can't do it on my own, and I am so scared. (more crying)
> T: So your mother's checking on you really infuriates you, *and* you are scared by what happened when she left you alone?
> C: Yes. I really need to prove that I can succeed on my own.
> T: Some kind of success is urgent.
> C: I *did* succeed with my article on the hockey team's season!
> My roomate, my English teacher, and some guys from the team all said they really liked it! … And I never told my mother about that. … And I'm not going to …

The preceding interchange includes examples of empathically acknowledging named aspects of experience, highlighting under-attended-to aspects of experience, and bringing together in attention conflicting aspects of experience. Through such processes, conditions are created in which clients can create novel synthetic transformations of their own experience. In the language of dialectics (Basseches, 1984), it is through the affirmation of theses, the discovery of antitheses, and the holding together of theses with their antitheses, that the creation of syntheses occurs.

Often, but not always, finding the dialectical tension involves "exploring differences between actual immediate experience and prior conceptually held views of how that experience should be" as Greenberg and Pascual-Leone (1995, p. 183) describe. This is done by attending to and symbolizing immediate experience and then holding it together with prior constructions. The interchange described in the preceding could easily lead to the client's acknowledging the belief that he or she should have achieved more sense of independence of mother than he or she frequently feels. In turn, accepting that conflict may lead to the client to construct fuller, more complex, more adaptive representations of experiences of both independence and dependence.

The terminology of "guiding and supporting" attention, which allows discovery of antitheses and creation of syntheses in experience, is broad enough to describe a fair amount of what therapists who identify themselves as psychodynamic, cognitive–behavioral, existential–humanistic, or in other ways (including "eclectic"), all do, and to relate this activity to the fundamental developmental processes of the client. Each therapeutic tradition has its own language, and each

---

[6] Different therapeutic schools use varying language to explain such processes of splitting (e.g., defenses, meeting conditions of worth, etc.). However, the explanations share a common element— that of contradiction or incompatibility among the split off components.

[7] In the examples that follow, "T" indicates hypothetical comments a therapist might make and "C" indicates hypothetical responses by a client.

therapist is guided by his or her own theories in guiding and supporting the client's attention.

The second category of forms of attentional support includes activities that are primarily reassuring rather than attention-directive. They support the client's deployment of attention by freeing it rather than by assisting or guiding it. Instead of the therapist joining the client in holding the many aspects of client's problems or conflicts in the attentional field, the therapist holds the client so that the client can more effectively hold the conflicts. If the client feels less alone in his or her experience, attending to even intensely painful feelings may be more manageable.

This type of activity was at the core of earlier formulations of "client-centered therapy" (e.g., Rogers, 1959). Rogers suggested that the therapist's primary job was to provide a climate in which the client experienced "unconditional positive regard," which Rogers believed would then free the client to optimally deploy his or her attention (or "awareness") in the service of his or her own development. It is questionable whether so-called "nondirective" responses merely echo the client's expression and communicate positive regard. Insofar as such responses inevitably *select* particular aspects of the client's communication to reflect, they must play a role in guiding the client's attention as well (Category 1). However, insofar as the principal effect is to reassure the client and to *encourage* the client's continued efforts to work toward growth, a "nondirective" empathic response may be said to fall into this second category.

Clearly, expressions of appreciation of the client's struggle made by therapists of all traditions also fall into this second category. When such appreciation is combined with therapist responses which (in any theoretical or diagnostic language) implicitly or explicitly express the therapist's faith that this form of therapy can help this particular client,[8] the client's panic may be relieved and sense of hopefulness fostered. When this occurs, the client's attentional resources are clearly freed to address the task of creating novel syntheses of meaning-making schemes, and therefore such reassuring responses also fall into the second category of forms of attentional support.

Attention-freeing responses may range from "mmhmm," to "I'm here," to "You've tried so hard to let him know how much you're hurting," to "You have a case of agoraphobia with panic attacks that can be treated successfully in any number of ways."

In sum, development can be understood as an attentional challenge. A wide range of emotion-laden action and experience schemes, constructed, organized, and reorganized throughout the client's entire history, and all in some way activated by various aspects of the client's current circumstances or deliberately employed by the client in response to that experience, must again be reorganized in novel and more adaptive ways to enable the client to function effectively and satisfactorily in these new circumstances. A client may be well aware of the functioning of many schemes, while other schemes may be functioning with little or no conscious attention, except when their negative consequences are noticed. The construction of successful novel syntheses will require paying sufficient attention over time to all of the activated schemes and to their relationships to

---

[8] Of course, this process depends on the client's attributing valid professional authority in the matter to the therapist, which does not follow necessarily from the choice to consult the therapist.

one another. Under conditions of crisis, the attentional challenge is even greater, as one's attention is inevitably drawn to those emotionally powerful schemes that are triggered by the experience of grave danger to the overall functioning of the entire organization of the self. While the forms of attentional support provided in psychotherapy vary considerably, they all involve the therapist sharing some of the client's attentional burden.

Offering the Therapist's Meaning-Making Schemes (Interpretation). Some approaches to therapy rely almost entirely on processes of attentional support (e.g., Rogers, 1951). They support the client in mining the developmental possibilities inherent in the conflicts that the client experiences, through increasing awareness and holding together of antithetical schemes and structures until novel syntheses are constructed. However, other approaches (e.g., traditional psychoanalysis) offer the therapist's meaning-schemes to the client as ingredients of the developmental process. This may facilitate development in a variety of ways.

The therapist may offer schemes which are useful to the client in explaining or reorganizing conflicting activated schemes of the client. For example, consider the following exchanges:

> T: it is quite common to long to be close to someone *and yet also* to very much fear closeness.
> C: Well, you know, now that you mention it, I think that is exactly what I feel when I have been going out with someone for quite some time. How do other people deal with that situation?

or

> T: My guess is that wanting to hide your feelings of how much you like her *has something to do with* fear of conflict with other men.
> C: I did have this really awful dream where I got beat up last night....and in the afternoon I had had this thought about buying flowers for Sandy.

The value of the interpretation is realized only once the client is able to make use of it in his or her own organization of experience. It is essential for the therapist, once offering the interpretation, to go on to collaborate in discovering whether or not it is useful and relevant to the client, and if it is useful, how?

Alternatively the therapist's schemes may function as the antitheses to meaning-making schemes of the client, and it is in integrating the client's schemes with those of the therapist that development may occur. For example,

> T: You told me that you concluded you were powerless to influence things at your daughter's school, *but* in the situation you just described, it sounds to me like you came across as pretty powerful, maybe even scary, to your daughter's teacher.
> C: Really? I seemed scary? I've never thought about myself that way.

or

> T: I can see why you might take his offering to do the presentation with you as reflecting a lack of confidence in you, *but* it occurs to me that he may actually see it as an opportunity to learn something from you, or maybe to get to know you better.
> C. Well, maybe, but you always have a more optimistic view about these things than I do.
> T: Why do you suppose that is?

Again, it is essential that the therapist go on from this point to collaborate in discovering whether or not, and how, the therapist's and client's perspectives can be integrated.

Within the generic developmental model of psychotherapy proposed here, the term "interpretation" is used in a very broad way, not in the particularly psychoanalytic sense of an account of unconscious dynamics which may underlie conscious experience. "Interpretation" describes any situation in which therapists offer clients their own ways of understanding phenomena—ways of understanding that differ from or extend beyond the clients' understanding. Because therapists may differ tremendously among themselves in the theories they use and in the idiosyncratic ways that they understand human experience, human suffering, the sources of human difficulties, and how they can be ameliorated, interpretations can take very many specific forms. From a dialectical–constructivist perspective, the therapeutic value of such interpretations lies in the contribution they make to the particular dialectical process that is the client's own evolving organization of his or her life.[9]

Dialogue is a rich source of individual learning and development, and it is the fact that the therapist has different ways of making meaning from the client that makes psychotherapeutic dialogue possible. Therapists' formal theories represent part of the therapists' different ways of making-meaning, and insofar as they contribute to the generation of interpretations that are useful to the client, they may play an important contributory role to effective psychotherapy.

Enactments.   A client's own past and present experience, and existing ways of organizing that experience, necessarily constitute the bulk of the ingredients out of which novel, more adaptive syntheses are created in therapy. The therapist's attentional support contributes to this creative work. At times, as we have seen in the previous section on interpretations, therapists communicate their understandings and these understandings also contribute to the creative work. The interpretations supplement the mix, serving as additional ingredients which facilitate the creation of new syntheses. However, often what is needed most in the mix of ingredients to increase the likelihood of the client's creating novel, more adaptive syntheses, is not only new understandings but also new experiences. Enactments provide these new experiences, and thereby constitute the third generic psychotherapeutic process. Consider the following example:

> A woman, Marla, who was sexually abused as a child, operates on schemes developed out of that experience, and repeatedly participates in new situations in which she is abused, devalued or exploited. She withdraws from relationships, becomes desperately lonely, and then, motivated by her desperation, enters new exploitative relationships. She views such exploitation as an inevitable aspect of intimate relationships, and has well-developed schemes for both creating and escaping from relationships. She describes her experiences in therapy. Her therapist formulates and offers interpretations of how she may have come

[9] This dialectical constructivist view of therapeutic value contrasts with the tendency of objectivist (see Neimeyer [1995], for the distinction between objectivism and constructivism in general) perspectives to derive the therapeutic value of the interpretation from some notion of accuracy of either the interpretation itself or the framework from which the interpretation is derived. It *also* contrasts with the dialectical–constructivist view of the *epistemological* value of both interpretations and frameworks, which lies in the role they play in all of the dialectical processes involved in the overall evolving social organization of people's interaction with each other and with their environment. For the dialectical–constructivist, individual meaning-making dialectics are contributing and interacting components out of which the larger processes of epistemological validation derive, whereas for the objectivist, therapeutic value derives from prior epistemological validity.

to limiting beliefs that abuse and exploitation are inevitable aspects of close relationships. These interpretations may be useful to her and she may become more aware of and able to conceptualize how her behaviors may contribute to cycles of abuse. She may even be able to entertain intellectually the possibility of escaping such cycles. However, she is likely to lack a repertoire of schemes for creating and participating in trusting and trustworthy relationships. Without these schemes, such relationships would likely remain in the realm of abstract possibility for her. As Herman (1992), in her work on *Trauma and Recovery* articulates, it is often through the creation and constant testing of a trusting relationship with the therapist, that victims of repeated exploitative relationships establish an initial experiential basis for alternative relational possibilities. It is through the therapist continually proving himself or herself trustworthy to not exploit the client, in the face of a variety of tests and challenges, that recovery is facilitated.

The process described in this example is one form of enactment, in that novel activity generated in the context of the therapy relationship itself becomes the basis of the development of new schemes, which then become key ingredients in novel syntheses.

There is a very wide range of other forms that enactment may take. These include, for example, (1) enactment of transference patterns in psychodynamic therapy, creating the opportunity to transform the relationship and alter the patterns through interpretation of the transference; (2) systematic desensitization in behavior therapy; (3) experiments in relational expression in gestalt therapy; (4) homework assignments in cognitive–behavioral therapy and other out-of-session experimentation (e.g., in family therapy), planned and reviewed within the therapy sessions; (5) developing new supportive relationships in group therapy; (6) enacting a new ending to an unresolved past relationship in psychodrama; and so forth. In all of these techniques, new experiences are produced, attended to, conceptualized, and ultimately conceptually and behaviorally integrated with the rest of the clients' experience.

Again, the diversity of forms of therapy suggest that there are a wide variety of theoretical frameworks that may guide therapists in creating novel experiences with their clients. There is also variation in the importance that particular theories of therapy and particular therapists may give to enactment of novel experience and creation of new activity schemes (relative to the offering of new interpretations and of support in attending to existing coactivated activity schemes in novel combinatorial ways). However, every course of psychotherapy will inevitably involve novel experiences, and what is most important, from a dialectical–constructivist perspective, is the therapist's ability to collaborate with the client's efforts to organize and make sense of that experience.

In sum, using lifespan developmental psychology as one's primary frame of reference reveals great diversity in the specific ways in which therapists (1) support and guide their clients' attention, (2) generate interpretations and offer them to their clients, and (3) contribute to clients' enactment of novel experiences. Therapists also vary in the relative importance that they place on each of these kinds of processes, and in how they conceptualize their integration. However, regardless of the extent to which the conceptual schemes and the immediate experiences reflected on are those the client brings from life outside the therapeutic context, or are supplemented within the therapeutic context, successful psychotherapy fosters development through three generic processes. Each process offers powerful resources for use in clients' own attentional challenges of creating novel, more adequate organizations of immediate experience and more abstract conceptualization.[10]

Relationships and Differences Among Theories and Techniques Employed by Therapists

Within clinical models of psychotherapy, techniques ideally cure disorders and theories ideally explain the disorders, as well as explain why the techniques work. Because extant psychotherapy techniques are so varied, and theoretical languages and assumptions so different, so incompatible and so difficult to integrate, the therapist primarily operating within a clinical frame of reference (as well as the student attempting to learn to be a therapist) faces extremely difficult challenges. These are challenges relating to choosing from among, or integrating, theoretical perspectives and choosing among, or integrating, techniques. If it is presumed to be the technique or the theory in which the key to cure or therapeutic success resides, how terrifying it is to be so unsure which theory or technique is correct, and how unsatisfying it is to be "eclectic" and to feel that one is trying all sorts of interventions because of one's own and one's profession's degree of uncertainty regarding the true explanation of a disorder.

The developmental model presented in the preceding offers a powerful alternative in several respects. First, in conceptualizing the goal of psychotherapy as developmental reorganization, and in conceptualizing the locus of the reorganization processes as in the client's activity and experience as supported by the therapeutic relationship, it recasts, and admittedly deemphasizes, the role of precisely what the therapist thinks and does.[11] What is crucial is that the therapist

---

[10] Greenberg and Pascual-Leone (1995) contrast the dialectical–constructivist view of change in therapy, with alternate views that emphasize "modifying cognition, ... intellectual insight, ... catharsis," and "'going with' one's feelings" in the following way:

> Rather change comes about through the construction of new personal meaning (i.e., affective and cognitive), which is based initially on the symbolization in awareness of truly novel dynamic syntheses occurring in the internal field of activation. In this process, the construction of new meaning is greatly facilitated by the vivid evocation in therapy of emotionally laden experience in order to bring emotional experience into contact with reflective processes. Dialectical syntheses of emotion and reflection are the key to therapeutic change, as opposed to catharsis or reasoning alone.
>
> As we have said, novel experience emerges by a process in which aspects of different (and sometimes even opposing) schemes are synthesized into new, higher level schemes. These new schemes incorporate compatible coactivated features (not just the common features) of original schemes into new unified structures with new capabilities. It is therefore important that the felt experience be activated in therapy so that clients can use it in new constructions. Purely conceptual or rational constructions will not produce enduring therapeutic change because they do not involve a synthesis of emotional experience with other elements ... opposing tendencies need to be simultaneously evoked and attended to with increased attentional effort to create a new synthesis. This ability to attend and synthesize is greatly facilitated by the safety of an empathic and respectful therapeutic environment (Greenberg, Rice, & Elliot, 1993). (Greenberg & Pascual-Leone, 1995, pp. 182–183)

[11] Again, this is entirely consistent with the current state of empirical research on psychotherapy outcomes, which suggests that the psychotherapy process produces helpful results but, outside very limited conditions, no particular theories or techniques demonstrate more effective results than others. (See *American Psychologist*, 1996).

offers resources that supplement the attentional, conceptual, and experiential resources that the client brings, and that the therapist musters the conceptual and interpersonal skills needed to collaborate effectively in the client's efforts to make developmental use of those resources. The therapist's theories and techniques are recast as resources, rather than as explanations of illness and methodologies of cure. Techniques and ideas may be perfectly compatible as resources, although they may be incompatible as explanations and methodologies.

Second, in conceptualizing the distinctions and interrelations among attentional, conceptual, and experiential processes, the developmental model provides a basis for understanding the relationships of various techniques and theories. Regarding techniques, they may be understood as efforts to create new experiences, to bring new conceptual tools into relation with experiences, or to support attention to the integration of conceptualization with experience. Techniques being considered to serve the same function within psychotherapeutic process can be compared in terms of their usefulness and compatibility for any given client and therapist. Techniques intended to serve different functions may be evaluated in relationship to each other by asking which kind of therapeutic resource is most needed by a client at any particular time, as well as by evaluating the effectiveness of the resources for the needed functions.

Within the developmental model, psychotherapists' theories appear to serve two functions in psychotherapeutic practice. One is as a resource that the therapist brings to the process of interpretation, in particular. The conceptualizations that the therapist offers to the client may be based on the therapist's preferred theories, and the discussion in the previous paragraph also applies to understanding the relationship of theoretical ideas which may be brought explicitly into therapeutic dialogue. The second function of theory is to help keep the therapist's activity as a whole organized, within the entire ongoing interactions between therapists and clients. Here then, the appropriate criterion of evaluation is whether the theory supports the therapist in achieving and maintaining a collaborative role relationship. This function is discussed more fully in the sections below.

Formulations of Clients' Psychological Functioning

In this section, discussion turns to the role of the field of adult development in providing alternative frames for assessment to the many diagnostic frameworks offered by traditions growing out of psychiatry and clinical psychology. As mentioned in previous sections, central to the clinical model is a formulation of the nature of client's mental status, adaptive style, symptoms, disorder, and in some cases, underlying disease process. This formulation then becomes the basis for a treatment plan—what the therapist intends to do to ameliorate the symptoms or the disorder. Thus, various approaches to psychotherapy have spawned various classification systems (e.g., oedipal vs. pre-oedipal issues, enmeshed vs. isolated family systems, borderline vs. narcissistic personality disorders, etc.), often along with specific formalized or standardized measurement tools to be used in conjunction with the classification systems. These classification systems make sense in the context of rationales for employing different techniques, or treatment approaches, depending on the formulation.

Generating classification systems—descriptions of steps, stages, phases, positions, and so forth within sequences—as a way of mapping common adult developmental processes is also a core activity in the field of adult development. These
systems may comprise stages in the development of very broad aspects of psychological functioning such as levels of cognitive organization (Piaget, 1972), levels
of ego development (Loevinger, 1976), or subject–object stages (Kegan, 1982), as
well as of much more specific aspects of adult functioning (e.g., parental awareness
levels—Newberger, 1980). Adult developmentalists have also developed various
specific formalized or standardized assessment tools (e.g., Loevinger, Wessler, &
Redmore, 1970) for use in accordance with their classification systems.

Insofar as all of these systems may have implications for treatment planning,
it is reasonable to expect that psychotherapists relying mainly on clinical frames
may incorporate developmental assessment tools and classification systems and
that those relying mainly on developmental frames may incorporate clinical
assessment tools and classification symptoms, in their ways of formulating clients'
difficulties and planning their responses. Thus one can find in the literature a variety of proposals or illustrations of ways of fruitfully bringing these conceptual
tools together in the psychotherapy process. For example, Noam (1986) has argued
that clients who share the diagnostic features that justify the label "borderline personality disorder" may vary dramatically in their levels of adult cognitive and ego
development, with very significant differential treatment implications.

For those psychotherapists for whom developmental frames are primary and
central in conceptualizing psychotherapeutic goals, it is not surprising that the
developmental aspects of the formulation most centrally influence the therapists'
understanding of the therapeutic work. So, for example, Kegan (1982) who clearly
opts for development as the central common goal of psychotherapy, argues that
psychotherapy, as any "holding environment" that adequately supports development, must most importantly provide a combination of "confirmation" (of a
client's form of meaning-making), "contradiction" (challenge to a client's form of
meaning-making), and "continuity" (opportunity to be reconstructed within a
client's emerging form of meaning-making). Yet Kegan illustrates the differences
in what the therapeutic environment must do to perform adequately these functions for different clients, depending on their current and emerging forms of
meaning-making.[12]

Ivey (1986) has perhaps taken furthest this approach of assessment, formulation, and treatment planning and evaluation based primarily on adult clients'
stages of developmental. Ivey's "developmental therapy suggests that it is indeed
possible to apply developmental theory directly to therapeutic and counseling
practice and to measure that process" (Ivey, 1986, p. xv). His book addresses these
"implications for therapeutic practice" in the form of a training manual to help
psychotherapists accomplish the following:

> First, it is possible to identify the cognitive–developmental level of the client. Second, we
> can match our verbal and non-verbal interventions to the specific cognitive level of the

---

[12] Kegan describes forms of meaning-making in terms of which aspects of the self and environment the
client experiences as objects of reflection and which aspects of self and environment the client experiences as inseparable from his or her subjectivity. In each new form, what was previously "subject"
is made "object."

> client, thus facilitating exploration and later cognitive-developmental processes. In effect, it is feasible to match counseling skills and theory to the observable developmental level of the client. (Ivey, 1986, p. xiv)

Ivey (1986, p. 144) goes on to elaborate examples of approaches that he believes create appropriate "therapeutic environments" for adults functioning at each of four successive developmental levels: "relaxation, gestalt, exercises, behavior modification" for the "Sensori-Motor" level; "assertiveness training, reality therapy" for the "concrete operations" level; "Rogerian, psychodynamic/reframing therapies" for the "formal operations" level; and "feminist therapy, Lacanian therapy, and family therapy" for the "dialectics" level.

In contrast, I have argued (Basseches, 1989a) for a much more modest use of stage classification systems in the context of psychotherapy. Although it does represent a developmental approach to assess a client's stage, and set as a therapeutic goal fostering development to a subsequent stage, I do not believe that such an approach adequately addresses clients' individuality. Although "stage structures" describe general abstract features of organizations of activity and meaning that may be recognized across groups of adults, I argue that the individual's unique psychological organization, and the unique conflicts that emerge within that organization, represents a more appropriate focus for therapeutic work than abstract stage structures. As discussed in the previous section, to focus on facilitating, through generic psychotherapeutic processes (entailed by all of Ivey's approaches as well as those of others), construction of unique novel syntheses that resolve unique conflicts through differentiation and integration, represents an equally developmental approach to psychotherapy.[13] Within this latter approach, the value of developmental stage theories for psychotherapy is that they identify both the power and the limitations of various general forms of meaning-making, and for any given form, specify the type of situations in which those forms tend to confront their limits. Such theories can therefore be used by therapists to help them recognize, and generate formulations of, those situations in which unique conflicts that a client reports correspond precisely to the limitations of a general form of meaning-making.

A second problem with an approach which locates every client at a particular stage of development and makes development of the next stage in the sequence the therapeutic goal is what I have called the problem of irrationality (Basseches, 1989a). Stage-change models describe idealized processes of complete reorganization and integration of psychic structure that are far from the reality that psychotherapists confront of how individuals' unique psychological organizations evolve. In the operation of any unique psychological organization, structures organized at different levels of complexity may be activated at different times, or even simultaneously (see Phillips, Basseches, & Lipson, 1998). What clients and therapists often experience as "irrationality," and appropriately make foci of therapy, is the activation and lack of adequate reciprocal assimilation of multiple

---

[13] This latter approach is more consistent with the Wernerian approach of adapting developmental modeling to a wide range of "microgenetic" processes (Wapner & Demick, 1998) as well as with Laske's proposal (this volume) that the related process of executive coaching focus on individuals' unique professional agenda, rather than structural stage.

levels of structures.[14] While stage-sequence models are woefully inadequate as *descriptions* of actual psychological processes of development, by describing ideal forms of psychological organization they offer very useful tools for recognizing the actual activation and conflictual functioning of multiple stage-structures within a particular client.

In sum, formulations and assessments of particular clients' functioning may perform the same two functions as therapists' theories in general, discussed in the section titled Relationships and Differences Among Theories and Techniques Employed by Therapists. They occasionally may be offered to clients as potentially useful interpretations. They may also organize the therapist's activity, insofar as within the therapist's frame of reference, differing classifications of different clients warrant different treatment approaches. Adult development frameworks have generated a broad range of systems of classification—quite different from those generated within clinical psychology—but that are also claimed to have treatment implications when used for formulation and assessments. Because it is relatively easy to mix and match elements of clinical and developmental assessment in therapists' practice of creating formulations, the field of adult development has the potential to inform the work of all therapists who work with adults. However, it is likely that therapists whose primary conceptual frameworks are developmental will prioritize promoting particular developmental tranformations, while therapists whose primary frameworks are clinical will prioritize ameliorating particular clinical disorders.

Regardless of whether a therapist's primary orientation is clinical or developmental, the question of the extent to which the therapist uses the formulation to guide the therapy, vs. the extent to the therapy is co-constructed by client and therapist, with both parties' formulations discussed, evolving, and negotiated as part of the process is a very important one. In considering the roles of both therapists' formulations and therapists' development in psychotherapeutic success and failure, the next two sections discuss the risks of therapists overvaluing their formulations. At this point, suffice it to say that a developmentally oriented therapist who believes he or she knows better than the client what's good for the client's development bears the same risks as a clinically oriented therapist who believes he or she knows better than the client what's healthy for the client. However, whereas the clinical model of psychotherapy-as-cure is fundamentally dependent in its operation on mental health experts' judgments of "health," the developmental model presented here relies on the therapy process, and any novel syntheses constructed through the psychotherapy process, to evaluate all the ideas, experiences, and formulations that contributed to the therapy process.

Developmental Perspectives on Therapists' Psychological Functioning

The relevance of adult development to professional functioning and the role of professional experience in adult development have been recognized in a variety

---

[14]Psychodynamic theories have at times posited various unconscious forces to explain such experiences, but viewing mental processing as multilayered and yet imperfectly organized recognizes the importance of such experience of intrapsychic conflict without necessitating the reification of eternally warring parties within the human psyche (see Basseches, 1989a).

of professional contexts (e.g., Drath & Palus, 1994; Kegan, 1994). Guided by a dialectical–constructivist perspective, some researchers have investigated the psychological development of psychotherapists, in particular (Bopp, 1984; Grigoriu, 1998; Pratt, 1993). A dialectical–constructivist perspective on the role of the therapist in the psychotherapy process differs considerably from traditional clinical perspectives. Yet these perspectives must be brought together to articulate the implications of adult development for thinking about the functioning of all psychotherapists in their practice, including those who operate primarily using clinical frameworks.

Within most frameworks derived from clinical perspectives, it is the clinician's knowledge about the nature of the client's disorder, and his or her ability to execute techniques that are theoretically or empirically associated with improvement of the client's condition, which will lead to therapeutic effectiveness. This view can be juxtaposed with my earlier claim that both therapists' general knowledge regarding psychological processes, as well as specific formulations (clinical and developmental) of the client's psychological functioning and associated conflicts, play two basic roles in the therapeutic processes. The first is as a source of interpretations that may be offered to the client as resources. The second role is to organize the therapist's activity within the therapeutic relationship.

Regarding the first role, a therapist who is more knowledgeable in areas related to the client's conflicts, and more capable of formulating accounts of the client's psychological functioning, has improved prospects for generating interpretations that may be useful to the client. Yet interpretation is only one of three generic psychotherapeutic processes, and the effectiveness of this process depends as much on the therapist's ability to collaborate effectively in the client's exploration of how he or she can use interpretations, as on the richness of the interpretations themselves. While psychological knowledge is clearly a resource for a therapist, overvaluation by a therapist of his or her own knowledge may be an impediment to effective collaboration.

With regard to the second function of organizing the therapist's activity, (1) training, experience, and comfort in the implementation of various techniques, as well as (2) adequate conceptual rationales for the choice of techniques for particular clients,[15] may play very crucial roles in keeping the therapist's activity well organized. But what exactly do we mean by "well organized?" From a dialectical–constructivist perspective, each new client, each new session, and each new moment in therapeutic process is an adult developmental challenge for the therapist to assimilate and accommodate with his or her therapeutic repertoire of schemes and structures. A relatively rigid, inflexible repertoire may nevertheless lead to effective collaboration with the client, and effective functioning of the generic therapeutic processes, if the repertoire is well matched to the challenges posed by the particular conflicts—the developmental curriculum—that the client brings. Such a result is likely to be experienced as confirming by the therapist. On the other hand, when the same relatively rigid repertoire is brought to work with

---

[15] Included in choice of techniques here is the idea of choice of focus. In providing attentional support, a therapist's formulations of a client's conflicts may influence to which aspect of the client's experience the therapist draws attention. In encouraging enactment, a therapist's formulation may influence what kinds of new experiences the therapist encourages the client to try outside of therapy, as well as to which experiences within the therapeutic relationship the therapist draws attention.

a client who is not able to make use of the resources that the therapist offers, the client's response is likely to be distressing and disorganizing for the therapist, who expects successful results. The therapist's distress and disorganization are likely to pose further obstacles to collaboration and challenges for the client.

I will take up this thread further in the next section on understanding failures of psychotherapy. At this point, let us compare the therapist with a relatively rigid repertoire to a therapist with a more developed repertoire. The more developed repertoire will be characterized by greater differentiation and integration, and it will provide a higher level of equilibrium (i.e., the ability to effectively assimilate and accommodate a broader range of client responses, without becoming disorganized in ways that pose obstacles to collaboration within the psychotherapy relationship). Because clients can be expected eventually to bring into the therapeutic relationship precisely those aspects of their repertoires that cause them most conflict and are likely most disruptive to clients' other social environments, it is ideal if the therapist's psychological organization can accommodate such typically disruptive behaviors. Since it is also likely that at times, a therapist's choice of techniques, and resources to offer, may be found unhelpful by a client, it is also ideal if the therapist's psychological organization can accommodate such experiences of frustration as well.

In my experience, this is where one encounters some vulnerabilities associated with the clinical assumption that it is the clinician's knowledge about the nature of the client's disorder, and his or her ability to execute techniques that are theoretically or empirically associated with improvement of the client's condition, which will lead to therapeutic effectiveness. This assumption often leads to the association of frustration with fears of incompetence, which lead to disorganization or efforts at self-maintenance and self-protection, which function as obstacles to maintaining effective collaboration with clients (see Basseches, 1997b). In contrast, a therapist's psychological organization is less likely to be disrupted if he or she understands psychotherapy developmentally as a process of inquiry into conflict, in which it is expectable and appropriate that (1) unanticipated events will provide opportunities for therapists as well as clients to construct novel syntheses and (2) conflicts between therapists' repertoires and clients' repertoires will emerge and require collaborative efforts to generate novel syntheses.

All of this suggests that two central aspects of therapists' ego development can have a powerful impact on his or her psychotherapy practice (see Basseches, 1997b). One aspect is the level of organization of a therapist's interpersonal functioning and related capacities to collaborate in inquiry into interpersonal conflict. The second is the level of organization of his or her epistemological reasoning regarding the relationships of received psychological knowledge, prior training and experience, and current and future experience in psychotherapy practice. Furthermore, therapeutic encounters provide powerful challenges that can stimulate therapists' ego development along both of these two crucial dimensions. Therapists at various levels of adult ego development can be effective participants in any of the three generic psychotherapeutic processes, provided that (1) there exists a good match between the resources the therapist offers and (2) the therapist's participation in the process is not disrupted by limitations in the repertoire of schemes which organize the therapist's activity. At the same time, increased differentiation and integration of therapists' psychological knowledge, interpersonal skills, and epistemological perspectives should increase their likelihood of effectiveness within a broader range of psychotherapeutic experiences.

In analyzing what is involved in the differentiation and integration of interpersonal skills, some authors have focused on the complex set of dialogue skills involved in what is commonly referred to by the single term "empathy" (see Bohart and Greenberg, 1997), while others have looked at the forms of language (see Havens, 1986) that a therapist needs to master to work with clients who require varying conditions of closeness/distance and encouragement to work effectively in psychotherapy. My own work on the epistemological dimension of therapists' development (Basseches, 1997b) illustrates how therapists' transforming universalistic epistemological assumptions to relativistic ones protects the therapists from risks of disorganization, but at the expense of making withdrawal from some kinds of therapeutic opportunities for clients more likely. In contrast, the subsequent transition from relativistic to dialectical epistemological perspectives facilitates fully embracing such therapeutic opportunities. Benack's (1981) study is unique in empirically considering the relationship between therapists' epistemological development and their capacities for empathy.

Understanding Psychotherapeutic Failure

From a clinical perspective, the potential effectiveness of psychotherapy for treating a range of disorders has been more clearly documented in the past decade than previously (*Consumer Reports*, 1995; Dawes, 1994). Yet the results of formal studies and most therapists' anecdotal knowledge converge on the conclusion that psychotherapy is still far from 100% effective. But having documented sufficient positive outcomes to recommend psychotherapy over no treatment for many "disorders," it is appropriate for psychotherapists to turn more attention to understanding negative outcomes. Of course, what is a negative outcome depends considerably on one's choice of outcome measures—an area where psychotherapists' with primarily clinical vs. primarily developmental orientations could be expected to part company. Developmentally oriented research might focus on whether or not novel syntheses of conflicts brought to therapy were created, while clinically oriented research might consider whether distress or abnormalities reflected in various symptoms were reduced. To illustrate how an adult development framework can help us understand failures of psychotherapy, this section considers one particular form of experience in psychotherapy that psychotherapists of all persuasions could probably agree is negative.

The majority of reports that I've heard regarding destructive or highly disappointing experiences[16] by former psychotherapy clients reflect the following common pattern:

> The client perceived that an otherwise competent and well-meaning therapist was unable to hear, understand, respond to, or tolerate some very important aspect of the

---

[16] I have also heard client reports of very destructive experiences of sexual abuse by therapists. Although such experiences have received far more public attention than other kinds of negative experiences, these experiences, as well as related but less damaging experiences of "boundary violations," represent a small fraction of the negative experiences about which I have heard.

client's experience, because it conflicted with some aspect of the therapist's conception of the client, the therapist's view of the nature of therapy, the therapist's view of human nature, or the therapist's own self-concept. The client then either gave up on trying to communicate that aspect of his or her experience or persevered and felt the therapist reacting increasingly defensively in response. Frequently, the therapy relationship ultimately ended with the client feeling disappointed, hurt, reinforced in limiting expectations regarding human relationships, or at worst, when the ending repeated a pattern of previous traumatic relationship endings, re-traumatized. During the process, clients' experiences varied from feeling irritation, frustration or rage toward the therapist, to feeling disempowered, devalued, shamed, or "annihilated" by the therapist, to making strenuous efforts to adopt the therapist's perspective at the expense of denying aspects of one's own experience, and only realizing the negative consequences of these efforts during or subsequent to the ending of the therapy. (Basseches, 1997b, p. 87)

Applying the developmental perspective proposed in this chapter, I have suggested (Basseches, 1997a, 1997b) that this pattern of negative experience reflects situations in which the three generic psychotherapeutic processes have been blocked by significant conflict between meaning-making needs of the therapists and the meaning-making needs of the client. For example, the therapist may not have been able to support the client's attention to aspects of the client's experience that made no sense to the therapist—that could not be assimilated and accommodated by the therapist's repertoire of meaning-making schemes.[17]

In the previous section we considered ways in which the psychological organization of the therapist influenced the likelihood of disruption to the therapist's effective functioning within the relationship. More differentiated and integrated and therefore more flexible repertoires, in the psychological organization of the therapist's activity, were associated with less vulnerability to disruption. Such development in the therapist probably also makes obstructions of therapeutic processes by meaning-making conflict less likely. However, based on my own experience and that of many other very experienced developmentally and clinically oriented therapists with whom I have talked, the occasional emergence of meaning-making conflicts that disrupt the smooth functioning of the generic therapeutic processes is an inevitable aspect of psychotherapeutic practice.[18] Therefore, in an attempt to prevent the type of negative experiences of therapy described above, I have applied my own prior research on adult cognitive and epistemological development (Basseches, 1984) to conceptualizing what is necessary to restore the generic therapeutic processes once they have been obstructed by meaning-making conflict:

When meaning-making conflict occurs in psychotherapy, 'a dialectical-constructivist perspective views the effectiveness of the therapy as depending on the degree of transcendence of the conflict through dialogue, via differentiation, mutual recognition and appreciation, and ideally integration of the client and therapist's meaning-making schemes.' (Basseches, 1997a, p. 29). Clients will vary greatly in their current meaning-making capacities, depending on such variable factors as age, level of cognitive and ego development, and degree of current distress. In some cases it may be appropriate for the

---

[17] See Basseches (1997a, 1997b) for explanations of how such conflicts subvert the therapeutic processes of interpretation and enactment as well.

[18] Furthermore, anecdotal evidence (e.g., Lipson, 1993) suggests that the emergence of such conflicts is more likely in longer term therapeutic relationships, and that such conflicts, when they emerge and are successfully transcended, often represent the most powerfully transformational learning experiences in therapy for clients, and for therapists as well.

therapist to take on himself or herself the bulk of the responsibility for identifying the meaning-making conflict, and working to differentiate and integrate the conflicting meaning-making schemes. In other cases ... greater degrees of explicitly sharing with the client the challenge of facing these tasks may be appropriate. However in all cases, the therapist's effectiveness in this process will depend on his or her own ability to simultaneously protect, maintain and advocate for the client's meaning-making structures *and* his or her own ... structures.

Furthermore, the therapist must also protect and advocate for the possibilities for greater mutual understanding and for synthesis of both parties' meaning-making structures. This protection and advocacy must be maintained until syntheses which differentiate and integrate the parties' structures and thereby transcend the conflict can be created. If a therapist is very good at the tasks of simultaneously protecting, maintaining, and advocating for both parties' structures, there is less of an urgency to synthesize. The mutual creation of syntheses can take place gradually over time. On the other hand, if the therapist is particularly skilled at helping to create adequate syntheses, this can also make more likely a transcendent ... playing out of the conflict. (Basseches, 1997b, p. 93)

I have argued that dialectical thinking plays a central role in a therapist's successfully facing these challenges. My argument is summarized in Table 28.2. The first two rows of the table present a developmental ordering of three epistemological perspectives that one may find among therapists, based on the complexity of the cognitive capacities underlying the perspective, and supported by my research on dialectical thinking (Basseches, 1984). The third row of the table describes contrasting views of the role of the therapist's theories within the psychotherapy process associated with the differing cognitive and epistemological structures. Finally, the fourth and fifth rows describe how therapists relying on the differing structures are likely to understand and respond to feedback that the therapist's activity, organized and guided by his or her theories and formulations of the client, have proved unhelpful to the client.

This argument illustrates how examining the adult development of the therapist may shed light on clients' negative experiences. Elaboration, including case material supporting this argument, may be found in Basseches (1997b).

Implications for Training and Supervision

Let us recapitulate the implications, stated thus far, of an approach to psychotherapy with adults which relies on lifespan developmental psychology as a primary conceptual framework. Psychotherapy can potentially support all adults who have encountered conflicts in employing their imperfectly organized repertoires of activity and meaning-making schemes in the contexts of their current lives. Through three generic processes, psychotherapy offers additional resources to adults' efforts to construct novel syntheses/reorganizations of their repertoire of schemes that transcend previous conflicts. Two of these generic processes involve supplementing clients repertoires with new conceptual schemes and new forms of experience discovered through dialogue with the therapist. The third involves support in optimally deploying attention to the synthesis of conceptualization with experience. The goal of psychotherapy is development, as reflected in creation of novel syntheses, and its justification is based on the epistemic adequacy of those syntheses.

The wide range of theories and techniques brought by therapists with different backgrounds and training to the psychotherapeutic endeavor may be understood as

Table 28.2. Epistemological and Cognitive-Structural Bases of Psychotherapists' Perspectives on Theory and Meaning-Making Conflict in Psychotherapy

Epistemological perspective:	Universalistic-Formal	Relativistic	Dialectical
Requisite cognitive capacities:	Formal operations	Formal operations, metasystematic schemes for differentiating, comparing, and contrasting systems	Formal operations, metasystematic schemes for differentiating, comparing, and contrasting systems and for modeling dynamic relations among systems and processes of system-transformation
Likely view of the therapist's theories:	Validity of the theory is the basis for the effectiveness of the therapist's interventions.	Theories reflect the specific ways in which the therapist and the communities with which s/he identifies create order in their experience.	Theories play significant roles in guiding the therapist in responding to the client and collaborating in the psychotherapeutic inquiry.
Likely understanding of client's not finding theory-guided responses useful:	Casts doubt on the therapist's competence by casting doubt on the foundations on which that competence is based: Validity of theory or therapist's ability to apply theory.	Provides "marketing" information regarding who is most likely to benefit from the therapist's services.	Provides useful data regarding the possible limitations of the organizations of meaning guiding the therapist's responses as well as the client's organizations of meaning
Likely approaches to meaning-making conflict:	1. Defending the adequacy of one's theories 2. Experiencing a "crisis of confidence"	1. Renewed effort to make a case for the potential usefulness of one's theoretical approach 2. Attempting to put aside concern with one's own meaning-making and focus attention on client's ways of making meaning of his/her experience 3. Referring the client to a new therapist	Continuing the inquiry while adding two supplementary focal questions to it: (1) What in the relationship between client's and therapist's organizations of meaning explains the obstruction to therapeutic processes? (2) How can the therapist accommodate to become more helpful to the client?

Adapted from Basseches (1997b), p. 98.

alternative forms of participating in and guiding the three generic processes. While clinical perspectives see theories and techniques as providing explanations of the nature of disorders and methods for their cure or amelioration, developmental perspectives see theories and techniques as mapping common features of different levels or forms of psychological organization, describing processes of reorganizing experience, and providing methods for facilitating such reorganization processes. Each set of theories entails classification systems which may provide tools for assessing and formulating specific clients' difficulties, and choosing among approaches to the therapy process. However, neither the use of specific theories nor techniques nor approaches to formulation is a necessary condition for effective psychotherapy. The basic therapeutic processes of attentional support, offering "interpretations" derived from the therapist's perspective, and the enactment of and collaborative reflection upon novel interaction, can be facilitated by therapists with a varied range of interpersonal skills, technical approaches, amounts of therapeutic experience, and theoretical perspectives—provided that the therapist's repertoire matches the client's needs.

The therapist's development plays important roles within the therapeutic process. The development on the part of the therapist of new skills and techniques, a rich experiential base, and increased depth, breadth and variety of theoretical thinking may make a therapist more effective with a broader range of clients. At the same time, the conflicts which clients bring to the therapy process may present developmental opportunities for the therapist, and the therapist's openness to new discoveries facilitates the success of their shared inquiry. Ultimately, however, for any therapist, the basic therapeutic processes may become blocked by meaning-making conflicts between therapist and client. Failure to transcend such conflicts may often account for clients' negative experiences of therapy. The keys to success in transcending such conflicts are not the therapist's specific theories, skills, techniques and experiences (which have come against their limits in the situation). Rather, the keys to success are *the epistemological perspectives and general cognitive structures within which the therapist organizes the tools and approaches which he or she brings.*

Many training programs focus on broadening therapists' experience, theoretical knowledge, and sets of tools for conceptualizing clients' experience as well as dynamics of the therapy relationship. However, the analysis above implies that dialectical thinking, and the capacity to apply it to situations of meaning-making conflict in psychotherapy, are also important aspects of therapeutic expertise. This leads to the question of how psychotherapy training can foster dialectical thinking. Based on analyses of general aspects of higher education (liberal arts or professional training) that foster the development of dialectical thinking (Basseches, 1989b). I would highlight the following factors within therapists' training:

1. Experience with multiple powerful frames of reference, or ways of understanding human experience and psychotherapy process
2. Exposure to the limitations that all the different frames of reference encounter when brought into contact with as wide a range as possible of (a) data from human experience, (b) cases of psychotherapy, and (c) alternative frames of reference

3. Exposure to role models of respectful and patient efforts to transform and integrate frames of reference when they come up against their limitations
4. Support and company in tolerating the ambiguity of being unable to easily transcend encounters with the limitations of one's meaning-making in one's professional work, yet not retreating from the challenge of doing so

The traditional structure of graduate education in clinical psychology provides opportunities for therapists to have precisely those experiences that will foster dialectical thinking about their work. Courses and research projects can be designed to maximize exposure to alternative frames of reference and sources of data. Planned training experiences provide the opportunity to work with a wide range of clients in a varied set of modalities. Clinical supervision provides opportunities for support and dialogue when encountering limitations in the context of one's own practice. However, in the absence of viewing psychotherapists' education from an adult development perspective, all these opportunities can be lost. Relying solely on the perspectives of clinical psychology, students could easily come out of their educational experiences equating their professional expertise with the explanatory and curative value of the theories and techniques that they have learned, rather than with their capacities for participation in processes of collaborative inquiry. An adult developmental perspective implies the value of being clear about the cognitive–developmental and epistemological intent of psychotherapists' education and of monitoring its effectiveness in realizing that intent. At the same time respect for the developmental nature of the psychotherapeutic process leads to recognition both (1) that a client might accomplish a great deal with a relatively inexperienced therapist, and (2) that a very experienced therapist can block the therapeutic process and cause harm to clients, by overvaluing what he or she knows and overlooking developmental possibilities for both therapist and client.

TOWARD PSYCHOTHERAPY INTEGRATION

Adult development offers psychotherapists a valuable set of theoretical lenses for viewing their clients and themselves, as well as an alternative metatheory for conceptualizing the potentials and processes of the interpersonal encounters which constitute "psychotherapy." Readers who see the value of what is offered may well ask how it can be integrated with traditional clinical perspectives.

Practitioners, researchers, and teachers of psychotherapy need not discard frameworks derived from clinical psychology, nor discontinue clinically oriented research, to make use of what lifespan developmental psychology has to offer. It is possible to adopt a developmental metatheory of psychotherapy as proposed here, within which various clinical theories are understood as important elements of the meaning-making structures that therapists bring to their encounters with clients. It is, of course, very appropriate to include within these meaning-making resources the wealth of models and received knowledge regarding the nature, etiology, and amelioration of syndromes generated by the systematic study of psychopathology. Such resources might be deemed most useful for that subset of psychotherapy cases in which a client clearly manifests the symptoms of an identified psychiatric disorder. However, as stated above, some psychotherapists have in practice tried to stretch the applicability of clinical models via the assumption that all people manifest psychiatric disorders to varying degrees. There is also a long tradition in psychiatry and clinical psychology dating back to Freud, which argues that the study of "abnormal" human psychological functioning sheds valuable light on understanding "normal" individuals. This latter perspective would imply that within a developmental model, knowledge of psychopathology still may be quite useful in developmentally oriented psychotherapy with clients for whom no disorder has been recognized.

This chapter also has acknowledged the possibility of an alternative form of integration in which a practitioner uses developmental assessments of adult clients' functioning within an overall clinical metatheory of psychotherapy. However, such a practitioner clearly would be making use of only a small portion of what the field of adult development has to offer to the practice of psychotherapy.

How would the integration of clinical knowledge within developmental metatheory affect psychotherapists' dealings with a social context in which they are viewed as health service providers? As stated previously, it is an implication of the developmental model that funding should be sought for psychotherapy as a form of adult education insofar as adult development research can demonstrate how such expenditures foster effective functioning of democracy. However, in the meantime, insofar as research also demonstrates that psychotherapy does lead to "improved" outcomes for clients with identifiable disorders, as well as to prevention of such disorders, psychotherapists need not deny that they provide very valuable health services.

However, from a developmental perspective, it is essential not to confuse such documented benefits of psychotherapy with its guiding principles or core processes. To view psychotherapy solely, primarily, or essentially as a method to promote psychological "normality" or to reduce pain and distress would reflect such confusion. If psychotherapists attempt to support their practices solely on clinical models and clinical psychological knowledge, either in their own understanding of their work or their presentation of their practices to clients or third party payers, they are vulnerable to Dawes' (1994) argument that their profession is built on a "house of cards."

REFERENCES

American Psychiatric Association (1968). *Diagnostic and statistical manual of mental disorders, 2nd edit.* Washington, DC: American Psychiatric Association.

American Psychiatric Association (1994). *Diagnostic and statistical manual of mental disorders, 4th edit.* Washington, DC: American Psychiatric Association.

American Psychologist (1996). Special issue: Outcome assessment of psychotherapy *51*(10) (October).

Basseches, M. (1984). *Dialectical thinking and adult development.* Norwood, NJ: Ablex.

Basseches, M. (1989a). Toward a constructive developmental understanding of the dialectics of individuality and irrationality. In D. N. A. Kramer & M. J. Bopp (Eds.), *Transformation in clinical and developmental psychology.* New York: Springer-Verlag.

Basseches, M. (1989b). Intellectual development: The development of dialectical thinking. In E. P. Maimon, B. F. Nodine, & F. W. O'Connor (Eds.), *Thinking, reasoning and writing.* White Plains: Longman.

Basseches, M. (1997a). A dialectical–constructivist view of human development, psychotherapy, and the dynamics of meaning-making conflict within therapeutic relationships (Part I of A developmental perspective on psychotherapy process, psychotherapists' expertise, and meaning-making conflict within therapeutic relationships). *Journal of Adult Development, 4*(1), 17–33.

Basseches, M. (1997b). Dialectical thinking and psychotherapeutic expertise: Implications for training therapists and protecting clients from "theoretical abuse." (Part II of: A developmental perspective on psychotherapy process, psychotherapists' expertise, and meaning-making conflict within therapeutic relationships). *Journal of Adult Development*, 4(2), 85–106.

Benack, S. (1981). The development of relativistic epistemological thought and the growth of empathy in late adolescence and early adulthood. Unpublished doctoral dissertation, Harvard University.

Bohart, A., & Greenberg, L. (Eds.) (1997). *Empathy reconsidered: New directions in psychotherapy.* Washington, DC: American Psychological Association.

Bopp, M. J. (1984). A study of dialectical metatheory in psychotherapy. Doctoral dissertation, Temple University. Ann Arbor, MI: University Microfilms No. 1416.

Consumer Reports (1995). Mental health: Does therapy help? November, 1995, 734–739.

Dawes, R. M. (1994). *House of cards: Psychology and psychotherapy built on myth.* New York: The Free Press.

Demick, J. (1996). Epilogue: What *are* clinical approaches to adult development? In M. L. Commons, J. Demick, and C. Goldberg (Eds.), *Clinical approaches to adult development.* Norwood, NJ: Ablex.

Dewey, J. (1916). *Democracy and education; an introduction to the philosophy of education.* New York: Macmillan.

Drath, W. H., & Palus, C. J. (1994). *Making common sense: Leadership as meaning-making in a community of practice.* Greensboro, NC: Center for Creative Leadership.

Greenberg, L., & Pascual-Leone, J. (1995). A dialectical–constructivist approach to experiential change. In R. A. Neimeyer & M. J. Mahoney (Eds.), *Constructivism in psychotherapy.* Washington, DC: American Psychological Association.

Greenberg, L. S., Rice, L. N., & Elliot, R. (1993). *Facilitating emotional change: The moment by moment process.* New York: Guilford Press.

Grigoriu, E. (1998). Mental health practitioners' use of emotion within Kegan's theory of constructive developmental psychology. Dissertation, Cornell University. *Dissertation Abstracts International*, Vol. 59-04b, p. 1883 (Bell & Howell microfilm no. aai9831094).

Habermas, J. (1971). *Knowledge and human interests.* Boston: Beacon Press.

Havens, L. (1986). *Making contact: Uses of language in psychotherapy.* Cambridge, MA: Harvard University Press.

Herman, J. L. (1992). *Trauma and recovery.* New York: Basic Books.

Ivey, A. E. (1986). *Developmental therapy: Theory into practice.* San Francisco: Jossey-Bass.

Kegan, R. G. (1982). *The evolving self: Problem and process in human development.* Cambridge, MA: Harvard University Press.

Kegan, R. G. (1994). *In over our heads: The mental demands of modern life.* Cambridge, MA: Harvard University Press.

Lipson, A. (Ed.) (1993). *Critical incidents in psychotherapy.* Unpublished manuscript, Harvard University.

Loevinger, J. (1976). *Ego development.* San Francisco: Jossey-Bass.

Loevinger, J., Wessler, R., & Redmore, C. (1970) *Measuring ego development.* San Francisco: Jossey-Bass.

McCarthy, T. (1978). *The critical theory of Jurgen Habermas.* Cambridge, MA: The MIT Press.

Neimeyer, R. A. (1995). Constructivist psychotherapies: Features, foundations and future directions. In R. A. Neimeyer & M. J. Mahoney (Eds.), *Constructivism in Psychotherapy*, pp. 11–38. Washington, DC: American Psychological Association.

Newberger, C. M. (1980). The cognitive structure of parenthood: Designing a descriptive measure. *New Directions for Child Development*, 7, 45–67.

Noam, G. G. (1986). The theory of biography and transformation and the borderline personality disorders (part II): A developmental typology. *McLean Hospital Journal*, XI(2), 79–105.

Pascual-Leone, J. (1990). Emotions, development and psychotherapy: A dialectical–constructivist perspective. In J. Safran & L. Greenberg (Eds.), *Emotion, psychotherapy and change* (pp. 302–335). New York: Guilford Press.

Phillips, A., Basseches, M., & Lipson, A. (1998). Meetings: A swampy terrain for adult development. *Journal of Adult Development*, 5(2), 85–104.

Piaget, J. (1972). *The principles of genetic epistemology.* New York: Basic Books.

Pratt, L. L. (1993). Becoming a psychotherapist: Implications of Kegan's model for counselor development and psychotherapy supervision. Unpublished doctoral dissertation. University of Massachusetts.

Rogers, C. R. (1951). *Client-centered therapy*. Boston: Houghton-Mifflin.

Rogers, C. R. (1959). A theory of therapy, personality and interpersonal relationships as developed in the client-centered framework. In S. Koch (Ed.), *Psychology: The study of a science, Vol. 3* (pp. 184–256). New York: McGraw-Hill.

Wapner, S., & Demick, J. (1998). Developmental analysis: A holistic, developmental, systems-oriented perspective. In W. Damon (Series Ed.) & R. M. Lerner (Vol. Ed.), *Handbook of child psychology, Vol. 1: Theoretical models of human development*, 5th edit., pp. 761–805.

CHAPTER 29

Executive Development as Adult Development

OTTO E. LASKE

*Few applications of adult development theories to the work
setting have, in fact, been reported.*

—CYTRYNBAUM & CRITES
(1989, p. 83)

The human resource function in organizations provides one of four perspectives in which to view human development in the workplace. In the framework conceived by Bolman and Deal (1991), this function gives rise to a perspective intersecting with three related but divergent perspectives the authors call structural, political, and symbolic. While the *human resource* perspective targets the creative potential and the needs of organization members, in a *structural* perspective organizations are seen as centered around a hierarchy or heterarchy of functions of power and control. By contrast, a *political* view sees organizational life as determined by coalitions competing for, and negotiating access to, scarce resources. Finally, the *symbolic* perspective regards organizational culture, a dimension in which values are created and shared that define an organization's raison d'être and mission (Schein, 1992).

It is sobering to think that the capacity to capture the complexity of organizations in terms of the four interrelated perspectives named is itself an adult developmental achievement. As shown by the organizational literature, this achievement is not attained by all, or even many, theorists and organization members (Senge, 1990; Senge, Kleiner, Roberts, Ross, Roth, & Smith, 1999). The fact that the literatures on executive development are bifurcated along either agentic or ontic lines is further testimony of the challenge involved in comprehending organizational reality. How is one to reconcile a focus on meeting the needs of an organizational task environment, that is, the *agentic imperative*, with a focus on the mental growth needs of

OTTO E. LASKE • Personnel Development Consultation, Inc., West Medford, Massachusetts, 02155.

Handbook of Adult Development, edited by J. Demick and C. Andreoletti. Plenum Press, New York, 2002.

individuals involving the epistemic adequacy of their construction of the world in them and around them, that is, the *ontic imperative*? And especially, how is one to do so under the "new career contract" which postulates that the career that matters is the "internal career," and that personal development is the worker's own business, not the responsibility of organizational task environments (Hall et al., 1996).

Apparently, there exist not only societal obstacles to harmonizing the agentic and ontic imperatives, but also adult developmental limits to organizational experiencing, learning, and acting (as well as the theory of these) that purely behavioral, and even adult developmental, discussions of organizations tend to gloss over or miss. In short, executive development practice as well as research, when viewed from the vantage point of epistemic adequacy (adult developmental maturity), exhibit a lack of epistemological realism as to the extent to which organization members can be free of the constraints of their own, developmentally rooted, ideological system (Laske, 1997). When one undertakes to tackle this lack exclusively from a human resource perspective (as is done, e.g., in coaching), one only documents the lack of epistemic realism one is trying to expell.

In this text, I hope to escape the human resource/symbolic tunnel vision of most writings on adult development in organizations, which typically omits two vital perspectives, the political and the structural one. (Most recently, this tunnel vision has been articulated by the slogan of a "learning organization.") In keeping with Bolman & Deal's (1991) attempt, to see organizational life as taking place at the intersection of the four dimensions introduced above, I discuss adult development in organizations from a synthetic adult developmental point of view. Since much of the recent literature on development stresses the *strategic* need to foster the development of executives, meaning the potential of their development to guarantee an organization's prospering, my focus is on notions pertaining to *strategic* executive development. I understand the strategic point of view expressed by this term as *a mix of structural and political interventions brought to bear on the human resource function*. In the structural view of human resources, the divisions of an organization are seen as different "schools of thought" that, aided by "catalysts" such as coaching, provide resources for the experiential learning of executives (McCall, 1998). In the political view, selecting individuals for opportunities of experiential learning is based on the competition for scarce resources within the organization (including those of attention), and is thus a compromise vis-à-vis other, more immediately advantageous, investments of capital. In a conceptual framework where the emphasis falls on replacing the a-developmental "survival of the fittest" by the agentic "development of the fittest," the issue of what human resources to foster utilizing which mechanisms creates its own peculiar dilemmas (McCall, 1998).

In short, while agentic theorists tend to favor the structural and political perspectives on the human resource function in organizations, ontic (adult developmental) theorists favor the human resource and symbolic perspectives on that function. Often, the split is one of short-term vs. long-term, or surface vs. deep structure, perspectives. The result is a schism responsible for two bifurcated, non-communicating sets of writings on executive development. As I show in the following, what makes matters more complicated is that each of the two universes of discourse internally struggles with its own peculiar dichotomies, often but dimly perceived. The result is an academic and real-world discourse on adult development in the workplace prone to reductionism and *simplification terrible*, and an adult developmental no-man's land pervading most organizations.

CONCEPTUAL CONTEXT

The State of the Art in Adult Development in the Workplace

As noted, executive development is an ambiguous term, as it conjoins two different, although intrinsically related, meanings of the term *development*. The first, agentic, meaning derives from the homo fabor metaphor of bringing about development by way of human change efforts. The second, ontic, meaning derives from the organismic metaphor of maturation over the lifespan (Werner, 1957). The term "executive development" evokes both metaphors in an uneasy mélange, often referred to as nature and nurture or, more atomistically, as "talent plus experience" (McCall, 1998).

There are a number of reasons why the discovery that executive development IS adult development is a recent and still novel one (Basseches, 1984; Laske, 1999b). These reasons are best explained with the aid of Fig. 29.1. As shown, not only are ontic and agentic notions of development presently not communicating with each other. There are also dichotomies within each of the two universes of discourse depicted in the diagram. On the ontic side, the split separates theories

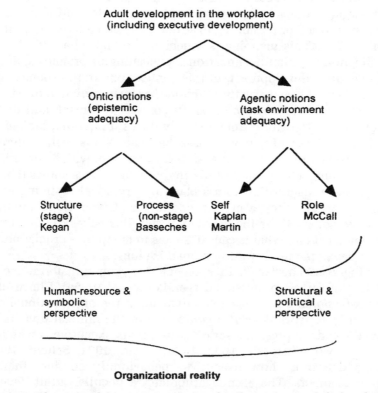

Figure 29.1. The state of the art in executive development as adult development.

paying primary attention to developmental structure, stage, level, or telos of development (e.g., Loevinger, 1976; Kegan, 1994), and those that are emphasizing the ("nonstage") processes (from brain processes to symbolic processes) that undergird such structure, stage, or level (e.g., Basseches, 1984). Thus, the Piagetian legacy of firmly associating structure/level with process has been lost. On the agentic side, the split is one between structural and political views of the executive as a bearer of functions and roles (McCall, 1998), on one hand, and of human resource and symbolic conceptions of the executive as character or self (Kaplan, 1991; Martin, 1996), on the other. As developmental structure by itself does not easily lend itself to a fruitful mapping into the organizational domain (leading, rather, to ideal typical character sketches of executives that are but a caricature of epistemological analysis), the structure/process dichotomy in developmental psychology intrinsically and perniciously supports that between self and role in the organizational literature. And since, moreover, it is easy to mistake the epistemological self on the ontic side for the behavioral self on the agentic side, and map them into each other at will (Drath, 1990), one ends up in a situation where adult developmental thinking has preciously little impact on actual executive development practice. This is true especially because that practice is largely carried out without psychological schooling.

The Notion of Professional Agenda

In my view, the greatest barrier to a higher profile of adult developmental thought in organizations is the unexplored epistemology of mapping developmental findings, whether pertaining to structure or process or both, into the organizational domain. This mapping is especially crucial when one is aiming not just for a diagnosis, but an organizationally meaningful prognosis, of individual and team executive development. In this context, one of the keenest analyses of ontic developmental stage/structure theories has been provided by M. Basseches, who characterizes them as *teleological* in contrast to causal, and distinguishes ontic developmental position from an individual's unique psychological organization and style (Basseches, 1989). Basseches' analysis is easily extended to the psychosocial profile of an individual in the workplace, which comprises organizational functioning. On account of his analysis, Basseches shows that it is futile to transform stage diagnostics from a classificatory, ideal-typifying method into an causally explanatory (or even ontogenetic) one. Given the temptation to mistake the epistemological for the behavioral self, Basseches' distinction between what is teleological and what is causal, seems to me to be a crucial one for creating an adult developmental culture in organizations.

Following Basseches lead, in a recent study on transformative effects of coaching on executives' professional agenda (Laske, 1999a), I have utilized the notion of *professional agenda* as an equivalent, in the organizational domain, of Basseches' *unique psychological organization* in the clinical domain (1989). A professional agenda expresses a set of assumptions executives make about their relationship to work (Argyris, 1992, 1993; Kegan, 1994; Schein, 1992). These assumptions determine how executives behaviorally conduct themselves as organization members. The agenda articulates a peculiar adult developmental status quo. *Importantly, the latter has both a structural and a procedural aspect.*

The structural aspect specifies ego level, and the procedural one, the processes supporting that level in any concrete professional situation.

One way to conceptualize executive development is to see it centered on changes to an executive's *professional agenda* (Laske, 1999a, 1999b). The agenda articulates a set of assumptions regarding an executive's relationship to work (Kegan, 1994; Schein, 1992), including how he/she uses formal status, communicates in the organizational environment, sets goals, approaches tasks, makes sense of personal experiences in the workplace, relates to the organization at large, and conceives of his/her self-developmental mandate. In cognitive science terms, the professional agenda has three levels, as depicted in Fig. 29.2. As shown, the assumptions made by an executive are the foundation for the executives behavior and verbal espousals. As a consequence, there are two kinds of potential changes to the agenda: structural–developmental changes effecting the basic assumption set, and behavioral or adaptational changes effecting observable behavior through "learning." In terms of assessment of the agenda, verbal espousals are used to decode the two lower levels. Because such espousals in most cases constitute "an espoused theory" that diverges from an individual's *theory in use* (Argyris, Putnam, & Smith, 1987), a *deep structure* analysis of the underlying assumptions articulated by the espousals is called for. From an adult developmental perspective, these assumptions change over a person's lifespan in accordance with what has variously been called ego level, developmental position, stage, or maturity (Kegan, 1994; Loevinger, 1976). These concepts refer to the structural aspect of development. The executive's assumption set is undergirded by a set of mental processes associated with a particular developmental level. These processes can be articulated *symbolically*, for example, by following Basseches' dialectical schema framework (Basseches, 1984). "Assessments for development" (Kaplan, 1998, p. 1) of the professional agenda comprise both a structure and a process statement. Together, these two assessments constitute the basis of evaluating an executives' behavior, learning, and adaptational changes to the agenda, as is topical, for example, in executive coaching. As Schein puts it:

> Most change processes emphasize the need for behavior change. Such change is important in laying the groundwork for cognitive redefinition but is not sufficient unless such redefinition (of some of the concepts in the assumption set, O.L.) takes place. (1992, p. 302)

In short, changing professional agenda is a type of cognitive restructuring that, taking place in the deep structure of the professional agenda, percolates upward to the behavioral and espousal levels. It is the task of developmental assessments

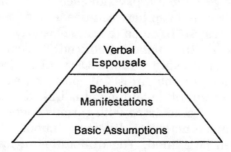

Figure 29.2. Three levels of the Professional Agenda.

to aid in the monitoring of restructuring. While the notion of self is thought to pertain to the assumption level, the notion of executive role or function, based on the notion of learning, is a behavioral one. Thus, when one speaks of the dialectic of self and role in executive functioning, what is involved is a complex interplay between two levels of the professional agenda, verbally articulated on a third, equally behavioral level. To gauge executive development, or development in the workplace more generally, it is therefore paramount methodologically, to have in place instruments that can gauge both the structural and procedural (process) aspect of human development in the workplace.

The 1990s Literature on Executive Development

In the recent literature on executive development, increasing attention has been paid to *self* in contrast to *role*. This is largely due to the increasing penetration of the "new, Protean, career contract," according to which personal development is a task of the employee, rather than the organization, the contract thus being a "contract with self" (Hall & Moss, 1998, p. 322). In the executive development literature, *self* has been conceptualized in various ways, as "character" (Kaplan, 1991, 1998), psychodynamic ego (Martin, 1996), even "talent" (McCall, 1998). Pervasively, self in the epistemological sense is confounded with (relational) "style" in the sense of feminist writing (Fletcher, 1996; Hodgetts, 1994; Kram & Hall, 1996) and, in behavioristic writing, with Myers-Briggs Type Indicator (Jung, 1971) and learning preferences (Kolb, 1984). Reinforced by how-to approaches in the burgeoning human services industry, these non- or a-developmental substitutes for the assumption set self pervade the popular as well as scholarly literature on executive development.

Among the writings most easily related to the deep structure notion of development of neo-Piagetian and -Kohlbergian vintage are those of M. W. McCall, Jr. (1998), I. Martin (1996), R. Kaplan (1991, 1998), and D. T. Hall (1996). These writers share a concern for two main issues: first, for executive development being *strategically* linked to business objectives; and second, *experiential (experience-based) learning* as the crux of executive development. These two concerns are not unrelated: experiential learning is what an organization can most easily provide its employees in a cost-effective and focused fashion without having to transcend its own realm. By strategic executive development is meant that business strategy should "logically" translate into "people strategy" by defining organizationally needed capabilities (e.g., by using psychological trait language), and then doing a means–ends analysis of the gap between needed and required capabilities (see Seibert, Hall, & Kram, 1995). To close the gap, executive developmental "mechanisms" and "catalysts" are then to be introduced. The commitment to such procedures is seen as confirming a shift from the Darwinian ideology of "survival of the fittest" to "development of the fittest" in enlightened organizations (McCall, 1998). This agentic concept of development is linked to a notion of adaptational learning as "experiential," associated with various degrees of self-transformation. Organizational divisions function as different "schools of thought" in which different experiences can be gained. The difficulties of learning from experience (Senge, 1990; Sims & Gioa, 1986) not to speak of ontic developmental preconditions, are not often taken into consideration in the literature.

I would characterize the present state of thinking about executive development by the partition separating writings on executive self (e.g., Kaplan, 1998; Martin, 1996) and executive role (McCall, 1998), although some intermediate positions exist that are poised to mitigate this separation (Hall et al., 1996). Predictably, this dichotomy is also one between the human resource and symbolic perspectives in the sense of Bolman and Deal (1991), on one hand, and structural and political perspectives, on the other. It is therefore pertinent to treat this dichotomy as one of *split organizational thinking* that, most likely for adult developmental reasons, fails to link together the four perspectives that cognitively define organizational reality. Because as Bolman and Deal (1991) convincingly show, each of the four perspectives is associated with different "action scenarios," the proposals in the literature for making executive development more "strategic," and learning more "experiential," are a logical outcome of the action scenarios associated with the respective writer's ideological persuasion and ontic developmental level. In what follows, I go into some more detail regarding the different approaches to executive development in the literature of the 1990s. I proceed from the structural/political to the human resource/symbolic pole of writings on executive development. Accordingly, I review in detail writings of M. W. McCall, Jr., I. Martin, R. Kaplan, and D. T. Hall.

FOUR APPROACHES TO EXECUTIVE DEVELOPMENT

A Model of Executive Development in Organizations

McCall (1998) approaches executive development from the structural/political perspective on organizations. He defines the lowest denominator for introducing adult developmental thinking into organizations. McCall's thinking is based on the notion of existing business divisions as "schools" for experience-based learning, and of scarce developmental resources competed for by antagonistic coalitions. McCall starts with what he perceives to be the lowest level of insight into development, where even development based on human agency (agentic development) is thought unnecessary due to a Darwinian belief in the survival of the fittest. By contrast, for McCall, "the right stuff," meaning "talent plus experience," can be generated only by active "development of the fittest."

McCall addresses executive development in the context of the immediate organizational task environment, largely ignoring the larger life context in which human development occurs. Despite this seemingly narrow vision, his writing convinces on account of systemic thinking and a fearless way of addressing intrinsic organizational antinomies, such as the conspiracies that support lopsided human development for the sake of immediate, short-term profit. Regarding the dialectic of executive self and role, McCall largely "leaves the person out of it" (Kaplan, 1991, p. 148), although he pays lip service to personal experience. Articulating the 1990s philosophy of "strategic" executive development, McCall wants the human resource function to be faultlessly integrated into the development of business strategy. (McCall is aware of the political issues this creates in the organization.) Conceptually, he follows Seibert et al.'s definition:

> Strategic executive development is the (1) implementation of explicit corporate and
> business strategies through the (2) identification and (3) growth of (4) wanted executive

skills, experiences, and motivations for the (5) intermediate and long-range future. (1995, p. 559)

This deceptively simple definition hides most of the antinomies of human development in organizations, especially the issue of what it takes, *in terms of the adult developmental status quo of those determining organizational strategy*, to define business strategy so that it can be "mapped into" developmentally productive (rather than arrestive or abortive) human resource goals and opportunities.

In terms of the mechanisms and catalysts that bring about human development in organizations, McCall is a believer in what he calls "experience." Because he neither distinguishes learning from development, nor experience from making sense of experiences in the adult developmental sense, his implicit formula is:

experience ⇒ learning (sometimes) ⇒ development,

where "experience" is closer in meaning to "organizational opportunities for making experiences" than to experience in a biographic, clinical, or epistemological sense. McCall's notion of experience is geared to *action learning*, in contrast to *classroom learning*. The notion focuses on the contingencies of learning, that is, the organizational, thus sociological, conditions under which learning and experience can be said to occur and relate to each other (Kolb, 1984). His approaoch is captured in Fig. 29.3 (McCall, 1998, p. 189).

What is called "mechanisms" in this diagram stands for structural opportunities existing inside an organization, while "catalysts" are supports such as coaching and development programs generally. The underlying notion is that (1) talent must be *found*, and (2) exposed to challenging situations inherent in organizational mechanisms, so that (3) talent is joined with experience which, when (4) supported by suitable catalysts, will render (5) the right stuff, viz., a transmutation of strategic into personal imperatives in the lives of talented executives. To arrive at this goal, McCall proposes to use preexisting business structures rather than learning opportunities removed from the actual task environment of executives:

> From a developmental perspective, business units or divisions can be thought of as "schools," each with a "curriculum" consisting of the experiences and exposures common to people who are successful within that part of the organization. (McCall, 1998, p. 84)

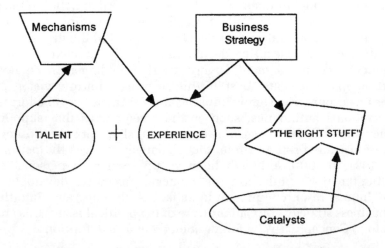

Figure 29.3. McCall's model of strategic executive development.

He underscores his belief in these schools as follows:

> Although the particular patterns are subject to change, the analytical approach assumes that the nature of the business and the structure of work in each of the organizations determines [sic!] the patterns of experience that talented people [sic!] will have. (McCall, 1998, pp. 84–85)

This statement has in mind executives' role, not self, given that it completely disregards what executives have to manage psychologically, as well as what is presupposed in terms of ontic-developmental status, for them to make *powerful experiences*. Suggesting that "executive development begins with experience and is driven by business strategy" (McCall, 1998, p. 18), McCall takes on the following issues:

- What experiences matter in shaping executives as leaders
- How important is the context in which development takes place
- How to choose among the valuable lessons many experiences teach
- How to think about talent other than as a static asset
- How to get the right people into the right experiences at the right time.

Throughout his discussion of these topics, the major issue for McCall is how the development of executives as leaders can be promoted by using the experiential resources available in an organization, which so far have not been optimally exploited for the purposes of leadership development. His polemic is directed against the "right stuff ideology" according to which fixed developmental sequences guarantee progressively more complex experiences (Dalton, 1989). McCall is thus critiquing notions deriving from the old career contract as a contract with an organization, in contrast to the "new career contract" which is a contract "with self" (Hall et al., 1996; see below). In this, he is motivated by his studies in corporate derailment, and in the organizational conspiracies giving rise to it. Aligning himself with the realities of the new career contract, he states:

> The bottom line for individuals is that no one cares as much about a person's development than the person. Whether the organization supports development Or inhibits it, individuals need to take responsibility for achieving their potential. (McCall, 1998, p. 59)

The fervor with which McCall endorses "development of the fittest" in organizations stems from the dichotomy between the situation targeted in the above statement, and his conviction that organizations comprise valuable, but un- or under-utilized, experiential resources. These resources comprise challenging job assignments, other people (especially supervisors), formal programs, but also "non-work experiences" and "hardships and setbacks" (McCall, 1998, pp. 65 f). However, it never occurs to McCall that both the organizational context of experiential learning as well as the powerful experiences that context allows for are not only a matter of factual existence, but of ontic developmental preparedness to "see" and take advantage of them. In short, McCall lacks an appreciation of ontic developmental limits of experiential learning. His argument thus ends in a prognosis without empirical evidence:

> People with the ability to learn from experience, (i.e., talented people, O.L.), when given [sic!] key experiences as determined by the business strategy, will learn the needed skills if given the right kind of support. (McCall, 1998, pp. 188)

According to McCall, human development in organizations is apt to encounter the following five "dilemmas" (McCall, 1998, p. 189 f):

Dilemma 1: How to think about talent.
Dilemma 2: Mechanisms controlling selection necessarily also control development.
Dilemma 3: Development is spurred by challenge and risk, which is contrary to organizational imperatives of predictability.
Dilemma 4: Learning from experience is not automatic.
Dilemma 5: Business strategy must address multiple possibilities (of using organizational resources).

Of these, dilemmas 2, 3, and 5 have a systemic, structural and political, aspect, while 1 and 4 are epistemological human resource concerns. The most important dilemma of McCall's model, in light of the adult developmental literature, may be his problematic relationship to conceptual complexity. Notions such as "talent," "experience," "the right stuff," and "strategic intent," and others are used by him in a developmentally as well as psychologically unreflected way that does not lend itself to epistemological subtlety. Nevertheless, McCall's systemic point of view and his spirited demonstration that agentic development is a strategical requirement for contemporary organizations are important assets. Among the unstated, and unresolved, dilemmas of McCall's model the following stand out:

- The *relationship of role to self*, and the issue of their "integration" in executives
- The *ontic developmental preconditions* of formulating business strategy and "translating" it into an executive development system
- The issue of *developmental catalysts* for learning from experience

The first issue is taken up by I. Martin, the remaining ones by Laske (1999a).

The Dialectic of Executive Role and Executive Self

A step toward conceptualizing the relationship of executive self and role is made by I. Martin who, in formulating a theory of corporate mentoring, unites the family systems and psychoanalytic traditions. Although Martin (1996) shares with McCall the systemic viewpoint, in contrast to him, she approaches executive development from a human resource and symbolic perspective. This entails that she focuses on the link connecting executives' personal development needs with the requirements of organizational development. Consequently, the system of importance to her is not the organization at large and its divisions per se, but the executive team and the family-of-origin reenactments its members are subject to in their organizational transactions. Seeing an organization's executive team as its culture bearer, she conceives of the self-transformation of each of its members as the enabling force by which to transform organizational cultures. In short, her theory of executive development is an ingredient of her theory of culture transformation.

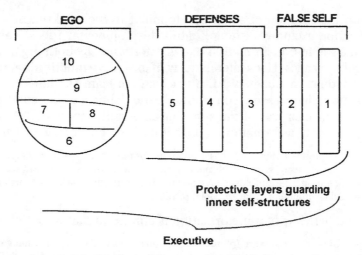

Figure 29.4. Martin's model of self-consciousness as a basis for executive mentoring aiming at culture transformation.

Just as the executive team, for Martin, forms a system of constituents bound together by a common dynamic, viz., that of individuals' family of origin, so does executives' individual psychological organization (Martin, 1996, p. 140).

As shown in Fig. 29.4, the executive self, identified by Martin with an individual's "unique psychological organization" (Basseches, 1989), is a system simultaneously operating on two sets of levels of consciousness "in which perception moves successively from an external to an internal focus" (Martin, 1996, pp. 140–141). The first set of levels makes up the individual's false self and defenses ("character"), while the second set of levels forms the ego. It is the task of the first five levels, to protect the inner self structures (levels 6–10) from feelings of shame and lack of self-worth. These levels constitute (social) role, self-illusions, defenses, developmental conflicts, and terror and rage (in the sense of A. Miller, 1984). In the second set of layers, 6 represents "basic self-love and eagerness for growth" (Martin, 1996, p. 145), followed by gender identification (7), triangulation tactics learned in the family of origin (8), and observing (9) and executive ego (10). Layer 9 represents "the ability to monitor one's own behavior with a realistic eye," that is, by taking it as object, while 10 is "the part of the ego that can oversee and direct an ongoing internal transformational process in which barriers (to growth) can be observed and then transmuted (Martin, 1996, p. 146). This "levels of self" model of an executive not only defines a framework for understanding the psychological dialectic of self (ego) and role; it also enables Martin to conceive of an executive's *professional agenda* as something determined by the executive's ability, to become aware of, and thereby transmute, the false-self and defense layers of the ego. This transmutation is the goal of mentoring as "corporate therapy." It simultaneously (and somewhat magically) aligns the executive with visionary strategic objectives of the organization, introduced by mentors (i.e., role models) external to the organization.

By joining executives' "unique psychological organization" (Basseches, 1989) to the sociological context of development outlined by McCall, and *defining executive role an an aspect of executive self*, Martin complexifies the set of parameters factoring into the equation of adult development in the workplace. Although her conceptualization excludes orthodox adult developmental ideas, she introduces the notion of a self-in-transformation that is determined by both organizational and psychological dynamics. Her systemic view of executive functioning, in the double sense of individuals' membership in the executive team as "family," and of each individual's multiple-level self, contributes to the richness of her conception. In her theory, there is no other way of transforming an organizational culture than by way of the personal self-transformation of individual members of the executive team:

> As culture transformation is viewed as a metaphor for the simultaneous transformation of a critical mass of executives through mentoring, it follows that systemic transformation will occur as mentoring unfolds. ...As people transform, they seek to transform their business and environment. (Martin, 1996, p. 156)

In many ways, her view of transformation is similar to that of E. Schein (1992):

> If any part of the core structure (of an organization's culture, O. L.) is to change in more than minor incremental ways, the system must first experience enough disequilibrium to force a coping process that goes beyond just reinforcing the assumptions that are already in place ... This is what Lewin called unfreezing, or creating a motivation to change.

except that she conceptualizes the obstacles to transformation in terms of systemic family therapy (Kirschner & Kirschner, 1986). In line with her multileveled model of self, she envisions a complex, 2-year course of mentoring that individual executives' pass through, to emerge as bearers of a new organizational culture firmly embracing strategically visionary objectives. Every mentee in the executive team engages in a kind of centrifugal mentoring, in that he or she takes on others to mentor who are not immediate supports. As a consequence, human development in the workplace and organizational development occur in close connection with each other. Both are based on personal self-transformation promoted by mentoring.

The Dialectic of Managerial Strengths and Weaknesses

A different approach to executive development from the vantage point of the dialectic of self and role is taken by R. Kaplan (1989, 1991, 1998) and W. H. Drath (1990), who adopt the method of *biographic action research*. This method combines ideas from action science (Argyris, 1987) with research into the biography of executives for the sake of coaching and mentoring them to provoke a self-transformation. The method has yielded important insights into executive derailment and, more generally, lopsided adult development in organizations. Kaplan, especially, is convinced that a behavioral approach to executive development does not suffice, and therefore suggests "getting personal" in executive development research and practice. For him, this entails a critique of all approaches that "leave the person out of it" (Kaplan, 1991, p. 148), an approach he demonstrates in his research on three types of "expansive" character in executives (Kaplan, 1991). In light of the foregoing, Kaplan takes on the ego-protective layers of Martin's multilevel self, both to explain expansiveness, and to bring about a "character shift" in executives. As for Martin, for Kaplan (1991, pp. 4–5) "character" is "a set of deep-seated strategies used to enhance or protect one's sense of self-worth." In his research, Kaplan comes upon two "styles" of executive functioning, "separate" and "relational." He points out that the organizational bias in favor of task mastery over relational competence (see also Fletcher, 1996) leads to incongruencies in adult development that ultimately, especially when secretly supported by an institutional "conspiracy," leads to derailment:

> There are organizational circumstances in which what is *required* is an executive who is clearly overbalanced on the side of results (emphasis O. L.). (Kaplan, 1998, p. 228)

Expansive character—whether in the form of the striver-builder, self-vindicator, or perfectionist-systematizer—is seen as a constellation of defenses that weakens executives' capacity to achieve a developmental balance. In addition, Kaplan shows that power and authority inhibit disconfirming criticism, thus further endangering the balance. To counteract these organizational conditions, mentoring is required:

> ... the class of interventions with which we have been principally concerned in this book is *deeply introspective self-development*. This is self-development precipitated by a concentrated dose of constructive criticism (emphasis O. L.). (Kaplan, 1991, p. 228)

Although he cautions that "deeply introspective self-development" is not a panacea for everyone, Kaplan sees such development as "a way of enhancing or accelerating the natural process of maturation" (Kaplan, 1991, p. 233), thus endorsing a notion of adult (ontic) development.

The process of *self-construction over the lifespan* has so far not been taken seriously in research on executive development. However, a first, shy step has nevertheless been made. Kaplan's exploration of the dialectics of executive self and role is extended by his colleague Drath, who emphasizes the intrinsic and constitutive relatedness, and thus the dialectic, of managerial strengths and weaknesses. To clarify this dialectic, Drath (1990), in a pioneering paper, adopts the early subject/object theory of R. Kegan (1982), without the benefit of its 1994 revisions. By "explaining" managerial weaknesses and strengths as a direct outcome of subject/object epistemologic (stage), thus by equating "unique psychological organization" and style with epistemologic, Drath demonstrates the stark reductionism that undialectical uses of the stage concept are likely to produce:

> Another prominent managerial strength arising from taking relationship as an object is toughness in decision making. This [toughness] is possible because of the way the institutional stage dramatically reduces the role of interpersonal feelings in decision making. Although a manager's "rational" approach to decisions can be explained in terms of learned skills [i.e., behaviorally], the objectification of feelings allows such a rational analysis to proceed without the manager's experiencing undue qualms. (Drath, 1990, p. 490)

This caricature of constructive developmental thought, in which a character trait such as "toughness" and a disposition of mind such as "undue qualms" straightforwardly follows from an individual's ontic developmental position, is likely to make adult developmental theory a bad name. It also explains the prevailing ineffectiveness of the theory. The caricature, which is not Drath's alone, shows an absence of epistemological know-how, one that has not benefited from Basseches' critique of epistemological reductionism in adult developmental psychology (Basseches, 1989). The caricature, while it introduces constructive developmental theory into executive development, demonstrates the difficulty of formulating a true "mapping" of teleological insight into adult development into concrete organizational contexts.

The Shift to the Protean ("Inner") Career

The schism of self and role in theories of executive development, demonstrated above, receives additional saliency in the context of what Hall has called "the new career contract" which, for him, is exemplified by the "Protean career" (Hall et al., 1996). The career contract regards the employer–employee relationship. It is ultimately a set of expectations regarding human development in organizations. As Hall explains:

> The idea of the psychological contract gained currency in the early 1960's when writers such as Chris Argyris, Harry Levinson, and Edgar Schein used the term to describe the employer-employee relationship. ... Later, Ian MacNeil discussed two forms of what he called the "social contract." (Hall & Moss, 1998, p. 23)

It is Hall's conjecture that MacNeil's *relational* social contract (if it was not a legend to begin with), has since the 1980s been replaced by a *transactional* contract. The latter essentially makes an employee's career contract one *with self*, and only secondarily one with an organization, as it no longer holds the employer responsible for an individual's adult development. Hall sees the shift as one from an "organizational" to a "Protean" career which is centered on an individual's *psychological success* as a basis for his or her development in the workplace:

> The protean career is a process which the person, not the organization, is managing. It consists of all the person's varied experiences in education, training, work in several organizations, changes in occupational field, etc. The protean person's own personal career choices and search for self-fulfillment are the unifying or integrative element in his or her life. The criterion of success is internal (psychological success), not external. (Hall & Moss, 1998, pp. 24–25)

In an even more emphatically constructivist way, Hall speaks of a shift toward the "internal career which describes the individual's perception and self-constructions of career phenomena" (Hall, Briscoe, & Kram, 1997, p. 321). This sociological change-of-scene clearly redefines the conditions and implications of adult development in the workplace. This shift is all the more noticeable since throughout the 1980's and 1990s, career theory was rather firmly wedded to theories of change (especially D. J. Levinson, Darrow, Klein, M. H. Levinson, & Mckee, 1978), rather than theories of development, in the sense of Fig. 29.5 (adapted from Demick, 1996, p. 118).

Not surprisingly, the saliency of stage theories of change ended with the old career contract, according to which an organizationally predefined sequence of career steps leads to increasingly more complex developmental opportunities and their psychological equivalent (e.g., Dalton, 1989, p. 100). As Kegan observed, the time for constructivist theories of development in the workplace was not ripe:

> What may be lacking is an understanding that the demand of work, the hidden curriculum of work, does not require that a new set of skills be "put in," but that a new *threshold of consciousness* be reached. (Kegan, 1994, p. 164)

Given the advent of the new career contract, has the time for constructivist theories of adult development in the organizational workplace arrived? Although

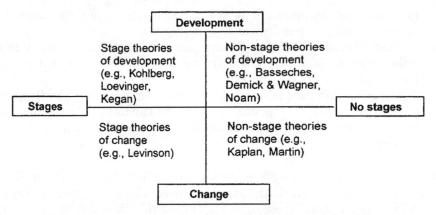

Figure 29.5. Stage and nonstage developmental theories.

career theory as well as executive development theories are poised to absorb and utilize adult developmental ideas, this potential still has to prove itself. Moreover, adult-developmental psychology must fulfil its side of the bargain. I briefly outline what this might entail in what follows.

LOOK TO THE FUTURE

Three Tasks of Constructivist Developmental Theory

If my hunch that there are adult developmental limits to comprehending organizational reality is correct (Laske, 1997), then no amount of preaching the good ontic developmental news, and no amount of scholarly reduction of organizational complexity to stages or derivatives thereof, will make the slightest dent in the state of the art of executive development research and practice. For adult developmental psychologists escaping the fascination with pathology, and concerned about creativity instead, there is a major self-critical task to accomplish before a cogent theory of *executive development as adult development* can emerge. This task, first seen in its full amplitude by Basseches (1984, 1989), in my view comprises three subtasks:

1. First, to firmly link developmental structure (stage, level, etc.) to the mental processes that undergird it (as Piaget taught us)
2. Second, to emerge from the human resource and/or symbolic tunnel vision that makes developmental theory incapable of taking structural and political perspectives on organizations (e.g., as outlined by McCall) into account
3. Third, to studiously scrutinize and avoid epistemological reductionism that portrays factual psychological or organizational content as a causal outcome of teleological principles.

Methodologically, these tasks entail, in the order followed above,

1. First, leaving stage vs. nonstage controversies behind, and researching what are the epistemological and neuropsychological processes that make

reaching and maintaining a given constructivist developmental level (however defined) possible in the first place

2. Second, furthering a theory of organizations as "thinking organizations," following the example of cognitive sociologists of the 1980s (e.g., Gioia & Sims, 1986)

3. Third, abstaining from false claims regarding what a teleologically grounded constructivist theory of human development can hope to "explain" that is of relevance to the theory of organizations, and to agentic practices in executive development.

In a research project undertaken to pursue the first subtask (and make a step toward the second one), entitled "Transformative effects of coaching on executives' professional agenda" (Laske, 1999a), I have designed and implemented an epistemological assessment tool, called the Developmental Structure/Process Tool (DSPT™). The instrument conjoins a structure description of developmental status quo, derived from Kegan's work (Kegan, 1994) and a process description of developmental level, deriving from Basseches' work on dialectical thinking (Basseches, 1984). Especially when used longitudinally (and with proper validity concerns regarding self-reports), the DSPT™ makes it possible to formulate a comprehensive assessment of an executive's developmental status quo (Laske, 2000). The instrument enables its user (psychologist or paraprofessional), to prognosticate developmental regression, stasis, and advance within a given developmental sequence. It requires the user to formulate a *nonreductionistic mapping of teleological findings into the organizational domain from which the executive's espousal has been taken*. By "mapping" is meant a confidential formulation of developmental findings that is grounded in the concrete details of an executive's organizational functioning at a particular time. Such a mapping becomes possible by utilizing, in addition to interview material, behavioral data deriving from, for example, 360-degree feedback procedures and related information about the organization's present strategic objectives.

This entails that rather than translating a specific "developmental stage" (e.g., the Kegan-stage 4(3); Lahey, Souvaine, Kegan, Goodman, & Felix, 1988) into a "character sketch" of the executive, thereby treating universalistic teleological data as causes of some unique psychological organization (Drath, 1990); or equating an epistemic profile dominated by the use of dialectical schemata of, for example, motion (Basseches, 1984) with behavioral processes that need to be "unlearned" or "improved," it is required to think through the epistemic (ontic) limits of the person assessed in terms of strategic, presently salient, agentic imperatives of the organizational task environment in question, formulating one's DSPT™ assessment accordingly. As a result of using the DSPT™, the user is enabled to build bridges from the domain of ontic developmental discourse to the agentic domain of coaching and mentoring, succession planning, and, when applied to groups of executives', the evaluation of entire corporate development programs. This is the case since developmental process descriptions, formulated symbolically by using Basseches' dialectical–schemata framework, when linked to associated structure (stage) descriptions, are prognostic of the movement of an individual's ego level within a teleological range of lower and higher stages. Linking stage and process descriptions has a high payoff, as prognostic assessments of executives' developmental status quo enable the DSPT™ user, to advise executive coaches and corporate development officers regarding the presence or lack of transformative effects of coaching on executives' professional agenda, with consequences for the reorientation of executive development efforts.

Based on a subject/object and a dialectical–schemata interview with each of six executives, I have been able to show that, and in what way, coaching outcome articulates the ontic developmental status quo of executives.

In columns A to E, Table 29.1 (Laske, 1999a, vol. 1, p. 239) relates modified subject/object stages (column B; Kegan, 1982) attained by six different executives (column A) to the number and quality of uses of metaform schemata in the sense of Basseches (1984, column D). The table compares six executives (column A), ranked in order of their ontic developmental position (column B), as to their reported uses of coaching (column E). The ranking of executives within subject/object-stage 4 is based on a "potential/clarity index" (column C) that quantitatively compares an executive's potential for transcending the present stage (column B) to the clarity and force by which the stage, and embeddedness in it, is expressed (column C). In the study, executives' self reports have been assessed by a subject/object (Lahey et al., 1988) and a dialectical–schemata analysis (Basseches, 1984), respectively. As the table shows, the use executives have been able to make of coaching is a reflection of their ontic developmental status quo. With a higher stage score, as well as quantitatively higher uses of metaform schemata, the ability to use coaching for more than a single purpose, and to make self-transformation a conscious telos of the coaching alliance, is strengthened. For example, using coaching strictly for skill building is a prerogative of executives who are firmly ensconced in a subject/object-stage 4 (i.e., clarity $>$ potential), where being at the particular ontic developmental level overshadows any potential to transcend it (e.g., S3, $p = 0 < c = 9$), which signals possible developmental arrest. On account of the process profile of the executives in question, the DSPT™ offers the capability to prognosticate movement within the teleological range of stages over the lifespan (i.e., regression, stasis/arrest, and transcendence), which has direct consequences for executives' resilience (Maddi, 1999). This leads me to the conclusion that having access to a symbolic representation of the mental

Table 29.1. Uses of Coaching as a Function of Ontic Developmental Status Quo

A Subject	B Stage & risk/ clarity/potential Index	C Potential to clarity (of stage)	D Meta-form (%)	E Type of coaching endorsed
S5	4(5){2:4:7}	$p > c$	44	Adult development, skill, performance, agenda
S6	4{2:9:4}	$p < c$	41	Adult development (inner agenda)
S2	4{1:8:5}	$p < c$	15	Performance/agenda
S4	4{0:5:3}	$p < c$	26	Performance
S1	4{3:9:2}	$p < c$	19	Skills/performance
S3	4{1:9:0}	$p < c$	1	Skills

processes that undergird "being at stage X," a developmental psychologist can formulate an ontic developmental prognosis of individual executives, both with regard to their concrete organizational functioning in the foreseeable future, and their ontic developmental resilience generally. As long as the error of epistemological reductionism is strenuously avoided (subtask 3), and the ontic developmental findings are responsibly mapped to organizational realities (e.g., the executive's functions, and existing strategic objectives of the organization), the developmental psychologist can provide coaches and corporate development officers with prognostic information, not only regarding individual executives-in-coaching, but equally regarding the outcome, over a longitudinal time span, of a corporate development effort in its entirety.

To conclude, while executive development has always been adult development—as much as we have always spoken prose, mostly without realizing it—the use theorists and organization members have made of self in thinking about, and promoting, that development, has not been commensurate with the complexities of organizational reality which includes adult development. For the same reason, developmental psychologists (as little as organization theorists and organization members) have so far been unable to gratify both the agentic and the ontic imperative of development in the workplace and larger society. However, if the new career contract in Western countries is indeed focused on the *internal career* (Hall et al., 1996), which implies attention to, and regard for, self (rather than just role), then there is a chance for practitioners of developmental psychology, to gain influence in, and become helpful to, organizations, but only if they eschew epistemological reductionism. Of course, this chance will depend on the extent to which these practitioners themselves can muster a commensurate ontic developmental status quo due to which they can epistemically cope with the complexities inherent in organizational reality. As I have shown, that reality is defined by the dialectic of agentic and ontic imperatives in the workplace.

REFERENCES

Argyris, C. (1960). *Understanding organizational behavior.* Homewood, IL: Irwin-Dorsey.

Argyris, C. (1992). *On organizational learning.* Malden, MA: Blackwell Business.

Argyris, C. (1993). *Knowledge for action: A guide to overcoming barriers to organizational change.* San Francisco: Jossey-Bass.

Argyris, C., Putnam, R., & Smith, D. M. (1987). *Action Science.* San Francisco: Jossey-Bass.

Basseches, M. (1984). *Dialectic thinking and adult development.* Norwood, NJ: Ablex.

Basseches, M. (1989). Toward a constructive-developmental understanding of the dialectics of individuality and irrationality. In D. A. Kramer & M. J. Bopp (Eds.), *Transformation in clinical and developmental psychology* (pp. 188–209). New York: Springer.

Bolman, L. G., & Deal, T. E. (1991). *Reframing organizations: Artistry, choice, and leadership.* San Francisco: Jossey-Bass.

Commons, M. L., Demick, J., & Goldberg, C. (1996). *Clinical approaches to adult development.* Norwood, NJ: Ablex.

Cook-Greuter, S. (1999). Postautonomous ego development: A study of its nature and measurement. Doctoral dissertation. Harvard Graduate School of Education.

Dalton, G. (1989). Developmental views of careers in organizations. In M. B. Arthur, D. T. Hall, & B. S. Lawrence (Eds.), *Handbook of career theory* (pp. 89–109). New York: Cambridge University Press.

Demick, J. (1996). Life transitions as a paradigm for the study of adult development. In Commons, M. L. et al. (Eds.), *Clinical approaches to adult development* (pp. 115–144). Norwood, NJ: Ablex.

Demick, J., & Miller, P. M. (Eds.) (1993). *Development in the work place*. Hillsdale, NJ: Lawrence Erlbaum.

Drath, W. H. (1990). Managerial strengths and weaknesses as functions of the development of personal meaning. *Journal of Applied Behavioral Science, 26*(4), 483–499.

Fletcher, J. C. (1996). A relational approach to the protean worker. In D. T. Hall, et al. (Eds.), *The career is dead—Long live the career*. San Francisco: Jossey-Bass.

Gioia, D. A., & Sims, H. P., Jr. (1986). Introduction: Social cognition in organizations. In H. P. Sims, Jr. & D. A. Gioia (Eds.), *The thinking organization* (pp. 1–19). San Francisco: Jossey-Bass.

Hall, D. T. (1996). Protean careers of the 21st century. *Academy of Management Executive, 10*(4), 7–16.

Hall, D. T., & Associates (1996). The career is dead — long live the career. San Francisco, CA: Jossey Bass.

Hall, D. T., Briscoe, J. P. & Kram, K. (1997). Identity, values and learning in the protean career. In C. L. Cooper & S. E. Jackson (Eds.), Creating tomorrow's organizations (pp. 321–335). New York: Wiley.

Hall, D. T., & Moss, J. E. (1998). The new protean career contract: Helping organizations and employees adapt. *Organizational Dynamics, 26*(3), 22–38.

Hodgetts, W. H. (1994). Coming of age: How male and female managers transform relationships with authority at midlife. Doctoral thesis, Harvard Graduate School of Education.

Jung, C. G. (1971). Psychological Types. *Collected works, Vol. 6*. Princeton, NJ: Princeton University Press.

Kaplan, R. E. (1989). *The expansive executive*. Report no. 135. Greensboro, NC: Center for Creative Leadership.

Kaplan, R. E. (1991). *Beyond ambition: How driven managers can lead better and live better*. San Francisco: Jossey-Bass.

Kaplan, R. E. (1998). Getting at character: The simplicity on the other side of complexity. In R. Jeanneret & R. Silzer (Eds.), *Individual assessment: The art and science of personal psychological evaluation in an organizational setting*. San Francisco: Jossey-Bass.

Kegan, R. (1982). *The evolving self*. Cambridge, MA: Harvard University Press.

Kegan, R. (1994). *In over our heads*. Cambridge, MA: Harvard University Press.

Kirschner, D., & Kirschner, S. (1986). *Comprehensive family therapy: An integration of systemic and psychodynamic treatment models*. New York: Bruner/Mazel.

Kolb, D. A. (1984). *Experiential learning: Experience as the source of learning and development*. Englewood Cliffs, NJ: Prentice-Hall.

Kram, K. E., & Hall, D. T. (1996). Mentoring in a context of diversity and turbulence. In E. Kossek & S. Lobel (Eds.), *Managing diversity: Human resource strategies for transforming the workplace* (pp. 108–136). Cambridge, MA: Blackwell.

Lahey, L., Souvaine, E., Kegan, R., Goodman, R., & Felix, S. (1988). *A guide to the subject–object interview: Its administration and interpretation*. Cambridge, MA: Harvard Graduate School of Education.

Laske, O. (1997). Four uses of self in cognitive science. In P. Pylkkänen, P. Pylkkö & A. Hautamäki (Eds.), *Brain, mind, and physics* (pp. 13–25). Amsterdam, The Netherlands: IOS Press.

Laske, O. (1999a). Transformative effects of coaching on executives' professional agenda. Doctoral dissertation. Massachusetts School of Professional Psychology. Ann Arbor, MI: Bell & Howell (order no. 9930438).

Laske, O. (1999b). An integrated model of developmental coaching. *Consulting Psychology Journal, 51*.3 (Fall), pp. 139–159.

Laske, O. (2000). Foundations of scholarly consulting: The developmental structure/process tool. *Consulting Psychology Journal, 52*.3, pp. 178–200.

Levinson, D. J., Darrow, C. N., Klein, E. B., Levinson, M. H., & McKee, B. (1978). *The Seasons of a man's life*. New York: Ballantine Books (Knopf).

Maddi, S. R. (1999). Comments on trends in hardiness research and theorizing. *Consulting Psychology Journal, 51*.2, 67–72.

Martin, I. (1996). *From couch to corporation: Becoming a successful corporate therapist*. New York: John Wiley & Sons.

McCall, M. W. (1998). *High Flyers*. Cambridge, MA: Harvard Business School Press.

McCall, M., Lombardo, M., & Morison, A. (1988). *The lessons of experience: How successful executives develop on the job.* Lexington, MA: Lexington Books.

Miller, A. (1984). *Thou shall not be aware.* New York: Basic Books.

Popp, N. (1996). Dimensions of psychological boundary development in adults. In M. L. Commons, D. J. Demick & C. Goldberg, (Eds.), *Clinical approaches to adult development* (pp. 145–174). Norwood, NJ: Ablex.

Schein, E. H. (1992). *Organizational culture and leadership,* 2nd edit. San Francisco: Jossey-Bass.

Seibert, K. W., Hall, D. T., & Kram, K. E. (1995). Strengthening the weak link in strategic executive development: Integrating individual development and global business strategy. *Human Resource Management 34*(4), 549–567.

Senge, P. M. (1990). *The fifth discipline: The art and practice of learning organizations.* New York: Doubleday Currency.

Senge, P. M., Kleiner, A., Roberts, C., Ross, R., Roth, G., & Smith, B. (1999). *The dance of change: The challenges of sustaining momentum in learning organizations.* New York: Doubleday Currency.

Sims, H. P., Jr., Gioia, D. A., & Associates (1986). *The thinking organization.* San Francisco: Jossey-Bass.

Werner, H. (1957). The concept of development from a comparative and organismic point of view. In D. Harris (Ed.), *The concept of development.* Minneapolis, MN: University of Minnesota Press.

CHAPTER 30

Community Service and Adult Development

Daniel Hart, Nancy Southerland, and Robert Atkins

Community service in adulthood is a topic of national significance and of considerable theoretical interest. In many countries, adults are involved in activities that are intended to benefit those who live in their communities, and perform these activities as volunteers—that is, without pay. The benefits that accrue to communities and societies as a result of these efforts are enormous, essential, and increasingly recognized; theoretical explanations to account for the participation of adults in volunteer activity abound but remain incomplete.

In this chapter, we explore community service and its connections to adult development. We begin by reviewing briefly some of the recent social policy debate concerning the importance of civic engagement, one form of which is community service. This overview helps to illustrate the societal significance of community service.

Most of the chapter is devoted to exploring various theoretical accounts for involvement in community service. We shall review recent theory and research, and to illustrate the claims from these works we shall present some findings from research underway that draws upon the National Survey of Midlife Development in the United States.

CIVIC ENGAGEMENT

Civic engagement, the extent of citizen participation in the social institutions constituting a society, has received considerable attention in the last 10 years from

Daniel Hart, Nancy Southerland, and Robert Atkins • Department of Psychology, Rutgers University, Camden, New Jersey, 08102.

Handbook of Adult Development, edited by J. Demick and C. Andreoletti. Plenum Press, New York, 2002.

political analysts and social scientists. One line of investigation has examined the role of civic engagement in supporting economic development. Fukayama (1995) has argued that the extent and forms of civic engagement are directly related to a country's economic success. According to Fukayama, civic engagement leads to economic success by fostering perceptions of trust among participants in the economy. For example, a society characterized by limited engagement in nonfamilial social institutions may produce adults who are reluctant to form alliances with others in the workplace, which in turn inhibits the growth of certain forms of corporate organization.

Voluntary association among adults varies within as well as between countries, and this intrasociety variation is of considerable significance. In a famous article entitled "Bowling Alone in America" Robert Putnam (1995) argued that participation in public life was declining in the United States. The title of Putnam's article was drawn from his observation that the decreasing rate of participation in bowling leagues over recent generations was counterpoised by an increasing number of Americans bowling. Marshalling together a range of similar trends led Putnam to conclude that Americans were increasingly detached from voluntary civic organizations. This detachment is forecasted to have undesirable consequences. To name just one, Kawachi, Kennedy, and Lochner (1997) report that states that have high per capita enrollment in bowling leagues—the illustrative example of civic participation prized by many of those engaged in the debate—have lower mortality rates than those states with low participation.

Still another way to illustrate the value of community service is to estimate the time that is volunteered by a country's populace. In the United States, the estimate for 1989 was that 15.7 billion hours of volunteer work was donated (Nonprofit Almanac, 1992–1993). The dollar value of this labor, if estimated at the average wage at the time, was well over 150 billion. While this sum is considerable, the economic value ought not overshadow the constructive contributions of community service to American life outlined in earlier paragraphs.

COMMITMENT TO COMMUNITY SERVICE

Why do adults enter into community service? What leads adults to commit to volunteer activities that have no instrumental benefit for themselves? What transforming effects does volunteer activity have on the individuals who become involved? There is no single theoretical approach that offers a comprehensive explanation to these questions. However, theoretical traditions can complement each other, and aligning them can provide an overview of the processes by which adults become involved and transformed by community service.

The first of these traditions is most common in sociological and economics research, and views volunteering by adults as the product of relatively static roles and resources. A second tradition, common among personality psychologists, attends to the traits, motives, and cognitive processes that are associated with committed community service. The third tradition, which has fostered relatively little research to date, focuses on the developmental, transforming effects of community service on involved adults.

ROLE/RESOURCE ACCOUNTS OF
VOLUNTEER ACTIVITY

One recent analysis has examined volunteer activity in the context of the many demands that characterize life in American families. Rossi (2001) examined how American families use their time, making use of data collected in the Midlife in the United States (MIDUS) study. Because we shall draw on the findings of this study at several points in this chapter, we describe the sample and methods in brief detail. The more than 3000 participants in the core sample constitute a representative slice of Americans between the ages of 25 and 74. The participants completed phone surveys and self-administered paper-and-pencil questionnaires that were returned by mail. The survey and questionnaire included a wide range of items concerning time use, volunteering and civic engagement, health, social relationships, educational and occupational histories, and psychosocial functioning. Ninety of the participants, generally representative of the larger group, were intensively interviewed about their lives and their understanding of their civic responsibilities. We shall describe Rossi's findings from the survey data and use some excerpts from the interviews to frame Rossi's claims in the complexities of human lives.

One of Rossi's most surprising findings was that participation in volunteer community service is essentially unrelated to the amount of free time that people have in their daily lives (for a similar finding from a study of older adults, see Warburton, Le Brocque, & Rosenman, 1998). Those who spend large amounts of time working are just as likely to volunteer as those who work far less; similarly, those with extensive family responsibilities volunteer as often as those with none. Apparently, then, community service is *not* undertaken to fill empty time.

What factors are associated with community service participation? First, like other investigators (Greeley, 1997), Rossi reports that adults who are connected to religious communities are more likely to volunteer than are others. We illustrate this association with a graph calculated from the data in the MIDUS survey. As is evident in the graph in Fig. 30.1, those who are actively involved in religious activities are indeed more likely to volunteer for community service.

Second, Rossi found that volunteer service is associated with educational attainment, a relation that is also quite typical in research. This association is depicted in Fig. 30.2, which indicates that those higher in educational attainment are more likely than those low in educational attainment to be volunteering time for community service activities.

What cannot be explained by these graphs—indeed, cannot be explained by the MIDUS data—are the reasons that religious involvement and education are associated with higher levels of volunteer activity. Rossi's interpretation for the first of these correlations is that participation in religious communities promotes a moral concern for social improvement that is reflected in direct participation. This inference is consistent with conclusions of other survey researchers (e.g., Greeley, 1997). However, because the connection of religious participation to volunteering could be the result of other factors (e.g., religious practice that shades into volunteering such as teaching Sunday school, etc.) it remains for future research to furnish the compelling evidence necessary to accept these interpretations as fact. A similar state of speculation and fact exists for the role of educational attainment

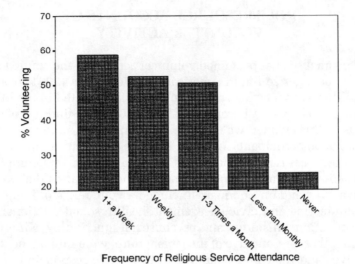

Figure 30.1. Religious service attendance and volunteerism. The figure shows the greater the level of religious service attendance, the higher the level of volunteerism.

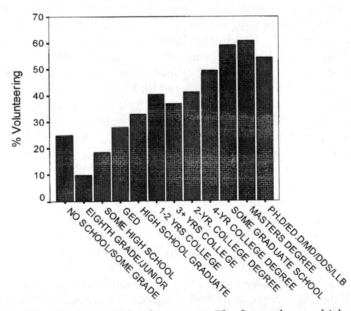

Figure 30.2. Level of education and volunteerism. The figure shows a higher level of education correlates with a higher level of volunteerism.

in fostering volunteer behavior. Wilson and Musick (1997) suggest that the consistent finding that persons with high educational attainment are more likely to volunteer than those low in educational attainment is a consequence of confidence in desirable credentials. That is, those with the cognitive and verbal skills developed through education believe that these skills will make them valuable to the causes they join, and their perceptions of their value make them likely to volunteer.

While plausible, this is more of a hypothesis than an interpretation that tracks closely research findings.

Wilson and Musick (1997) find additional evidence for the importance of religion and education in volunteer in their analyses of two waves of data from a longitudinal panel study with a large sample of American adults. Wilson and Musick formulate the answer to the question of why people participate in economic terms: volunteering is a form of work that is "paid for" with different forms of "capital." *Human capital* includes individual resources such as education, physical health, and income. *Social capital* refers to the network of relationships upon which a person can draw for support and information. *Cultural capital* is

> acquired, sometimes unwittingly. Like any other form of capital, it can be "invested" to yield "social profits" in the form of symbolic goods, such as titles, honors, and club memberships. These "social profits," in turn, yield social esteem, which is denied to those who lack cultural capital. (p. 696)

Cultural capital as defined by the authors encompasses a number of variables measured in the data set they use in their study, including religious engagement.

Wilson and Musick used these forms of capital—human, social, cultural—to predict *change* in their index of volunteering (a composite of number of volunteer positions and total hours volunteered) in Americans surveyed in both 1986 and 1989. The findings indicated that Americans high in human and social capital were more likely than were Americans low in these resources to increase their involvement in volunteering over the 3-year span. Particularly influential forms of human and cultural capital were, respectively, educational attainment and religious involvement. Variables tapping attitudes about help, health status, and frequency of social interaction with other people were much weaker predictors of change in volunteering.

Rick Munson: Volunteering in the Context of a Life

Characterizing participation in community service in terms of roles and resources—as these are described in the preceding—strips volunteering of much of the detail that makes prosocial action understandable. In this section, we use the life of one man to illustrate the intertwining of community service with the strands of life. Rick Munson, an alias for one of the interviewed participants in the MIDUS study,[1] is a 34-year-old male who was born and raised in the southwest. His mother was a social worker and his father was a minister. Munson is in the top half of the sample in educational attainment, with a degree in accounting, and his self-report of frequency of church attendance is in the top quarter.

Munson's entry into community service was through participation in a foster child program. After involvement with several children, he and his wife were asked to care for two physically handicapped children. Munson and his wife quickly became attached to the children, and adopted them. The challenges of obtaining appropriate care for his children led Munson to become a dedicated advocate on behalf of children with special needs.

---

[1] Minor details have been altered to protect the anonymity of the participant.

Munson's account of his involvement in care and advocacy on behalf of special needs children refers to his religious faith, a faith for which he saw little use as a child but as an adult finds central to his life:

> When you go to church as much as I did when I was growing up, it's a tough tough thing, a tough habit, it's not a habit, because I was going to church because my father told me I had to go, but now I go to church because I want to ... my father would walk through the house, he would quote you a scripture out of the Bible thirty times an hour...but, come to find out that he was giving me the basics that I would need, you know, to deal with life.

One of the reasons that religion is central to Munson is that it sustains him in times of hardship; with God in life, Munson feels able to remain optimistic even when others are deeply pessimistic. For example, Munson discusses his decision to transfer his children from one school to another, a transfer that deepened his own resolve to advance the causes of handicapped children, in terms of his refusal to accept the gloomy prognosis offered by a school official:

> She [the school official] always kept at this issue of how my children would never walk, they would never do this, or they would never do that, you know? But with my God, all things are possible.

Rick Munson's involvement in community service does not seem to be a direct result of his religious faith. However, his commitment to community service is deepened by his belief that his own hard work, the efforts of others, are noticed, appreciated, and rewarded by a deity that intervenes in the functioning of the world. The excerpts from an interview with Rick Munson cannot sustain claims of fact; however, they do suggest that the relation of religion to community service may have complexities that escape from the abstractions revealed by analyses of survey data.

TRAIT/MOTIVATIONAL/ATTITUDINAL EXPLANATIONS FOR VOLUNTEERING

Generativity

Community service has also been examined from the perspective of individual differences in traits, motivation, and developmental stage. Erikson's (1982) theory has been very influential in one line of this research. According to Erikson, a central task of adulthood is the development of generativity, characterized by a

> widening commitment to *take care of* the persons, products, and the ideas one has learned to *care for*. All of the strengths arising from earlier developments in the ascending order from infancy to young adulthood (hope and will, purpose and skill, fidelity and love) now prove, on closer study, to be essential for the generational task of cultivating strength in the next generation. (Erikson, 1982, p. 67)

Community service is one facet of generativity; by contributing to the welfare of others in community service one is demonstrating concern and care for those who benefit. While community service is not the only manifestation of generativity, it is nonetheless plausible to presume that individuals successfully engaged in generativity are more likely to be involved in community service than those best characterized as failing at the task.

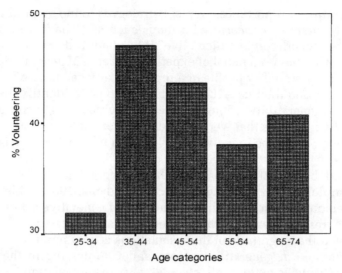

Figure 30.3. Age and volunteerism. The figure shows a sharp increase in volunteerism from the age interval of 25–34 to 35–44. A steady decrease in the level of volunteerism occurs following the 35–44 age interval up until the age interval of 65–74 where the level of volunteerism begins to increase once again.

Two implications following from Erikson's work are relevant for this chapter. The first is that a weak association between age and generative action—of which community service is one form—might be expected. This is because Erikson gave greatest emphasis to generativity in his account of middle age; consequently, one might predict that generative action would be highest in this period. Second, across adulthood, there are likely to be individual differences in generativity. Those most engaged in this task (who therefore can be characterized as high in generativity) are most likely to be involved in generative behavior. We discuss findings relevant to both implications below.

Figure 30.3 depicts the relationship between age and volunteering in the MIDUS study.

As might be predicted from Erikson's theory, volunteering increases between very early adulthood and middle adulthood, and then declines slightly in late middle age. While there is a trend, it is important to acknowledge that the relation of age and volunteering in middle and late adulthood is quite weak (Wilson & Musick [1997] find a similarly small association). This is consistent with theory: although Erikson associated the challenge of developing generativity with middle adulthood, MacDermid, Franz, and De Reus (1998) point out that Erikson believed that the task was apiece of the entire life span.

Among adults of any age, there are likely to be individual differences in the extent of engagement in the generativity task. Those deeply engaged are more likely to be involved in generative tasks—such as community service and volunteer activity—than those for whom generativity concerns are peripheral. A number of different strategies have been used to assess individual differences in generativity. One approach is to ask persons to report on a questionnaire which of a number of prototypical generative actions they have recently performed (e.g., McAdams & de St. Aubin, 1992). This is best viewed as a measure of *generative action.* A second self-report measurement strategy is to tap an orientation toward generativity (*generative concern*) with items such as "I have a responsibility to improve the neighborhood in which I live" (McAdams & de St. Aubin, 1992). Still a third approach relies on a panel of experts in the area of generativity research to characterize the personality profile of a prototypically generative person (*generative personality*), and then uses the experts' prototype to identify the key personality traits for generativity. Using this latter approach, Peterson and Klohnen (1995) identified 13 traits that were judged to be very characteristic of the generative person (e.g., "behaves in a giving way toward others") and 13 believed to be uncharacteristic ("distrustful of people in general").

As should be expected, scores on these different measures are associated (see, for example, Himsel, Hart, Diamond, & McAdams, 1997). This makes sense: we expect persons who score high on a measure of generative action to also score high on measures of generative concern and generative personality. In the MIDUS study, however, the magnitude of the associations is very modest. The correlation between a measure of generative concern and volunteering in the core sample (3000+ participants) is quite small, $r = -.21$, as is the association between generative concern and the indicators of positive and negative generative personality (the latter assessed via the measure discussed by Peterson and Klohnen (1995), described above, and using the interviewed subsample of 90 participants), $rs < .22$.

Personality Traits

Attempts to identify personality traits associated with volunteering and community service have flourished without the theoretical underpinnings of Erikson's work. For example, Clary and Snyder (1999) have proposed that volunteering can be understood as resulting from one of six broad types of motivation. These types are: values ("I feel it is important to help others"), understanding ("Volunteering lets me learn through direct, hands-on experience"), enhancement ("Volunteering makes me feel better about myself"), career ("Volunteering can help me to get my foot in the door at a place where I would like to work"), social ("People I know share an interest in community service"), and protective ("Volunteering is a good escape from my own troubles"). In a range of studies, Clary and Snyder marshal evidence (1999) that suggests that the types of motivation are relatively independent of each other and that satisfaction with volunteering is partially a function of whether the qualities of the experience match the motives that led to volunteering. In other words, a person might be motivated principally by the motive to advance one's career, and volunteering is likely to be satisfying for such a person only if it is consistent with this motive.

A related approach to volunteering focuses on personality traits. Traits refer to relatively stable, relatively consistent behavioral, affective, and cognitive tendencies. Some recent research has focused on identifying the traits that are associated with volunteering, and, more broadly, prosocial behavior. For example, Penner and Finkelstein (1998) claimed that the trait of *other-oriented empathy* is an important determinant of volunteer behavior. Persons who are high in other-oriented empathy report experiencing high levels of empathy with, and responsibility for the welfare of others. Penner and Finkelstein tested their claim by administering a measure of other-oriented empathy (and a number of other measures) to a sample of volunteers working for an AIDS service organization who were tracked for nearly a year. The analyses indicated that persons who were high in other-oriented empathy were active volunteers for longer periods of time than persons low in other-oriented empathy (for similar findings, see Unger & Thumuluri, 1997). In a related line of research, Eisenberg and Okun (1996) measured sympathy and affect experienced while volunteering in a hospital-setting among older adults. They found that participants who were high in sympathy, a tendency to experience concern for another's plight, were more likely than those low in sympathy to experience positive emotions while volunteering. It seems likely that those who experience positive emotions—the individuals high in sympathy—will persist in their volunteer activities.

Attitudes

Finally, there is evidence to indicate that stable attitudes influence volunteering. In her analyses of the MIDUS data, Rossi (2001) found that adults who felt an obligation to fulfill civic obligations and who held generally benevolent and altruistic attitudes were more likely to volunteer than those without these qualities were. Janoski, Musick, and Wilson (1998) also assessed the effects of attitudes on volunteering. In their study of more than 1000 young adults following longitudinally for 17 years, they focused on the contributions of prosocial attitudes to the prediction of volunteering. Prosocial attitudes was a composite of items measuring commitment to citizenship, tolerance for the opinions and perspectives of others, and a sense of political efficacy. Janoski and his colleagues found that those high on the measure of prosocial attitudes were more likely to volunteer than were adults who were low on the measure. More importantly, the researchers found that highly prosocial attitudes in high school or in early adulthood predicted to volunteer activity 9 years later, even among those who were not involved in volunteer activity at the time prosocial attitudes were measured. This suggests that prosocial attitudes can precede, and perhaps lead to, volunteer activity at a later date.

To summarize, the studies reviewed in this section found that volunteering and community service are related to developmental tasks (e.g., generativity), motives (e.g., values, occupation), traits (e.g., other-oriented empathy), and attitudes. However, the magnitude of the associations suggests that volunteering and community service are *not* well predicted by personality variables.

COMMUNITY SERVICE AND THE TRANSFORMATION OF IDENTITY

In this final section, we explore the possibility that community service can effect change in important components of adult identity. This reflects a change in direction from the previous two sections. In the two previous sections, we suggested that relatively stable features of persons—their roles, motives, personalities, and attitudes—might lead individuals to become engaged in community service. In this section, we suggest that community service can lead to transformations in roles and personalities.

The Transformation of Rick Munson

Before considering the research evidence for the transformation of identity through community service, we return to the life of Rick Munson. Munson, it will be remembered, had adopted physically challenged children, which in turn led to his entry into community service. The community service had a profound effect on Munson, transforming his orientation toward life from a focus on economic success to a concern with providing assistance to children in need. As he described it:

> I volunteered work at the school where my kids go. I have volunteered a lot more of my time since I have had the kids then I did before the kids. Before I had kids I was basically running fast in my [real estate] business. You know, I was trying to hook into these mortgages and financing to try and buy me a nice house and stuff. But lately, I picked up more and more volunteer stuff.

Munson provided direct services to children, he served on the board of directors for several charitable organizations, and he acquired a bus driver's license so that he could transport children from place to place. Eventually, the range and depth of experience led Munson to dream of developing his own facility that could provide integrated services to physically challenged children:

> It really started with us having trouble finding decent facilities for the kids to go to. So we said why not let's take the background we have and open up a center where we can offer kids physical therapy … occupational therapies, hearing specialists, speech, you know, we're gonna have a nurse on staff, they will all be there so … the kids can get a full round of services.

This dream became so powerful that Munson and his wife agreed that he should leave the real estate business and devote all of his time to the development of this new agency. For Munson then, involvement in community service had transformed his life, shifting his focus from business and material success to educational and human services.

Empirical Research

There is some research indicating that community service can have the transforming effect on adulthood that seems to be true of Munson. Some of the best of this work has focused on the long-term effects of involvement in the civil rights on adult development. For example, McAdam (1989) examined the influence of participation in the 1964 Mississippi Freedom Summer Project on life paths. The Freedom Summer Project used college student volunteers to register black voters to vote in Mississippi (there were other activities as well). College students applied to join the project, and as part of the application procedure completed fairly extensive questionnaires.

McAdam was able to locate twenty years later more than 300 of the original applicants. To assess the effect of participation on development, McAdam compared the life outcomes of two groups formed from the initial pool of applicants: those who were accepted but did not actually participate, or the "no-shows," and those who participated. The analytical strategy was to statistically control for initial attitudinal, educational, political, and economic differences between the groups in order to test the effect of participation in Freedom Summer on the lives of individuals in the twenty years that followed. Differences between the two groups were substantial. Even after statistically controlling for differences that existed between the two groups prior to Freedom Summer, those who participated were much more likely to be politically active in adulthood than were those who did not participate. Moreover, participation appeared to have profound effects on occupational attainment and social relationships. For example, the participants were making substantially *less* money than were the no-shows almost twenty years after Freedom Summer. The income differential seems attributable to the tendencies for participants, in comparison to no-shows, to attend fewer years of school following freedom summer and to enter the job market at an older age. Participants were also less likely than no-shows to marry.

What does this pattern of results mean? McAdam concludes that participation in Freedom Summer resulted in the "alternation" of identity. An alternation in identity refers to changes in an individual's relationships, self-conception, and worldview. In contrast to a "conversion" in identity in which these facets of life are essentially unconnected to the individual's prior life, an alternation in identity is characterized by change that retains continuity with the previous identity. In other words, the "altered" identity can be seen to have developed out of the previous one. Returning to the study, McAdam interpreted the community service experience as one that amplified and elaborated facets of the identity that were already present in the individuals. Prior to 1964, participants had been politically active and involved in community service, however Freedom Summer altered their identities by providing a context in which political and community service commitments could become central in defining the sense of self, relationships, and a worldview. The no-shows had no such experience, and consequently their identities—so McAdam presumes—were not elaborated as committed activists.

Colby and Damon (1995) found a broadly similar trend in their study of moral exemplars, many of whom were notable for their extensive careers in community service. Like the individuals studied by McAdam (1989), the moral exemplars often found that their initial commitments to community service resulted in changes in their values, jobs, and relationships that in turn deepened their commitments to their community service.

There is evidence that community service of more modest sorts can effect change in facets of identity. Janoski, Musick, and Wilson (1998) used a panel study of a representative sample of high school seniors followed for 17 years to test the hypothesis that volunteering can effect changes in attitudes (as noted in an earlier section, these researchers also found that attitudes predict volunteering). Janoski and his colleagues found evidence for a reciprocal relationship between attitudes and volunteering, with volunteering apparently causing an increase in prosocial attitudes as well as being a result of prosocial attitudes.

While these studies (and research in related areas, e.g., the investigation of blood donation by Piliavin and her colleagues [Piliavin & Callero, 1991]) suggest that community service can produce change in identity, the processes through which this change occurs remain unclear. In our judgment, this is an important topic for future research.

CONCLUSIONS

Community service is an essential constituent of civic life. The value of the work that adults donate to service activities is enormous. Moreover, volunteer activity and community service contribute in fundamental ways to the effective functioning of society.

While there is widespread recognition of the importance of community service to civic life, relatively little is known about the psychology of volunteering. We do know something about role and trait correlates of volunteering. For example, our review indicated that religiosity and educational attainment, two roles or resources, are consistently and positively associated with the tendency to volunteer. Similarly, generative tendencies, prosocial personality traits, and prosocial attitudes also are predictive of community service involvement. All of these variables show only weak relationships to volunteer behavior, a pattern that indicates that our account of community service is very incomplete.

Where the research literature is particularly thin is on the topic of this handbook: adult development. There is little research to indicate how and why involvement in community service has an effect on the life course. That there is such an effect is suggested by the studies reviewed above and our anecdotes drawn from the life of Rick Munson; none of these sources provides a research-based account of the developmental processes that culminate in the transformation of a life. In our view, this is a particularly important area for future research.

REFERENCES

Clary, E. G., & Snyder, M. (1999). The motivations to volunteer: Theoretical and practical considerations. *Current Directions in Psychological Science, 8*(5), 156–159.

Colby, A., & Damon, W. (1995). The development of extraordinary moral commitment. In M. Killen & D. Hart (Eds.), *Morality in everyday life: Developmental perspectives* (pp. 342–370). New York: Cambridge University Press.

Eisenberg, N., & Okun, M. A. (1996). The relations of dispositional regulation and emotionality to elders' empathy-related responding and affect while volunteering. *Journal of Personality, 64*(1), 157–183.

Erikson, E. (1982). *The life cycle completed.* New York: W. W. Norton.

Fukuyama, F. (1995). *Trust: The social virtues and the creation of prosperity.* New York: The Free Press.

Greeley, A. (1997). The other civic America: Religion and social capital. *The American Prospect, 32,* 68–73.

Himsel, A. J., Hart, H., Diamond, A., & McAdams, D. P. (1997). Personality characteristics of highly generative adults as assessed in *q*-sort ratings of life stories. *Journal of Adult Development, 4*(3), 149–161.

Janoski, T., Musick, M., & Wilson, J. (1998). Being volunteered? The impact of social participation and prosocial attitudes on volunteering. *Sociological Forum, 13*(3), 495–519.

Kawachi, I., Kennedy, B. P., & Lochner, K. (1997). Long live community: Social capital as public health. *The American Prospect, 35,* 56–59.

MacDermid, S., Franz, C., & De Reus, L. (1998). Generativity. At the crossroads of social roles and personality. In D. McAdams & E. de St. Aubin (Eds.), *Generativity and adult Development: How and why we care for the next generation.* Washington D.C.: American Psychological Association.

McAdam, D. (1989). The biographical consequences of activism. *American Sociological Review, 54,* 744–760.

McAdams, D. P., & de St. Aubin, E. (1992). A theory of generativity and its assessment through self-report, behavioral acts, and narrative themes in autobiography. *Journal of Personality and Social Psychology, 62*(6), 1003–1015.

Penner, L. A., & Finkelstein, M. A. (1998). Dispositional and structural determinants of volunteerism. *Journal of Personality and Social Psychology, 74*(2), 525–537.

Peterson, B. E., & Klohnen, E. C. (1995). Realization of generativity in two samples of women at midlife. *Psychology and Aging, 10*(1), 20–29.

Piliavin, A. J., & Callero, P. L. (1991). *Giving blood: The development of an altruistic identity.* Baltimore, MD: John Hopkins University Press.

Putnam, R. (1995). Bowling alone in America. *Journal of Democracy, 6*, 65–78.

Rossi, A. S. (2001). Caring and doing for others. Chicago, IL: University of Chicago Press.

Unger, L. S., & Thumuluri, l. K. (1997). Trait empathy and continuous helping: The case of volunteerism. *Journal of Social Behavior and Personality, 12*(3), 785–800.

Warburton, J., LeBrocque, R., & Rosenman, L. (1998). Older people—The reserve army of volunteers?: An analysis of volunteerism among older Australians. *International Journal of Aging and Human Development, 46*(3), 229–245.

Wilson, J., & Musick, M. (1997). Who cares? Toward an integrated theory of volunteer work. *American Sociology Review, 62*, 694–713.

Future Research Directions in Adult Development

JACK DEMICK AND CARRIE ANDREOLETTI

Only 25 years ago, a majority of developmental psychologists considered development to end in adolescence with adults largely restricted to living out roles and scripts (cf. Baltes, Reese, & Nesselroade, 1977; Commons, Richards, & Armon, 1984; Levinson, 1978). However, recent demographic changes and events have forced us all to stop and reassess. For example, consider the following statistics (cf. Lemme, 1999). The largest generation in American history—*baby boomers* born between 1946 and 1964—is now in its 50s. The average life expectancy has risen significantly to 75 years. The fastest growing segment of our population is now over 85 years. More people are now changing jobs, divorcing, and/or moving than ever before. The average married couple currently has more living parents than children. Children either staying home or returning home after college has also become a relatively frequent occurrence. The average retirement age, 62 years, is higher than ever before. The recent events of September 11, 2001 have presented us with new changes and challenges, which require reorganization of our person-in-environment systems.

In light of these changes, it has become almost hackneyed to assert that development continues across the lifespan. However, as an academic discipline, the field of adult development has only more recently come into its own, yet there is still much more to be done. For example, less than 15 years ago, Baltes (1987), one of the progenitors of the related field of lifespan human development, proposed six key postulates on lifespan development. These postulates included the notions that development is: a lifelong process; multidimensional; multidirectional; multicausal; plastic; embedded in historical, cultural, and social contexts; and multidisciplinary. Thus, against this backdrop, the present volume begins to contribute significantly to the burgeoning field of adult development. As mentioned in the

JACK DEMICK • Center for Adoption Research, University of Massachusetts Medical School, Worcester, Massachusetts, 01605-2397.   CARRIE ANDREOLETTI • Department of Psychology, Brandeis University, Waltham, Massachusetts, 02454-9110.

Handbook of Adult Development, edited by J. Demick and C. Andreoletti. Plenum Press, New York, 2002.

preface, represented herein are proponents of both positive adult development (cf. Sheldon & King, 2001, on positive psychology) and more mainstream approaches to adult development. The volume also contains selections by academics as well as by clinicians and by those who adhere to quantitative methodologies as well as more qualitative ones.

One strategy for integrating these disparate selections and for systematically identifying directions for future research on adult development has come from our (Wapner & Demick, 1998, 1999, 2000, 2002, in press) perspective on person-in-environment functioning across the lifespan. Before proceeding with the following analysis that is based on our perspective, we emphasize that this choice was clearly an arbitrary and convenient one. While our approach (see Wapner & Demick, Chapter 4) represents one application of a grand theory to the field of adult development, there are other grand theories (e.g., see related to Piaget's theory, Commons & Richards, Chapter 11; related to Kohlberg's theory, Armon & Dawson, Chapter 15; related to Riegel's theory, Basseches, Chapter 28 and Fisher & Pruyne, Chapter 10) as well as more specific theories (e.g., see Moshman, Chapter 3) represented in the volume. Parallel comparisons may be made using these other perspectives as the basis for comparison. Indeed, we would encourage such undertakings because, as we have asserted elsewhere (e.g., Wapner & Demick, 1998; Wapner, Demick, Yamamoto, & Minami, 2000), they would lead to better understanding among the contributors of the large variety of perspectives represented herein.

From our perspective, we begin by acknowledging that there is no process of *neutral* observation, inquiry, or conclusion in any science. This assertion is based on the notion of *perspectivism* (cf. Lavine, 1950a, 1950b, on *interpretationism*), which in its most general form assumes that any object, event, or phenomenon is always mentally viewed from a particular standpoint, or worldview, that is capable of definition (see also the Heisenberg Uncertainty Principle in physics). This then leads us to the corollary assertion that inquiry and knowledge are biased insofar as problem, theory, method, and praxis in science are interrelated. That is, one's theoretical orientation at least partly determines one's problem (what one studies), method (how one studies it), and praxis (how one translates findings into practical action). Thus, in light of these interrelations, we will organize our following discussion of research implications with respect to problem, theory, method, and practice, respectively.

PROBLEM

The chapters in this volume cover a range of contexts relevant to the study of adult development. Only 10 years ago, many authors in this volume contributed chapters to our innovative volumes on parental development (Demick, Bursik, & DiBiase, 1993) and development in the workplace (Demick & Miller, 1993). Since that time, the notion of *context (system/environment)* has permeated all developmental science and has been expanded routinely to consider the overlapping historical, socioeconomic, cultural, and ethnic contexts. Further, in contrast to many approaches that have equated context with situational moderator variables, we (Wapner & Demick, 2002) have recently proposed an even broader concept of *context* (in line with our elaborated concepts of person, of environment, and of development). That is, on the general level, our approach has suggested six contexts, namely, the *physical, psychological (intrapersonal), and sociocultural contexts of the person* and, analogously, the *physical, interpersonal, and sociocultural contexts of the environment.* However, on the specific level, we have also proposed that there are *an infinite number of specific situations or contexts within each of the six more general contexts.* This broader conceptualization has the potential to provide a more systematic means for attacking open research problems on the wide range of contextual variation relevant to all aspects of adult development.

Employing this elaborated system, the current volume treats the majority of these six general contexts. For example, several selections address the *physical context of the developing adult* (see Brabeck & Shore, Chapter 18, on sex differences or the lack thereof in intellectual and moral development; LoCicero, Chapter 17, on a technologically driven model of women's health care that integrates experiential and procedural perspectives). A significant number of chapters examine the *intrapersonal context of the developing individual* (e.g., see Hoyer & Touron, Chapter 2, Berg & Sternberg, Chapter 6, and O'Connor & Kaplan, Chapter 7, on cognitive functioning; Kramer, Chapter 8, and Shedlock & Cornelius, Chapter 9, on wisdom; Fisher & Pruyne, Chapter 10, Commons & Richards, Chapter 11, and Sinnott, Chapter 12, on postformal thinking; Tahir & Gruber, Chapter 13, on creativity; Armon & Dawson, Chapter 14, Schrader, Chapter 5, on moral thinking; and Day & Youngman, Chapter 27, on religious development). Several treat the *sociocultural context of the developing adult* (e.g., see Azar, Chapter 20, and Wells, Chapter 21, on parenthood and Brown and Roodin, Chapter 24, on grandparenthood).

With respect to *environmental contexts,* Kenny and Barton (Chapter 19) and Climo and Stewart (Chapter 23) explore the developing adult in relation to the *interpersonal context* of family relationships (attachment and eldercare, respectively). Several authors also address adult development in the *sociocultural context* of psychotherapy (e.g., see Basseches, Chapter 28, and Goldberg, Chapter 26), the workplace (see Laske, Chapter 29), the community (see Hart, Southerland, & Atkins, Chapter 30), and contemporary culture (see Josselson, Chapter, 22).

Virtually no chapters have addressed the developing adult in the *physical context of the environment.* This is striking in light of the well documented notion (e.g., Butler, Lewis, & Sunderland, 1991) that, over the course of adult development, many people come to appreciate nature and aesthetic experiences in a deeper way. Moreover, research within our own approach (e.g., Capobianco, 1992) has attested to the notion that, during certain periods of adult life (e.g., the so-called empty nest period), parents may become focused on the physical aspect of the home (e.g., actively changing or preserving their children's rooms and/or possessions). While several contributors have focused on adult development in the sociocultural context of the environment, cross-cultural studies on aspects of adult development (with potential to elaborate universal vs. culturally specific processes) are missing at least from this volume. Thus, as we have argued elsewhere (Wapner & Demick, 2002), a broader conceptualization of context has the potential to open up new research problems and to help us conceptualize problems that are more in line with the complex character of everyday life and that cut across various aspects of developing adults and the various aspects of their environments.

Nonetheless, the volume opens up problems relevant to adult development that have previously been unexplored. For example, Bybee and Wells (Chapter 14) expand classic research on real vs. ideal selves in adulthood to include the notions of the nightmare self (the self as one does not want to be), the moral (or ought) self, and the fantasy self (the self as one would like to be if anything were possible). LoCicero (Chapter 17) explores the types of constructivist activities (integration of experiential and procedural knowledge) that are required for women as a function of a technologically driven (Internet) model of health care. Demick (Chapter 25) raises the issue of whether the world of adoption may be seen as a microcosm of adult life and development. Further, although the problem of parental development (see, e.g., Demick, 2002; Demick et al., 1993) has been treated in the literature, Azar (Chapter 20) presents a relatively holistic and innovative social cognitive perspective on parenting. These and related problems have the potential to contribute to adult development and related fields for some time to come.

THEORY

From our perspective, we believe that there is considerable value in articulating one's underlying theoretical assumptions (since they determine, at least in part, aspects of one's problem, method, and practice). We agree with Werner and Altman (2000), who have stated that: "...we have found articulating research assumptions to be quite liberating. By putting traditional assumptions in perspective, we recognize they are just one of several ways of doing research. We feel comfortable trying out alternative approaches and exploring new ways of thinking about and studying phenomena...Indeed, being aware of alternative ways of knowing has helped us see limitations in traditional psychological approaches. In our own work, it helps us to see where we have been, where we could go, as well as enabling us to see what we have overlooked" (p. 22). Further, these investigators have added that: "Emerging opportunities for...researchers also argue for a more careful articulation of research assumptions. More and more funding agencies expect research proposals to adopt a multidisciplinary approach...In order for researchers to communicate effectively across disciplinary boundaries, it is essential that we be aware of our fundamental research assumptions, know how to select methodologies most appropriate for different problems" (p. 36).

Our holistic, developmental, systems-oriented approach to person-in-environment functioning across the lifespan has been based on Werner's (1957; Werner & Kaplan, 1963) comparative–developmental theory. Both the original theory and our elaborated approach have been termed *organismic* (insofar as psychological part-processes, e.g., cognition, affect, valuation, behavior, are considered in relation to the total context of human activity) and *developmental* (in that it provides a systematic principle governing developmental progression and regression so that living systems may be compared with respect to their formal, organizational features).

On the most general level, our elaborated approach is:

1. *Holistic* insofar as we assume that the person-in-environment system is an integrated system, whose parts may be considered in relation to the functioning whole

2. *Developmental* insofar as we assume that progression and regression may be assessed against the ideal of development embodied in the *orthogenetic principle* (change from dedifferentiated to differentiated and isolated or in conflict to differentiated and hierarchically integrated person-in-environment functioning) and that development encompasses not only *ontogenesis*, but also additional processes such as *phylogenesis* (e.g., adaptation manifest by different species), *microgenesis* (e.g., development of a percept or idea), pathogenesis (e.g., development of both functional and organic pathology), and *ethnogenesis* (e.g., changes during the history of humankind); and

3. *Systems-oriented* insofar as we assume that the person-in-environment system, which includes three aspects of the *person* (*biological*, e.g., health; *intrapersonal*, e.g., stress; *sociocultural*, e.g., role) and three analogous aspects of the *environment* (*physical*, e.g., natural or built; *interorganismic*, e.g., friends, relatives, pets; *sociocultural*, e.g., rules, laws of society).

Corollary notions include the assumptions of:

4. *Transactionalism* (The person and the environment mutually define, and cannot be considered independent of, one another: similarly, the person-in-environment system's experience—consisting of cognitive, affective, and valuative processes—and action are inseparable and operate contemporaneously under normal conditions.)

5. *Multiple modes of analysis* including *structural analysis* (part–whole relationships) and *dynamic analysis* (means–ends relationships)

6. *Constructivism* (The person-in-environment system actively constructs or construes his or her experience of the environment.)

7. *Multiple intentionality* (The person-in-environment system adopts different intentions with respect to self-world relationships, i.e., toward self or world-out-there.)

8. *Directedness and planning* (The person-in-environment system is directed toward both long- and short-term goals related to the capacity to plan.)

9. *Multiple worlds* (The person-in-environment system operates in different spheres of existence, e.g., home, work, recreation.)

10. Preference for *process rather than achievement* analysis.

Against this backdrop, we (e.g., Wapner & Demick, 1999, 2000; Wapner et al., 2000) have delineated extensive categories of comparison between our approach and others. Here, however, we will restrict our discussion to similarities and differences in contributors': *worldviews; use of levels of organization; concepts of person and of environment*; and *concept of development*.

World Views

There is a major difference between the underlying *worldview* (Altman & Rogoff, 1987) or *world hypothesis* (Pepper, 1942, 1967) of our approach and some of the others. Specifically, *our approach adopts elements of both organismic (organicist) and transactional (contextual) worldviews.* The organismic worldview is embodied in an attempt to understand the world through the use of synthesis (i.e., by putting its parts together into a unified whole). Such a view highlights the relationships among the parts, but the relations are viewed as part of an integrated process rather than as unidirectional chains of cause–effect relationships. The major feature of a transactional worldview is that the person and the environment are considered parts of a whole so that one cannot deal with one aspect of the whole without treating the other (cf. Cantril, 1950; Ittelson, 1973; Lewin, 1935). The transactional view specifically treats the "... person's behaving, including his most advanced knowings as activities not of himself alone, nor even primarily his, but as processes of the full situation of organism-environment" (Dewey & Bentley, 1949, p. 104). These worldviews have figured prominently within our approach by impacting our choice of paradigmatic problems (e.g., critical person-in-environment transitions across the lifespan) and of methodological flexibility.

On the most general level, Stevens-Long and Michaud (Chapter 1) have concluded that theoretical advances in adult development are converging on dynamic systems models (e.g., Wapner & Demick, Chapter 4; Fisher & Pruyne, Chapter 10), "... which appear to hold promise for many kinds of integration... [since they] encourage us to analyze human activities in all their complexity and variability and yet maintain our interest in order and patterning within that variation" (p. 19). While we concur, we also note that dynamic system models in and of themselves may have differing underlying worldviews. For example, our (Wapner & Demick, Chapter 4) approach is most sympathetic with a transactional worldview in which the person and the environment are seen as mutually defining entities (cf. person-in-environment as unit of analysis). In contrast, other approaches represented in this volume (e.g., Fisher & Pruyne, Chapter 10; Basseches, Chapter 28; Laske, Chapter 29) appear most in tune with contextualist (Pepper, 1942) and/or dialectical (Riegel, 1973) worldviews. In the former (contextualism), the root metaphor or main comparison is the historic event (e.g., constantly changing sets of circumstances), while in the latter (dialecticism), the root metaphor is the dialectic (or dialogue in which theses and antitheses are forged into syntheses).

With respect to the other theories represented in this volume that differ from ours, those such as Tahir and Gruber (Chapter 13), Josselson (Chapter 22), Climo and Stewart (Chapter 23), and Goldberg (Chapter 26) appear *at least partially* based on the more recently identified narrative as a root metaphor for psychology. As Sarbin (1986) has argued, such approaches believe that "our lives are ceaselessly intertwined with narrative, with the stories that we tell and hear told, with the stories that we dream or imagine or would like to tell. All these stories are reworked in the story of our own life which we narrate to ourselves in an episodic, sometimes semiconscious, virtually uninterrupted monologue. We live immersed in narrative, recounting and reassessing the meanings of our past actions, anticipating the outcomes of our future projects, situating ourselves at the intersection of several stories not yet completed. We explain our actions in terms of plots, and often no other form of explanation can produce sensible statements" (p. 160; cf. Bruner, 1986).

Finally, Moshman (Chapter 3) presents a postmodern metatheory of adult development, *pluralist rational constructivism*, that "does not rule out the possibility of universal sequences and outcomes but does not assume them either. It leaves open the possibility that change may proceed in more than one justifiable direction" (p. 56). This perspective, which "enables us to see how development continues into adulthood and to understand how developmental change in adulthood both resembles and differs from developmental change in early childhood" (p. 59), portends the inevitable and necessary development of new world views for the 21st century.

Levels of Organization

This assumption states that organism-in-environment processes may be categorized in terms of levels of integration (Feibelman, 1954; Herrick, 1949; Novikoff, 1945a, 1945b; Schneirla, 1949), that is, biological (e.g., breathing), psychological (e.g., thinking), and sociocultural (e.g., living by a moral code). There is a contingency relationship among these levels: functioning at the sociocultural level requires functioning at the psychological and biological levels, and functioning at the psychological level requires functioning at the biological level. The levels differ qualitatively and functioning at one level is not reducible to functioning on the prior, less complex level because we assume that higher level functioning does not substitute for, but rather integrates and transforms, lower level functioning. This assumption has played a very important role within our research and highlights a significant driving force, namely, rejection of reductionism.

Numerous authors within this volume have either explicitly or implicitly ascribed to this notion. Some have treated relationships among multiple levels, while others have focused on the relationships between and/or among processes within a given level (most notably, the psychological level). For example, with respect to the former, LoCicero (Chapter 17) has discussed the relationships among women's health (biological level), cognitive processing strategies (psychological level), and modern technology (sociocultural level). Tahir and Gruber (Chapter 13) have addressed relationships between the psychological and sociocultural levels in their evolving systems approach to creativity. A similar focus has appeared in the work by Sinnott (Chapter 12), Basseches (Chapter 28), Laske (Chapter 29), and others. Within the psychological level per se, Bybee and Wells (Chapter 14) have, for example, dealt with the interrelationships between cognitive and affect (self-esteem), while Armon and Dawson (Chapter 15) and Schrader (Chapter 16) have treated relationships among cognition, valuation (morality), and action.

As we (Wapner & Demick, 1998, 1999, 2000, 2002) have noted, there is a need for psychologists routinely to consider the relationships between and among aspects of functioning at the various levels of integration. Because psychological processes operate contemporaneously in the normal functioning adult, we feel that it is increasingly important for scientists to eschew the notion that psychological part-processes (cognition, affect, valuation, action) exist in a vacuum and/or that the individual operates, at any given time, at only one level of organization (biological, psychological, sociocultural). We further believe that future advances in theory in both the fields of adult development and psychology more generally would do well to consider and to assess the structural organization of processes both among and within levels of organization at different points in development (see Wapner & Demick, 1998, for an elaborated discussion).

Concepts of Person and of Environment

As mentioned, we define the person aspect of the person-in-environment system with respect to levels of integration and so assume that the person is comprised of mutually defining physical/biological (e.g., health), intra-personal/ psychological (e.g., self-esteem), and sociocultural (e.g., role as worker, family member) aspects. In line with this, we assume that the person is characterized by constructivism, a rage for order (Kuntz, 1968), and multiple intentionality (i.e., the ability to focus on different objects of experience such as self, environment, and their relationships).

Analogous to our conceptualization of person, we assume that the environmental aspect of the person-in-environment system is comprised of mutually defining physical (e.g., natural and built objects), living organism (e.g., spouse, friend, pet), and sociocultural (e.g., rules and mores of the home, community, and other cultural contexts) aspects. Again, we do not focus on the person or on the environment per se but rather consider the person and the environment relationally as parts of one whole (cf. person-in-environment system as unit of analysis).

While many of the authors in this volume share a multifaceted conception of the person (e.g., see Tahir & Gruber, Chapter 13; Bybee & Wells, Chapter 14; Armon & Dawson, Chapter 15; Shrader, Chapter 16; Wells, Chapter 21; Josselson, Chapter 22; Climo & Stewart, Chapter 23), the work of Sinnott (Chapter 12) is particularly noteworthy. That is, in her discussion of postformal thought and adult development, she has linked the adult's cognitive development (intra-personal) to emotion (intrapersonal), spirituality (intrapersonal, sociocultural), community (interpersonal, sociocultural), and existential meaning (intra-personal, sociocultural). In contrast, fewer authors share a multidimensional view of the environment. However, those who have discussed development in context (e.g., Day & Youngman, Chapter 27; Basseches, Chapter 28, Laske, Chapter 29) appear to consider the physical, interpersonal, and/or sociocultural aspects of the environment. Most notably, in their discussion of community service and adult development, Hart, Southerland, and Atkins (Chapter 30) have integrated the physical (e.g., specific activities in specific locales), interpersonal (e.g., civic engagement), and sociocultural (e.g., economic) aspects of the environment.

Concept of Development

Our view of development transcends the boundaries within which the concept of development is ordinarily applied and not restricted to child growth, to ontogenesis. In contrast, we view development more broadly as a mode of analysis of diverse aspects of person-in-environment functioning (including, e.g., ontogenesis, microgenesis, pathogenesis, phylogenesis, ethnogenesis).

Components (person, environment), relations among components (e.g., means-ends), and part-processes (e.g., cognition) of person-in-environment systems are assumed to be developmentally orderable in terms of the degree of organization attained by a system. Development is seen as a change from a dedifferentiated to a differentiated and isolated or in conflict to a differentiated and integrated system vis-à-vis its parts, its means, and its ends. Thus, optimal development entails a differentiated and hierarchically integrated person-in-environment system with flexibility, freedom, self-mastery, and the capacity to shift from one mode of person-in-environment relationship to another as required by goals, by demands of the situation, and by the instrumentalities available.

There are some clear differences among the contributors with respect to their general concept of development and its endpoints or lack thereof. In their discussion of recent theory building in adult development, Stevens-Long and Michaud (Chapter 1) have concluded that "Dynamic systems models... leave open the question of direction. They may predict movement toward greater complexity, but they don't tell us whether that complexity is good or mature or wise..." (p. 19) While their characterization appears apt for some dynamic systems models (e.g., Fisher & Pruyne, Chapter 10; cf. Basseches, Chapter 28), others such as our own (Wapner & Demick, Chapter 4) have posited flexibility, freedom, self-mastery, and multiple intentionality as possible endpoints of development. Further, many theories of (postformal) cognitive development (e.g., Commons & Richards, Chapter 11; cf. Sinnott, Chapter 12) and moral development (e.g., Armon & Dawson, Chapter 15) have posited a telos to development.

Finally, most related to our concept of development as a mode of analysis has been Kramer's (Chapter 8) discussion of the ontogeny of wisdom in its variations. Thus, future work in adult development might do well both to integrate developmental and nondevelopmental notions within theory on the complexity of adult functioning (e.g., Moshman, Chapter 3) and to go beyond the discussion of age differences in adulthood toward considering the development of various processes (e.g., cognition, affect, valuation) and their interrelations over the course of adult life (e.g., see Wapner & Demick, 1998, 2002).

METHOD

From our point of view, research on adult development should be open to a range of methodologies from both the natural science and the human science perspectives (cf. Cavanaugh & Whitbourne, Chapter 5). For the most part, the former is concerned with prediction, which is linked to: a focus on observable behavior and explanation in terms of cause–effect relationships; an analytic analysis that begins with parts and assumes that the whole cannot be understood through addition of those elements; and the adoption of scientific experimentation as the appropriate methodology. In contrast, the latter (e.g., Giorgi, 1970): "specifies as its goal the understanding of experience or the explication of structural relationships, pattern, or organization that specifies meaning... adopts the descriptive method... seeks detailed analysis of limited numbers of cases that are presumably prototypic of larger classes of events... carries out qualitative analysis through naturalistic observation, empirical, and phenomenological methods" (Wapner, 1987, p. 1434).

Thus, as we see it, since explication makes focal the meaning of a phenomenon or what is explained and causal explanation focuses on underlying process (e.g., the question of how and under what conditions developmental transformation is reversed, arrested, or advanced), both approaches have advantages and limitations. While the natural science approach may be characterized by precision and reliability, it may also suffer from lack of validity. In contrast, while the human science approach may be characterized by validity, it may suffer from lack of precision and reliability. Accordingly, both of these approaches to understanding complement each other and should be fostered.

A central issue for us then concerns when these methods should be used. This depends on the level of complexity of the phenomenon under investigation. Phenomena on the level of complex features of human experience—where manipulation and control of conditions are not possible—may more appropriately be analyzed by human science methods; in contrast, less complex and more simplified phenomena where conditions can be manipulated and controlled may be appropriately addressed by the methods of natural science (cf. Maslow, 1946 on problem- vs. means-centering in psychology).

Several contributors to this volume share our methodological eclecticism. In this regard, most notable is the work of: O'Connor and Kaplan (Chapter 7), who have typically focused on both process and achievement analyses of neuropsychological functioning (cf. Hoyer & Touron, Chapter 2); Fisher and Pruyne (Chapter 10), who have considered nomothetic and idiographic approaches to adult cognitive development; and Armon and Dawson (Chapter 15), who have provided both quantitative and qualitative analyses. In line with our reasoning, still others who are well versed in both methods (e.g., Azar, Chapter 20; Wells, Chapter 21; Climo & Stewart, Chapter 23) appear to use a particular method depending on the level of organization in which they are interested.

PRACTICE

Strongly believing that theory and praxis are *flip sides of the same coin*, we are very glad to see that, unlike many reference texts in the field, the *Handbook of Adult Development* contains a significant number of contributions that focus on practice. While some chapters (e.g., Kenny & Barton, Chapter 19; Hart, Southerland, & Atkins, Chapter 30) have strong implications for practice (e.g., college students' adaptation to college, community service in adulthood, respectively), others have dealt directly with the issue of practice. For example, in his discussion of adult development and the practice of psychotherapy, Basseches (Chapter 28) begins to provide specific suggestions for both the training and supervision of developing psychotherapists. In a similar manner, Laske (Chapter 29) provides a view of executive development that emphasizes constructivist–developmental theory in general and the notion of human agency in particular. Not only do we feel that the relationships between experience and action (e.g., see Wapner & Demick, 1998, 1999, 2000, 2002) constitute an important theoretical and empirical problem, but we also think that such theory-driven practical application has the potential to optimize the transactions (experience and action) of developing adults in their diverse environments.

SUMMARY AND CONCLUSIONS

The above comparison of similarities and differences in the problem, theories, methods, and practices of leading researchers in the field of adult development has suggested that this field has come a long way since its inception over 25 years ago. That is, in its infancy, adult developmentalists consisted largely of a disparate group of researchers, problems, theories, and methodologies. Currently, there appear to be more similarities than dissimilarities in the theoretical orientations at least among this current cohort of researchers. That is, common assumptions that unify this group revolve around the notions of: constructivism; contextualism; transactionalism; relations among levels of integration; a multifaceted conception of person; relationships between experience and action; and methodological eclecticism. Less common assumptions among this group included: the person-in-environment system as the unit of analysis; a multidimensional conception of the environment; and a broader concept of development.

Thus, theory and research on adult development such as (but not limited to) ours has implications for the problem, theory, method, and practice of adult developmental science as well as for the larger field of psychology in the following ways.

1. *Problem*: A focus such as ours, namely, on the person-in-environment system with mutually defining physical/biological, intra-personal, and sociocultural aspects of the person and physical, living organisms, and sociocultural aspects of the environment has the potential to broaden the scope of research in adult development to include the study of problems that are not limited to the person and that are more in line with the complex character of everyday life.

2. *Theory*: Our work and that of others suggest that advances in adult developmental research as well as in psychological research more generally can be made through the adoption of overarching theories that utilize both normative and causal analyses. That is, the complementarity of normal explication (description) and causal explanation (conditions under which cause–effect relationships occur) has the potential to shed light on both adult experience and action rather than focusing on one or the other as has heretofore been the case.

3. *Method*: Research such as ours and that of others represented in this volume suggest that holistic, ecologically oriented research may be a necessary complement to more traditional laboratory work and that it might be conducted through reducing the number of focal individuals studied rather than the number and kinds of interrelationships among aspects of the person, of the environment, and of the systems to which they belong.

4. *Practice*: Particularly in light of recent demographic changes, theoretical and practical considerations in the study of adult development must go hand in hand. With more people living longer than ever before, it is incumbent on psychological scientists to work toward a unified theory-driven science of psychology and of adult development that maximizes adults' transactions with their multifaceted environments.

Such reframing as embodied in the aforementioned four areas may also help the larger field of psychology see itself as well as be seen by others as a unified—or in our terminology, a differentiated and integrated—science concerned not only with the study of isolated aspects of human functioning but with the rigorous treatment of a wide range of problems that cut across the various aspects of developing adults and the various aspects of their environments (cf. Demick & Andreoletti, 1995). This message takes on added significance in light of the recent events of September 11, 2001. It is now more imperative than ever before for psychological scientists to uncover similarities and differences in our assumptions, research problems, and methodologies to serve as the basis for accelerated growth of the science and practice of adult development in the 21st century. We sincerely hope that such endeavors will optimize adults' transactions with their physical, interpersonal, and sociocultural environments as well as integrate their relations with other adults living in different physical and sociocultural contexts.

Finally, it should again be emphasized that the commonalities and differences reported here are based on the assumptions ingredient in our holistic, developmental, systems-oriented perspective. We strongly encourage others, both contributors and readers of this volume, to make similar comparisons utilizing their own perspective as its basis. Such endeavors will bring us closer to understanding similarities and differences among researchers and thereby serve to advance the study of adult development even further.

REFERENCES

Altman, I., & Rogoff, B. (1987). World views in psychology: Trait, interactional, organismic and transactional perspectives. In D. Stokols & I. Altman (Eds.), *Handbook of environmental psychology* (pp. 7–40). New York: John Wiley & Sons.

Baltes, P. B. (1987). Theoretical propositions of life span developmental psychology: Some converging observations on history and theory. In P. B. Baltes & O. G. Brim (Eds.), *Life span development and behavior, Vol. 2* (pp. 255–279). New York: Academic Press.

Baltes, P. B., Reese, H. W., & Nesselroade, J. R. (1977). *Life span developmental Psychology: Introduction to research methods.* Monterey, CA: Brooks/Cole.

Bruner, J. (1986). *Actual minds, possible worlds.* Cambridge, MA: Harvard University Press.

Butler, R. N., Lewis, M., & Sunderland, T. (1991). *Aging and mental health: Positive psychosocial and biomedical holdings,* 4th edit. New York: Merrill.

Cantril, H. (Ed.) (1950). *The why of man's experience.* New York: Macmillan.

Capobianco, N. (1992). *Changes in the home setting following the departure of adult children.* Unpublished manuscript. Worcester, MA: Clark University.

Commons, M. L., Richards, F. A., & Armon, C. (1984). *Beyond formal operations: Late adolescent and adult cognitive development.* New York: Praeger.

Demick, J. (2002). Stages of parental development. In M. Bornstein (Ed.), *Handbook of parenting,* Vol. 3. Being and becoming a parent (2nd ed., pp. 389–413). Mahwah, NJ: Lawrence Erlbaum.

Demick, J., & Andreoletti, C. L. (1995). Some relations between clinical and environmental psychology. *Environment & Behavior, 27*(1), 56–72.

Demick, J., Bursik, K., & DiBiase, R. (Eds.) (1993). *Parental development.* Hillsdale, NJ: Lawrence Erlbaum.

Demick, J., & Miller, P. (Eds.) (1993). *Development in the workplace.* Hillsdale, NJ: Lawrence Erlbaum.

Dewey, J., & Bentley, A. F. (1949). *Knowing and the known.* Boston, MA: Beacon Press.

Feibelman, J. K. (1954). Theory of integrative levels. *British Journal of Philosophy of Science, 5,* 59–66.

Giorgi, A. (1970). Towards phenomenologically based research in psychology. *Journal of Phenomenological Psychology, 1,* 75–98.

Herrick, C. J. (1949). A biological survey of integrative levels. In R. W. Sellars, V. J. McGill, & M. Farber (Eds.), *Philosophy for the future* (pp. 222–242). New York: Macmillan.

Ittelson, W. H. (1973). Environmental perception and contemporary perceptual theory. In W. H. Ittelson (Ed.), *Environment and cognition* (pp. 1–19). New York: Seminar Press.

Kuntz, P. G. (1968). *The concept of order.* Seattle: University of Washington Press.

Lavine, T. (1950a). Knowledge as interpretation: An historical survey. *Philosophy and Phenomenological Research, 10,* 526–540.

Lavine, T. (1950b). Knowledge as interpretation: An historical survey. *Philosophy and Phenomenological Research, 11,* 80–103.

Lemme, B. H. (1999). *Development in adulthood,* 2nd ed. Boston, MA: Allyn & Bacon.

Levinson, D. J. (1978). *The seasons of a man's life.* New York: Ballentine.

Lewin, K. (1935). *A dynamic theory of personality.* New York: McGraw-Hill.

Maslow, A. H. (1946). Problem-centering vs. means-centering in science. *Philosophy of Science, 13,* 326–331.

Novikoff, A. B. (1945a). The concept of integrative levels and biology. *Science, 101,* 209–215.

Novikoff, A. B. (1945b). Continuity and discontinuity in evolution. *Science, 101,* 405–406.

Pepper, S. C. (1942). *World hypotheses.* Berkeley: University of California Press.

Pepper, S. C. (1967). *Concept and quality: A world hypothesis.* LaSalle, IL: Open Court.

Riegel, K. (1973). Dialectical operations: The final period of cognitive development. *Human Development, 16,* 346–370.

Sarbin, T. R. (1986). The narrative as a root metaphor for psychology. In T. R. Sarbin (Ed.), *Narrative psychology: The storied nature of human conduct* (pp. 3–21). New York: Praeger.

Schneirla, T. C. (1949). Levels in the psychological capacities of animals. In R. W. Sellars, V. J. McGill, & M. Farber (Eds.), *Philosophy for the future* (pp. 243–286). New York: Macmillan.

Sheldon, K. M., & King, L. (2001). Why positive psychology is necessary. *American Psychologist, 56*(3), 216–217.

Wapner, S. (1987). A holistic, developmental, systems-oriented psychology: Some beginnings. In D. Stokols & I. Altman (Eds.), *Handbook of environmental psychology* (pp. 1433–1465). New York: John Wiley & Sons.

Wapner, S., & Demick, J. (1998). Developmental analysis: A holistic, developmental, systems-oriented perspective. In W. Damon (Series Ed.) & R. M. Lerner (Vol. Ed.) *Handbook of child psychology, Vol. 1: Theoretical models of human development,* 5th edit. (pp. 761–805). New York: John Wiley & Sons.

Wapner, S., & Demick, J. (1999). Developmental theory and clinical practice: A holistic, developmental, systems-oriented approach. In W. K. Silverman & T. H. Ollendick (Eds.), *Developmental issues in the clinical treatment of children* (pp. 3–30). Boston, MA: Allyn & Bacon.

Wapner, S., & Demick, J. (2000). Person-in-environment psychology: A holistic, developmental, systems-oriented perspective. In W. B. Walsh, K. H. Craik, & R. H. Price (Eds.), *Person-environment psychology: New directions and perspectives,* 2nd edit. (pp. 25–60). Hillsdale, NJ: Lawrence Erlbaum.

Wapner, S., & Demick, J. (2002). The increasing *contexts* of *context* in the study of environment-behavior relations. In R. B. Bechtel & A. Churchman (Eds.), *Handbook of environmental psychology* (pp. 3–14). New York: John Wiley & Sons.

Wapner, S., & Demick, J. (in press). Critical person-in-environment transitions across the life span: A holistic, developmental, systems-oriented program of research. In J. Valsiner (Ed.), *Differentiation and integration of a developmentalist: Heinz Werner's ideas in Europe and America.*

Wapner, S., Demick, J., Yamamoto, T., & Minami, H. (Eds.). (2000). *Theoretical perspectives in environment-behavior research: Underlying assumptions, research problems, and methodologies.* New York: Kluwer Academic/Plenum.

Werner, C. M., & Altman, I. (2000). Humans and nature: Insights from a transactional view. In S. Wapner, J. Demick, T. Yamamoto, & H. Minami (Eds.), *Theoretical perspectives in environment-behavior research: Underlying assumptions, research problems, and methodologies* (pp. 21–37). New York: Kluwer Academic/Plenum.

Werner, H. (1957). *Comparative psychology of mental development* (rev. ed., originally published 1940). New York: International Universities Press.

Werner, H., & Kaplan, B. (1963). *Symbol formation.* New York: John Wiley & Sons.

Index